Space Sciences Series of ISSI

Volume 89

The Space Sciences Series of ISSI books are coherent reports of the findings, discussions, and ideas that result from international scientific workshops regularly held at the International Space Science Institute (ISSI) in Bern, Switzerland. ISSI's main task is to contribute to the achievement of a deeper understanding of the results from space-research missions, adding value to those results through multi-disciplinary research in an atmosphere of international cooperation. The books are reprints of special issues in the Space Science Reviews journal and occasionally of special issues in the Surveys in Geophysics journal.

Valéry Lainey • Michel Blanc • Aurélien Crida •
Jeffrey Cuzzi • Maryame El Moutamid •
Gianrico Filacchione • Carly Howett •
Tilman Spohn
Editors

New Vision of the Saturnian System in the Context of a Highly Dissipative Saturn

Previously published in *Space Science Reviews, New Vision of the Saturnian System in the Context of a Highly Dissipative Saturn, Vol. 219–222, 2023–2026*

Springer

Editors
Valéry Lainey
Paris Observator
Paris, France

Michel Blanc
Research Institute in Astrophysics
and Planetology
Toulouse, France

Aurélien Crida
Université Côte d'Azur
Nice, France

Jeffrey Cuzzi
NASA - Ames Research Center
Moffet Field, CA, USA

Maryame El Moutamid
Southwest Research Institute
Boulder, CO, USA

Gianrico Filacchione
Institute for Space Astrophysics
and Pl
Rome, Italy

Carly Howett
Atmospheric, Oceanic and Planetary
Physics Department
Clarendon Laboratory
Oxford, UK

Tilman Spohn
German Aerospace Center (DLR)
Berlin, Germany

ISSN 1385-7525
Space Sciences Series of ISSI
ISBN 978-94-024-2386-0

Cover illustration: IMCCE/CNRS

This Springer imprint is published by the registered company Springer Nature B.V.
The registered company address is: Van Godewijckstraat 30, 3311 GX Dordrecht, The Netherlands

Contents

Space Science Reviews (2026) 222:38
https://doi.org/10.1007/s11214-026-01294-7

New Vision of the Saturnian System in the Context of a Highly Dissipative Saturn – Editorial

Valéry Lainey[1] · Michel Blanc[2,3] · Aurélien Crida[4] · Jeffrey N. Cuzzi[5] · Maryame El Moutamid[6] · Gianrico Filacchione[7] · Carly Howett[8] · Alyssa Rhoden[6] · Tilman Spohn[9]

Accepted: 2 April 2026 / Published online: 17 April 2026

Over the past fifteen years, our understanding of the Saturnian system has undergone a profound transformation. Long considered a relatively stable and slowly evolving system, Saturn and its moons are now recognized as a highly dynamic environment shaped by unexpectedly strong tidal dissipation within the planet. This paradigm shift has far-reaching consequences, from the orbital evolution of the satellites to their origin and long-term history.

A key breakthrough came from the ability to measure the current orbital evolution of Saturn's moons with unprecedented precision. These observations revealed that tidal interactions within Saturn are far more efficient than previously thought, driving rapid outward migration of the satellites. This result opens new possibilities for interpreting long-standing puzzles, such as the intense internal activity of Enceladus, but also challenges classical assumptions about the age and formation of the Saturnian moons and rings.

The wealth of high-quality data provided by the Cassini spacecraft during its 13 years-long mission at Saturn has played a central role in this transformation. Long-term, in-situ observations have enabled a detailed reconstruction of the dynamical evolution of the system, while also highlighting inconsistencies with traditional tidal models. In response, new

✉ V. Lainey
valery.lainey@obspm.fr

1 LTE, Observatoire de Paris, Université PSL, Sorbonne Université, Université de Lille, LNE, CNRS, 61 avenue de l'Observatoire, 75014, Paris, France

2 Institut de Recherche en Astrophysique et Planétologie, 9 Av. du Colonel Roche, 31400, Toulouse, France

3 Weihai Institute for Interdisciplinary Research, Shandong University, Weihai, 264209, China

4 Lagrange (UMR 7293), Université Côte d'Azur / CNRS / Observatoire de la Côte d'Azur, Boulevard de l'Observatoire, CS 34229, 06304 NICE Cedex 4, France

5 Ames Research Center, NASA MS 245-3, Moffet Field, CA, 94035-1000, USA

6 Southwest Research Institute, 1301 Walnut St Suite 400, Boulder, CO, 80302, USA

7 IAPS-INAF, Via del Fosso del Cavaliere, 100, 00133, Rome, Italy

8 University of Oxford, Clarendon Laboratory, Parks Road, Oxford, OX1 3PU, UK

9 International Space Science Institute, Hallerstrasse 6, 3012 Bern, Switzerland

theoretical frameworks have emerged, in which tidal dissipation is governed by complex interactions between the satellites and the evolving internal structure of Saturn. These mechanisms naturally account for the observed diversity of orbital behaviors and rapid migration rates.

Beyond their dynamical implications, these results provide a unique window into the deep interior of Saturn. The efficiency and apparent frequency dependence of tidal dissipation suggest that internal processes (such as stratification, layered convection, or resonant oscillation modes) play a central role in shaping the system's evolution. In this sense, the orbital architecture of the moons becomes a diagnostic tool, offering indirect but powerful constraints enlightening the physical properties and long-term evolution of the planet itself.

At the same time, the current orbital state of the Saturnian system (marked by numerous resonances and ongoing dynamical interactions) points to a history of significant evolution. These constraints suggest that some moons may be much younger than previously assumed, raising fundamental questions about their origin. In this context, alternative formation scenarios have been developed. In one class of models, moons form from the spreading of Saturn's rings and subsequently migrate outward under the influence of strong tidal forces. In another, large tidal expansions may lead to resonance crossings, triggering collisions, satellite re-accretion, and possibly the formation of rings. Together, these scenarios highlight a coupled evolution of rings and satellites, providing a coherent framework that naturally explains several key features of the system.

Taken together, these advances outline a new, consistent picture of the Saturnian system, in which rings, moons, and planetary interiors are intimately connected through coupled dynamical and physical processes. This emerging vision not only reshapes our understanding of Saturn but also provides a broader framework for interpreting the origin and evolution of giant planet systems, both within and beyond our solar system.

More broadly, the Saturnian system now appears as a natural laboratory for studying tidal processes under extreme conditions. The lessons learned here are likely to guide our understanding of a wide range of astrophysical systems, from exoplanetary systems hosting close-in satellites or planets, to the evolution of giant planets' internal structure. Future observational efforts and potential new space missions will be essential to collect better data necessary for testing and refining this new paradigm.

This collection originates from an International Space Science Institute (ISSI) workshop entitled "New Vision of the Saturnian System in the Context of a Highly Dissipative Saturn", which brought together experts in planetary dynamics, interior physics, ring science, and satellite geophysics. The workshop provided the opportunity to confront recent observational constraints with emerging theoretical models, with a particular focus on the implications of strong tidal dissipation for the coupled evolution of Saturn, its rings, and its satellites. These discussions led to the identification of key open questions and helped structure the contributions presented in this volume. The articles are organized around the major themes addressed during the workshop, providing complementary perspectives on the origin, evolution, and present-day dynamics of the Saturnian system. Together, they offer a coherent synthesis of recent progress while outlining the main challenges that remain to be addressed. The Saturnian system thus stands as a cornerstone for future studies of planetary dynamics and evolution.

Declarations

Competing Interests The authors declare no competing interests.

Space Science Reviews (2024) 220:22
https://doi.org/10.1007/s11214-024-01055-4

Tidal Dissipation in Giant Planets

Jim Fuller[1] · Tristan Guillot[2] · Stephane Mathis[3] · Carl Murray[4]

Received: 8 October 2023 / Accepted: 7 February 2024 / Published online: 20 February 2024

Abstract
Tidal interactions between moons and planets can have major effects on the orbits, spins, and thermal evolution of the moons. In the Saturn system, tidal dissipation in the planet transfers angular momentum from Saturn to the moons, causing them to migrate outwards. The rate of migration is determined by the mechanism of dissipation within the planet, which is closely tied to the planet's uncertain structure. We review current knowledge of giant planet internal structure and evolution, which has improved thanks to data from the *Juno* and *Cassini* missions. We discuss general principles of tidal dissipation, describing both equilibrium and dynamical tides, and how dissipation can occur in a solid core or a fluid envelope. Finally, we discuss the possibility of resonance locking, whereby a moon can lock into resonance with a planetary oscillation mode, producing enhanced tidal migration relative to classical theories, and possibly explaining recent measurements of moon migration rates.

Keywords Saturn · Tidal interaction · Moons · Enceladus · Titan

1 Introduction

During the last decade, new discoveries in the Jupiter and Saturn systems have compelled us to revisit the estimates of tidal dissipation in giant gaseous planets of our solar system. Figure 1 illustrates how dissipation leads to a phase lag between the tidal bulge and the satellite, creating a torque that drags the satellite forward or backward. When the planet rotates faster than the moon orbits ($\Omega_\mathrm{p} > \Omega_\mathrm{m}$, as in the case of solar system moon systems), angular momentum is transferred from the planet to the moon, dragging the moon forward and making it migrate outwards. The initial predictions for the tidal quality factor of Jupiter and Saturn were computed by Goldreich and Soter (1966), who solved the evolution equation for the semi-major axis of the moons assuming a frequency-independent tidal dissipation rate, constant over the evolution, in order to place them at their current positions at the age of the

✉ C. Murray
c.d.murray@qmul.ac.uk

1 TAPIR, Walter Burke Institute for Theoretical Physics, Caltech, Mailcode 350-17, Pasadena, 91125, CA, USA

2 Observatoire de la Côte d'Azur, Université Côte d'Azur, CNRS, Nice, 06304, France

3 Université Paris-Saclay, Université Paris Cité, CEA, CNRS, AIM, Gif-sur-Yvette, F-91191, France

4 School of Physical and Chemical Sciences, Queen Mary University of London, E1 4NS, London, UK

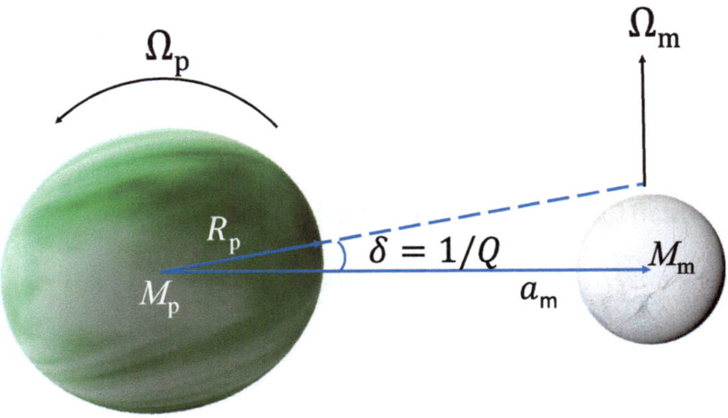

Fig. 1 Cartoon showing a tidally distorted planet of mass M_p and radius R_p, due to a moon of mass M_m at semi-major axis a_m. The planet's spin frequency is Ω_p and the moon's orbital frequency is Ω_m. The tidal quality factor of the planet, Q, can be visualized as a lag angle of $\delta = 1/Q$ between the tidal bulge and the moon's position. In reality, the effective value of Q can involve more complicated tidal distortion within the planet due to dynamical tides

solar system. This procedure predicts tidal quality factors of $10^4 \lesssim Q \lesssim 10^6$ for Saturn and Jupiter.

Using high precision astrometry, a series of measurements (Lainey et al. 2009, 2012, 2017; Polycarpe et al. 2018) demonstrated that tidal dissipation in Jupiter and Saturn can be more than an order of magnitude higher than what was computed initially by Goldreich and Soter (1966) (see also Sinclair 1983), i.e., a tidal quality factor $Q \lesssim 10^3$ instead of $Q \sim 10^4$ for Saturn. Moreover, the dissipation in Saturn has a complex variation as a function of the tidal frequency: we observe a weak (and potentially smooth) dependence on the frequency for Enceladus, Tethys and Dione, and a much stronger dissipation for Rhea and Titan. Observations of faster moon migration than expected (and hence stronger dissipation) may require revisions of our understanding of the formation and evolution of giant planets systems (Charnoz et al. 2011), and of tidal dissipation in planetary interiors.

These results are also of major interest for exoplanetary systems since giant gaseous planets have been the first type of discovered exoplanets. Indeed, the discovery of 51 Pegasi b by Mayor and Queloz (1995) opened the path to the exploration of giant planets outside our Solar system. The occurrence of giant gaseous planets close to their host stars, the so-called "hot" Jupiters, demonstrated that the orbital architecture of our solar system may not be common. It also spawned new challenging questions about the formation, the dynamical evolution, and the stability of exoplanetary systems. In this context, the study of tidal (and magnetic) star-planet interactions between low-mass stars and close giant planets came to the forefront (e.g., Ahuir et al. 2021).

For example, tidal interactions between a host star and orbiting giant exoplanets were proposed as a candidate to explain the "bloated" hot Jupiter phenomenon (Bodenheimer et al. 2001; Guillot and Showman 2002; Baraffe 2005). The proposed idea is that tidal dissipation can sustain intense internal heating in giant exoplanets, modifying their internal structure with an increase of their radii. Another important but unsolved question is the effects of tides on the orbital distribution of hot Jupiters around their host stars, for which dissipation in both the stars and the planets can play a key role. Let us finally point that these major questions are among the key objectives of ongoing and forthcoming major space missions like CHEOPS, TESS, JWST, PLATO, and ARIEL (Benz et al. 2017; Ricker et al.

2015; Lagage 2015; Rauer et al. 2014; Tinetti et al. 2018) and ground-based instruments like SPIRou (Moutou et al. 2015).

While tidal dissipation is crucial for orbital evolution of both giant planet moon systems and exoplanet systems, the underlying physics is complex. In this chapter, we review our current understanding of giant planet structure, and we discuss how planetary structure and internal dynamics affect various tidal dissipation mechanisms. Comparison between models and measured migration rates are now allowing us to uncover the complex and interrelated histories of giant planet interiors and their moon systems.

2 Giant Planet Structure

2.1 Interiors of Jupiter, Saturn, Uranus and Neptune

In the solar system, four planets have acquired enough hydrogen and helium during the early formation phase of the system to grow rapidly in size and mass and earn the qualification of "giant". With radii of about 10 times our Earth, Jupiter and Saturn are commonly known as "gas giants" while Uranus and Neptune, which are about 4 times larger than Earth, are usually called "ice giants". Both denominations are convenient, but we should be aware of their limitations: most of Jupiter and Saturn's interiors behave as a liquid rather than a gas (they are in fact *fluid*, e.g., Guillot 2005), while the compositions of Uranus and Neptune are highly uncertain: they could well be more rocky than icy (Helled and Fortney 2020; Kunitomo et al. 2018).

An important property of giant planets in our solar system is that they radiate about twice as much energy as they receive from the Sun (Pearl and Conrath 1991, and references therein). The difference is due to their progressive contraction and cooling (Hubbard 1968). A notable exception is Uranus, whose intrinsic heat luminosity is about 5 to 10 times lower than the other three giant planets and compatible with zero (Pearl and Conrath 1991). Several hypotheses have been put forward (see Helled and Fortney 2020), but the recent 30% upward revision of Jupiter's intrinsic luminosity (Li et al. 2018) calls for a reanalysis of available data, and ultimately, in situ measurements at Uranus (e.g., Fletcher et al. 2020; Guillot 2022).

In most of the interiors of these four planets, the high radiative opacities and low heat conductivities imply that any non-negligible intrinsic luminosity should ensure convection (Stevenson 1976; Guillot et al. 1994, 2004). The atmospheres of all planets, including Uranus, indeed exhibit significant convective activity (Hueso et al. 2020). At photospheric levels, their temperatures are remarkably uniform in spite of the strongly varying insulation, also a sign of efficient heat transport and hence convection (Ingersoll and Porco 1978). In planets, only very slightly super-adiabatic temperature gradients are required for convection to carry the planetary heat flux, so convective regions are thought to have a nearly adiabatic density profile. These facts motivate the widely used hypothesis that interiors are convective and hence nearly adiabatically stratified.

The structures of our four giant planets as obtained from interior models using the latest gravity measurements by *Juno* for Jupiter (Durante et al. 2020), gravity and seismology measurements by *Cassini* for Saturn (Iess et al. 2019; Hedman et al. 2019), and by much less accurate gravity measurements for Uranus (Jacobson 2014) and Neptune (Jacobson 2009), are shown in Fig. 2. For Jupiter, the models predict the existence of a fluid envelope separated into a helium-poor upper part and a helium-rich lower part, and a deep dilute core with a higher proportion of heavy elements mixed with hydrogen (Wahl et al. 2017; Debras and Chabrier 2019; Miguel et al. 2022; Militzer et al. 2022).

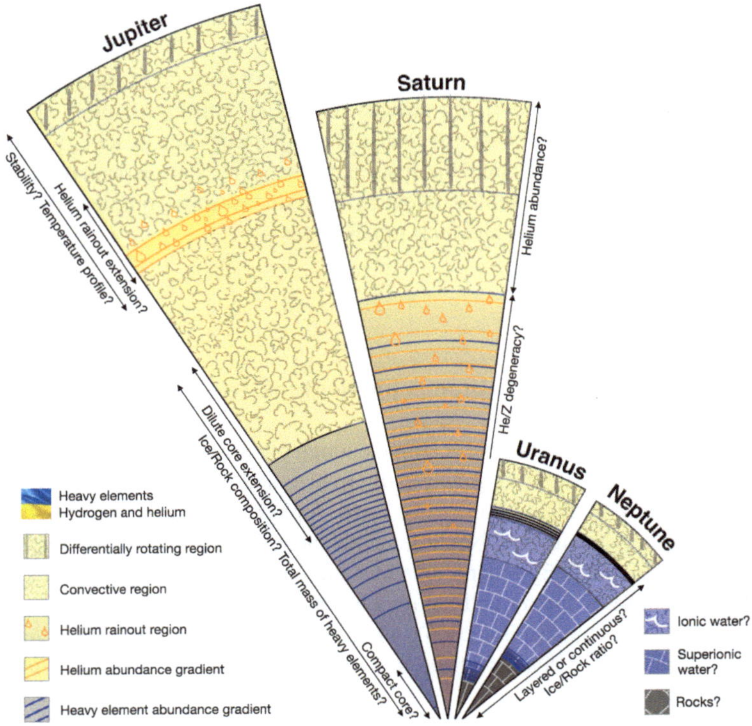

Fig. 2 Slices of the internal structures of Jupiter, Saturn, Uranus and Neptune highlighting compositional gradients in helium (orange) and heavy elements (blue). Jupiter and Saturn are characterized by a phase separation of helium in metallic hydrogen starting near Mbar pressures. *Juno* and *Cassini* measurements also indicate the presence of a dilute heavy element core in these planets, implying that heavy elements are partially mixed with hydrogen and helium rather forming separate pure layers. In this diagram, each layer corresponds to a 2% increase of the helium/hydrogen ratio (orange) or of the heavy element mass fraction Z (blue). For Uranus and Neptune, a structure dominated by well-separated layers, including a solid superionic water layer, is presented. However, solutions with compositions that evolve more continuously are also highly plausible. Image reproduced with permission from Guillot et al. (2023), copyright by ASP

For Saturn, models predict that the transition from helium-poor to helium-rich region occurs at much larger depths, and combines with a progressive increase in heavy elements (Mankovich and Fuller 2021). Models of Uranus and Neptune have considerable uncertainties (Nettelmann et al. 2013; Helled et al. 2020) and point to a small, 0.5 to 4 $M_\oplus$ fluid hydrogen-helium envelope overlaying a core made of ices and rocks which could either be layered or mixed, possibly including hydrogen and helium as well. In these models, the envelopes are believed to be largely convective and of uniform composition, except for Jupiter and Saturn, in a region in which helium separates from hydrogen and rains out (e.g., Schöttler and Redmer 2018; Brygoo et al. 2021). The cores, however, may be highly stratified and exhibit large composition gradients with depth (see Sect. 2.3).

Although hydrogen and helium are definitely fluid, the state of other elements is more difficult to assess. In Jupiter and Saturn, the temperature profile lies above the estimated solidification curves for most elements shown in Fig. 3, except possibly iron and magnesium oxide (Mazevet et al. 2019; Guillot et al. 2023). Even in this case, these elements are believed to be soluble in metallic hydrogen (González-Cataldo et al. 2014). Whether a solid compact core remains will depend on the availability of energy to mix elements upward (Guillot et al.

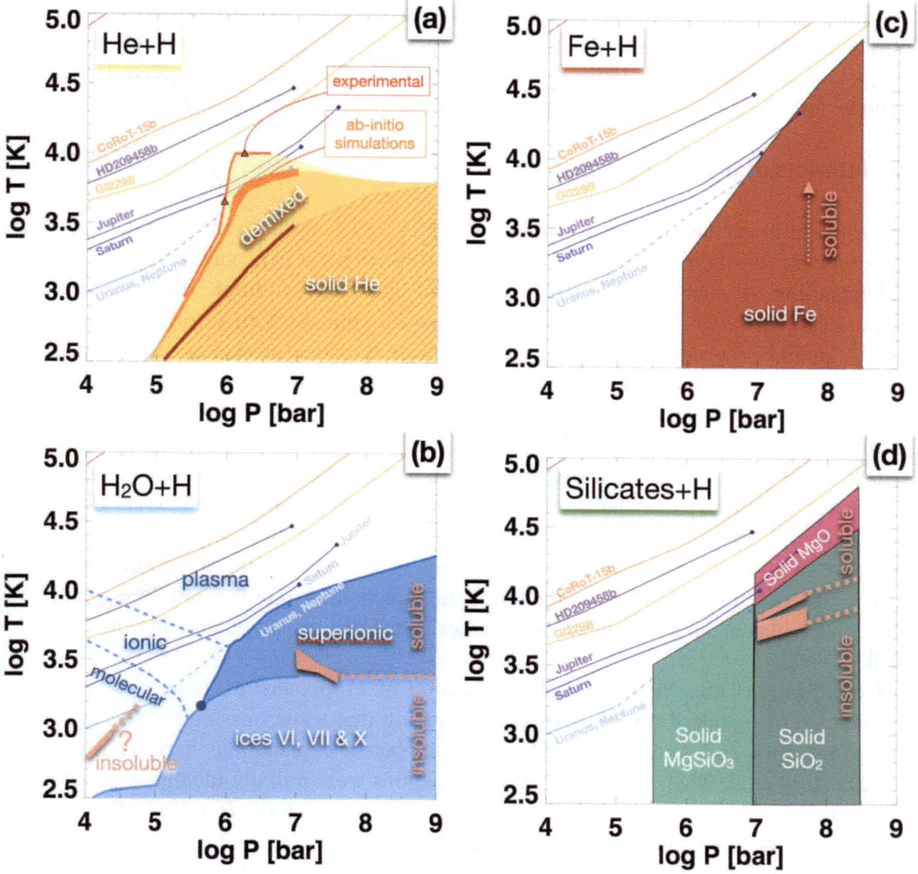

Fig. 3 Phase diagrams of key elements with pressure-temperature profiles of relevant astrophysical objects (the Sun, brown dwarfs CoRoT-15b and Gl229 B, giant exoplanet HD 209458 b, and Jupiter, Saturn, Uranus and Neptune). Pressures are in bars (1 bar = 10^5 Pa = 10^6 dyn/cm^2). (a): Location of the hydrogen-helium phase separation leading to helium rain-out in Jupiter and Saturn (yellow). The critical demixing temperatures obtained from ab-initio simulations are shown in orange. Those obtained from high-pressure experiments are shown as two triangles connected by a red curve. (b): Phase diagram of H_2O, including the fluid molecular, ionic and plasma phases, the solid ice phases and the superionic phase, as labelled. (c): Region of iron solidification, relevant for the central regions of Jupiter and possibly Saturn. (d): Regions of silicate solidification, relevant in the interiors of Uranus, Neptune, Saturn and Jupiter. Note that $MgSiO_3$ decomposes at high pressures into MgO (which should be solid in Jupiter and Saturn) and SiO_2 (which may still be liquid in the deep interior of Jupiter and Saturn). Image reproduced with permission from Guillot et al. (2023), copyright by ASP; see references therein

2004; Vazan et al. 2018) and on how high-pressure elemental interactions between different species affect phase diagrams.

In Uranus and Neptune, gravitational moments indicate that the hydrogen-helium envelope should transition to a denser fluid (Helled et al. 2020). For models assuming a three layer structure (hydrogen-helium overlying ices overlying rocks), water in the middle layer forms an ionic fluid which transitions to being superionic at Mbar pressures (French et al. 2009; Redmer et al. 2011, see Fig. 3). Ab-initio calculations indicate that superionic water should behave as a solid (Millot et al. 2019), with strong consequences for the planetary evolution (Stixrude et al. 2021). This situation corresponds to the structure shown in Fig. 2. However, in the case of a non-layered structure, the mixing of other elements may affect the picture, perhaps considerably (Guarguaglini et al. 2019), so that whether Uranus and Neptune are partially solid remains an open question (see Guillot et al. 2023, for details).

2.2 Interiors of Giant Exoplanets

So far, the primary measurements that we can rely on to constrain the global composition of exoplanets are their size and mass, coupled to theoretical expectations drawn from our understanding of solar system giant planets and planetary evolution. The known sample of giant exoplanets is extremely diverse, and includes planets with surprisingly large radii which cannot be reproduced by conventional evolution models as well as very dense planets requiring the presence of tens to hundreds of Earth masses of heavy elements in their interior (e.g., Guillot et al. 2006; Thorngren et al. 2016).

Many of the giant exoplanets known today orbit much closer to their parent star than planets in our solar system. Because of observational biases, some of the planets that can be best characterized (in particular from transit photometry and radial velocimetry) also endure extremely high level of irradiation, orders of magnitude higher than Jupiter. This implies the existence of an outer radiative region that grows as the planet cools and contracts (Guillot et al. 1996), potentially to kbar depths for mature hot Jupiters (Guillot and Showman 2002). Thus, these planets could be made of an inner convective envelope and an outer radiative shell.

Complications arise however: a large fraction of exoplanets are bloated beyond what is predicted by standard evolution models, implying that some missing physics is slowing their contraction (see Guillot et al. 2006). A number of explanations have been proposed (e.g., Fortney et al. 2021). Although tidal interactions were initially identified as a possible explanation (Bodenheimer et al. 2001; Guillot and Showman 2002; Baraffe 2005), the small eccentricities of hot Jupiters indicate tidal heating could only account for a small fraction of the anomalously large planets (Fortney et al. 2021). More likely mechanisms such as Ohmic dissipation or downward energy transport (Batygin and Stevenson 2010; Fortney et al. 2021, and references therein) alter the simple picture of a thick radiative shell. It could be much narrower (Sarkis et al. 2021), or more generally, several radiative/convective zones could be present. In the deeper interior, the probable existence of compositional gradients would also affect static stability in the interior, as well as the cooling of the giant planets themselves (Chabrier and Baraffe 2007).

Finally, giant exoplanets which are more irradiated than those in the solar system (i.e., most of the ones which are currently observed with large instruments) should be warmer and therefore even less likely of being partially solid in their interior. This is particularly obvious when considering the case of HD 209458 b in Fig. 3.

2.3 Static Stability in Giant Planets

While simple models of giant planets are fully convective and adiabatically stratified, several signs point toward stable (non-adiabatic) stratification due to compositional gradients at large depths. The existence of a largely stable deep interior in Saturn had been previously proposed in order to account for Saturn's high luminosity relative to predictions of adiabatic models (e.g., Leconte and Chabrier 2012, 2013). Fuller (2014) also appealed to stable stratification in order to account for the splitting of normal modes as detected in the rings by Cassini (Hedman and Nicholson 2013). The mode splitting is interpreted as the interaction between f-modes present in the whole planet and g-modes which are present only in stably stratified regions. The presence of a dilute core both in Jupiter and in Saturn (see Fig. 2) seems to indicate that the existence of such a stable region may be the norm rather than the exception. However, for giant planets more massive than Jupiter, more energy is available to mix any primordial core more efficiently (see Guillot et al. 2004), raising the possibility that these massive giant planets might have fully eroded their core and be fully convective, erasing any primordial composition gradients.

The question of stability also arises in other contexts: in the hydrogen-helium phase separation region (see Fig. 3), the formation of helium-rich droplets may or may not lead to the inhibition of convection (Stevenson and Salpeter 1977; Mankovich et al. 2016). In the presence of compositional gradients, double-diffusive convection may arise, possibly leading to a small-scale staircase structure of convective regions sandwiched between conductive interfaces (Rosenblum et al. 2011; Leconte and Chabrier 2012). However, double-diffusive staircases may not always form for realistic conditions in giant planets (Rosenblum et al. 2011; Mirouh et al. 2012; Fuentes et al. 2022), and they may become unstable over long time scales (Moore and Garaud 2016).

In Jupiter, the existence of stable regions deep in the molecular envelope (between 95% and 80% of the planet's total radius) have been invoked to account for the inferred zonal wind decay (Christensen et al. 2020) and measured magnetic field properties (Moore et al. 2022; Connerney et al. 2022; Sharan et al. 2022). But even features in the convective part of the atmosphere require that the atmosphere is at least partially stable. This is shown in Fig. 4 by measurements of the Brünt-Väisälä frequency, N, defined by

$$N^2 = \frac{g}{T}\left(\frac{dT}{dz} + \Gamma_{\rm ad}\right), \tag{1}$$

where g is the gravitational acceleration, T temperature, z altitude, $\Gamma_{\rm ad} \equiv -(dT/dz)_S$ is the dry adiabatic lapse rate (at constant entropy S), and a perfect gas of uniform composition has been assumed for simplicity (Wong et al. 2011). Thus, $N^2 > 0$ only in stable regions. The values of N shown in Fig. 4 directly measured range from high values of around 0.02 s^{-1}

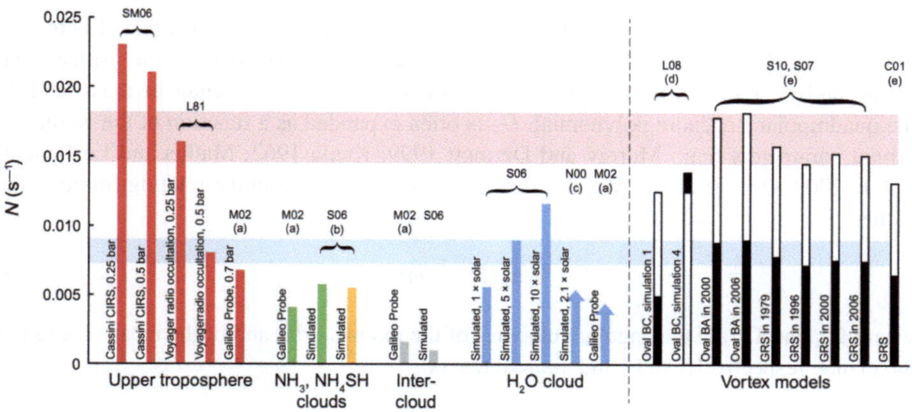

Fig. 4 Estimates of static stability in Jupiter's atmosphere as measured from its Brünt-Väisälä frequency N. Shaded bars (to the left of the dashed line) are from measurements and models of Jupiter's vertical structure; black and white bars (to the right of the dashed line) are for vortex surroundings as determined by numerical models of the Great Red Spot, and of Jupiter's anticyclones. Image reproduced with permission from Wong et al. (2011), copyright by Elsevier

in the radiative upper atmosphere, to about $2 \times 10^{-3}\ \mathrm{s}^{-1}$ in the troposphere near 10 bars, as measured by the Galileo probe. Intermediate values are required to model the observed vortexes. In these regions, latent heat release due to the condensation of H_2O, NH_4SH and NH_3 can lead to subadiabatic temperature gradients accounting for the observed static stability. However, the penetration of cyclones and anticyclones to great depths as seen by *Juno* (Bolton et al. 2021; Guillot et al. 2023, and references therein) indicates that the stable region is not limited solely to the water condensation zone near 5 bar.

3 Tidal Dissipation Mechanisms

3.1 Tides: General Principles

3.1.1 Tidal Force and Potential

Gaseous giant planets of our solar system have strong tidal interactions with their numerous moons, with tides raised on the moons by the host planet, and vice versa. Tides are for instance responsible for the volcanism of Io (Johnson et al. 1984; Ojakangas and Stevenson 1986) and for the heating of Enceladus (Lainey et al. 2012). The same is hypothesized to occur between close-in giant exoplanets and their host stars (e.g. Winn and Fabrycky 2015, and references therein). In this section, we focus on tidal interactions with moons, but many of the same processes could happen in gas giant exoplanet systems where the tidal force is produced by the planet's host star.

Giant planets (hereafter the primary) are large, so the gravitational force exerted by the companions (e.g., a moon, hereafter the secondary) varies non-negligibly between the center and the surface. The resulting differential force is the tidal force ($\mathbf{f}_\mathrm{T}$), whose dominant quadrupolar component is derived from the tidal potential (U_T) as:

$$\mathbf{f}_\mathrm{T} = -\boldsymbol{\nabla} U_\mathrm{T} \quad \text{with} \quad U_\mathrm{T} = -\frac{G M_\mathrm{m} r^2}{d^3} P_2 \left[\frac{\mathbf{d} \cdot \mathbf{r}}{d\, r}\right], \tag{2}$$

where G is the gravitational constant, M_m the mass of the companion (treated here as a point mass), d ($\mathbf{d}$) the distance (vector) between the centres of mass of the primary and the secondary, $\mathbf{r}$ the position vector in the primary relative to its center of mass, and P_2 the quadrupolar Legendre polynomial. U_T is often expanded as a function of the Keplerian orbital parameters (e.g., Murray and Dermott 1999; Kaula 1962; Mathis and Le Poncin-Lafitte 2009; Ogilvie 2014). We also define the tidal frequency in the rotating frame of the planet:

$$\omega \equiv l\Omega_\mathrm{m} - m\Omega_\mathrm{p}, \tag{3}$$

where Ω_m and Ω_p are the angular velocities of the moon's orbit and of the mean rotation of the planet, respectively, and l and m are integers.

3.1.2 Equilibrium and Dynamical Tides and Their Dissipation

The perturbation of the hydrostatic balance of celestial bodies by tidal forces results in mass redistribution and gravity and pressure perturbations. This mass redistribution creates a tidal bulge approximately in the direction of the companion (Fig. 1) and a large-scale velocity or

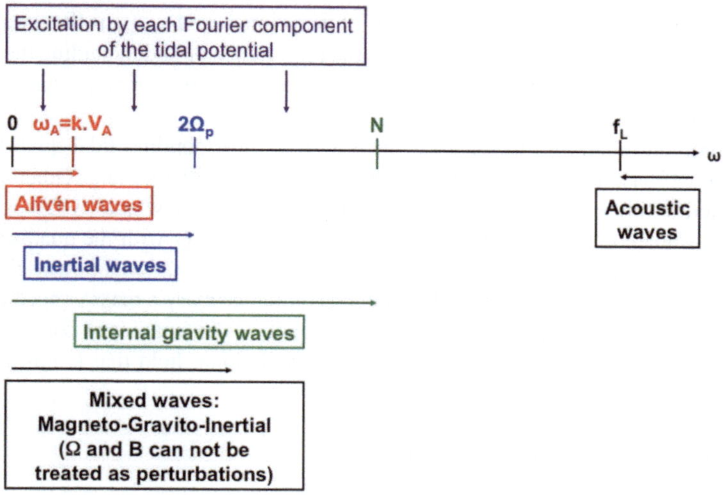

Fig. 5 Type of waves constituting the dynamical tide in fluid planetary layers. We introduce their characteristic frequencies: the Alfvén frequency $\omega_{\rm A}$, where $\vec{k}$ is the wave vector and $\vec{V}_{\rm A} = \vec{B}/\sqrt{\mu_0 \rho}$ is the Alfvén velocity ($\vec{B}$ being the magnetic field, ρ the density, μ_0 the vacuum permeability), the inertial frequency $2\Omega_{\rm p}$ ($\Omega_{\rm p}$ being the angular velocity of the planet), the Brünt-Väisälä frequency N defined in Eq. (1), and the Lamb frequency $f_{\rm L}$. For tidally excited waves in stably stratified regions of giant planets, we typically expect $\omega_{\rm A} < 2\Omega_{\rm p} < N < f_{\rm L}$, but this may not always be true. Image adapted from Mathis and de Brye (2011)

an elastic displacement in fluid/solid layers, respectively, the so-called fluid/elastic equilibrium tide (e.g. Zahn 1966; Remus et al. 2012; Love 1911; Tobie et al. 2005; Remus et al. 2012). Because the equilibrium tide is not a solution of the full momentum equation, it is completed by the so-called dynamical tide. In fluid bodies, the dynamical tide is composed of inertial, gravity, Alfvén, and acoustic waves excited by the tidal force (e.g. Mathis et al. 2013). Their restoring forces are the Coriolis acceleration, buoyancy (in convectively stable layers or at the free surface), magnetic tension, and pressure, respectively (e.g. Rieutord 2015, and Fig. 5). In solid bodies, the dynamical tide includes tidally-excited shear waves (e.g. Alterman et al. 1959; Tobie et al. 2005).

In an ideal case without any friction in the primary, the equilibrium and dynamical tides would be in phase with the tidal potential, with the resulting tidal bulges (one per Fourier mode of $U_{\rm T}$) being aligned along the line of centres. In this adiabatic situation, the time-averaged tidal torques on the orbit would vanish (e.g. Zahn 1966). However, the equilibrium and dynamical tides are submitted to dissipation because of viscous or turbulent friction, the diffusion of heat and chemicals, Ohmic heating from magnetic fields, and viscoelastic dissipation in solid regions. Because of the induced losses of energy, the direction of the tidal bulge lags behind the secondary's position (Fig. 1), resulting in a tidal torque. In addition, the conversion of the kinetic and potential energies of tides into heat impacts the structure and the evolution of the primary.

In the simplest case of a circular co-planar system with semi-major axis $a_{\rm m}$, the torque applied to the spin of the primary can be approximated as (Zahn 2013)

$$\Gamma = -\frac{(\Omega_{\rm p} - \Omega_{\rm m})}{t_{\rm friction}} \left(\frac{M_{\rm m}}{M_{\rm p}}\right)^2 M_{\rm p} R_{\rm p}^2 \left(\frac{R_{\rm p}}{a_{\rm m}}\right)^6, \tag{4}$$

where M_p and R_p are the mass and the radius of the primary and $t_{friction}$ is the time characterising tidal dissipation. This expression shows the importance of understanding the physics of tidal dissipation from first principles to obtain a robust evaluation of $t_{friction}$ and a prediction of the evolution of the system.

It is common to express $t_{friction}$ as a function of a corresponding tidal quality factor Q (MacDonald 1964), in analogy with the theory of forced oscillators (e.g. Greenberg 2009). A short friction time corresponds to a strong dissipation and a low quality factor (and vice-versa). To evaluate tidal dissipation, we compute the ratio of Q with the quadrupolar Love number k_2, defined as the ratio of the primary's perturbed gravitational potential to the tidal potential at the primary's surface, which depends on the primary's mass concentration. The tidal torque is proportional to k_2/Q (which is equal to the imaginary component of k_2, $\mathrm{Im}[k_2]$), i.e., the component of the tidal bulge's gravitational field that is misaligned with respect to the moon.

3.1.3 Tidal Orbital and Rotational Evolution

For moon orbits that are nearly circular and aligned with the planet's rotation (as is the case for most of the solar system's major moons), tidal evolution is relatively simple. In Eq. (4), one can identify the crucial importance of the so-called co-rotation radius at which $\Omega_p = \Omega_m$. If $\Omega_m > \Omega_p$, then the companion migrates inwards and the primary rotation accelerates. If $\Omega_m < \Omega_p$, the companion migrates outwards and the primary rotation slows down. The latter is the configuration of the Earth-Moon system where the Moon migrates outwards by $\sim$3.8 centimetres per year. When $\Omega_m > \Omega_p$, the Darwin instability results if $L_{orb} \leq 3L_p$, where L_{orb} and L_p are the angular momentum of the orbit and of the primary's rotation respectively. In this case, the companion spirals towards the primary until it is tidally disrupted at the Roche limit (Hut 1980). If $L_{orb} \geq 3L_p$, the system evolves towards an equilibrium state where the orbit is circular and the primary's rotation is aligned and synchronized with the orbit.

We can see that the time characterising dissipation ($t_{friction}$) allows one to predict orbital migration and circularisation time scales, and spin alignment and synchronisation time scales. Each dissipation mechanism leads to a specific frequency-dependence of $t_{friction}$ (i.e. smooth or resonant; we refer to Mathis et al. 2013; Ogilvie 2014, for mathematical formalisms), with very sensitive frequency-dependence for resonant fluid dynamical tides. This has major consequences for the dynamical evolution of planetary systems (Zahn and Bouchet 1989; Witte and Savonije 2002; Efroimsky and Lainey 2007; Auclair-Desrotour et al. 2014).

3.2 Dissipation in the Solid Core

In the standard core-accretion scenario for the formation of gaseous giant planets, a rocky/icy core is needed to ensure the accretion and the stability of a deep gaseous envelope (Pollack et al. 1996). If it solidifies, this core constitutes a potentially strong source of tidal dissipation, because of the friction applied through its rheology on the solid equilibrium tide (e.g. Tobie et al. 2005; Efroimsky and Makarov 2013; Tobie et al. 2019, and Fig. 6 left panel), and a boundary for tidal flows studied in the next section.

In this framework, Dermott (1979) studied the equilibrium tide in homogeneous solid cores. He showed how its amplitude is boosted thanks to the hydrostatic pressure exerted by a friction-less homogeneous deep gaseous envelope when compared to the case of a sole solid core without any surrounding fluid layer. Remus et al. (2012, 2015) built on this first

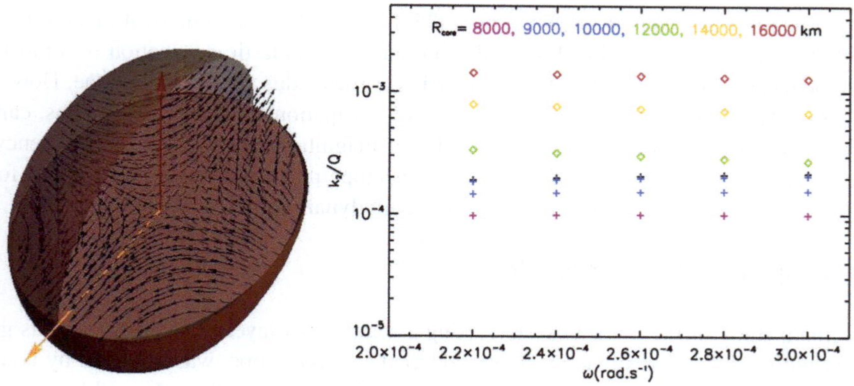

Fig. 6 *Left*: elastic equilibrium tide in the rocky/icy core of a giant planet. The red and orange arrows are the rotation axis and the direction of the companion respectively. *Right*: Tidal dissipation in the solid core of Saturn, computed for different values of its radius as a function of the tidal frequency, taking into account its radial density stratification, for a viscosity $\eta = 10^{15}$ Pa s. The dissipation is expressed as a function of the second-order Love number k_2, which evaluates the amplitude of the hydrostatic tidal deformation, and the tidal quality factor Q. Images reproduced from [left] Remus et al. (2012), copyright by ESO; and [right] Lainey et al. (2017), copyright by Elsevier

work and provided a coherent model of the dissipation of solid tides in a simplified bi-layer model with an homogeneous viscoelastic solid core surrounded by a deep homogeneous non-dissipative fluid envelope. Using complex Love numbers and the correspondence principle (e.g. Tobie et al. 2005), they computed the elastic deformation of solid layers and the viscous dissipation in its bulk. Because of our lack of knowledge of the rheology of the core of giant planets in their extreme conditions of pressure and temperature, they considered the case of the simplest Maxwell rheology and they explored plausible values of the core's rigidity and viscosity.

These studies showed that friction can simultaneously explain the high amplitude of the tidal dissipation observed for Jupiter and Saturn by Lainey et al. (2009) and Lainey et al. (2012) and its smooth frequency-dependence in the Saturnian system for Enceladus, Tethys and Dione. The moon measurements, made using high-precision ground-based astrometric measurements of the orbital migration, seemed to favour at that time the hypothesis of the dissipation of a solid tide. Indeed, the dissipation of tidal waves propagating in the gaseous envelope would have a resonant frequency-dependence with possible sharp variations of Q by several orders of magnitude (Ogilvie and Lin 2004; Mathis et al. 2016).

The measured values of Q were an order of magnitude smaller than what was predicted assuming a constant value of Q and simultaneous formation of the planets and of their satellites. This opened the possibility of having young moons as proposed in Charnoz et al. (2011) and in Crida and Charnoz (2012). However, Shoji and Hussmann (2017) took into account the frequency-dependence of the tidal dissipation predicted in the case of a visco-elastic core when they solved the equations for the orbital migration of the moons. They showed that a strong dissipation in Saturn does not necessarily imply young moons. Storch and Lai (2014) also studied this mechanism to provide us an explanation of the formation of hot Jupiters through circularization in a high-eccentricity migration scenario. Finally, Storch and Lai (2015) improved this bi-layer model by considering a set-up with an homogeneous core surrounded by a polytropic gas envelope.

These works were generalized in Lainey et al. (2017) by taking into account the radial profiles of the density and of the incompressibility parameter of the assumed Maxwell rheology (we refer the reader to Tobie et al. 2005, for details of the numerical modelling). As illustrated in the right panel of Fig. 6, the computed viscoelastic dissipation is compatible with the measured dissipation at the frequencies of Enceladus, Tethys and Dione. However, this viscoelastic dissipation, if it is the source of dissipation for these three moons, cannot explain the stronger dissipation by one order of magnitude observed at the frequency of Rhea. Additional dissipation within the fluid envelope must be invoked: we have to focus our efforts on modelling the dissipation of the fluid dynamical tide.

3.3 Dissipation in the Fluid Envelope

A seminal study of the dynamical tide, using a standard tri-layer model of a gaseous giant planet with a deep convective envelope modelled as a polytrope, was achieved by Ogilvie and Lin (2004). Assuming the presence of a central solid core, they showed how tidally-excited inertial waves can be efficiently dissipated assuming an effective (turbulent) viscous friction. The restoring force governing the dynamics of these waves is the Coriolis acceleration. The dissipation is primary localised in intense shear layers called wave attractors launched at the critical co-latitude $\theta_c = \arccos\left(\omega/2\Omega_p\right)$ (Ogilvie 2005; Goodman and Lackner 2009; Rieutord and Valdettaro 2010). It varies by several orders of magnitude with the tidal frequency and the value of the effective viscosity. The resonances in the dissipation frequency spectrum are stronger and sharper for low viscosity.

In this framework, Mathis et al. (2016) studied the efficiency of the turbulent friction applied on tidal inertial waves by rapidly rotating convection. Using scaling laws of the rotating mixing-length theory derived by Stevenson (1979), which have been confirmed in recent numerical simulations (Barker et al. 2014; Currie et al. 2020; Vasil et al. 2021; Fuentes et al. 2023; de Vries et al. 2023), they showed that rapid rotation leads to less efficient turbulent viscous friction applied to tidally-excited inertial modes. The associated sharper and higher resonant peaks in the dissipation frequency-spectrum (e.g. Auclair Desrotour et al. 2015) may provide a possible explanation for rapid moon migration as observed in the case of Rhea.

The efficiency of inertial wave dissipation also depends on the structure of the planet. In simple planetary models with a uniform density envelope surrounding a solid core, the frequency-average inertial wave dissipation rate scales approximately as the core radius to the fifth power (Ogilvie 2013). For more realistic planetary models with barotropic structures, significant inertial wave dissipation can occur even in the absence of a solid core (Ogilvie 2013; Lazovik et al. 2024). Inertial wave dissipation has also been investigated for highly eccentric orbits applicable to exoplanet systems (Ivanov and Papaloizou 2004, 2007; Papaloizou and Ivanov 2010). These studies investigate the more complex dynamics that can occur in the case of eccentric/misaligned orbits.

In the case where a solid (denser) core was assumed, it is also important to compare the relative strengths of the dissipation of tidal inertial waves and of the viscoelastic dissipation. This was achieved by Guenel et al. (2014), who considered a simplified bi-layer model with a homogeneous solid core and a deep convective envelope with tidally excited inertial modes. Using the method developed by Ogilvie (2013), they computed the frequency-averaged values of both types of dissipation. They showed that fluid and solid tidal friction might have comparable strength when assuming plausible values for the core rigidity. This demonstrated the importance of computing dissipation in each fluid (gaseous or liquid) and solid planetary layer. In the case of a fully convective core-less gaseous giant planet, Wu (2005a,b) demonstrated that the dissipation of tidal inertial modes will be weakened, as in the case where a solid core is replaced by a denser motionless fluid core (Mathis 2015).

Our picture of the tidal dissipation mechanism has been greatly changed by the revolution of our understanding of gaseous giant planets interiors thanks to the *Juno* and *Cassini* space missions as explained in Sect. 2. Indeed, the newly obtained constraints show the importance of possible stably stratified regions within giant planets. As illustrated in Fig. 2, state-of-the-art models of the internal structure of giant planets are all constituted by a combination of convective and stably-stratified regions. These convectively stable layers are triggered by heavy element and helium abundance gradients, which stabilise the entropy gradient.

Stably stratified layers allow for dissipation of tidally excited gravito-inertial waves which propagate within these layers. The restoring forces that govern their dynamics are

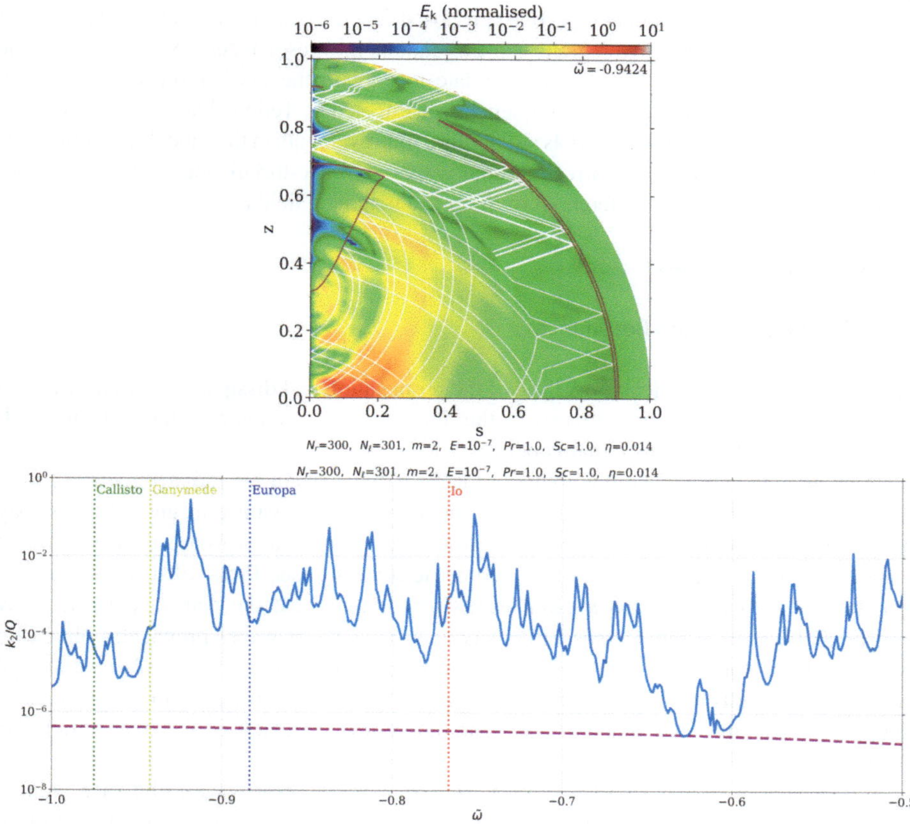

Fig. 7 *Top*: Kinetic energy density of tidal waves propagating in the interior of Jupiter at the tidal frequency of Ganymede. We use a model of the internal structure of the planet computed by Debras and Chabrier 2019 with (from core to surface) a dilute stably stratified core, a convective metallic hydrogen shell, a very thin stably stratified layer induced by the immiscibility of Helium into Hydrogen, and an external convective molecular envelope. We recognise the propagation of gravito-inertial waves in the central stable core and the one of inertial waves in convective layers. We assume a solid core with a radius $R_c = 0.014 R_p$ with R_p the radius of the planet. The ratio of the (turbulent) viscous force to the Coriolis acceleration (the Ekman number E) is 10^{-7} and viscous, heat and chemical diffusions are equal (i.e. the Prandtl (P_r) and the Schmidt (S_c) numbers are set to unity). Image reproduced from Dhouib et al. (2024). *Bottom*: The value of k_2/Q as a function of the tidal frequency corresponding to the dissipation of tidal waves as computed in the left panel (blue solid line). The purple line corresponds to the value obtained in the case of tidal inertial waves computed in a sole deep convective envelope surrounding the small core as this was expected before the constraints obtained by the *Juno* mission

buoyancy and the Coriolis acceleration. Ioannou and Lindzen (1993a,b) were pioneers in envisioning the possible importance of these waves to predict tidal dissipation within Jupiter. The development of a new generation of models of the tidal dissipation in giant planets, with the dissipation of tidal inertial waves in convective zones and of tidal gravito-inertial waves in stable layers has been undertaken first in local Cartesian geometry (André et al. 2017, 2019) and now in global spherical (Pontin et al. 2020, 2023; Lin 2023; Lazovik et al. 2024; Dhouib et al. 2024, Pontin et al. submitted) and spheroidal geometries (Dewberry 2023). These models show how gravito-inertial waves within stably stratified layers enhance the tidal dissipation when compared to the case of a "standard" fully convective envelope (see Fig. 7). They also show that a stably stratified layer increases the amplitude of inertial wave excitation and associated dissipation within convective regions.

Fuller et al. (2016), Luan et al. (2018) proposed that the gravito-inertial modes in either stably stratified or convective regions can trap moons into resonances (Witte and Savonije 2002). This mechanism is a natural candidate to explain the dissipation rates measured by astrometric observations at the frequencies of Enceladus, Tethys, Dione, and Rhea (see Sect. 4). However, all of these models still require improvements in the modelling of possible nonlinearities and instabilities affecting tidal waves, of wave-turbulence nonlinear interactions, and of the impact of differential rotation and magnetic fields.

3.4 Current Challenges

3.4.1 Nonlinearities and Resonances

As of today, hydrodynamical modelling of the propagation and dissipation of tidal waves in giant planets has been done primarily in the linear regime. As a consequence, many challenging important questions remain open.

First, as in the case of high-amplitude tidal gravity waves in the radiative cores of low-mass stars (e.g. Barker and Ogilvie 2010), tidal waves propagating in giant planets may break. This may be the case for gravito-inertial waves propagating towards the center of the planet in absence of a core because of geometric focusing (see Press 1981, in the case of gravity waves). Tidal (gravito-)inertial waves propagating near the planetary surface may also break because of the decreasing density as in the case of waves propagating in early-type stars (e.g. Rogers et al. 2013). Inertial waves have very large wave numbers at critical latitudes, potentially driving non-linear wave breaking. Even when waves do not break, nonlinear wave coupling may increase wave damping rates (Weinberg et al. 2012; Essick and Weinberg 2016).

Second, future studies should examine the complex nonlinear wave-turbulence interactions that mediate dissipation. The key questions to answer are: can we model the action of the convective turbulence on tidal inertial waves as an effective turbulent viscous friction, and what is its strength (see, e.g., Terquem 2021; Barker and Astoul 2021)? How do the attractors predicted in the linear regime evolve in the presence of turbulent convection or in a nonlinear regime? Do they survive or are they destroyed? Are we entering in the so-called wave turbulence regime? How do tidal inertial waves interact with the numerous convective vortices observed at the surface of giant planets with a broad diversity of characteristic scales?

Studies have begun to examine those crucial questions (e.g. Jouve and Ogilvie 2014; Le Reun et al. 2017; Astoul and Barker 2022; Dandoy et al. 2023; Astoul and Barker 2023; Terquem 2023) but this research area in the specific framework of the prediction of tidal dissipation is still in its infancy. In the case where the modelling of the turbulent friction can be done using an effective eddy viscosity, important progress has been made in predicting its strength and frequency-dependence thanks to systematic studies using direct nonlinear numerical simulations of the interactions between turbulent convection and oscillatory flows (Duguid et al. 2020a,b; Vidal and Barker 2020a,b; Barker and Astoul 2021), and more realistic tidal flows (de Vries et al. 2023).

When tidal deformation is large (which can occur in hot Jupiters but not in solar system planets), global-scale tidal flows can become unstable because of the elliptic instability (Kerswell 2002). In this case, the oscillatory tidal flow non-linearly excites a pair of gravity waves. This triggers the instability, turbulence, and dissipation (for a complete review, we refer the reader to Le Bars et al. 2015, and references therein). Using hydrodynamical simulations, Barker and Lithwick (2013) and Barker (2016) have identified the important role of the differential rotation triggered by the instability, which acts to saturate it. This leads to an intermittent dissipation. In the presence of a weak magnetic field, differential rotation is damped and turbulence is sustained that triggers a stronger dissipation (Barker and Lithwick 2014). The elliptic instability may be most important in exoplanet systems where the tidal distortion from the host star can be much larger than that produced by moons.

3.4.2 Zonal Flows and Magnetic Fields

The observations provided by *Juno* and *Cassini* (during its "grand finale") have revealed that the convective molecular envelopes of Jupiter and Saturn are the seat of the differential rotation observed at their surface (Guillot et al. 2018; Galanti et al. 2019). Below $\sim$3000 km for Jupiter and $\sim$8000 km for Saturn, the differential rotation weakens as the electrical conductivity increases with depth. That means that the impact of zonal flows on tidal (gravito-)inertial waves excited in these external layers must be taken into account.

First, differential rotation can modify the cavities where (gravito-)inertial waves propagate and also change their frequencies (Mathis 2009; Baruteau and Rieutord 2013; Guenel et al. 2016; Mirouh et al. 2016). Next, critical layers appear where the local Doppler-shifted frequency of the fluid oscillation vanishes. They are of crucial importance since they are the place of strong wave-mean zonal flows interactions that can constitute the dominant component of the tidal torque both in stably-stratified (Goldreich and Nicholson 1989) and in convective regions (e.g. Favier et al. 2014; Guenel et al. 2016; Astoul et al. 2021; Astoul and Barker 2022, Baruteau et al., in prep). Finally, an interesting and necessary perspective will be to compare the relative strengths of zonal flows induced by tidal waves (e.g. Morize et al. 2010; Cébron et al. 2021) and those triggered by the turbulent convection.

Gaseous giant planets are also magnetic celestial bodies (Connerney et al. 2018; Duarte et al. 2018; Dougherty et al. 2018; Yadav et al. 2022). As a consequence, it will be necessary to consider the impact of magnetic fields on tidal waves (e.g. Wei 2016; Lin and Ogilvie 2018; Wei 2018). Tidal (gravito-)inertial waves become tidal magneto-(gravito-)inertial waves while the Ohmic heating related to their dissipation adds to the viscous and heat diffusion. Finally, it will be important to study carefully the tidally-induced modification of planetary magnetic fields (e.g. Cébron and Hollerbach 2014).

4 Resonance Locking

As discussed above, the tidal torque due to dynamical tides can easily dominate over equilibrium tides, and is usually very sensitive to the tidal forcing frequency. Unfortunately, our understanding of giant planet structures is not detailed enough to accurately calculate

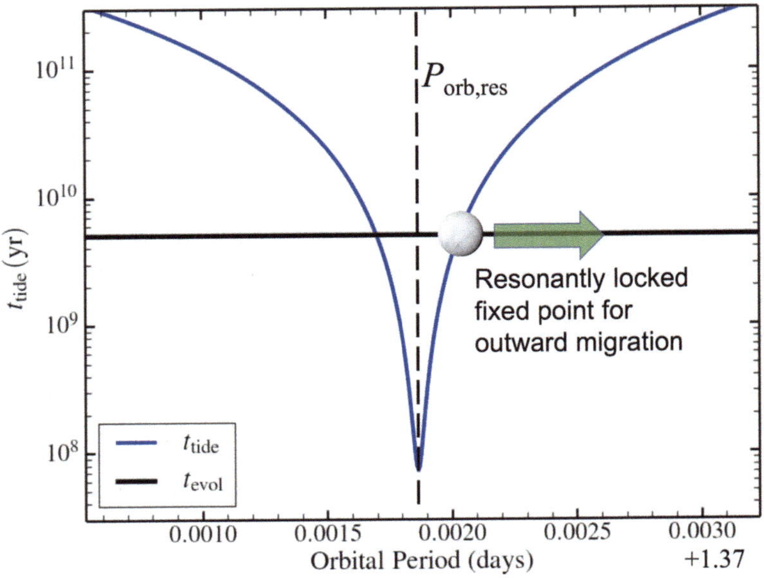

Fig. 8 Example of the tidal migration time scale (blue line) as a function of orbital period, near a resonance with a Saturn oscillation mode. A moon with orbital period $P_{\rm orb,res}$ and semi-major axis $a_{\rm res}$ would lie at exact resonance with the oscillation mode (vertical dashed line). The horizontal black line denotes the time scale $t_{\rm evol}$ at which $P_{\rm orb,res}$ moves outward. At the resonantly locked fixed point (where the blue and black lines intersect), the moon would migrate out at the same rate as the resonance location, allowing for a stable resonance lock

the frequencies of gravito-inertial modes at which tidal dissipation peaks. Hence, we cannot determine whether the observed moons will resonate with planetary oscillation modes. A resonance occurs when the tidal forcing frequency in the inertial frame, $m\Omega_{\rm m}$, is nearly equal to one of the planetary oscillation mode frequencies, σ_α. Since the resonances are typically narrow in frequency, it is tempting to conclude that the moons are unlikely to be in resonance, and that tidal dissipation due to dynamical tides should be weak.

However, just as there are physical processes that allow moons to be stably locked into mean-motion resonances with one another, there are also processes that allow moons to be stably locked into resonances with planetary oscillations. This phenomenon is known as "tidal resonance locking", as was first described by Witte and Savonije (1999) and first considered for moons by Fuller et al. (2016). If resonance locking can act in the Saturn system, it is much more likely (perhaps even probable) that a moon will be in resonance with one of Saturn's oscillations, allowing it to migrate much faster than expected otherwise. While it is not currently clear whether resonance locking is required to explain the observed moon migration rates, it is an appealing mechanism to increase tidal dissipation rates.

Resonance locking arises from the fact that planets and stars are not static objects, but are evolving in time. Because a planet's internal structure is slowly changing, so too are the oscillation mode frequencies that can become resonantly excited by one of the moons. Since the moons migrate outward, their tidal forcing frequencies $m\Omega_{\rm m}$ decrease with time. Crucially, if the oscillation modes that resonate with the moons also have decreasing frequencies σ_α in the inertial frame (corresponding to resonant semi-major axes that move outward in time), then resonance locking can potentially occur, as illustrated in Fig. 8.

In many cases, the moon migration rates are fast at exact resonance, and slow away from resonance. Just outward of the resonant location, there is a "fixed point" at which the moon will tidally migrate at the same rate that the resonance location is moving outwards (see Fig. 8). This fixed point is stable: an inward perturbation will move the moon deeper into resonance, so that it migrates outward faster, and returns to the fixed point. An outward perturbation will move the moon out of resonance, so that it migrates slower, allowing the resonance location to catch up to the moon. Hence, moons can stably "surf" these resonance locations outwards as the planet evolves, so that their migration rates are determined by the rate at which the planetary structure is evolving. Moreover, planetary evolution may allow the resonant locations to sweep past the moons, such that moons will naturally be caught up in these resonances, regardless of when and where the moons formed.

4.1 When Resonance Locking Will Not Occur

There are several reasons why some moons may not be caught in resonance locks. First, if the damping rates of Saturn's oscillation modes are too high, the resonant migration time scales (blue line in Fig. 8) will be broader and shallower, such that the blue line no longer intercepts the black line, and the resonance locking fixed point no longer exists. When this occurs, the moon will not be able to migrate outward as fast as the resonance location is moving, so resonance locks cannot be sustained. This can also occur if Saturn evolves too quickly (i.e., the black line moves downward in Fig. 8), or if a moon has a large semi-major axis (causing the blue line to shift upward). Hence, we expect resonance locks to eventually break as the moons migrate outwards, though it is not clear when exactly this will occur.

Resonance locks also cannot occur if the resonant locations are moving inward, because then there is no resonance locking fixed point. In this scenario, the moons simply pass through the resonances on short time scales, so we do not expect them to be in resonance today.

To determine whether resonance locking can occur, we must examine the resonance condition,

$$\sigma_\alpha \simeq m\Omega_{\rm m} \,. \tag{5}$$

Differentiating the resonance criterion of equation (5) with respect to time leads to the locking criterion

$$\dot{\sigma}_\alpha \simeq m\dot{\Omega}_{\rm m} \,. \tag{6}$$

Since the moons migrate outward, $\dot{\Omega}_{\rm m}$ is negative, so the mode frequency must decrease in the inertial frame in order for resonance locking to occur.

In Saturn's rotating frame the mode frequency is

$$\omega_\alpha = \sigma_\alpha - m\Omega_{\rm p}, \tag{7}$$

which is negative for all of Saturn's moons (using $m \geq 2$), such that resonant modes have negative frequencies and are retrograde modes. Resonance locking thus requires

$$\dot{\omega}_\alpha \simeq m(\dot{\Omega}_{\rm m} - \dot{\Omega}_{\rm p}) \,. \tag{8}$$

For a constant rotation rate of Saturn, or if Saturn's rotation rate increases with time, the mode frequency ω_α must decrease in the rotating frame, i.e., become more negative and such that $|\omega_\alpha|$ increases. Increasing rotation frequency and $|\omega_\alpha|$ are what we expect if Saturn is slowly contracting, but explicit verification of this requirement has not been demonstrated with evolutionary models of Saturn.

Finally, resonance locking may not occur if non-linear effects (which are neglected in most calculations of tidally excited oscillations) become important. A tidally driven parent mode can non-linearly couple to two daughter modes, as described in detail in works such as Weinberg et al. (2012), Essick and Weinberg (2016). The transfer of energy to daughter modes (and subsequently to granddaughter modes, etc.) effectively acts as damping that may saturate the resonance and prevent the fixed point from existing, as described above. This likely prevents resonance locking from occurring in many Sun-like stars (see Ma and Fuller 2021), but non-linear coupling calculations have not been performed for tidally excited modes of giant planets. While we can be sure that resonance locks will eventually break as moons move to larger semi-major axis, it is not clear when this non-linear saturation will occur.

4.2 Consequences of Resonance Locking

If a moon becomes caught in a resonance lock, its tidal migration time scale is dictated by the evolution of the resonant mode frequency, which is determined by the structure of the planet. Hence, the moon's tidal migration time scale is directly linked to the time scale on which the planet's structure changes. For gas giant planets undergoing slow contraction, we expect this time scale to be comparable to (or somewhat larger than) the age of the solar system. Hence, to order of magnitude, we expect moon migration time scales of billions to tens of billions of years.

Given a model for the evolution of Saturn's internal structure, one can calculate the resonance locking-driven migration rates exactly. We defined the mode frequency evolution time scale $t_\alpha = \omega_\alpha / \dot{\omega}_\alpha$, and the planetary spin evolution timescale $t_\mathrm{p} = \Omega_\mathrm{p}/\dot{\Omega}_\mathrm{p}$. Equation (8) becomes

$$\dot{\Omega}_\mathrm{m} = \frac{\Omega_\mathrm{p}}{t_\mathrm{p}} + \frac{\omega_\alpha}{m t_\alpha} . \tag{9}$$

Since the moon's orbital frequency changes with semi-major axis as $\dot{a}_\mathrm{m}/a_\mathrm{m} = -(2/3)\dot{\Omega}_\mathrm{m}/\Omega_\mathrm{m}$, we have

$$\frac{\dot{a}_\mathrm{m}}{a_\mathrm{m}} = -\frac{2}{3}\left[\frac{\omega_\alpha}{m\Omega_\mathrm{m} t_\alpha} + \frac{\Omega_\mathrm{p}}{\Omega_\mathrm{m} t_\mathrm{p}}\right]. \tag{10}$$

Since ω_α is negative, we see that outward migration requires that ω_α be decreasing with time (t_α is positive) or that the planetary spin frequency be decreasing with time (t_p is negative).

For moons at large semi-major axis, resonant modes have $\omega_\alpha \simeq -m\Omega_\mathrm{p}$, in which case we obtain

$$\frac{\dot{a}_\mathrm{m}}{a_\mathrm{m}} \approx \frac{2}{3}\frac{\Omega_\mathrm{p}}{\Omega_\mathrm{m}}\left[\frac{1}{t_\alpha} - \frac{1}{t_\mathrm{p}}\right]. \tag{11}$$

Hence, if t_α is not a strong function of mode frequency, we expect outer moons (which have smaller Ω_m) to migrate faster than inner moons. This is obviously a very different prediction from viscoelastic dissipation or constant Q models. Using the definition

$$t_\mathrm{tide}^{-1} = \frac{\dot{a}_\mathrm{m}}{a_\mathrm{m}} , \tag{12}$$

we can also write

$$t_{\rm tide} \approx \frac{3}{2}\frac{\Omega_{\rm m}}{\Omega_{\rm p}}\left[\frac{1}{t_\alpha} - \frac{1}{t_{\rm p}}\right]^{-1}. \tag{13}$$

If resonance locks are driven by planetary spin evolution such that the $t_{\rm p}$ term dominates, outward migration requires that $t_p < 0$, i.e., the planet's spin frequency must be decreasing with time. Hence, if giant planets are gradually spinning up as they contract, resonance locking cannot occur unless mode frequencies evolve such that $t_\alpha < t_{\rm p}$.

Using the definition of the tidal quality factor Q, the moon migration rate is given by

$$\frac{\dot{a}_{\rm m}}{a_{\rm m}} = \frac{3k_2}{Q}\frac{M_m}{M_{\rm p}}\left(\frac{R_{\rm p}}{a_{\rm m}}\right)^5 \Omega_{\rm m}. \tag{14}$$

Equating this with equation (10) yields the effective quality factor due to a resonance lock, $Q_{\rm RL}$

$$Q_{\rm RL} = \frac{9k_2}{2}\frac{M_{\rm m}}{M_{\rm p}}\left(\frac{R}{a}\right)^5\left[\frac{\omega_\alpha}{m\Omega_{\rm m}^2 t_\alpha} - \frac{\Omega_{\rm p}}{\Omega_{\rm m}^2 t_{\rm p}}\right]^{-1}. \tag{15}$$

The effective tidal quality factor therefore depends on many parameters of the system and is expected to be different for each moon.

4.3 Flavors of Resonance Locking

Depending on the type of oscillation mode that is resonantly locked with a moon, and depending on how a planet is evolving, migration driven by resonance locking could exhibit different behaviors. Let us examine equation (13) above.

Let us start with the limit of constant planetary spin such that $t_p \to \infty$. In this case, $t_{\rm tide} \approx (3\Omega_{\rm m}/2\Omega_{\rm p})t_\alpha$. If the mode evolution time scale t_α is the same for different moons locked in resonance, then this entails that $t_{\rm tide} \propto \Omega_{\rm m}$. Outer moons with smaller orbital frequencies would migrate *faster* than inner moons, as mentioned above. This also entails that pairs of moons would cross mean-motion resonances divergently, if both moons are driven outwards by resonance locking. Obviously this cannot always be the case for the Saturn system, because Tethys and Dione are in mean-motion resonance with Mimas and Enceladus, respectively. It is possible that Mimas and Enceladus are in resonance locks with Saturn, while Tethys and Dione are not.

Another possibility is that outer moons (e.g., Rhea) were previously in a mean-motion resonance with one of the inner moons. If Rhea then encountered a resonance lock with Saturn, it could start migrating outwards faster than the inner moon, escaping the mean-motion resonance and migrating outwards as we see it today.

However, it is important to remember that the value of t_α is not the same for each oscillation mode. If we consider pure gravity modes trapped in a stably stratified region, then we expect the g mode frequencies to scale as $\omega_\alpha \propto \int N dr/r$. For a contracting planet with increasing N, we would expect the value of t_α to be positive (allowing for resonance locking) and similar for each mode. However, if modes are partially trapped in different regions of the planet, their values of t_α could be different. For instance, in models of evolving stars (Fuller et al. 2017), the value of t_α changes by a factor of several (and can even change sign) because some g modes are trapped near the core of the star, while others are trapped in the envelope. Detailed models of Saturn and its evolving structure are needed to quantify the expected values of t_α.

Another flavor of resonance locking discussed in Lainey et al. (2020) is with predominantly inertial "modes." It is well known (e.g., Ogilvie and Lin 2004) that inertial waves confined to a spherical shell can be reflected onto paths at critical latitudes where their wavenumber diverges, internal shear layers are created, and large amounts of dissipation are produced. This creates a stronger tidal torque at certain tidal forcing frequencies. While these frequencies are not normal oscillation mode frequencies, recent work (Lin and Ogilvie 2021) shows that this pattern is related to resonances with an underlying spectrum of pure global-scale inertial modes. In any case, an evolving planetary structure and/or spin rate would cause these "mode" frequencies to evolve with time, possibly allowing for resonance locking.

Unfortunately, the associated values of t_α (and even their sign) for inertial "modes" are unclear, making it difficult to predict the associated trends in resonance locking migration rates. Lainey et al. (2020) posited that inertial "mode" frequencies may evolve such that $t_\alpha \propto \Omega_{\rm m}^{-1}$. If we again let $t_{\rm p} \to \infty$, this would mean that $t_{\rm tide}$ would be nearly the same for each moon caught in a resonance lock, consistent with observations of Saturn's moons. However, this possibility needs to be investigated with detailed calculations of inertial wave dissipation in a sequence of evolutionary models, such that the realistic values of t_α can be estimated.

4.4 Long-Term Orbital Evolution

The long-term orbital evolution resulting from resonance locking is very difficult to predict. It depends on the long-term evolution of Saturn's structure and oscillation modes, whereas even the current structure is poorly understood. Moreover, the relevant oscillation modes have frequencies (in Saturn's rotating frame) $|\omega_\alpha| < 2\Omega_{\rm p}$ and hence lie in the gravito-inertial regime. Even computing the relevant oscillation modes (frequencies, eigenfunctions, and damping rates) for a single planetary model is challenging due to the non-separability of the oscillation equations and the presence of critical latitudes where inertial waves develop extremely short wavelengths. Quantitative predictions await improved planetary structure models, evolution models, and mode calculations.

Nonetheless, we can predict some general properties of tidal migration driven by resonance locking, and how it differs from constant Q or viscoelastic models. As mentioned above, the resonance locking time scale is proportional to the planetary evolution time scale, regardless of the planet's mass and orbital period. Hence, whereas constant Q models predict that higher mass moons migrate otward faster, the resonance locking migration timescale is independent of mass. Most importantly, whereas constant Q models predict a very strong dependence on semi-major axis (with migration time scale proportional to $a_{\rm m}^{13/2}$, equation (14)), the resonance lock migration time scale is only weakly dependent on semi-major axis. Hence, resonance locking typically predicts much faster migration for outer moons (and correspondingly low effective Q values) relative to constant Q models.

Figure 9 shows an example of the qualitative differences between migration driven by resonance locking vs. migration driven by constant Q models. The biggest difference is that with resonance locking, outer moons migrate relatively faster, and inner moons migrate slower. In Saturn, the consequence is that Rhea and Titan may have migrated by a large fraction of their current semi-major axes, whereas constant Q models predict almost no migration for these moons. Additionally, Mimas, Enceladus, and Tethys migrate slower and hence may be much older than inferred from constant Q models.

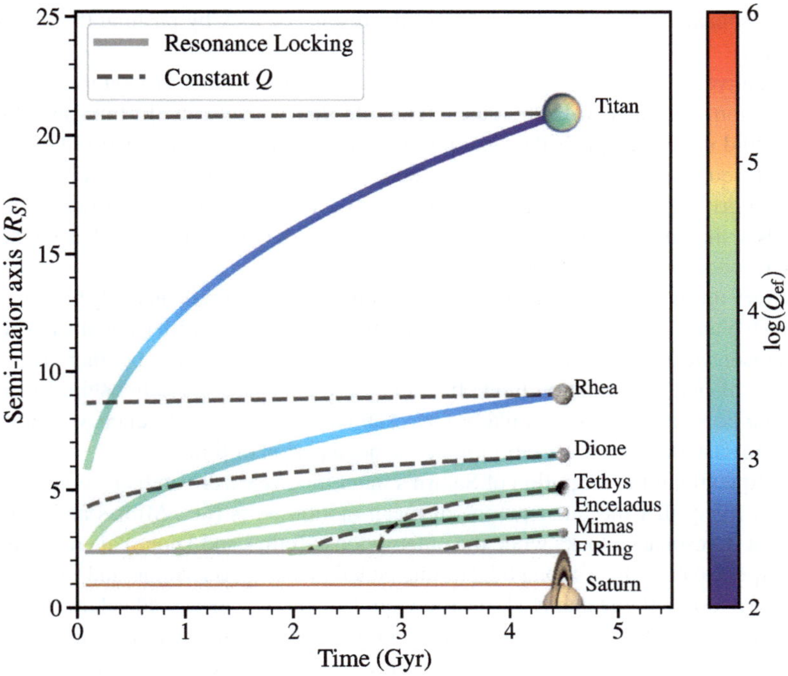

Fig. 9 An example of the differing orbital evolution between resonance locking (colored lines) and constant Q models (dashed lines). For the latter, Q values for each moon of Saturn are taken from Lainey et al. (2020) and assumed to be constant in time. For resonance locking, we use $t_{\rm tide} = 3t$, which matches measured moon migration rates fairly well, and the color of the lines corresponds to the effective quality factor (Eq. (15)). Neither evolution accounts for mean-motion resonances, which will significantly affect the orbital evolution of the inner moons. Image reproduced from Lainey et al. (2020)

However, detailed predictions are plagued by the uncertainties mentioned above. While present day migration rates and hence t_α values can be inferred, it is not clear how those values have evolved over time. Figure 9 assumes that $t_{\rm tide} \sim t$, where t is the age of Saturn, but the constant of proportionality has not been predicted from theoretical models. Current moon migration rates require $t_{\rm tide} \sim 3t$ (Lainey et al. 2020) In reality, the form of $t_{\rm tide}(t)$ and $t_\alpha(t)$ is determined by the very uncertain evolution of Saturn's structure and oscillation mode frequencies. Even the sign of t_α has not been calculated with realistic models. Extrapolation from the present day to small values of t becomes increasingly uncertain.

Another major uncertainty is whether moons pass through mean-motion resonances, and what happens when they do so. If two moons are each caught in a resonance lock, their relative migration rates are determined by the value of t_α for each resonantly locked mode, which could be quite different from one another. Hence it is certainly possible for either the inner or outer moon to migrate faster, i.e., the migration may be either convergent or divergent. In the latter case (the outer moon migrates faster), the system will simply pass through the mean-motion resonance, and each moon will presumably remain resonantly locked with its respective planetary oscillation.

In the case of convergent migration, the moons may get locked into mean motion resonance, but the details of this process are highly uncertain. At present, nobody has ever computed an orbital integration for this case, including the changing tidal torque, which is an extremely sensitive function of orbital period near resonance (Fig. 8). Hence, the mean-motion resonance could potentially disrupt the resonance lock by pushing the moon out of resonance.

Alternatively, the system may be able to adjust so that the inner moon stably locks into mean motion resonance with the outer moon, driving it outwards. In this case, the inner moon would need to be pushed deeper into resonance with the planetary oscillation, so that the tidal torque on it can be stronger to account for the extra energy and angular momentum needed to drive out the outer moon. If this can occur, the entire system would be driven outwards at the same resonance locking time scale that the inner moon had before the mean-motion resonance was established.

It is very unlikely that both an inner and outer moon could simultaneously be in resonance locks with planetary oscillations, and in mean-motion resonance with each other. The mean-motion resonances occur near integer ratios of the orbital frequencies, but there is no reason to suspect planetary oscillation mode frequencies to obey such a relationship. Even if they did at one point in time, it is unlikely that each oscillation mode frequency would evolve together to sustain this integer ratio for an extended period of time.

Consequently, if the migration of Saturn's inner moons are driven by resonance locking, it is most likely that the inner moons of mean-motion pairs (i.e., Mimas and Enceladus) are caught in resonance locks while the outer moons (Tethys and Dione) are not. Hence, we would expect to measure a strong tidal torque (small planet Q) on Mimas and Enceladus, and a weak tidal torque (large planet Q) on Tethys and Dione. The measured values (Lainey et al. 2020) may indeed be larger for those outer moons, but current measurement uncertainties are very large so it is difficult to claim anything with confidence. Continued monitoring of moon migration rates will help distinguish between different tidal dissipation mechanisms and hence differentiate between possible evolutionary histories.

5 Conclusions

Our understanding of tidal dissipation in giant planets and its consequences for moon migration remain incomplete, but they have improved greatly over the last decade. A major advance has been in our picture of giant planet structure. Both Jupiter and Saturn show evidence for a large and "diffuse" core, likely containing a smooth and gradual compositional change from ice/rock in the core to the gaseous envelope. In Jupiter, this diffuse core helps explain gravity data from *Juno*, while in Saturn it is revealed in part by ring seismology data from *Cassini*. A diffuse core is also predicted by modern planetary formation models.

Stably stratified regions between the core and envelope have major implications for tidal dissipation in giant planets. They allow for the excitation of gravity waves in these regions, which can contribute greatly to tidal dissipation. Additionally, thick stably stratified shells increase the excitation rate of inertial waves in the overlying convective envelope, enhancing their contribution to tidal dissipation. Recent models accounting for these effects may explain the rapid moon migration rates in Saturn that have been measured using long-baseline astrometry.

Finally, the continually evolving structures of giant planets may feedback onto tidal migration through a resonance locking process. As a planet's structure evolves (e.g., due to cooling, helium rain, compositional settling, or core erosion), the frequencies of gravito-inertial modes evolve as well. This may allow moons to lock into resonance with these modes, such that the energy dissipation rate is enhanced over long periods of time. This process could be necessary to explain the rapid migration of Saturn's outer moons Rhea and Titan.

Although recent progress is promising, there is plenty of room for improvement in future work. More sophisticated planetary formation and evolution models matched to thermal/gravity/seismology data can be developed to provide more realistic planetary internal structures. These will allow for more reliable calculations of tidal energy dissipation rates, including the feedback between the evolving planetary structure and expanding orbits of the moons. Such work will also shed light on the physics of ice giant systems, exoplanet systems, tidal heating of moons, and even the possibility of life in the oceans of Enceladus and Europa.

Acknowledgements This work was inspired by a workshop hosted by the International Space Science Institute in Bern, Switzerland.

Author Contribution Authors Fuller, Guillot, and Mathis contributed equally to the writing of this manuscript. Author Murray contributed expertise on moon orbital evolution.

Funding We gratefully acknowledge support from the International Space Science Institute. S. M. acknowledges support from PNP (CNRS/INSU) and PLATO CNES grant at CEA/DAp.

Data Availability No new data has been used in this work. Data and analysis scripts can be found in referenced materials.

Declarations

Competing Interests The authors have no competing interests to declare that are relevant to the content of this article.

References

Ahuir J, Strugarek A, Brun A-S, Mathis S (2021) Magnetic and tidal migration of close-in planets. Influence of secular evolution on their population. Astron Astrophys 650:126. https://doi.org/10.1051/0004-6361/202040173. arXiv:2104.01004 [astro-ph.EP]

Alterman Z, Jarosch H, Pekeris CL (1959) Oscillations of the Earth. Proc R Soc Lond Ser A 252:80–95. https://doi.org/10.1098/rspa.1959.0138

André Q, Barker AJ, Mathis S (2017) Layered semi-convection and tides in giant planet interiors – I. Propagation of internal waves. arXiv e-prints. arXiv:1704.08974 [astro-ph.EP]

André Q, Mathis S, Barker AJ (2019) Layered semi-convection and tides in giant planet interiors. II. Tidal dissipation. Astron Astrophys 626:82. https://doi.org/10.1051/0004-6361/201833674. arXiv:1902.04848 [astro-ph.EP]

Astoul A, Barker AJ (2022) The effects of non-linearities on tidal flows in the convective envelopes of rotating stars and planets in exoplanetary systems. Mon Not R Astron Soc 516(2):2913–2935. https://doi.org/10.1093/mnras/stac2117. arXiv:2207.12780 [astro-ph.SR]

Astoul A, Barker AJ (2023) Tidally-excited inertial waves in stars and planets: exploring the frequency-dependent and averaged dissipation with nonlinear simulations. arXiv e-prints. https://doi.org/10.48550/arXiv.2309.02520. arXiv:2309.02520 [astro-ph.SR]

Astoul A, Park J, Mathis S, Baruteau C, Gallet F (2021) The complex interplay between tidal inertial waves and zonal flows in differentially rotating stellar and planetary convective regions. I. Free waves. Astron Astrophys 647:144. https://doi.org/10.1051/0004-6361/202039148. arXiv:2101.04656 [astro-ph.SR]

Auclair Desrotour P, Mathis S, Le Poncin-Lafitte C (2015) Scaling laws to understand tidal dissipation in fluid planetary regions and stars I. Rotation, stratification and thermal diffusivity. Astron Astrophys 581:118. https://doi.org/10.1051/0004-6361/201526246. arXiv:1506.07705 [astro-ph.EP]

Auclair-Desrotour P, Le Poncin-Lafitte C, Mathis S (2014) Impact of the frequency dependence of tidal Q on the evolution of planetary systems. Astron Astrophys 561:7. https://doi.org/10.1051/0004-6361/201322782. arXiv:1311.4810 [astro-ph.EP]

Baraffe I (2005) Structure and evolution of giant planets. Space Sci Rev 116:67–76. https://doi.org/10.1007/s11214-005-1948-0

Barker AJ (2016) Non-linear tides in a homogeneous rotating planet or star: global simulations of the elliptical instability. Mon Not R Astron Soc 459:939–956. https://doi.org/10.1093/mnras/stw702. arXiv:1603.06840 [astro-ph.EP]

Barker AJ, Astoul AAV (2021) On the interaction between fast tides and convection. Mon Not R Astron Soc 506(1):69–73. https://doi.org/10.1093/mnrasl/slab077. arXiv:2105.00757 [astro-ph.SR]

Barker AJ, Lithwick Y (2013) Non-linear evolution of the tidal elliptical instability in gaseous planets and stars. Mon Not R Astron Soc 435:3614–3626. https://doi.org/10.1093/mnras/stt1561. arXiv:1309.0107 [astro-ph.EP]

Barker AJ, Lithwick Y (2014) Non-linear evolution of the elliptical instability in the presence of weak magnetic fields. Mon Not R Astron Soc 437:305–315. https://doi.org/10.1093/mnras/stt1884. arXiv:1309.0108 [astro-ph.EP]

Barker AJ, Ogilvie GI (2010) On internal wave breaking and tidal dissipation near the centre of a solar-type star. Mon Not R Astron Soc 404:1849–1868. https://doi.org/10.1111/j.1365-2966.2010.16400.x. arXiv:1001.4009 [astro-ph.EP]

Barker AJ, Dempsey AM, Lithwick Y (2014) Theory and simulations of rotating convection. Astrophys J 791:13. https://doi.org/10.1088/0004-637X/791/1/13. arXiv:1403.7207 [astro-ph.SR]

Baruteau C, Rieutord M (2013) Inertial waves in a differentially rotating spherical shell. J Fluid Mech 719:47–81. https://doi.org/10.1017/jfm.2012.605. arXiv:1203.4347 [astro-ph.SR]

Batygin K, Stevenson DJ (2010) Inflating hot Jupiters with Ohmic dissipation. Astrophys J Lett 714(2):238–243. https://doi.org/10.1088/2041-8205/714/2/L238. arXiv:1002.3650 [astro-ph.EP]

Benz W, Ehrenreich D, Isaak K (2017) CHEOPS: CHaracterizing ExOPlanets satellite, p 84. https://doi.org/10.1007/978-3-319-30648-3_84-1

Bodenheimer P, Lin DNC, Mardling RA (2001) On the tidal inflation of short-period extrasolar planets. Astrophys J 548(1):466–472. https://doi.org/10.1086/318667

Bolton SJ, Levin SM, Guillot T, Li C, Kaspi Y, Orton G, Wong MH, Oyafuso F, Allison M, Arballo J, Atreya S, Becker HN, Bloxham J, Brown ST, Fletcher LN, Galanti E, Gulkis S, Janssen M, Ingersoll A, Lunine JL, Misra S, Steffes P, Stevenson D, Waite JH, Yadav RK, Zhang Z (2021) Microwave observations reveal the deep extent and structure of Jupiter's atmospheric vortices. Science 374(6570):968–972. https://doi.org/10.1126/science.abf1015

Brygoo S, Loubeyre P, Millot M, Rygg JR, Celliers PM, Eggert JH, Jeanloz R, Collins GW (2021) Evidence of hydrogen-helium immiscibility at Jupiter-interior conditions. Nature 593(7860):517–521. https://doi.org/10.1038/s41586-021-03516-0

Cébron D, Hollerbach R (2014) Tidally driven dynamos in a rotating sphere. Astrophys J Lett 789:25. https://doi.org/10.1088/2041-8205/789/1/L25. arXiv:1406.3431 [astro-ph.SR]

Cébron D, Vidal J, Schaeffer N, Borderies A, Sauret A (2021) Mean zonal flows induced by weak mechanical forcings in rotating spheroids. J Fluid Mech 916:39. https://doi.org/10.1017/jfm.2021.220. arXiv:2103.10260 [physics.flu-dyn]

Chabrier G, Baraffe I (2007) Heat transport in giant (exo)planets: a new perspective. Astrophys J Lett 661(1):81–84. https://doi.org/10.1086/518473. arXiv:astro-ph/0703755 [astro-ph]

Charnoz S, Crida A, Castillo-Rogez JC, Lainey V, Dones L, Karatekin Ö, Tobie G (2011) Accretion of Saturn's mid-sized moons during the viscous spreading of young massive rings: solving the paradox of silicate-poor rings versus silicate-rich moons. Icarus 216:535–550. https://doi.org/10.1016/j.icarus.2011.09.017. arXiv:1109.3360 [astro-ph.EP]

Christensen UR, Wicht J, Dietrich W (2020) Mechanisms for limiting the depth of zonal winds in the gas giant planets. Astrophys J 890(1):61. https://doi.org/10.3847/1538-4357/ab698c

Connerney JEP, Kotsiaros S, Oliversen RJ, Espley JR, Joergensen JL, Joergensen PS, Merayo JMG, Herceg M, Bloxham J, Moore KM, Bolton SJ, Levin SM (2018) A new model of Jupiter's magnetic field from Juno's first nine orbits. Geophys Res Lett 45(6):2590–2596. https://doi.org/10.1002/2018GL077312

Connerney JEP, Timmins S, Oliversen RJ, Espley JR, Joergensen JL, Kotsiaros S, Joergensen PS, Merayo JMG, Herceg M, Bloxham J, Moore KM, Mura A, Moirano A, Bolton SJ, Levin SM (2022) A new model of Jupiter's magnetic field at the completion of Juno's prime mission. J Geophys Res, Planets 127(2):07055. https://doi.org/10.1029/2021JE007055

Crida A, Charnoz S (2012) Formation of regular satellites from ancient massive rings in the solar system. Science 338:1196. https://doi.org/10.1126/science.1226477. arXiv:1301.3808 [astro-ph.EP]
Currie LK, Barker AJ, Lithwick Y, Browning MK (2020) Convection with misaligned gravity and rotation: simulations and rotating mixing length theory. Mon Not R Astron Soc 493(4):5233–5256. https://doi.org/10.1093/mnras/staa372. arXiv:2002.02461 [astro-ph.SR]
Dandoy V, Park J, Augustson K, Astoul A, Mathis S (2023) How tidal waves interact with convective vortices in rapidly rotating planets and stars. Astron Astrophys 673:6. https://doi.org/10.1051/0004-6361/202243586. arXiv:2211.05900 [astro-ph.EP]
de Vries NB, Barker AJ, Hollerbach R (2023) Tidal dissipation due to the elliptical instability and turbulent viscosity in convection zones in rotating giant planets and stars. Mon Not R Astron Soc 524(2):2661–2683. https://doi.org/10.1093/mnras/stad1990. arXiv:2306.17622 [astro-ph.EP]
Debras F, Chabrier G (2019) New models of Jupiter in the context of Juno and Galileo. Astrophys J 872(1):100. https://doi.org/10.3847/1538-4357/aaff65. arXiv:1901.05697 [astro-ph.EP]
Dermott SF (1979) Tidal dissipation in the solid cores of the major planets. Icarus 37:310–321. https://doi.org/10.1016/0019-1035(79)90137-4
Dewberry JW (2023) Dynamical tides in Jupiter and other rotationally flattened planets and stars with stable stratification. Mon Not R Astron Soc 521(4):5991–6004. https://doi.org/10.1093/mnras/stad546. arXiv:2301.07097 [astro-ph.EP]
Dhouib H, Baruteau C, Mathis S, Debras F, Astoul A, Rieutord M (2024) Hydrodynamic modelling of dynamical tide dissipation in Jupiter's interior as revealed by Juno. Astron Astrophys 682:A85. https://doi.org/10.1051/0004-6361/202347703
Dougherty MK, Cao H, Khurana KK, Hunt GJ, Provan G, Kellock S, Burton ME, Burk TA, Bunce EJ, Cowley SWH, Kivelson MG, Russell CT, Southwood DJ (2018) Saturn's magnetic field revealed by the Cassini grand finale. Science 362(6410):5434. https://doi.org/10.1126/science.aat5434
Duarte LDV, Wicht J, Gastine T (2018) Physical conditions for Jupiter-like dynamo models. Icarus 299:206–221. https://doi.org/10.1016/j.icarus.2017.07.016
Duguid CD, Barker AJ, Jones CA (2020a) Tidal flows with convection: frequency dependence of the effective viscosity and evidence for antidissipation. Mon Not R Astron Soc 491(1):923–943. https://doi.org/10.1093/mnras/stz2899. arXiv:1910.06034 [astro-ph.SR]
Duguid CD, Barker AJ, Jones CA (2020b) Convective turbulent viscosity acting on equilibrium tidal flows: new frequency scaling of the effective viscosity. Mon Not R Astron Soc 497(3):3400–3417. https://doi.org/10.1093/mnras/staa2216. arXiv:2007.12624 [astro-ph.EP]
Durante D, Parisi M, Serra D, Zannoni M, Notaro V, Racioppa P, Buccino DR, Lari G, Gomez Casajus L, Iess L, Folkner WM, Tommei G, Tortora P, Bolton SJ (2020) Jupiter's gravity field halfway through the Juno mission. Geophys Res Lett 47(4):86572. https://doi.org/10.1029/2019GL086572
Efroimsky M, Lainey V (2007) Physics of bodily tides in terrestrial planets and the appropriate scales of dynamical evolution. J Geophys Res, Planets 112(E11):12003. https://doi.org/10.1029/2007JE002908. arXiv:0709.1995
Efroimsky M, Makarov VV (2013) Tidal friction and tidal lagging. Applicability limitations of a popular formula for the tidal torque. Astron Astrophys 764(1):26. https://doi.org/10.1088/0004-637X/764/1/26. arXiv:1209.1615
Essick R, Weinberg NN (2016) Orbital decay of hot jupiters due to nonlinear tidal dissipation within solar-type hosts. Astrophys J 816:18. https://doi.org/10.3847/0004-637X/816/1/18. arXiv:1508.02763 [astro-ph.EP]
Favier B, Barker AJ, Baruteau C, Ogilvie GI (2014) Non-linear evolution of tidally forced inertial waves in rotating fluid bodies. Mon Not R Astron Soc 439:845–860. https://doi.org/10.1093/mnras/stu003. arXiv:1401.0643 [astro-ph.EP]
Fletcher LN, Helled R, Roussos E, Jones G, Charnoz S, André N, Andrews D, Bannister M, Bunce E, Cavalié T, Ferri F, Fortney J, Grassi D, Griton L, Hartogh P, Hueso R, Kaspi Y, Lamy L, Masters A, Melin H, Moses J, Mousis O, Nettleman N, Plainaki C, Schmidt J, Simon A, Tobie G, Tortora P, Tosi F, Turrini D (2020) Ice giant systems: the scientific potential of orbital missions to Uranus and Neptune. Planet Space Sci 191:105030. https://doi.org/10.1016/j.pss.2020.105030. arXiv:1907.02963 [astro-ph.EP]
Fortney JJ, Dawson RI, Komacek TD (2021) Hot jupiters: origins, structure, atmospheres. J Geophys Res, Planets 126(3):06629. https://doi.org/10.1029/2020JE006629. arXiv:2102.05064 [astro-ph.EP]
French M, Mattsson TR, Nettelmann N, Redmer R (2009) Equation of state and phase diagram of water at ultrahigh pressures as in planetary interiors. Phys Rev B 79(5):054107. https://doi.org/10.1103/PhysRevB.79.054107
Fuentes JR, Cumming A, Anders EH (2022) Layer formation in a stably stratified fluid cooled from above: towards an analog for Jupiter and other gas giants. Phys Rev Fluids 7(12):124501. https://doi.org/10.1103/PhysRevFluids.7.124501. arXiv:2204.12643 [astro-ph.EP]

Fuentes JR, Anders EH, Cumming A, Hindman BW (2023) Rotation reduces convective mixing in Jupiter and other gas giants. Astrophys J Lett 950(1):4. https://doi.org/10.3847/2041-8213/acd774. arXiv:2305.09921 [astro-ph.EP]
Fuller J (2014) Saturn ring seismology: evidence for stable stratification in the deep interior of Saturn. Icarus 242:283–296. https://doi.org/10.1016/j.icarus.2014.08.006. arXiv:1406.3343 [astro-ph.EP]
Fuller J, Luan J, Quataert E (2016) Resonance locking as the source of rapid tidal migration in the Jupiter and Saturn moon systems. Mon Not R Astron Soc 458(4):3867–3879. https://doi.org/10.1093/mnras/stw609. arXiv:1601.05804 [astro-ph.EP]
Fuller J, Hambleton K, Shporer A, Isaacson H, Thompson S (2017) Accelerated tidal circularization via resonance locking in KIC 8164262. Mon Not R Astron Soc 472(1):25–29. https://doi.org/10.1093/mnrasl/slx130. arXiv:1706.05053 [astro-ph.SR]
Galanti E, Kaspi Y, Miguel Y, Guillot T, Durante D, Racioppa P, Iess L (2019) Saturn's deep atmospheric flows revealed by the Cassini grand finale gravity measurements. Geophys Res Lett 46(2):616–624. https://doi.org/10.1029/2018GL078087. arXiv:1902.04268 [astro-ph.EP]
Goldreich P, Nicholson PD (1989) Tidal friction in early-type stars. Astrophys J 342:1079–1084. https://doi.org/10.1086/167665
Goldreich P, Soter S (1966) Q in the solar system. Icarus 5:375–389. https://doi.org/10.1016/0019-1035(66)90051-0
González-Cataldo F, Wilson HF, Militzer B (2014) Ab initio free energy calculations of the solubility of silica in metallic hydrogen and application to giant planet cores. Astrophys J 787(1):79. https://doi.org/10.1088/0004-637X/787/1/79
Goodman J, Lackner C (2009) Dynamical tides in rotating planets and stars. Astrophys J 696:2054–2067. https://doi.org/10.1088/0004-637X/696/2/2054. arXiv:0812.1028
Greenberg R (2009) Frequency dependence of tidal q. Astrophys J Lett 698:42–45. https://doi.org/10.1088/0004-637X/698/1/L42
Guarguaglini M, Hernandez J-A, Okuchi T, Barroso P, Benuzzi-Mounaix A, Bethkenhagen M, Bolis R, Brambrink E, French M, Fujimoto Y, Kodama R, Koenig M, Lefevre F, Miyanishi K, Ozaki N, Redmer R, Sano T, Umeda Y, Vinci T, Ravasio A (2019) Laser-driven shock compression of "synthetic planetary mixtures" of water, ethanol, and ammonia. Sci Rep 9:10155. https://doi.org/10.1038/s41598-019-46561-6
Guenel M, Mathis S, Remus F (2014) Unravelling tidal dissipation in gaseous giant planets. Astron Astrophys 566:9. https://doi.org/10.1051/0004-6361/201424010. arXiv:1406.1672 [astro-ph.EP]
Guenel M, Baruteau C, Mathis S, Rieutord M (2016) Tidal inertial waves in differentially rotating convective envelopes of low-mass stars. I. Free oscillation modes. Astron Astrophys 589:22. https://doi.org/10.1051/0004-6361/201527621. arXiv:1601.04617 [astro-ph.SR]
Guillot T (2005) The interiors of giant planets: models and outstanding questions. Annu Rev Earth Planet Sci 33:493–530. https://doi.org/10.1146/annurev.earth.32.101802.120325. arXiv:astro-ph/0502068 [astro-ph]
Guillot T (2022) Uranus and Neptune are key to understand planets with hydrogen atmospheres. Exp Astron 54(2–3):1027–1049. https://doi.org/10.1007/s10686-021-09812-x. arXiv:1908.02092 [astro-ph.EP]
Guillot T, Showman AP (2002) Evolution of "51 Pegasus b-like" planets. Astron Astrophys 385:156–165. https://doi.org/10.1051/0004-6361:20011624. arXiv:astro-ph/0202234 [astro-ph]
Guillot T, Gautier D, Chabrier G, Mosser B (1994) Are the giant planets fully convective? Icarus 112(2):337–353. https://doi.org/10.1006/icar.1994.1188
Guillot T, Burrows A, Hubbard WB, Lunine JI, Saumon D (1996) Giant planets at small orbital distances. Astrophys J Lett 459:35. https://doi.org/10.1086/309935. arXiv:astro-ph/9511109 [astro-ph]
Guillot T, Stevenson DJ, Hubbard WB, Saumon D (2004) The interior of Jupiter. In: Bagenal F, Dowling TE, McKinnon WB (eds) Jupiter. The planet, satellites and magnetosphere, vol 1, pp 35–57
Guillot T, Santos NC, Pont F, Iro N, Melo C, Ribas I (2006) A correlation between the heavy element content of transiting extrasolar planets and the metallicity of their parent stars. Astron Astrophys 453(2):21–24. https://doi.org/10.1051/0004-6361:20065476. arXiv:astro-ph/0605751 [astro-ph]
Guillot T, Miguel Y, Militzer B, Hubbard WB, Kaspi Y, Galanti E, Cao H, Helled R, Wahl SM, Iess L, Folkner WM, Stevenson DJ, Lunine JI, Reese DR, Biekman A, Parisi M, Durante D, Connerney JEP, Levin SM, Bolton SJ (2018) A suppression of differential rotation in Jupiter's deep interior. Nature 555:227–230. https://doi.org/10.1038/nature25775
Guillot T, Fletcher LN, Helled R, Ikoma M, Line MR, Parmentier V (2023) Giant planets from the inside-out. In: Inutsuka S et al (eds) Protostars and Protoplanets VII. Astronomical Society of the Pacific, pp 947–991
Hedman MM, Nicholson PD (2013) Kronoseismology: using density waves in Saturn's C ring to probe the planet's interior. Astron J 146(1):12. https://doi.org/10.1088/0004-6256/146/1/12. arXiv:1304.3735 [astro-ph.EP]

Hedman MM, Nicholson PD, French RG (2019) Kronoseismology. IV. Six previously unidentified waves in Saturn's middle C ring. Astron J 157(1):18. https://doi.org/10.3847/1538-3881/aaf0a6. arXiv:1811.04796 [astro-ph.EP]

Helled R, Fortney JJ (2020) The interiors of Uranus and Neptune: current understanding and open questions. Philos Trans R Soc Lond Ser A 378(2187):20190474. https://doi.org/10.1098/rsta.2019.0474. arXiv:2007.10783 [astro-ph.EP]

Helled R, Nettelmann N, Guillot T (2020) Uranus and Neptune origin, evolution and internal structure. Space Sci Rev 216(3):38. https://doi.org/10.1007/s11214-020-00660-3. arXiv:1909.04891 [astro-ph.EP]

Hubbard WB (1968) Thermal structure of Jupiter. Astrophys J 152:745–754. https://doi.org/10.1086/149591

Hueso R, Guillot T, Sánchez-Lavega A (2020) Convective storms and atmospheric vertical structure in Uranus and Neptune. Philos Trans R Soc Lond Ser A 378(2187):20190476. https://doi.org/10.1098/rsta.2019.0476. arXiv:2111.15494 [astro-ph.EP]

Hut P (1980) Stability of tidal equilibrium. Astron Astrophys 92:167–170

Iess L, Militzer B, Kaspi Y, Nicholson P, Durante D, Racioppa P, Anabtawi A, Galanti E, Hubbard W, Mariani MJ, Tortora P, Wahl S, Zannoni M (2019) Measurement and implications of Saturn's gravity field and ring mass. Science 364(6445):2965. https://doi.org/10.1126/science.aat2965

Ingersoll AP, Porco CC (1978) Solar heating and internal heat flow on Jupiter. Icarus 35(1):27–43

Ioannou PJ, Lindzen RS (1993a) Gravitational tides in the outer planets. I – implications of classical tidal theory. II – interior calculations and estimation of the tidal dissipation factor. Astrophys J 406:252–278. https://doi.org/10.1086/172437

Ioannou PJ, Lindzen RS (1993b) Gravitational tides in the outer planets. II. Interior calculations and estimation of the tidal dissipation factor. Astrophys J 406:266. https://doi.org/10.1086/172438

Ivanov PB, Papaloizou JCB (2004) On the tidal interaction of massive extrasolar planets on highly eccentric orbits. Mon Not R Astron Soc 347(2):437–453. https://doi.org/10.1111/j.1365-2966.2004.07238.x. arXiv:astro-ph/0303669 [astro-ph]

Ivanov PB, Papaloizou JCB (2007) Dynamic tides in rotating objects: orbital circularization of extrasolar planets for realistic planet models. Mon Not R Astron Soc 376:682–704. https://doi.org/10.1111/j.1365-2966.2007.11463.x. arXiv:astro-ph/0512150

Jacobson RA (2009) The orbits of the neptunian satellites and the orientation of the pole of Neptune. Astron J 137(5):4322–4329. https://doi.org/10.1088/0004-6256/137/5/4322

Jacobson RA (2014) The orbits of the uranian satellites and rings, the gravity field of the uranian system, and the orientation of the pole of Uranus. Astron J 148(5):76. https://doi.org/10.1088/0004-6256/148/5/76

Johnson TV, Morrison D, Matson DL, Veeder GJ, Brown RH, Nelson RM (1984) Volcanic hotspots on Io – stability and longitudinal distribution. Science 226:134–137. https://doi.org/10.1126/science.226.4671.134

Jouve L, Ogilvie GI (2014) Direct numerical simulations of an inertial wave attractor in linear and nonlinear regimes. J Fluid Mech 745:223–250. https://doi.org/10.1017/jfm.2014.63. arXiv:1402.2769 [astro-ph.SR]

Kaula WM (1962) Development of the lunar and solar disturbing functions for a close satellite. Astron J 67:300. https://doi.org/10.1086/108729

Kerswell RR (2002) Elliptical instability. Annu Rev Fluid Mech 34:83–113. https://doi.org/10.1146/annurev.fluid.34.081701.171829

Kunitomo M, Guillot T, Ida S, Takeuchi T (2018) Revisiting the pre-main-sequence evolution of stars. II. Consequences of planet formation on stellar surface composition. Astron Astrophys 618:132. https://doi.org/10.1051/0004-6361/201833127. arXiv:1808.07396 [astro-ph.SR]

Lagage P-O (2015) Exoplanets characterisation with the JWST and particularly MIRI. In: European planetary science congress 2015, vol 10, p 757

Lainey V, Arlot J-E, Karatekin Ö, van Hoolst T (2009) Strong tidal dissipation in Io and Jupiter from astrometric observations. Nature 459:957–959. https://doi.org/10.1038/nature08108

Lainey V, Karatekin Ö, Desmars J, Charnoz S, Arlot J-E, Emelyanov N, Le Poncin-Lafitte C, Mathis S, Remus F, Tobie G, Zahn J-P (2012) Strong tidal dissipation in Saturn and constraints on Enceladus' thermal state from astrometry. Astrophys J 752:14. https://doi.org/10.1088/0004-637X/752/1/14. arXiv:1204.0895 [astro-ph.EP]

Lainey V, Jacobson RA, Tajeddine R, Cooper NJ, Murray C, Robert V, Tobie G, Guillot T, Mathis S, Remus F, Desmars J, Arlot J-E, De Cuyper J-P, Dehant V, Pascu D, Thuillot W, Le Poncin-Lafitte C, Zahn J-P (2017) New constraints on Saturn's interior from Cassini astrometric data. Icarus 281:286–296. https://doi.org/10.1016/j.icarus.2016.07.014. arXiv:1510.05870 [astro-ph.EP]

Lainey V, Casajus LG, Fuller J, Zannoni M, Tortora P, Cooper N, Murray C, Modenini D, Park RS, Robert V, Zhang Q (2020) Resonance locking in giant planets indicated by the rapid orbital expansion of Titan. Nat Astron 4:1053–1058. https://doi.org/10.1038/s41550-020-1120-5. arXiv:2006.06854 [astro-ph.EP]

Lazovik YA, Barker AJ, de Vries NB, Astoul A (2024) Tidal dissipation in rotating and evolving giant planets with application to exoplanet systems. Mon Not R Astron Soc 527(3):8245–8256. https://doi.org/10.1093/mnras/stad3689. arXiv:2311.15815 [astro-ph.EP]

Le Bars M, Cébron D, Le Gal P (2015) Flows driven by libration, precession, and tides. Annu Rev Fluid Mech 47:163–193. https://doi.org/10.1146/annurev-fluid-010814-014556

Le Reun T, Favier B, Barker AJ, Le Bars M (2017) Inertial wave turbulence driven by elliptical instability. Phys Rev Lett 119(3):034502. https://doi.org/10.1103/PhysRevLett.119.034502. arXiv:1706.07378 [physics.flu-dyn]

Leconte J, Chabrier G (2012) A new vision of giant planet interiors: impact of double diffusive convection. Astron Astrophys 540:20. https://doi.org/10.1051/0004-6361/201117595. arXiv:1201.4483 [astro-ph.EP]

Leconte J, Chabrier G (2013) Layered convection as the origin of Saturn's luminosity anomaly. Nat Geosci 6:347–350. https://doi.org/10.1038/ngeo1791. arXiv:1304.6184 [astro-ph.EP]

Li L, Jiang X, West RA, Gierasch PJ, Perez-Hoyos S, Sanchez-Lavega A, Fletcher LN, Fortney JJ, Knowles B, Porco CC, Baines KH, Fry PM, Mallama A, Achterberg RK, Simon AA, Nixon CA, Orton GS, Dyudina UA, Ewald SP, Schmude RW (2018) Less absorbed solar energy and more internal heat for Jupiter. Nat Commun 9(1):3709. https://doi.org/10.1038/s41467-018-06107-2

Lin Y (2023) Dynamical tides in Jupiter and the role of interior structure. Astron Astrophys 671:37. https://doi.org/10.1051/0004-6361/202245112. arXiv:2301.02418 [astro-ph.EP]

Lin Y, Ogilvie GI (2018) Tidal dissipation in rotating fluid bodies: the presence of a magnetic field. Mon Not R Astron Soc 474:1644–1656. https://doi.org/10.1093/mnras/stx2764. arXiv:1710.07690 [astro-ph.EP]

Lin Y, Ogilvie GI (2021) Resonant tidal responses in rotating fluid bodies: global modes hidden beneath localized wave beams. Astrophys J Lett 918(1):21. https://doi.org/10.3847/2041-8213/ac1f23. arXiv:2108.08515 [physics.flu-dyn]

Love AEH (1911) Some problems of geodynamics. Cambridge University Press

Luan J, Fuller J, Quataert E (2018) How Cassini can constrain tidal dissipation in Saturn. Mon Not R Astron Soc 473:5002–5014. https://doi.org/10.1093/mnras/stx2714. arXiv:1707.02519 [astro-ph.EP]

Ma L, Fuller J (2021) Orbital decay of short-period exoplanets via tidal resonance locking. Astrophys J 918(1):16. https://doi.org/10.3847/1538-4357/ac088e. arXiv:2105.09335 [astro-ph.EP]

MacDonald GJF (1964) Tidal friction. Rev Geophys Space Phys 2:467–541. https://doi.org/10.1029/RG002i003p00467

Mankovich CR, Fuller J (2021) A diffuse core in Saturn revealed by ring seismology. Nat Astron. https://doi.org/10.1038/s41550-021-01448-3. arXiv:2104.13385 [astro-ph.EP]

Mankovich C, Fortney JJ, Moore KL (2016) Bayesian evolution models for Jupiter with helium rain and double-diffusive convection. Astrophys J 832(2):113. https://doi.org/10.3847/0004-637X/832/2/113. arXiv:1609.09070 [astro-ph.EP]

Mathis S (2009) Transport by gravito-inertial waves in differentially rotating stellar radiation zones. I – theoretical formulation. Astron Astrophys 506:811–828. https://doi.org/10.1051/0004-6361/200810544

Mathis S (2015) Variation of tidal dissipation in the convective envelope of low-mass stars along their evolution. Astron Astrophys 580:3. https://doi.org/10.1051/0004-6361/201526472. arXiv:1507.00165 [astro-ph.SR]

Mathis S, de Brye N (2011) Low-frequency internal waves in magnetized rotating stellar radiation zones. I. Wave structure modification by a toroidal field. Astron Astrophys 526:65. https://doi.org/10.1051/0004-6361/201015571

Mathis S, Le Poncin-Lafitte C (2009) Tidal dynamics of extended bodies in planetary systems and multiple stars. Astron Astrophys 497:889–910. https://doi.org/10.1051/0004-6361/20079054

Mathis S, Le Poncin-Lafitte C, Remus F (2013) Tides in planetary systems. In: Souchay J, Mathis S, Tokieda T (eds) Tides in astronomy and astrophysics. Lecture notes in physics, vol 861. Springer, Berlin, p 255. https://doi.org/10.1007/978-3-642-32961-6_7

Mathis S, Auclair-Desrotour P, Guenel M, Gallet F, Le Poncin-Lafitte C (2016) The impact of rotation on turbulent tidal friction in stellar and planetary convective regions. Astron Astrophys 592:33. https://doi.org/10.1051/0004-6361/201527545. arXiv:1604.08570 [astro-ph.SR]

Mayor M, Queloz D (1995) A Jupiter-mass companion to a solar-type star. Nature 378:355–359. https://doi.org/10.1038/378355a0

Mazevet S, Musella R, Guyot F (2019) The fate of planetary cores in giant and ice-giant planets. Astron Astrophys 631:4. https://doi.org/10.1051/0004-6361/201936288. arXiv:1909.07640 [astro-ph.EP]

Miguel Y, Bazot M, Guillot T, Howard S, Galanti E, Kaspi Y, Hubbard WB, Militzer B, Helled R, Atreya SK, Connerney JEP, Durante D, Kulowski L, Lunine JI, Stevenson D, Bolton S (2022) Jupiter's inhomogeneous envelope. Astron Astrophys 662:18. https://doi.org/10.1051/0004-6361/202243207. arXiv:2203.01866 [astro-ph.EP]

Militzer B, Hubbard WB, Wahl S, Lunine JI, Galanti E, Kaspi Y, Miguel Y, Guillot T, Moore KM, Parisi M, Connerney JEP, Helled R, Cao H, Mankovich C, Stevenson DJ, Park RS, Wong M, Atreya SK, Anderson J, Bolton SJ (2022) Juno spacecraft measurements of Jupiter's gravity imply a dilute core. Planet Sci J 3(8):185. https://doi.org/10.3847/PSJ/ac7ec8

Millot M, Coppari F, Rygg JR, Correa Barrios A, Hamel S, Swift DC, Eggert JH (2019) Nanosecond X-ray diffraction of shock-compressed superionic water ice. Nature 569(7755):251–255. https://doi.org/10.1038/s41586-019-1114-6

Mirouh GM, Garaud P, Stellmach S, Traxler AL, Wood TS (2012) A new model for mixing by double-diffusive convection (semi-convection). I. The conditions for layer formation. Astrophys J 750(1):61. https://doi.org/10.1088/0004-637X/750/1/61. arXiv:1112.4819 [astro-ph.SR]

Mirouh GM, Baruteau C, Rieutord M, Ballot J (2016) Gravito-inertial waves in a differentially rotating spherical shell. J Fluid Mech 800:213–247. https://doi.org/10.1017/jfm.2016.382. arXiv:1511.05832 [astro-ph.SR]

Moore K, Garaud P (2016) Main sequence evolution with layered semiconvection. Astrophys J 817(1):54. https://doi.org/10.3847/0004-637X/817/1/54. arXiv:1506.01034 [astro-ph.SR]

Moore KM, Barik A, Stanley S, Stevenson DJ, Nettelmann N, Helled R, Guillot T, Militzer B, Bolton S (2022) Dynamo simulations of Jupiter's magnetic field: the role of stable stratification and a dilute core. J Geophys Res, Planets 127(11):2022–007479. https://doi.org/10.1029/2022JE007479

Morize C, Le Bars M, Le Gal P, Tilgner A (2010) Experimental determination of zonal winds driven by tides. Phys Rev Lett 104(21):214501. https://doi.org/10.1103/PhysRevLett.104.214501

Moutou C, Boisse I, Hébrard G, Hébrard E, Donati J-F, Delfosse X, Kouach D (2015) SPIRou: a spectropolarimeter for the CFHT. In: Martins F, Boissier S, Buat V, Cambrésy L, Petit P (eds) SF2A-2015: proceedings of the annual meeting of the French Society of Astron. Astrophys., pp 205–212

Murray CD, Dermott SF (1999) Solar system dynamics. Cambridge University Press, Cambridge

Nettelmann N, Helled R, Fortney JJ, Redmer R (2013) New indication for a dichotomy in the interior structure of Uranus and Neptune from the application of modified shape and rotation data. Planet Space Sci 77:143–151. https://doi.org/10.1016/j.pss.2012.06.019. arXiv:1207.2309 [astro-ph.EP]

Ogilvie GI (2005) Wave attractors and the asymptotic dissipation rate of tidal disturbances. J Fluid Mech 543:19–44. https://doi.org/10.1017/S0022112005006580. arXiv:astro-ph/0506450

Ogilvie GI (2013) Tides in rotating barotropic fluid bodies: the contribution of inertial waves and the role of internal structure. Mon Not R Astron Soc 429:613–632. https://doi.org/10.1093/mnras/sts362. arXiv:1211.0837 [astro-ph.EP]

Ogilvie GI (2014) Tidal dissipation in stars and giant planets. Annu Rev Astron Astrophys 52:171–210. https://doi.org/10.1146/annurev-astro-081913-035941. arXiv:1406.2207 [astro-ph.SR]

Ogilvie GI, Lin DNC (2004) Tidal dissipation in rotating giant planets. Astrophys J 610(1):477–509. https://doi.org/10.1086/421454

Ojakangas GW, Stevenson DJ (1986) Episodic volcanism of tidally heated satellites with application to Io. Icarus 66(2):341–358. https://doi.org/10.1016/0019-1035(86)90163-6

Papaloizou JCB, Ivanov PB (2010) Dynamic tides in rotating objects: a numerical investigation of inertial waves in fully convective or barotropic stars and planets. Mon Not R Astron Soc 407:1631–1656. https://doi.org/10.1111/j.1365-2966.2010.17011.x. arXiv:1005.2397 [astro-ph.SR]

Pearl JC, Conrath BJ (1991) The albedo, effective temperature, and energy balance of Neptune, as determined from Voyager data. J Geophys Res 96:18921–18930. https://doi.org/10.1029/91JA01087

Pollack JB, Hubickyj O, Bodenheimer P, Lissauer JJ, Podolak M, Greenzweig Y (1996) Formation of the giant planets by concurrent accretion of solids and gas. Icarus 124:62–85. https://doi.org/10.1006/icar.1996.0190

Polycarpe W, Saillenfest M, Lainey V, Vienne A, Noyelles B, Rambaux N (2018) Strong tidal energy dissipation in Saturn at Titan's frequency as an explanation for Iapetus orbit. Astron Astrophys 619:133. https://doi.org/10.1051/0004-6361/201833930. arXiv:1809.11065 [astro-ph.EP]

Pontin CM, Barker AJ, Hollerbach R, André Q, Mathis S (2020) Wave propagation in semiconvective regions of giant planets. Mon Not R Astron Soc 493(4):5788–5806. https://doi.org/10.1093/mnras/staa664. arXiv:2003.02595 [astro-ph.EP]

Pontin CM, Barker AJ, Hollerbach R (2023) Tidal dissipation in stratified and semi-convective regions of giant planets. Astrophys J 950(2):176. https://doi.org/10.3847/1538-4357/accd67. arXiv:2304.11898 [astro-ph.EP]

Press WH (1981) Radiative and other effects from internal waves in solar and stellar interiors. Astrophys J 245:286–303. https://doi.org/10.1086/158809

Rauer H, Catala C, Aerts C, Appourchaux T, Benz W, Brandeker A, Christensen-Dalsgaard J, Deleuil M, Gizon L, Goupil M-J, Güdel M, Janot-Pacheco E, Mas-Hesse M, Pagano I, Piotto G, Pollacco D, Santos C, Smith A, Suárez J-C, Szabó R, Udry S, Adibekyan V, Alibert Y, Almenara J-M, Amaro-Seoane P, Eiff MA-V, Asplund M, Antonello E, Barnes S, Baudin F, Belkacem K, Bergemann M, Bihain G,

Birch AC, Bonfils X, Boisse I, Bonomo AS, Borsa F, Brandão IM, Brocato E, Brun S, Burleigh M, Burston R, Cabrera J, Cassisi S, Chaplin W, Charpinet S, Chiappini C, Church RP, Csizmadia S, Cunha M, Damasso M, Davies MB, Deeg HJ, Díaz RF, Dreizler S, Dreyer C, Eggenberger P, Ehrenreich D, Eigmüller P, Erikson A, Farmer R, Feltzing S, de Oliveira Fialho F, Figueira P, Forveille T, Fridlund M, García RA, Giommi P, Giuffrida G, Godolt M, Gomes da Silva J, Granzer T, Grenfell JL, Grotsch-Noels A, Günther E, Haswell CA, Hatzes AP, Hébrard G, Hekker S, Helled R, Heng K, Jenkins JM, Johansen A, Khodachenko ML, Kislyakova KG, Kley W, Kolb U, Krivova N, Kupka F, Lammer H, Lanza AF, Lebreton Y, Magrin D, Marcos-Arenal P, Marrese PM, Marques JP, Martins J, Mathis S, Mathur S, Messina S, Miglio A, Montalban J, Montalto M, Monteiro MJPFG, Moradi H, Moravveji E, Mordasini C, Morel T, Mortier A, Nascimbeni V, Nelson RP, Nielsen MB, Noack L, Norton AJ, Ofir A, Oshagh M, Ouazzani R-M, Pápics P, Parro VC, Petit P, Plez B, Poretti E, Quirrenbach A, Ragazzoni R, Raimondo G, Rainer M, Reese DR, Redmer R, Reffert S, Rojas-Ayala B, Roxburgh IW, Salmon S, Santerne A, Schneider J, Schou J, Schuh S, Schunker H, Silva-Valio A, Silvotti R, Skillen I, Snellen I, Sohl F, Sousa SG, Sozzetti A, Stello D, Strassmeier KG, Švanda M, Szabó GM, Tkachenko A, Valencia D, Van Grootel V, Vauclair SD, Ventura P, Wagner FW, Walton NA, Weingrill J, Werner SC, Wheatley PJ, Zwintz K (2014) The PLATO 2.0 mission. Exp Astron 38:249–330. https://doi.org/10.1007/s10686-014-9383-4. arXiv:1310.0696 [astro-ph.EP]

Redmer R, Mattsson TR, Nettelmann N, French M (2011) The phase diagram of water and the magnetic fields of Uranus and Neptune. Icarus 211(1):798–803. https://doi.org/10.1016/j.icarus.2010.08.008

Remus F, Mathis S, Zahn J-P (2012) The equilibrium tide in stars and giant planets. I. The coplanar case. Astron Astrophys 544:132. https://doi.org/10.1051/0004-6361/201118160. arXiv:1205.3536 [astro-ph.SR]

Remus F, Mathis S, Zahn J-P, Lainey V (2012) Anelastic tidal dissipation in multi-layer planets. Astron Astrophys 541:165. https://doi.org/10.1051/0004-6361/201118595. arXiv:1204.1468 [astro-ph.EP]

Remus F, Mathis S, Zahn J-P, Lainey V (2015) The surface signature of the tidal dissipation of the core in a two-layer planet. Astron Astrophys 573:23. https://doi.org/10.1051/0004-6361/201424472. arXiv:1409.8343 [astro-ph.EP]

Ricker GR, Winn JN, Vanderspek R, Latham DW, Bakos GÁ, Bean JL, Berta-Thompson ZK, Brown TM, Buchhave L, Butler NR, Butler RP, Chaplin WJ, Charbonneau D, Christensen-Dalsgaard J, Clampin M, Deming D, Doty J, De Lee N, Dressing C, Dunham EW, Endl M, Fressin F, Ge J, Henning T, Holman MJ, Howard AW, Ida S, Jenkins JM, Jernigan G, Johnson JA, Kaltenegger L, Kawai N, Kjeldsen H, Laughlin G, Levine AM, Lin D, Lissauer JJ, MacQueen P, Marcy G, McCullough PR, Morton TD, Narita N, Paegert M, Palle E, Pepe F, Pepper J, Quirrenbach A, Rinehart SA, Sasselov D, Sato B, Seager S, Sozzetti A, Stassun KG, Sullivan P, Szentgyorgyi A, Torres G, Udry S, Villasenor J (2015) Transiting Exoplanet Survey Satellite (TESS). J Astron Telesc Instrum Syst 1(1):014003. https://doi.org/10.1117/1.JATIS.1.1.014003

Rieutord M (2015) Fluid dynamics: an introduction. Springer, Berlin. https://doi.org/10.1007/978-3-319-09351-2

Rieutord M, Valdettaro L (2010) Viscous dissipation by tidally forced inertial modes in a rotating spherical shell. J Fluid Mech 643:363–394. https://doi.org/10.1017/S002211200999214X. arXiv:0909.3922 [astro-ph.SR]

Rogers TM, Lin DNC, McElwaine JN, Lau HHB (2013) Internal gravity waves in massive stars: angular momentum transport. Astrophys J 772(1):21. https://doi.org/10.1088/0004-637X/772/1/21. arXiv:1306.3262 [astro-ph.SR]

Rosenblum E, Garaud P, Traxler A, Stellmach S (2011) Turbulent mixing and layer formation in double-diffusive convection: three-dimensional numerical simulations and theory. Astrophys J 731(1):66. https://doi.org/10.1088/0004-637X/731/1/66. arXiv:1012.0617 [astro-ph.SR]

Sarkis P, Mordasini C, Henning T, Marleau GD, Mollière P (2021) Evidence of three mechanisms explaining the radius anomaly of hot Jupiters. Astron Astrophys 645:79. https://doi.org/10.1051/0004-6361/202038361. arXiv:2009.04291 [astro-ph.EP]

Schöttler M, Redmer R (2018) Ab initio calculation of the miscibility diagram for hydrogen-helium mixtures. Phys Rev Lett 120(11):115703. https://doi.org/10.1103/PhysRevLett.120.115703

Sharan S, Langlais B, Amit H, Thébault E, Pinceloup M, Verhoeven O (2022) The internal structure and dynamics of Jupiter unveiled by a high-resolution magnetic field and secular variation model. Geophys Res Lett 49(15):98839. https://doi.org/10.1029/2022GL098839

Shoji D, Hussmann H (2017) Frequency-dependent tidal dissipation in a viscoelastic Saturnian core and expansion of Mimas' semi-major axis. Astron Astrophys 599:10. https://doi.org/10.1051/0004-6361/201630230. arXiv:1612.03664 [astro-ph.EP]

Sinclair AT (1983) A re-consideration of the evolution hypothesis of the origin of the resonances among Saturn's satellites. In: Markellos VV, Kozai Y (eds) IAU colloq. 74: dynamical trapping and evolution in the solar system. Astrophysics and space science library, vol 106, pp 19–25. https://doi.org/10.1007/978-94-009-7214-8_2

Stevenson DJ (1976) The interior of Jupiter. PhD thesis, Cornell University, New York
Stevenson DJ (1979) Turbulent thermal convection in the presence of rotation and a magnetic field – a heuristic theory. Geophys Astrophys Fluid Dyn 12:139–169. https://doi.org/10.1080/03091927908242681
Stevenson DJ, Salpeter EE (1977) The phase diagram and transport properties for hydrogen-helium fluid planets. Astrophys J Suppl Ser 35:221–237
Stixrude L, Baroni S, Grasselli F (2021) Thermal and tidal evolution of Uranus with a growing frozen core. Planet Sci J 2(6):222. https://doi.org/10.3847/PSJ/ac2a47
Storch NI, Lai D (2014) Viscoelastic tidal dissipation in giant planets and formation of hot Jupiters through high-eccentricity migration. Mon Not R Astron Soc 438:1526–1534. https://doi.org/10.1093/mnras/stt2292. arXiv:1308.4968 [astro-ph.EP]
Storch NI, Lai D (2015) Analytical model of tidal distortion and dissipation for a giant planet with a viscoelastic core. Mon Not R Astron Soc 450:3952–3957. https://doi.org/10.1093/mnras/stv904. arXiv:1502.06550 [astro-ph.EP]
Terquem C (2021) On a new formulation for energy transfer between convection and fast tides with application to giant planets and solar type stars. Mon Not R Astron Soc 503(4):5789–5806. https://doi.org/10.1093/mnras/stab224. arXiv:2101.10047 [astro-ph.SR]
Terquem C (2023) On the energetics of a tidally oscillating convective flow. Mon Not R Astron Soc 525(1):508–526. https://doi.org/10.1093/mnras/stad2163. arXiv:2309.01450 [astro-ph.SR]
Thorngren DP, Fortney JJ, Murray-Clay RA, Lopez ED (2016) The mass-metallicity relation for giant planets. Astrophys J 831(1):64. https://doi.org/10.3847/0004-637X/831/1/64. arXiv:1511.07854 [astro-ph.EP]
Tinetti G, Drossart P, Eccleston P, Hartogh P, Heske A, Leconte J, Micela G, Ollivier M, Pilbratt G, Puig L et al (2018) A chemical survey of exoplanets with ARIEL. Exp Astron 46:135–209. https://doi.org/10.1007/s10686-018-9598-x
Tobie G, Grasset O, Lunine JI, Mocquet A, Sotin C (2005) Titan's internal structure inferred from a coupled thermal-orbital model. Icarus 175:496–502. https://doi.org/10.1016/j.icarus.2004.12.007
Tobie G, Grasset O, Dumoulin C, Mocquet A (2019) Tidal response of rocky and ice-rich exoplanets. Astron Astrophys 630:70. https://doi.org/10.1051/0004-6361/201935297
Vasil GM, Julien K, Featherstone NA (2021) Rotation suppresses giant-scale solar convection. Proc Natl Acad Sci 118(31):2022518118. https://doi.org/10.1073/pnas.2022518118
Vazan A, Helled R, Guillot T (2018) Jupiter's evolution with primordial composition gradients. Astron Astrophys 610:14. https://doi.org/10.1051/0004-6361/201732522. arXiv:1801.08149 [astro-ph.EP]
Vidal J, Barker AJ (2020a) Turbulent viscosity acting on the equilibrium tidal flow in convective stars. Astrophys J Lett 888(2):31. https://doi.org/10.3847/2041-8213/ab6219. arXiv:1912.07910 [astro-ph.SR]
Vidal J, Barker AJ (2020b) Efficiency of tidal dissipation in slowly rotating fully convective stars or planets. Mon Not R Astron Soc 497(4):4472–4485. https://doi.org/10.1093/mnras/staa2239. arXiv:2007.13392 [astro-ph.SR]
Wahl SM, Hubbard WB, Militzer B, Guillot T, Miguel Y, Movshovitz N, Kaspi Y, Helled R, Reese D, Galanti E, Levin S, Connerney JE, Bolton SJ (2017) Comparing Jupiter interior structure models to Juno gravity measurements and the role of a dilute core. Geophys Res Lett 44(10):4649–4659. https://doi.org/10.1002/2017GL073160. arXiv:1707.01997 [astro-ph.EP]
Wei X (2016) Calculating rotating hydrodynamic and magnetohydrodynamic waves to understand magnetic effects on dynamical tides. Astrophys J 828:30. https://doi.org/10.3847/0004-637X/828/1/30. arXiv:1606.06232 [astro-ph.SR]
Wei X (2018) The magnetic effect on dynamical tide in rapidly rotating astronomical objects. Astrophys J 854:34. https://doi.org/10.3847/1538-4357/aaa54d. arXiv:1801.05552 [astro-ph.SR]
Weinberg NN, Arras P, Quataert E, Burkart J (2012) Nonlinear tides in close binary systems. Astrophys J 751:136. https://doi.org/10.1088/0004-637X/751/2/136. arXiv:1107.0946 [astro-ph.SR]
Winn JN, Fabrycky DC (2015) The occurrence and architecture of exoplanetary systems. Annu Rev Astron Astrophys 53:409–447. https://doi.org/10.1146/annurev-astro-082214-122246. arXiv:1410.4199 [astro-ph.EP]
Witte MG, Savonije GJ (1999) Tidal evolution of eccentric orbits in massive binary systems. A study of resonance locking. Astron Astrophys 350:129–147. arXiv:astro-ph/9909073
Witte MG, Savonije GJ (2002) Orbital evolution by dynamical tides in solar type stars. Application to binary stars and planetary orbits. Astron Astrophys 386:222–236. https://doi.org/10.1051/0004-6361:20020155
Wong MH, de Pater I, Asay-Davis X, Marcus PS, Go CY (2011) Vertical structure of Jupiter's Oval BA before and after it reddened: what changed? Icarus 215(1):211–225. https://doi.org/10.1016/j.icarus.2011.06.032
Wu Y (2005a) Origin of tidal dissipation in Jupiter. I. Properties of inertial modes. Astrophys J 635:674–687. https://doi.org/10.1086/497354. arXiv:astro-ph/0407627

Wu Y (2005b) Origin of tidal dissipation in Jupiter. II. The value of Q. Astrophys J 635:688–710. https://doi.org/10.1086/497355. arXiv:astro-ph/0407628

Yadav RK, Cao H, Bloxham J (2022) A global simulation of the dynamo, zonal jets, and vortices on Saturn. Astrophys J 940(2):185. https://doi.org/10.3847/1538-4357/ac9d94. arXiv:2212.10617 [astro-ph.EP]

Zahn JP (1966) Les marées dans une étoile double serrée. Ann Astrophys 29:313

Zahn J-P (2013) Stellar tides. In: Souchay J, Mathis S, Tokieda T (eds) Tides in astronomy and astrophysics. Lecture notes in physics, vol 861. Springer, Berlin, p 301. https://doi.org/10.1007/978-3-642-32961-6_8

Zahn J-P, Bouchet L (1989) Tidal evolution of close binary stars. II – orbital circularization of late-type binaries. Astron Astrophys 223:112–118

Space Science Reviews (2025) 221:93
https://doi.org/10.1007/s11214-025-01209-y

Quantification of Tides in Giant Planets from Observations

Valéry Lainey[1] · Marco Zannoni[2] · Vincent Robert[1,3] · Tristan Guillot[4]

Received: 23 April 2024 / Accepted: 6 August 2025 / Published online: 13 October 2025

Abstract
Quantifying tidal effects on giant planets has recently made significant advances, thanks in particular to the Cassini space probe. During its thirteen-year orbit around Saturn, numerous measurements from different instruments made it possible to characterize fundamental parameters such as Saturn's Love number k_2 and quality factor Q at different frequencies. In this article, we summarize the various measurements and methods that have allowed to arrive at such a result, as well as the extrapolations that can be deduced for other systems. More generally, the state of the art concerning the four giant planets of the Solar System is presented, as well as the case of exoplanets.

Keywords Astrometry · Radio-science data · Tides

1 Introduction

Knowledge of moon dynamics has significantly improve over the last 150 years. The advent of new, ever more precise instruments, including heliometers and the use of photographic plates, meant that by the end of the 19th century, precision had rapidly increased from 1 arcsecond to 200 milliarcsecond (mas). These observations were then used by analytical theories to produce precise ephemerides for these moons. In the case of the Jovian system, Sampson's analytical theory (Sampson 1921) was the reference throughout the first part of the 20th century. It took eight years to perform the numerous calculations required. These ephemerides were used unchanged until the beginning of the space era. In 1973, J.H. Lieske took up Sampson's theory and made many improvements, enabling the theory to be adapted to new observations (Lieske 1974). This modified theory, known as the Sampson-Lieske theory, was used by NASA for the Voyager and Galileo missions (Lieske 1996). It was eventually replaced by purely numerical methods on the Galileo mission (Peters 1981).

✉ M. Zannoni
m.zannoni@unibo.it

1 IMCCE, Observatoire de Paris, PSL Research University, Sorbonne Université, CNRS, Univ. Lille; 77 avenue Denfert Rochereau, 75014 Paris, France

2 Dipartimento di Ingegneria Industriale, Alma Mater Studiorum - Università di Bologna, Forlì, Italy

3 Institut Polytechnique des Sciences Avancées IPSA, 63 bis Boulevard de Brandebourg, 94200 Ivry-sur-Seine, France

4 Observatoire de la Côte d'Azur, UniCA, Laboratoire Lagrange CNRS UMR 7293, Bd de l'Observatoire, 06304 Nice, France

The study of the orbital motion of the Saturnian moons followed a somewhat similar treatment. The presence of three independent mean motion resonances within the system (Mimas-Tethys, Enceladus-Dione and Titan-Hyperion) led to the development of dynamical models specific to different satellites subsets (Kozai 1957; Sinclair 1977; Harper et al. 1988; Harper and Taylor 1993). A global analytical treatment was eventually given by Duriez and Vienne (1991), Vienne and Duriez (1991, 1992). Analytical developments were then superseded by numerical integration (Jacobson 2004). In the case of the Uranian and Neptunian system, the absence of mean motion resonances made the use of analytical theories more efficient (Laskar 1986; Emelyanov and Samorodov 2015), although numerical integration is mostly used today (Jacobson et al. 1991; Lainey 2008; Emelyanov and Nikonchuk 2013; Jacobson 2014).

Quantifying tidal effects in systems of giant planets from astrometric observations is nothing new. It began in the early 20th century with the work of de Sitter (1928). He targeted the four Galilean moons of Jupiter, which had been observed continuously since their discovery by Galileo in 1610. To do so, de Sitter benefited from a number of observations, many of which he had reduced himself (de Sitter 1916). His early work showed notable accelerations, which we now know to be erroneous. de Sitter's theory was too succinct, given the complexity of the dynamic effects to be modelled. In this respect, the Laplace resonance linking energy and angular momentum of the three moons Io, Europa, and Ganymede will remain a major difficulty in the development of any analytical theory of these satellites. The search for secular tidal accelerations in the moons' motion resumed in the 1970s with the work of Lieske. With Sampson-Lieske's theory in the public domain, researchers began to make their own, sometimes contradictory, measurements (Lieske 1987; Vasundhara et al. 1996). In 2003, Lainey et al. (2004) and Avdyushev (2004) independently found that the Sampson-Lieske theory suffered from a lack of long-period terms, which could often be confused with secular accelerations. Hence, depending on the observations used, former researchers would fit different tidal accelerations. In 2009, the first measurements from purely numerical models appeared (Lainey et al. 2009). The underlying method was already used to numerically quantify the orbital acceleration of Phobos. The main difference is that tidal effects in at least one of the satellites, Io, needed to be introduced. Practically, the dynamical models introduce tides by mean of the ratio k_2/Q, where k_2 is the degree-2 Love number and Q is the quality factor characterizing the dissipation of tidal energy in the planet.

The extension of this work to the Saturn system was quite natural in the context of NASA's Cassini mission. Although initial estimates of secular accelerations had been found by Kozai (1957) and Dourneau (1987), it would appear that, as with the Jupiter system, it was the absence of a long period in the analytical theory that was lacking (Vienne et al. 1992). The first measurements from a numerical model were published in 2012 by Lainey et al. This work was then enhanced by the introduction of the numerous measurements made by the ISS/Cassini instrument (Porco et al. 2004) to produce the first measurement of Saturn's k_2 (Lainey et al. 2017), as well as the frequency dependence of the dissipative factor Q. Last but not least, radio science data revealed to be of high importance for constraining the tidal expansion of Titan's orbit (Lainey et al. 2020).

In Sect. 2 we present the available astrometric methods that have been used to determine tidal parameters in giant planets. Section 3 focuses on radio science data. The next section details the various measurements made for the Jupiter system. Section 5 presents the latest results for the Saturn system. In Sect. 6, we discuss the first measurements for the Uranus system and possible future ones for Neptune system. Section 7 focuses on a few extra-solar systems.

2 Astrometry

2.1 Classical Astrometry

Astrometry – referring to positioning astronomy – is the first science deriving from observational astronomy. It allows the determination of spherical coordinates in right ascension and declination of the observed objects, to assess their motion and dynamics, and to determine parameters of their physical environment (masses, internal mass distribution, tidal parameters…), in particular. As the methods of observation and data acquisition have benefited from technological advances with the transition from analog (astrophotographic plates) to digital (CCD and CMOS sensors), measurement methods have also evolved with the development of new techniques, which now complement the classical approach of plate constants calibration. Finally, these data are used in conjunction with a dynamic model to quantify physical parameters, including tides.

2.1.1 Photographic Plates

The astrophotographic plates were to the telescope what the photographic films were to the camera: analog supports to acquire and save shots. They were produced by the manufacturers of the epoch (Kodak, Agfa-Gevaert…), used since the end of the 19th century to take pictures of scenes of life and by extension, astronomical observations (Fig. 1). They have been progressively replaced by numerical CCD sensors since the end of the 1990's, and the production has stopped.

Photographic plate formats could be very different depending on their use for astrometry and photometry (example of a 35x35 cm Schmidt plate), or spectroscopy (example of a 2x25 cm star spectra). One side was covered by a photosensitive emulsion whose silver halides blackened under a light beam. This was followed by the classical photographic development phases in a darkroom to fix the emulsion and clean the support. Exposure times range from several seconds to a few hours depending on the observed target. The size of the field, depending on the instrumentation and the size of the plates, varies from a few arcminutes to several degrees on the celestial sphere.

Due to their analog nature, it was possible to filter the plates by direct apposition, partial or complete, with the huge advantage to allow for simultaneous observation objects with very different magnitudes. Another particularity, due to the relatively high cost of this equipment: it was very common to carry out several exposures on the same support by shifting the plate in right ascension or declination according to the mode of mounting at the focus of the telescope, between exposures.

Firstly analyzed by hand, photographic plates could now be digitized using specific machines to provide numerical images to be re-studied with new numerical tools, providing new accurate information of past observations (Robert et al. 2021). Moreover, because relatively few photographic plates have already been analyzed, mainly for the needs of the first space missions (Mariner, Viking…), we there find a huge observation tank with original data allowing the adjustment periods to be extended in dynamical models.

2.1.2 CCDs

Charge-Coupled Devices, CCDs are digital photographic sensors developed since the 1970's, fully by the end of the 20th, and now commonly used for industry, scientific imagery, and general public (smartphones, reflex cameras…).

Fig. 1 Example of a USNO photographic plate (center part) of Saturn in 1974, showing stars and four of the main satellites

CCDs are generally composed with semiconductor elements able to convert photons into electrons that are stored in a memory cell before decoding. They are mounted as pixel lines or pixel matrices. They are characterized by their quantum efficiency (the ratio between the number of incoming photons and the number of generated electrons), their transmission (wavelength response), resolution, pixel size and number, information encoding... Depending on their architecture, they could generate monochrome or RGB colored images.

CCD cameras provide numerical images that can be used in astrometry, photometry, and spectroscopy (Fig. 2). They directly benefit from numerical tools for image transfer, treatment, analysis and archiving. Apart from their intrinsic analogic-numeric difference, CCDs have a quantum efficiency that can peak at 90% compared to 2-10% for photographic plates (Howell 2006), arguing in favour of the replacement of these former materials. Moreover, CCDs allow to observe objects by considerably decreasing the exposure time, without sensibility loss. When observing natural satellites, exposure times range typically from some tenths of second to several minutes depending on the observed target. The size of the field, depending on the instrumentation and the size of the sensor is up to some tens of arcminutes on the celestial sphere.

Fig. 2 Example of a monochrome OHP-T120 CCD observation of Pluto in 2023, July

2.1.3 Precision Premium

Precision premium is a complement to the classical astrometric techniques, first presented by Pascu (1994) within an application to the Jovian system. It consists of observing a pair of close moons using the other moons present all around in the field for calibration. Indeed, a major source of error often comes from an inaccurate image scale factor. Since relative positions on the tangential plane depend essentially linearly on the scale factor, the use of close pair of moons is extremely beneficial.

Historically, before the arrival of the first accurate astrometric star catalogs, astronomers could not link the reference stars on the celestial sphere to calculate the observation plate constants, which are essential to determine spherical coordinates in right ascension and declination. Thus, only relative (intersatellite) positions could be used for the adjustments of dynamical models. This has the side advantage to provide measurements independent on the planetary ephemeris error.

Precision Premium is an observation concept that constraints the realization of the observations to provide the most precise astrometric results. One could also consider Precision Premium as a condition of observation sampling since it can refer to very specific parts in large data set. Within his application, Pascu (1994) wrote that the most accurate relative positions of two objects should be obtained when their apparent angular separation is small. It

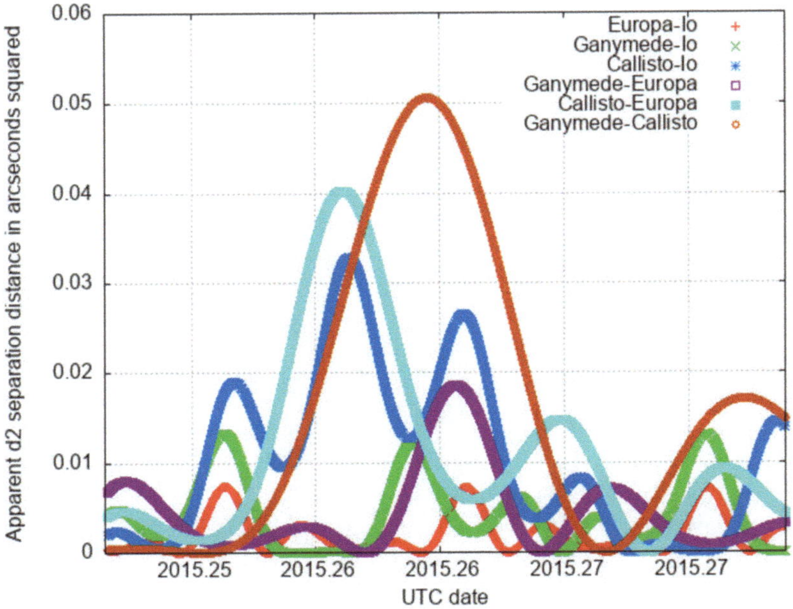

Fig. 3 Occurrences of minimum and maximum intersatellite distances over 10 days

was demonstrated with observations of the Galileans with separations below 50 arcseconds. More recently, Lin et al. (2019) showed that the range of separation for which Precision Premium could be used is up to 100 arcseconds, allowing to decrease the object positioning errors up to several milliarcseconds.

Precision Premium has been used for intersatellite measurements, but it can also be generalized to all inter-object positioning with the same accuracy. Moreover, it allows the decorrelation of the motion of the measured bodies even if they are rotating around a primary.

2.1.4 Mutual Approximations

Mutual approximations (Morgado et al. 2016) were somewhat inspired from the mutual events observation (e.g. mutual occultation or eclipses of one moon by another). The apparent relative motion of satellites allows the observation of the approaches and distances. We no longer fit magnitude drops but distance drops. Thus, minimum and maximum distance times can be used to constrain dynamical models. The geometric parameters of central instant, impact parameter and relative velocity that describe mutual events are the same for mutual approximations.

Mutual approximations occur several times each day, and can be continuously observed to provide regular accurate positioning (Fig. 3).

2.2 Astrometry Derived from Photometry

Although not strictly speaking an astrometric measurement, the observation of celestial phenomena involving moons is often one of the most reliable means of doing an astrometric measurement. The longest observed phenomena are eclipses of moons, particularly Galilean moons. For a long time, these have served as a reliable time reference for deducing longitude measurements of places, thus improving our knowledge of global cartography (Lieske

1986a). These easy-to-make measurements continued to be used until the beginning of the 21st century (Mallama et al. 2010). However, the observation of so-called mutual phenomena, involving the eclipse or occultation of one moon by another, made it possible in the 1970s to considerably increase the precision of astrometric measurements of the moons of Jupiter (first campaign in 1973) and Saturn (first campaign in 1979).

The clear advantage of observing phenomena is that they are not an angular measurement, but a photometric one. This means that, provided the signal-to-noise ratio is sufficiently high, the astrometric measurement (deduced from the photometric light curve) is much less dependent on the distance of the observed system than conventional astrometric measurements, for which a precision of a tenth of an arcsecond will not give the same kilometric precision within the system, depending on whether it is near or far from the Earth. In this respect, the campaign to observe the mutual phenomena of Uranus' satellites enabled us to obtain measurements of a quality unattainable from the ground using conventional methods (Arlot et al. 2013). Today, the processing of observations of mutual phenomena is limited by two main factors. The first is the lack of albedo maps of objects at the observed wavelengths (Vasundhara et al. 2003; Emelyanov 2017). The second is the difficulty often encountered in obtaining detailed information on the observing conditions of many observers like detector characteristics, bandwidth of filters used, etc.

However, even the observation of mutual phenomena has now been superseded by an even more powerful technique: the observation of star occultations. While this technique is by no means new in itself, it is the presence of the Gaia stellar catalog that now enables this observational technique to take the lead in moon astrometry (Morgado et al. 2019). Thanks to observations made at various locations on Earth, it is possible not only to determine the shape of the occulting object, but also its precise position on the celestial sphere. Accuracies of just a few milliarcseconds can easily be achieved, which is an order of magnitude more precise than the observation of mutual phenomena.

2.3 Space Astrometry

Astrometric processing of space images is not fundamentally different from that used for ground-based images (see Sect. 2.1). When the conditions are right, we can generally find enough stars in the image background to perform a field calibration. The presence of star trackers on probes often means that only a minor correction has to be made to the center of the field, although this is absolutely essential for a quality measurement. The camera's optical parameters are generally obtained by pointing at star clusters (Owen 2003). The essential differences therefore concern the treatment of star photocentres (Zhang et al. 2021) and the determination of the geometric center of resolved celestial objects (Tajeddine et al. 2013).

The Caviar software (Cooper et al. 2018), available under an open source license, enables precisely this type of measurement (Fig. 4). It was used throughout the Cassini mission to provide astrometric data for the moons of the Saturn system. A typical accuracy of a few tenths of a pixel is frequently obtained, which amounts to an accuracy of between a hundred meters and a few kilometers, depending on the probe's distance from its target. It should be noted that the accuracy of these measurements is sometimes worse than the precision. Among the sources of error that may explain these measurement biases, the difficulty of precisely defining the position of the edge of objects has been reported (Cooper et al. 2014), as has the relative lack of precision of the ellipsoidal models used for images taken at close range to the target (Cooper et al. 2018).

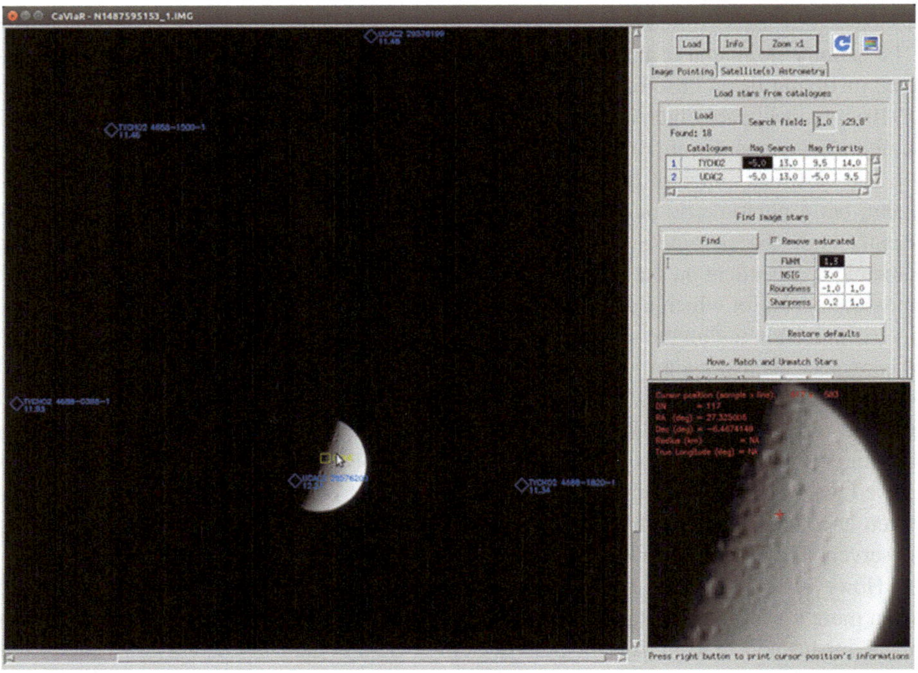

Fig. 4 Screenshot of Caviar software available under open source license (Cooper et al. 2018). Dione can be seen in the image used N1487595153_1 (available on https://pds.nasa.gov/)

2.4 Gaia

Although Gaia is a space mission, the astrometric processing carried out is somewhat different from the standard procedure. In fact, the Gaia probe scans the sky according to the evolution of its axis of rotation. Gaia's astrometric measurements are so precise because they are based on a measurement of time, not space. As the probe rotates, the stars move across the focal plane. Since Gaia's rotation is perfectly well known, measuring the time taken for the various objects to pass over the nine CCDs enables to obtain an extremely precise relative position of the stars in relation to each other in the direction of Gaia's movement. Measurements of a few tenths of a milliarcsecond are common for moons in the solar system (Tanga et al. 2023). Measurements are therefore essentially made in a single direction, although a measurement perpendicular to the axis of rotation is also made, albeit with much lower precision (typically 0.6 arcseconds).

A number of technical difficulties somewhat limit Gaia's contribution to the solar system's moons motion. Firstly, moons that are too close to their planet are not detectable. Secondly, measurement of very large objects such as Jupiter's four Galilean satellites or Titan are not reliable. Last, objects of magnitude below 20.9 are observed by Gaia, only. Nonetheless, for outer moons and a good number of main moons, Gaia data are exceptionally accurate. So, while the first measurements of the solar system's moons have not been provided until DR3 (Gaia Collaboration 2023), it's pretty clear that Gaia astrometric data is becoming a remarkable asset for the possible quantification of tidal dissipation in Uranus and Neptune, especially after the publication of Gaia DR5 release that will provide 10 years of high accuracy astrometric measurements.

2.5 Radar Measurements

Radar measurements are another way of obtaining direct astrometric information about the position of moons. The principle is to measure the round trip made by a series of radar shots between the transmitting antenna and one of the moons. This can be done from an antenna on the ground or from a space probe. In the first case, the great distance separating the Earth from the giant planets makes it necessary to use a very large antenna. Until now, it has only been possible to make observations of the Galilean satellites from the Arecibo radar station (Harmon et al. 1994; Brozovic et al. 2020). The first measurements were made of the moons Ganymede and Callisto in 1992, then Europa in 1999 and finally Io in 2015. A set of 22 measurements taken over the period 1992-2016 is now available. The precision obtained ranges from 1.5 km for the best observations to several tens of kilometres for the least precise (Brozovic et al. 2020).

More recently, radar shots taken by the MARSIS instrument on board the European Mars Express probe have also been analysed for astrometric purposes. A measurement accuracy of a few hundred metres was obtained for the position of Phobos (Desage et al. 2024). It is therefore likely that similar measurements will be possible with the arrival of the Juice probe in the Jovian system in 2031 (Van Hoolst et al. 2024).

2.6 Orbital Dynamics

All these astrometric data can be compared with the predictions of orbital models for the moons concerned. These models, formerly analytical and now numerical, consist of an integration of the equations of motion. Starting from approximate initial conditions, it is then possible to study the discrepancy between observed and calculated positions. From these deviations, a correction can be made to the physical parameters modelled within the dynamical system, as well as to the initial conditions. To do this, one also needs to know the derivative of the position and velocity vectors with respect to these quantities at the observation dates. In practice, it is often necessary to integrate an additional system of differential equations, also known as variational equations. Ultimately, the inversion of a linear system is performed using least-squares techniques. More details on these aspects can be found in Lainey (2016).

3 Radio Science

Radio science measurements are fundamentally different from astrometric ones because they do not refer directly to the target bodies and, therefore, they have a different information content. The main radio science observables are two-way range and range-rate (Doppler) measurements between the spacecraft and the ground station's antennas, enabled by an onboard transponder (Fig. 5). Range-rate measurements are obtained from the Doppler shift of a highly stable microwave carrier, while range is obtained from the round-trip light-time of a ranging code modulated onto the radio signal. The radio link provides direct information on the state of the spacecraft. However, since its trajectory is affected by the gravity of close celestial bodies, it is possible to derive the 3-d position of those bodies through an orbit determination process. The core of this process is the comparison between the measurements obtained from the spacecraft and the ones computed by means of mathematical models (Milani and Gronchi 2009). To better fit the observations, the trajectory of the spacecraft and the orbits of close celestial bodies must be precisely reconstructed, so a set of parameters that

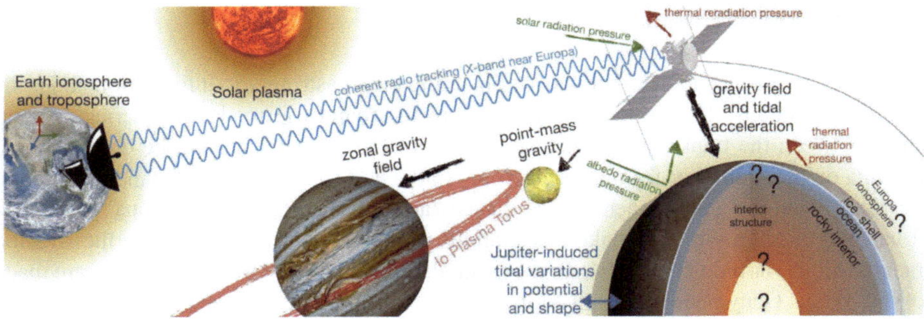

Fig. 5 Illustration of radio science investigations of a deep space mission (from Mazarico et al. 2023). The main radio science measurements are derived from the coherent radio link with Earth ground station antennas. They provide information on the spacecraft trajectory, influenced by both gravitational and non-gravitational accelerations

affect their dynamics must be jointly estimated. The dynamical model must include all the non-negligible forces that, for missions at giant planets, typically include: the point-mass gravity of the Sun, the planets of the Solar System, and close moons, including relativistic effects; the extended gravity of the close celestial bodies, usually in terms of spherical harmonics coefficients; rotational models of the celestial bodies that have an extended gravity field; time-varying gravity due to tidal interactions; non-gravitational forces, such as the solar radiation pressure and atmospheric drag (e.g. Lainey et al. 2020; Gomez Casajus et al. 2022).

The quality of radiometric measurements depends on many elements of both the space and ground segments, but the main driving factor is the carrier frequency adopted in the radio link. Higher frequencies are less affected by dispersive media, among which the solar plasma is typically the dominant source of noise in the data (Asmar et al. 2005; Iess et al. 2014). With time, space exploration missions adopted higher and higher carrier frequencies. The S-band link (2.1-2.3 GHz), used by the Pioneer, Voyager, and Galileo probes, allowed accuracies of the order of 1 mm/s (Doppler, 60 s integration time) and 10 m (ranging). Moving to X-band (7.1-8.4 GHz), used for example by the standard Cassini radio link, improved the accuracies to about 0.05 mm/s (Doppler, 60 s integration time) and 2 m (ranging). Finally, the Ka-band link (32-34 GHz), adopted by Juno (Doppler only), BepiColombo (Doppler and ranging), and Juice (Doppler and ranging), allows reaching accuracies of better than 0.03 mm/s (Doppler, 60 s integration time) and 1 m (ranging). In addition, the availability of multiple radio links at different frequencies, guaranteed by state-of-the-art radio science instruments such as the ones carried by Cassini (up to 2003), Juno, BepiColombo, and Juice, allows the removal of almost completely the dispersive noises (Bertotti et al. 1993; Mariotti and Tortora 2013), leaving the Earth troposphere as the residual leading noise source. State-of-the-art tropospheric calibrations are produced through water-vapour radiometers, currently available only at NASA's DSS 25 and ESA's Malargue stations (Buccino et al. 2021; Lasagni Manghi et al. 2021). When multi-frequency links and water-vapour radiometers are combined, accuracies of the order of 0.01 mm/s (Doppler, 60 s integration time) and 20 cm (ranging) can be reached, as demonstrated by Juno (Durante et al. 2020) and BepiColombo (Lasagni Manghi et al. 2023), opening the way to next-generation radio science investigations (Iess et al. 2021, 2024).

When a spacecraft is in orbit around a body, range-rate measurements allow to reconstruct the relative orbit of the spacecraft with respect to that body. Complementary range measurements provide strong information on the body's orbit, making them the main observables for the development of the planetary ephemerides (Folkner et al. 2014; Park et al. 2021).

When the spacecraft performs a flyby of a celestial body, most of the information content comes from range-rate measurements. During each encounter, these measurements provide the relative position of the spacecraft with respect to the body. Before and after the encounter, when the spacecraft is outside the sphere of influence of the body, range-rate measurements provide the relative position of the spacecraft with respect to the central body (the planet, in case of moon flybys). As a result, precisely reconstructing the trajectory during the entire encounter allows us to get the relative position of the body with respect to the central body. Typical X-band accuracies in the retrieved position of moons with a flyby are in the order of few hundreds of meters on the orbital plane, tens of km normal to the orbital plane. However, because flybys provide high-accuracy but rather sparse data, to correctly reconstruct the satellite ephemerides radio science data must be properly combined with measurements providing better coverage, such as astrometry.

3.1 Voyager

In 1979, Pioneer 11 became the first space mission to collect radio science data of the natural satellites of Saturn, allowing to estimate the gravity field of Saturn, together with the masses of Rhea, Titan, and Iapetus (Null et al. 1981). However, the satellite ephemerides were taken directly from analytical theories based on astrometric measurements. Later, in 1980 and 1981 the twin Voyager probes performed a flyby of Saturn. A joint analysis of Pioneer and Voyager radio science data, together with Voyager astrometry allowed to improve the estimation of the gravity of the Saturn system (Campbell and Anderson 1989). In addition, parameters of the analytical theory describing the orbits of Tethys, Rhea, and Titan were estimated.

Later, in preparation for the Cassini mission, the addition of Earth-based astrometry to radio science and space-based astrometry allowed an improved and consistent estimate of the gravity field of Saturn, and the masses and orbits of the Saturnian moons, through numerical integration of a full dynamical model (Jacobson 2004). The resulting uncertainty in the moons' positions during the analysed timespan was $\sim$50 km in the radial and normal directions and 150-500 km in the transverse direction, not sufficient to detect tidal dissipation effects in Saturn.

3.2 Cassini

The Voyager probes returned a huge impact in science and helped to understand better the satellite systems of the gas giants. However, they left many unanswered questions about Saturn and its moons until 2004, when the Cassini spacecraft entered into orbit around the planet. During its 13-year mission, it performed many close flybys of the moons of Saturn, including 127 flybys of Titan. Radio science data were collected during the closest approach of 3 flybys of Enceladus, 2 of Rhea, 3 of Dione, and 10 of Titan. During the mission, the satellite ephemerides of the Saturn systems were periodically updated by the Cassini team to support the navigation and science activities. These analyses made use of an extensive data set comprising all available radio tracking data of Cassini, Pioneer 11, and Voyager, space-based and Earth-based astrometric data, including stellar and ring occultations (Jacobson et al. 2006; Jacobson 2022).

In 2017, the Cassini mission ended with a "Grand Finale", a series of inclined, highly eccentric orbits characterized by a very low pericenter, with altitudes 2600-3900 km above the

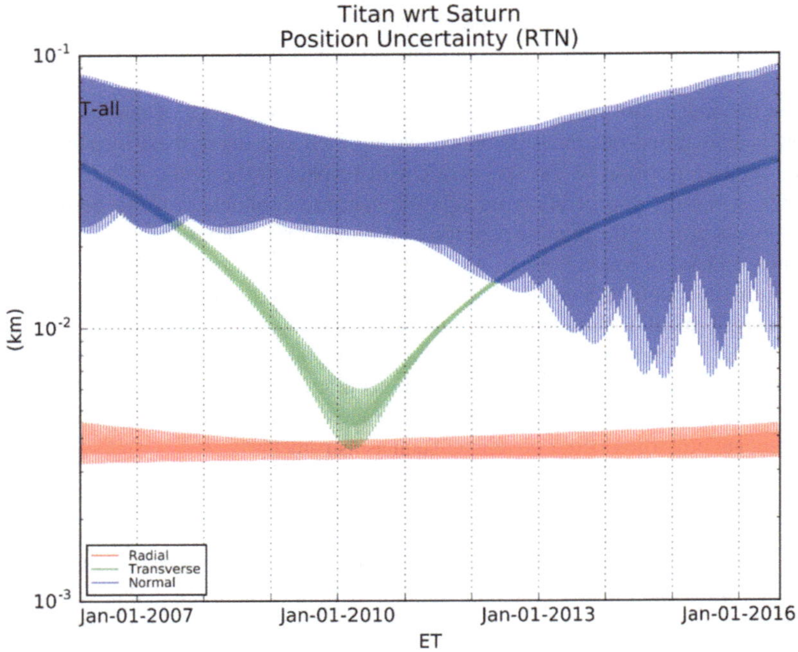

Fig. 6 Uncertainties of Titan trajectory with respect to Saturn in the radial, transverse and normal to the orbital plane directions, obtained from Cassini radio science data (from Gomez Casajus 2019)

cloud tops. Radio science data were collected during five of these pericenter arcs, allowing to estimate the gravity field of Saturn at an unprecedented level of accuracy, including for the first time the mass of its rings (Iess et al. 2019). A time-variable component of the Saturn gravity field was also detected, possibly related to the planet's normal modes (Markham et al. 2020), similar to what was later observed in Jupiter by the Juno mission (Durante et al. 2022).

The gravity field of Saturn observed during the Grand Finale also allowed to better characterize the orbital motion of the main Saturnian satellites. In particular, given the overall good quality of radio tracking data and the good time coverage of Titan flybys, Lainey et al. (2020) precisely reconstructed a coherent orbit of Titan for the entire timespan of the Cassini mission, using only radiometric data (see Sect. 5.2). The obtained formal uncertainties in the Titan's position are 3 m in the radial direction, almost constant during the timespan of the Cassini mission. In the transverse direction, the uncertainty increases almost linearly from a minimum of 4 m in January 2010 to a maximum of 40 m. The sensitivity in the orbit normal direction is lower, with uncertainties up to 50-100 m, during the entire mission (Fig. 6).

3.3 Galileo

The Galileo mission studied the Jovian system from December 1995 to September 2003, dramatically increasing our knowledge of the system, and especially of its Galilean moons, thanks to more than 30 close flybys. Radio science data were acquired during most of these close encounters. Unfortunately, the overall scientific return of the mission was significantly reduced by the missing deployment of the umbrella-like high-gain antenna, forcing to communicate with the spacecraft using only the low-gain antenna. The quality of radio science data was heavily affected because the low-gain antenna supported only a low signal-to-noise ratio radio link at S-band, resulting in high noise levels due to solar plasma and station electronics. In addition, the unavailability of a multi-frequency link at both S- and X- bands, as originally envisioned by the high-gain antenna radio link, prevented calibration of the Doppler shift caused by the ionospheres of Jupiter, its moons, and the Io plasma torus, possibly introducing biases in the orbit determination results.

As a result, using only Galileo radio tracking data acquired during the moons' close encounters, the obtainable formal uncertainty in the moons' positions with respect to the central body was a few kilometers in the radial and transverse directions and tens of kilometers in the orbit normal direction.

Nevertheless, during and after the Galileo mission, the orbits of the Galilean moons, together with their gravity and the gravity of Jupiter, were periodically updated jointly analysing Galileo and Ulysses radiometric data, space-based and Earth-based astrometry, including mutual events and eclipses (Jacobson et al. 2000; Jacobson 2001).

3.4 Juno

The Juno mission, dedicated to the study of the origin and evolution of Jupiter, reached the planet in July 2016, when it entered into a highly eccentric, polar orbit characterized by very low pericenter altitudes (less than 5000 km above the cloud tops). The orbit was specifically designed to avoid Jupiter's intense radiation belts and, at the same time, to provide a detailed characterization of the planet. In particular, Juno measured the gravity field of Jupiter at an unprecedented level of detail (Durante et al. 2020), including the possible detection of normal modes (Durante et al. 2022).

Because of Jupiter's oblateness perturbation, Juno's orbit periapsis precesses northward, allowing the inclusion of 4 close encounters of the inner Galilean satellites after the end of the prime mission in 2021: one of Ganymede (June 2021), one of Europa (September 2022) and two of Io (December 2023 and February 2024). During the flybys, radio tracking measurements have been acquired, offering the opportunity to update the moons' gravity and orbits for the first time since the end of the Galileo mission (Gomez Casajus et al. 2022). This opportunity was particularly interesting for Io, which will not be visited by any of the currently approved missions to the Jupiter system, namely Juice and Europa Clipper (Park et al. 2025).

Moreover, as opposed to Galileo, Juno is equipped with a state-of-the-art radio system that supports the acquisition of Doppler measurements at both X and Ka-band, allowing to reach noise levels of 0.01-0.03 mm/s (60 s integration time) and to remove the Doppler signatures caused by the ionospheres. However, Juno's orbit is not optimized for moon investigations, and the very high velocities relative to the moons decrease the sensitivity of radio science measurements. As a result, the formal uncertainty in the Ganymede's position obtained during the G34 close flyby was 100 m in the radial direction, 200 m in the transverse direction, and 5 km in the orbit normal direction (Gomez Casajus et al. 2022). The Juno radio science measurements of the moons, jointly analysed with Galileo data, astrometric data, together with the radio data of the future Juice and Europa Clipper missions, have the potential to significantly improve our knowledge of the orbit of the moons and the tidal dissipation of the Jupiter system (Magnanini et al. 2024; Fayolle et al. 2023).

Table 1 Tidal acceleration under the form of $\dot{n}/n(10^{-10}\ yr^{-1})$ where n stands for the mean motion and determined from observations. Even though Io is expected to suffer tides more than Europa and Ganymede, the presence of the Laplace resonance dispatches the change of Io's orbital energy into the three innermost Galilean moons. It is noteworthy that Lieske (1987), Lieske (1986a) and Aksnes and Franklin (2001) used the Sampson-Lieske theory to fit their data

	Io	Europa	Ganymede
de Sitter (1928)	+3.3 +/− 0.5	+2.7 +/− 0.7	+1.5 +/− 0.6
Lieske (1987)	−0.074 +/− 0.087	−0.082 +/− 0.097	−0.098 +/− 0.153
Goldstein and Jacobs (1995)	+4.54 +/− 0.95	+5.6 +/− 5.7	+2.8 +/− 2.0
Lieske (1986a)	+2.27 +/− 0.70	−0.67 +/− 0.80	+1.06 +/− 1.00
Aksnes and Franklin (2001)	+3.6 +/− 1.0	-	-
Lainey et al. (2009)	+0.14 +/− 0.01	-0.43 +/− 0.10	-1.57 +/− 0.27

4 The Jovian System

4.1 The Early Measurements

De Sitter was probably the first to measure the expected tidal acceleration of the Galilean moons. Benefiting from heliometric instruments and photographic plates, he was able to envisage the detection of tidal accelerations. Unfortunately, the complexity of the dynamics involved in the large masses of the Galilean moons made it extremely difficult to develop an analytical solution for orbital motions. This question remained topical throughout the twentieth century, notably with the famous work of J.H. Lieske. Lieske put a great deal of effort into demonstrating the expected tidal accelerations. Interestingly, he collected 16,842 of eclipses data, some dating back to the mid-17th century (Lieske 1986a,b). If the connection of the time scale used at the time with the modern time scale was a problem, so too was the absence of long periodic terms in the so-called Sampson-Lieske theory.

It was only at the beginning of the 21st century that the dynamic inconsistencies of this orbital theory came to light (Lainey et al. 2004; Avdyushev 2004). In particular, the absence of long periodic terms could easily be confused with tidal accelerations. As many observers used Sampson-Lieske theory to adjust tidal accelerations, they obtained inaccurate results. Interestingly, depending on the period covered by the astrometric data used by the observers, their results simply seemed to contradict each other (Table 1).

4.2 Using Numerical Models

Benefiting from numerical simulations, the first accurate estimation of tides inside Jupiter and Io was given by Lainey et al. (2009). Introducing the Jovian k_2 from Gavrilov and Zharkov (1977) and assuming a constant quality factor Q for Jupiter, they found that Io's orbit is decaying, while Ganymede's and Europa's orbits are expanding. The fitted dissipation values correspond to orbital acceleration values (dn/dt)/n of $+0.14+/-0.01$ x 10^{-10} yr^{-1}; $-(0.43+/-0.10)$ x 10^{-10} yr^{-1}; and $-(1.57+/-0.27)$ x 10^{-10} yr^{-1}; (formal error bars, 1-sigma) for Io, Europa and Ganymede, respectively. These accelerations represent a shift in the satellite orbital positions of respectively 55 km, 2125 km and 2365 km over the 116 years considered. Using such numerical values, the study of the Laplace resonance angle suggests that the Laplace resonance is evolving, which could be the consequence of tidal cycles within Io (Ojakangas and Stevenson 1986). More recently, Park et al. (2025) re-estimated the tidal dissipation within Io and Jupiter. They found a solution consistent with

Lainey et al. (2009). However, in the continuation of former works, such results assumed a constant Jovian Q value for all tidal frequencies. How far such an approximation is reliable remains an open question. In particular, computation of tidal dissipation inside planets with a fluid envelope and solid core predicts a strongly frequency dependent dissipation (e.g. Ogilvie and Lin 2004). This suggests that the error bar on the Jupiter and Io tidal ratio k_2/Q could be underestimated.

4.3 Love Numbers Estimation

The knowledge of the gravity field of Jupiter was revolutionized by the Juno spacecraft, which has been in orbit around the gas giant since 2016. Its polar orbit, characterized by a very low pericenter radius, allows a strong sensitivity to many of the components in which the planet's gravity can be decomposed. Given an isolated non-rotating body in space, its internal mass distribution is expected to be spherically symmetric, and the corresponding external gravity field is equivalent to that of a simple point mass. Because of the rotation around its spin axis, the shape and mass distribution change under the combined effects of internal pressure and the centrifugal force. If the body is in hydrostatic equilibrium, which is a good assumption over long time scales, especially for a gaseous planet, the shape of the planet will follow an equipotential surface. For a gaseous planet rotating rigidly around its principal polar axis of inertia, the mass distribution is expected to be symmetric around the spin axis and between the North and South hemispheres. The resulting gravity field can be represented only by zonal harmonics (order $m=0$) of even degree ($l=2n$), whose magnitude decreases with the degree (solid body rotation). Thanks to the favourable orbit and the state-of-the-art radio science instrumentation, Juno was able to measure the even zonal harmonics coefficients up to degree 10 (Serra et al. 2019; Durante et al. 2020), allowing to characterize its interior structure to an unprecedented level of detail (e.g. Stevenson 2020; Helled et al. 2022; Miguel et al. 2022; Militzer et al. 2022).

In addition to the dominating solid body rotation, many smaller effects perturbing the mass distribution symmetry are at play. First, the dissipation of internal energy sustains the atmospheric dynamics visible on the upper layer of the gas giants, causing variations in the interior pressure and density and, thus, in the gravity field. The largest effect is represented by the zonal winds, which exhibits a symmetry around the spin axis but are asymmetric between the North and South hemispheres. As a result, the gravity field also presents smaller odd zonal harmonics. Juno detected for the first time the odd zonal harmonics coefficients up to degree 9 (Iess et al. 2018), helping constraining the depth of the zonal winds to about 3000 km (Kaspi et al. 2018), after which the planet rotates nearly as a rigid body (Guillot et al. 2018). In addition to the zonal winds, smaller-scale vortices introduce non-zonal, time-varying terms in the gravity field expansion, whose amplitude is expected to be smaller than the Juno sensitivity. However, Juno data allowed to detect the gravity signature of the great red spot, the biggest atmospheric vortex in Jupiter's atmosphere, constraining its depth to less than 500 km (Parisi et al. 2021).

Besides atmosphere dynamics, normal modes of giant planets are, in principle, able to perturb the interior density profile, producing a time-varying gravity field. Accumulating the data of many perijove passes, Juno experienced a time-varying gravity field that could be explained by normal modes, in particular by p-modes with a peak radial velocity at the surface of 10-50 cm/s at 900-1200 μHz (Durante et al. 2022).

Finally, Jupiter's shape and distribution of mass are perturbed by the tidal potential generated by its natural satellites, among which the dominant effect is due to the Galilean satellites and especially Io, because of its combination of mass and distance. Each satellite produces a tidal bulge on the planet, oriented mainly along the direction of the satellite itself, where a small misalignment is expected due to the planet's internal dissipation, equivalent to time and phase lags. The resulting effect on the planet's gravity field can be modeled through a time variation of the spherical harmonics coefficients, parameterized by a complex Love number $k_{l,m}$ for each degree l and order m, where the imaginary part is related to the tidal dissipation. Up to its mid-mission in December 2018, Juno was able to detect the variation of the gravity field due to the different orientations of the tidal bulge between the perijove passes, providing a statistically significant measure of the real part of the Love numbers $k_{2,2} = 0.565 + / - 0.018$, $k_{3,1} = 0.248 + / - 0.046$, $k_{3,3} = 0.34 + / - 0.12$, $k_{4,2} = 1.29 + / - 0.19$, and marginally $k_{4,4} = 0.55 + / - 0.41$, assumed constant for all satellites (Durante et al. 2020). In particular, the Love numbers $k_{2,2}$ and $k_{4,2}$ were found to be different than the values coming from interior model predictions for a static tidal response to Io (0.590 and 1.74, respectively; Wahl et al. 2020), possibly caused by a dynamical response (Idini and Stevenson 2021, 2022a,b; Dewberry and Lai 2022; Dewberry 2023; Lin 2023). However, interior models predict that Jupiter's static tidal response varies with the distance to the perturber because of its large oblateness (Wahl et al. 2020), meaning that a different Love number has to be expected for each satellite. Considering satellite-dependent tides, Juno estimation capability degrades, only the Love numbers at Io's frequency are observable, and $k_{2,2}$ and $k_{4,2}$ become statistically compatible with theory without the need for dynamical tides (Durante et al. 2020). The smaller sensitivity of Juno data to satellite-dependent tides is caused by the fact that Juno's orbital period ($\sim$53 d) is close to resonance with Io and Europa, so the longitudinal separation between the two moons during the perijove passes is almost constant, making difficult to distinguish the different tidal bulges (Notaro et al. 2019). However, during the extended mission, Juno performed several close encounters with the Galilean satellites (see Sect. 3.4), which decreased its orbital period to 43 days after Ganymede's flyby, to 38 days after the one Europa, and finally to 33 days after the two flybys of Io. This would disentangle the tidal effects of Io and Europa, potentially allowing to obtain a reliable frequency-dependent tidal estimation and detect deviations from the static theory.

Finally, Juno does not allow to characterize the tidal dissipation within Jupiter. In fact, the data are directly sensitive to the gravity field variation caused by the tidal bulges, so that a comparable estimation accuracy can be achieved on both the real and imaginary parts of Jupiter's Love numbers. If the obtained accuracy provides a good estimation of the real parts, the imaginary parts are expected to be much smaller in value, preventing a statistically significant estimation.

5 The Saturnian System

Although further from Earth than Jupiter, the Saturn system has a number of features that make it ideal for observational determination of tidal effects within this system. First of all, the number of moons is much greater, making it easier to retrieve information and obtain a global solution. But above all, there is no resonance involving at once the majority of the system's moons, as is the case with the Laplace resonance in the Jovian system. Irrespective of the crucial importance of the Cassini mission, these assets have certainly helped a great deal in obtaining the tidal parameters within this system.

5.1 Mid-Sized Moon Orbital Expansion

As mentioned in the introduction, Kozai (1957) and Dourneau (1987) determined some first estimations of tidal secular accelerations in the Saturnian system. But similarly to the Galilean system, long periodic terms were missing in their dynamical model making their observation irrelevant to tides (Vienne et al. 1992). The first accurate quantification of Saturn's k_2/Q factor was published in 2012 (Lainey et al. 2012). At the time of their analysis, the authors did not have access to astrometric data from the ISS camera. A set of observations covering the period 1886-2009 was required to obtain a reliable signal from the astrometric data. Lainey et al. (2012) found an intense tidal dissipation ratio ($k_2/Q = (2.3+/-0.7)$ x $10-4$), which was about 10 times higher than the usual value estimated from theoretical arguments (Goldreich and Soter 1966). To support their measurement, the authors noticed that eccentricity equilibrium for Enceladus could now account for the huge heat emitted from Enceladus' south pole (Spencer et al. 2009). Moreover, the measured k_2/Q was found to be poorly sensitive to the tidal frequency, within the uncertainty of the measurement, on the short frequency interval considered. Such a large tidal dissipation within Saturn suggested that the moon current orbits would be inconsistent with the moon formations 4.5 Byr ago above the synchronous orbit in the Saturnian subnebulae. But it would be compatible with a new model of satellite formation in which the Saturnian satellites formed possibly over a longer timescale at the outer edge of the main rings (Charnoz et al. 2011).

A second estimation of Saturnian tides was performed several years later, which benefitted from ISS data and presented an independent analysis of the Saturnian system from both IMCCE-Paris Observatory and JPL (Lainey et al. 2017). The use of ISS-Cassini data significantly reduced the uncertainty on the measurements and revealed a strong dissipative ratio at Rhea's tidal frequency. Such result was interpreted by Fuller et al. (2016) as the sign of a resonance locking mechanism between the moons and internal oscillation modes of the planet which can produce rapid tidal migration (Fuller et al. 2024). Using Rhea's orbital expansion, Fuller et al. (2016) predicted that Titan's orbit could similarly be involved in tidal locking, allowing its orbit to expand extremely fast over time.

5.2 Titan's Orbital Expansion

The hypothesis of a fast orbital expansion was tackled by Lainey et al. (2020) using two different approaches. The first one relied on radio-science data acquired during ten of the Titan's flyby by Cassini. Solving for both Cassini and Titan's orbit over a decade, the authors found a peculiar feature on Titan's orbital motion that suggested a strong orbital expansion corresponding to a tidal Saturnian Q of 124 (105-150 at 3-sigma level). In particular, as discussed in Sect. 3.2, the Cassini radio science data accuracy proved fundamental to obtain Titan's orbital acceleration over the period of Cassini mission. The second method relied on classical astrometry data (both from ground and space), adding new images involving Hyperion and Titan. This solution provided a Q estimation of 61 (30-301 at 3-sigma level). Both solutions were found consistent within 2-sigma uncertainties.

In their work, the authors provided a new estimation of tidal dissipation at mid-sized moons frequency, also. The roughly constant migration time t_{tide} for several of Saturn's moons suggested the authors that resonance locking with inertial waves, rather than gravity modes (which predicts smaller t_{tide} values for outer moons), is the most probable explanation for the moons' migration. This also helped to explain why mean-motion resonances between moons have survived, because resonance locking with gravity modes typically results in divergent migration that can disrupt mean-motion resonances between moons, whereas

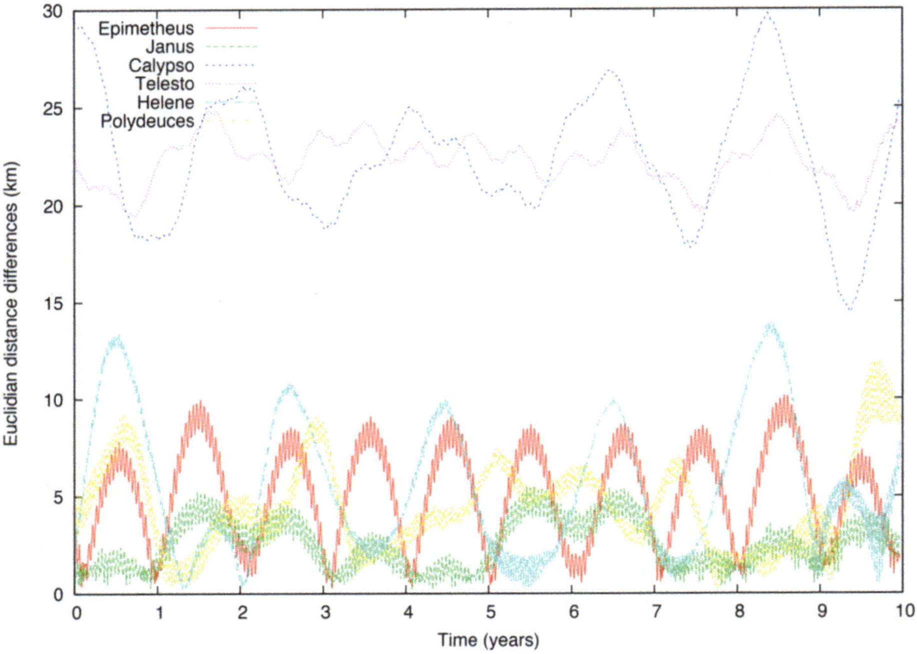

Fig. 7 Postfit residuals associated with cross-tidal effects from Lainey et al. (2017). Kilometric signals appear associated with the Saturn's tidal bulge at Janus/Epimetheus', Mimas', Enceladus', Tethys' and Dione's tidal frequencies

resonance locking with inertial waves does not necessarily produce divergent migration and allows mean-motion resonances to survive.

More recently, the large tidal expansion of Titan was called into question by Jacobson (2022). Using similar dataset, the author could not reproduce the results obtained from radio-science data by Lainey et al. (2020). But the author found that a Voyager 1 flyby in addition to Cassini's flyby data can measure Titan's orbital expansion. Jacobson (2022) obtained a value about ten times larger for the Saturnian tidal Q parameter. Future analysis will have to be done to understand the source of such a discrepancy.

5.3 Love Number Estimations

The large number and accuracy of the ISS-Cassini dataset concerned also the small coorbital moons of Tethys and Dione, called Telesto, Calypso, Helen and Polydeuces. In particular, (Lainey et al. 2017) could use the orbital motion of these moons to determine, for the first time from observations, the Love number k_2 of Saturn. This determination allowed the authors to discriminate significantly the interior modelling of Saturn. In particular, the Love number appeared to be extremely complementary in term of interior constraints to the usual gravitational coefficients J_2, J_4 and J_6 determined from radio-science data (Lainey et al. 2017).

Figure 7 shows the incompressible dynamical effect of Saturn's k_2 on the orbital motion of the system's small moons after fitting the moons' initial conditions, masses, Saturn's J_2, polar orientation and precession and Saturn's tidal Q. This represents up to several tens of kilometers on the motion of Telesto and Calypso. Such signal appears extremely clearly in the post-fit residuals. Even more recently, it has been possible to quantify Saturn's k_2 at several frequencies. To do this, Lainey et al. (2024) took advantage of a number of observations covering the whole thirteen years of the Cassini mission. They were able to determine the number of Love k_2 of Saturn at the frequency of Janus/Epimetheus, Mimas, Tethys and Dione. Within the error bars, Saturn's k_2 solution appears consistent with both a static value and predictions from numerical models of the tidal response of Saturn, including contributions from fundamental modes, gravito-inertial modes, and inertial waves (Lainey et al. 2024).

6 The Icy Giants: Uranus and Neptune

The first accelerations reported for icy giants is the work of Emelyanov and Nikonchuk (2016). But the large and negative sign makes these estimations extremely suspicious. Considering the large distance of Uranus and Neptune, it is challenging to make any reliable constraints before a dedicated space mission or the release of more Gaia data. More recently, Jacobson and Park (2025) determined orbital expansion of the main Uranian moons implying $Q = 678 \pm 231$ (assuming $k_2 = 0.3$). While their values (including the sign) are significantly different from Emelyanov and Nikonchuk (2016), they still partly use the same dataset. The question of potential biases within the old astrometric data, particularly less accurate than for the Jovian and Saturnian system is raised. Digitisation of old plates may help in the future to solve this question (Sect. 2.1.1).

For the moment, the Gaia data available covers the period 2014-2020. With a magnitude of between 14 and 15 (with the exception of Miranda, which is not observable by Gaia), the main Uranian satellites correspond perfectly to Gaia's capabilities. Along track measurements guarantee fantastic precision on the orbits of these moons, with residuals of typically 1 milliarcsecond (Fig. 8). Recent theoretical estimation of the Uranian tidal parameter Q expect possible values as low as 10^3 (Nimmo 2023). Lainey (2016) studied the tidal signal amplitude to look for assuming $Q = 100$ for Uranus. The author found a post-fit signal of about 40 mas on Ariel over 100 years of data. Assuming linearity in Q, the Nimmo (2023) value implies a post-fit amplitude associated to tides of about few to several mas, only. Limiting the analysis to the currently available Gaia interval data, this would induce a signal even few hundred times smaller, since tides are a quadratic effect over time, way below the Gaia precision. Nevertheless, a Q of several hundred might be detectable with the arrival of the last Gaia astrometric catalogue in late 2020s and the use of the most accurate astrometric data gathered over more than a century, including Voyager (Jacobson 1992) and mutual events data (Arlot et al. 2013).

For the Neptune system, Lainey (2016) provided a similar study, assuming Q=100 for the planet. A signal of 5 mas over 100 years was found. This is almost an order of magnitude smaller than for the Uranian system, due to the largest distance to the Earth and the significant semi-major axis of Triton. A quantification of tides in Neptune may well depend on the existence of a future space mission.

7 Exoplanets

With the discovery of 51 Peg b (Mayor and Queloz 1995), a so-called hot-Jupiter on an extremely tight 4-day orbit around its star, tides have been recognized early as crucial in defining the orbital properties of exoplanets: close-in planets, with orbital periods less than about

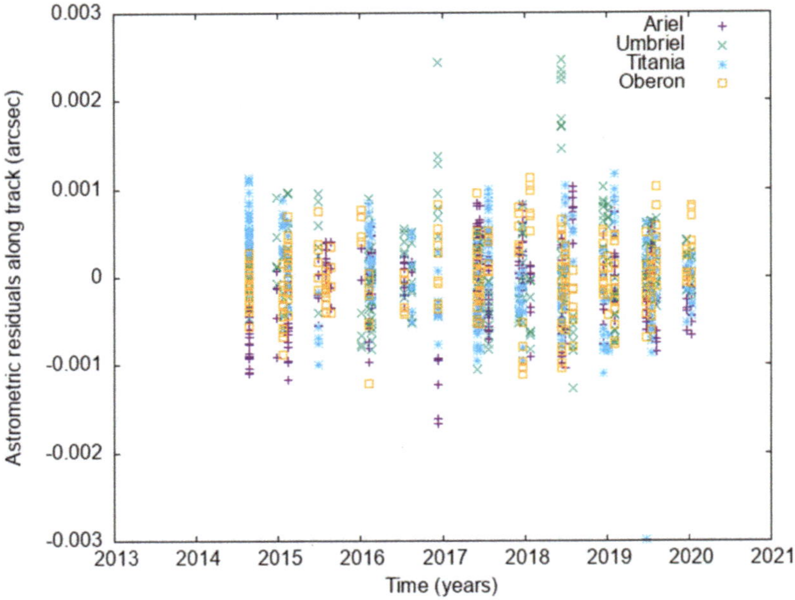

Fig. 8 Posfit residuals for the main Uranian moons with the Gaia data. Being too close to Uranus, Miranda's data are not available

10 days, should be synchronized with their parent star (Guillot et al. 1996) and circularized (Rasio et al. 1996). Tides have been shown to potentially affect the sizes of hot Jupiters (e.g., Bodenheimer et al. 2001; Guillot and Showman 2002; Guillot and Havel 2011). It was later found that Hot Jupiters generally orbit in the equatorial plane of their star except when orbiting hot stars with no outer convective zone (see Winn and Fabrycky 2015), showing the importance of tides in realigning the orbital plane and of convection in ensuring an efficient dissipation (see Zahn 1977; Ogilvie 2014).

The tidal evolution timescales were derived by Hut (1981), Eggleton et al. (1998) on the basis of an assumed quality factor inversely proportional to the efficiency of dissipation of the tidal energy per orbit. It is useful to consider simple configurations, with a star+planet (« hot Jupiter ») system defined by their masses $M_{1,2}$, radii $R_{1,2}$, modified quality factors $Q'_{1,2}$, semi-major axis a, eccentricity i, orbital period P, stellar spin period $P_\star$. In that case the mean orbital evolution timescale is (Barker and Ogilvie 2009):

$$\tau_a \equiv -\frac{2}{13}\frac{a}{\dot{a}} \approx 12.0\,\mathrm{Myr}\left(\frac{Q'_1}{10^6}\right)\left(\frac{M_1}{\mathrm{M}_\odot}\right)^{8/3}\left(\frac{M_2}{\mathrm{M_J}}\right)^{-1}\left(\frac{R_1}{\mathrm{R}_\odot}\right)^{-5}\left(\frac{P}{1\,\mathrm{d}}\right)^{13/3}\left(1-\frac{P}{P_\star}\right)^{-1}.$$

This implies that if, as is the case for most hot Jupiters $P > P_\star$, the planet's orbit will progressively shrink. The $P^{13/3}$ dependence however implies that only the planets with the shortest orbits will be affected in a measurable way.

The orbit is also circularized on a timescale:

$$\tau_e \equiv -\frac{e}{\dot{e}} \approx 16.8\,\mathrm{Myr}\left(\frac{Q'_1}{10^6}\right)\left(\frac{M_1}{\mathrm{M}_\odot}\right)^{8/3}\left(\frac{M_2}{\mathrm{M_J}}\right)^{-1}\left(\frac{R_1}{\mathrm{R}_\odot}\right)^{-5}\left(\frac{P}{1\,\mathrm{d}}\right)^{13/3}$$
$$\times\left[\left(f_1\left(e^2\right)-\frac{11}{18}\frac{P}{P_\star}f_2\left(e^2\right)\right)+\beta\left(f_1\left(e^2\right)-\frac{11}{18}f_2\left(e^2\right)\right)\right]^{-1},$$

where $f_{1,2}(e^2)$ are functions defined in Barker and Ogilvie (2009) with values close to unity and β defines the ratio of the importance of the planetary to the stellar tides:

$$\beta = \frac{Q_1'}{Q_2'}\left(\frac{M_1}{M_2}\right)^{1/3}\left(\frac{\overline{\rho}_1}{\overline{\rho}_2}\right)^{5/3},$$

and $\overline{\rho}_{1,2}$ are the stellar and planetary mean densities, respectively.[1] Hot Jupiters have mean densities that are generally between 1 and 0.1 g/cm^3 for the most inflated ones implying that even though they are ~1000 times less massive than their parent star, their low density and low quality factor can lead to $\beta > 1$, i.e., planetary tides dominating the circularization over the stellar tides.

Lastly, it the orbital and stellar equatorial planes are misaligned, dissipation of the stellar tide (raised on the star by the planet) would align them on a timescale

$$\tau_i \equiv -\frac{i}{di/dt} \approx 70.0\,\mathrm{Myr}\left(\frac{Q_1'}{10^6}\right)\left(\frac{M_1}{\mathrm{M}_\odot}\right)\left(\frac{M_2}{\mathrm{M_J}}\right)^{-2}\left(\frac{R_1}{\mathrm{R}_\odot}\right)^{-3}\left(\frac{P}{1\,\mathrm{d}}\right)^{4}\left(\frac{\Omega_1}{\Omega_0}\right)$$
$$\times\left[1-\frac{P}{2P_\star}\left(1-\frac{1}{\alpha}\right)\right]^{-1},$$

where a circular orbit and small i are assumed, Ω_1 is the star spin frequency, $\Omega_0 = 5.8 \times 10^{-6}\ \mathrm{s}^{-1}$ corresponds to a ~12.5 d spin period and α is the ratio of orbital to stellar spin angular momentum.

From ultra-precise timing of transits over a long time, at least two planets have been measured to be falling into their star. WASP-12b, an inflated 1.9 $\mathrm{R_J}$, 1.4 $\mathrm{M_J}$ hot Jupiter with an orbital period of about P=26 hours around a solar-type star, has been measured to have a shrinking orbit with $\dot{P} = -29.81 \pm 0.94$ ms/yr (Maciejewski et al. 2016a,b; Wong et al. 2022), compatible with a stellar quality factor $Q_\star' = 1.5 \times 10^5$, on the lower end of expectations from stellar models. Kepler-1658b, a massive hot Jupiter (1.1 $\mathrm{R_J}$, 1.4 $\mathrm{M_J}$) with a P=3.85 day orbit around an evolved F-type star (2.9 $\mathrm{R}_\odot$, 1.5 $\mathrm{M}_\odot$), was found to also have a decreasing orbital period $\dot{P} = -131^{+20}_{-22}$ ms/yr, corresponding to an infall timescale $\tau_a \sim 2.5$ Myr and a stellar quality factor $Q_\star' = 2.5 \times 10^4$ (Vissapragada et al. 2022). This quality factor appears much lower than expectations, except during a very limited ~100 yrs evolution phase during which inertial waves may be driven in the stellar interior indicating that other explanations should be found (Barker et al. 2024).

Studies of orbital circularization have also been able to put constraints on the tidal quality factors for hot Jupiters as shown in Fig. 8. Studies of individual systems have the potential to provide some constraints: For example, the unique configuration of HAT-P-13b, an eccentric hot Jupiter in resonance with an outer eccentric companion puts constraints on tidal dissipation in the inner planet (Batygin et al. 2009). This is also the case if WASP-12b's rapid inward migration is caused by obliquity tides (Millholland and Laughlin 2018). Lower limits can be obtained from the period decay of XO-3b (Yang and Wei 2022), and from detailed analyses using asteroseimic information for the Kepler-91 system (Fellay et al. 2023).

Studies of ensemble of systems provide powerful statistical constraints. Studies of orbit circularization in open clusters are promising because the known stellar ages lift at least one degeneracy. Figure 8 shows that these studies applied to 0.7 Gyr M67 and 4 Gyr Hyades and Praesepe lead to values of Q_p' ranging from about 2×10^6 to 2×10^7 (O'Connor and

[1] This equation corrects a typo in Barker and Ogilvie (2009) in which had been inverted $\beta\, Q_1'/Q_2'$.

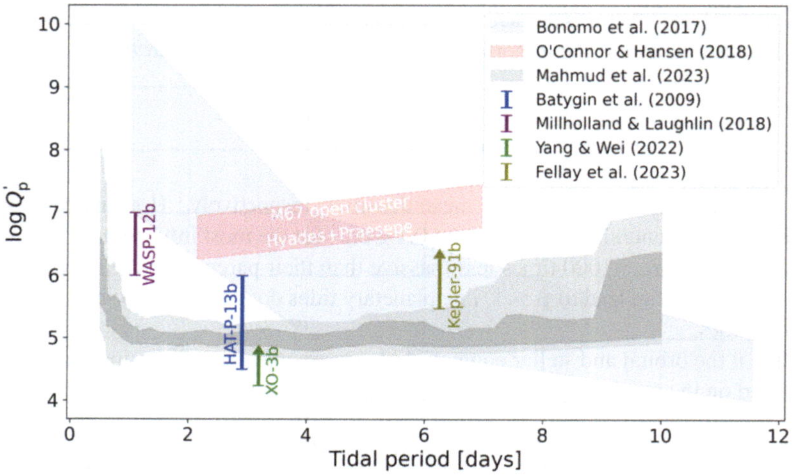

Fig. 9 Constraints obtained on the modified planetary quality factor $log Q'_p$ as a function of the tidal period obtained from different works as labelled. Constraints on individual systems (WASP-12b, HAT-P-13b) and lower limits (XO-3b, Kepler-91b) are indicated with error bars, and arrows, respectively. Upper limits obtained from the circularization of planets supposed to be initially eccentric coupled to lower limits for systems with a non-zero eccentricity are shown in pale blue and approximated from the results of Bonomo et al. (2017). Constraints obtained from 6 planets in the open clusters M67, Hyades and Praesepe by O'Connor & Hansen and recalculated as a function of $log Q'_p$ are shown as a pink area. Constraints derived from tidal evolution calculations of 78 planets analyzed with a Bayesian approach (Mahmud et al. 2023) are shown in gray

Hansen 2018), but the small number of planets (6) in these clusters limits the robustness of the estimation. With a selected sample of 211 planets, Bonomo et al. (2017) focused on deriving upper limits on Q'_p for planets with circular orbits and lower limits for planets with known eccentricites yielding constraints that appear to favor higher values of Q'_p for the shortest orbits and lower ones, down to about 104 for periods of 10 days and higher. Perhaps the most involved study so far is that of Mahmud et al. (2023) which includes a detailed study of the combined tidal evolution of the star and planet for a sample of 78 single-planet systems with the hypothesis that the planet quality factor only depends on the tidal period. The Bayesian analysis results in rather well constrained values $Q'_p = 10^{5\pm0.5}$, independently of the tidal period between 1 and 10 days. The discrepancies between the different methods seen in Fig. 8 however shows that more work is needed.

8 Conclusion

The determination of tidal parameters within giant planet systems has taken a major leap forward in the last fifteen years, thanks in particular to data from the Cassini and Juno space probes. The imminent arrival of the new generation of probes such as Juice (ESA) and Europa Clipper (NASA) should take us a step further in our understanding of the dissipative phenomena at work within these systems. The question of the origin of the high energy dissipation under the effect of tidal phenomena, as well as its dependence on the excitation frequency, remains a subject for further study if we ultimately wish to understand the past dynamic evolution of these systems.

Beyond the Solar system, data from future ground- and space-based programs such as PLATO (ESA) should enable us to balance what we observe today in the Solar system with the dynamic richness of the thousands of exoplanetary systems known today. It is to be hoped that the decade to come will enable us to integrate the dynamic study of all these systems into a global context, clearly revealing the physical mechanisms underlying tidal dissipation in these objects.

Acknowledgements This paper resulted from the Workshop "New Vision of the Saturnian System in the Context of a Highly Dissipative Saturn" held at the International Space Science Institute (ISSI) in Bern (Switzerland).

Funding Information Open access funding provided by Alma Mater Studiorum - Università di Bologna within the CRUI-CARE Agreement. M. Zannoni is grateful to the Italian Space Agency (ASI) for financial support through Agreement No. 2023-6-HH.O.

Declarations

Competing Interests The authors have no conflicts of interest to report.

References

Aksnes K, Franklin FA (2001) Secular acceleration of Io derived from mutual satellite events. Astron J 122:2734–2739

Arlot J-E, et al (2013) Astrometric results of observations of mutual occultations and eclipses of the Uranian satellites in 2007. Astron Astrophys 557

Asmar SW, Armstrong JW, Iess L, Tortora P (2005) Spacecraft Doppler tracking: noise budget and accuracy achievable in precision radio science observations. Radio Sci 40(2):1–9. https://doi.org/10.1029/2004RS003101

Avdyushev VA (2004) A new system of initial parameters for numerical simulation of the motion of Jupiter's Galilean satellites. Sol Syst Res 38:238–240

Barker AJ, Ogilvie GI (2009) On the tidal evolution of hot Jupiters on inclined orbits. Mon Not R Astron Soc 395:2268

Barker AJ, Efroimsky M, Makarov VV, Veras D (2024) On the orbital decay of the gas giant Kepler-1658b. Mon Not R Astron Soc 527:5131

Batygin K, Bodenheimer P, Laughlin G (2009) Determination of the interior structure of transiting planets in multiple-planet systems. Astrophys J 704:L49

Bertotti B, Comoretto G, Iess L (1993) Doppler tracking of spacecraft with multi-frequency links. Astron Astrophys 269:608–616

Bodenheimer P, Lin DNC, Mardling RA (2001) On the tidal inflation of short-period extrasolar planets. Astrophys J 548:466

Bonomo AS, Desidera S, Benatti S, Borsa F, Crespi S, et al (2017) The GAPS programme with HARPS-N at TNG. XIV. Investigating giant planet migration history via improved eccentricity and mass determination for 231 transiting planets. Astron Astrophys 602:A107

Brozovic M, et ak (2020) Arecibo Radar Astrometry of the Galilean Satellites from 1999 to 2016. Astron J 159

Buccino DR, Kahan DS, Parisi M, Paik M, Barbinis E, Yang O, Park RS, Tanner A, Bryant SH, Jongeling A (2021) Performance of Earth troposphere calibration measurements with the advanced water vapor radiometer for the Juno gravity science investigation. Radio Sci 56(12):1–9. https://doi.org/10.1029/2021RS007387

Campbell JK, Anderson JD (1989) Gravity field of the Saturnian system from Pioneer and Voyager tracking data. Astron J 97:1485. https://doi.org/10.1086/115088

Charnoz S, et al (2011) Accretion of Saturn's mid-sized moons during the viscous spreading of young massive rings: solving the paradox of silicate-poor rings versus silicate-rich moons. Icarus 216(2):535–550. https://doi.org/10.1016/j.icarus.2011.09.017

Cooper NJ, Murray CD, Lainey V, Tajeddine R, Evans MW, Williams GA (2014) Cassini ISS mutual event astrometry of the mid-sized Saturnian satellites 2005-2012. Astron Astrophys 572

Cooper NJ, et al (2018) The Caviar software package for the astrometric reduction of Cassini ISS images: description and examples. Astron Astrophys 610

de Sitter W (1916) Jupiter, the longitude of the satellites of, from photographs taken at the Cape Observatory in 1913. Mon Not R Astron Soc 76:448

de Sitter W (1928) Orbital elements determining the longitudes of Jupiter's satellites, derived from observations. Leiden Ann 16(2):1–92

Desage L, Hérique A, Lainey V, Kofman W, Cicchetti A, Orosei R (2024) MARSIS data as a new constraint for the orbit of Phobos. Astron Astrophys 686:A136. https://doi.org/10.1051/0004-6361/202348655

Dewberry JW (2023) Dynamical tides in Jupiter and other rotationally flattened planets and stars with stable stratification. Mon Not R Astron Soc 521(4):5991–6004. https://doi.org/10.1093/mnras/stad546

Dewberry JW, Lai D (2022) Dynamical tidal Love numbers of rapidly rotating planets and stars. Astrophys J 925(2):124. https://doi.org/10.3847/1538-4357/ac3ede

Dourneau G (1987) Observations et études du mouvement des huit principaux satellites de Saturne. Thèse, Bordeaux

Durante D, Parisi M, Serra D, Zannoni M, Notaro V, Racioppa P, Buccino DR, Lari G, Gomez Casajus L, Iess L, Folkner WM, Tommei G, Tortora P, Bolton SJ (2020) Jupiter's gravity field halfway through the Juno mission. Geophys Res Lett 47(4):1–8. https://doi.org/10.1029/2019GL086572

Durante D, Guillot T, Iess L, Stevenson DJ, Mankovich CR, Markham S, Galanti E, Kaspi Y, Zannoni M, Gomez Casajus L, Lari G, Parisi M, Buccino DR, Park RS, Bolton SJ (2022) Juno spacecraft gravity measurements provide evidence for normal modes of Jupiter. Nat Commun 13(1):4632. https://doi.org/10.1038/s41467-022-32299-9

Duriez L, Vienne A (1991) A general theory of motion for the eight major satellites of Saturn. I. Equations and method of resolution. Astron Astrophys 243:263–275

Eggleton PP, Kiseleva LG, Hut P (1998) The equilibrium tide model for tidal friction. Astrophys J 499:853

Emelyanov NV (2017) Current problems of dynamics of moons of planets and binary asteroids based on observations. Sol Syst Res 51:20–37

Emelyanov NV, Nikonchuk DV (2013) Ephemerides of the main Uranian satellites. Mon Not R Astron Soc 436:3668–3679. https://doi.org/10.1093/mnras/stt1851

Emelyanov NV, Nikonchuk DV (2016) Erratum: Ephemerides of the main Uranian satellites: Table 1. Mon Not R Astron Soc 461:1718

Emelyanov NV, Samorodov MY (2015) Analytical theory of motion and new ephemeris of Triton from observations. Mon Not R Astron Soc 454:2205–2215. https://doi.org/10.1093/mnras/stv2116

Fayolle M, Magnanini A, Lainey V, Dirkx D, Zannoni M, Tortora P (2023) Combining astrometry and JUICE – Europa Clipper radio science to improve the ephemerides of the Galilean moons. Astron Astrophys 677:A42. https://doi.org/10.1051/0004-6361/202347065

Fellay L, Pezzotti C, Buldgen G, Eggenberger P, Bolmont E (2023) Constraints on planetary tidal dissipation from a detailed study of Kepler 91b. Astron Astrophys 669:A2

Folkner WM, Williams JG, Boggs DH, Park RS, Kuchynka P (2014) The Planetary and Lunar Ephemerides DE430 and DE431. The Interplanetary Network Progress Report, 1–81. http://tmo.jpl.nasa.gov/progress_report/42-196/196C.pdf

Fuller J, Luan J, Quataert E (2016) Resonance locking as the source of rapid tidal migration in the Jupiter and Saturn moon systems. Mon Not R Astron Soc 458:3867–3879

Fuller J, Guillot T, Mathis S, Murray C (2024) Tidal Dissipation in Giant Planets. Space Sci Rev 220

Gaia Collaboration (2023) Gaia data release 3: summary of the content and survey properties. Astron Astrophys 674:A1. https://doi.org/10.1051/0004-6361/202243940

Gavrilov SV, Zharkov VN (1977) Love numbers of the giant planets. Icarus 32:443–449. https://doi.org/10.1016/0019-1035(77)90015-X

Goldreich P, Soter S (1966) Q in the Solar System. Icarus 5(1–6):375–389. https://doi.org/10.1016/0019-1035(66)90051-0

Goldstein SJ Jr, Jacobs KC (1995) A recalculation of the secular acceleration of Io. Astron J 110:3054–3057

Gomez Casajus L (2019) Development of methods for the global ephemerides estimation of the gas giant satellite systems. PhD thesis, Alma Mater Studiorum Università di Bologna
Gomez Casajus L, Ermakov AI, Zannoni M, Keane JT, Stevenson DJ, Buccino DR, Durante D, Parisi M, Park RS, Tortora P, Bolton SJ (2022) Gravity field of Ganymede after the Juno extended mission. Geophys Res Lett 49(24):1–10. https://doi.org/10.1029/2022GL099475
Guillot T, Havel M (2011) An analysis of the CoRoT-2 system: a young spotted star and its inflated giant planet. Astron Astrophys 527:A20
Guillot T, Showman AP (2002) Evolution of "51 Pegasus b-like" planets. Astron Astrophys 385:156
Guillot T, Burrows A, Hubbard WB, Lunine JI, Saumon D (1996) Giant planets at small orbital distances. Astrophys J 459:L35
Guillot T, Miguel Y, Militzer B, Hubbard WB, Kaspi Y, Galanti E, Cao H, Helled R, Wahl SM, Iess L, Folkner WM, Stevenson DJ, Lunine JI, Reese DR, Biekman A, Parisi M, Durante D, Connerney JEP, Levin SM, Bolton SJ (2018) A suppression of differential rotation in Jupiter's deep interior. Nature 555(7695):227–230. https://doi.org/10.1038/nature25775
Harmon JK, Ostro SJ, Chandler JF, Hudson RS (1994) Radar ranging to Ganymede and Callisto. Astron J 107:1175
Harper D, Taylor DB (1993) The orbits of the major satellites of Saturn. Astron Astrophys 268:326–349
Harper D, Taylor DB, Sinclair AT, Kaixian S (1988) The theory of the motion of Iapetus. Astron Astrophys 191:381–384
Helled R, Stevenson DJ, Lunine JI, Bolton SJ, Nettelmann N, Atreya S, Guillot T, Militzer B, Miguel Y, Hubbard WB (2022) Revelations on Jupiter's formation, evolution and interior: challenges from Juno results. In: Icarus, vol 378. Academic Press, San Diego. https://doi.org/10.1016/j.icarus.2022.114937
Howell SB (2006) Handbook of CCD astronomy. Cambridge University Press, Cambridge
Hut P (1981) Tidal evolution in close binary systems. Astron Astrophys 99:126
Idini B, Stevenson DJ (2021) Dynamical tides in Jupiter as revealed by Juno. Planet Sci J 2(2):69. https://doi.org/10.3847/psj/abe715
Idini B, Stevenson DJ (2022a) The Lost Meaning of Jupiter's High-degree Love Numbers. Planet Sci J 3(1). https://doi.org/10.3847/PSJ/ac4248
Idini B, Stevenson DJ (2022b) The Gravitational Imprint of an Interior–Orbital Resonance in Jupiter–Io. Planet Sci J 3(4). https://doi.org/10.3847/PSJ/ac6179
Iess L, di Benedetto M, James N, Mercolino M, Simone L, Tortora P (2014) Astra: interdisciplinary study on enhancement of the end-to-end accuracy for spacecraft tracking techniques. Acta Astronaut 94(2):699–707. https://doi.org/10.1016/j.actaastro.2013.06.011
Iess L, Folkner WM, Durante D, Parisi M, Kaspi Y, Galanti E, Guillot T, Hubbard WB, Stevenson DJ, Anderson JD, Buccino DR, Gomez Casajus L, Milani A, Park RS, Racioppa P, Serra D, Tortora P, Zannoni M, Cao H, Bolton SJ, et al (2018) Measurement of Jupiter's asymmetric gravity field. Nature 555(7695). https://doi.org/10.1038/nature25776
Iess L, Militzer B, Kaspi Y, Nicholson P, Durante D, Racioppa P, Anabtawi A, Galanti E, Hubbard WB, Mariani MJ, Tortora P, Wahl SM, Zannoni M (2019) Measurement and implications of Saturn's gravity field and ring mass. Science 364(6445):eaat2965. https://doi.org/10.1126/science.aat2965
Iess L, Asmar SW, Cappuccio P, Cascioli G, de Marchi F, di Stefano I, Genova A, Ashby N, Barriot J, Bender P, Benedetto C, Border JS, Budnik F, Ciarcia S, Damour T, Dehant V, di Achille G, di Ruscio A, Fienga A, Zannoni M, et al (2021) Gravity, geodesy and fundamental physics with BepiColombo's MORE investigation. Space Sci Rev 217(1):21. https://doi.org/10.1007/s11214-021-00800-3
Iess L, et al (2024) The 3GM investigation of ESA's JUICE mission: Geodesy and Geophysics of Jupiter and the Galilean Moons. Space Sci Rev. Under submission
Jacobson RA (1992) Astrographic observations of the major Uranian satellites from Voyager 2. Astron Astrophys Suppl Ser 96:549–563
Jacobson RA (2001) The gravity field of the Jovian system and the orbits of the regular Jovian satellites. In: AAS/Division for Planetary Sciences Meeting, vol 33
Jacobson RA (2004) The orbits of the major saturnian satellites and the gravity field of Saturn from spacecraft and Earth-based observations. Astron J 128(1):492–501. https://doi.org/10.1086/421738
Jacobson RA (2014) The orbits of the Uranian satellites and rings, the gravity field of the Uranian system, and the orientation of the pole of Uranus. Astron J 148:76. https://doi.org/10.1088/0004-6256/148/5/76
Jacobson RA (2022) The orbits of the main Saturnian satellites, the Saturnian system gravity field, and the orientation of Saturn's pole. Astron J 164:199. https://doi.org/10.3847/1538-3881/ac90c9
Jacobson RA, Park RS (2025) The orbits of Uranus, its satellites and rings, the gravity field of the Uranian system, and the orientation of the poles of Uranus and its satellites. Astron J 169:65. https://doi.org/10.3847/1538-3881/ad99d1
Jacobson RA, Riedel JE, Taylor AH (1991) The orbits of Triton and Nereid from spacecraft and earthbased observations. Astron Astrophys 247:565–575

Jacobson RA, Haw RJ, McElrath TP, Antreasian PG (2000) A comprehensive orbit reconstruction for the Galileo prime mission in the J2000 system. J Astronaut Sci 48(4):495–516. https://doi.org/10.1007/bf03546268

Jacobson RA, Antreasian PG, Bordi JJ, Criddle KE, Ionasescu R, Jones JB, Mackenzie RA, Meek MC, Parcher DW, Pelletier FJ, Owen WM, Roth DC, Roundhill IM, Stauch JR (2006) The gravity field of the saturnian system from satellite observations and spacecraft tracking data. Astron J 132(6):2520–2526. https://doi.org/10.1086/508812

Kaspi Y, Galanti E, Hubbard WB, Stevenson DJ, Bolton SJ, Iess L, Guillot T, Bloxham J, Connerney JEP, Cao H, Durante D, Folkner WM, Helled R, Ingersoll AP, Levin SM, Lunine JI, Miguel Y, Militzer B, Parisi M, Wahl SM (2018) Jupiter's atmospheric jet streams extend thousands of kilometres deep. Nature 555(7695):223–226. https://doi.org/10.1038/nature25793

Kozai Y (1957) On the astronomical constants of Saturnian satellites system. Ann Tokyo Astron Obs 5:73–106

Lainey V (2008) A new dynamical model for the Uranian satellites. Planet Space Sci 56:1766–1772. https://doi.org/10.1016/j.pss.2008.02.015

Lainey V (2016) Quantification of tidal parameters from Solar System data. Celest Mech Dyn Astron 126:145–156. https://doi.org/10.1007/s10569-016-9695-y

Lainey V, Duriez L, Vienne A (2004) New accurate ephemerides for the Galilean satellites of Jupiter. I. Numerical integration of elaborated equations of motion. Astron Astrophys 420:1171–1183

Lainey V, Arlot J-E, Karatekin Ö, van Hoolst T (2009) Strong tidal dissipation in Io and Jupiter from astrometric observations. Nature 459:957–959

Lainey V, et al (2012) Strong tidal dissipation in Saturn and constraints on Enceladus' thermal state from astrometry. Astrophys J 752:14. https://doi.org/10.1088/0004-637X/752/1/14

Lainey V, et al (2017) New constraints on Saturn's interior from Cassini astrometric data. Icarus 281:286–296

Lainey V, et al (2020) Resonance locking in giant planets indicated by the rapid orbital expansion of Titan. Nat Astron 4:1053–1058

Lainey V, Dewberry W, Fuller J, Cooper N, Rambaux N, Zhang Q (2024) Tidal frequency dependence of the Saturnian k_2 Love number. Astron Astrophys 684:L3. https://doi.org/10.1051/0004-6361/202449639

Lasagni Manghi R, Zannoni M, Tortora P, Martellucci A, de Vicente J, Villalvilla J, Mercolino M, Maschwitz G, Rose T (2021) Performance Characterization of ESA's Tropospheric Delay Calibration System for Advanced Radio Science Experiments. Radio Sci 56(10). https://doi.org/10.1029/2021RS007330

Lasagni Manghi R, Bernacchia D, Gomez Casajus L, Zannoni M, Tortora P, Martellucci A, de Vicente J, Villalvilla J, Maschwitz G, Cappuccio P, Iess L (2023) Tropospheric delay calibration system performance during the first two BepiColombo solar conjunctions. Radio Sci 58(2):1–15. https://doi.org/10.1029/2022RS007614

Laskar J (1986) A general theory for the Uranian satellites. Astron Astrophys 166:349–358

Lieske JH (1974) A method of revitalizing Sampson's theory of the Galilean satellites. Astron Astrophys 31:137

Lieske JH (1986b) A collection of Galilean satellite eclipse observations, 1652-1983. II. Astron Astrophys Suppl Ser 63:143–202

Lieske JH (1986a) A collection of Galilean satellite eclipse observations, 1652-1983. I. Astron Astrophys 154:61–76

Lieske JH (1987) Galilean satellite evolution - observational evidence for secular changes in mean motions. Astron Astrophys 176:146–158

Lieske JH (1996) Galilean satellites and the Galileo space mission. Celest Mech Dyn Astron 66:13–20

Lin FR, Peng JH, Zheng ZJ, Peng QY (2019) Characterization of the precision premium in astrometry. Mon Not R Astron Soc 490(3):4382–4387. https://doi.org/10.1093/mnras/stz2871

Lin Y (2023) Dynamical tides in Jupiter and the role of interior structure. Astron Astrophys 671:A37. https://doi.org/10.1051/0004-6361/202245112

Maciejewski G, Dimitrov D, Fernández M, Sota A, Nowak G, et al (2016a) Departure from the constant-period ephemeris for the transiting exoplanet WASP-12. Astron Astrophys 588:L6

Maciejewski G, Dimitrov D, Mancini L, Southworth J, Ciceri S, et al (2016b) New transit observations for HAT-P-30 b, HAT-P-37 b, TrES-5 b, WASP-28 b, WASP-36 b and WASP-39 b. Acta Astron 66:55

Magnanini A, Zannoni M, Casajus LG, Tortora P, Lainey V, Mazarico E, Park RS, Iess L (2024) Joint analysis of JUICE and Europa Clipper tracking data to study the Jovian system ephemerides and dissipative parameters. Astron Astrophys 687:A132. https://doi.org/10.1051/0004-6361/202347616

Mahmud MM, Penev KM, Schussler JA (2023) Measuring tidal dissipation in giant planets from tidal circularization. Mon Not R Astron Soc 525:876

Mallama A, Stockdale C, Krobusek BA, Nelson P (2010) Assessment of the resonant perturbation errors in Galilean satellite ephemerides using precisely measured eclipse times. Icarus 210:346–357

Mariotti G, Tortora P (2013) Experimental validation of a dual-uplink multifrequency dispersive noise calibration scheme for deep space tracking. Radio Sci 48. https://doi.org/10.1002/rds.20024
Markham S, Durante D, Iess L, Stevenson D (2020) Possible evidence of p-modes in Cassini measurements of Saturn's gravity field. Planet Sci J 1:27. https://doi.org/10.3847/PSJ/ab9f21
Mayor M, Queloz D (1995) A Jupiter-mass companion to a solar-type star. Nature 378:355
Mazarico E, Buccino DR, Castillo-Rogez JC, Dombard AJ, Genova A, Hussmann H, Kiefer WS, Lunine JI, McKinnon WB, Nimmo F, Park RS, Roberts JH, Srinivasan DK, Steinbrügge G, Tortora P, Withers P (2023) The Europa Clipper gravity and radio science investigation. Space Sci Rev 219(4):30. https://doi.org/10.1007/s11214-023-00972-0
Miguel Y, Bazot M, Guillot T, Howard S, Galanti E, Kaspi Y, Hubbard WB, Militzer B, Helled R, Atreya SK, Connerney JEP, Durante D, Kulowski L, Lunine JI, Stevenson D, Bolton S (2022) Jupiter's inhomogeneous envelope. Astron Astrophys 662:A18. https://doi.org/10.1051/0004-6361/202243207
Milani A, Gronchi G (2009) Theory of orbit determination, vol 1. Cambridge University Press, Cambridge. https://doi.org/10.1017/CBO9781139175371
Militzer B, Hubbard WB, Wahl S, Lunine JI, Galanti E, Kaspi Y, Miguel Y, Guillot T, Moore KM, Parisi M, Connerney JEP, Helled R, Cao H, Mankovich C, Stevenson DJ, Park RS, Wong M, Atreya SK, Anderson J, Bolton SJ (2022) Juno spacecraft measurements of Jupiter's gravity imply a dilute core. Planet Sci J 3(8):185. https://doi.org/10.3847/PSJ/ac7ec8
Millholland S, Laughlin G (2018) Obliquity tides may drive WASP-12b's rapid orbital decay. Astrophys J 869:L15
Morgado B, Assafin M, Vieira-Martins R, Camargo JIB, Dias-Oliveira A, Gomes-Jùnior AR (2016) Astrometry of mutual approximations between natural satellites. Application to the Galilean moons. Mon Not R Astron Soc 460:4086–4097
Morgado B, et al (2019) First stellar occultation by the Galilean moon Europa and upcoming events between 2019 and 2021. Astron Astrophys 626:L4. https://doi.org/10.1051/0004-6361/201935500
Nimmo F (2023) Strong tidal dissipation at Uranus? Planet Sci J 4:241. https://doi.org/10.3847/PSJ/ad0cfb
Notaro V, Durante D, Iess L (2019) On the determination of Jupiter's satellite-dependent Love numbers from Juno gravity data. Planet Space Sci 175:34–40. https://doi.org/10.1016/j.pss.2019.06.001
Null GW, Lau EL, Biller ED, Anderson JD (1981) Saturn gravity results obtained from Pioneer 11 tracking data and Earth-based Saturn satellite data. Astron J 86:456. https://doi.org/10.1086/112905
O'Connor CE, Hansen BMS (2018) Constraining planetary migration and tidal dissipation with coeval hot Jupiters. Mon Not R Astron Soc 477:175
Ogilvie GI (2014) Tidal dissipation in stars and giant planets. Annu Rev Astron Astrophys 52:171
Ogilvie GI, Lin DNC (2004) Tidal dissipation in rotating giant planets. Astrophys J 610:477–509. https://doi.org/10.1086/421454
Ojakangas GW, Stevenson DJ (1986) Episodic volcanism of tidally heated satellites with application to Io. Icarus 66:341–358
Owen WM Jr (2003) Cassini ISS Geometric Calibration of April 2003. JPL IOM 312.E-2003
Parisi M, Kaspi Y, Galanti E, Durante D, Bolton SJ, Levin SM, Buccino DR, Fletcher LN, Folkner WM, Guillot T, Helled R, Iess L, Li C, Oudrhiri K, Wong MH (2021) The depth of Jupiter's Great Red Spot constrained by Juno gravity overflights. Science 374(6570):964–968. https://doi.org/10.1126/science.abf1396
Park RS, Folkner WM, Williams JG, Boggs DH (2021) The JPL planetary and lunar ephemerides DE440 and DE441. Astron J 161(3):105. https://doi.org/10.3847/1538-3881/abd414
Park RS, Jacobson RA, Gomez Casajus L, Nimmo F, Ermakov AI, Keane JT, et al (2025) Io's tidal response precludes a shallow magma ocean. Nature 638(8049):69–73. https://doi.org/10.1038/s41586-024-08442-5
Pascu D (1994) An appraisal of the USNO program for photographic astrometry of bright planetary satellites. Galactic and Solar System Optical Astrometry, Proceedings of the RGO / Institute of Astronomy Workshop, Cambridge, 21. Cambridge Univ Press, Cambridge
Peters CF (1981) Numerical integration of the satellites of the outer planets. Astron Astrophys 104:37–41
Porco CC, West RA, Squyres S, et al (2004) Cassini imaging science: instrument characteristics and anticipated scientific investigations at Saturn. Space Sci Rev 115:363–497. https://doi.org/10.1007/s11214-004-1456-7
Rasio FA, Tout CA, Lubow SH, Livio M (1996) Tidal decay of close planetary orbits. Astrophys J 470:1187
Robert V, Desmars J, Lainey V, Arlot JE, Perlbarg AC, Horville D, Aboudarham J, Etienne C, Gérard J, Ilovaisky S, Khovritchev MY, Le Poncin-Lafitte C, Le Van Suu A, Neiner C, Pascu D, Poirier L, Schneider J, Tanga P, Valls-Gabaud D (2021) The NAROO digitization center - overview and scientific program. Astron Astrophys 652:A3. https://doi.org/10.1051/0004-6361/202140472
Sampson RA (1921) Theory of the four great satellites of Jupiter. Mem R Astron Soc 63:1

Serra D, Lari G, Tommei G, Durante D, Casajus LG, Notaro V, Zannoni M, Iess L, Tortora P, Bolton SJ (2019) A solution of Jupiter's gravitational field from Juno data with the ORBIT14 software. Mon Not R Astron Soc 490(1):766–772. https://doi.org/10.1093/mnras/stz2657
Sinclair AT (1977) The orbits of Tethys, Dione, Rhea, Titan and Iapetus. Mon Not R Astron Soc 180:447–459. https://doi.org/10.1093/mnras/180.3.447
Stevenson DJ (2020) Jupiter's interior as revealed by Juno. Annu Rev Earth Planet Sci 48(1):465–489. https://doi.org/10.1146/annurev-earth-081619-052855
Tajeddine R, Cooper NJ, Lainey V, Charnoz S, Murray CD (2013) Astrometric reduction of Cassini ISS images of the Saturnian satellites Mimas and Enceladus. Astron Astrophys 551:A129
Tanga P, et al (2023) Gaia data release 3. The Solar System survey. Astron Astrophys 674:A12
Van Hoolst T, Tobie G, Vallat C, et al (2024) Geophysical characterization of the interiors of Ganymede, Callisto and Europa by ESA's JUpiter ICy moons Explorer. Space Sci Rev 220:54. https://doi.org/10.1007/s11214-024-01085-y
Vasundhara R, Arlot J-E, Descamps P (1996) Dynamics, ephemerides and astrometry of the Solar System. In: Ferraz-Mello S, Morando B, Arlot J-E (eds) Proc. 172nd symp. Int. Astron. Union, IAU, pp 145–149
Vasundhara R, Arlot J-E, Lainey V, Thuillot W (2003) Astrometry from mutual events of Jovian satellites in 1997. Astron Astrophys 410:337–341
Vienne A, Duriez L (1991) A general theory of motion for the eight major satellites of Saturn. II. Short-period perturbations. Astron Astrophys 246:619–633
Vienne A, Duriez L (1992) A general theory of motion for the eight major satellites of Saturn. III. Long-period perturbations. Astron Astrophys 257:331–352
Vienne A, Sarlat JM, Duriez L (1992) About the secular acceleration of mimas. In: Chaos, resonance, and collective dynamical phenomena in the Solar System, IAU Symposium, vol 152. Kluwer Academic, Dordrecht, p 219
Vissapragada S, Chontos A, Greklek-McKeon M, Knutson HA, Dai F, et al (2022) The possible tidal demise of Kepler's first planetary system. Astrophys J 941:L31
Wahl SM, Parisi M, Folkner WM, Hubbard WB, Militzer B (2020) Equilibrium tidal response of Jupiter: detectability by the Juno spacecraft. Astrophys J 891(1):42. https://doi.org/10.3847/1538-4357/ab6cf9
Winn JN, Fabrycky DC (2015) The occurrence and architecture of exoplanetary systems. Annu Rev Astron Astrophys 53:409
Wong I, Shporer A, Vissapragada S, Greklek-McKeon M, Knutson HA, et al (2022) TESS revisits WASP-12: updated orbital decay rate and constraints on atmospheric variability. Astron J 163:175
Yang F, Wei X (2022) Transit timing variation of XO-3b: evidence for tidal evolution of hot Jupiter with high eccentricity. Publ Astron Soc Pac 134:024401
Zahn J (1977) Tidal friction in close binary systems. Astron Astrophys 57:383
Zhang QF, et al (2021) A comparison of centring algorithms in the astrometry of Cassini imaging science subsystem images and Anthe's astrometric reduction. Mon Not R Astron Soc 505:5253–5259

Space Science Reviews (2024) 220:20
https://doi.org/10.1007/s11214-024-01049-2

Long-Term Evolution of the Saturnian System

Matija Ćuk[1] · Maryame El Moutamid[2] · Giacomo Lari[3] · Marc Neveu[4,5] · Francis Nimmo[6] · Benoît Noyelles[7] · Alyssa Rhoden[8] · Melaine Saillenfest[9]

Received: 1 May 2023 / Accepted: 18 January 2024 / Published online: 19 February 2024

Abstract
Here we present the current state of knowledge on the long-term evolution of Saturn's moon system due to tides within Saturn. First we provide some background on tidal evolution, orbital resonances and satellite tides. Then we address in detail some of the present and past orbital resonances between Saturn's moons (including the Enceladus-Dione and Titan-Hyperion resonances) and what they can tell us about the evolution of the system. We also present the current state of knowledge on the spin-axis dynamics of Saturn: we discuss arguments for a (past or current) secular resonance of Saturn's spin precession with planetary orbits, and explain the links of this resonance to the tidal evolution of Titan and a possible recent cataclysm in the Saturnian system. We also address how the moons' orbital evolution, including resonances, affects the evolution of their interiors. Finally, we summarize the state of knowledge about the Saturnian system's long-term evolution and discuss prospects for future progress.

Keywords Satellites of Saturn · Tidal evolution · Orbital resonances · Obliquity of Saturn

1 Tidal Dissipation Within the Planet and Orbital Evolution

Satellites raise a tidal bulge on the planet they orbit. If there is any kind of lag in the planet's response, the tidal bulge will not point directly towards the satellite and as a result, the latter will experience a torque. Generally, planets spin faster than satellites orbit and, as a result, the torque pushes the satellites outwards. In the limit of low-eccentricity and inclination, a spin-synchronous satellite, and static solid tidal dissipation, for a tidal bulge averaged over an orbital cycle and apsidal precession, the semi-major axis increases as (Murray and Dermott 1999)

$$\dot{a} = 3\frac{k_{2,p}}{Q_p}\sqrt{\frac{G}{M_p}}R_p^5 m_s a^{-11/2}, \tag{1}$$

where G is the gravitational constant, M_p, R_p and $k_{2,p}$ are the planet's mass, radius and tidal Love number, respectively, while m_s and a are the mass and semi-major axis of the satellite that raises the tides. Traditionally, the lag has been assumed to be proportional to Q_p^{-1}, where Q_p is the so-called dissipation factor of the planet and is often taken to be constant.

Extended author information available on the last page of the article

In this theory of "equilibrium tides", the semi-major axis evolves according to $t^{2/13}$, where t is time (Murray and Dermott 1999) and more distant satellites tend to evolve outwards more slowly. The result is converging orbits, which raises the possibility of capture into mean-motion resonance (see Sect. 2). Tides on the planet also cause the eccentricity of satellites to increase, even though this effect is generally smaller than the eccentricity damping due to the tidal dissipation within the moons (see e.g. Goldreich and Gold 1963).

More recently, a different mechanism, termed "resonance locking", has been proposed (Fuller et al. 2016). This mechanism is described in more detail in Fuller et al. (2024, this collection). The key results, however, are as follows: 1) orbital migration can be significantly more rapid than previously thought; 2) outward evolution is exponential or power-law in time, rather than going as $t^{2/13}$, meaning that the rate of outwards motion accelerates; and 3) more distant satellites tend to evolve outwards faster. All three statements are true only if a satellite is in a resonance lock, which is not assured and depends on the internal structure of the planet.

Point (1) has received some observational confirmation from astrometry and radio science (Lainey et al. (2020) but cf. Jacobson (2022)). Point (2) means that an observed rapid present-day outward migration does *not* necessarily imply that the satellite formed recently. Point (3) means that the opportunities for capture into resonance are more limited than in the traditional picture, though the results are sensitive to very poorly known parameters.

Another important consequence of the resonance-locking hypothesis is that for satellites in a mean-motion resonance, the tidal heat production rate corresponding to constant (equilibrium) orbital eccentricities may be much higher than previously thought, because the effective Q_p of the planet is lower (e.g. Meyer and Wisdom 2007). This issue is discussed in more detail in Nimmo et al. (2023, this collection), and in Sects. 3 and 4.2 below.

2 Mean-Motion Resonances

A mean-motion resonance (MMR) arises when two orbiting bodies (in this case satellites of a planet) have a ratio of orbital mean motions that is close to a simple fraction, i.e. 1/2, 2/3, 3/5 etc. MMRs are usually locations of strong gravitational interaction between the moons, and depending on many factors, this interaction could be temporary or the bodies may be permanently *captured into the resonance*, meaning that their period ratio stays fixed as their orbits evolve.

More rigorously, two bodies are in resonance if a relevant angle called the *resonant argument* is librating rather than circulating. For a MMR, the general form of the resonant argument is $j_1\lambda_1 + j_2\lambda_2 + j_3\varpi_1 + j_4\varpi_2 + j_5\Omega_1 + j_6\Omega_2$, where λ, ϖ and Ω respectively designate the mean longitude, the longitude of pericenter and the longitude of the ascending node (Murray and Dermott 1999), and the subscripts 1 and 2 refer to the inner and the outer satellite. The coefficients j_{1-6} are integers, with the condition that $j_1 + j_2 + j_3 + j_4 + j_5 + j_6 = 0$ (D'Alembert's rule, arising because the geometry of MMR does not depend on the origin of longitude; Murray and Dermott 1999). The order of the resonant argument is $|j_3| + |j_4| + |j_5| + |j_6|$, and the corresponding resonant term in the Hamiltonian (i.e. interaction potential) will be multiplied by $e_1^{|j_3|}e_2^{|j_4|}s_1^{|j_5|}s_2^{|j_6|}$, where e and $s = \sin(i/2)$ refer to orbital eccentricity and inclination. As $e < 1$ and $s < 1$, and especially because the orbits of the major satellites of the giant planets tend to have small eccentricities and inclinations, lower-order resonances are typically stronger than higher-order ones. Note that each period commensurability ($j_2 : j_1$) contains many formal resonances (often called "sub-resonances") with different resonant arguments, which may affect either inclination or eccentricity. For example, within the 2:1 commensurability between Mimas ("M") and Tethys ("Θ"), we can have 1st order resonance (e.g. $2\lambda_{\Theta} - \lambda_M - \varpi_M$), 2nd order resonance ($4\lambda_{\Theta} - 2\lambda_M - \Omega_M - \Omega_{\Theta}$, which is currently librating) as well as 3rd order resonance (e. g. $2\lambda_{\Theta} - \lambda_M + \varpi_M - \Omega_M - \Omega_{\Theta}$). In general, $|j_1 + j_2|$ is the lowest order that a resonance within the commensurability can have, but higher order sub-resonances are always also present. Note that $j_5 + j_6$ has to be an even number (due to inclinations being relative, rather than absolute, and the geometry of the MMR not depending on the choice of reference plane; Murray and Dermott 1999), with the implication that resonances affecting inclination have to be at least of second order, while eccentricity-affecting resonances can be of first order.

Satellite systems of the giant planets introduce two elements in the resonant dynamics that are not as prominent for heliocentric orbits. Major satellites experience significant tidal torques from the primary, which makes moons migrate, potentially encountering resonances with each other (see Sect. 1). Additionally, the highly oblate shapes of the giant planets strongly perturb the orbits of the satellites, inducing fast orbital precession. Different sub-resonances of the same commensurability are separated in frequency space by secular (i.e. precessional) frequencies, so fast orbital precession leads to well-separated sub-resonances that do not overlap in frequency (e.g. Meyer and Wisdom 2008). This non-overlap is important in order for a stable resonance capture to occur, as resonance overlap produces dynamical chaos (Wisdom 1980), which sometimes breaks resonances and sometimes leads to a much more complex evolution (as seen in the Uranian system; Dermott et al. 1988; Ćuk et al. 2020).

One fundamental rule of resonant encounters is that convergent orbital evolution between two bodies can lead to resonance capture, while divergent crossings of resonances always lead to "jumps", i.e. passages through resonance during which relevant orbital elements experience discontinuous "kicks" but no long-term resonant evolution (e.g. Dermott et al. 1988). As the resonances within the same commensurability are separated by precession frequencies, it is always the case that inclination-type resonances (i.e. those with resonant arguments including longitudes of the node) are encountered first during convergent migration, as the precession of orbital nodes is generally retrograde (against the direction of orbital motion). Eccentricity-type resonances (which include the longitudes of pericenter in the resonant argument) tend to be crossed by converging satellites later as the pericenters precess in the direction of orbital motion. Whether convergent migration results in capture into a specific resonance depends on the parameters of the bodies and the resonance (lower-order resonances and higher masses make capture more likely), as well as migration rates (slower migration enhances capture probability) and pre-resonance orbital elements (low eccentricity or inclination are required for capture; Murray and Dermott 1999). While dissipation within satellites does not directly affect the capture into resonances, it can indirectly enable capture by damping previously existing eccentricity (or inclination).

The above discussion assumes that a MMR is only affecting the two satellites whose orbits are in near-commensurability. Close proximity of several sizeable satellites in the Saturnian system means that three-body resonances are sometimes present. Most relevant three-body resonances are semi-secular ones involving Titan, in which the longitudes of the apse or node of Titan (the most massive moon and the most important perturber) appear in a resonant argument of a MMR between two moons other than Titan (Ćuk and El Moutamid 2022). Three-body resonances independent of two-body resonances are also possible, but tend to be relatively weak and any capture tends to be temporary (Ćuk and El Moutamid 2022).

Once two moons are captured into a resonance, their semi-major axes evolve in unison in order to preserve the resonant relationship. In a stable MMR, eccentricity or inclination of at least one of the moons has to increase over time. This eccentricity or inclination is determined by the resonant argument, with the increasing orbital element being a conjugate of the secular angle in the resonant argument (Murray and Dermott 1999). For example, in the current Titan-Hyperion resonance with the resonant argument $4\lambda_H - 3\lambda_T - \varpi_H$ the secularly increasing element is e_H, conjugate to ϖ_H. Higher-order resonances can have orbital elements of both moons increase over time, as is the case with the inclinations of both Mimas and Tethys in the current Mimas-Tethys 4:2 resonance. The eccentricities and/or inclinations of bodies in MMR are expected to grow indefinitely, unless balanced by dissipation in the satellite or more usually until some other dynamical mechanism arrests further evolution or even breaks the resonance. Often this is due to the onset of resonance overlap. Even if closely-spaced sub-resonances are separate at the moment of resonant capture, their widths are proportional to eccentricities and inclinations and become wider over time. Eventually this can produce resonance overlap, resulting in dynamical chaos and likely breaking of the resonance relation.

One special case in which the orbital eccentricity can be constant indefinitely in the resonance is when the tidal dissipation within the satellite (described in the next subsection) acts to damp the eccentricity. Depending on the masses of the moons, their rates of tidal evolution and the strength of dissipation within the affected moon, the system can settle in an equilibrium in which the increase of eccentricity within the resonance is exactly counteracted by the damping by satellite tides (Meyer and Wisdom 2007). It is widely thought that Enceladus is currently in such equilibrium, in which eccentricity pumping by 2:1 MMR with Dione is counteracted by tidal dissipation within Enceladus, keeping eccentricity constant and providing energy for the geothermal activity on Enceladus (Lainey et al. 2012). It is important to note that this situation can be stable only for some combinations of tidal dissipation parameters that damp the librations of the resonant argument, as in other cases the equilibrium may be temporary as librations grow and the moons evolve out of the resonance (Meyer and Wisdom 2008).

3 Tidal Dissipation and Heating Within Satellites

A synchronous satellite (i.e. one that has identical rotational and orbital periods) will experience a permanent tidal bulge as a consequence of its proximity to its parent planet. However, if the satellite's orbit is eccentric, the size and orientation of this tidal bulge will fluctuate; the latter is also true if the satellite has an axial tilt (obliquity). If there is any friction in the system, some of the mechanical energy deforming the satellite will be converted to thermal energy. This is the origin of tidal heating.

Quantitatively, the heat production rate $\dot{E}$ is given by (e.g. Wisdom 2008)

$$\dot{E} = \frac{3}{2}\frac{n^5 R^5}{G}\frac{k_2}{Q}\left(7e^2 + \sin^2\theta\right). \tag{2}$$

Here n is the satellite mean motion, R is its radius, e its eccentricity and θ its obliquity. This expression is correct in the limit of small e and θ. The quantity k_2 is the satellite's tidal Love number, which describes the response of its gravity field to the perturbing tidal potential; the maximum value (for a uniform, fluid body) is 1.5 and it is reduced if the body has appreciable rigidity or central concentration of mass. The quantity Q describes the friction in the satellite; Q^{-1} is proportional to the lag angle between the applied potential and the response, so that a low Q implies a large lag angle and correspondingly higher dissipation.

Calculation of k_2 and (especially) Q for satellites[1] is not straightforward. In the simplest case, a solid viscoelastic (Maxwellian) body, the density, rigidity and viscosity at each point in the satellite need to be specified to derive k_2/Q (e.g. Moore and Schubert 2000). Since viscosity is strongly temperature-dependent, and the heat production rate depends on k_2/Q, even this very simple case raises the possibility of feedbacks between heat production and internal structure (Ojakangas and Stevenson 1986; Hussmann and Spohn 2004).

The Maxwell description of viscoelasticity, though often used, fails to adequately describe how real materials respond to forcing at different frequencies, so more complicated models (e.g. Andrade, Burgers) are needed (Renaud and Henning 2018; Renaud et al. 2021). If oceans are present, other modes of tidal heating are possible, such as turbulent dissipation in the oceans themselves (Chen et al. 2014), or tidal flushing of water through a permeable crust (Rovira-Navarro et al. 2022). However, the contribution of these mechanisms to the overall tidal heating budget appears in general to be small.

Because tidal heating represents an energy loss, the semi-major axis of the satellite must contract. If it is not experiencing external torques, for angular momentum to be conserved the eccentricity must also decrease. Thus, for an isolated satellite any eccentricity will be damped relatively rapidly. The damping rate for small e is given by (Murray and Dermott 1999)

$$\dot{e} = -\frac{a\dot{E}}{eGMm}, \tag{3}$$

where a is the semi-major axis, m and M the mass of the satellite and primary, respectively, and $\dot{E}$ is the tidal heating rate (equation (2)).

If eccentricity tides dominate then Eq. (3) can be used to derive the eccentricity damping timescale $\tau_e = e/\dot{e}$:

$$\tau_e = \frac{GMme^2}{a\dot{E}}. \tag{4}$$

Since $\dot{E}$ depends on e^2 (Eq. (2)), the damping timescale is independent of eccentricity.

If dissipation is dominated by obliquity rather than eccentricity tides, the inclination (i) damping timescale τ_i for small inclination is given by Chyba et al. (1989)

$$\tau_i = \frac{GMm\sin^2 i}{a\dot{E}}. \tag{5}$$

Because the satellites are expected to be in a Cassini state,[2] the obliquity θ will scale with inclination and thus τ_i is independent of inclination. Other things being equal, inclination damping is slower than eccentricity damping because of the factor of $7(\sin i/\sin\theta)^2$ from Eqs. (2), (4) and (5). Note that the factor $(\sin i/\sin\theta)$ is typically $\gg 1$ for relatively close-in satellites of giant planets that are in the Cassini State 1 (Chen et al. 2014). However, for satellites with oceans, obliquity tidal heating can be much more effective than eccentricity

[1] Not to be confused with the analogous parameter Q_P for the planet; see Sect. 1.

[2] Cassini states are the end-points of energy damping within a rotating satellite, in which the obliquity of the satellite is determined solely by its shape and orbital inclination, and any initial free obliquity is damped (Ward 1975).

tidal heating (Tyler 2008; Chen et al. 2014), and so inclination damping may be more rapid than eccentricity damping.

Equation (3) is the reason that orbital resonances (see Sect. 2) are so important; although the eccentricity of an isolated satellite (like Triton) will swiftly damp to zero, if the eccentricity is being excited by an orbital resonance, then long-lasting tidal heating can arise. In this case the ultimate source of energy is the rotational kinetic energy of the planet. Because the damping rate depends on the tidal heating rate, and the tidal heating rate depends on the thermal structure, complicated and non-monotonic orbital and thermal histories can result (e.g. Ojakangas and Stevenson 1986; Hussmann and Spohn 2004).

Because torques from the planet tend to increase the eccentricity and semi-major axis (Sect. 1), while dissipation in the satellite decreases both, for satellites in a MMR an equilibrium can result in which e is constant (see Sect. 2). Meyer and Wisdom (2007) show that the power of tidal heating of the inner moons of a resonant pair, assuming no tidal torque on the outer moon and all eccentricities being in equilibrium, is:

$$H = n_E T_E - \frac{T_E}{L_E + L_D}\left(\frac{G m_S m_E}{a_E} + \frac{G m_S m_D}{a_D}\right), \tag{6}$$

where T_E is the tidal torque on Enceladus, while L is angular momentum ($= ma^2 n \times \sqrt{(1-e^2)}$), n is mean motion, a is semi-major axis, m is mass, and G is gravitational constant (subscripts E, D and S refer to Enceladus, Dione and Saturn, respectively). If we ignore terms of order e^2 (Meyer and Wisdom 2007):

$$H = n_E T_E \left(1 - \frac{1 + m_D a_E/(m_E a_D)}{1 + (m_D/m_E)\sqrt{a_D/a_E}}\right). \tag{7}$$

The torque on Enceladus is given by:

$$T_E = \frac{3}{2}\frac{G m_E^2 R_p^5 k_{2p}}{a_E^6 Q_p}. \tag{8}$$

In this special case, the dissipation rate depends on the Q_p of the planet but *not* Q of the satellite, which affords a very significant simplification. Since the effective Q_p of Saturn has been measured at various frequencies (Lainey et al. 2020), the equilibrium tidal heat production rate in satellites in MMRs (either current or ancient) can in principle be calculated (e.g. Meyer and Wisdom 2007). Thus, for instance, equilibrium tidal heating rates of 11 GW and 4.8 GW are obtained for the 2:1 present-day Enceladus-Dione resonance and a putative earlier 3:2 Mimas-Enceladus resonance, assuming a Saturn Q_p of 1800 as suggested by astrometry (Lainey et al. 2020). A higher Saturn Q_p would yield a correspondingly lower heat flow. Note that a tidal equilibrium for Enceladus requires it to be quite dissipative (with $k_{2E}/Q_E \simeq 0.01$ or even larger), likely due to the presence of an internal ocean, making solid-body dissipation estimates (e.g. Murray and Dermott 1999) invalid. Extensive further discussion of this issue is provided in Sect. 4.2 and Nimmo et al. (2023, this collection).

4 Important Resonance Passages in Saturn's Moon System

4.1 Mimas-Tethys 4:2 MMR

Mimas and Tethys are locked in a 4:2 inclination-type MMR, associated with the resonant argument $2\lambda_M - 4\lambda_\Theta + \Omega_M + \Omega_\Theta$. This argument slowly librates around 0° with a large amplitude of 95° and a period of 70 years. Such a resonance is expected to raise the inclinations of the two involved bodies over time. However, since Tethys is about 16 times more massive than Mimas, only the inclination of Mimas is significantly affected. Allan (1969) derived the time evolution of these two inclinations since the trapping and has shown that, while the current inclination of Mimas (1.6°) could be due to the excitation by this MMR, the one of Tethys (1.1°) cannot, and was probably nearly as high as it is today when the capture happened. Ćuk et al. (2016) propose an explanation for this inclination, involving a past 5:3 MMR between Dione and Rhea, followed immediately by a secular resonance between Tethys and Dione.

The trapping into this current resonance and its stability require that the orbits of Mimas and Tethys converge. Since the trapping, assuming no other perturbations, the libration amplitude has decreased from 180° at the onset of resonance to the currently observed 95°, while the inclination of Mimas increased from the initial value to 1.6°. Using a linear model, Sinclair (1972) found that when reaching the separatrix delimiting the resonance, the satellites had only a 4% chance to be trapped. This low probability of the capture is due to the large current libration amplitude of the resonant argument, which implies an initial inclination of 0.4° for Mimas. Champenois and Vienne (1999a,b) solved this paradox by showing that a non-linear model, which is more accurate, allows for secondary resonances to affect the libration amplitude in the main resonance. Allowing for much lower resonant libration amplitudes post-capture leads to much higher capture probabilities. These probabilities may almost reach 100% for a null initial free inclination of Mimas, which means that when Mimas and Tethys encountered the resonance during their convergent migration, the trapping into the currently observed configuration was the only plausible outcome.

Before the current 4:2 inclination resonance was reached, Mimas and Tethys had to cross the sub-resonance with the argument $2\lambda_M - 4\lambda_\Theta + 2\Omega_M$ ("i_M^2" resonance). In the simplest case in which the pre-resonance inclination of Mimas was 0.4° the probability of capture into the i_M^2 resonance was only 7% (Luan 2014), explaining why the capture did not happen. However, scenarios in which the initial inclination of Mimas was small and the libration amplitude of the current ("$i_M i_\Theta$") resonance was subsequently changed (such as Champenois and Vienne 1999b) would also make the capture into the i_M^2 resonance highly likely. Therefore any hypothesis on the origin of the current 4:2 $i_M i_\Theta$ Mimas-Tethys resonance would also need to account for the prior crossing of the i_M^2 resonance not resulting in capture.

None of these analytical and numerical studies actually constrain the dissipation rate inside Saturn, which is modelled as a scale temporal factor. The capture process and the evolution into resonance are assumed to be slow enough, i.e. adiabatic, to not be affected by the dissipation timescale. In assuming a Love number $k_{2,p} = 0.341$ for Saturn and that the dissipation function Q_S is constant over the whole orbital evolution, Champenois and Vienne (1999a) estimate that the capture happened some 17,000 Q_S years ago. This implies that the resonance was assembled only about 30 Myr ago using the Q_S value from Lainey et al. (2012). Note that in equilibrium tidal theory (Eq. (1)) orbital evolution timescales of Mimas and Tethys are relatively similar. On the other hand, if the tidal response of Saturn depends on the frequency (Fuller et al. 2016), the estimated age of the resonance could greatly differ from the value given above.

Recent results, which imply non-equilibrium tides in Saturn, complicate the issue of the history and age of this resonance. Capture into resonance requires convergent evolution of Mimas and Tethys, which contradicts the end-member case in which all moon orbits are expanding with constant semi-major axis ratios (Lainey et al. 2020). While the directly measured tidal accelerations of Mimas and Tethys still have large uncertainties (Lainey et al. 2020; Jacobson 2022), the relatively fast observed evolution makes it unlikely that the age of the Mimas-Tethys resonance is greater than a few tens of Myr. Better determination of the moons' tidal accelerations is necessary to put firmer limits on the age of this resonance.

4.2 Enceladus-Dione 2:1 MMR

Enceladus and Dione are currently in a 2:1 MMR with an argument $2\lambda_D - \lambda_E - \varpi_E$ that acts to increase the eccentricity of Enceladus over time. The libration amplitude of this resonant argument is below one degree (Murray and Dermott 1999), implying long-term damping of the libration amplitude. Given that the eccentricity of Enceladus is modest ($e_E = 0.0047$), it was always considered likely that the tides within Enceladus act to decrease this eccentricity (Sect. 3). This eccentricity damping releases heat within Enceladus, affecting its geophysics and geology. As the eccentricity of Enceladus is modified over time by satellite tides, it is not possible to use the current orbits of the moons to determine the age of this resonance.

After the *Voyager* mission, it was known that parts of Enceladus's surface were very young, but current tidal heating was considered insufficient for resurfacing due to low eccentricity and then-prevailing estimates of the rate of Enceladus's orbital evolution (Squyres et al. 1983). As shown by Meyer and Wisdom (2007), if we assume that the Dione-Enceladus resonance is in a long-term equilibrium, the resonance produces $(18{,}000/Q_S) \times 1.1$ GW of heat within Enceladus. $Q_S = 18{,}000$ is the smallest value for which Mimas stays outside Saturn's Roche limit over the age of the Solar System (assuming frequency-independent equilibrium tides; Eq. (1)). As it was widely accepted before the *Cassini* mission that Saturn's major moons were primordial, long-term average tidal heating of Enceladus was thought to be restricted to 1 GW or less.

Cassini observations have shown that the heat flux from the south pole of Enceladus is 10-15 GW (Porco et al. 2006; Howett et al. 2011, 2024, this collection), an order of magnitude larger than predicted. Early solutions to this discrepancy suggested non-equilibrium scenarios (e.g. O'Neill and Nimmo 2010), until Lainey et al. (2012) found that fast tidal evolution of Saturn's moons suggested by astrometric observations can naturally explain the tidal heating in Enceladus being in equilibrium. Equilibrium heating of 15 GW implies $Q_E/k_{2E} = 100$ for Enceladus (Sect. 3), giving us an eccentricity-damping timescale (in absence of excitation) of 0.5 Myr. Therefore Enceladus would need only a few Myr to settle into a dynamical steady state, and an assumption of equilibrium does not preclude geophysical evolution on longer timescales (Nimmo et al. 2023, this collection).

More recent findings of non-equilibrium, likely resonant, tides dominating the evolution of the Saturnian system (Lainey et al. 2020) complicate this picture. Just like with the Mimas-Tethys resonance (previous subsection), the existence of the Enceladus-Dione resonance requires that the orbits of those two moons converge. Furthermore, unlike in the case of Mimas-Tethys resonance which only requires past net convergent evolution, tidal heating within the Enceladus-Dione resonance requires ongoing orbital convergence. This is consistent with the small libration amplitude which would be growing if the two moons' orbits were diverging, making the resonance evolve toward dissolution.

Convergent evolution is not compatible with the end-member orbital evolution model in which all moons are locked to resonant modes that all evolve on the same timescale (Lainey et al. 2020), or with a model in which the modes are divergent (Fuller et al. 2016). However, a situation in which Enceladus is in a resonance lock, but Dione is not, can result in a MMR. Using the approach of Meyer and Wisdom (2007) and adding the assumption about Enceladus being much less massive than Dione, we can roughly estimate the (equilibrium) tidal heating of Enceladus in the resonance as:

$$H \approx \frac{G m_S m_D}{2 t_a a_D} = \frac{5 \times 10^{28}\ \mathrm{J}}{t_a}, \tag{9}$$

where $t_a = (a_D/\dot{a}_D)_{MMR}$ is the timescale for the evolution of Dione's orbit forced by resonance with Enceladus. In the case of equilibrium tides, t_a would be the timescale of the two moons' orbital convergence multiplied by the moons' mass ratio m_D/m_E, so $t_a \approx 100$ Gyr if $Q_S = 1700$, yielding $H \approx 16$ GW. If Enceladus is evolving through a resonant lock, then t_a is simply the timescale of the resonant mode's evolution, as Dione would be forced to evolve at the same rate. If we assume $t_a = 9$ Gyr, as proposed by Lainey et al. (2020) as being a typical value for orbital evolution due to resonance locking in the Saturnian system, $H \approx 180$ GW, more than an order of magnitude in excess of the observed value (Howett et al. 2011).

On the other hand, if the resonance lock evolves uniformly in a reference frame rotating with Saturn and the satellites' orbits evolve divergently (Fuller et al. 2016), steady-state heating of Enceladus will be much lower. The Saturn-frame resonance lock would produce $a/\dot{a} \approx 200$ Gyr for Enceladus, based on Titan's evolution timescale of 11 Gyr (Lainey et al. 2020) and $a/\dot{a} \propto n/(\Omega_p - n)$, where Ω_p is the planet's rotation rate (Fuller et al. 2016). This rate of evolution gives us a steady-state heat flow of about 8 GW for Enceladus, relatively close to the measured value, but almost certainly less than the true heat production rate. However, if we use Rhea's observed evolution ($a/\dot{a} = 6$ Gyr; Lainey et al. 2020) to estimate Enceladus's tidal heating in the same model, we get $t_a \simeq 27$ Gyr and $H \approx 60$ GW. This last estimate is motivated by the main result of Jacobson (2022), who finds that Titan does not exhibit non-equilibrium tidal evolution but Rhea does. If we use the observational results for the secular acceleration of Rhea from Jacobson (2022), the timescale for Enceladus's evolution through resonant lock (assuming divergent modes) could be in the 30-40 Gyr range and the associated tidal heating in the 40-55 GW range. Note that the total tidal heating rate of Enceladus must exceed the measured value, as the thermal observations are not sensitive to distributed tidal heating outside the South Polar Terrain. Models of Enceladus's ice shell suggest a heat loss rate in the range 25-40 GW (Hemingway and Mittal 2019), so there may not be a discrepancy between the heating rates predicted here and the actual heat production rate (see Nimmo et al. 2023, this collection).

A separate constraint on the possible resonant-lock driven heating of Enceladus comes from the observational limits on the orbital deceleration of Dione. The minimum possible effective $k_{2,p}/Q_p$ for Dione found by Lainey et al. (2020) is 0.6×10^{-4}, which implies a slowest evolution of about 30 Gyr. This means that 55 GW is the minimum tidal heating in the Enceladus-Dione system allowed by the astrometric observations, if all of Dione's orbital evolution is due to Enceladus. Jacobson (2022) has larger error bars on the observed evolution of Dione, and allows for minimum heating rates that are about twice smaller. Note that these "lower limits" apply only to pure evolution of Enceladus by resonance lock that includes zero tidal torque on Dione. Models in which equilibrium tides are significant are not constrained by these limits.

We conclude that Enceladus's observed heat flux (which is a lower bound on the actual heat production rate) may be consistent with the resonance lock model in which the resonant modes are divergent, but is well short of that expected from the simple picture in which all resonant modes evolve on the same timescale. A picture in which modes affecting the inner satellites evolve more slowly (Fuller et al. 2016) can explain the current Enceladus-Dione resonance, but requires Dione to have avoided direct resonance locking to an internal mode within Saturn. Further refinement of our understanding of Enceladus's tidal heating will require more precise measurement of the current orbital evolution of the Saturnian moons, as well as more complete constraints on the thermal flux from Enceladus (Howett et al. 2024, this collection). Further discussion of Enceladus's tidal heat budget and thermal evolution may be found in Nimmo et al. (2023).

Unlike in the case of the Mimas-Tethys resonance, less attention has been given to the formation of the Dione-Enceladus resonance than to its continuing maintenance. In order for Enceladus and Dione to reach the current sub-resonance of their 2:1 MMR through convergent orbital evolution, numerous other sub-resonances must be crossed first. Past work (Meyer and Wisdom 2008; Zhang and Nimmo 2009) has concentrated on the sub-resonances that can be modelled assuming two bodies on planar orbits, including the second-order 4:2 $e_E e_D$ mixed-eccentricity resonance. However, inclination-type resonances (equivalent to the one Mimas and Tethys are currently in) would be encountered first, and their effect on the (currently very low) inclination of Enceladus has yet to be fully explored. Additionally, using numerical integrations featuring all major moons, Ćuk and El Moutamid (2022) have recently shown that two-body resonances also contain additional three-body sub-resonances that can result in capture. Clearly more work is needed to understand the history of the Enceladus-Dione 2:1 resonance.

4.3 Past Resonances Between Mid-Sized Moons?

In addition to currently active resonances, researchers have long suspected that Saturn's moons may have experienced additional resonances in the past which have since been broken (Fig. 1). Here we will address several proposed past resonances among the inner moons (out to Rhea), while those involving Titan and the Sun (i.e. the evection resonance) are discussed in separate subsections. For an introduction to the topics of orbital migration, resonance capture and evolution within the resonance, we refer the reader to Sect. 2.

Among the inner moons, past resonances are often considered responsible for excited eccentricities and inclinations that cannot be accounted for with observed interactions. For example, while the inclination of Mimas is likely a product of its current 4:2 MMR with Tethys, the eccentricity of Mimas is unaffected by this resonance and must predate it. Similarly, the large inclination of Tethys mostly precedes the resonance with Mimas, requiring an origin mechanism (Sect. 4.1). Here it is assumed that the moons formed on coplanar,

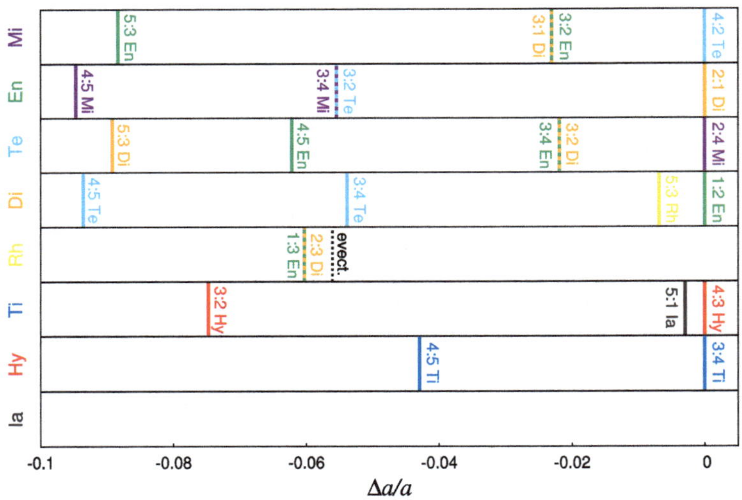

Fig. 1 Current location of nominal MMR for each satellite in terms of variation of its semi-major axis, assuming that the other moons do not move. Within a 10% variation in a, we reported all first-order resonances up to 5:4, all second-order resonances up to 5:3, the evection resonance for Rhea and the 5:1 resonance between Titan and Iapetus. Overlapping between resonances is due to currently active MMR

circular orbits, as both eccentricity and inclination are damped by expected collisions with satellitesimals (Peale 1999). Apart from current orbital elements, the record of tectonic activity, such as Ithaca Chasma on Tethys, could indicate past tidal heating driven by high eccentricity (Chen and Nimmo 2008). Tethys's orbital eccentricity is presently very low, but past high eccentricity implies that Tethys experienced one or more orbital resonances during its history.

Reconstruction of past resonance crossings among the inner moons is highly dependent on the assumed strength and nature of tidal dissipation within Saturn. The potential past resonances that were studied in past work and are therefore discussed here are expected to have happened in the context of equilibrium tides. Equilibrium tidal theory predicts that all adjacent inner moon pairs are on converging paths, with the sole exception of Enceladus and Tethys, which were thought to be on divergent orbits (Murray and Dermott 1999). Frequency-dependent tidal dissipation within Saturn certainly changes this picture and may rule out some past resonances while introducing new ones. Here we will review past work on these resonances and follow with a brief discussion on how the latest results on Saturn's tidal response (Lainey et al. 2020; Jacobson 2022) change our assessment of these proposed resonant encounters.

Mimas-Enceladus 3:2 Resonance In the classical picture of equilibrium tides, one of the most recent resonance passages was that of the Mimas-Enceladus 3:2 MMR. This is expected to have been a convergent encounter, and has been proposed by Meyer and Wisdom (2008) as the source of Mimas's eccentricity. In particular, Meyer and Wisdom (2008) have suggested capture into e_M and $e_M e_E$ (i.e. the 6:4 Mimas-Enceladus MMR) as possible sources of Mimas's eccentricity, while ruling out past capture into the e_E sub-resonance as it would have been impossible to break. Tian and Nimmo (2020) later revisited this work using a wider range of tidal parameters, and found regions of tidal parameter space in which the hypothesis of Meyer and Wisdom (2008) is likely. Both of these studies relied on semi-analytical treatment of the resonant Hamiltonian, and assumed a planar system. The last issue is crucial, as the inclination of Enceladus is presently very low ($i_E \approx 0.01°$), putting strong constraints on past resonance crossings involving Enceladus. El Moutamid et al. (2019) proposed that a long-lived Mimas-Enceladus 3:2 e_E resonance was broken when the pair encountered respective 3:1 and 2:1 resonances with Dione. However, ongoing work by some of this work's authors (El Moutamid and Ćuk) using direct numerical integrations suggests that this triple resonance usually produces dynamical chaos, exciting the inclination of Enceladus beyond the observed value. In principle, a high inclination could have been damped by fluid tides in Enceladus's ocean (Sect. 3) but no quantitative study has been attempted. Clearly more work is needed to explore the dynamical viability of a past Mimas-Enceladus 3:2 MMR crossing, especially as the tidal evolution rates of these two moons (which may also not be constant over time) are hard to constrain by other means.

Mimas-Dione 3:1 Resonance The fact that Enceladus and Dione are currently in a 2:1 resonance implies that their orbital evolution (at least recently) has been convergent. Therefore, before this resonance was established Enceladus must have been interior to the resonant location. A clear implication is that if Mimas ever crossed the 3:2 MMR with Enceladus, it must have also crossed the 3:1 resonance with Dione (either at the same time or later). The reverse is not true: depending on the relative tidal evolution rates and the age of the system, Mimas could have in principle crossed the 3:1 resonance with Dione and not the 3:2 MMR with Enceladus. Given the large mass difference between Dione and Mimas, Mimas is likely to be affected by their 3:1 MMR much more than Dione. Meyer and Wisdom (2008) suggested the Mimas-Dione 3:1 MMR as another possible source of the eccentricity of Mimas.

They prefer capture into the $e_M e_D$ sub-resonance as it would also excite the eccentricity of Dione in line with constraints from Dione's later resonance passage with Enceladus. More recently Ćuk and El Moutamid (2022) simulated the Mimas-Dione 3:1 resonance and found that three-body resonances involving the eccentricity of Titan can be important. This three-body sub-resonance, not included in the semi-analytical model of Meyer and Wisdom (2008), usually leads to a short-lived capture that would excite the eccentricity of Mimas, consistent with the present state. Since Mimas's eccentricity is not being excited at present, survival of its eccentricity from an earlier time places limits on how dissipative it could be. At present, the 3:1 Mimas-Dione resonance appears to be a viable candidate for the source of Mimas's eccentricity, but more work is needed to confirm this.

Tethys-Dione 3:2 Resonance This resonance should have taken place relatively recently if Tethys and Dione were to follow equilibrium tidal theory in their orbital evolution. In the classical picture in which Saturn's tidal $Q_S = 18{,}000$ (Murray and Dermott 1999) and is constant, this resonance would have happened about 1 Gyr ago, while $Q_S \approx 1800$ proposed by Lainey et al. (2012) would put this resonance at 100 Myr ago. This resonance was proposed by Chen and Nimmo (2008) as the source of tidal heating of Tethys, and this hypothesis was explored in more detail by Zhang and Nimmo (2012) using numerical integrations. Zhang and Nimmo (2012) found that the Tethys-Dione 3:2 resonance was impossible to break once established, and proposed a large impact on Tethys (that formed the Odysseus basin) as a mechanism of breaking this resonance. However, Zhang and Nimmo (2012) assumed a planar system, therefore ignoring all constraints from the satellites' inclinations, and also underestimating the amount of chaos in the system (as all inclination-type resonances were absent in their simulations). Ćuk et al. (2016) studied this resonance using direct integrations with non-zero inclinations, and found that the inclination of Tethys is excited up to several degrees, after which a chaotic phase excites both the eccentricities and inclinations of both moons. The chaotic interactions ultimately break the resonance, but leave Tethys and Dione with inclinations well in excess of the observed ones. Unless both bodies experienced efficient inclination damping, these results imply that the Tethys-Dione 3:2 MMR never happened and therefore that the present inner satellite system of Saturn must be younger than 100 Myr (Ćuk et al. 2016, assuming $Q_p = 1800$). However, the recent astrometric results implying that the tidal response of Saturn is frequency-dependent suggest that Saturn's response to Tethys is significantly *weaker* than to other moons (Lainey et al. 2020; Jacobson 2022). A slowly-migrating Tethys would have either converged with Dione much more slowly or not at all (observations are still too uncertain to distinguish between these possibilities), so this resonance clearly cannot be used to put an upper limit on the age of the system. In summary, a passage through the Tethys-Dione 3:2 resonance would have produced inclinations inconsistent with the present values (Ćuk et al. 2016, unless the inclinations of both Tethys and Dione were greatly damped by some process), and its past occurrence is also disfavored by recent observational results on the moon's orbital evolution (Lainey et al. 2020; Jacobson 2022), so this resonance is unlikely to have much bearing on the history of the system.

Dione-Rhea 5:3 Resonance In the equilibrium tides-based relative chronology, the Dione-Rhea 5:3 resonance happens soon after the Tethys-Dione 3:2 resonance. Ćuk et al. (2016) studied the former using numerical simulations and unexpectedly found that this resonance between Dione and Rhea is a good candidate for the origin of Tethys's inclination. They find that following a largely chaotic Dione-Rhea 5:3 MMR passage, Dione and Tethys encounter a secular resonance with the argument $(\varpi_\Theta + \Omega_\Theta) - (\varpi_D - \Omega_D)$ (see Sect. 2 for definitions). The two expressions within parentheses designate very slow-moving angles, as the apsidal and nodal precession rates for each moon are approximately equal and in opposite directions. However, precession rates are perturbed in proximity of MMRs, and in this case Rhea's perturbations just outside the Dione-Rhea 5:3 MMR modify the sum of Dione's apsidal and nodal precession rates just enough to produce the secular resonance. In the secular resonance, the eccentricity and inclination of Dione are transferred to Tethys in equal amounts (i.e. $\sin(i_\Theta) \approx e_\Theta$). This sequence of resonances (Dione-Rhea 5:3 MMR followed by Tethys-Dione secular resonance) can explain the high inclinations of Tethys ($i_\Theta = 1°$ before resonance with Mimas) and Rhea ($i_R = 0.33°$) despite the low inclination of Dione ($i_D = 0.02°$, decreased by the secular resonance). These resonances also produce past large eccentricity of Tethys ($e_\Theta \simeq 0.01$) and the current eccentricity of Dione, assuming reasonable tidal dissipation. A past excited eccentricity of Tethys could in principle explain a past heating episode on Tethys inferred from Ithaca Chasma (Chen and Nimmo 2008). The free eccentricity of Rhea, however, is currently just $e_R \approx 2 \times 10^{-4}$ requiring either strong dissipation or some other mechanism for damping.

The biggest challenge to the past passage through the Dione-Rhea 5:3 MMR is the currently observed tidal evolution of Rhea (Lainey et al. 2020), which is five times faster than equilibrium tides would predict and would make the orbits of Dione and Rhea divergent. Two obvious solutions are that either the inclinations of Tethys and Rhea require different explanations, or Rhea's fast evolution is a phenomenon that started relatively recently. More work is urgently needed, both on exploring alternative possible sources of Tethys's inclination, and on the nature and operation of tidal dissipation within Saturn.

4.4 Rhea and the Evection Resonance

The evection resonance is a semi-secular resonance in which a moon's precession period is equal to the parent planet's orbital period (Touma and Wisdom 1998). While no moons in the solar system are currently in the evection resonance, studies of the early orbital history of Earth's Moon have identified the evection resonance as an important dynamical mechanism (Touma and Wisdom 1998; Ćuk and Stewart 2012; Tian et al. 2017; Rufu and Canup 2020). Evection is a "semi-secular" resonance in the sense that the (apparent) mean motion of the perturber (the Sun) is in resonance with the apsidal precession of the satellite's orbit. As the moons' orbital precession are overwhelmingly driven by Saturn's oblateness, the dynamics of the evection resonance does not depend on the relative arrangement of moons, as MMR do. The evection resonance affects any moon at a certain distance from Saturn, which is given by:

$$a_{eve} = R_S \left(\frac{3}{2} J_2 \frac{\Omega_0}{n_s} \right)^{2/7}, \tag{10}$$

where J_2 is Saturn's oblateness moment, $\Omega_0 = \sqrt{GM/R_S^3}$ is the orbital frequency at $a = R_S$ (Saturn's radius), and n_S is Saturn's heliocentric orbital mean motion. Equation (10) gives $a_{eve} = 8.1 R_S$, but Titan's perturbations shift the location of the evection resonance to $a_{eve} = 8.2 R_S$. This distance is somewhat smaller than the current orbital distance of Rhea (which has $a = 8.7 R_S$).

Estimates of the past orbital evolution of Rhea based on extrapolation from observations (Lainey et al. 2020; Jacobson 2022) suggest that Rhea should have crossed the evection resonance about 300-400 Myr ago. The evection resonance, with the resonant argument $2\lambda_S - 2\varpi$ (λ_S is Saturn's mean longitude, ϖ Rhea's longitude of pericenter), is expected to

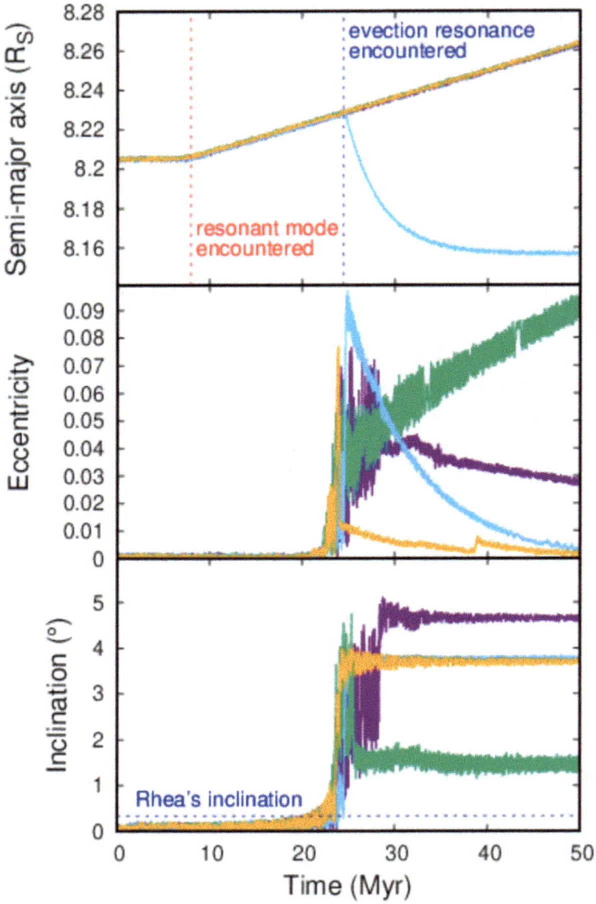

Fig. 2 Simulations of Rhea's encounter with the solar evection resonance using different parameters for the width of Saturn's resonant mode and Rhea's tidal dissipation. From top to bottom, the panels plot Rhea's semi-major axis, eccentricity and inclination. The purple and the green lines had $Q/k_2 = 1000$ for Rhea, and the light blue and the orange had $Q/k_2 = 100$ (i.e. likely value for Enceladus). The purple and the light blue lines had effective tidal $Q = Q_0 + (\frac{n-\nu}{\nu\sigma})^2$ (where ν is the mode's frequency and $\sigma = 10^{-6}$) due to the resonant mode, while the green and the orange lines used the same Q profile with $\sigma = 10^{-5}$. The evolution timescale is $a/\dot{a} = 6$ Gyr, as found for Rhea by Lainey et al. (2020). The dotted line on the bottom panel shows the present-day inclination of Rhea. This simulation included the Sun, Jupiter, Saturn, Titan, and Rhea. From work in preparation by MĆ

significantly excite Rhea's eccentricity. One possible outcome is that Rhea's satellite tides become strong enough to push the satellite inwards through the resonant mode, leaving it stranded interior to the mode. Another possibility would be Rhea evolving with the mode, while captured in the evection resonance, producing a continuously increasing eccentricity. Yet another outcome would be a Rhea that continues evolving through the mode, but moves beyond the resonance, with Rhea's excited eccentricity slowly damping. Only the last of these possibilities would be consistent with the present orbital distance, small eccentricity (≈ 0.001) and observed evolution rate of Rhea.

Figure 2 shows four preliminary simulations done by author M.Ć. using a modified version of the numerical integrator SIMPL (Ćuk et al. 2016). One of the simulations shows a long-term resonance capture (green line), one "drops out" from the resonant mode (light blue line), while the other two simulations have Rhea staying with the mode but moving beyond the evection resonance. However, the bottom panel that plots inclination, indicates that the evection and associated resonances also significantly excite the *inclination* of Rhea, well in excess of the observed value ($i_R = 0.33°$).

The excitation of inclination appears to be chaotic and occurs concurrently with the evection resonance. This is consistent with results obtained by Ćuk et al. (2016) assuming that Rhea migrated by equilibrium tides. The best candidate for the excitement of the inclination is the resonance as $\lambda_S - \varpi_S + \Omega - \Omega_S$ where Ω refers to the longitude of the ascending node and the subscript S refers to Saturn's heliocentric orbit. As the reference plane is the Laplace plane[3] of Rhea's orbit, which is close to Saturn's equatorial plane, Ω_S is identical to the longitude of Saturn's equinox. Both Ω_S and ϖ_S are relatively slow-precessing (with periods of about 1.8 Myr and 50 kyr), so the resonance happens because $\dot{\lambda}_S = n_S \approx -\dot{\Omega}$. One can think of this resonant term as a combination of the Sun's "annual equation" (a term in classical lunar theory; Brouwer and Clemence 1961) and the Sun's main secular perturbation on the inclination of Rhea. In other words, the Sun induces perturbations in Rhea's inclination once per precession period, and solar perturbations vary in strength during one orbit of Saturn because of the planet's orbital eccentricity. When these two cycles are commensurable, this strong resonant perturbation occurs. This resonance is chaotic, as its strength depends on the eccentricity of Saturn's orbit, which varies in the $0.01 < e_S < 0.09$ range with a period of 5×10^4 yr due to Jupiter's perturbations. Jumps in inclination also affect the apsidal precession, making the main evection resonance also chaotic. Direct numerical integrations are clearly the only practical and accurate way of studying this problem.

In some cases, the orbital inclination of a moon can be damped by obliquity tides (Chyba et al. 1989), but this is unlikely to be applicable to Rhea. Large-scale inclination damping requires large forced obliquities, either through a Cassini State transition like the one experienced by the Moon (Chen and Nimmo 2016), or spin-orbit resonance, as proposed for Uranian moons (Ćuk et al. 2020). The forced obliquity of Rhea is currently $\theta = 0.03°$ (Chen et al. 2014), and the timescale for inclination damping is longer than that for eccentricity damping by a factor of $7(\sin\theta/\sin i)^2$ (Chyba et al. 1989), which in the case of Rhea is on the order of 10^3. If Rhea is as dissipative as "Enceladus in equilibrium" with $Q/k_2 = 100$, the eccentricity damping timescale is on the order of 10 Myr (Fig. 2), which puts the timescale for inclination damping to many Gyr. Resonant tides excited in satellite oceans offer a different mechanism of inclination damping (Tyler 2008), but Chen et al. (2014) find that ocean obliquity tides would produce less dissipation in Rhea than classic obliquity tides that assume $Q/k_2 \simeq 100$. The above upper limits on dissipation may be overestimates, as Rhea currently does not appear to be in hydrostatic equilibrium (Tortora et al. 2016), and a solid tidal Love number $k_2 = 0.01$ (Chen et al. 2014) may be more applicable. Therefore, it appears that the damping of multi-degree inclination of Rhea within the last 0.5 Gyr is implausible.

It appears that Rhea should have crossed the evection resonance about 300-400 Myr ago and acquired a large inclination that could not have been subsequently damped. This dynamical history is clearly in conflict with Rhea's modest inclination ($i_R = 0.33°$), and this disagreement suggests that one or more assumptions that produced this result are incorrect. While it cannot be completely ruled out that Rhea somehow avoided acquiring high inclination during resonance crossing, numerical results available so far suggest that it is highly unlikely that Rhea crossed the evection resonance, at least in the present dynamical environment. If the obliquity of Saturn is only 1-2 Gyr old (Saillenfest et al. 2021a), it is possible that Rhea did cross the evection resonance in the distant past when the out-of-plane perturbations by the Sun were much weaker. This would still require a separate solution for the timescale problem. One possible explanation is that Rhea became locked to a normal mode (Fuller et al. 2016) more recently than 400 Myr ago, with its orbital evolution before that being much slower. Another explanation would be that Rhea re-accreted outside the evection resonance distance, as a part of a dynamical cataclysm (e.g. Ćuk et al. 2016). In order

[3] See Sect. 5.1 for a definition.

to distinguish between these possibilities a better understanding of tidal dissipation within Saturn, both past and present, is likely needed.

The only firm conclusions we can draw from the proximity of Rhea to the evection resonance is that we can exclude the global picture proposed by Lainey et al. (2020) in which the moons evolve by resonance lock practically since the system's formation. Unlike the tidal evolution of Titan where the results of Lainey et al. (2020) and Jacobson (2022) are in disagreement, both studies agree on the fast orbital evolution of Rhea, corresponding to a timescale of $(a/\dot{a}) \approx 6$ Gyr. Large-scale migration of Rhea, if it is a long-term steady state, assures relatively recent (300-400 Myr) crossing of the evection resonance which would overexcite Rhea's inclination. Therefore, within the last few hundred Myr, Rhea must have experienced either a dramatic speeding up of its tidal evolution, or possibly a disruption and re-accretion.

4.5 Titan-Hyperion 4:3 MMR

Titan and Hyperion are currently in the 4:3 MMR, which keeps their orbital periods fixed in a 3:4 ratio. As Hyperion's orbit is relatively eccentric ($e_H = 0.104$), the resonance is important for protecting Hyperion from having close encounters and colliding with Titan. It is established that the eccentricity of Hyperion in the resonance grows as Titan migrates outward (Murray and Dermott 1999); in the context of equilibrium tides, the orbital evolution of Hyperion on its own has a negligible effect on the resonance. Therefore, the eccentricity of Hyperion could be used to constrain the past orbital evolution of Titan.

According to Eq. 8.242 in Murray and Dermott (1999), the increase in the eccentricity of Hyperion is given by:

$$\frac{\dot{e}_H}{e_H} = \frac{1}{e_H^2} \frac{m_T}{m_S} n_H a_H \frac{F}{3g}, \tag{11}$$

where m, a, e and n are masses, semi-major axes, eccentricities and mean motions as defined before, and subscripts S, T, and H refer to Saturn, Titan and Hyperion. The other variables are defined as $F = 4\dot{n}_H - 3\dot{n}_T \approx -3\dot{n}_T$, and $g = 16Gm_T/a_H^2 + 9Gm_H/a_T^2 \approx 16Gm_T/a_H^2$. Expanding F and g into Eq. (11), and using $Gm_S/a_H^3 = n_H^2$, $n_H = (3/4)n_T$ and $\dot{n}/n = -(2/3)\dot{a}/a$, we get:

$$2e_H\dot{e}_H = \frac{\mathrm{d}(e^2)}{\mathrm{d}t} = \frac{1}{4}\frac{\dot{a}_T}{a_T}. \tag{12}$$

Assuming that Hyperion started with a low eccentricity, and evolved to $e_H = 0.1$, Titan's semi-major axis must have evolved by about 4%, assuming a constant rate of migration. This is a very rough estimate, but it gives us the correct order of magnitude of how much Titan could have migrated since Hyperion was captured into the resonance.

Before 2012, when Saturn was estimated to have $Q_p > 18{,}000$ (Murray and Dermott 1999), Titan was thought to have migrated less than 1% of its orbital distance, and therefore the resonance was thought to have been established and largely evolved by non-tidal means, such as gas drag in protosatellite nebula (Peale 1999). Greenberg (1973) proposed an alternative view that the Titan-Hyperion resonance was evolved by tides alone, and implied a much faster global evolution rate of Saturn's moons.

When Lainey et al. (2012) presented evidence for the fast tidal evolution of Saturn's moons, Ćuk et al. (2013) noted that the new value of Saturn's tidal quality factor $Q_p \approx 1700$ would make the Titan-Hyperion resonance about as old as the Solar System. The more recent direct measurement of Titan's orbital evolution puts the effective tidal Q_p of Saturn at Titan's frequency as $Q_p \approx 120$, equivalent to an evolution timescale of $a/\dot{a} = 11$ Gyr (Lainey et al. 2020). Such a rapid expansion of Titan's orbit would place the origin of the Titan-Hyperion resonance at about 400-500 Myr ago.

The above calculation relies on two assumptions. One is that Hyperion's own rate of tidal evolution is negligible, and the other is that Hyperion did not damp its eccentricity. The first assumption about negligible tidal evolution is straightforward in the theory of equilibrium tides, in which a moon's tidal evolution rate is proportional to its mass. In the context of resonant mode locking (Fuller et al. 2016), the issue shifts from the rate of evolution to the question of whether Hyperion can lock to a mode within Saturn. Some quick calculations assuming $\dot{a}/a = 11$ Gyr (i.e. the observed rate of evolution of Titan) and using standard definition of tidal parameters (Murray and Dermott 1999) yield that Saturn would need to have a tidal quality factor $Q_p \simeq 10^{-3}$ at Hyperion's frequency, which is nonphysical. Therefore, Hyperion is too small and too distant to lock to a resonant mode by itself.

The second issue of eccentricity damping is somewhat more difficult to model as Hyperion is in chaotic rotation (Wisdom et al. 1984). One way to estimate the amount of energy dissipation is to treat it like never-ending tidal spindown (Murray and Dermott 1999). This approach gives an eccentricity damping timescale of 10^{11} yr (assuming $Q_H = 10$ and $k_{2H} = 0.01$). Alternatively, we can estimate the energy dissipation within chaotic rotation as if Hyperion was a wobbling asteroid (Sharma et al. 2005), and this approach gives us an eccentricity damping timescale of 10^{12} yr. Unless another source of dissipation is identified, it appears unlikely that the eccentricity damping within Hyperion contributed significantly to the evolution of its resonance with Titan.

While the age of the Titan-Hyperion resonance is not the same as the age of Hyperion, there are strong reasons to think Hyperion cannot be much older than its 3:4 resonance with Titan. As Titan's and Hyperion's orbits converged, a primordial Hyperion would have likely been captured into other resonances that were crossed before the 4:3 resonance (notably the 3:2 and 7:5 resonances). Capture of Hyperion into those resonances is clearly inconsistent with Hyperion being on a relatively low-e, low-i orbit before capture into the 4:3 resonance. While more work is needed to directly confirm the likelihood of capture into these outer resonances, we would expect at least the capture into the first-order 3:2 resonance to be robust. If Hyperion formed just interior to Titan's 3:2 resonance, Hyperion cannot be older than 1.5 Gyr, assuming Titan maintained the migration rate measured by Lainey et al. (2020). If capture into the mutual 7:5 resonance is also found to be a certain outcome, then Hyperion must be younger than about 1 Gyr.

Hyperion is a relatively small moon, but its late formation would very likely imply a wider cataclysm at that time. Hamilton (2013) has proposed that Titan was a late merger between multiple satellites, with Hyperion being an unaccreted fragment. Asphaug and Reufer (2013) also proposed a late formation mechanism for the Saturnian moons that involved a major impact on Titan. Confirming that Hyperion is not primordial would support the hypothesis of a late cataclysm involving Titan. On the other hand, if the slower migration of Titan implied by the orbital solutions of Jacobson (2022) is correct, Hyperion is likely to be primordial or almost primordial.

4.6 Past Titan-Iapetus 5:1 MMR Crossing?

Iapetus is the third-largest moon of Saturn, as well as the major moon that is the most distant from the planet. Iapetus is notable for its albedo dichotomy (Buratti and Mosher 1995; Porco et al. 2005), oblate shape (Thomas et al. 2007; Castillo-Rogez et al. 2011), and equatorial ridge (Levison et al. 2011; Dombard et al. 2012; Stickle and Roberts 2018), but here we will restrict ourselves to studying its orbital motion. Like other regular satellites, Iapetus has a relatively low orbital eccentricity ($e_I = 0.028$), but it also has a substantial orbital inclination ($i_I = 8°$ with respect to its Laplace plane[4]), the origin of which has been a long-standing problem (Ward 1981; Nesvorný et al. 2014). As the solar perturbations on Iapetus's orbit are comparable to those arising from Saturn's oblateness and the inner moons (chiefly Titan), the Laplace plane of Iapetus is significantly tilted to Saturn's equator ($i_L = 14°$). As Iapetus's orbit precesses around its Laplace plane, the instantaneous inclination of Iapetus to Saturn's equator varies approximately over a $5° - 21°$ range over Iapetus's nodal precession period of about 3400 yr.

While in the classical picture (e.g. Murray and Dermott 1999) Iapetus does not take part in any resonances with other satellites, faster tidal evolution (Lainey et al. 2012, 2017) would make Titan and Iapetus cross their mutual 5:1 MMR in the past. This crossing should have happened about 500 Myr ago if we assume a uniform tidal quality factor $Q = 1500 - 2000$ for all satellites (Ćuk et al. 2013), or could have happened at a very different epoch if the tidal evolution of Saturn's moons is driven by resonant modes inside the planet (Fuller et al. 2016).

Ćuk et al. (2013) have modeled the Titan-Iapetus 5:1 MMR crossing and found that the orbits of both bodies are chaotic during the crossing of the resonant region, which consists of numerous sub-resonances of the 5:1 resonance (they assumed $Q_p/k_{2,p} = 4000$ for Saturn). If Titan was as eccentric as it is now, the most likely outcome is eccentricity growth for Iapetus, followed by orbit crossing. If the eccentricity of Titan was low (about 0.005), Iapetus typically survives the resonance, acquiring an eccentricity of a few percent, consistent with its present orbit ($e_I = 0.03$). Ćuk et al. (2013) conclude that if the Titan-Hyperion 5:1 resonance was crossed in the past (which is unavoidable for $Q_p/k_{2,p} = 4000$), the current large eccentricity of Titan must postdate this resonance crossing. This suggestion is consistent with the later excitation of Titan's eccentricity (and inclination) by an encounter with a large moon, now lost (Wisdom et al. 2022) (see Sect. 5).

Ćuk et al. (2013) also find that the inclination of Iapetus is affected only weakly by the 5:1 resonance crossing, with the typical change being only a degree or so, compared to the current 8°. Therefore, this resonance cannot constrain the timing or the source of Iapetus's inclination excitation. Interestingly, changes to the inclination of Titan can be comparable to the free inclination itself (which is 0.3°), making it possible that Titan's inclination was significantly modified by this resonance.

Polycarpe et al. (2018) have carried out numerical simulations of the resonance crossing using an N-body code as well as using averaged equations of motion. A large span of migration rates were explored for Titan and Iapetus was started on its local Laplace plane (14° with respect to the equatorial plane) with a circular orbit. Polycarpe et al. (2018) find that the resonance crossing can trigger a chaotic evolution of the eccentricity and the inclination of Iapetus. The outcome of the resonance is highly dependent on the migration rate (or equivalently on Q_p). For a quality factor Q_p of over around 2000, the chaotic evolution of Iapetus in the resonance leads in most cases to its ejection, while simulations with a quality factor between 100 and 2000 show a departure from the resonance with post-resonant eccentricities spanning from 0 up to 15%, and free inclinations capable of reaching 11°. Usually high inclinations come with high eccentricities but some simulations (less than 1%) show elements compatible with Iapetus' current orbit. Polycarpe et al. (2018) conclude a

[4]The instantaneous Laplace plane can be defined for every perturbed orbit as plane normal to the vector around which the orbit normal is precessing. See Sect. 5.1 for a formal definition.

quality factor between 100 and 2000 at the frequency of Titan would bring Titan and Iapetus into a 5:1 resonance, which could in some cases perturb Iapetus' eccentricity and inclination to values observed today. Such rapid tidal migration would have avoided Iapetus' ejection around 40-800 Myr ago.

Ćuk et al. (2018) find that Iapetus is currently in a secular resonance with an argument $\varpi - \Omega + \varpi_J - \Omega_{eq}$ librating around 180°, where ϖ and ϖ_J are the longitudes of pericenter of Iapetus and Jupiter, while Ω and Ω_{eq} are the longitudes of Iapetus's ascending node and Saturn's vernal equinox. The libration period is several Myr and the libration is likely to persist for several tens of Myr. Longer-term stability of this resonance is tied to the precession of Saturn's spin axis, and more definite predictions await better determinations of Saturn's precession rate (see Sect. 5). Most allowable solutions for Saturn's pole precession lead to eventual breaking of the Iapetus-g_5 secular resonance,[5] but some solutions preserve the secular resonance for at least 100 Myr. The Iapetus-g_5 secular resonance was almost certainly established more recently than the proposed 5:1 MMR crossing between Titan and Iapetus (500-50 Myr ago, depending on the Titan's unknown tidal evolution rate). While Ćuk et al. (2018) find cases when the secular resonance was established in the aftermath of this MMR (with the more rapidly evolving Titan offering promising results), they did not find a high-probability mechanism for establishing the secular resonance.

There are a number of profound mysteries regarding Iapetus's orbital history. There is no agreed-upon high-probability mechanism of producing Iapetus's very large inclination. The history of its secular resonance is also unclear. The eccentricity of Iapetus is probably the orbital parameter that is most closely coupled to the overall evolution of the system. Survival of Iapetus with a low eccentricity after the 5:1 Titan-Iapetus resonance is consistent with rapid evolution found by Lainey et al. (2020), with no restrictions on Titan's own eccentricity at the time. However, it is also consistent with a slower evolution of Titan (Jacobson 2022), but this requires that Titan acquired its eccentricity in the last 500 Myr (Ćuk et al. 2013), requiring some kind of catastrophic event (cf. Asphaug and Reufer 2013; Hamilton 2013; Ćuk et al. 2016). It is still not clear whether the low eccentricity of Iapetus can be consistent with a recent loss of a moon exterior to Titan (Wisdom et al. 2022), as such a moon would strongly interact with Iapetus before its disruption.

5 Obliquity of Saturn

In the previous sections, specific features of the satellite dynamics were reviewed in the context of strong tidal dissipation within Saturn. On a gigayear timescale, however, satellites cannot be considered in isolation; they are part of a vast coupled system that comprises the spin-axis dynamics of their host planet. The fast migration of Saturn's satellites, and of Titan in particular, has been shown to strongly affect the motion of Saturn's spin axis (see Saillenfest et al. 2021a,b; Saillenfest and Lari 2021; Wisdom et al. 2022). In this section, we review the main implications of these findings.

5.1 Basic Effect of Saturn's Satellites

From a satellite's perspective, the orientation of a planet's spin axis has direct dynamical consequences. During their formation in a disc, regular moons are naturally damped towards an equilibrium configuration with near-zero eccentricity. Their orbital inclination at

[5]The eigenmode with fundamental frequency g_5 dominates the apsidal precession of Jupiter, enabling us to use ϖ_J as its proxy.

Fig. 3 Equilibrium orbital inclination of a small satellite as a function of its semi-major axis. The orbital inclination of the satellite is measured with respect to the planet's equator. ε is the planet's obliquity

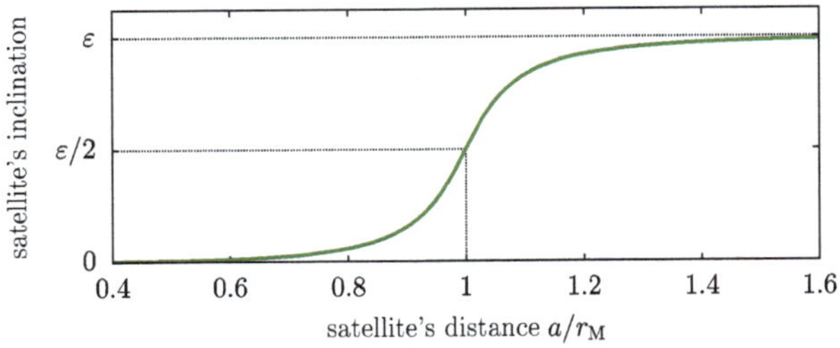

equilibrium defines the "Laplace plane", which results from a balance between the attraction of the equatorial bulge of the planet and the attraction of the star. For a massless satellite, the inclination of the Laplace plane measured from the planet's equator is:

$$I_{\mathrm{L}} = \frac{\pi}{2} + \frac{1}{2}\mathrm{atan2}\big[-\sin(2\varepsilon), -r_{\mathrm{M}}^5/a^5 - \cos(2\varepsilon)\big] \tag{13}$$

(see Tremaine et al. 2009; Saillenfest and Lari 2021), where ε is the planet's obliquity. As shown in Fig. 3, the characteristic length r_{M} is the distance at which the satellite's Laplace plane lies exactly halfway between the equator and the orbital plane of the planet (the index M stands for 'midpoint'). It can be written

$$r_{\mathrm{M}}^5 = 2\frac{M}{m_\odot} J_2 R_{\mathrm{eq}}^2 a_{\mathrm{P}}^3 (1 - e_{\mathrm{P}}^2)^{3/2}. \tag{14}$$

Here, $m_\odot$ is the mass of the star, and M, J_2, and R_{eq} are the mass, second zonal gravity coefficient, and equatorial radius of the planet. The orbit of the planet around the star is described by its semi-major axis a_{P} and eccentricity e_{P}. The value of r_{M} for Saturn is about 42 R_{eq}, so that the Laplace planes of most satellites are close to Saturn's equator. After their formation, regular moons keep oscillating about their local Laplace planes. Among the regular satellites of Saturn, only Iapetus presents today a substantial deviation from the exact equilibrium, about which it oscillates with an offset of 8° (see Sect. 4.6). From Eq. (13), we deduce that the spin-axis orientation of the planet directly sets the mean orbital inclination of its moons. Saturn's obliquity change resulting from high dissipation within Saturn (see below) has therefore an indirect effect on Saturn's whole satellite system.

From a planet's perspective, a system of satellites has a long-term contribution to its spin-axis motion. In absence of satellites, the average motion of a planet's spin axis is merely due to the stellar torque applied on its equatorial bulge. Noting **s** as the unit spin-axis vector and **n** the unit orbit normal, this motion can be written

$$\dot{\mathbf{s}} = -\alpha(\mathbf{n}\cdot\mathbf{s})(\mathbf{n}\times\mathbf{s}) \tag{15}$$

(see e.g. Goldreich 1966; Tremaine 1991), where $\mathbf{n}\cdot\mathbf{s} = \cos\varepsilon$. This equation produces a precession of the spin axis at a rate $\alpha\cos\varepsilon$ around the instantaneous orbit normal **n** of the planet. The timescale of this precession is set by the "precession constant" α, which can be expressed as

$$\alpha = \frac{3}{2}\frac{Gm_\odot}{a_{\mathrm{P}}^3(1-e_{\mathrm{P}}^2)^{3/2}}\frac{J_2}{\lambda\omega}, \tag{16}$$

where λ is the normalised polar moment of inertia of the planet, and ω is its spin rate. Satellites alter this motion in several ways: they contribute to the angular momentum of the system, they apply a torque on the planet's equatorial bulge, and their orbits are themselves torqued by the stellar attraction. The relative magnitude of these effects depends on the satellite distances. The moon-planet coupling is most obvious for the Moon around the Earth, as the lunar mass is about 1% of the Earth's mass (for comparison, Titan's mass is only 2×10^{-4} of Saturn's mass). As such, the first formulas that took into account the coupled motion of a planet's spin axis and the orbit of its moons were developed for the Earth-Moon system (see e.g. Goldreich 1966). From these formulas, we know that satellites enhance the mean spin-axis precession rate of their host planet in a way that is intimately linked to their local Laplace plane. Analytical expressions were obtained by Ward (1975) in the two extreme cases of close-in satellites (whose mean orbital plane is the planet's equator) and far-away satellites (whose mean orbital plane is the planet's own orbital plane; see Fig. 3). The first closed-formed expressions valid for any distance of the satellites were introduced by Tremaine (1991) and French et al. (1993). These expressions can be written as a modified precession constant α' for the planet, obtained by replacing J_2 and λ in Eq. (16) by their effective values:

$$
\begin{aligned}
J_2' &= J_2 + \frac{1}{2}\sum_k \frac{m_k}{M}\frac{a_k^2}{R_{\mathrm{eq}}^2}\frac{\sin(2\varepsilon - 2I_k)}{\sin(2\varepsilon)}, \\
\lambda' &= \lambda + \sum_k \frac{m_k}{M}\frac{a_k^2}{R_{\mathrm{eq}}^2}\frac{n_k}{\omega}\frac{\sin(\varepsilon - I_k)}{\sin\varepsilon}.
\end{aligned} \tag{17}
$$

In this expression, the sum runs over all regular satellites (assumed to have circular orbits); m_k is the mass of the kth satellite, a_k is its semi-major axis, and I_k is the inclination of its local Laplace plane with respect to the planet's equator. Equation (17) can be retrieved as a particular case of the self-consistent expressions of Boué and Laskar (2006): here, the motion of satellites is averaged over their precession around the Laplace plane. This approach is valid as long as they precess much faster than the planet's spin axis. This condition is well verified in practice for any distance of the satellites. As a result, the planet and its satellites rigidly precess as a whole (Goldreich 1965).

Equation (17) implies that the boost imparted by a moon on a planet's spin-axis precession rate strongly depends on the moon's distance. As an example, Fig. 4 shows the enhancement factor of Saturn's spin-axis precession rate as a function of Titan's distance. Despite the small mass of Titan as compared to Saturn's, the spin-axis precession rate of Saturn is currently multiplied by roughly a factor 4 due to the presence of Titan. This precession boost strongly depends on Titan's distance. The high dissipation within Saturn and

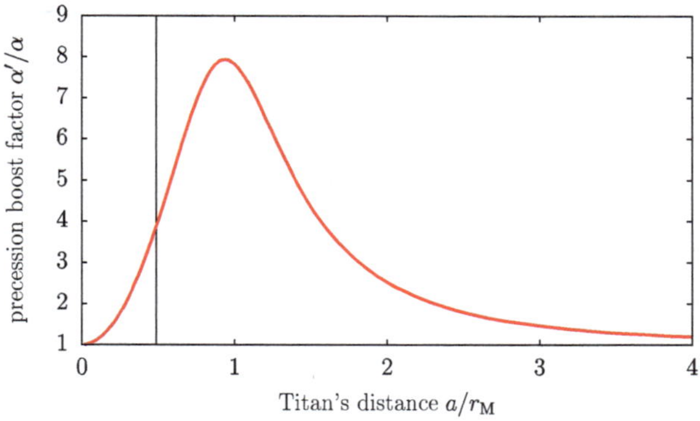

Fig. 4 Enhancement factor of Saturn's spin-axis precession rate due to its most massive satellite, Titan. The current semi-major axis of Titan is shown by a vertical line. The obliquity of Saturn is fixed to its current value $\varepsilon \approx 27°$. The curve reaches its maximum at $(a/r_M)^5 \approx (\sqrt{\cos^2(2\varepsilon) + 24} - \cos(2\varepsilon))/6$, see Saillenfest and Lari (2021)

the resulting fast migration of Titan therefore have strong implications for Saturn's spin-axis dynamics.

5.2 Secular Spin-Orbit Resonances

Because of the existence of secular spin-orbit resonances, a variation in a planet's spin-axis precession rate can drastically change its spin-axis dynamics. To understand this mechanism, we must notice that the orbit pole **n** appearing in Eq. (15) is actually not fixed, but it precesses itself as a result of mutual planetary perturbations. As the orbital motion of the planet remains largely unaffected by its spin state, the vector **n** behaves in Eq. (15) as an autonomous forcing term. Building on the early works by Lagrange and Laplace, very accurate theories now exist for the long-term orbital motion of the Solar System planets. As an example, Table 1 gives the dominant terms of the secular motion of Saturn's orbit pole computed by Laskar (1990). The solution is written through the complex variable

$$\zeta = \sin\frac{I}{2}\exp(i\Omega) = \sum_k S_k \exp[i\phi_k(t)], \tag{18}$$

in which the amplitudes S_k are real constants and the sum runs over all terms with non-negligible amplitude. The angles ϕ_k evolves linearly over time as $\phi_k(t) = \nu_k t + \phi_k^{(0)}$, where ν_k is a fixed frequency and $\phi_k^{(0)}$ is the phase at $t = 0$.

Each term in the motion of the orbit pole **n** in Eq. (15) can generate a resonance for the motion of the spin axis **s**. These resonances are called "secular spin-orbit resonances", because they involve the precession of the spin axis and the precession of the orbit. The geometry of the lowest-order resonances is given by Colombo's top Hamiltonian (see Colombo 1966; Henrard and Murigande 1987; Saillenfest et al. 2019; Su and Lai 2022). As a function of the planet's precession constant α, the system near a resonance can have up to four equilibrium points which are called the "Cassini states" (Peale 1969). For a critical value of α, a separatrix appears around the Cassini state 2; this separatrix divides the inside from the outside of the resonance. At this point, it becomes clear how even a set of small moons

Table 1 First ten terms in the series decomposition of Saturn's inclination and longitude of ascending node, from Laskar (1990). The orbital elements of Saturn are measured in the J2000 ecliptic and equinox reference frame. Terms are sorted by decreasing amplitude and identified in terms of the fundamental frequencies g_j and s_j of the Solar System (using standard definitions, see e.g. Murray and Dermott 1999). The period associated with each frequency ν_k is given in the last column

k	Identification	ν_k ($''\,\mathrm{yr}^{-1}$)	$S_k \times 10^6$	$\phi_k^{(0)}$ (°)	P_k (kyr)
1	s_5	0.00000	13,774	107.59	
2	s_6	−26.33023	7850	127.29	49
3	s_8	−0.69189	560	23.96	1873
4	s_7	−3.00557	391	140.33	431
5	$g_5 - g_6 + s_7$	−26.97744	59	43.05	48
6	$2g_6 - s_6$	82.77163	34	128.95	16
7	$g_5 + g_6 - s_6$	58.80017	20	212.90	22
8	$2g_5 - s_6$	34.82788	16	294.12	37
9	s_1	−5.61755	14	168.70	231
10	s_4	−17.74818	13	123.28	73

can dramatically affect a planet's spin dynamics: because of their effective enhancement of the precession constant α', the presence of moons may result in the planet being inside instead of outside a given resonance. On top of that, the moons' tidal migration may slowly drive the value of α' past the critical value and produce a resonance capture. In case of resonance capture, the spin-axis precession rate of the planet becomes roughly constant, as it remains close to a given frequency ν_k appearing in the orbital precession (see Table 1). In other words, $\alpha' \cos\varepsilon \approx -\nu_k$. As the moons continue to migrate and produce large changes in the value of α' (see Fig. 4), the planet's obliquity ε is forced to evolve so as to maintain the equality. This phenomenon is likely to happen for Jupiter in the future and produce an increase in its obliquity (Lari et al. 2020; Saillenfest et al. 2020).

5.3 Saturn Is Close to a Resonance...

By taking into the account the effect of Saturn's satellites, Ward eand Hamilton (2004) and Hamilton and Ward (2004) have shown that Saturn is today very close to a strong secular spin-orbit resonance with s_8 (third term in Table 1). It was immediately recognised that this striking match and Saturn's large obliquity ($\varepsilon \approx 27°$) are unlikely to be coincidental. Indeed, due to the process of gas accretion, giant gaseous planets are expected to form with a near-zero obliquity ($\varepsilon \approx 0°$); some mechanism should therefore have tilted Saturn after its formation. As the strong resonance with s_8 has necessarily affected Saturn's obliquity in the past, it offers a natural explanation for its large obliquity (see Ward eand Hamilton 2004; Hamilton and Ward 2004). Yet, in order to produce a resonance capture and tilting, the two frequencies involved (i.e. Saturn's spin-axis precession rate and Neptune's precession mode s_8) must have evolved and crossed the 1 to 1 commensurability. The question is when this crossing has happened in the past, and under which circumstances.

It is known today that shortly after the planetary formation, the orbits of the giant planets have been reshaped, as planets were migrating through a swarm of planetesimals (Malhotra 1993). The last large-scale changes in the orbits of the Solar System planets probably happened during this stage, more than four gigayears ago, when planetary migration triggered a phase of orbital instability (see e.g. Tsiganis et al. 2005; Nesvorný and Morbidelli 2012). The dominant orbital precession modes of the planets are a direct function of their semi-major axes (see e.g. Murray and Dermott 1999); for this reason, planetary migration and instability will have produced a variation in the frequency s_8. By virtue of Eq. (16), Saturn's spin-axis precession rate also varied during this period of time. Hence, planetary migration first appeared to astronomers as a natural explanation for the crossing of Saturn's secular spin-orbit resonance with s_8. More than that, the current obliquity of Saturn was viewed as a practical constraint for the planetary migration process itself (Boué et al. 2009; Vokrouhlický and Nesvorný 2015; Brasser and Lee 2015).

However, the high tidal dissipation inside Saturn measured by Lainey et al. (2020) has changed our understanding of Saturn's tilting process. The migration rate measured for Titan is so fast that Titan has most probably migrated over a distance of several radii of Saturn in the past (and possibly even more than that; see Lainey et al. 2024, this collection, but also the contrary opinion of Jacobson 2022). Due to the steepness of its enhancement factor on Saturn's precession constant (see Fig. 4), this migration for Titan must have greatly altered Saturn's precession rate over time. As a result, it appears very unlikely that Saturn crossed the s_8 resonance during late planetary migration, more than four gigayears ago, because its spin-axis precession motion was too slow at that time. Instead, Titan's orbital expansion offers a new explanation for Saturn's capture into resonance and tilting to its current 27° obliquity (Saillenfest et al. 2021a; Wisdom et al. 2022).

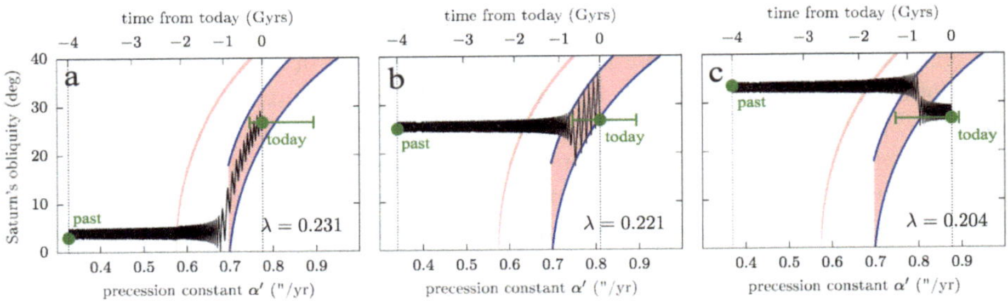

Fig. 5 The three types of possible obliquity evolution for Saturn (adapted from Saillenfest et al. 2021a). Each panel shows an example of simulation for a given value of λ (see labels). The explored range of values lies in the horizontal green bar. The first and last point of each trajectory are shown by a green spot. First-order secular spin-orbit resonances are shown in pink: the large area on the right is the resonance with s_8 and the thin area on the left is the resonance with $g_7 - g_8 + s_7$ (not shown in Table 1; 19th term). The separatrices of the s_8 resonance are highlighted in blue. Panels a and b show resonant capture while panel c does not

The actual dynamical path followed by Saturn depends on its current location with respect to the resonance. As described by Eq. (16) and (17), the precise precession rate of Saturn depends on several physical parameters. Each of these parameters are well known today from direct measurements, except Saturn's normalised polar moment of inertia λ. Indeed, the gravitational potential measured by spacecraft only provides differences between the moments of inertia (J_2, J_4, etc.), but no measure of the individual moments of inertia. In order to obtain the value of Saturn's polar moment of inertia, authors usually use semi-empirical models of Saturn's interior structure that are fitted to Saturn's gravitational moments (see e.g. Hubbard and Marley 1989; Nettelmann et al. 2013; Movshovitz et al. 2020). Such estimates, although formally accurate, are model-dependent and generally do not agree with each other. Moreover, it is not clear how to convert the moment of inertia derived from state-of-the-art interior models to that used in simplified models of rotation dynamics (which use the gyroscopic approximation, a purely rigid rotation, no internal dynamical processes, etc.); the small discrepancy between these two concepts of moment of inertia may lead to a slight shift of Saturn's precession constant in the neighbourhood of the s_8 secular spin-orbit resonance. For these reasons, previous authors did not rely on a given value of λ, but they explored a range of possible values spanned by the estimates obtained through interior modelling. Using this approach, Saillenfest et al. (2021a) showed that Saturn could have followed three possible types of trajectories as a result of Titan's fast migration (see Fig. 5). For a given range of λ values, one obtains very small past obliquities for Saturn (panel a) and a resonance capture with 100% certainty without crossing the separatrix. For slightly lower values of λ, one obtains a recent resonance capture by crossing the separatrix (panel b) or a resonance crossing without capture (panel c). By conducting large-scale Monte Carlo experiments, Saillenfest et al. (2021b) investigated the likelihood for each pathway as a function of Saturn's initial obliquity. The result is shown in Fig. 6 for Titan's nominal migration rate measured by Lainey et al. (2020). Trajectories labelled "b" are by far the less likely, because separatrix crossings are probabilistic events and in this case they rarely lead to a capture (see Saillenfest et al. 2021b). Trajectories labelled "a" are about ten times more likely that those labelled "c". This higher likelihood is mostly due to the precession phase of Saturn's spin axis, which points today roughly in the direction of the

Fig. 6 Likelihood of reproducing Saturn's current state as a function of its past obliquity (adapted from Saillenfest et al. 2021b). Each point is made of 240 numerical simulations with random initial precession angle for Saturn, started 4 Gyrs ago and propagated until today. The colour scale measures the relative likelihood of obtaining a final obliquity and precession angle within 0.5° and 5° of Saturn's current values, respectively. A small grey dot means that no successful trajectory was found. The labels a, b, and c refer to the trajectory types illustrated in Fig. 5

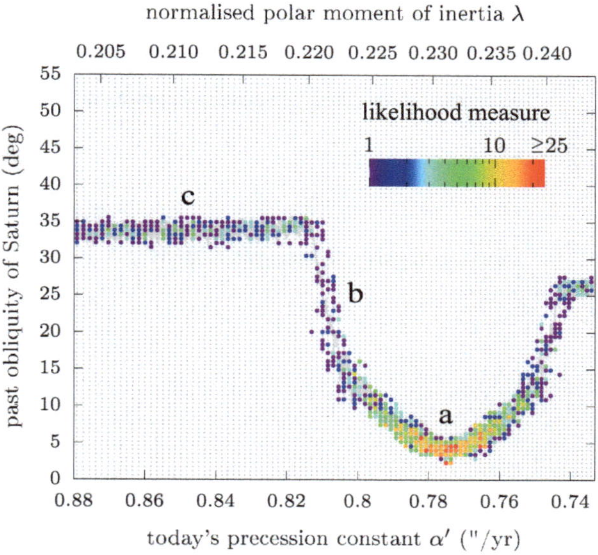

centre of the resonance.[6] Hence, the tilting of Saturn from a small obliquity – and Saturn still being in resonance today – is by far the most likely scenario from a dynamical point of view.

In this situation, Saturn's obliquity would continue to increase in the future as a result of the still ongoing migration of Titan; it could even reach values as high as $\varepsilon \gtrsim 75^\circ$ within 5 Gyrs, at which point Titan would be strongly destabilised (see Saillenfest and Lari 2021).

5.4 ...but Saturn Is Just Out of Sync

The dynamical pathway followed by Saturn depends on its current location with respect to the s_8 resonance. If Saturn is inside the resonance today, then it would have been captured from a small obliquity and it would continue to follow the resonance as Titan migrates (Fig. 5a). If instead Saturn is outside the resonance today, then it would have recently crossed the resonance without capture, from a higher obliquity (Fig. 5c). In order to determine whether Saturn is currently inside or outside the resonance, Wisdom et al. (2022) developed new models of Saturn's interior structure that were fitted to the latest estimates of Saturn's gravitational moments measured by the Cassini spacecraft. From these models, they deduced refined values of Saturn's polar moment of inertia λ (estimated to be equal to $0.2182^{+0.0006}_{-0.0003}$) that they used in long-term numerical integrations of Saturn's spin axis.

According to their results, Saturn is not in resonance today, but just outside its edge, at the level of 1% on the value of λ. As for the ambiguity and model dependency inherent to this approach (see above), Wisdom et al. (2022) argue that the four models that they consider cover the entirety of possible solutions that one could obtain from realistic models of Saturn's interior. The more recent interior models of Mankovich et al. (2023), which include a fit to ring seismology, also suggest that Saturn is out of resonance at present (see

[6]The current value of the resonance angle is $\sigma \approx 30^\circ$, which is relatively close to zero (Ward eand Hamilton 2004). If Saturn is not locked in resonance today (as in case c), then σ would circulate between 0° and 360° and have no particular reason to be small today; it would rather be a coincidence. If instead Saturn is locked in resonance today (as in case a), then σ would oscillate around 0° and always keep a relatively small value.

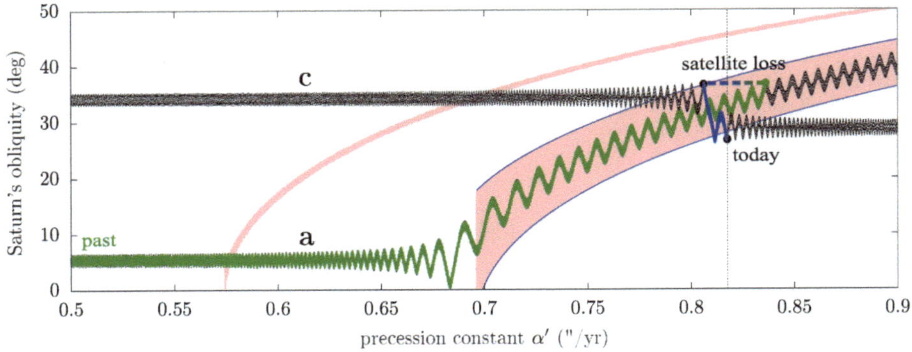

Fig. 7 Variant scenario proposed by Wisdom et al. (2022). The small black dots show the types of trajectories "a" and "c" depicted in Fig. 5. The two portions of pathway followed by Saturn are superimposed in green and blue, respectively. When Saturn's ancient satellite is lost, the curve instantly jumps from the green to the blue curve (dashed segment)

below). On this basis, Saturn should definitely be considered out of the s_8 resonance today. Saturn should therefore follow a pathway similar to that in Fig. 5c.

In this case, we are back to the question of the origin of Saturn's large obliquity. As recalled by Saillenfest et al. (2021a), an early giant impact could be a possibility. Instead of a giant impact, Wisdom et al. (2022) propose to connect the two possible types of pathways ("a" and "c") by a timely reorganisation in Saturn's moon system. Indeed, if a moon of Saturn is suddenly ejected or engulfed into Saturn, then Saturn's effective precession constant α' would jump to a lower value (see Eq. (17)). If this moon is massive enough, this would instantly kick Saturn out of the resonance. Wisdom et al. (2022) show that a relatively small ancient moon would be enough to produce this effect, for instance with a mass of the order of Iapetus's. Then, the still ongoing migration of Titan would make Saturn cross the resonance again, but without capture. This scenario is illustrated in Fig. 7. Ejection of the moon is hypothesized to occur when its orbit is destabilized during a 3:1 MMR with Titan (Wisdom et al. 2022).

As Saturn is still very close to the resonance today, and Titan's migration drives Saturn farther and farther away as time goes by, the removal of Saturn's ancient moon must have occurred recently in the history of the Solar System. For a former moon with the mass of Iapetus (that would not kick Saturn very far off the resonance), Wisdom et al. (2022) estimate that this event should have happened about 100 Myrs ago. Such a recent event reminds us of the controversial young age of Saturn's ring (see Crida et al., this collection). Wisdom et al. (2022) thus suggest to link these two phenomena: the lost moon would have been ripped apart within Saturn's Roche limit, which would have supplied the ring material, similarly to what has been proposed by Canup (2010). A recent instability in Saturn's system may also explain some peculiarities in the moons' orbits, such as Titan's eccentricity or Iapetus's inclination (see Sect. 4.6). It is still unclear, however, how such a catastrophic event would relate to the existence of Saturn's mid-sized moons, which all have rather unexcited orbits today and are connected through a fragile web of MMR (see Sect. 4).

The scenario of Wisdom et al. (2022) has the merit of preserving Titan's migration as the explanation for Saturn's large obliquity (see Saillenfest et al. 2021a), while allowing Saturn to be out of the resonance today. However, the likelihood analysis shows that, if one remains somewhat sceptical about the exact value of Saturn's polar moment of inertia, Saturn's current dynamical state is about ten times more likely to result from Saturn still being in resonance today (mostly due to the current small value of the resonance angle σ; see footnote 6 above). It is also more probable in a cosmogonic point of view: if Saturn had freshly crossed the resonance, then humanity would be witnessing today a transient event that could have happened any time over the history of the Solar System. That being said, Saturn's ring does seem exceptional in the Solar System, so one may need to accept some exceptional level of chance to account for its existence (see the discussion by Maryame El Moutamid 2022). Furthermore, the Saturn system seems to have possessed a population of planetocentric impactors which are not observed elsewhere (Ferguson et al. 2022) and suggest that something unusual happened.

The debate is revived by the recent article by Jacobson (2022). His self-consistent adjustment of Saturn's spin axis and satellite dynamics to all available data yields a lower migration rate for Titan than previously measured (see Lainey et al. 2024, this collection). If we extrapolate the results of Jacobson (2022) into the past, we find that Titan would have migrated just the right amount to explain Hyperion's eccentricity (Ćuk et al. 2013; see Sect. 4.5) and Saturn's tilt (Saillenfest et al. 2021b; see Sect. 5.3) over the age of the Solar System. Moreover, Jacobson (2022) obtains a value for Saturn's moment of inertia[7] that would put Saturn near the centre of the most probable region in Fig. 6; this would imply that Saturn's obliquity has evolved as in Fig. 5a, rather than the scenario proposed by Wisdom et al. (2022) and displayed in Fig. 7. If so, then the origin of Saturn's rings, Titan's eccentricity and the population of planetocentric impactors all remain to be explained. In the next few years, we can expect different teams of astronomers to revisit the studies of Wisdom et al. (2022) and Jacobson (2022); the future will tell which scenario will reach scientific consensus. Similarly, future work will hopefully yield agreement on the correct value for Saturn's polar moment of inertia λ.

A recent step in this direction has been made by Mankovich et al. (2023). The authors fitted interior models not only to the measured gravity harmonics of Saturn, but also on the frequencies of Saturn's internal oscillation modes imprinted in the rings. This approach allows the deep rotation profile of Saturn to be probed. From their models of internal structure, Mankovich et al. (2023) computed the expected value of Saturn's moment of inertia: they note that their results are consistent with the upper end of estimates of Wisdom et al. (2022), and approximately 0.5% smaller than the value required for the s_8 resonance to be active today. This supports the notion that Saturn is today outside the resonance, though close to its edge. Mankovich et al. (2023) also stress that large residuals in the frequencies of several modes (mostly for Saturn's deep structure, but also for outer layers) reflect the fact that current models for Saturn's interior are still not complete. According to them, a parallel can be drawn between this mismatch and the persistent difficulty encountered in fitting Jupiter and Saturn interior models to gravity while retaining the supersolar envelope heavy element enrichments indicated by spectroscopic measurements. These discussions reveal the richness and complexity of planetary interior modelling, and new avenues to better match the available observational constraints. Mankovich et al. (2023) advocate giving more freedom to the composition, temperature, and rotation profiles used as working interior models for their adjustments. It will be interesting to know whether such future refined interior models could change estimates for Saturn's moment of inertia by a non-negligible amount, and whether the strongly non-rigid rotation of Saturn that they reveal – still never taken into account in simulations of Saturn's long-term spin-axis dynamics – can alter in some way the precession rate of Saturn's spin axis.

[7] When converted to the same normalising radius R_{eq} and spin rate ω as those used in Sect. 5.3, the values obtained by Jacobson (2022) for Saturn's normalised moment of inertia are $\lambda = 0.2283$ and 0.2313 (see Eq. (16)), with a 1σ uncertainty of about 1%. The two values correspond to two different sets of observations.

6 Interior Evolution of Moons

The interior evolution of moons is driven by heating processes that can affect the state (e.g., solid or liquid), location (e.g., homogeneous or segregated), and composition of their materials. These materials appear to be primarily ices (chiefly water ice) and, as inferred from meteorites and interplanetary dust particles, silicate rock and refractory organic material. The proportion of ice is inferred from the moons' bulk densities (985 $kg\,m^{-3}$ for Tethys to 1879 $kg\,m^{-3}$ for Titan) to be at least 50 vol.%, with some uncertainty arising from porosity (for smaller moons such as Mimas) and compression (for Titan). Heating predominantly arises from the long-term decay of radioactive isotopes sourced in silicate rock and from tidal dissipation (Eq. (2)) with additional contributions, likely limited in time, from the radioactivity of short-lived isotopes, release of potential energy during accretion or separation of materials by density (differentiation), and chemical reactions between materials (e.g., rock hydration). Radiogenic heating drives larger, more silicate-rich moons, with a lower ratio of surface area (heat loss) to volume (heat production), to have warmer, more evolved interiors. Tidal heating drives closer-in, more eccentric or inclined moons to be warmer and thus more evolved (Eq. (2)). Tidal dissipation also couples the interior and orbital evolution of a moon and, because of gravitational interactions (e.g., resonances) within the moon system, to those of the other moons, Saturn, and its rings as well. While interior and orbital evolution aspects can and have been studied separately for specific satellites, or a subset of the moon system, system-level studies of the coupled interior-orbital evolution are challenging to carry out owing to the diversity of timescales involved (from orbital periods of days to radioactive decay timescales of billions of years) and the larger space of uncertain or unknown parameters.

A number of observational constraints provide a starting point. In addition to the dynamical constraints on ongoing and past orbital evolution described in previous sections, these include geological observations (see Rhoden et al. 2024, this collection, for an in-depth discussion). The moons' ice-rich surfaces and (assuming hydrostatic equilibrium) moments of inertia indicate some degree of ice-rock separation, but this separation is incomplete inside at least Titan (Iess et al. 2010) and perhaps Enceladus (Hemingway and Mittal 2019) and Rhea (Tortora et al. 2016). The data at other moons are too uncertain to tell (Tajeddine et al. 2014; Beuthe et al. 2016). Enceladus (see Rhoden et al. 2024, this collection) and, perhaps, Titan (Nixon et al. 2018) show evidence of ongoing endogenic geological activity, and Tethys and Dione show evidence of past activity linked to past high heat fluxes (Chen and Nimmo 2008; Hammond et al. 2013; White et al. 2017). The sizes of the moons increase with increasing semi-major axis from Mimas to Titan, but their bulk densities (rock content, including heat-producing radioisotopes) do not.

An especially noteworthy observation, the 'Mimas paradox', is the puzzling contrast in geological activity between inactive Mimas and (cryo)volcanic Enceladus. From Eq. (2), the dissipation of solid tides raised by Saturn, which dominates radiogenic heating at these small, close-in moons, is roughly 40 times stronger at Mimas than at Enceladus, assuming a similar interior structure and material response (i.e., k_2/Q). Why, then, is Enceladus so active and Mimas so quiescent? Proposed solutions to this paradox involve different interior material responses between Mimas and Enceladus, i.e., different loci in the positive feedback loop in which increasing interior temperatures lead to decreasing material viscosities, which lead to increasing tidal dissipation and more heating. Malamud and Prialnik (2013) proposed that Enceladus was nudged into this positive feedback loop by a spike of chemical heating during the hydration of the core, possible at Enceladus but not Mimas due to the former's higher radiogenic heat supply (Enceladus is denser and thus has a higher rock fraction than Mimas). Czechowski and Witek (2015) suggested that this nudging could have also been caused by higher short-lived radiogenic heating at Enceladus. Neveu and Rhoden (2019) invoked more radical differences: a young age for Mimas and earlier encounter of MMR by Enceladus which would not be today in equilibrium heating. Rhoden and Walker (2022) suggested that Mimas may have an ocean just like Enceladus, which could be compatible with observations of its libration (Tajeddine et al. 2014). Whether this suggestion is compatible with the lack of obvious indications of extensive relaxation of Mimas's large impact basins remains to be established. An ocean-bearing Mimas would also have to have avoided damping of Mimas's eccentricity (Sect. 4.3).

The long-term evolution of the moons depends on how and when they formed. This is largely an open question for the Saturn system. While Titan and Iapetus are likely primordial, the age of Rhea and closer-in moons is unclear (Rhoden et al. 2024, this collection). They too could have formed concurrently with Saturn at the dawn of the Solar System (Mosqueira et al. 2010, and references therein), possibly coexisting and interacting with a long-lived massive ring (Nakajima et al. 2019), or from the reaccretion of moon fragments from prior collisions with Sun-orbiting impactors (Movshovitz et al. 2015), or from accretion at the outer edge of Saturn's rings followed by tidally driven orbital expansion (Charnoz et al. 2010). Each formation scenario is compatible only with a subset of possible orbital expansion histories (Sect. 1): roughly, the end-members are accelerating expansion for primordial or old moons and decelerating expansion for young moons. Astrometric observations of fast orbital expansion (Lainey et al. 2020) currently lean toward the former (but see Jacobson 2022). If moons formed from rings, rock (silicates and organics) and ice may have been separated inside moons at the time of formation: the lesser propensity of rock to undergo tidal disruption would have led first to the formation of a rock-rich core, onto which ice would then have accreted as tidally driven orbital expansion led to decreased tidal forcing (Charnoz et al. 2011). This appears to be the case for Mimas, whose predicted fast orbital expansion, dominated by interactions with Saturn's nearby rings, suggests a young age if Mimas postdates the rings (Rhoden et al. 2024, this collection), and whose libration necessitates some degree of internal differentiation (Tajeddine et al. 2014). However, there is no independent evidence that Mimas postdates the rings. The higher propensity of close-in moons (Enceladus, Tethys, Dione) to encounter MMR implies a greater potential for interior evolution via episodes of high heat flow or collisions, especially if the moons are old. This is reflected in the broader diversity of geological features at their surfaces relative to Mimas, which may be younger, and Rhea, which is farther out.

In contrast, the interior evolution of Rhea, Titan, and Iapetus appears dominated by radiogenic heating and thus less tumultuous (see Rhoden et al. 2024, this collection). Dating events in the moons' history, such as past periods of geological activity, can constrain or test hypotheses on their internal evolution (Zahnle et al. 2003); progress on crater-based dating accounting for planetocentric impactor populations (Ferguson et al. 2020, 2022) and future in situ measurements could decrease the currently considerable uncertainties on surface ages.

7 Conclusions

The long-term evolution of the Saturnian system has probably never been more contentious. The proposed views of the system have changed over the years, from a primordial one driven by relatively slow tides (Murray and Dermott 1999; Meyer and Wisdom 2008), through a fast-evolving system (Lainey et al. 2012) that may not be primordial (Ćuk et al. 2016), to a resonance-lock driven one that could either be primordial (Fuller et al. 2016; Lainey et al. 2020) or recently reset (Wisdom et al. 2022). Our models of the system's long-term evolution are closely related to the observational results on the Saturnian rings, satellite astrometry and crater statistics which are covered in other articles of this collection. Some of the past controversies have been resolved, as we now know that the rings have a relatively low mass (Iess et al. 2019), even if their age is still in question (Crida et al. 2019), and the impact craters on the satellites are at least partially caused by planetocentric debris (Ferguson et al. 2020, 2022). New controversies include the current migration rate of Titan (Lainey et al. 2020; Jacobson 2022) and whether Saturn's spin precession is still in a resonance with Neptune's orbit (Saillenfest et al. 2021a; Wisdom et al. 2022).

One firm data point is that the observed tidal evolution of Rhea is well constrained (Lainey et al. 2017, 2020; Jacobson 2022) and is incompatible with equilibrium tides, indicating that highly frequency-dependent (probably resonant) tides must be operating in the Saturnian system (Lainey et al. 2024, this collection). While this is technically just an instantaneous migration rate, the existence of resonant tides greatly influences our understanding of the past evolution of the system. In this article we reviewed past work on the long-term dynamical evolution of Saturn's satellites, and attempted to show how various resonances and other dynamical features constrain the current theories about the system's age and tidal evolution. Our findings can be summarized as follows:

1. The currently observed Mimas-Tethys 4:2 and Enceladus-Dione 2:1 resonances indicate that in each of these two pairs, the satellites are on converging orbits. This is incompatible with all satellites being on diverging (Fuller et al. 2016) or parallelly-evolving orbits (Lainey et al. 2020). If the system is primarily driven by resonance locking, these resonances indicate that not all satellites are currently locked to Saturn's internal modes.
2. The current heat flux of Enceladus (if it is in equilibrium) is consistent both with strong equilibrium tides (Lainey et al. 2012) or diverging resonant modes (Fuller et al. 2016), but is likely significantly smaller than the expected tidal heating if Enceladus is locked to one of the parallelly-evolving modes (as illustrated in Fig. 3 of Lainey et al. 2020).
3. The excitation of Tethys's 1° inclination is an important constraint on the evolution of the system before the current resonances. It is possible that Tethys's inclination is a consequence of Dione-Rhea 5:3 resonance crossing followed by Tethys-Dione secular resonance (Ćuk et al. 2016), but that scenario would require the current fast evolution of Rhea to be a very recent phenomenon. If Rhea's current rate of evolution is a long-term value, then Dione and Rhea never crossed their 5:3 resonance.
4. Another "fossil" feature of the inner system is the excited eccentricity of Mimas, which could have been excited by the past crossing of the Mimas-Dione 3:1 resonance (Ćuk and El Moutamid 2022).
5. In every modern theory of its tidal evolution, Rhea should have crossed the semi-secular evection resonance with the Sun which should have excited Rhea's eccentricity and inclination well beyond the observed values. The damping of Rhea's inclination would be difficult to achieve, possibly indicating that Rhea may not be primordial but that it could have reaccreted close to its present orbit.
6. The current Titan-Hyperion 4:3 resonance has been evolved by tides and, depending on which estimate of Titan's tidal evolution rate we use, may be either 500 Myr old (using Lainey et al. 2020) or 4 Gyr old (using Jacobson 2022). In the former case Hyperion is not primordial and must be a product of a cataclysm of some kind within the last 1-1.5 Gyr.
7. Titan and Iapetus must have crossed their mutual 5:1 resonance at some point in the past. Once again, different estimates of Titan's migration rate place this crossing at either 50 Myr ago (Lainey et al. 2020) or 500 Myr ago (Jacobson 2022). The latter case of slower

evolution through the resonance leads to more dynamical excitation, so in this case Titan must have acquired its eccentricity after the 5:1 resonance with Iapetus, again indicating some kind of possibly catastrophic, recent event.

8. Titan's relatively high eccentricity and mechanically weak interior have long been thought to be problematic, indicating that the eccentricity had to be generated relative recently or it would have been damped. Mechanisms for this excitation usually involve some sort of catastrophe in which at least some moons are disrupted (Asphaug and Reufer 2013; Hamilton 2013; Ćuk et al. 2016; Wisdom et al. 2022)
9. A resonance between the precession of Saturn's spin axis and the orbit of Neptune almost certainly generated Saturn's obliquity (Ward eand Hamilton 2004; Hamilton and Ward 2004; Saillenfest et al. 2021a,b). The substantial orbital expansion of Titan offers a natural origin for this resonance capture. Wisdom et al. (2022) find that Saturn is currently not in this spin-orbit resonance, but this claim depends on the value of Saturn's moment of inertia derived from interior models, which are not all in agreement. However, if the resonance is broken, this would strongly indicate a recent cataclysm in the outer part of the satellite system (beyond Titan), although the viability of the specific scenario of Wisdom et al. (2022) and the connection to the formation of the rings may need further confirmation.

The above points demonstrate that there are no easy answers to the questions posed by the Saturnian system. It appears that there is no simple unified explanation for the orbital evolution of all moons (past and observed), and while some moons may not be primordial, the age and extent of any disruption events is highly uncertain.

Fortunately, there are two observable quantities that may be constrained enough in the near future to help us narrow down the range of the system's possible histories. One is the precession rate of Saturn's spin axis. Current measured values lean toward Saturn being in the spin-orbit resonance with Neptune (Jacobson 2022), though non-resonant values are also permitted. If Saturn is determined to be in the spin-orbit resonance, that would remove the need for (but not fully rule out) the proposed cataclysm in the outer part of the system (which is more important for Saturn's precession than the inner satellites). The second observable quantity is the current rate of tidal evolution of Titan. At the time of writing there are two contradictory results in the literature (Lainey et al. 2020; Jacobson 2022) that are largely based on the same data. If a consensus could be reached on this issue, we would have better understanding of the physics of tidal evolution in the system. Titan is the largest moon and is relatively far from Saturn. As such it is most likely to be ancient and most likely to be locked to a resonant mode within Saturn. Confirmation of a fast-evolving Titan would demonstrate the reality of resonant locking; a slower migrating Titan could indicate that fast migration driven by resonant modes may not be permanent.

Apart from the two above mentioned observable quantities, progress may come from sources of data other than the satellite dynamics. Better understanding of Saturn's interior (Mankovich and Fuller 2021), the rings (Iess et al. 2019) and satellite surfaces (Ferguson et al. 2022, also see Rhoden et al. 2024, this collection) could all help us understand the history of the system. Furthermore, more theoretical work on the moons' past orbital evolution, while not assumption-free, could still produce valuable new insights on the possible histories of the system, especially if combined with constraints from other lines of evidence.

Acknowledgements The authors thank the International Space Science Institute and Dr. Valery Lainey for organizing the workshop "New Vision of the Saturnian System in the Context of a Highly Dissipative Saturn" that inspired this collection.

Author Contribution All authors have contributed to this review article.

Funding Open access funding provided by Università di Pisa within the CRUI-CARE Agreement. MĆ is supported by NASA Solar System Workings award 80NSSC22K0979. ME and MĆ are supported by NASA-SSW award 80NSSC23K0681. FN is supported by NASA-SSW award 80NSSC21K1802. MN is supported by NASA-CDAP award 80NSSC21K0531 and by the CRESST II agreement between NASA GSFC and Univ. Maryland, College Park (award number 80GSFC17M0002). AR is supported by NASA Cassini Data Analysis Program award 80NSSC21K0535.

Data Availability Not applicable.

Code Availability Not applicable.

Declarations

Competing Interests The authors have no relevant financial or non-financial interests to disclose.

References

Allan RR (1969) Evolution of Mimas-Tethys commensurability. Astron J 74:497. https://doi.org/10.1086/110827

Asphaug E, Reufer A (2013) Late origin of the Saturn system. Icarus 223(1):544–565. https://doi.org/10.1016/j.icarus.2012.12.009

Beuthe M, Rivoldini A, Trinh A (2016) Enceladus's and dione's floating ice shells supported by minimum stress isostasy. Geophys Res Lett 43(19):10–088

Boué G, Laskar J (2006) Precession of a planet with a satellite. Icarus 185(2):312–330

Boué G, Laskar J, Kuchynka P (2009) Speed limit on Neptune migration imposed by Saturn tilting. Astrophys J 702(1):L19–L22

Brasser R, Lee MH (2015) Tilting Saturn without tilting Jupiter: constraints on giant planet migration. Astron J 150(5):157

Brouwer D, Clemence GM (1961) Methods of celestial mechanics. Academic Press, San Diego

Buratti BJ, Mosher JA (1995) The dark side of Iapetus: additional evidence for an exogenous origin. Icarus 115(2):219–227. https://doi.org/10.1006/icar.1995.1093

Canup RM (2010) Origin of Saturn's rings and inner moons by mass removal from a lost Titan-sized satellite. Nature 468(7326):943–946

Castillo-Rogez JC, Efroimsky M, Lainey V (2011) The tidal history of Iapetus: spin dynamics in the light of a refined dissipation model. J Geophys Res, Planets 116(E9):E09008. https://doi.org/10.1029/2010JE003664

Champenois S, Vienne A (1999a) Chaos and secondary resonances in the Mimas-Tethys system. Celest Mech Dyn Astron 74(2):111–146

Champenois S, Vienne A (1999b) The role of secondary resonances in the evolution of the Mimas-Tethys system. Icarus 140(1):106–121. https://doi.org/10.1006/icar.1999.6115

Charnoz S, Salmon J, Crida A (2010) The recent formation of Saturn's moonlets from viscous spreading of the main rings. Nature 465(7299):752–754

Charnoz S, Crida A, Castillo-Rogez JC, Lainey V, Dones L, Karatekin Ö, Tobie G, Mathis S, Le Poncin-Lafitte C, Salmon J (2011) Accretion of Saturn's mid-sized moons during the viscous spreading of young massive rings: solving the paradox of silicate-poor rings versus silicate-rich moons. Icarus 216(2):535–550

Chen EMA, Nimmo F (2008) Implications from Ithaca Chasma for the thermal and orbital history of Tethys. Geophys Res Lett 35(19):L19203. https://doi.org/10.1029/2008GL035402

Chen EMA, Nimmo F (2016) Tidal dissipation in the lunar magma ocean and its effect on the early evolution of the Earth-Moon system. Icarus 275:132–142. https://doi.org/10.1016/j.icarus.2016.04.012
Chen EMA, Nimmo F, Glatzmaier GA (2014) Tidal heating in icy satellite oceans. Icarus 229:11–30. https://doi.org/10.1016/j.icarus.2013.10.024
Chyba CF, Jankowski DG, Nicholson PD (1989) Tidal evolution in the Neptune-Triton system. Astron Astrophys 219(1–2):L23–L26
Colombo G (1966) Cassini's second and third laws. Astron J 71:891
Crida A, Charnoz S, Hsu H-W, Dones L (2019) Are Saturn's rings actually young? Nat Astron 3:967–970. https://doi.org/10.1038/s41550-019-0876-y
Ćuk M, El Moutamid M (2022) Three-body resonances in the Saturnian system. Astrophys J Lett 926(2):L18. https://doi.org/10.3847/2041-8213/ac501c
Ćuk M, Stewart ST (2012) Making the Moon from a fast-spinning Earth: a giant impact followed by resonant despinning. Science 338(6110):1047. https://doi.org/10.1126/science.1225542
Ćuk M, Dones L, Nesvorný D (2013) Titan-hyperion resonance and the tidal Q of Saturn. arXiv e-prints. arXiv:1311.6780
Ćuk M, Dones L, Nesvorný D (2016) Dynamical evidence for a late formation of Saturn's Moons. Astrophys J 820(2):97. https://doi.org/10.3847/0004-637X/820/2/97
Ćuk M, Dones L, Nesvorný D, Walsh KJ (2018) Secular resonance between iapetus and the giant planets. Mon Not R Astron Soc 481(4):5411–5421
Ćuk M, El Moutamid M, Tiscareno MS (2020) Dynamical history of the Uranian system. Planet Sci J 1(1):22. https://doi.org/10.3847/PSJ/ab9748
Czechowski L, Witek P (2015) Comparison of early evolutions of Mimas and Enceladus. Acta Geophys 63(3):900–921
Dermott SF, Malhotra R, Murray CD (1988) Dynamics of the Uranian and Saturnian satelite systems: a chaotic route to melting Miranda? Icarus 76(2):295–334
Dombard AJ, Cheng AF, McKinnon WB, Kay JP (2012) Delayed formation of the equatorial ridge on Iapetus from a subsatellite created in a giant impact. J Geophys Res, Planets 117(E3):E03002. https://doi.org/10.1029/2011JE004010
El Moutamid M (2022) How Saturn got its tilt and its rings. Science 377(6612):1264–1265
El Moutamid M, Ćuk M, Tiscareno M (2019) The orbital connection between Mimas and Enceladus. In: AAS/Division of Dynamical Astronomy meeting. AAS/Division of Dynamical Astronomy meeting, vol 51, p 102.04
Ferguson SN, Rhoden AR, Kirchoff MR (2020) Small impact crater populations on Saturn's moon Tethys and implications for source impactors in the system. J Geophys Res, Planets 125(9):e2020JE006400
Ferguson SN, Rhoden AR, Kirchoff MR (2022) Regional impact crater mapping and analysis on Saturn's moon dione and the relation to source impactors. J Geophys Res, Planets 127(6):e2022JE007204
French RG, Nicholson PD, Cooke ML, Elliot JL, Matthews K, Perkovic O, Tollestrup E, Harvey P, Chanover NJ, Clark MA, Dunham EW, Forrest W, Harrington J, Pipher J, Brahic A, Grenier I, Roques F, Arndt M (1993) Geometry of the Saturn system from the 3 July 1989 occultation of 28 Sgr and Voyager observations. Icarus 103(2):163–214.
Fuller J, Luan J, Quataert E (2016) Resonance locking as the source of rapid tidal migration in the Jupiter and Saturn moon systems. Mon Not R Astron Soc 458:3867–3879
Fuller J, Guillot T, Mathis S, Murray C (2024) Tidal dissipation in giant planets. Space Sci Rev 220
Goldreich P (1965) Inclination of satellite orbits about an oblate precessing planet. Astron J 70:5
Goldreich P (1966) History of the lunar orbit. Rev Geophys Space Phys 4:411–439
Goldreich P, Gold T (1963) On the eccentricity of satellite orbits in the solar system. Mon Not R Astron Soc 126(3):257–268
Greenberg R (1973) Evolution of satellite resonances by tidal dissipation. Astron J 78:338. https://doi.org/10.1086/111423
Hamilton DP (2013) Late major merger at Saturn: consequences for Titan and iapetus. In: AAS/Division for Planetary Sciences meeting abstracts. AAS/Division for Planetary Sciences meeting abstracts, vol 45, p 302.01
Hamilton DP, Ward WR (2004) Tilting Saturn. II. Numer Model Astron J 128(5):2510–2517
Hammond NP, Phillips CB, Nimmo F, Kattenhorn SA (2013) Flexure on dione: investigating subsurface structure and thermal history. Icarus 223(1):418–422
Hemingway DJ, Mittal T (2019) Enceladus's ice shell structure as a window on internal heat production. Icarus 332:111–131
Henrard J, Murigande C (1987) Colombo's top. Celest Mech 40:345–366
Howett CJA, Spencer JR, Pearl J, Segura M (2011) High heat flow from Enceladus' south polar region measured using 10-600 cm^{-1} Cassini/CIRS data. J Geophys Res, Planets 116(E3):E03003. https://doi.org/10.1029/2010JE003718

Howett C, Miles G, Quick LC (2024) Constraining the mass and energy of dissipation systems. Space Sci Rev 220
Hubbard WB, Marley MS (1989) Optimized Jupiter, Saturn, and Uranus interior models. Icarus 78(1):102–118
Hussmann H, Spohn T (2004) Thermal-orbital evolution of Io and Europa. Icarus 171(2):391–410
Iess L, Rappaport NJ, Jacobson RA, Racioppa P, Stevenson DJ, Tortora P, Armstrong JW, Asmar SW (2010) Gravity field, shape, and moment of inertia of Titan. Science 327(5971):1367–1369
Iess L, Militzer B, Kaspi Y, Nicholson P, Durante D, Racioppa P, Anabtawi A, Galanti E, Hubbard W, Mariani MJ, Tortora P, Wahl S, Zannoni M (2019) Measurement and implications of Saturn's gravity field and ring mass. Science 364(6445):aat2965. https://doi.org/10.1126/science.aat2965
Jacobson RobertA (2022) The orbits of the main Saturnian satellites, the Saturnian system gravity field, and the orientation of Saturn's pole. Astron J 164(5):199
Lainey V, Karatekin Ö, Desmars J, Charnoz S, Arlot J-E, Emelyanov N, Le Poncin-Lafitte C, Mathis S, Remus F, Tobie G, Zahn J-P (2012) Strong tidal dissipation in Saturn and constraints on Enceladus' thermal state from astrometry. Astrophys J 752(1):14. https://doi.org/10.1088/0004-637X/752/1/14
Lainey V, Jacobson RA, Tajeddine R, Cooper NJ, Murray C, Robert V, Tobie G, Guillot T, Mathis S, Remus F, Desmars J, Arlot J-E, De Cuyper J-P, Dehant V, Pascu D, Thuillot W, Le Poncin-Lafitte C, Zahn J-P (2017) New constraints on Saturn's interior from Cassini astrometric data. Icarus 281:286–296. https://doi.org/10.1016/j.icarus.2016.07.014
Lainey V, Gomez Casajus L, Fuller J, Zannoni M, Tortora P, Cooper N, Murray C, Modenini D, Park R, Robert V, Zhang Q (2020) Resonance locking in giant planets indicated by the rapid orbital expansion of Titan. Nat Astron 4:1053–1058
Lainey V, Zannoni M, Iess L, Robert V, Fuller J, Mathis S, Guillot T (2024) Quantification of tides in giant planets from observations. Submitted to Space Sci Rev
Lari G, Saillenfest M, Fenucci M (2020) Long-term evolution of the Galilean satellites: the capture of Callisto into resonance. Astron Astrophys 639:A40
Laskar J (1990) The chaotic motion of the solar system: a numerical estimate of the size of the chaotic zones. Icarus 88(2):266–291
Levison HF, Walsh KJ, Barr AC, Dones L (2011) Ridge formation and de-spinning of Iapetus via an impact-generated satellite. Icarus 214(2):773–778. https://doi.org/10.1016/j.icarus.2011.05.031
Luan J (2014) Special mean motion resonance pairs: Mimas-Tethys and Titan-Hyperion. arXiv e-prints. https://doi.org/10.48550/arXiv.1410.2648. arXiv:1410.2648
Malamud U, Prialnik D (2013) Modeling serpentinization: applied to the early evolution of Enceladus and Mimas. Icarus 225(1):763–774
Malhotra R (1993) The origin of Pluto's peculiar orbit. Nature 365(6449):819–821
Mankovich CR, Fuller J (2021) A diffuse core in Saturn revealed by ring seismology. Nat Astron 5:1103–1109. https://doi.org/10.1038/s41550-021-01448-3
Mankovich CR, Dewberry JW, Fuller J (2023) Saturn's seismic rotation revisited. Planet Sci J 4(4):59
Meyer J, Wisdom J (2007) Tidal heating in Enceladus. Icarus 188(2):535–539. https://doi.org/10.1016/j.icarus.2007.03.001
Meyer J, Wisdom J (2008) Tidal evolution of Mimas, Enceladus, and Dione. Icarus 193(1):213–223
Moore WB, Schubert G (2000) The tidal response of Europa. Icarus 147(1):317–319
Mosqueira I, Estrada P, Turrini D (2010) Planetesimals and satellitesimals: formation of the satellite systems. Space Sci Rev 153(1):431–446. https://doi.org/10.1007/s11214-009-9614-6
Movshovitz N, Nimmo F, Korycansky DG, Asphaug E, Owen JM (2015) Disruption and reaccretion of mid-sized moons during an outer solar system late heavy bombardment. Geophys Res Lett 42(2):256–263
Movshovitz N, Fortney JJ, Mankovich C, Thorngren D, Helled R (2020) Saturn's probable interior: an exploration of Saturn's potential interior density structures. Astrophys J 891(2):109
Murray CD, Dermott SF (1999) Solar system dynamics. Cambridge University Press, Cambridge
Nakajima A, Ida S, Kimura J, Brasser R (2019) Orbital evolution of Saturn's mid-sized moons and the tidal heating of Enceladus. Icarus 317:570–582
Nesvorný D, Morbidelli A (2012) Statistical study of the early solar system's instability with four, five, and six giant planets. Astron J 144:117
Nesvorný D, Vokrouhlický D, Deienno R, Walsh KJ (2014) Excitation of the orbital inclination of Iapetus during planetary encounters. Astron J 148(3):52. https://doi.org/10.1088/0004-6256/148/3/52
Nettelmann N, Püstow R, Redmer R (2013) Saturn layered structure and homogeneous evolution models with different EOSs. Icarus 225(1):548–557
Neveu M, Rhoden AR (2019) Evolution of Saturn's mid-sized moons. Nat Astron 3(6):543–552
Nimmo F, Neveu M, Howett C (2023) Origin and evolution of Enceladus's tidal dissipation. Space Sci Rev 219(7):57. https://doi.org/10.1007/s11214-023-01007-4

Nixon CA, Lorenz RD, Achterberg RK, Buch A, Coll P, Clark RN, Courtin R, Hayes A, Iess L, Johnson RE et al (2018) Titan's cold case files-outstanding questions after Cassini-Huygens. Planet Space Sci 155:50–72

Ojakangas GW, Stevenson DJ (1986) Episodic volcanism of tidally heated satellites with application to Io. Icarus 66(2):341–358

O'Neill C, Nimmo F (2010) The role of episodic overturn in generating the surface geology and heat flow on Enceladus. Nat Geosci 3(2):88–91. https://doi.org/10.1038/ngeo731

Peale SJ (1969) Generalized Cassini's laws. Astron J 74:483

Peale SJ (1999) Origin and evolution of the natural satellites. Annu Rev Astron Astrophys 37:533–602. https://doi.org/10.1146/annurev.astro.37.1.533

Polycarpe W, Saillenfest M, Lainey V, Vienne A, Noyelles B, Rambaux N (2018) Strong tidal energy dissipation in Saturn at Titan's frequency as an explanation for iapetus orbit. Astron Astrophys 619:A133

Porco CC, Baker E, Barbara J, Beurle K, Brahic A, Burns JA, Charnoz S, Cooper N, Dawson DD, Del Genio AD, Denk T, Dones L, Dyudina U, Evans MW, Giese B, Grazier K, Helfenstein P, Ingersoll AP, Jacobson RA, Johnson TV, McEwen A, Murray CD, Neukum G, Owen WM, Perry J, Roatsch T, Spitale J, Squyres S, Thomas PC, Tiscareno M, Turtle E, Vasavada AR, Veverka J, Wagner R, West R (2005) Cassini imaging science: initial results on Phoebe and Iapetus. Science 307(5713):1237–1242. https://doi.org/10.1126/science.1107981

Porco CC, Helfenstein P, Thomas PC, Ingersoll AP, Wisdom J, West R, Neukum G, Denk T, Wagner R, Roatsch T, Kieffer S, Turtle E, McEwen A, Johnson TV, Rathbun J, Veverka J, Wilson D, Perry J, Spitale J, Brahic A, Burns JA, Del Genio AD, Dones L, Murray CD, Squyres S (2006) Cassini observes the active south pole of Enceladus. Science 311(5766):1393–1401. https://doi.org/10.1126/science.1123013

Renaud JP, Henning WG (2018) Increased tidal dissipation using advanced rheological models: implications for io and tidally active exoplanets. Astrophys J 857(2):98

Renaud JP, Henning WG, Saxena P, Neveu M, Bagheri A, Mandell A, Hurford T (2021) Tidal dissipation in dual-body, highly eccentric, and nonsynchronously rotating systems: applications to Pluto–Charon and the exoplanet TRAPPIST-1e. Planet Sci J 2(1):4

Rhoden AR, Walker ME (2022) The case for an ocean-bearing Mimas from tidal heating analysis. Icarus 376:114872

Rhoden AR, Ferguson SN, Castillo-Rogez JC, Martin E, Bland M, Zannoni M, Rambaux N, Salmon J (2024) Geologic constraints on the formation and evolution of Saturn's mid-sized moons. Submitted to Space Sci Rev

Rovira-Navarro M, Katz RF, Liao Y, van der Wal W, Nimmo F (2022) The tides of Enceladus' porous core. J Geophys Res, Planets 127(5):e2021JE007117

Rufu R, Canup RM (2020) Tidal evolution of the evection resonance/quasi-resonance and the angular momentum of the Earth-Moon system. J Geophys Res, Planets 125(8):e06312. https://doi.org/10.1029/2019JE006312

Saillenfest M, Lari G (2021) Future destabilisation of Titan as a result of Saturn's tilting. Astron Astrophys 654:A83

Saillenfest M, Laskar J, Boué G (2019) Secular spin-axis dynamics of exoplanets. Astron Astrophys 623:A4

Saillenfest M, Lari G, Courtot A (2020) The future large obliquity of Jupiter. Astron Astrophys 640:A11

Saillenfest M, Lari G, Boué G (2021a) The large obliquity of Saturn explained by the fast migration of Titan. Nat Astron 5:345–349

Saillenfest M, Lari G, Boué G, Courtot A (2021b) The past and future obliquity of Saturn as Titan migrates. Astron Astrophys 647:A92

Sharma I, Burns JA, Hui CY (2005) Nutational damping times in solids of revolution. Mon Not R Astron Soc 359(1):79–92. https://doi.org/10.1111/j.1365-2966.2005.08864.x

Sinclair AT (1972) On the origin of the commensurabilities amongst the satellites of Saturn. Mon Not R Astron Soc 160:169. https://doi.org/10.1093/mnras/160.2.169

Squyres SW, Reynolds RT, Cassen PM, Peale SJ (1983) The evolution of Enceladus. Icarus 53(2):319–331. https://doi.org/10.1016/0019-1035(83)90152-5

Stickle AM, Roberts JH (2018) Modeling an exogenic origin for the equatorial ridge on Iapetus. Icarus 307:197–206. https://doi.org/10.1016/j.icarus.2018.01.017

Su Y, Lai D (2022) Dynamics of Colombo's top: tidal dissipation and resonance capture, with applications to oblique super-earths, ultra-short-period planets and inspiraling hot Jupiters. Mon Not R Astron Soc 509(3):3301–3320

Tajeddine R, Rambaux N, Lainey V, Charnoz S, Richard A, Rivoldini A, Noyelles B (2014) Constraints on Mimas' interior from Cassini ISS libration measurements. Science 346(6207):322–324

Thomas PC, Burns JA, Helfenstein P, Squyres S, Veverka J, Porco C, Turtle EP, McEwen A, Denk T, Giese B, Roatsch T, Johnson TV, Jacobson RA (2007) Shapes of the Saturnian icy satellites and their significance. Icarus 190(2):573–584. https://doi.org/10.1016/j.icarus.2007.03.012

Tian Z, Nimmo F (2020) Implications of second-order resonance for the thermal and orbital evolution of Mimas. Mon Not R Astron Soc 492(1):369–376. https://doi.org/10.1093/mnras/stz3427
Tian Z, Wisdom J, Elkins-Tanton L (2017) Coupled orbital-thermal evolution of the early Earth-Moon system with a fast-spinning Earth. Icarus 281:90–102. https://doi.org/10.1016/j.icarus.2016.08.030
Tortora P, Zannoni M, Hemingway D, Nimmo F, Jacobson RA, Iess L, Parisi M (2016) Rhea gravity field and interior modeling from Cassini data analysis. Icarus 264:264–273
Touma J, Wisdom J (1998) Resonances in the early evolution of the Earth-Moon system. Astron J 115(4):1653–1663. https://doi.org/10.1086/300312
Tremaine S (1991) On the origin of the obliquities of the outer planets. Icarus 89(1):85–92
Tremaine S, Touma J, Namouni F (2009) Satellite dynamics on the Laplace surface. Astron J 137(3):3706–3717
Tsiganis K, Gomes R, Morbidelli A, Levison HF (2005) Origin of the orbital architecture of the giant planets of the solar system. Nature 435(7041):459–461
Tyler RH (2008) Strong ocean tidal flow and heating on moons of the outer planets. Nature 456(7223):770–772. https://doi.org/10.1038/nature07571
Vokrouhlický D, Nesvorný D (2015) Tilting Jupiter (a bit) and Saturn (a lot) during planetary migration. Astrophys J 806(1):143
Ward WR (1975) Tidal friction and generalized Cassini's laws in the solar system. Astron J 80:64–70
Ward WR (1981) Orbital inclination of Iapetus and the rotation of the Laplacian plane. Icarus 46(1):97–107. https://doi.org/10.1016/0019-1035(81)90079-8
Ward WR, Hamilton DP (2004) Tilting Saturn. Astron J 128(5):2501–2509
White OL, Schenk PM, Bellagamba AW, Grimm AM, Dombard AJ, Bray VJ (2017) Impact crater relaxation on dione and Tethys and relation to past heat flow. Icarus 288:37–52
Wisdom J (1980) The resonance overlap criterion and the onset of stochastic behavior in the restricted three-body problem. Astron J 85:1122–1133. https://doi.org/10.1086/112778
Wisdom J (2008) Tidal dissipation at arbitrary eccentricity and obliquity. Icarus 193(2):637–640
Wisdom J, Peale SJ, Mignard F (1984) The chaotic rotation of Hyperion. Icarus 58(2):137–152. https://doi.org/10.1016/0019-1035(84)90032-0
Wisdom J, Dbouk R, Militzer B, Hubbard WB, Nimmo F, Downey BG, French RG (2022) Loss of a satellite could explain Saturn's obliquity and young rings. Science 377(6612):1285–1289
Zahnle K, Schenk P, Levison H, Dones L (2003) Cratering rates in the outer solar system. Icarus 163(2):263–289. https://doi.org/10.1016/S0019-1035(03)00048-4
Zhang K, Nimmo F (2009) Recent orbital evolution and the internal structures of Enceladus and Dione. Icarus 204(2):597–609. https://doi.org/10.1016/j.icarus.2009.07.007
Zhang K, Nimmo F (2012) Late-stage impacts and the orbital and thermal evolution of Tethys. Icarus 218(1):348–355. https://doi.org/10.1016/j.icarus.2011.12.013

Authors and Affiliations

Matija Ćuk[1] · Maryame El Moutamid[2] · Giacomo Lari[3] · Marc Neveu[4,5] · Francis Nimmo[6] · Benoît Noyelles[7] · Alyssa Rhoden[8] · Melaine Saillenfest[9]

✉ G. Lari
giacomo.lari@unipi.it

M. Ćuk
mcuk@seti.org

M. El Moutamid
maryame@astro.cornell.edu

M. Neveu
marc.f.neveu@nasa.gov

F. Nimmo
fnimmo@ucsc.edu

B. Noyelles
benoit.noyelles@univ-fcomte.fr

A. Rhoden
alyssa@boulder.swri.edu

M. Saillenfest
melaine.saillenfest@obspm.fr

[1] Carl Sagan Center, SETI Institute, 339 N Bernardo Ave, Mountain View, 94043, CA, USA

[2] Cornell Center of Astronomy and Planetary Sciences, Cornell University, Space Sciences Building, Ithaca, 14850, NY, USA

[3] Dipartimento di Matematica, Università di Pisa, Largo Bruno Pontecorvo 5, 56127, Pisa, Italy

[4] Department of Astronomy, University of Maryland, 4296 Stadium Dr., College Park, 20742, MD, USA

[5] Planetary Environments Laboratory, NASA Goddard Space Flight Center, 8800 Greenbelt Rd., Greenbelt, 20771, MD, USA

[6] Department of Earth and Planetary Sciences, University of California Santa Cruz, 1156 High St, Santa Cruz, CA 95064, USA

[7] Institut UTINAM UMR 6213 / CNRS, Univ. of Franche-Comté, OSU THETA, BP 1615, 25010, Besançon Cedex, France

[8] Department of Space Studies, Southwest Research Institute – Boulder, 1050 Walnut St., Boulder, 80302, CO, USA

[9] IMCCE, Observatoire de Paris, PSL Research University, CNRS, Sorbonne Université, Université de Lille, 77 av. Denfert-Rochereau, 75014, Paris, France

Space Science Reviews (2024) 220:55
https://doi.org/10.1007/s11214-024-01084-z

Geologic Constraints on the Formation and Evolution of Saturn's Mid-Sized Moons

Alyssa Rose Rhoden[1] · Sierra N. Ferguson[1] · William Bottke[1] · Julie C. Castillo-Rogez[2] · Emily Martin[3] · Michael Bland[4] · Michelle Kirchoff[1] · Marco Zannoni[5] · Nicolas Rambaux[6] · Julien Salmon[1]

Received: 5 April 2023 / Accepted: 13 June 2024 / Published online: 17 July 2024

Abstract
Saturn's mid-sized icy moons have complex relationships with Saturn's interior, the rings, and with each other, which can be expressed in their shapes, interiors, and geology. Observations of their physical states can, thus, provide important constraints on the ages and formation mechanism(s) of the moons, which in turn informs our understanding of the formation and evolution of Saturn and its rings. Here, we describe the cratering records of the mid-sized moons and the value and limitations of their use for constraining the histories of the moons. We also discuss observational constraints on the interior structures of the moons and geologically-derived inferences on their thermal budgets through time. Overall, the geologic records of the moons (with the exception of Mimas) include evidence of epochs of high heat flows, short- and long-lived subsurface oceans, extensional tectonics, and considerable cratering. Curiously, Mimas presents no clear evidence of an ocean within its surface geology, but its rotation and orbit indicate a present-day ocean. While the moons need not be primordial to produce the observed levels of interior evolution and geologic activity, there is likely a minimum age associated with their development that has yet to be determined. Uncertainties in the populations impacting the moons makes it challenging to further constrain their formation timeframes using craters, whereas the characteristics of their cores and other geologic inferences of their thermal evolutions may help narrow down their potential histories. Disruptive collisions may have also played an important role in the formation and evolution of Saturn's mid-sized moons, and even the rings of Saturn, although more sophisticated modeling is needed to determine the collision conditions that produce rings and moons that fit the observational constraints. Overall, the existence and physical characteristics of Saturn's mid-sized moons provide critical benchmarks for the development of formation theories.

Keywords Saturnian satellites · Icy moons · Geology · Tectonics · Cratering · Interiors · Ocean worlds · Mimas · Enceladus · Tethys · Dione · Rhea

1 Introduction

Saturn possesses a diverse system of satellites, including moonlets embedded within the rings, a suite of mid-sized icy moons (Mimas, Enceladus, Tethys, Dione, and Rhea), one

Extended author information available on the last page of the article

Properties	Mimas	Enceladus	Tethys	Dione	Rhea
Mean Radius, *R* (km) (1-σ)	198.2±0.4	252.1±0.2	531.1±0.6	561.4±0.4	763.5±0.6
GM (km^3/s^2)	2.50349 ± 0.00042	7.21037 ± 0.00027	41.21353 ± 0.00094	73.11607 ± 0.00016	153.94175 ± 0.00123
Mass, *M* (10^{17} kg)	375.094 ± 0.023	1080.318 ± 0.028	6174.959 ± 0.146	10,954.868 ± 0.246	23,064.854 ± 0.522
Mean Density (kg/m^3)	1.1501± 0.0070	1.6097± 0.0038	0.9840± 0.0033	1.4781± 0.0032	1.2372± 0.0029
Sub-saturnian Equatorial Radius, *a* (km)	207.8±0.5	256.6±0.6	538.4±0.3	563.4±0.6	765.0±0.6
Along-Orbit Equatorial Radius, *b* (km)	196.7±0.5	251.4±0.2	528.3±1.1	561.3±0.5	763.1±0.6
Polar Radius, *c* (km)	190.6±0.3	248.3±0.2	526.3±0.6	559.6±0.4	762.4±0.6

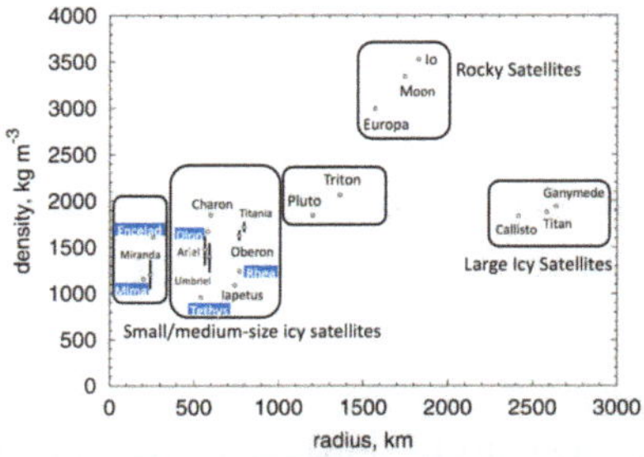

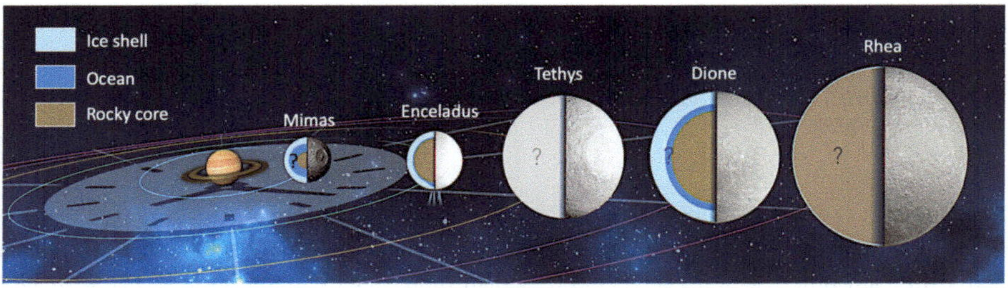

Fig. 1 (Top left) Key physical properties of Mimas, Enceladus, Tethys, Dione, and Rhea (error bars are 1-sigma); (Top right) comparison of their densities and mean radii to those of other moons in the outer solar system. References for radii and densities are Thomas (2010) and Archinal et al. (2018). (Bottom) Summary of our knowledge of their internal structures. There are still major uncertainties about the state of differentiation of Rhea and Tethys. Modified from Neveu and Rhoden (2019)

Table 1 Key orbital properties of Saturn's large moons (Source: ssd.jpl.nasa.gov)

Properties	Mimas	Enceladus	Tethys	Dione	Rhea
Semi-Major Axis, *a* (km)	186,000	238,400	295,000	377,700	527,200
Orbital Eccentricity, *e*	0.020	0.005	0.0001	0.002	0.001
Orbital Inclination, *i* (°)	1.6	0.0	1.1	0.0	0.3
Orbital Period (days)	0.942422	1.370218	1.887802	2.736916	4.517503

large moon that possesses a thick atmosphere and a global hydrological cycle (Titan), distant moons with puzzling properties (Hyperion, Iapetus, Phoebe), and several populations of distant irregular satellites. In addition, two of the mid-sized moons share their orbits with co-orbital moons (Calypso/Telesto at Tethys and Helene/Polydeuces at Dione). The *Cassini-Huygens* mission, building upon the data sets acquired by *Voyager*, revealed many of the physical, dynamical, geological, and compositional properties of these moons (Fig. 1; Table 1), although many questions regarding their origins, interiors, and habitability still remain.

A peculiarity of Saturn's mid-sized moons is that they do not display a density gradient with distance from the planet (Fig. 1), unlike the regular moons of Jupiter. The moon with the highest density is Enceladus, which orbits between the two lowest density moons: Mimas (interior) and Tethys (exterior). Enceladus' density corresponds to a rock mass fraction of 60%, assuming a dry rock composition akin to a CI chondrite (Castillo-Rogez et al. 2018). In contrast, Tethys' rock fraction is limited to only ∼7% by mass, although it could be as much as 15–20% if Tethys maintains substantial porosity (e.g., Thomas 2010; Castillo-Rogez et al. 2018), and Mimas' density implies only ∼9% rock, by mass. While the densities and inferred rock fractions vary wildly, the moons increase in mass and radius with distance from Saturn (Fig. 1), which may be a clue as to the processes that led Saturn's moons to display such different characteristics than the moons of Jupiter. In addition, our new understanding of dissipation within Saturn has dramatically altered our interpretations of the system, opening up new evolutionary pathways not previously considered.

Dissipation within Saturn, described by the parameter Q, exerts a strong control on the orbital migration rates of its moons. The "classical" value of Saturn's Q is ~18,000 (e.g., Meyer and Wisdom 2008), which was derived assuming that the moons evolved to their current orbital distances over the age of the solar system. However, analyses of a combination of historical astrometry measurements and contemporary *Cassini* data (Lainey et al. 2012, 2017, 2020) have revealed much higher orbital migration rates for the moons, from Mimas to Titan, which imply that Saturn's Q is much lower than the classical value. In fact, the orbital migration rates of the mid-sized moons each indicate a different tidal Q (Lainey et al. 2017, 2020), ranging from ~300 to ~2500 (Lainey et al. 2012; Nimmo et al. 2018; Lainey et al. 2017, 2020). It may be the case that Saturn's Q changes with time and/or that Saturn's Q is frequency-dependent, such that the moons experience a different effective tidal Q as their orbits expand (e.g., Fuller et al. 2016; Nimmo et al. 2018; Lainey et al. 2020). This new view of Saturn's Q allows for the possibility that the mid-sized moons can be much younger than the age of the Solar System and still have evolved to their present-day orbits (Lainey et al. 2020). However, ancient moons are not ruled out by a lower Q of Saturn.

In parallel with these findings, a plethora of new hypotheses have been put forth to explain the formation of the mid-sized moons inward of Titan that do not invoke contemporaneous formation with Saturn from its circumplanetary disk. In particular, it has been suggested that at least one, and possibly all, of these moons were spawned from Saturn's rings (Canup 2010; Charnoz et al. 2011; Crida and Charnoz 2012; Salmon and Canup 2017; Dubinski 2019). In that case, the age of the rings and the ages of the moons are inextricably linked. A second suite of hypotheses invoke disruptive collisions into or among a previous generation of moons, with the debris coalescing to become the present-day mid-sized moons (e.g., Ćuk et al. 2016; Asphaug and Reufer 2013; Teodoro et al. 2023). We discuss these models in more detail in the following section.

The new vision of Saturn's Q, along with the breadth of formation models for the mid-sized moons, has resulted in a range of allowable ages for the moons from only ~100 Myr to nearly the age of Saturn itself. Here, we turn to the geophysical characteristics of Mimas, Enceladus, Tethys, Dione, and Rhea, to obtain additional constraints on both the timing and mechanism of their formation and subsequent histories. This chapter focuses on the mid-sized icy moons found within Titan's orbit, in part because they are more diagnostic of processes within Saturn and its ring-moon system than more distant moons. These bodies are also particularly intriguing because they may all have had subsurface oceans, some of which persist to the present day, despite their small sizes (radii between ~200 and 750 km) and relatively rock-poor interiors (Fig. 1); we discuss the evidence for past and contemporary oceans in the following sections. Taken together, Saturn's mid-sized moons provide an opportunity to explore a wide range of conditions that may promote ocean development and the development of habitable worlds. In addition, the mid-sized moons can inform our understanding of satellite formation and the potential links between dense rings and mid-sized moons, which are also observed at Uranus.

We summarize the present hypotheses for the formation of Saturn's mid-sized moons (Sect. 2) and then describe the present-day geophysical observations that provide constraints on the histories of the moons (Sects. 3–5). Specifically, we explore the cratering records of the mid-sized moons, as well as their surface geology, interior structures, and thermal budgets. We conclude with a summary intended to guide the development and testing of theories for the moons' formation and evolution (Sect. 6). Because this review covers a wide breadth of topics, where applicable, we point to reviews already in the literature to provide the reader a starting point from which to delve deeper into these topics. In particular, we refer to the reader to Schenk et al. (2018), Kirchoff et al. (2018), and Patterson et al. (2018) for surface geology of the mid-sized moons and to Castillo-Rogez et al. (2018) and Hemingway et al. (2018) for interior structures of the mid-sized moons; the latter also includes a thorough description of the methodology used to infer interior structure (e.g., through gravity measurements).

2 Hypotheses for the Origins of Saturn's Mid-Sized Moons

Moon formation within a circumplanetary disk (CPD) is a well-established idea (e.g., Canup and Ward 2009; Mosqueira et al. 2010 and references therein). However, Saturn's mid-sized moons grow in both mass and radius with distance from Saturn, whilst displaying variable densities (Fig. 1), which is not the pattern observed among the regular moons of Jupiter nor the expected outcome of CPD formation (e.g., Canup and Ward 2009; Mosqueira et al. 2010). Hence, as we discuss in this section, many studies have investigated whether the present mid-sized moons (Mimas through Rhea) initially formed by a different mechanism and/or manifested from disruptions into or among precursor moons.

One suite of models posits that the mid-sized moons accreted in Saturn's rings (Charnoz et al. 2010; Canup 2010; Charnoz et al. 2011; Crida and Charnoz 2012; Salmon and Canup 2017). In that case, one moon forms, grows, and migrates out of the rings before the next. A natural outcome of the model is that the sizes of the newly formed moons would decrease over time, as the remaining mass in the ring decreases, which could explain the observed mass gradient of the moons. In particular, Crida and Charnoz (2012) showed that the relative masses of the moons would naturally result from the merging of similarly-sized moons during formation from the rings. The densities of the moons, which are set by their rock fractions, are determined somewhat stochastically in these models; either rock "seeds" present in the rings become the cores of the moon (e.g., Charnoz et al. 2011) or the rocky material is brought in by impacts over time (Salmon and Canup 2017).

Ring-born moon scenarios require that a massive ring around Saturn predates any mid-sized moons that emerged from it, and imply that the moons are different ages. Some models invoke ring formation for all of the moons (Charnoz et al. 2011; Crida and Charnoz 2012) – with Rhea being as much as a billion years older than Mimas (Charnoz et al. 2011) – while others suggest that only Mimas, Enceladus, and Tethys formed in this manner (Salmon and Canup 2017). In either case, the ages of the mid-sized moons would be tied to the age of the rings, which we will now briefly discuss.

Canup (2010) proposed that Saturn's rings formed from a Titan-sized body that accreted in the CPD, migrated inward due to interactions with the gas disk, and was disrupted when crossing Saturn's Roche limit. The study assumed that the moon's outer ice layers were tidally stripped to form massive icy rings while the bulk of the moon's rocky core was lost to the planet, resulting in a low rock abundance in the rings. Reliance on gas-driven migration means this process must have occurred early in Saturn's history. Alternatively Saturn's rings may have formed from tidal disruption of a large transneptunian object (e.g., Centaur) that crossed Saturn's Roche's limit (e.g., Dones 1991; Hyodo and Charnoz 2017). It has not been possible to discriminate between these two scenarios based on the *Cassini* data.

The Canup (2010) model for ring formation, coupled with ring-born moon models (Charnoz et al. 2010; Canup 2010; Charnoz et al. 2011; Crida and Charnoz 2012; Salmon and Canup 2017), implies that the mid-sized moon formation began within ~1 Gyr after the dissipation of the gas disk around Saturn, roughly 4.5 Gyr ago (see Canup 2010, *Supplementary Information*). Invoking the disruption of an interloper (e.g., Dones 1991; Hyodo and Charnoz 2017) may allow for somewhat younger rings, and hence, younger ring-born moons. Although there is still some uncertainty in the availability of these large objects over time, ring formation via tidal disruption of a passing object would have been much more likely billions of years ago (e.g., Nesvorny 2018; Nesvorny et al. 2023). In either case, the amount of time needed to assemble the mid-sized moons depends on the details of the model and how many of the moons are thought to have emerged from the rings, which also places a lower limit on the initial ring mass. Both options for ring formation produce rings early enough in Saturn's history that the moons would have billions of years to emerge from the rings, undergo interior evolution, and develop their geologic records.

Cassini measurements have shown that the mass of Saturn's rings is consistent with that of ancient rings. Over time, a disk of material around a planet will lose mass, with ~80% of the disk's mass falling inward onto the planet and ~20% of the mass escaping outward, through the Roche limit, where it can assemble into moons (e.g., Salmon and Canup 2017). As the disk mass decreases, the process of material loss slows, leading to an asymptotic mass over time. The current mass of Saturn's rings (Iess et al. 2019) is consistent with the predicted asymptotic mass for an evolved disk, a process that takes ~2 Gyr (Salmon et al. 2010; Crida et al. 2019). Therefore, the rings are at least 2 billion years old or they just happened to form with a mass close to that of an ancient, evolved ring system. Even though an ancient ring system seems to provide an explanation for the current ring mass and the emergence of mid-sized moons, the current ring system at Saturn may not be ancient. While the idea of young rings is not new (e.g., Esposito 1986; Cuzzi and Durisen 1990; Cuzzi and Estrada 1998, and references therein), *Cassini* measurements have provided robust data that any ring formation model must now address. In particular, the brightness and composition of the rings suggest that they may be only ~100 Myr old (Zhang et al. 2017; Iess et al. 2019).

Very young rings present a challenge for the ring-born moon scenario, particularly if all five mid-sized moons are assumed to form in this manner, because 100 Myr may not be sufficient time for the orbits to expand to their present locations (e.g., Salmon and Canup 2017), and undergo any geologic or geophysical modification of the moons that must take place. In addition, the initial mass of the ring must have been close to the present-day mass because of the limited time available for viscous spreading and mass loss, which further reduces the likelihood of moon formation and requires that the rings coincidentally formed with about the same mass as long-lived rings. Hence, the conflicting age estimates of the rings inferred by different data sets (e.g., Salmon et al. 2010; Zhang et al. 2017; Iess et al. 2019; Crida et al. 2019 and references therein) creates uncertainty as to whether emergence from rings is a viable scenario for the initial formation of the mid-sized moons.

An alternative scenario, that does not invoke rings to form the mid-sized moons, comes from Asphaug and Reufer (2013), who suggest that Saturn had an initial set of larger moons – more similar to the Galilean moons of Jupiter – that formed in the CPD. The moons underwent a series of mergers, with the collisions creating ejecta that became the mid-sized moons and the final merger producing Titan. They posit that the mergers occurred relatively early in the system's history (roughly 3.5 Gyr ago), and that the variations in rock content across the mid-sized moons reflect different proportions of rocky core and icy mantle material that were incorporated into each moon. This model does not directly provide an explanation of the mass gradient of the mid-sized moons or how Saturn's rings formed.

Whether mid-sized moons formed directly from the CPD, were born from ancient rings, or formed as ejecta from mergers early in Saturn system history, there would be billions of years in which the moons could then be subjected to potentially-disruptive collisions (e.g., Nesvorny et al. 2023). Several models suggest that some or all of the current mid-sized moons reassembled after collisions into or among preexisting moons (Ćuk et al. 2016; Dubinski 2019; Teodoro et al. 2023); these models are typically agnostic as to the origins of the preexisting moons. The collision and reassembly is assumed to occur within the past 100 Myr either to help explain the evolution of mean motion resonances (Ćuk et al. 2016; Teodoro et al. 2023) or to produce rings from collisional debris that are consistent with the presumed young age of the rings (Dubinski 2019; Teodoro et al. 2023). Initially, numerical models of such an event suggested that limited debris would escape from the precursor moons (Hyodo and Charnoz 2017), but more recent simulations show that both rings and substantial planetocentric debris can be generated after a collision between precursor moons (Teodoro et al. 2023). Separately, Wisdom et al. (2022) suggested that the rings formed recently via tidal disruption of a highly eccentric hypothetical moon, nicknamed Chrysalis. The model does not detail the interactions between Chrysalis and the current suite of mid-sized moons in the simulations of its migration, so we do not consider it further here.

Models that invoke disruption and reassembly imply that the entire cratering histories of the reassembled moons would have to be emplaced since the collision. The accretion of debris after the impact may account for some of the craters, as suggested by Dubinski (2019) for Mimas, but the effectiveness and geophysical implications of this mechanism have not been fully investigated. A major open question is whether reassembled moons, whether formed out of the ejecta from early mergers (Asphaug and Reufer 2013) or late collisions (Ćuk et al. 2016; Dubinski 2019; Teodoro et al. 2023), can develop interiors and geologic records consistent with the observations of Saturn's mid-sized moons.

To summarize: 1) The ring-born moon model can explain the mass gradient of the mid-sized moons (e.g., Crida and Charnoz 2012) and relies upon a material transport mechanism that must occur (e.g., Salmon and Canup 2017), making it an appealing option for the initial assembly of moons at Saturn. Producing the mid-sized moons from the rings is more tenable if the rings are ancient (as in Charnoz et al. 2011). An important consideration is whether moons that formed billions of years ago can avoid disruptive collisions (cf., Charnoz et al. 2009; Movshovitz et al. 2015) and how collisions might have altered the very characteristics of the moons that these formation models reproduce. Although, because the moons form in sequence in these models (e.g., Crida and Charnoz 2012), with time between the emergence of each moon, it is possible that the innermost mid-sized moons were produced after the heaviest bombardment, lessening this concern. If the present rings are very young, it would be challenging to have produced the mid-sized moons via this mechanism. However, spawning moons from a previous generation of rings, an older age for the current rings, or a variant of the ring-born moon model that can produce moons and expand their orbits within $\sim$100 Myr may be compatible with a young ring age.

2) Models that invoke late collisions and reassembly (Ćuk et al. 2016; Dubinski 2019; Teodoro et al. 2023) may provide a pathway to generate both rings and mid-sized moons that are consistent with the inferred young age of the rings (e.g., Zhang et al. 2017; Iess et al. 2019) while relying on a process that is likely to have occurred at some point within the lifetime of the Saturn system (i.e., large collisions). Reassembly after recent disruptive collisions does not independently explain the mass gradient of the mid-sized moons. If collisional debris was preferentially reaccreted by the same precursor body (e.g., Charnoz et al. 2009; Hyodo and Charnoz 2017), characteristics of the moon that stem from its formation, such as a mass set by formation in the rings, may be preserved. However, collisional debris has also been suggested as a source of ring material and impactors that cratered the moons (Teodoro et al. 2023), implying a substantial amount of debris is lost from the precursor

moon. Alternatively, a model that produces mid-sized moons from early disruptive collisions must then invoke another mechanism to generate rings (e.g., combining Asphaug and Reufer 2013 with Wisdom et al. 2022), particularly if the rings are young. Most importantly, reassembled moons must still be old enough to have developed their rich geologic records and high crater densities.

At present, models for the formation of the mid-sized moons imply ages of either billions of years or $\sim$100 Myr. Hence, in the following sections, we focus on geologic constraints that can help differentiate between these two end-members. It is worth noting that both ring-born moons and reassembled moons could have produced moons with intermediate ages if the processes invoked occur at different times than the present studies have explored.

3 Cratering of Saturn's Mid-Sized Moons, Insights and Limitations

3.1 Crater-Based Ages

One of the primary ways of estimating planetary surface ages is by counting impact craters and estimating the time required to produce them, which relies upon knowledge of the impactor populations responsible for cratering. The outer planets and their satellites are predominantly cratered by Kuiper Belt objects (KBOs), Jupiter-family comets (JFCs), Centaurs, and planetocentric debris. Studies of outer planet cratering rates and/or impactor sources include: Shoemaker and Wolfe (1981), Nakamura and Yoshikawa (1995), Chapman and McKinnon (1986), Levison et al. (2000), Di Sisto et al. (2009), Di Sisto and Brunini (2007, 2011), Di Sisto and Rossignoli (2020), Di Sisto and Zanardi (2013, 2016), Rossignoli et al. (2019), Volk and Malhotra (2008), Zahnle et al. (2003), Alvarellos et al. (2005, 2017), Dones et al. (2009), Schenk and Zahnle (2007), Kirchoff and Schenk (2010, 2015), Greenstreet et al. (2015), Hirata (2016), Kirchoff et al. (2018), Singer et al. (2019), Spencer et al. (2020), Bell (2020), Ferguson et al. (2020, 2022a, 2022b, 2024), Kirchoff et al. (2022), Bottke et al. (2023, 2024), Robbins et al. (2024). It has also been proposed that ejected main belt asteroids are the primary source of impactors on the Saturn satellites (e.g., Horedt and Neukum 1984; Wagner et al. 2006; Neukum et al. 2006; Schmedemann et al. 2009), but this scenario has many challenges to overcome (see Bottke et al. 2023). Unlike the Earth's Moon, there are no known absolute ages for the outer solar system, so crater-derived chronologies will have substantial uncertainties.

Zahnle et al. (2003), using the knowledge gained in previous works, developed two production functions and chronologies for outer solar system moons – Case A and Case B – that became the primary means, for many years, of dating cratered surfaces that were more affected by outer solar system material than by material originating in the asteroid belt. Case A was based on estimates of the contemporary impact rate of comets onto Jupiter and was tailored to match crater counts on Europa and Ganymede (Zahnle et al. 2003). Case B was formulated to provide a better match to crater counts on the surface of Neptune's moon, Triton, under the assumption that Triton's size frequency distribution of craters differed from that at Jupiter because the impactor flux changed with distance from the Sun (Zahnle et al. 2003). The resolution of *Voyager* images at Triton limited crater identification to $D > 3$ km, so the Case B production function is only tied to observations for crater sizes above that limit (Schenk and Zahnle 2007).

Overall, the impactor size frequency distribution (SFD) represented by Case A matches expectations for a heliocentric population coming from the primordial Kuiper belt that has experienced substantial collisional evolution (e.g., Bottke et al. 2023, 2024). In addition, the *New Horizons* mission acquired crater data at Pluto and Charon that is a much closer match to Case A than Case B (Singer et al. 2019; Robbins and Singer 2021; Bottke et al. 2023), casting doubt on the initial hypothesis that Triton's SFD represented a heliocentric population that changed with distance from the Sun (Zahnle et al. 2003). Work by Schenk and Zahnle (2007) and Mah and Brasser (2019) suggested that Triton's craters were mainly produced by planetocentric debris and possibly secondaries from unseen large craters on Triton, instead of heliocentric impactors, providing an alternative interpretation of Case B. However, the impact velocities, timing and duration of impacts, and other details of the chronology were never updated to be consistent with planetocentric impactors, so Case B cannot be used to derive ages on Triton or other bodies where planetocentric impactors dominate the cratering record. Attempts have been made to recover the characteristics of planetocentric impactor populations from crater counts on Saturn's moons (e.g., Bell 2020), but more work is needed to attach robust ages to these empirical production functions.

We describe the crater counting studies thus far performed for the mid-sized moons of Saturn in the next section. Many of these studies, particularly those that focus on elliptical craters, support the idea that planetocentric impactors have played an important role in cratering the moons (e.g., Kirchoff and Schenk 2010; Hirata 2016; Kirchoff et al. 2018 and references therein; Ferguson et al. 2020, 2022a, 2022b, 2024; Robbins et al. 2024), even when considering updated estimates of the heliocentric impactor production function and flux (e.g., Dones et al. 2009; Di Sisto and Zanardi 2013, 2016). Hence, we refrain from presenting age estimates based on fits to existing production functions, particularly Case B from Zahnle et al. (2003), even where the cited literature does.

The real value in these crater counts is that they provide an observable population against which we can test models of possible events that produce and distribute planetocentric material to the mid-sized moons. Once the sources of planetocentric material are better constrained, we can begin estimating the timescales over which the craters were emplaced. It is important to note, however, that some studies report fitting the broad characteristics of the crater populations on Saturn's mid-sized moons using models of exclusively heliocentric impactors (Wong et al. 2019, 2021, 2023; Bottke et al. 2024). The models estimate the surface ages of the moons to be >4 Gyr old. We consider these ages to be upper limits as the presence of planetocentric sources would add to the crater populations contemporaneously with heliocentric sources.

In addition to primary cratering, secondary and sesquinary cratering may contribute to the populations on icy moons, particularly on the lower-gravity moons around Saturn. Secondary cratering is the process in which material ejected from a primary impact onto a body is launched outwards and re-impacts the same surface as the primary crater, creating an additional crater (Bierhaus et al. 2012, 2018; Alvarellos et al. 2005, 2017; McEwen and Bierhaus 2006; Singer et al. 2013). Impact velocities for the ejecta that form secondaries are, necessarily, slower than the target's escape velocity (McEwen and Bierhaus 2006). Mimas and Enceladus are not expected to have significant secondary crater populations due to their very low escape velocities, which approach the threshold velocity required to make a secondary crater (Bierhaus et al. 2012). Rhea, Tethys, and Dione have high enough surface gravities to retain ejected material and form secondaries (Bierhaus et al. 2012; Schenk et al. 2020).

Ejecta from a primary impact that does exceed the escape velocity of the impacted body can eventually impact either the same body or another body, depending on the dynamics of the system. Craters created by this ejecta are called sesquinaries (Alvarellos et al. 2005, 2017; Zahnle et al. 2008; Bierhaus et al. 2012, 2018), and the ejecta that forms them can be considered a type of planetocentric impactor. Sesquinaries may remain in orbit around the original moon or travel between moons, where they can contaminate the primary crater population. Alvarellos et al. (2005, 2017) used dynamical models to estimate the fraction of debris that should be present on the surfaces of the mid-sized Saturnian moons due to sesquinaries from large primary impacts onto a different moon. For the impact events that created craters such as Odysseus and Penelope on Tethys, 82–96% of ejected material came back to Tethys, while the remaining fragments were likely captured into an orbit around Saturn (Alvarellos et al. 2005). Similarly for Rhea, 91–96% of the material ejected from the Tirawa impact event would have re-impacted Rhea (Alvarellos et al. 2005). For the Herschel basin on Mimas, ~99% of all ejected material returned to Mimas in the form of sesquinary impacts (Alvarellos et al. 2005, 2017). Most sesquinary cratering on worlds like Enceladus and Dione will have average crater diameters on a scale of 200 m or smaller (Alvarellos et al. 2017), rendering these craters far below the current image resolutions for the data from *Cassini*.

An important consideration in the interpretation of cratered surfaces is the extent of saturation. A surface is in crater saturation equilibrium when the crater density becomes high enough that, on average, a newly-formed crater erases another previously formed crater, such that the crater spatial density reaches a steady-state (e.g., Gault 1970; Hartmann 1984; Chapman and McKinnon 1986; Richardson 2009). Saturation primarily affects the computation and comparison of surface ages because a saturated surface is older than can be derived from either absolute or model-based ages. In particular, if the heavily-cratered terrains of the satellites are saturated, the surfaces could be different ages even if their relative crater spatial densities are similar. Whether the surfaces of the mid-sized moons are saturated has been debated since *Voyager* (e.g., Hartmann 1984; Chapman and McKinnon 1986; Lissauer et al. 1988; Squyres et al. 1997; Kirchoff and Schenk 2010; Kirchoff et al. 2018). Until the saturation issue is resolved, no definitive conclusions can be drawn as to the relative ages of heavily cratered regions across different mid-sized moons.

The other effect saturation equilibrium can have is modifying the shape of the crater size-frequency distribution (SFD) from its production function (Gault 1970; Chapman and McKinnon 1986; Richardson 2009). Specifically, in cases where the slope of the production function is steeper than approximately −2.5, the crater SFD would become shallower (closer to −2) when the distribution enters saturation. However, if the production slope is shallower than approximately −3, the crater SFD generally does not change significantly as a result of saturation and still represents the production function (e.g., Woronow 1978, 1985; Chapman and McKinnon 1986; Richardson 2009). Fortunately, the crater SFDs of Saturn's mid-sized moons appear to have overall shallower slopes (Kirchoff and Schenk 2010; Bell 2020; Ferguson et al. 2020, 2022a, 2024; Robbins et al. 2024), making it likely that they are still representative of the impactor population(s) that created them.

Finally, several models have investigated how the mid-sized moons would have been affected by the intense heliocentric bombardment produced by the clearing of the primordial Kuiper belt (e.g., Gomes et al. 2005; Nesvorny 2018; Nesvorny et al. 2023; Bottke et al. 2023). For example, Nimmo and Korycansky (2012) determined that early bombardment would cause Mimas and Enceladus to lose all of their volatiles (as well as Miranda at Uranus), which does not match the high mass fractions of ice in the moons (e.g., Castillo-Rogez et al. 2018). The favored explanation of Nimmo and Korycansky (2012) was that their modeled bombardment population was roughly ten times too large. The bombardment predictions were revised downward by Bottke et al. (2024), who found the expected impact flux was a factor of 5 smaller than previously suggested, enough that volatiles on the icy worlds could survive early bombardment.

Using the original estimated mass of the early bombardment population, Charnoz et al. (2009) showed that Mimas would have been disrupted, Enceladus had a roughly 50% chance

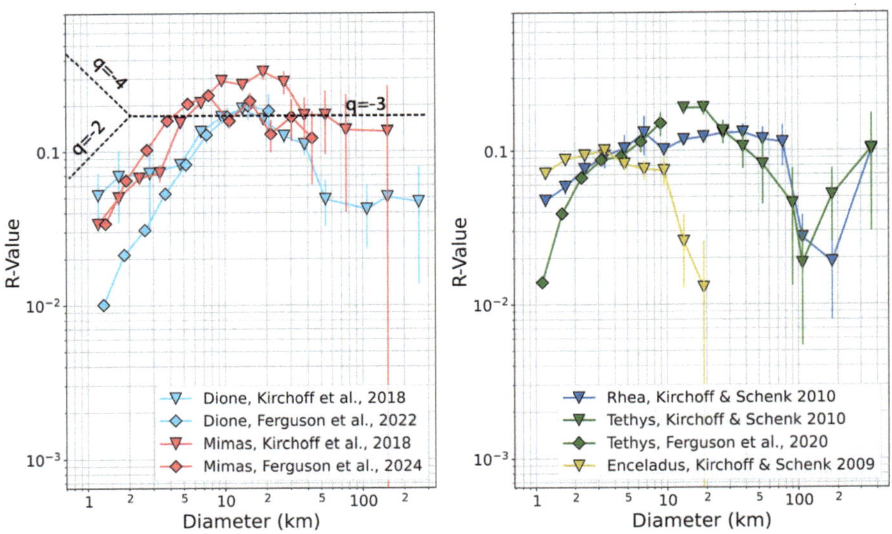

Fig. 2 Crater distributions shown on an R-plot, on which a horizontal line would correspond to a -3 slope on a log-log SFD plot (see dotted lines in the left panel). Data for each moon is shown in a different color, with symbols representing the crater counting study from which the data was taken. For Tethys, we show results from Kirchoff and Schenk (2010) only for craters larger than $\sim$10 km because the size and locations of the areas mapped make the counts less complete than the Ferguson et al. (2020) data set

of disruption, and Tethys, Dione, and Rhea likely survived early heavy bombardment. In contrast, Movshovitz et al. (2015) found that Mimas, Enceladus, and Tethys experienced at least one catastrophic impact in every one of their bombardment simulations. Furthermore, they determined that the delivered mass of impactors would need to be a factor of 100 lower for Mimas to avoid disruption and about 30 times lower to save Enceladus. Comparable results were also found by Wong et al. (2019, 2021). Regardless of the early impactor mass, Mimas and/or Enceladus may have avoided disruption if their final assembly occurred late enough to miss peak bombardment, such as via ring formation (e.g., Salmon and Canup 2017). Given the various ways in which impacts can affect mid-sized moons, it is plausible that the current suite of mid-sized moons experienced major changes in their bulk properties since their formation, whether they formed in Saturn's CPD, emerged from ancient rings, or formed from early mergers.

3.2 Overview of Crater Counts in the Saturn System

The study of impact cratering on the mid-sized moons has a long history, dating back to the earliest exploration of the outer solar system. *Voyager*-era studies, and their implications, are described in detail in Kirchoff et al. (2018), along with the post-*Cassini* view up to that point. As impact cratering is only one of the geologic feature types we discuss in this work, we summarize the major characteristics of the cratering records of the moons, with an emphasis on work that post-dates Kirchoff et al. (2018). Examples of crater data for the moons are shown on the relative plots (R-plots) in Fig. 2, in which the colors represent the moons, and the symbols represent the different studies where the data was presented. The data shown from Kirchoff et al. (2018) combines several different datasets (Kirchoff and Schenk 2009, 2010, 2015; Robbins et al. 2015).

The value of an R-plot is that it more clearly separates data with log-log SFD slopes (labeled q on Fig. 2, left panel) that fall between -2 and -4, which applies to many surfaces throughout the Solar System (Crater Analysis Techniques Working Group, 1979). As indicated in Fig. 2 (left panel), data that plots as a horizontal line on an R-plot would have a slope of -3 on a log-log SFD plot. When comparing data on R-plots, the vertical placement of the curve indicates the overall number of craters normalized by area, in which higher R values correspond to a higher spatial density of craters and, in the simplest case, indicate a longer exposure age. Because the data is Fig. 2 has not been scaled to account for differences in surface gravity, impactor speed, or the expected increase in the number of impactors with proximity to Saturn across the mid-sized moons, the vertical placement of the data cannot be used to infer anything about the relative ages of the moons. The slopes of distributions on an R-plot represent the shapes of the impact populations that created them. Breaks in slope (i.e., going from a shallow slope to a steeper slope) may indicate a change in the impactor population at the diameter value where the break in slope occurred. If the slope of the impactor production function is well-known, deviations could indicate that multiple impact sources have affected the surface and/or the surface lost craters due to other types of geologic resurfacing.

The shapes of the crater distributions on Mimas and Dione are similar between the data sets shown in Fig. 2 (left panel; pink and blue, respectively), with a sharp drop in crater density at small sizes (D < 5 km) and a modest drop for larger craters (D > 40 km). For Tethys (right panel; green), we show crater data from Ferguson et al. (2020) for smaller craters and Kirchoff and Schenk (2010) for larger craters, based on the estimated completeness diameters of each data set. In particular, Kirchoff and Schenk (2010) also reported data for the smaller size range of craters on Tethys, which differ from the Ferguson et al. (2020) results due to differences in the sizes and locations of the count areas, but those counts are not as complete. For craters with diameters of 1–10 km, Tethys' craters follow trends fairly similar to those on Mimas and Dione, while for the larger craters (20–100 km), Tethys' distribution shows a more significant drop in crater density. The crater distribution on Rhea (right panel; blue) is flatter than those on Mimas, Tethys, and Dione until ~70 km where the crater density drops in a similar fashion to the Tethys data. Enceladus shows a flatter distribution for small craters (<10 km), but drops off rapidly for craters larger than 10 km (right panel, yellow).

Comparisons between the moons show that the heavily cratered regions of Mimas, Tethys, and Dione have size frequency distributions more similar to each other than to Rhea or more distant Iapetus (Kirchoff et al. 2018). For smaller impact craters (D < 5 km), Mimas has a higher overall crater density than Tethys, which has a higher crater density than Dione (Ferguson et al. 2020, 2022a, 2024). In contrast, there are some indications that Mimas lacks impact craters with diameters ~30–80 km relative to the other mid-sized moons (Kirchoff and Schenk 2010). Using relative spatial densities of impact craters ~10 km and larger, which are harder to remove by subsequent resurfacing than smaller craters, Ferguson et al. (2024) found that Dione has preserved the oldest surfaces, followed by Tethys, and then Mimas. However, all three moons appear to be within about a factor of two of one another in age. For the largest craters sizes (D > 100 km), the cumulative number of impact craters per square kilometer on the most ancient terrains of Mimas, Tethys, Dione, and Rhea are fairly comparable to one another (Kirchoff and Schenk 2010). Although there are sometimes variations in the results of crater counting studies, there is good agreement across the Mimas data sets, and Rhea consistently shows a relatively flat slope across a wide range of impact crater sizes and some variation between the leading and trailing hemispheres (Robbins et al. 2024).

On Enceladus, crater spatial densities are highly variable due to geologic resurfacing (e.g., Crow-Willard and Pappalardo 2015; Kinczyk et al. 2022). Overall the crater density is lower than the other moons, particularly for impact craters smaller than 2 km, and Enceladus lacks larger impact craters (~6 km and larger) when compared to the other satellites in the system (Kirchoff and Schenk 2009). The greatest numbers of impact craters are on the sub- and anti-Saturn hemispheres (e.g. Kirchoff and Schenk 2009) rather than the leading and trailing hemispheres where tectonic resurfacing has erased most of the early cratering record. Impact craters may have also been removed through a combination of viscous relaxation and mantling by plume fall-back (Bland et al. 2012) as evidenced by heavily modified craters, especially in the northern hemisphere (e.g., Bland et al. 2024). The sparseness of impact craters on Enceladus' surface is further highlighted when compared against the heavily cratered terrains of Rhea and Tethys (Fig. 2, right panel, yellow). Enceladus' craters show a different slope from the other two moons, and the crater density drops for craters larger than ~4 km, which suggests erasure by various resurfacing and surface modification mechanisms that are active on Enceladus (Kirchoff and Schenk 2009).

Impact basins provide additional benchmarks with which to assess surface ages of a moon. Here, we focus on the Odysseus basin ($D = 445$ km) on Tethys and the Herschel basin ($D = 139$ km) on Mimas. By assessing the crater distributions both within and outside of the basins, the Odysseus impact basin was shown to be ~4–9 times younger than heavily-cratered terrains on Tethys (Kirchoff and Schenk 2010). For the Herschel basin on Mimas, Ferguson et al. (2024) showed that the ejecta blanket and interior have far lower crater densities than the most heavily cratered regions (i.e., around Mimas' north pole), suggesting that Herschel formed within the most recent ~10-20% of Mimas' surface age. Hence, large impacts can occur even relatively recently in the lifetimes of the mid-sized moons.

Based on the cratering records across Saturn's moons, Kirchoff and Schenk (2010) hypothesized that Mimas, Tethys, and Dione record mainly a planetocentric population of impactors, as evidenced by their steeper-sloped SFDs that indicate an abundance of small craters relative to larger ones. More recently, Bell (2020) compared the crater populations across Saturn's moons, using existing crater catalogs and scaling laws, and found similar crater densities from Mimas through Rhea, suggesting a common impactor source. When compared with predictions using heliocentric (using the model of Di Sisto and Zanardi 2013) or planetocentric impactors, Bell (2020) concluded that the moons experienced predominantly planetocentric cratering, even when crater saturation was considered.

At Mimas, the Case B production function provides an excellent fit to the SFDs in all mapped regions (Ferguson et al. 2024), for impact craters larger than a few km, supporting a planetocentric origin. Case A can match the slope of the SFDs for smaller impact craters within one heavily cratered region, but it vastly overpredicts the number of larger impact craters as compared with the observations. Within the Herschel basin and the ejecta blanket, which have a lower spatial density of craters overall, Case A and Case B can both match large portions of the SFD. However, surface ages derived from Case A fits (which are tied to the more robust heliocentric chronology) produce unrealistic results, in which the interior and ejecta blanket of Herschel are estimated to be billions of years older than the heavily cratered regions of Mimas. While it's possible that the improvement in the fit of Case A within regions of Mimas with lower crater densities indicates a change to a more balanced mix of heliocentric and planetocentric impactors, the unphysical ages add skepticism (Ferguson et al. 2024).

Similarly, at Tethys and Dione, Case B provides a much better fit to the data at impact crater diameters ~3–10 km and larger (depending on region) than Case A, although the fits are not as good as those at Mimas, perhaps indicating differences in the planetocentric flux across the mid-sized moons (Ferguson et al. 2020, 2022a, 2024). Again, Case A only fits the SFD slopes at small diameters and overpredicts the number of larger impact craters due to the slope of the production function. At Rhea and Iapetus, Kirchoff and Schenk (2010) posit that the impact craters are more representative of the heliocentric population, although Hirata (2016) suggests that – for Rhea – only $D > 20$ km craters are predominantly heliocentric in origin, due to the presence of an asymmetry in the crater distribution at that size between hemispheres.

3.3 The Distribution and Implications of Elliptical Craters

Recent work mapping and analyzing elliptical impact craters on the mid-sized moons may provide some additional constraints on the scope and nature of planetocentric debris in the Saturn system. Elliptical impact craters (Fig. 3) are particularly useful for characterizing an impactor population because the long axis of an elliptical crater records the impact direction, and the formation of an elliptical, rather than circular, crater implies a relatively low impact angle (Gault and Wedekind 1978; Bottke et al. 2000; Collins et al. 2011; Elbeshausen et al. 2013; Holo et al. 2018).

On Mimas, Tethys, and Dione, elliptical impact craters with diameters between 2 and 70 km have been mapped as globally as possible within the existing image data (Ferguson et al. 2022b, 2024). Mimas and Tethys have a similar spatial density of elliptical craters, with the more heavily resurfaced Dione displaying a somewhat lower spatial density (Ferguson et al. 2024). Tethys and Dione display a remarkable pattern in which the majority of elliptical craters have their long axes oriented east-west (Fig. 4) and are concentrated in an equatorial band between 30°N and 30°S (Ferguson et al. 2022b). At Mimas, the equatorial east-west oriented population is not as apparent (Fig. 4), perhaps because the Herschel-forming impact erased existing elliptical craters along the equator (Ferguson et al. 2024). Tethys and Dione also display smaller populations of elliptical craters with long axis orientations that are isotropic, which appear to be globally-distributed; this population is not observed on Mimas. Instead, Mimas' elliptical craters become more azimuthally clustered far from the equator, and the dominant orientation is north-south rather than east-west.

One potential source for the elliptical impact craters on Tethys and Dione is a planetocentric debris disk that flattened over time. A debris disk could form after a large impact,

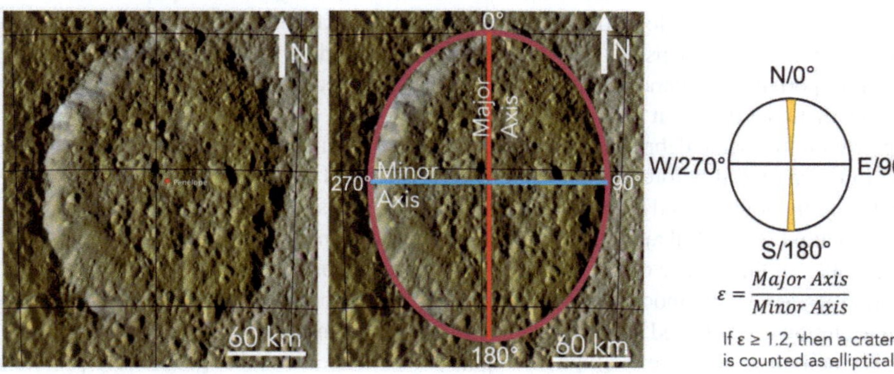

Fig. 3 Penelope crater on Tethys (left) is elliptical rather than circular, which indicates a relatively slow, oblique impact. Here, we show the major and minor axes of Penelope (center), the relationship between "orientation" and the azimuth of the major axis (right), and the equation for determining the ellipticity of a crater.

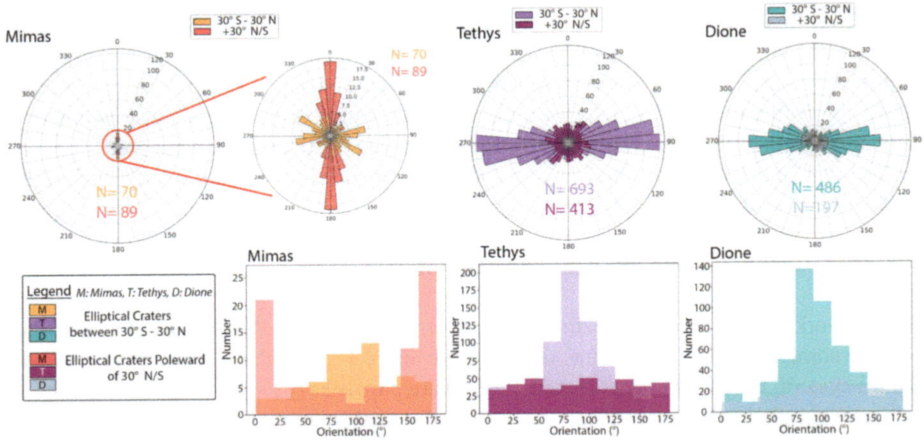

Fig. 4 Two styles of histogram show the orientations of the long-axes of elliptical craters on Mimas, Tethys, and Dione, in which the populations are split by latitude. Across all three moons, regions within 30° of the equator contain elliptical craters that are predominantly oriented east-west, although the signal is less apparent on Mimas. On Tethys and Dione, this group makes up the majority of the elliptical craters, with a smaller, more isotropically-oriented group spanning all mapped latitudes. On Mimas, elliptical craters above 30°N are predominantly oriented north-south; this population appears to be unique among the three moons. Image credit: Ferguson et al. (2024)

as explored by Hyodo and Charnoz (2017), although more recent models suggest a significant fraction of collisional debris may be lost to Saturn instead of impacting the moons (Teodoro et al. 2023). Because the trajectories of debris tend to become more uniform as a disk flattens (e.g., Brahic 1977), this mechanism may explain both the isotropically-oriented global population and the east-west equatorial population (Ferguson et al. 2022b). Currently unknown is whether the moons were impacted with material from a common debris disk or experienced impacts that generated independent debris disks. Tethys and Dione have similar diameters and similar latitudinal extents to their east-west elliptical crater populations, suggesting that the source of impacting material had the same thickness at the locations of each moon. Curiously, Mimas has a similar band of east-west elliptical craters as Tethys and Dione, despite its smaller diameter, and it lacks an isotropically-oriented population. Either Mimas was exposed to a local debris disk with different characteristics, or a common debris disk affected all the moons but had already flattened considerably when it reached Mimas.

An important consideration is that both Mimas and Tethys currently have orbital inclinations that take them out of Dione's orbital plane (Table 1). In order for the moons to be subjected to the same debris disk and have similar latitudinal patterns, the elliptical craters had to be emplaced before the inclinations of Mimas and Tethys were raised. The uncertainties in the long-term dynamical evolution of the moons (e.g., Ćuk et al. 2016; Ćuk and El Moutamid 2022; Nakajima et al. 2019) make this scenario plausible and may provide constraints on the history of mean motion resonances among the moons. Alternatively, the similarities across the moons may indicate a common process of impact-generated debris and accretion that affected each of the three moons independently (Ferguson et al. 2022b).

The non-circular nature of elliptical impact craters makes it challenging to interpret their SFDs with the standard crater analysis tools that were developed for circular craters. First, there is the issue of crater diameter, which varies over the crater's perimeter. If only elliptical craters are being included in the SFD, it may be reasonable to simply apply a consistent rule, such as using the geometric mean of the major and minor axes of each elliptical crater (as in Ferguson et al. 2022b). However, mixing elliptical and circular craters on the same SFD is problematic as the scaling laws that connect impact parameters with elliptical crater dimensions on icy surfaces are still being developed (e.g., Elbeshausen et al. 2013). While we know that the impact angle controls whether the impact forms a circular or elliptical crater, how the dimensions of the elliptical crater change with impact angle is still under investigation. Thus, we cannot tell which circular and elliptical craters were formed by impacts with similar impact velocities and impactor sizes.

The second challenge in interpreting the SFDs of elliptical craters is that the production functions were developed assuming impact angles that create circular craters. Again, our lack of detailed knowledge as to how lower impact angle events manifest in the dimensions of elliptical craters means that we cannot simply compare the predicted crater sizes from a production function to the crater populations on the moons. More modeling and/or experimental work that fully connects the shapes of elliptical craters on icy bodies with the dynamics and physical characteristics of the impactor (e.g., Collins et al. 2011; Elbeshausen et al. 2009, 2013) would be highly valuable.

3.4 Linking Craters with Thermal History

In addition to the spatial distribution of impact craters, their morphologies can be used to record the thermal history of a moon. Impact crater topography induces deviatoric (non-hydrostatic) stresses in the lithosphere (e.g. Dombard and McKinnon 2006). If the viscosity of the ice is low enough, these stresses can drive viscous flow that reduces the crater topography (e.g., Scott 1967; Johnson and McGetchin 1973). As the topography is reduced, the driving stress is removed and viscous flow ceases, resulting in a preserved crater rim and flat or even up-bowed floor (Fig. 5; Scott 1967; Parmentier and Head 1981). Elastic flexure enhances relaxation but is a secondary effect (Dombard and McKinnon 2006).

The viscosity of rock and ice is strongly temperature dependent (e.g., Karato and Wu 1993; Durham and Stern 2001). Thus, determining whether crater topography has been modified by viscous relaxation provides a constraint on the thermal conditions the crater has experienced. Furthermore, because the viscous relaxation process also depends on the wavelength of the topography (long wavelengths are reduced faster than short ones; Parmentier and Head 1981), determining the diameter range over which craters have viscously relaxed provides a lower bound on the thermal conditions (e.g., small craters require warmer conditions to viscously relax). Thus, the thermal history of an icy satellite can be constrained by combining robust numerical modeling of viscous relaxation (e.g., Dombard and McKinnon 2006) with observations of crater depths (e.g., Passey 1983; Bland et al. 2012, 2017, 2023; White et al. 2013, 2017; Singer et al. 2018; Bland and Bray 2024).

While viscous relaxation provides critical insight into the thermal history of icy moons, it also faces several challenges. The observation that an impact crater's topography is subdued only indicates that the viscosity of the lithosphere was low enough for relaxation to occur. The viscosity depends on both temperature and composition and is therefore non-unique. Although pure H_2O ice is often assumed for simplicity, the lithospheres of icy moons may include other materials, including salts, dusty particulates, or ice clathrates that increase its viscosity (e.g., Friedson and Stevenson 1983; Durham et al. 1992; Mangold et al. 2002; Durham et al. 2010; Qi et al. 2018), or ammonia and ammonia hydrates that reduce it (Durham et al. 1992). The latter species has been tentatively detected from ground-based observations of Enceladus (Emery et al. 2005; Verbiscer et al. 2006) and Tethys (Verbiscer et al. 2008) and *Cassini* Visible and Infrared Mapping Spectrometer observations of Dione (Clark et al. 2008) and Iapetus (Clark et al. 2012).

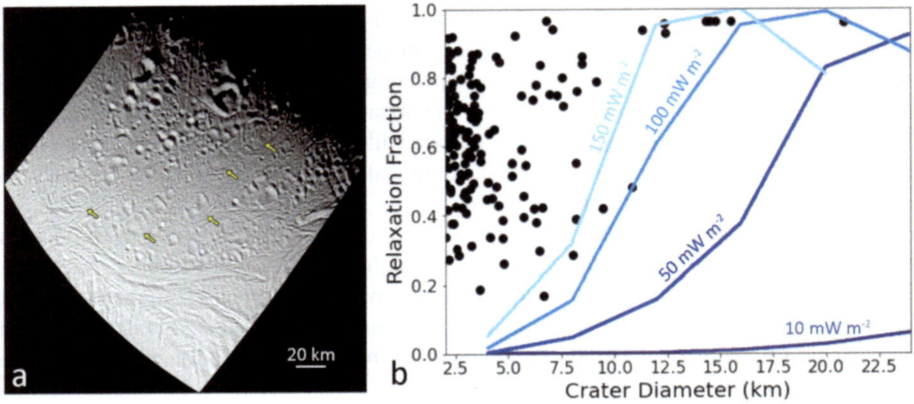

Fig. 5 a) *Cassini* ISS image of Enceladus showing numerous viscously relaxed impact craters (yellow arrows, but many more occur throughout the area). Image N1487299402_1 in a local orthographic projection with north up. b) An example of combining observations (black points are measured crater depths on Enceladus relative to their expected depth) with numerical modeling (blue curves) to constrain the heat flux. The simulations shown here assumed a surface temperature of 120 K and a pure ice (non-porous) rheology and thermal conductivity. The modeling suggests a flux in excess of 150 $\mathrm{mW\,m^{-2}}$ was necessary to viscously relax Enceladus' craters. Panel 'b' modified from Bland et al. (2012)

Even if the composition can be assumed, the temperature structure of the lithosphere is a function of surface temperature (T_s), heat flux (F), and thermal conductivity (k). Assuming a temperature dependent thermal conductivity of the form $k = C/T$, where T is the temperature and C is a constant (e.g., Petrenko and Whitworth 1999, state $C = 651\ \mathrm{Wm^{-1}}$, whereas Carnahan et al. 2021, find $C = 612\ \mathrm{Wm^{-1}}$), the temperature as a function of depth, z, is given by

$$T(z) = T_s \exp\left(\frac{Fz}{C}\right).$$

The surface temperature, T_s, is a function of surface albedo and latitude, can be modified by an insulating regolith (Passey and Shoemaker 1982) or solid-state greenhouse (Brown and Matson 1987; Matson and Brown 1989), and may have changed over the moon's history due to substantial changes in surface albedo, especially for low-albedo moons like Callisto or Iapetus, and a weaker young Sun (Gough 1981; Bahcall et al. 2001). The thermal conductivity of pure ice is well known, but it can be modified by the inclusion of other species. For example, clathrates (English and Tse 2009) and ammonia-doped water ice (Lorenz and Shandera 2001) have thermal conductivity 2–3 times lower than water ice. Even for pure water ice, if the upper portion of the lithosphere is porous (e.g., Eluszkiewicz 1990; Kossacki and Leliwa-Kopystyński 1993; Leliwa-Kopystyński and Kossacki 2000; Besserer et al. 2013), the thermal conductivity will be substantially reduced (Shoshany et al. 2002). Thus, the bounds on the lithospheric heat flux, F, itself must be carefully qualified because the temperature structure $T(z)$ is a non-unique combination of F and k. Furthermore, in many cases the same subdued topography can result from either a high heat flux imposed over a short timescale (e.g., a pulse of heat) or a lower heat flux imposed over a long timescale (e.g., Bland et al. 2017), so additional information must be used, such as the moon's dynamical history, to reduce the uncertainty.

Inferring heat flows from crater shapes also requires knowledge of the original dimensions of the crater, which can vary based on the material properties at the surface, the thermal profile of the ice at the time of impact, whether the ice shell is conductive or convective, and whether there is an ocean under the ice (Turtle and Pierazzo 2001; Senft and Stewart 2011; Bray et al. 2014; Silber and Johnson 2017; Denton and Rhoden 2022; Bjonnes et al. 2022). Although the original crater dimensions are often inferred by measuring fresh-appearing craters on the moon's surface, in some cases relaxation occurs rapidly enough under relatively quiescent conditions that even these depths are too shallow (Bland and Bray 2024). Incorrect inference of the original crater depth leads to an inaccurate estimate of both the degree to which the crater has relaxed (e.g., the crater may simply have formed shallow), and the timescale over which relaxation has occurred. For a stress-independent rheology, the relaxation timescale (i.e., the e-folding time) is independent of crater depth; although relaxation occurs faster for a deep crater than a shallow one due to the larger stresses involved, greater vertical displacement of the crater floor is also required (Turcotte and Schubert 2002). However, the viscosity of icy material is highly stress-dependent (for a summary, see e.g., Durham et al. 2010) and deep initial craters in ice therefore relax more rapidly than shallow ones. In other words, under the same conditions, a deeper initial crater will have a greater relaxation fraction than an initially shallower crater.

Despite these challenges, crater shapes can still be used to draw inferences as to the past heat flows on icy moons, as described for each moon within Sect. 5. In addition, numerical tools to model the formation and modification of craters – particularly for the size of Herschel and for moons with subsurface oceans – have also greatly improved in recent years (e.g., Silber and Johnson 2017; Denton et al. 2021; Denton and Rhoden 2022; Bjonnes et al. 2022). These tools provide an opportunity to better predict initial crater shapes and determine more precise relaxation fractions in future studies.

4 Interior Structure Characterization, Approaches and Observations

4.1 Rotational Dynamics

Here, we briefly describe a valuable observational tool, in which the rotational motion of a moon is used to infer its internal mass distribution. Details of the theoretical approach are provided by Hemingway et al. (2018); see also Rambaux et al. (2010), Rambaux (2014), and Nimmo (2018). Results obtained for the mid-sized moons are summarized, here, while the interpretations in context with their overall geologic histories are covered in Sect. 5.

Most moons of the giant planets maintain a 1:1 spin-orbit resonance, in which the moon completes one rotation about its spin axis in the same amount of time that it completes one orbit around the planet (e.g., Peale 1976). In this case, averaged over an orbit, the same face of the moon is always pointed toward the planet, which leads to an elongation of the moons' physical shape in the direction of the planet (i.e., the tidal axis). Moons in this configuration are described as being in synchronous rotation. Although long-period non-synchronous rotation (NSR) has been suggested for several ocean-bearing icy moons from geological interpretations, it typically refers only to the ice shell's rotation relative to the interior, which may still be tidally locked (e.g., Greenberg et al. 1998; Patthoff and Kattenhorn 2011; Collins et al. 2010). Whether NSR is likely from a dynamical perspective is still debated (e.g., Greenberg and Weidenschilling 1984; Bills et al. 2009; Goldreich and Mitchell 2010). For simplicity, we neglect this type of NSR in the present discussion.

When a synchronously-rotating moon has an eccentric orbit, the moon's orbital speed will increase and decrease as the moon moves through the closer (pericenter) and farther (apocenter) parts of its orbit, respectively. The moon's spin rate and orbital rate will, therefore, be equal *on average* but not instantaneously throughout the orbit. As a result, the long axis of the moon will oscillate back and forth relative to the direction of the planet. Such motion is referred to as the optical libration because it is related to the relative velocity of spin and orbit. The misalignments between the long axis of the moon and the planet, associated with the optical libration, will lead to torques that also generate oscillations; these oscillations are called physical librations, and they are the dynamical response of the moon to the gravitational torque of the planet. The amplitudes of the short period physical librations are tied to the interior structure of the moon through the moment of inertia. The largest libration is the diurnal libration (libration at orbital period); its amplitude can reveal whether a moon is differentiated and, in some cases, reveal the presence of an ocean.

Libration can be measured by tracking the subtle longitudinal motion of surface features within images taken at different times in a moon's orbit. The changes in location resulting from libration can be small – of order 100s of meters for Enceladus (Thomas et al. 2016), for example. Hence, identifying the libration signature requires (at a minimum) precise knowledge of the moon's orbit and rotation, the spacecraft's position/pointing, and the locations of surface features. Once the libration has been measured, its value is compared with models of the librations, derived using different plausible interior structures and the measured shape of the moon. These models typically use simplified structures in which the body is represented as nested, uniform layers; the number of layers, the layer radii, and the layer densities are then varied to capture the parameter space.

Cassini images provided sufficient coverage to measure librations of Mimas (Tajeddine et al. 2014) and Enceladus (Thomas et al. 2016; van Hoolst et al. 2016; Park et al. 2024). The Mimas results require that the interior is differentiated, but the data could be fit by either a subsurface ocean or a frozen outer layer over an elongated core (Tajeddine et al. 2014). As described in Sect. 5.2, additional studies of Mimas' librations and other dynamical properties favor the ocean model (Caudal 2017; Noyelles et al. 2019; Lainey et al. 2024). Enceladus' libration is only compatible with a differentiated interior and global sub-surface ocean (Thomas et al. 2016). Complications of the data collection and interpretations using *Cassini* data are described in Hemingway et al. (2018) as well as in the papers presenting libration measurements (Tajeddine et al. 2014; Thomas et al. 2016).

4.2 Shape and Gravity

Here, we summarize the approach of gravity and shape measurements for inferring the interior structure of a moon. For a more detailed treatment, including the main equations, we refer the reader to Hemingway et al. (2018). We also provide a summary of *Cassini* gravity measurements of the mid-sized moons, and their interpretations. Additional discussion of these findings is provided within the individual moon sections (Sect. 5).

For a fluid moon in synchronous rotation around its parent planet, the moon will adopt a triaxial ellipsoidal shape with the associated gravity field. However, because a moon is likely to increase in density with depth, its physical responsiveness to deformation deviates from that of a uniform fluid sphere, so the triaxial shape and gravity field include the scaling factors, h_{2f} and k_{2f}, which are the respective degree-2 fluid Love numbers. The scaled shape and gravity field can then be described using weighted mathematical functions (i.e., spherical harmonics), where the weights are H_{lm} and C_{lm}, respectively, and l and m relate the weight to its function. This approach is qualitatively similar to describing a color in terms of the percentage of red, yellow, and blue that – when mixed – can produce it; the percentages act as weights while the primary colors represent orthogonal "basis" functions. By matching observational data of a moon's shape and gravity field, the values of the weights can be constrained.

If there has been sufficient time and/or heat, the shape of a moon will conform to an equipotential surface determined by the body's gravity, rotation, and tidal deformation. In this case, the moon is considered to be in hydrostatic equilibrium, and the ratios of the degree-2 shape and gravity weights, $-H_{20}/H_{22}$ and $-C_{20}/C_{22}$, will be 10/3. By convention, $-C_{20}$ is referred to as J_2, such that $J_2/C_{22} = 10/3$ for a hydrostatic gravity field, although slight corrections may also be required for fast rotating bodies (see Tricarico 2014). These ratios provide a measure of how close (or far) the moon is to a hydrostatic shape or hydrostatic gravity field. Critically, if a moon is in hydrostatic equilibrium, H_{20}, H_{22}, J_2 and C_{22} can be directly related to the polar moment of inertia, C, which constrains the density structure of the moon. In particular, $C/MR^2 = 0.4$ for an undifferentiated moon, where M is the mass and R is the mean equatorial radius. Lower values indicate a more centrally-condensed interior, suggesting a differentiated moon.

If the moon is not in hydrostatic equilibrium, more complicated strategies must be employed to constrain the interior. Typically, the hydrostatic and non-hydrostatic components of gravity and shape are assumed to be separable from one another. Then, using multiple models of the interior structure of the body, the model parameters (such as radius and density of each layer) that best fit the measured gravity and shape can be determined (e.g., Iess et al. 2014; Tortora et al. 2016; Zannoni et al. 2020). Non-hydrostatic components may be related to uncompensated surface topography, which would suggest a relatively cold and rigid exterior. They could also be the signature of density variations within the body such that the assumption of uniform layers is invalid, or a core shape that is not hydrostatic. This last option may be common to moons that have low central pressures or may be an indication that the interior experienced insufficient heating to relax the core shape (e.g., Castillo-Rogez et al. 2018; Hemingway et al. 2018).

For the mid-sized moons, the Cassini spacecraft collected sufficient limb images (for shape) and Doppler tracking data (for gravity) to determine H_{20}/H_{22} and J_2/C_{22} for Enceladus (Iess et al. 2014; Hemingway et al. 2018 and references therein; Park et al. 2024), Dione (Thomas 2010; Nimmo et al. 2011; Zannoni et al. 2020), and Rhea (Tortora et al. 2016). Detailed descriptions of the results are included in the individual moon section. These studies show that the shapes of Enceladus and Dione are strongly non-hydrostatic, while the larger uncertainty at Rhea makes the results inconclusive. In terms of gravity, Dione and Rhea are significantly non-hydrostatic, while Enceladus has a more modest deviation from hydrostaticity. For both Enceladus and Dione, the gravity field is closer to hydrostatic equilibrium than their shape, which is interpreted as the result of a sub-surface ocean enabling compensation of overlying topography (e.g., Iess et al. 2014; Beuthe et al. 2016; Zannoni et al. 2020). In contrast, the large uncertainty in Rhea's shape makes it compatible with hydrostatic equilibrium, so the non-hydrostatic gravity is interpreted as a limited excess oblateness of a core, which implies differentiation of the moon (Tortora et al. 2016). There is no gravity data at Mimas or Tethys; estimates from their global shapes suggest that neither moon is in hydrostatic equilibrium (Thomas 2010).

5 The Geology of the Mid-Sized Moons

5.1 Mimas

5.1.1 Overview of Surface Geology

Mimas' geologic record is dominated by craters (Fig. 6), including the Herschel impact basin (D = 139 km). As discussed in Sect. 3.2, the Case B production function from Zahnle et al.

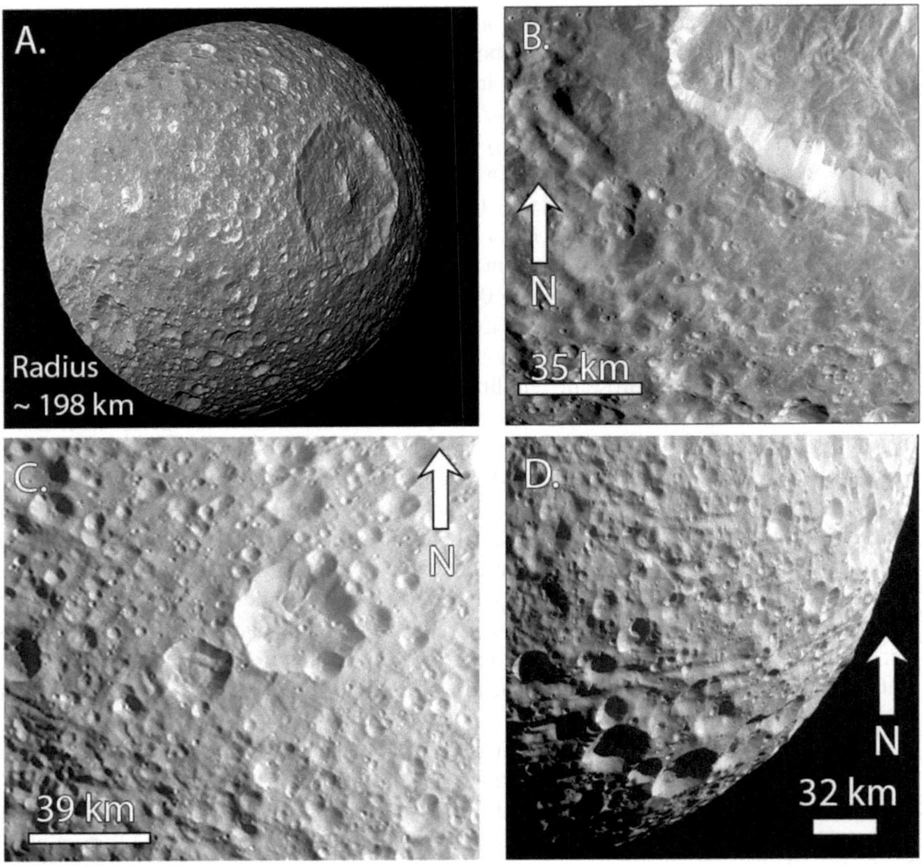

Fig. 6 Views of Mimas across its leading and trailing hemispheres. A) Full disk view of Mimas with a focus on the Herschel impact basin (D = 139 km) and its central peak. Image PIA12570. B) Close-up of Herschel's ejecta blanket. The lower density of craters on the ejecta blanket suggests relatively recent formation of the basin (Ferguson et al. 2024). Image N1644778567_1. C) Trailing hemisphere cratered terrain along with some grooves. Image N1831441742_1. D) Oblique view of the grooves on Mimas' trailing hemisphere, which are among the only tectonic features so far identified on Mimas. Image N1831443018_1

(2003) provides an excellent fit to much of the crater data at Mimas (Ferguson et al. 2024), providing further support for planetocentric impactors (Kirchoff and Schenk 2010; Hirata 2016; Kirchoff et al. 2018; Bell 2020). Cratering is more sparse within the Herschel basin and on its eject blanket than on other regions of Mimas, implying a young relative age for the basin (Ferguson et al. 2024). Using the relative crater spatial densitites at Mimas, Tethys, and Dione, Ferguson et al. (2024) determined that heavily-cratered regions of Mimas are younger than those on Tethys and Dione. However, the relative ages are all within a factor of 2–3 of each other. Hence, Mimas cannot be of order 100 Myr old unless Tethys and Dione are as well, but if Mimas is at least a billion years old, Dione could be 2–3 billion years older.

As described in Sect. 3.4, crater morphologies can record the extent of heating over time. Using Voyager data, Schenk (1989) concluded that Herschel is unrelaxed. Because large craters relax more quickly than small craters, this result implied limited crater relaxation at Mimas, overall. Post-*Cassini* analyses of Mimas' crater relaxation have been sparse. Using photoclinometry, White and Schenk (2011) analyzed the depth-to-diameter ratios of nine craters on Mimas, which were consistent with limited relaxation. Unpublished digital elevation models (DEMs) show that many deep craters, including Herschel, exist on Mimas (Schenk, pers comm), but the extent of relaxation that is compatible with the observed depths is not yet known. In addition, moons can display both heavily relaxed and unrelaxed craters due to differences in crater ages or location-based variations in heat flow, as seen on Enceladus, Tethys, and Dione (Bland et al. 2012; White et al. 2017; see also, Sects. 5.3 and 5.4), motivating a more global study. Revisiting the crater morphologies of Mimas, using the full *Cassini* data set and our improved understanding of the effects of a moon's thermal profile and interior structure on the initial crater shape used in such studies (see Sect. 3.4), is critical for evaluating the extent of modification of Mimas' craters.

Mimas displays little tectonic activity, particularly when compared with the other mid-sized moons, and shows no evidence of plumes or cryovolcanic flows (Schenk et al. 2018, and references therein). Arcuate grooves of uncertain origin have been identified (Fig. 6D), extending 10s of km across the surface, which may be ~10 km deep (Schenk 2011; Schenk et al. 2018). Given that they are sometimes overprinted by craters, the grooves are probably not the youngest features on the surface. Mimas displays no evidence of tidally-driven activity akin to that observed on Enceladus (Rhoden et al. 2017; Rhoden 2023), despite a high eccentricity and close orbit around Saturn.

5.1.2 Interior Structure and Evolution

Mimas' low density of 1.150 g/cm^3 suggests an interior dominated by ice, as much as ~91% in volume or ~73% in mass, if there is no porosity (assuming a dry rock density of 3.1 g/cm^3 consistent with a CI chondrite composition, Castillo-Rogez et al. 2018). The fraction of porosity is unconstrained, but it would be ~30% if Mimas has an average grain density similar to that of the Uranian moons (1.6 g/cm^3; ssd.jpl.nasa.gov). A better estimate of Mimas' porosity could provide constraints on its formation and evolution, such as the extent of global-scale melting (e.g., Kossacki and Leliwa-Kopystyński 1993; Leliwa-Kopystyński and Kossacki 2000).

Simultaneously matching *Cassini* measurements of Mimas' librations and precession of its orbit requires a differentiated interior and a present-day subsurface ocean (Tajeddine et al. 2014; Noyelles 2017; Caudal 2017; Lainey et al. 2024). The overlying ice shell thickness is constrained to be 20 to 30 km thick at present (Tajeddine et al. 2014; Lainey et al. 2024). If the ice shell is currently thicker than ~29 km, it should be actively melting because there is more tidal heat than can be conducted out of the shell (Rhoden and Walker 2022). Even the thinnest shells allowed by the libration and precession measurements may be actively melting, depending on the heat flow entering the ice shell from below. Heat sources include (likely low) radiogenic heating (e.g., Hussmann et al. 2006) and tidal dissipation in Mimas' core, which could be substantial if the core is porous (Roberts 2015; Choblet et al. 2017; Rovira-Navarro et al. 2022; Rhoden et al. 2024). Tidal dissipation within Mimas would act to circularize its orbit, particularly in the absence of a mean motion resonance (cf., Io-Europa; Peale et al. 1979), providing constraints on Mimas' orbital evolution.

Lainey et al. (2024) and Rhoden et al. (2024) both modeled the co-evolution of Mimas' eccentricity and interior, assuming a recent eccentricity-pumping event led to the onset of melting and ocean generation. Both investigations determined that the initial eccentricity was between 2 and 3 times the present value. Over time, tidal dissipation caused both eccentricity decay and thinning of the ice shell, eventually reaching today's conditions, in which the shell is between 20 and 30 km thick (Tajeddine et al. 2014; Lainey et al. 2024). Based on these results, both studies concluded that Mimas' ocean has likely emerged within the last

~10 Myr (Lainey et al. 2024; Rhoden et al. 2024). As described in the following section, a young ocean, in which the ice shell has thinned over Mimas' recent past, is compatible with Mimas' geology (Rhoden 2023; Rhoden et al. 2024), although it does not provide as much insight as to Mimas' initial origin or age.

In a pair of companion papers, Baillié et al. (2019) and Noyelles et al. (2019) investigated a potential history for Mimas in which a recent resonance passage with either Tethys or Enceladus resulted in Mimas migrating inward, increasing its eccentricity. As a result of its migration, Mimas would have opened a gap in Saturn's rings, perhaps explaining the formation of the Cassini Division (Baillié et al. 2019). The changes in Mimas' orbit would also lead to enhanced tidal heating within Mimas (Baillié et al. 2019; Noyelles et al. 2019). The scenarios investigated so far either induce too much melting within Mimas to be compatible with its geology or do not reproduce the current orbital elements of the moons (Noyelles et al. 2019). In any case, once melting occurs, and dissipation is further enhanced, the eccentricity drops, eventually causing the moons to exit the resonance (Baillié et al. 2019; Noyelles et al. 2019). At that point, Mimas would return to a state of outward migration, with its eccentricity continuing to decrease (Baillié et al. 2019; Noyelles et al. 2019).

Based on estimated rates of infilling of the Cassini Division, the entire inward and outward migration of Mimas is presumed to have occurred within the past 10 Myr (Baillié et al. 2019; Noyelles et al. 2019). While this proposed history fits with the timescale and sequence of events suggested by Lainey et al. (2024) and Rhoden et al. (2024), the implied eccentricity of Mimas is too high by an order of magnitude, and there is a mismatch in the time since Mimas resumed outward migration (~1 Myr in Baillié et al. 2019 vs 3–25 Myr in Lainey et al. 2024). It remains to be seen whether these models can be reconciled, or whether Mimas' emerging ocean is entirely disconnected from the process that opened the Cassini Division.

Modeling of Mimas' coupled thermal-orbital evolution over billion year timescales has confirmed that, if a primordial Mimas were heated sufficiently to differentiate, it would also lose its eccentricity (Neveu and Rhoden 2017). Any ocean would be lost as a result of eccentricty decay and the subsequent drop in tidal heating, and any geologic features formed in response would need to be obscured by later cratering to match observations. In contrast, a ring-born Mimas could have formed layered (as in Charnoz et al. 2011) while preserving any initial eccentricity and avoiding geologic activity associated with an ocean (Neveu and Rhoden 2019). While the ability to preserve a primordial high eccentricity initially seemed like a valuable attribute of the ring-formation model, the likelihood of a recent eccentricity-pumping event at Mimas means neither model can be ruled out. In this way, the presence of a recently-formed ocean within Mimas creates a disconnect between the present conditions and Mimas' early origins. The current eccentricity cannot be primordial, and the higher eccentricity at the onset of melting must be a recent event or the entire lifecycle of ocean generation, eccentricity decay, and ocean freezing would have already occurred (Lainey et al. 2024; Rhoden et al. 2024). We do not yet have constraints on Mimas' eccentricity before the recent eccentricity-pumping event that triggered melting within Mimas.

Reassembly of Mimas as the result of a disruptive impact (e.g., Asphaug and Reufer 2013; Ćuk et al. 2016; Dubinski 2019) could also be compatible with the emergence of an ocean and Mimas' current high eccentricity, provided that there is sufficient time for Mimas to accumulate craters, including Herschel (see next section). An important open question is whether large impacts, involved in either disrupting precursor moons or generating observed basins on the current moons, would significantly increase Mimas eccentricity (as investigated for Tethys by Zhang and Nimmo 2012) and whether they would be ocean-promoting or ocean-limiting events.

5.1.3 Geologic Features and a Young Mimean Ocean

Mimas' most distinctive geologic feature, the Herschel impact basin, has been used to probe the ice shell thickness at the time of its formation. Numerical modeling of the impact that formed Herschel showed that the ice shell had to be at least 55 km thick in order to match the overall shape and morphology of the basin (Denton and Rhoden 2022). Given that the libration and precession measurements indicate a present-day ice shell no thicker than 30 km (Tajeddine et al. 2014; Lainey et al. 2024), these results support the hypothesis that Mimas' ice shell has thinned considerably since the formation of Herschel (Denton and Rhoden 2022).

Tidal heating within an ocean-bearing Mimas would generate a globally-averaged surface heat flow of 10s of mW/m^2 at the present ice shell thickness (Rhoden and Walker 2022). Estimates of historic heat flows for plausible evolutions of a developing ocean and ice shell within Mimas are similar in magnitude (Rhoden et al. 2024). Rhoden et al. (2024) modeled relaxation of a 26-km diameter crater using the highest heat flows identified in their ocean evolution models and determined that the low gravity and cold surface temperatures on Mimas would generate only $\sim$10 m of topographic change, which is far below the detection limit enabled by *Cassini* image data. Hence, limited crater relaxation is compatible with the development and presence of an ocean.

Mimas lacks the tectonic structures and processes that have been linked to tidal stresses on confirmed ocean worlds, such as Europa and Enceladus (e.g., Greenberg et al. 1998; Rhoden et al. 2010, 2020). Mimas also lacks large canyons, such as those observed on Tethys and Charon, which have been attributed to stresses caused from a past ocean freezing out (Chen and Nimmo 2008; Spencer et al. 2021 and references therein; Rhoden et al. 2023; see also Sect. 5.3), although it has also been suggested that the formation of Ithaca Chasma on Tethys was caused or exacerbated by the Odysseus forming impact (e.g., Moore et al. 2004). When compared with moons for which fracture formation from ocean freezing has been investigated (Nimmo 2004; Rudolph and Manga 2009; Rudolph et al. 2022; Rhoden et al. 2023), cooling cracks on Mimas should be able to transit even thicker ice shells due to Mimas' lower gravity. In fact, Rhoden et al. (2024) showed that ice shell thickening at Mimas leads to crack penetration to the ocean and surface eruptions of ocean material within all of the evolution models they studied.

The limited tectonic record, and lack of eruptive activity, suggests that Mimas' ice shell has not thickened enough to form fractures from cooling, and that tidal stresses alone are insufficient to fracture the ice (Rhoden et al. 2017; Rhoden 2023; Rhoden et al. 2024). A recently emerging ocean, and thinning ice shell, are thus compatible with the surface geology of Mimas. In addition, the continued eccentricity decay caused by tidal dissipation within Mimas is likely to cause Mimas to enter a state of ice shell thickening within the next $\sim$10 Myr, which ought to generate similar tectonic and eruptive activity as we currently observe on Enceladus (Rhoden et al. 2024).

5.1.4 Summary of Constraints

Mimas' geologic record, including the lack of tectonic activity (Rhoden et al. 2017; Schenk et al. 2018), morphology of Herschel (Denton and Rhoden 2022), and limited crater relaxation (Schenk 1989; White and Schenk 2011), is compatible with a young ocean and recently thinning ice shell (e.g., Rhoden et al. 2024). The emergence of an ocean points to an eccentricity-pumping event that triggered melting within the last $\sim$10 Myr of Mimas' history (Lainey et al. 2024; Rhoden et al. 2024), although the exact mechanism that raised the eccentricity is still under investigation (e.g., Baillié et al. 2019; Noyelles et al. 2019). Such a recent change in Mimas' orbit and interior somewhat decouples its present state from its initial formation. In particular, retaining a high eccentricity over long timescales is no longer a strict constraint on Mimas' evolution as the ocean implies a recent eccentricity-pumping event (Lainey et al. 2024; Rhoden et al. 2024).

Mimas' differentiated interior (Tajeddine et al. 2014; Noyelles et al. 2019; Caudal 2019; Lainey et al. 2024) requires either global-scale melting (e.g., Neveu and Rhoden 2017), a layered origin such as from ring formation (Charnoz et al. 2011), or some form of impact-induced differentiation (e.g., Barr and Canup 2010). Models that invoke reassembly of Mimas after a disruptive collision or mergers of precursor moons (Asphaug and Reufer 2013; Ćuk et al. 2016; Dubinski 2019) must also provide a pathway to promote or preserve internal layering. Given the ease with which cracks and eruptions form as a result of ocean freezing (Rhoden et al. 2024), any past ocean that formed during differentiation or from an older epoch of high tidal heating, would need to have been lost before the majority of craters were emplaced to erase the evidence. Mimas' substantial crater population also requires adequate time and impactor sources to develop (e.g., Kirchoff and Schenk 2010; Ferguson et al. 2024) and a surface solid and cold enough to retain crater shapes over the surface age. However, the lower spatial densities of larger craters on Mimas as compared with Tethys and Dione suggests that Mimas is overall younger, and thus, unlikely to be primordial (Ferguson et al. 2024). Finally, Mimas must be able to participate in an eccentricity-pumping resonance within the past $\sim$10 Myr in order to develop the ocean that exists today (Tajeddine et al. 2014; Lainey et al. 2024; Rhoden et al. 2024).

5.2 Enceladus

5.2.1 Overview of Surface Geology

Enceladus' surface (Fig. 7) is broadly arranged into four terrain types: 1) the cratered terrains co-located with the sub- and anti-Saturn points, 2) the tectonized terrains on the leading and trailing hemispheres, 3) the more heavily cratered, but tectonically-modified, north polar terrain and 4) the crater-free and heavily tectonized south polar terrain (e.g., Crow-Willard and Pappalardo 2015; see also Patterson et al. 2018 and references therein). Despite their appearance, there is evidence that the cratered terrains on the sub- and anti-Saturn points are experiencing recent or current tectonic dissection by tectonic structures called pit chains (Martin et al. 2017; Whitten and Martin 2019; Martin et al. 2023). Pit chains are also found at the boundaries between cratered terrains and tectonized terrains, often in en echelon patterns suggesting shear motions, extending into and across north polar terrains (Martin 2016). Given the heavy focus on the geologically/tectonically active south polar terrain (see below), it is intriguing that terrains along the equator (Martin et al. 2023) and at/near the north pole (Martin 2016) may also be undergoing current (or recent) tectonism.

The south polar terrain and the tectonized leading/trailing terrains are thought to be relatively young, as evidenced by their overall lack of impact craters (Kirchoff and Schenk 2009, 2010). In addition, tectonic terrains of the leading and trailing hemispheres differ in character (Crow-Willard and Pappalardo 2015; Patterson et al. 2018; Patthoff et al. 2022). The leading hemisphere terrain is dominated by ridge and trough structures of varying size and morphology that likely formed under compressional stress (Patthoff et al. 2015). The trailing hemisphere terrain includes two vast regions of tectonically striated plains (Sarandib and Diyar planitia) separated by a complex of ridges called dorsa (Crow-Willard and Pappalardo 2015). The morphology of the dorsa suggests they formed as thrust blocks and implies lithospheric shortening (compressive stress), consistent with the sense of strain inferred

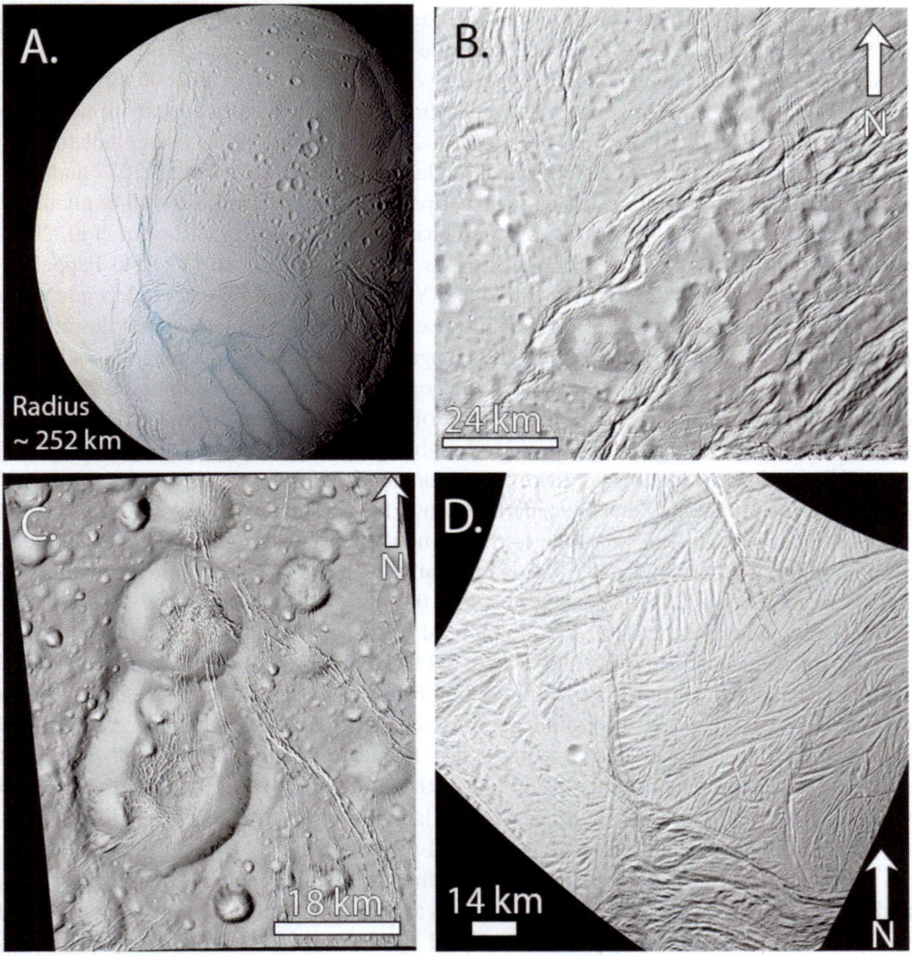

Fig. 7 A) View of Enceladus' south polar terrain, with a focus on the Tiger Stripes fractures. Material from within Enceladus can erupt into space via these fractures. Image PIA07800. B) A transitional region between tectonized and cratered terrains. While craters are present on the surface, they are often overprinted by tectonics and appear to have undergone viscous relaxation. Image N1637465264_1. C) "Snowman" craters near the North Pole of Enceladus. While the cratered terrains in the north are likely representative of older surfaces on Enceladus, these craters are often cross-cut by other fractures. Image N1823513163_1. D) Fractured terrain on Enceladus. Due to the lack of impact craters on this surface, it's inferred that fracture formation occurred relatively recently in geologic history. Image N1604168315_3

for the leading hemisphere (Patthoff et al. 2022). In contrast, the striated plains appear to have formed through extension (Bland et al. 2007), with a deep, graben-like trough (Harran Sulci) separating the tectonic terrains from cratered terrain to the east (Giese et al. 2008). The central portions of the leading and trailing hemispheres are surrounded by complex sets of curvilinear ridges and troughs (e.g., Hamah and Samarkand Sulci) that likely record multiple episodes of tectonism (Crow-Willard and Pappalardo 2015; Patterson et al. 2018). The presence of clear indicators of both extension (the plains and graben) and contraction (the dorsa) is rare on icy moons, and suggests a complex tectonic history in Enceladus' mid-latitudes.

The south polar terrain (SPT) is characterized by four long, roughly-parallel fissures (dubbed Tiger Stripes) from which emanate the on-going plume activity identified by the *Cassini* mission (Hansen et al. 2006; Dougherty et al. 2006; Porco et al. 2006; Spencer et al. 2006). The volume of plume material emitted from the Tiger Stripes appears to be tidally-modulated (Hurford et al. 2007; Hedman et al. 2013), although it has been challenging to match the variations with tidal-mechanical models (e.g., Běhounková et al. 2017 and references therein). The orientations of the Tiger Stripes, on the other hand, are well-explained by eccentricity-driven tidal stresses in an ocean-bearing Enceladus (e.g., Rhoden et al. 2020), although non-tidal origins for the Tiger Stripes have been proposed (Yin and Pappalardo 2015; Hemingway et al. 2020; Schoenfeld and Yin 2024), which may act independently or in concert with tidal stresses. Variations in Enceladus' ice shell thickness, as described in the following section, will affect tidal stress magnitudes, perhaps explaining the dichotomy in tidally-driven tectonic activity between the north and south poles (Beuthe 2018; Rhoden et al. 2020). However, the overall magnitudes of tidal stresses are still an order of magnitude lower than the failure strength of ice in laboratory studies (e.g., Collins et al. 2009), suggesting additional stresses are involved in fracture formation.

Stresses due to cooling and freezing of a subsurface ocean are often invoked for moons with tidally-driven tectonics in order to overcome the tensile strength of the ice shell (e.g., Nimmo 2004; Rudolph et al. 2022). In particular, thickening of Enceladus' ice shell would induce large, isotropic, extensional stresses in the upper portion of the ice shell (Nimmo 2004; Rudolph and Manga 2009; Rudolph et al. 2022), resulting in cracks that propagate radially through shell (Rudolph and Manga 2009; Hemingway et al. 2020; Rudolph et al. 2022). In shells that are thin enough for cooling fractures to transit the entire shell ($\sim$10 km on Enceladus), it takes $\sim$10 Myr of thickening to connect the surface and the ocean (Rudolph et al. 2022). It seems likely that cooling cracks formed the initiation points of Enceladus' south polar plumes (Hemingway et al. 2020; Rhoden et al. 2020). However, the extent to which radial cooling cracks are related to the laterally-propagating Tiger Stripe fractures from which the plumes now emanate is not well-understood.

The Tiger Stripes lie atop a region of dense fractures, which includes many crosscutting and overlapping fractures of various orientations and a set of parallel fractures that have been interpreted as paleo Tiger Stripes (Patthoff and Kattenhorn 2011). The history of fracture formation, activation, and decommission at the south pole implies changes in the stress field controlling fracture orientations (Patthoff and Kattenhorn 2011). As discussed more in the following section, non-synchronous rotation (NSR) of Enceladus' ice shell could plausibly cause the requisite changes in stress over time to explain the orientations of older fractures in the SPT, and the relatively large magnitude of NSR stress may also facilitate fracture formation (Patthoff and Kattenhorn 2011). Whether the past generations of fractures also served as conduits for plumes is presently unknown.

Many craters on Enceladus are heavily relaxed, even near the equator, which is well outside the active south polar region (e.g., Fig. 5). Bland et al. (2012) found that even with 10s of mW/m^2, applied over millions of years, the measured relaxation could not be reproduced. Only by adding a brief pulse of high heat flow (150 mW/m^2), along with an insulating regolith that could keep the near-surface temperature higher than the measured value at the surface, could the modeling results begin to approach the levels of relaxation on Enceladus. Muted crater morphologies on Enceladus have been attributed to a combination of viscous relaxation and infilling from plume fallback (Bland et al. 2012). These processes can erase many of the craters with diameters <2 km and >6 km (Kirchoff and Schenk 2009) and help match the observed depth-to-diameter ratios observed on Enceladus. While crater infilling by plume fallback would contribute to muting the short wavelength topography of craters on Enceladus (Bland et al. 2012), thick sequences of plume fallback, on the order of 100s of meters, would take millions or billions of years to deposit across the surface given current plume mass fluxes (Martin et al. 2023).

Composition measurements of material erupting from Enceladus' plumes indicates on-going hydrothermal alteration, which may be indicative of geologically recent interactions between rock and liquid water (e.g., Hsu et al. 2015). In particular, the detection of H_2 in Enceladus' plumes (Waite et al. 2017) suggests that water-rock interactions are on-going. Since rock hydration operates on short timescales (i.e., million years, e.g., Martin and Fyfe 1970), the presence of H_2 in the plume might indicate that Enceladus' ocean is young.

5.2.2 Interior Structure and Evolution

Enceladus has the highest bulk density of the mid-sized moons (1.6 g/m^3; Rappaport et al. 2007) suggesting a rock fraction of ~60% by mass. Enceladus' shape shows significant deviation from hydrostatic equilibrium in the form of excess flattening, particularly at the south pole (Thomas et al. 2007; Nimmo et al. 2011; Hemingway et al. 2018 and references therein). The shape and gravity data obtained by *Cassini* result in a J_2/C_{22} of 3.51 ± 0.05 (Iess et al. 2014). When the rapid rotation of Enceladus is accounted for, the J_2/C_{22} for a hydrostatic Enceladus would be 3.25 rather than 10/3 (Tricarico 2014; McKinnon 2015). Either way, the results suggest that Enceladus' gravity field is not hydrostatic (Iess et al. 2014; McKinnon 2015). That Enceladus' shape deviates more from hydrostatic than its gravity implies some amount of compensation of excess topography, which a subsurface ocean could provide.

As discussed in Sect. 4, the lack of hydrostatic equilibrium complicates inferences of Enceladus' moment of inertia and interior structure. Therefore, many studies have attempted to further constrain Enceladus' interior from gravity and shape measurements using different approaches to isolate non-hydrostatic components and/or incorporate isostatic compensation (Iess et al. 2014; McKinnon 2015; Čadek et al. 2016; Beuthe et al. 2016; Hemingway and Mittal 2019; Park et al. 2024). These models have sufficient complexity to produce a variety of solutions. However, there is general agreement across these studies that 1) Enceladus is differentiated, 2) the rocky core is ~190 km in radius and has a low density of 2.3–2.45 g/cm^3, 3) there is a global ocean that enables compensation of topography, and 4) the mean global ice shell thickness is most likely 20–30 km, with a total hydrospheric thickness of ~60 km. Enceladus' low core density may be the result of high porosity, with ice or brines in the pores (Roberts 2015; Choblet et al. 2017), although the porosity of the core is not well-constrained. Assuming uniform porosity, a grain density of 3.1 g/cm^3 and a brine density of 1.03 g/cm^3, the average porosity of the core would be ~35%.

Enceladus has a large physical libration; the original study using *Cassini* imagery found a libration amplitude of $0.12° \pm 0.014°$ (2σ) that was 180° out of phase with the eccentricity libration (Thomas et al. 2016). The libration amplitude was revised down in a later analysis to $0.091° \pm 0.009°$ (3σ; Park et al. 2024). With either value, the large libration amplitude requires a global ocean to mechanically decouple the ice shell from the interior (Thomas et al. 2016; van Hoolst et al. 2016; Park et al. 2024), consistent with the inferences from the gravity studies (e.g., Hemingway and Mittal 2019). However, the compensation depth of the ice shell (i.e., its thickness) depends on the assumed libration amplitude. Hence, using the value from Thomas et al. (2016), the ice shell would have a global average thickness of 21–26 km if it behaves entirely elastically or 14–19 km if it is viscoelastic. Park et al. (2024) find a global average ice shell thickness of 27 to 33 km based on their libration analysis.

Enceladus' ice shell varies in thickness with latitude, with the south pole being the thinnest, then the north pole, and then equatorial regions (Iess et al. 2014; McKinnon 2015;

Beuthe et al. 2016; Thomas et al. 2016; Čadek et al. 2016, 2019). Because the estimates are model-dependent, there is disagreement as to just how thick the shell currently is at the south pole, from less than 5 km to at least 10 km. Hypotheses for the source of the thickness variations include, but are not limited to, stochastic effects of interior evolution leading to one hemisphere experiencing more heating/melting than the other, patterns of convection (e.g., Běhounková et al. 2012 and references therein), or a giant impact that led to reorientation (Peale and Greenberg 2007; Roberts and Stickle 2021). In addition, gravity data suggests that the ice shell has thickened with time (Čadek et al. 2019), which could have facilitated formation of the conduits that enable Enceladus' plumes (Hemingway et al. 2020; Rudolph et al. 2022).

The evolution of Enceladus' interior may provide an explanation for the changes in fracture orientations at the south pole and the current locations of tectonized terrains near the equator (Rhoden et al. 2020). If Enceladus were frozen, but separated into an ice layer over a rocky interior, the pattern of tidal heating would be nearly inverted from the pattern with an ocean (e.g., Roberts 2015). In that case, tidal heating would be concentrated at the equator, centered on the leading and trailing points. These are the locations at which we observe older tectonized terrains (Crow-Willard and Pappalardo 2015). If Enceladus then transitioned from a frozen world to one with a subsurface ocean, these locations would have experienced tidal stress that may have led to fracturing. Once the ocean became global in extent, the pattern of tidal heating and stress would change such that maximum tidal heating and stress were generated at the poles (Roberts 2015; Rhoden et al. 2020). If the Tiger Stripes, which are the youngest geologic features at the south pole (Crow-Willard and Pappalardo 2015; Patterson et al. 2018), formed as a result of ice shell thickening (as in Hemingway et al. 2020; Rudolph et al. 2022), it would imply that Enceladus has since transitioned from a growing ocean to a shrinking one, which is consistent with the gravity measurements (Čadek et al. 2019).

Because the subsurface ocean decouples Enceladus' ice shell from its interior, the shell can spin via a process called non-synchronous rotation (NSR). As a result of the ice shell's motion, regions would move to new locations, relative to the direction of Saturn, thus exposing them to a different tidal stress field. In addition, NSR can impart large stresses on the shell that can facilitate fracture formation (e.g., Kattenhorn and Hurford 2009), although the relative importance of NSR versus diurnal tidal stress at Enceladus is not yet known. Slow NSR has been suggested from an analysis of fracture orientations at Enceladus' south pole (Patthoff and Kattenhorn 2011) and the systematic rotation of pit chain orientations within Enceladus' cratered terrains (e.g. Martin et al. 2017, 2023). If NSR has been acting to reorient the ice shell in longitude, it would imply that the tectonized terrains near the equator, which are currently aligned with the sub- and anti-Saturn points, have also moved since they formed and are at their current locations by happenstance. In addition, a minimum of millions of years would be required to create the observed fracture population within the south polar region, depending on the rotation period (Patthoff and Kattenhorn 2011), which limits the youth of tectonic activity on Enceladus overall.

Cassini measurements revealed that Enceladus is producing a tremendous amount of heat, particularly in the south polar region (Spencer et al. 2006; Howett et al. 2011), and high heat flows are recorded in heavily relaxed craters even well outside the SPT (Bland et al. 2012). Although tidal heating can be produced in Enceladus' ice shell (e.g., Soucek et al. 2019), and to lesser extent in its ocean (Hay and Matsuyama 2019; Rovira-Navarro et al. 2019), such high heat flows may require tidal dissipation in the core of Enceladus. Several studies have shown that a porous core within Enceladus would generate a significant amount of tidal dissipation (e.g., Roberts 2015; Choblet et al. 2017; Rovira-Navarro et al.

2022). The link between tidal dissipation, core characteristics, and formation is discussed more in Sect. 5.6.

Enceladus maintains a mean motion resonance with Dione (e.g., Meyer and Wisdom 2008; Ćuk and El Moutamid 2022), which enables its eccentricity to remain relatively high (0.0047) in spite of its extreme dissipation. Whether Enceladus is currently in a state of equilibrium tidal heating or oscillating around an equilibrium depends on how dissipation is partitioned between Enceladus and Dione, as well as the evolution of Saturn's Q, as discussed in detail by Nimmo et al. (2023). In addition, Enceladus may have experienced past epochs of higher eccentricity during the complex dynamical evolution of all five mid-sized moons (e.g., Meyer and Wisdom 2008; Běhounková et al. 2012; Neveu and Rhoden 2019; Noyelles et al. 2019; Nakajima et al. 2019; Ćuk and El Moutamid 2022). For a more complete review of dissipation within Enceladus, see Nimmo et al. (2023).

5.2.3 Summary of Geologic Constraints

Models of Enceladus' formation and evolution need to account for the following observations. First, Enceladus displays differences in geologic activity that vary both spatially and temporally (Crow-Willard and Pappalardo 2015; Patterson et al. 2018). Tectonic features at the south pole have been linked to tidal stress and stresses from a freezing ocean (e.g., Beuthe 2018; Hemingway et al. 2020; Rhoden et al. 2020; cf., Yin and Pappalardo 2015), although stress from non-synchronous rotation may have also played a role (Patthoff and Kattenhorn 2011). The origins of older fractures both within the south polar terrain and in equatorial regions are less certain (e.g., Patterson et al. 2018). One potential explanation for Enceladus' tectonic history is the development of an ocean that has since entered an epoch of freezing (Rhoden et al. 2020). That we observe the most tectonic activity where the ice shell is thinnest is not surprising given that tidal stresses are enhanced at locally thin regions (e.g., Beuthe 2018), although the process(es) that led to the south pole being thin relative to the global average is still debated (e.g., Běhounková et al. 2012; Roberts and Stickle 2021). Overall, the evolution of Enceladus' interior, orbit, and rotation state must account for the observed geology and how features have changed with time.

Second, the present-day heat flux is quite high, particularly at the south pole (Spencer et al. 2006; Howett et al. 2011), and craters even well outside the south polar terrain imply high heat flows (Bland et al. 2012). These results require high dissipation in the present as well as in the past (see reviews by Nimmo et al. 2018, 2023). Enhanced tidal dissipation within a porous (also called "fluffy") core may be critical to explaining such high heat flows (Roberts 2015; Choblet et al. 2017; Rovira-Navarro et al. 2022). Hence, both the assembly and thermal evolution of Enceladus must be compatible with a core that has significant pore space within the rock, to generate the observed dissipation while matching the low core density.

5.3 Tethys

5.3.1 Overview of Surface Geology

Tethys' surface is heavily cratered (Fig. 8), including the large (D = 445 km) impact basin, Odysseus, which displays a central depression along with interior mounds and other topographic features (Schenk et al. 2018). Crater counts within Odysseus suggest that it is ~4–9 times younger than the heavily cratered terrains on Tethys (Kirchoff and Schenk 2010). Crater SFDs for the most heavily cratered regions of Tethys are similar to those on Dione and

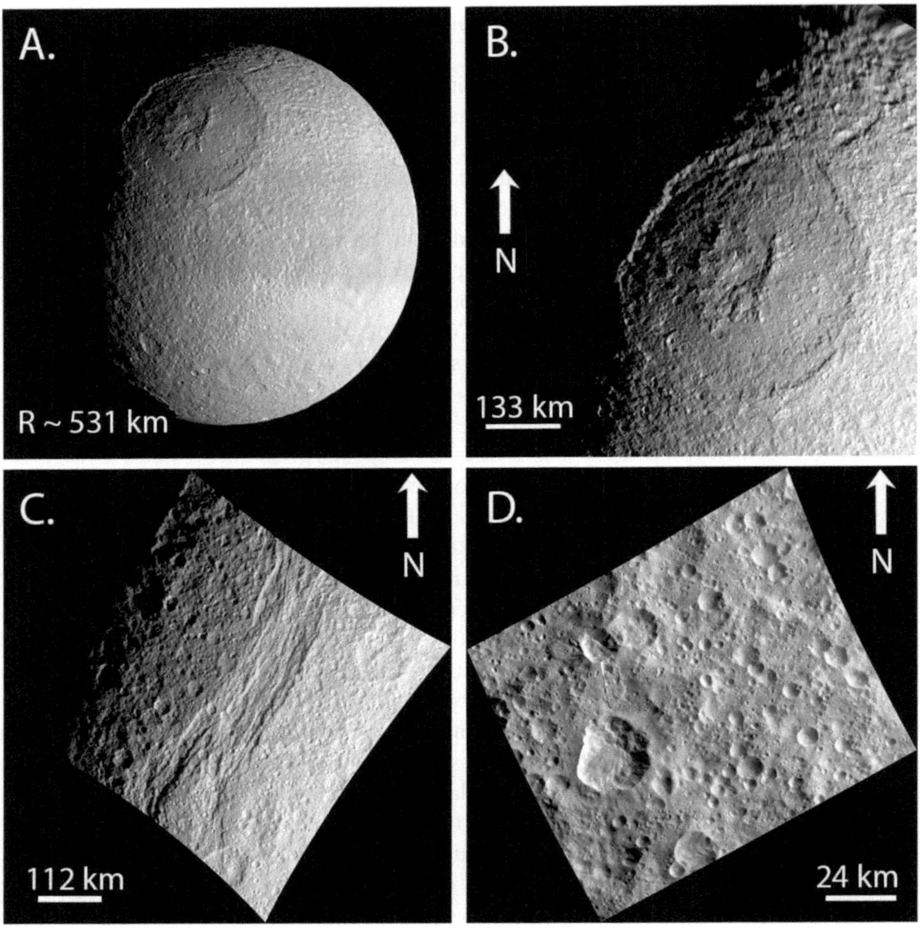

Fig. 8 A) Full disk image of the leading hemisphere of Tethys, with a focus on the Odysseus impact basin (D = 445 km). Image PIA08400. B) Closer view of the Odysseus basin and its cratered interior in which the central pit of the basin is clearly visible. Image N1567098978_1. C) Portion of the Ithaca Chasma canyon system, which stretches ~1800 km across the surface of Tethys. Image N1489061272_1. D) Close up of cratered terrain on Tethys, including some instances of mass wasting. Image N1713137226_1

Mimas, but Tethys shows less variation in the slope of the crater SFD (i.e., flatter R-values) particularly between roughly 6–20 km in diameter, lower crater density from roughly 7 to 100 km diameters (i.e., lower R-values), and the strong dip in crater density occurs at D ~80–110 km rather than at D ~100–120 km as observed on Dione (Kirchoff and Schenk 2010; Kirchoff et al. 2018); these trends are apparent in Fig. 2.

While there are some variations in crater density across different regions of Tethys, there are no regions in which cratering is sparse (Kirchoff and Schenk 2010; Kirchoff et al. 2018; Ferguson et al. 2020). For example, there is a ~600 km wide region on Tethys' leading hemisphere that displays a dearth of craters with D > 50 km but is still densely cratered at smaller sizes (Ferguson et al. 2020) and is relatively smooth as compared with the very rugged terrain that is observed across most of Tethys (Moore and Ahern 1983; Schenk et al. 2018). Whether this region was affected by the formation of Odysseus (e.g., Moore et al.

2004; Bruesch and Asphaug 2004), which is close to antipodal, or by an ancient cryovolcanic flow (Moore and Ahern 1983), is still undetermined. There are no other features thus far associated with cryovolcanism on Tethys, and a search for plume activity at Tethys with the *Cassini* Ultraviolet Imaging Spectrograph and Visible and Infrared Mapping Spectrometer yielded negative results (Buratti et al. 2018).

Based on regional crater maps of Tethys, the Case B production function (Zahnle et al. 2003; see also Sect. 3.1) deviates more from the SFDs of craters larger than D = 4–8 km (Ferguson et al. 2020) than at Dione (Ferguson et al. 2022a) or Mimas (Ferguson et al. 2024). For smaller craters, Case A provides a better fit to the data than Case B in most regions, but it then vastly overpredicts the number of craters with $D > 4$ km (Ferguson et al. 2020). The larger mismatch between Case B and craters on Tethys suggests differences between the planetocentric impactor population at Tethys as compared with Dione and Mimas (e.g., Ferguson et al. 2024).

As discussed in Sect. 3.3, Tethys also has a substantial population of elliptical craters, the majority of which lie within 30° of the equator and are oriented with their long axes pointed east-west (Ferguson et al. 2022b). There is a smaller population of elliptical craters with isotropic orientations that are more globally distributed. Elliptical craters are more likely to form in slow, oblique impacts (e.g., Collins et al. 2011; Elbeshausen et al. 2013), suggesting a planetocentric origin due to their significantly lower impact speeds (Alvarellos et al. 2017). Polygonal craters have also been identified on Tethys (Ferguson et al. 2020); this crater morphology is thought to indicate interactions with fractures, which may not be visible at the surface (Beddingfield et al. 2016).

Tethys displays a global-scale (~1800 km) rift zone called Ithaca Chasma, which roughly follows a great circle that runs through the sub-Saturn point (Moore and Ahern 1983; Moore et al. 2004). There are several hypotheses for the formation of Ithaca Chasma, including a response to the Odysseus-forming impact (Moore et al. 2004), tidal stresses from orbital recession after an eccentricity-pumping resonance passage (e.g., Hussmann et al. 2019), and at least partial freezing of a subsurface ocean (e.g., Chen and Nimmo 2008). One reason the impact origin has been strongly considered is the apparent geographical correlation between Odysseus and Ithaca Chasma. Specifically, the canyon system roughly follows a great circle that is concentric to Odysseus (Smith et al. 1981; Schenk et al. 2018). Updated crater counts suggest that Ithaca Chasma predates Odysseus, which is the main argument against the impact origin, but there is enough uncertainty in the ages that this hypothesis cannot be fully ruled out (Kirchoff and Schenk 2010; Stephan et al. 2016; Ferguson et al. 2020, 2022b, 2022c).

Complicating the origin of Ithaca Chasma is the source of stress that fractured the ice shell. One option is orbital recession of Tethys, which can lead to a collapse of the tidal bulge, generating large stresses (e.g., Hussmann et al. 2019). However, the orientation of Ithaca Chasma is inconsistent with that stress pattern, leading Hussmann et al. (2019) to suggest that additional stresses were involved in the canyon's formation. Other sources of tidal stresses have not been fully vetted against the characteristics of Ithaca Chasma. Tidal stresses from eccentricity or obliquity, alone, may be too small to overcome the tensile strength of ice (see discussion in Collins et al. 2009, for example); tidal stresses induced from rotations of the ice shell (e.g., non-synchronous rotation or polar wander) can be much larger, but these stresses have not yet been investigated as sources for Ithaca Chasma's formation. It is also possible that freezing of a subsurface ocean facilitated fracture formation, as it induces large tensile stresses in the ice shell (e.g., Nimmo 2004) and may act in concert with other stress sources to generate tectonic patterns (as suggested by Hussmann et al. 2019). Models that invoke tidal stress from orbital recession and/or stresses from a freezing ocean also require passage through a mean motion resonance that pumped Tethys' eccentricity (e.g., Chen and Nimmo 2008; Hussmann et al. 2019).

In addition to Ithaca Chasma, Tethys displays somewhat sparse, globally-distributed tectonic features, generally in the form of troughs, grooves, or pit chains (Moore and Ahern 1983; Schenk et al. 2018; Ferguson et al. 2020). In general, Tethys' tectonic features have been interpreted as extensional in origin, although most features are not well-explained by any particular formation model (e.g., Schenk et al. 2018). One region near Odysseus, in which higher-resolution *Cassini* imagery allowed for detailed mapping, showed a higher density of grooves and significantly more small craters ($D = 1$–4 km) than were identified in similar image data sets in other regions (Ferguson et al. 2020), perhaps suggesting a causal link between impacts and grooves in the region.

Craters on Tethys record variable heat flows, ranging from a few mW/m^2 to ~ 100 mW/m^2 (White et al. 2017), which suggests either spatial or temporal changes in heat flow. Shallow angles of normal faults also indicate extensive viscous relaxation of the fault scarps, implying high heat flows (Beddingfield et al. 2015). Analysis of the flexural uplift of the margin of the Ithaca Chasma tectonic structure (Fig. 8) indicates 18–30 mW/m^2 along the northern limb of the canyon system (Giese et al. 2007) and 12–39 mW/m^2 across the south limb (Beddingfield et al. 2023), suggesting that large regions of the satellite experienced similar heat fluxes. However, measurements for Ithaca Chasma's heat flux are lower than what is expected for the nearby Telemus basin (Beddingfield et al. 2023). Ithaca Chasma is thought to be older than other regions of Tethys (e.g., Giese et al. 2007), whereas relaxed craters may have a variety of ages, so one interpretation of the heat flow results is that Tethys has undergone multiple epochs of high heat flow (Beddingfield et al. 2023).

5.3.2 Interior Structure and Evolution

Cassini measurements provided few constraints on the present-day interior structure of Tethys. The bulk density of 0.984 g/cm^3 limits the amount of rock present, even if there is substantial porosity in Tethys' outer layer. The rock fraction is about 7 wt.%, although we do not yet know whether the rock is globally distributed or concentrated in a core (Castillo-Rogez et al. 2018, and references therein).

Given Tethys' low rock fraction and relatively small size ($r = 531$ km), it is unlikely that radiogenic or primordial accretionary heating contributed to the high heat flows preserved in Tethys' geologic record, leaving tidal heating as the most likely candidate (e.g., Castillo-Rogez et al. 2018; Tian and Nimmo 2020). Presently, tidal effects from orbital eccentricity are negligible due to Tethys' nearly circular orbit. However, the eccentricity may have been higher in the past (see below). In addition, if Tethys' spin pole is tilted, which may result from its inclined orbit, obliquity tides may be able to generate substantial tidal heating and stress (e.g., Rhoden et al. 2010; Chen et al. 2014). The extent of tidal heating will depend on whether the interior of Tethys is readily deformable. For example, tidal heating from obliquity-raised tides in the ocean, which could be significant on Tethys (Chen et al. 2014), would not be available if Tethys has remained frozen.

Past tidal heating within Tethys has been inferred from the moon's global shape, under the simplifying assumption that ice shell thickness variations mimic the tidal heating pattern (Gyalay and Nimmo 2023). The best fits to the global shape support the idea that obliquity tides have played a key role in shaping Tethys. However, tidal heating in the best fit model would produce less than 2 mW/m^2, which is far less heat than is inferred from the other geology (e.g., White et al. 2017), and less than would be required to create the differentiated (but presently frozen) interior structure used in the fits. Therefore, the model still relies upon a past high eccentricity that generated high heat flows, with the shape emplaced later, after the orbit circularized but the obliquity was high (Gyalay and Nimmo 2023).

Thermal-orbital evolution models of the five mid-sized moons, for a variety of assumptions about Saturn's Q, revealed that Tethys' history of mean motion resonance passages could have generated an ocean that persists to the present day (Neveu and Rhoden 2019). Whether the simulations assume a primordial origin in the CPD or a late, layered origin within the rings, the ocean within Tethys could have been 100s of km thick in the past, with a much thicker ice shell today (Neveu and Rhoden 2019). The models also reproduce Tethys' nearly circular present-day orbit. Circularization results from tidal dissipation caused by an initially high eccentricity, and the decrease in eccentricity leads to rapid thickening of the ice shell (Neveu and Rhoden 2019). Obliquity was not included in the model. Additional studies of past mean motions resonances between Tethys and other Saturnian moons (e.g., Chen and Nimmo 2008; Hussmann et al. 2019) can also produce enhanced tidal dissipation that could produce the high heat flows inferred from Tethys' geologic features (e.g., Giese et al. 2007; Beddingfield et al. 2015; White et al. 2017), although ocean evolution within Tethys was not tracked. More work is needed to quantify the combinations of eccentricity, obliquity, and interior structure that can produce 10s of mW/m^2 of heat within Tethys.

Partial or complete freezing of an interior ocean is a predicted outcome of the thermal-orbital evolution models (Neveu and Rhoden 2019) and generates extensional stresses in the upper half of the overlying ice shell (e.g., Nimmo 2004). Within the Saturn system, radially-propagating fractures that form as a result of ocean freezing have been modeled for Enceladus and Mimas (Rudolph et al. 2022; Rhoden et al. 2024). Tethys' surface gravity is larger than that on either moon, so we can infer that fractures formed from ocean freezing will be less able to fully penetrate ice shells of comparable thickness (e.g., $\sim$10 km on Enceladus and up to $\sim$50 km on Mimas; Rudolph et al. 2022; Rhoden et al. 2024). The limited evidence of flows or eruptions on Tethys suggests that the hypothesized ocean remained thin enough that fractures could not fully penetrate the thick ice shell and tap the ocean. However, if the smooth region on Tethys leading hemisphere is shown to be the result of cryovolcanic flows (e.g., Moore and Ahern 1983), it may argue for at least one fully penetrating fracture in Tethys' history. A study of the conditions under which freezing of a subsurface ocean within Tethys could have formed deep fractures, enabling the formation of Ithaca Chasma, while avoiding widespread eruptions, may help constrain the range of thermal histories that can account for Tethys' geology and the inferred heat flows through time.

Impact formation models have shown that crater size and morphology are sensitive to the structure and thermal profile of an ice shell, potentially even revealing the presence of a subsurface ocean (e.g.,Turtle and Pierazzo 2001; Senft and Stewart 2011; Bray et al. 2014; Silber and Johnson 2017; Denton et al. 2021; Denton and Rhoden 2022; Bjonnes et al. 2022). A detailed study of the Odysseus-forming impact, with varying ice shell thicknesses (including a frozen Tethys) may provide additional constraints on the timing of the impact with respect to the evolution of Tethys' interior and orbit. Moreover, the formation and subsequent relaxation of Odysseus may help constrain whether and when Tethys had a subsurface ocean and at what depth, as was recently done for Herschel on Mimas (Denton and Rhoden 2022).

5.3.3 Summary of Geologic Constraints

Hypotheses regarding the formation and evolution of Tethys need to satisfy the following geologic constraints. First, as with Mimas, there must be adequate time and impactors to generate Tethys' considerable cratering record, including formation of the Odysseus impact basin. The age of Odysseus is not well-constrained, in part due to the uncertainties in impactor source populations described in Sect. 3; estimates range from 200 Myr to 3.9 Gyr (Giese et al. 2007; Dones et al. 2009; Kirchoff and Schenk 2010; Kirchoff et al. 2018). Tethys also has a substantial elliptical crater population, with a curious spatial distribution that requires an impactor source that is not heliocentric (see Sect. 3; Ferguson et al. 2022b). Hence, Tethys' bombardment history must account for the formation and dispersal of this additional population of planetocentric impactors.

Second, Tethys must have experienced at least one epoch of high heat flow, which can most easily be produced by tidal heating after an eccentricity-pumping resonance passage, in order to account for the extensive crater relaxation (White et al. 2017) and the heat flow implied by the topography of Ithaca Chasma (Giese et al. 2007; Beddingfield et al. 2015). However, the interior structure of Tethys during that time period will have a major control on the extent of tidal dissipation and heating. Coupled thermal-orbital evolution models suggest that Tethys could have a long-lived subsurface ocean (e.g., Neveu and Rhoden 2019), which may provide a heat source for crater relaxation via eccentricity and/or obliquity tidal heating (e.g., Chen et al. 2014), although the link between tidal heating and Tethys' varying states of crater relaxation has not been fully investigated. Whether these heat sources are simultaneously compatible with Tethys' global shape (e.g., Gyalay and Nimmo 2023) and other geologic features is still unknown.

Third, a mechanism is needed for the formation of Ithaca Chasma that can satisfy the following constraints: the elastic thickness of the ice shell was 5–7 km at the time of formation (Giese et al. 2007), there was a source of stress high enough to induce tensile failure in the ice such as freezing of a subsurface ocean (e.g., Chen and Nimmo 2008), and there was a heat source sufficient to explain the associated relaxation and shallow fault angles (Giese et al. 2007; Beddingfield et al. 2015). A mechanism that can also explain the location and orientation of Ithaca Chasma would be particularly compelling (e.g., discussion in Schenk et al. 2018).

Finally, Tethys has a substantially lower density than the other mid-sized moons, particularly its neighbors, Enceladus (interior) and Dione (exterior). None of the current formation models provide a compelling explanation for Tethys' very low density; CPD formation would imply a density gradient that is not observed (e.g., Canup and Ward 2006, 2009; Mosqueira et al. 2010), whereas ring-based formation (e.g., Charnoz et al. 2011) and early mergers (Asphaug and Reufer 2013) appeal to stochastic effects to reproduce the variable rock contents of the moons. Reassembly of a precursor moon would imply that either more rock than ice was lost in the collision or that the low rock fraction predated the collision, which would still require explanation. In the former case, Tethys' density could provide a key constraint on the conditions of the impact that led to its final assembly.

5.4 Dione

5.4.1 Overview of Surface Geology

After Enceladus, Dione is the most heavily tectonized of the mid-sized moons (Fig. 9; Schenk et al. 2018 and references therein). The majority of Dione's tectonism is preserved in the heavily dissected "wispy" terrains concentrated within the trailing hemisphere (Wagner et al. 2006; Moore and Schenk 2007). With few overprinted impact craters, the fractures within the wispy terrains are likely some of the most recent geologic structures preserved on Dione's surface (Kirchoff and Schenk 2015). Older tectonic activity on Dione includes wide troughs that may be graben and at least one large ridge, Janiculum Dorsa, on the leading

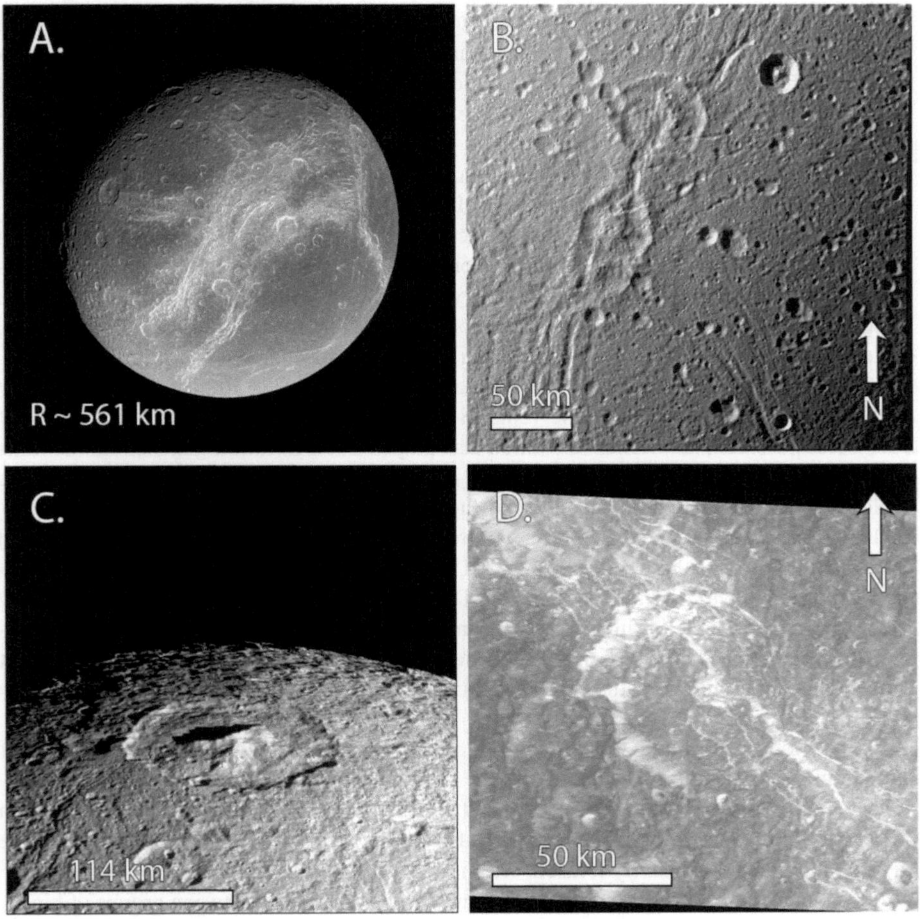

Fig. 9 A) Full disc view of Dione's trailing hemisphere showcasing the wispy terrain (PIA08526). B) Cratered terrain on Dione's leading hemisphere. Pictured off-center are the heavily degraded and modified craters of Murranus (56.8 km) and Metiscus (43.8 km). Image N1820417749_1. C) Oblique view of Erulus crater (120 km), which is thought to be heavily relaxed (White et al. 2017), with a focus on the central peak complex. Image N1665975031_1. D) Amastrus crater (62.4 km) located within the wispy terrain on Dione's trailing hemisphere, illustrating the combined effects of cratering and tectonic resurfacing on Dione. Image N1507743880_2

hemisphere (Hammond et al. 2013). Inferences of early tectonism are supported by numerous polygonal craters found across terrains on Dione, both in and outside the wispy terrain (Beddingfield et al. 2016), implying that widespread fracturing has been occurring throughout Dione's history. The relative ages of these periods of tectonism are supported by crater densities, where the terrains around Janiculum have a greater crater density than those within the wispy terrains (Kirchoff and Schenk 2015).

Resurfacing obscures older tectonic features, making it difficult to determine their original distribution and, thus, test likely dominant stress mechanisms. However, the graben and normal fault scarps within the wispy terrains show good matches with diurnal tidal and non-synchronous rotation stress fields (Collins et al. 2010; Martin et al. 2016). The magnitude of stresses from either of these sources would be amplified in the presence of a global subsurface ocean. In fact, an ocean is necessary to enable NSR by mechanically decoupling the ice shell from the interior (e.g., Helfenstein and Parmentier 1985). Although this first order result requires further analysis, the distribution and orientation of fractures within Dione's wispy terrains suggest a global ocean at some point in Dione's evolution. In total, the extensional features on Dione are consistent with $\sim$1% surface expansion (Collins et al. 2010), such as from partial freezing of a subsurface ocean, although fracture formation from ocean freezing has not been explicitly modeled for Dione.

In addition to Dione's extensive tectonic activity, evidence for cryovolcanism is preserved in the geologic record. Smooth plains that dominate the trailing hemisphere have been suggested to have formed by some amount of cryovolcanic resurfacing (e.g. Smith et al. 1981; Plescia and Boyce 1982; Plescia 1983; Moore 1984; Morrison et al. 1984). Furthermore, centered within the smooth terrains are Murranus and Metiscus craters within Fidena Fossae (Fig. 9B). The morphology of these craters is not indicative of impact craters, and they are hypothesized to be cryovolcanic vents (Schenk and Moore 2009; Kirchoff and Schenk 2015). It is possible that viscous relaxation also played a role in forming the smooth terrains (Hammond et al. 2013). Overall, more work is needed to further understand the endogenic geologic processes that have shaped the surface of Dione.

The *Cassini* mission revealed plasma streams originating separately from Dione and Tethys (Burch et al. 2007; Buratti et al. 2011), as well as the possibility of a tenuous atmosphere around Dione (Clark et al. 2008). These results suggest some amount of on-going endogenic activity, which could be cryovolcanic. Despite detailed searches for active plumes over the course of the *Cassini* mission (Buratti et al. 2011, 2018), none were detected at Dione. The possibility remains that Dione is still geologically active, but future mission work will be required to solve this mystery.

Dione's long-wavelength topography (Nimmo et al. 2011), along with the topography of Janiculum Dorsa (Hammond et al. 2013), the shallow dip angles of normal faults (Beddingfield et al. 2015), and Dione's relaxed craters (White et al. 2017) all indicate a history of high heat flows. Craters record variable heat flows from only a few mW/m^2 to $\sim$100 mW/m^2. The large Evander basin is heavily relaxed, with models suggesting heat flows over 60 mW/m^2 (White et al. 2017). A rough geographical boundary can be drawn between groups of highly relaxed and minimally relaxed craters (White et al. 2017). As shown in Fig. 10, the pattern correlates with the spatial variations in tidal heating from eccentricity, although the present-day eccentricity (0.0022), along with Dione's large orbital distance, is unlikely to generate high heat flows at present. Furthermore, there are only a handful of highly relaxed craters on Dione, so the pattern may not be significant (White et al. 2017).

5.4.2 Interior Structure and Evolution

Dione's average density is about 1.478 g/cm^3, which corresponds to an ice fraction of about 48 wt.% or 75 vol.% assuming no porosity (Castillo-Rogez et al. 2018). Gravity data from multiple *Cassini* flybys indicate a J_2/C_{22} of 4.102 ± 0.044, which is far from the hydrostatic value (Zannoni et al. 2020). Dione's shape also deviates from hydrostatic, with a $-H_{20}/H_{22}$ of 4.9 ± 0.4 (Zannoni et al. 2020). The fact that the shape deviates more than the gravity suggests partial compensation of excess topography (e.g., Hemingway et al. 2016, 2018; Zannoni et al. 2020). Studies combining gravity, shape, compensation models, and plausible interior structures suggest a differentiated interior with a present-day ocean, likely buried under $\sim$100 km of ice, which provides some compensation of topography (Beuthe et al. 2016; Hemingway et al. 2016; Zannoni et al. 2020). However, more data is needed before this interpretation can be considered robust (e.g., Castillo-Rogez et al. 2018; Zannoni et al. 2020).

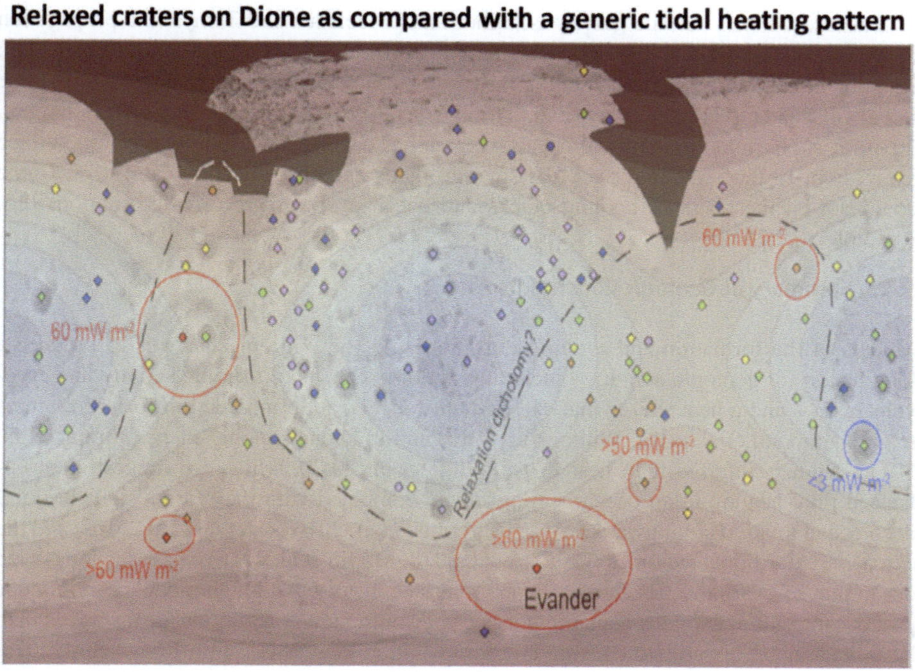

Fig. 10 Dione's craters record variable rates of relaxation, which could mean they are different ages or that the heat flow that led to relaxation was spatially-variable. Here, we overlay the relaxation results from White et al. (2017) on a generic contour map of eccentricity-driven tidal heating (similar to the one shown in Roberts 2015). Warm colors represent higher heat flows relative to the cooler colors. There is good correlation between areas of higher relaxation (and inferred heat flow) and the regions of higher tidal heating, but the fit is not perfect. Generating these high heat flows from tides alone is challenging without both a subsurface ocean and a higher past eccentricity, owing to Dione's distance from Saturn. It is also possible that the initial depths assumed for the craters on Dione are too large, leading to an overestimate of the past heat flow (e.g., Sect. 3.4)

Thermal-orbital evolution models do not predict a present or recent ocean within Dione (Neveu and Rhoden 2019). Formation from the rings seems to generate a longer-lived past ocean than primordial formation (e.g., Neveu and Rhoden 2019). It is possible that the models have neglected an important source of heat, such as core dissipation, an additional unidentified resonance passage, or fluctuations in heating caused by a change in dissipation within Saturn.

Despite the challenges of producing an ocean within Dione, interpretations of its geology generally rely upon an ocean's existence to enhance stresses and/or heat flows (Collins et al. 2010; Martin et al. 2016; White et al. 2017). For example, a past epoch of higher eccentricity has been invoked to explain the high heat flows recorded by relaxation of Dione's craters (e.g., White et al. 2017). In addition, Dione displays globally-distributed extensional tectonic features attributed to ocean freezing (Collins et al. 2010), while also showing evidence of cryovolcanic flows (e.g. Smith et al. 1981; Plescia and Boyce 1982; Plescia 1983; Moore 1984; Morrison et al. 1984). Ocean freezing has also been suggested as a key mechanism in creating Enceladus' south polar tectonic features and enabling eruptions (Hemingway et al. 2020; Rudolph et al. 2022) and for facilitating the creation of the Ithaca Chasma canyon system on Tethys (Moore and Ahern 1983; Schenk et al. 2018). These observations lead to the question of why the same geologic process (i.e., ocean freezing) manifests so differently across the moons.

Because Dione's surface gravity is higher than Enceladus and Tethys, the propagation depth of cooling cracks will be limited in comparison (e.g., Rudolph and Manga 2009; Rudolph et al. 2022), perhaps affecting the resulting tectonic features. In addition, if Dione has always had a relatively thin ocean, buried beneath ∼100 km of ice (e.g., Zannoni et al. 2020), it is possible that minimal ocean freezing has occurred, limiting the overall surface expression. In that case, however, the smooth plains and cryovolcanic features on Dione could not be the result of cooling cracks and ocean overpressure driving eruptions (cf., Rudolph et al. 2022).

5.4.3 Summary of Geologic Constraints

Models of the formation and evolution of Dione must provide mechanisms to produce a long history of geologic activity – including multiple epochs of tectonic activity and cryovolcanism – and a heat source that can account for relaxation recorded in surface features (e.g., Hammond et al. 2013; White et al. 2017) and in Dione's global shape (Nimmo et al. 2011). Dione's inferred high heat flows (White et al. 2017), which are suggestive of enhanced tidal heating in an ocean-bearing moon, combined with the gravity signature of a deep ocean today (Hemingway et al. 2016; Beuthe et al. 2016; Zannoni et al. 2020) and the extensive extensional tectonics (Collins et al. 2010; Martin et al. 2016), suggest a long-lived ocean that has partially frozen over Dione's surface history. Although models can produce past oceans within Dione, a contemporary ocean would require additional heat sources than have been incorporated into existing interior evolution models (Castillo-Rogez et al. 2018; Neveu and Rhoden 2019). Studies of fractures that would arise from ocean freezing may provide some constraints on the extent of freezing and/or the minimum thickness of Dione's ice shell throughout its recorded history.

Finally, Dione's surface displays polygonal, elliptical, and circular craters, including many large craters and many relaxed craters (Beddingfield et al. 2016; White et al. 2017; Ferguson et al. 2020, 2022b). These craters need both time and sufficient source populations to accumulate. As described in Sect. 3.3, Dione's elliptical craters have the same geographical signatures as those on Tethys (Ferguson et al. 2022b), suggesting a common origin or process for their formation.

5.5 Rhea

5.5.1 Overview of Surface Geology

Rhea's surface (Fig. 11) is heavily cratered (e.g., Kirchoff and Schenk 2010; Schenk et al. 2020) and displays large, degraded basins such as Mamaldi (D = 480 km) and Powehiwehi (D = 271 km). These basins are, themselves, heavily cratered, supporting the idea that Rhea's surface is much older than, for example, the cratered terrains of Enceladus (Kirchoff and Schenk 2009). There is some debate in the literature as to whether Rhea's smaller craters record the same impactor population as the other mid-sized moons (e.g., Kirchoff and Schenk 2010 versus Bell 2020) and which craters are most likely to be heliocentric in origin (e.g., Kirchoff and Schenk 2010 versus Hirata 2016). Rhea has a dearth of large (D > 100 km) craters, as compared with Iapetus, leading to the suggestion that Rhea had such high heat flows early in its history that many early craters were relaxed beyond recognition (White et al. 2013).

As summarized by Schenk et al. (2018), tectonic features on Rhea include long (>1000 km), north-south trending canyons and a series of ridges and troughs. Many of these features are thought to be extensional, suggesting a formation mechanism related to a

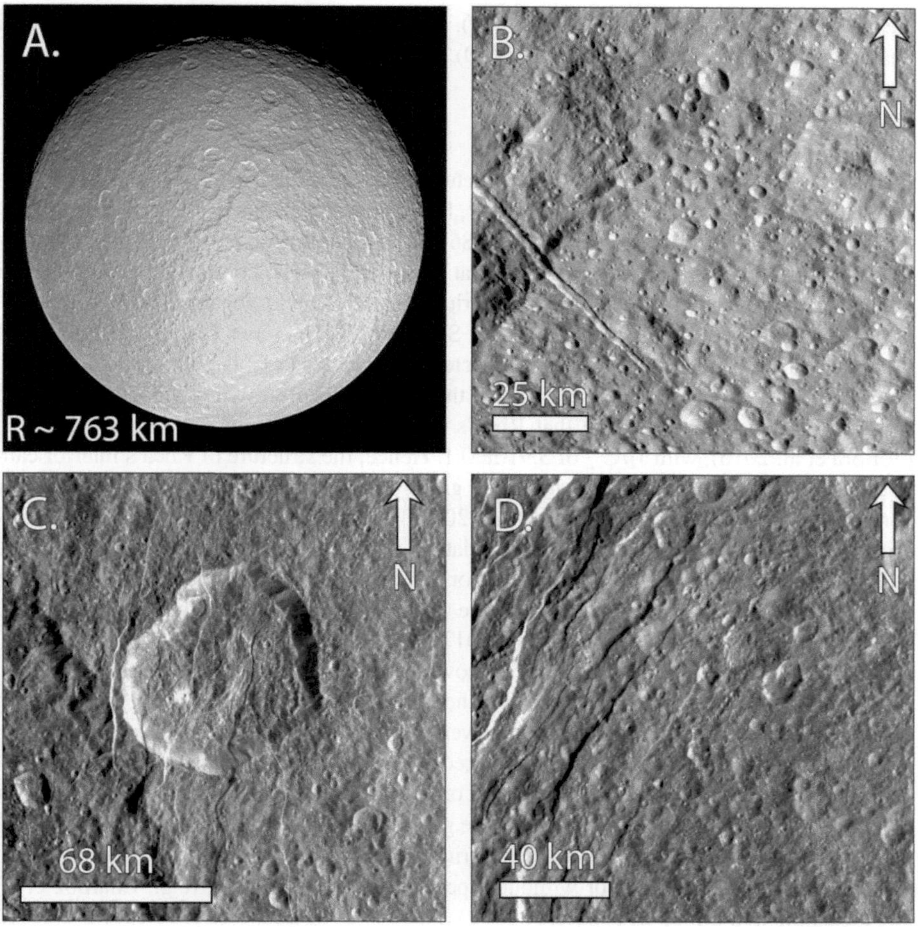

Fig. 11 A) Full disk image of Rhea with the Tirawa basin located near the center. Image PIA07763. B) Cratered terrain with a linear feature. Image N1567132880_1. C) Obatala crater (62.5 km) overprinted by tectonics. Image N1637518901_1. D) Tectonic features alongside cratered terrain. Image N1637519610_1

volume increase within the moon, although an older period of global contraction has also been suggested (e.g., Moore et al. 1985). Rhea's heavily cratered surface obscures some older tectonic features, making it challenging to discern their origins (Schenk et al. 2018 and references therein). However, there are also several craters crosscut by more recent tectonic features, indicating different epochs of tectonic activity (Beddingfield et al. 2015; Aponte-Hernández et al. 2021).

Craters on Rhea with diameters greater than 100 km appear to all record similar amounts of crater relaxation, which require 10s of mW/m^2 of heat (White et al. 2013). Regardless of the magnitude of the heat flow, White et al. (2013) find that the majority of resulting relaxation takes place within ~10 Myr, with little additional change in crater depth even if the crater is heated for 1 Gyr. This result suggests that even a relatively short-lived pulse of enhanced heat flow could substantially relax craters on Rhea, although craters could have been exposed to high heat flows for much longer than 10 Myr without leaving a clear signature. Shallow slopes of normal faults on Rhea are also suggestive of viscous relaxation, implying elevated heat flows (Beddingfield et al. 2015), and Rhea's global shape also indicates higher heat flows than expected (Nimmo et al. 2010).

5.5.2 Interior Structure and Evolution

Even with *Cassini* measurements, the structure of Rhea's interior is still somewhat ambiguous. Rhea' density is about 1.237 g/cm^3, lower than Enceladus and Dione. The density corresponds to an ice fraction of 65 wt.% or 86 vol.%, assuming no porosity (Castillo-Rogez et al. 2018). The single initial gravity pass at Rhea led to some controversy as to whether it supported or excluded hydrostatic equilibrium (Anderson and Schubert, 2007; Iess et al. 2007; Mackenzie et al. 2008; Anderson and Schubert 2010), in part because the motion of the spacecraft must be very precisely characterized in order to isolate the gravitational effects of the single coefficients J_2 and C_{22} of the moon. However, with a second flyby, there was sufficient data to determine that Rhea's gravity field is significantly non-hydrostatic (Tortora et al. 2016), with J_2/C_{22} of 3.91 ± 0.1. Hence, the structure of Rhea's interior cannot be directly constrained. Combining the gravity data, plausible interior structures, and constraints on Rhea's shape (Thomas et al. 2007; Thomas 2010; Nimmo et al. 2011), Tortora et al. (2016) showed that both differentiated and undifferentiated interiors are allowed within the current data set. Overall, they favor a model in which Rhea is differentiated with a low density, modestly non-hydrostatic core - perhaps similar to what has been suggested for Mimas (Tajeddine et al. 2014; Caudal 2017; Noyelles 2017).

Thermal-orbital evolution models have shown that a primordial Rhea, formed in the CPD, could have differentiated and maintained a short-lived ocean early in its history (Hussmann et al. 2006; Kamata and Nimmo 2015; Neveu and Rhoden 2019). The thermal evolution of a ring-born Rhea has not yet been explored as that scenario requires a Q of Saturn that can enable Rhea to migrate out to its current orbital distance within the lifetime of the Solar System. As our understanding of Q evolves (e.g., Lainey et al. 2020), there may be additional pathways for Rhea's formation via Saturn's rings. In that case, Rhea may have formed with a rocky interior and ice exterior without experiencing global melting (e.g., Charnoz et al. 2011; Salmon and Canup 2017).

As with Tethys and Dione, Rhea's geology records higher heat flows than thermal models would have suggested (Passey 1983; Schenk 1989; Moore et al. 2004; Nimmo et al. 2010; White et al. 2013; Aponte-Hernández et al. 2021). The source of Rhea's internal heating remains somewhat mysterious. Radiogenic heating alone is insufficient to explain the high heat flows recorded by craters (White et al. 2013; Kamata and Nimmo 2015). Rhea's potential for tidal heating is limited due to its large distance from Saturn and low eccentricity, and Rhea is not currently in any mean motion resonances. If Rhea is differentiated, its core is likely non-hydrostatic (e.g., Tortora et al. 2016) and low density ($\sim$2.4 g/cm^3), perhaps similar to the "fluffy" core proposed for Enceladus (e.g., Roberts 2015). However, dissipation within the core would still require some kind of tidal forcing, which is again challenging due to the moon's distance from Saturn.

In contrast to geologic inferences that Rhea is old, its orbital inclination is challenging to explain if Rhea is primordial (Ćuk et al. 2016), particularly given its current rapid orbital migration rate (Lainey et al. 2020). Passing through the so-called evection resonance, which is located within Rhea's current orbital distance, should have pumped the moon's inclination beyond what is currently observed, with no clear mechanism to subsequently damp it (Ćuk et al. 2016). The timing of Rhea's passage through the resonance (100s of Myr ago versus billions of years ago) depends sensitively on how long Rhea's orbital migration has been as rapid as observed today (e.g., Ćuk et al. 2016; Teodoro et al. 2023).

To address the issue of Rhea's low inclination, Teodoro et al. (2023) suggested that a precursor moon evolved through the evection resonance, which resulted in its eccentricity and inclination being pumped and leading to a disruptive collision with a Dione-sized moon. Present-day Rhea then assembled from the collisional debris, with a low eccentricity and inclination, outside the evection resonance. Teodoro et al. (2023) assumed the event occurred very recently, such that the debris from the collision could supply material to Saturn's rings. A challenge with this model, and any model of very recent formation or reassembly of Rhea, is how to produce the geologic history preserved in its surface, which implies multiple tectonic episodes, a significant crater population, and sufficiently high heat flows to relax craters and other features. Alternatively, if Rhea (or its precursor moon) passed through the evection resonance earlier in Saturn system history, and then disrupted and reassembled, there would be more time for it to accumulate its geologic record.

5.5.3 Summary of Constraints

Models of Rhea's formation and evolution need to account for the high heat flows recorded by Rhea's craters (White et al. 2013), shallow faults (Beddingfield et al. 2015), and long-wavelength topography (Kamata and Nimmo 2015). As with Tethys and Dione, the presence of extensional tectonics suggests a volume increase, which could be attributed to ocean freezing, although global contraction has not been ruled out either (see discussion in Schenk et al. 2018). In addition, the heat source to generate an ocean, and the likely timing of its existence, is not well constrained. There appear to be multiple epochs of tectonic activity on Rhea (e.g., Beddingfield et al. 2015; Aponte-Hernández et al. 2021), which could indicate long-term cooling, additional drivers of tectonic activity, or variability in heat flow, internal structure, and stress related to a complex orbital history. Rhea's geologic record may provide additional constraints on its evolution, but more work is needed to analyze its surface features and their implications. Furthermore, there is a challenge in reconciling the relatively low inclination of Rhea with its heavily cratered, and presumed ancient, surface (e.g., Ćuk et al. 2016; Teodoro et al. 2023).

Finally, characterizing the population of elliptical craters on Rhea, similar to studies conducted for Mimas, Tethys, and Dione (Ferguson et al. 2022b, 2024), may help constrain the characteristics of planetocentric debris, which could help reduce uncertainty in crater-based ages. The lack of co-orbital moons at Rhea also provides a means of testing the role of co-orbital debris in cratering the mid-sized moons. In addition, an examination of polygonal craters may help identify older tectonics, as has been determined at Dione (Beddingfield et al. 2016).

5.6 Summary of Geologic Constraints Across All Five Mid-Sized Moons

The interior structures of Mimas, Enceladus, and Dione are layered and have possessed internal heating sufficient to generate oceans that likely persist to the present day. At Mimas, observations of its librations and precession can only be explained by a differentiated interior, are most consistent with a subsurface ocean, and also support an elongated core (Tajeddine et al. 2014; Caudal 2019; Noyelles et al. 2019; Lainey et al. 2024). At Enceladus, there is definitive evidence of a global ocean and a low-density core (e.g., Thomas et al. 2016; Hemingway et al. 2018 and reference therein; Hemingway and Mittal 2019; Park et al. 2024), which has been interpreted as a porous core subject to hydrothermal circulation (e.g., Roberts 2015; Choblet et al. 2017; Rovira-Navarro et al. 2022). At Dione, where libration measurements are not available, interpretations of gravity and shape data suggest a deep ocean under a thick ice shell (Hemingway et al. 2016; Beuthe et al. 2016; Zannoni et al. 2020). Hence, models for the formation of Mimas, Enceladus, and Dione, need to account for their layered interiors and provide a pathway to develop their present-day oceans.

The data for Rhea is suggestive of a differentiated moon with some core topography, although a largely undifferentiated interior is also possible (Tortora et al. 2016). Unfortunately, the interior structure of Tethys is not well-constrained by direct measurements (Thomas 2010; Castillo-Rogez et al. 2018). However, as discussed below, there are inferences from geologic features on both Tethys and Rhea that suggest they once possessed oceans, implying compositional layering within their interiors.

There is evidence of either sustained or recurring epochs of tectonic activity on Enceladus (e.g., Patthoff and Kattenhorn 2011; Crow-Willard and Pappalardo 2015), Dione (e.g., Collins et al. 2010; Hammond et al. 2013; Beddingfield et al. 2016; Martin et al. 2016), and Rhea (e.g., Beddingfield et al. 2015; Aponte-Hernández et al. 2021). Diurnal tidal stresses have been linked to the formation of young fractures on Enceladus (e.g., Rhoden et al. 2020), while non-synchronous rotation (NSR) stresses have been suggested to explain older fracture sets on Enceladus (Patthoff and Kattenhorn 2011) and features on Dione (Collins et al. 2010; Martin et al. 2016). NSR would require a subsurface ocean to decouple the ice shell from the rocky interior (e.g., Helfenstein and Parmentier 1985). The extent to which diurnal tidal stress or stress from NSR has produced tectonic activity on Tethys or Rhea is not well-known.

Freezing of subsurface oceans has been invoked to facilitate the formation of extensional tectonic features observed on Enceladus, Tethys, Dione, and Rhea (e.g., Moore et al. 1985, 2004; Collins et al. 2010; Hemingway et al. 2020). Hence, these features indicate a heat source capable of generating an ocean, particularly if tidal stresses are involved in their formation, and subsequent reduction of the total heat budget in order to transition to cooling and freezing. Tectonic activity is minimal at Mimas, which is consistent with an emerging ocean or a frozen interior (Rhoden et al. 2017, 2024). Formation models for these moons must provide sufficient time, internal heating, and stress to drive thermal evolution of their interiors and the associated tectonic activity preserved at the surface.

Enceladus, Tethys, Dione, and Rhea all display geologic evidence of high heat flows (Spencer et al. 2006; Chen and Nimmo 2008; White et al. 2013; Hammond et al. 2013; White et al. 2017; Beddingfield et al. 2023; Gyalay and Nimmo 2023). There are comparatively fewer constraints on Mimas' surface heat flow. The $\sim$10 craters so far examined show no clear signature of relaxation (Schenk 1989; White and Schenk 2011; Schenk et al. 2018), but Mimas' low gravity leads to less topographic change for a given heat flow than other mid-sized moons, making Mimas' craters less diagnostic than craters elsewhere (Rhoden et al. 2024; cf., Bland et al. 2012 for Enceladus).

Achieving high heat flows from tidal dissipation within these moons implies the presence of subsurface oceans, warm ice that responds to tides, and/or highly dissipative cores. In addition, past epochs of higher eccentricity (or obliquity, e.g., Chen et al. 2014) may be required in order to generate high enough tidal dissipation to match observations. This is particularly true for Tethys, which currently has negligible eccentricity. Orbital evolution models have shown that eccentricity-pumping resonance crossings can be produced under a wide variety of conditions, including high, low, or variable Q of Saturn and in cases assuming that the moons formed contemporaneously with Saturn or were spawned from the rings within the last billion years (e.g., Ćuk et al. 2016; Neveu and Rhoden 2019; Nakajima et al. 2019; Tian and Nimmo 2020; Nakajima et al. 2019; Ćuk and El Moutamid 2022).

Dissipation within Enceladus appears to be dominated by tidal heating in its core (Roberts 2015; Choblet et al. 2017; Rovira-Navarro et al. 2022). If the other mid-sized moons also have dissipative cores, it could affect their heat budgets, perhaps providing a source for the heat that relaxed craters on Tethys, Dione, and Rhea (White et al. 2013, 2017). However, the orbits of the moons must also be conducive to tidal forcing. The enhanced core dissipation comes from the assumption that the rocky core is highly porous, and that the pores are filled with ice (Roberts 2015) or liquid water that flows through the core (Rovira-Navarro et al. 2022). Preserving ice and/or porosity in the core may be a discriminator in testing models of the formation and thermal evolutions of the moons. Similarly, the cores of Mimas and Rhea are unlikely to be hydrostatic (Tajeddine et al. 2014; Caudal 2017; Tortora et al. 2016), which may help constrain their formation.

All of the mid-sized moons record extensive bombardment histories - even Enceladus, outside of the south polar terrain (e.g., Kirchoff and Schenk 2009, 2010). The crater SFDs indicate that planetocentric material played an important role in cratering Mimas, Enceladus, Tethys, and Dione, although the crater populations vary even across these moons (Kirchoff et al. 2018; Bell 2020; see also, Sect. 3). Although some studies suggest that all five mid-sized moons have been subjected to the same population of impactors (e.g., Bell 2020), differences in Rhea's crater population have been interpreted as a clearer signature of heliocentric impactors relative to the other moons (e.g., Kirchoff and Schenk 2010; Hirata 2016).

The elliptical crater populations on Mimas, Tethys, and Dione display some key similarities, such as clustering of east-west oriented craters within 30° of the equator of each moon (Ferguson et al. 2022b, 2024). Given the present-day orbital inclinations of the moons, the similar location and extent of the elliptical craters suggests either individual impactor populations with similar dynamics or a common impactor population that predated the increased inclinations of the moons (e.g., Ferguson et al. 2022b). The observed elliptical crater populations provide constraints on the distribution of planetocentric material within the Saturn system.

In addition to the geophysical properties of the moons, their relationships with other components of the Saturn system may provide important constraints. Tethys and Dione each have co-orbital moons within their Lagrange points. Analyses of stability regions at Saturn suggest that Rhea, if it ever had co-orbitals, would have lost them upon crossing the evection resonance (Giuppone et al. 2018), while overlapping resonances may reduce stability of Lagrange points at Mimas and Enceladus (Christou et al. 2007). Hence, it is possible that more of the mid-sized moons had co-orbitals in the past. Although it may be tempting to dismiss these small moons, accounting for their existence can provide insight as to the formation and evolution of the mid-sized moons. Similarly, the on-going evolution of the Cassini Division may constrain the recent orbital changes at Mimas (Baillié et al. 2019; Noyelles et al. 2019), providing a potential pathway to form a young ocean within the moon (Lainey et al. 2024; Rhoden et al. 2024). And finally, characterizing the nature of Saturn's interior and how it has evolved through time would enable more robust models of Saturn's Q, which controls the orbital expansion of the mid-sized moons, resonances passages, and the orbital properties that control tides. Narrowing down the possible thermal-orbital pathways for the moons can help pinpoint where additional mechanisms are needed to explain their geophysical characteristics.

6 Synthesis and Conclusions

With the geologic constraints in mind (i.e., Sects. 3, 4, and 5), we now connect them – to the extent possible – to the proposed formation mechanisms for the mid-sized moons (Sect. 2). Specifically, we consider the mass gradient of the moons, their global shapes and interior structures, inferences as to their thermal histories, their craters, and clues as to the dynamical evolution of the Saturn system.

Mass gradient. Although accretion within Saturn's CPD (Canup and Ward 2006, 2009; Mosqueira et al. 2010) or formation via mergers of larger planetesimals (Asphaug and Reufer 2013) cannot be ruled out, these models do not provide a clear pathway to explain the mass gradient of the moons. In contrast, ring-based formation of Saturn's mid-sized moons can explain the mass, rather than density, variation of the moons with distance from Saturn (e.g., Crida and Charnoz 2012). If the moons formed from rings billions of years ago, they would either need to survive disruptive impacts or reassemble in a manner that preserves the mass gradient we currently observe. If the rings formed during a heavy bombardment period, such as by tidal disruption of a passing Centaur (e.g., Dones 1991; Hyodo and Charnoz 2017), the moons would emerge later in Solar System history and the likelihood of avoiding disruptive collisions with heliocentric impactors increases.

If any or all of the mid-sized moons emerged from the rings, their formation age is tied to the ring age. While the mass of the present-day rings suggests it has evolved over billions of years (Salmon et al. 2010), other observations imply an age on the order of 100 Myr (e.g., Zhang et al. 2017; Iess et al. 2019). If confirmed, such a young age for the rings would all but rule out direct assembly of the present-day moons from the present-day rings. In that case, we suggest that the Saturn system may have experienced cycles of ring formation, moon emergence, and ring replenishment via moon disruption, which could help explain the signature of ancient rings within the current ring mass (Salmon et al. 2010; see also Sect. 2).

Global shape and interior structure. Where *Cassini* data was sufficient to assess the interiors of the mid-sized moons, all of them show deviations from hydrostatic equilibrium in global shape, core shape, and/or gravity (e.g., Tajeddine et al. 2014; Thomas et al. 2016; Tortora et al. 2016; Zannoni et al. 2020), which may be an indication as to the conditions in which they formed. Moons that formed in the CPD and differentiated due to internal heating may produce a different interior structure than reassembly after mergers or ring-based formation models. For example, ring-based models may produce layered moons without invoking global internal melting and hydrothermal alteration (e.g., Neveu and Rhoden 2019), which may result in different shapes and interior structures than are typically assumed for differentiated moons (e.g., Sect. 4). The formation models also differ in the source of rock within the moons and how it would evolve into a core. Even the ring-based formation models have multiple approaches to introducing rocky material (e.g., Charnoz et al. 2011 versus Salmon and Canup 2010). Studies that track the thermal and physical evolution of the cores or/or global shapes of the moons across different moon formation models would be highly valuable. In addition, testing whether non-hydrostatic cores can be retained through disruptive collisions could help determine whether the moons have avoided such collisions.

Inferred thermal histories. There is evidence that Mimas, Enceladus, and Dione possess present-day oceans (Hemingway and Mittal 2019; Tajeddine et al. 2014; Thomas et al. 2016; Zannoni et al. 2020; Lainey et al. 2024). In addition, Enceladus, Tethys, Dione, and Rhea all display relaxed craters that indicate epochs of high heat flows (Bland et al. 2012; White et al. 2013, 2017). Freezing of ocean material has been suggested as a mechanism to drive tectonic activity on Enceladus (e.g., Hemingway et al. 2020; Rudolph et al. 2022), may have contributed to the formation of Ithaca Chasma on Tethys (Moore and Ahern 1983), and has been suggested as a source of extensional tectonics on Dione (Collins et al. 2010). At Rhea, there is no unambiguous evidence of a differentiated interior (e.g., Tortora et al. 2016), although its extensional tectonic features and highly relaxed craters have been linked to a past ocean, which would imply compositional layering in its interior (e.g., White et al. 2013).

If all of the mid-sized moons have/had oceans, it implies that the moons either underwent global differentiation or assembled layered (e.g., from rings). Oceans may form as part of the early interior evolution of a moon and/or during a later epoch of enhanced internal heating. Using combinations of accretionary, radiogenic, and tidal heating under different formation scenarios, thermal-orbital co-evolution models of all five moons can produce oceans at some point within the lifetimes of Enceladus, Tethys, Dione, and Rhea (Neveu and Rhoden 2019). Whether born from the CPD or Saturn's rings, the simulations begin billions of years ago (e.g., Neveu and Rhoden 2019). Separate studies that focus only on the past $\sim$10 Myr of Mimas' evolution have shown that a young ocean can be generated within Mimas by invoking a relatively recent increase in its eccentricity (Rhoden et al. 2024; Lainey et al. 2024). A critical open question is whether the development and, in some cases loss, of oceans can be reproduced in very young moons, particularly after disruptive collisions and reassembly.

The effectiveness of tidal heating to grow or preserve oceans and to relax topography depends on the responsiveness of the interior to deformation and the eccentricity, obliquity, and physical librations of the moon, which can be forced by mean motion resonances. In order to match the observed heat flows at Enceladus, some models incorporate a tidally dissipative core (e.g., Roberts 2015). We do not yet know whether other mid-sized moons possess dissipative cores or how porous cores may be formed and retained across different formation models. At Mimas, Tethys, and Dione, past epochs of higher eccentricity or obliquity that promote tidal heating have been suggested to account for the presence of an ocean and/or high heat flows (e.g., Mimas: Lainey et al. 2024; Rhoden et al. 2024; Tethys: Chen and Nimmo 2008; Dione: White et al. 2017). Hence, models for the formation of the mid-sized moons, and their subsequent thermal-orbital evolution, must provide opportunities for tidal dissipation consistent with these observations and inferences.

Given the currently available models, we cannot yet tie the geologic records and thermal evolutions they imply directly to an initial formation model. With additional studies, the extent of geologic activity could be used to place constraints on the ages of the moons. For example, some histories may be ruled out by the resonance crossings required to match the heat flows recorded on the moons. Additional models are needed to connect dynamical evolution of the moons – accounting for things like the addition or loss of moons over time or different initial interior structures based on the proposed formation models – in order to fully utilize the geologic constraints. These models will also depend on, and may thus help to constrain, the historic Q of Saturn and its potential evolution through time.

Craters. Mimas, Tethys, Dione, and Rhea are heavily-cratered, as is Enceladus outside the south polar terrain (e.g., Kirchoff and Schenk 2009, 2010). These crater populations need time and source material to develop, which may be challenging if the moons are extremely young. However, the lack of craters at the largest sizes, as compared with Iapetus and model predictions for early heliocentric bombardment, suggest that the surfaces are not primordial either (Kirchoff and Schenk 2010; Bottke et al. 2024).

It may be difficult to preserve primordial mid-sized moons through early heavy bombardment (e.g., Movshovitz et al. 2015; Bottke et al. 2023, 2024), particularly if all of the moons formed contemporaneously in the CPD. Moons that emerged from the rings (e.g., Charnoz et al. 2011; Crida and Charnoz 2012; Salmon and Canup 2017) would have entered the system later than CPD moons, although most ring-based formation models still have the process begin billions of years ago. Whether that delay is sufficient to "save" the mid-sized moons from disruptive collisions by external impactors may depend on exactly when the rings formed and how many moons are assumed to emerge. Alternatively, one or more of the mid-sized moons may have reassembled from a disruptive collision due to early heliocentric bombardment, regardless of how it initially formed.

There is strong evidence of planetocentric debris in the Saturn system (e.g., Kirchoff et al. 2018), particularly as recorded in the distribution of elliptical craters (Ferguson et al. 2022b, 2024), for which collisions into or among existing moons seems the most likely source (e.g., Ćuk et al. 2016; Teodoro et al. 2023). An important consideration is the extent to which these collisions reset the surface geologic records of the moons, as well as their internal structures and eccentricities, and how much material is reaccreted versus lost to Saturn (e.g., Hyodo and Charnoz 2017, versus Teodoro et al. 2023). Additional modeling is needed to quantify the outcomes of hypothesized collisions for comparison with the moons.

Dynamical evolution. Looking at the Saturn system more broadly, there appears to be evidence of relatively recent dynamical changes. Dynamic destabilization of one or more moons, leading to collisions destructive enough to generate a planetocentric population of impactors, is the current prevailing hypothesis for the source of the material that cratered the mid-sized moons (e.g., Wisdom et al. 2022; Teodoro et al. 2023). However, the timing and longevity of planetocentric bombardment is somewhat uncertain. Elliptical craters, which are thought to be planetocentric in origin, seem to be less prevalent in the youngest regions of Tethys and Dione, as compared with more heavily cratered regions (Ferguson et al. 2022b), perhaps because the impactor population became depleted within the time frame recorded on their surfaces. Furthermore, the similarities in elliptical crater populations at Tethys and Dione suggest the craters predate inclination pumping of Tethys, although other options are plausible (Ferguson et al. 2022b). Therefore, it seems likely that whatever generated planetocentric material at Saturn predated at least some of the geological and dynamical evolution of the moons.

The fact that Rhea currently orbits outside the evection resonance, while having a relatively low inclination, suggests either dynamically-driven destruction and reassembly to produce the present-day Rhea or a complex series of resonances that can accommodate the dynamical effects of evection (e.g., Ćuk et al. 2016; Teodoro et al. 2023). In either case, Rhea's heavily cratered surface is more consistent with an ancient surface than a young one (e.g., Kirchoff and Schenk 2010). Identifying pathways that reproduce Rhea's present-day low inclination may provide useful constraints on the overall evolution of the mid-sized moons.

If Mimas' ocean is confirmed by additional measurements, it implies recent ($\sim$10 Myr) eccentricity pumping of Mimas (Lainey et al. 2024; Rhoden et al. 2024). Furthermore, Mimas' recent orbital evolution may be related to the on-going evolution of the Cassini Division (e.g., Baillié et al. 2019), and Mimas is in an inclination-type resonance with Tethys. Pinning down the timing and conditions over which Saturn's rings, Mimas, and Tethys co-evolved would be useful for understanding the evolution of the moons and may also help constrain the lifetime of Saturn's present ring system. Additional measurements or experiments that can date the rings would also be helpful in constraining formation models of the rings and mid-sized moons.

Conclusions. The *Cassini* mission enabled characterization of the surfaces of the mid-sized moons and, to varying extent, their interior structures and thermal histories. As more sophisticated models for the initial formation of the mid-sized moons, and subsequent collisions and reassembly, are developed, we can better test and identify the scenarios most consistent with these constraints. Of particular value would be identifying the starting locations of the mid-sized moons, such that their past resonance passages and overall tidal dissipation can be constrained. Better estimates of the present orbital migration rates of the moons, as well as information about Saturn's interior that governs its dissipation over time, can aid in this endeavor. While the ages of the mid-sized moons cannot yet be firmly determined by their cratering records or inferred thermal evolution, neither extreme in age (primordial or $\sim$100 Myr) appears fully consistent with the observations. Hence, we consider middle-aged moons to be most plausible and encourage continued investigation into formation and/or reassembly models that would allow retention of geologic features over billion year timescales.

Acknowledgements The authors wish to acknowledge the organizers and attendees of the ISSI workshop, particularly V. Lainey, for stimulating discussion that contributed to the ideas presented in this chapter. The authors also wish to thank S. Charnoz for valuable comments and an anonymous reviewer for their thorough evaluation that led to improvements in this manuscript. A. Rhoden and S. Ferguson were partially supported by a NASA CDAP Grant 80NSSC21K0535. M. Zannoni is grateful to the Italian Space Agency (ASI) for financial support through Agreement No. 2018-25-HH.O. Any use of trade, firm, or product names is for descriptive purposes only and does not imply endorsement by the U.S. Government.

Funding Open access funding provided by Alma Mater Studiorum - Università di Bologna within the CRUI-CARE Agreement.

Declarations

Competing Interests The authors have no conflicts of interest to report.

References

Alvarellos JL, Zahnle KJ, Dobrovolskis AR, Hamill P (2005) Fates of satellite ejecta in the Saturn system. Icarus 178:104–123

Alvarellos JL, Dobrovolskis AR, Zahnle KJ, Hamill P, Dones L, Robbins S (2017) Fates of satellite ejecta in the Saturn system, II. Icarus 284:70–89

Anderson JD, Schubert G (2010) Rhea's gravitational field and interior structure inferred from archival data files of the 2005 Cassini flyby. Phys Earth Planet Inter 178:176–182

Aponte-Hernández B, Rivera-Valentin EG, Kirchoff MR, Schenk P (2021) Morphometric study of craters on Saturn's moon Rhea. Planet Sci J 2:235

Archinal BA, Acton CH, A'Hearn MF et al (2018) Report of the IAU working group on cartographic coordinates and rotational elements: 2015. Celest Mech Dyn Astron 130:22. https://doi.org/10.1007/s10569-017-9805-5

Asphaug E, Reufer A (2013) Late origin of the Saturn system. Icarus 223:544–565

Bahcall JN, Pinsonneault MH, Basu S (2001) Solar models: current epoch and time dependences, neutrinos, and helioseismological properties. Astrophys J 555:990–1012

Baillié K, Noyelles B, Lainey V, Charnoz S, Tobie G (2019) Formation of the Cassini division – I. Shaping the rings by Mimas inward migration. Mon Not R Astron Soc 486(2):2933–2946. https://doi.org/10.1093/mnras/stz548

Barr AC, Canup RM (2010) Origin of the Ganymede-Callisto dichotomy by impacts during the late heavy bombardment. Nat Geosci 3:164–167

Beddingfield CB, Burr DM, Dunne WM (2015) Shallow normal fault slopes on Saturnian icy satellites. J Geophys Res, Planets 120:2053–2083

Beddingfield CB, Burr DM, Tran LT (2016) Polygonal impact craters on Dione: evidence for tectonic structures outside the wispy terrain. Icarus 274:163–194. https://doi.org/10.1016/j.icarus.2016.03.020

Beddingfield CB, Cartwright RJ, Ferguson SN, Leonard EJ (2023) Tethys's heat fluxes varied with time in the Ithaca Chasma and Telemus Basin region. Planet Sci J 4(3):57

Běhounková M, Tobie G, Choblet G, Čadek O (2012) Tidally-induced melting events as the origin of south-pole activity on Enceladus. Icarus 219(2):655–664
Běhounková M, Souček O, Hron J, Čadek O (2017) Plume activity and tidal deformation on Enceladus influenced by faults and variable ice shell thickness. Astrobiology 17:941–954. https://doi.org/10.1089/ast.2016.1629
Bell SW (2020) Relative crater scaling between the major moons of Saturn: implications for planetocentric cratering and the surface age of Titan. J Geophys Res, Planets 125:1–18. https://doi.org/10.1029/2020JE006392
Besserer J, Nimmo F, Roberts JH, Pappalardo RT (2013) Convection-driven compaction as a possible origin of Enceladus's long wavelength topography. J Geophys Res, Planets 118(5):908–915. https://doi.org/10.1002/jgre.20079
Beuthe M (2018) Enceladus's crust as a non-uniform thin shell: I tidal deformations. Icarus 302:145–174
Beuthe M, Rivoldini A, Trihn A (2016) Enceladus's and Dione's floating ice shells supported by minimum stress isostasy. Geophys Res Lett 43:10088–10096
Bierhaus EB, Dones L, Alvarellos JL, Zahnle K (2012) The role of ejecta in the small crater populations on the mid-sized saturnian satellites. Icarus 218:602–621
Bierhaus EB, McEwen AS, Robbins SJ, Singer KN, Dones L, Kirchoff MR, Williams JP (2018) Secondary craters and ejecta across the Solar System: populations and effects on impact-crater-based chronologies. Meteorit Planet Sci 53:638–671
Bills BG, Nimmo F, Karatekin O, Van Hoolst T, Rambaux N, Levrard B, Laskar J (2009) Rotational dynamics of Europa. In: Pappalardo RT, McKinnon WB, Khurana KK (eds) Europa. The University of Arizona space science series. Arizona Press, p 119. ISBN: 9780816528448
Bjonnes E, Johnson BC, Silber EA, Singer KN, Evans AJ (2022) Ice shell structure of Ganymede and Callisto based on impact crater morphology. J Geophys Res, Planets 127:e2021JE007028
Bland MT, Bray VJ (2024) The inevitability of large shallow craters on Callisto and Ganymede: implications for crater depth-diameter trends. Icarus 408:115811. https://doi.org/10.01016/j.icarus.2023.115811
Bland MT, Beyer RA, Showman AP (2007) Unstable extension of Enceladus' lithosphere. Icarus 192:92–105. https://doi.org/10.1016/j.icarus.2007.06.011
Bland MT, Singer KN, McKinnon WB, Schenk PM (2012) Enceladus' extreme heat flux as revealed by its relaxed craters. Geophys Res Lett 39:L17204
Bland MT, Singer KN, McKinnon WB, Schenk PM (2017) Viscous relaxation of Ganymede's impact craters: constraints on heat flux. Icarus 296:275–288. https://doi.org/10.1016/j.icarus.2017.06.012
Bland MT, Beddingfield CB, Nordheim TA, Patthoff DA, Vance SD (2023) Constraints on the composition and thermal structure of Ariel's icy crust as inferred from its largest observed impact crater. Icarus 395:115452. https://doi.org/10.1016/j.icarus.2023.115452
Bland MT, Kirchoff MR, Hoover RH, Singer KN (2024) More evidence for extreme regionally localized heat fluxes in Enceladus' past. 55th LPSC, abstract 1944
Bottke WF, Love SG, Tytell D, Glotch T (2000) Interpreting the elliptical crater populations on Mars, Venus, and the Moon. Icarus 145:108–121. https://doi.org/10.1006/icar.1999.6323
Bottke WF, Vokrouhlicky D, Marschall R, Nesvorny D, Morbidelli A, Deienno R, Marchi S, Dones L, Levison HF (2023) The collisional evolution of the primordial Kuiper Belt, its destabilized population, and the Trojan asteroids. Planet Sci J 4:168
Bottke WF, Vokrouhlický D, Nesvorný D, Marshall R, Morbidelli A, Deienno R, Marchi S, Kirchoff M, Dones L, Levison HF (2024) The bombardment history of the giant planet satellites. Planet Sci J 5:88
Brahic A (1977) Systems of colliding bodies in a gravitational field: I – numerical simulation of the standard model. Astron Astrophys 54:895–907
Bray VJ, Collins GS, Morgan JV, Melosh HJ, Schenk PM (2014) Hydrocode simulation of Ganymede and Europa cratering trends – how thick is Europa's crust? Icarus 231:394–406
Brown RH, Matson DL (1987) Thermal effects of insolation propagation into the regoliths of airless bodies. Icarus 72:84–94
Bruesch S, Asphaug E (2004) Modeling global impact effects on middle-sized icy bodies: applications to Saturn's moons. Icarus 168:457–466
Buratti BJ, Faulk SP, Mosher J, Baines KH, Brown RH, Clark RN, Nicholson PD (2011) Search for and limits on plume activity on Mimas, Tethys, and Dione with the Cassini Visual Infrared Mapping Spectrometer (VIMS). Icarus 214:534–540. https://doi.org/10.1016/j.icarus.2011.04.030
Buratti BJ, Hansen CJ, Hendrix AR, Esposito LW, Mosher JA, Brown RH, Clark RN, Baines KH, Nicholson PD (2018) The search for activity on Dione and Tethys with Cassini VIMS and UVIS. Geophys Res Lett 45:5860–5866. https://doi.org/10.1029/2018GL078165
Burch JL, Goldstein J, Lewis WS, Young DT, Coats AJ, Dougherty MK, André N (2007) Tethys and Dione as sources of outward-flowing plasma in Saturn's magnetosphere. Nature 447:833–835

Čadek O, Tobie G, Van Hoolst T, Massé M, Choblet G, Lefèvre A et al (2016) Enceladus's internal ocean and ice shell constrained from Cassini gravity, shape, and libration data. Geophys Res Lett 43(11):5653–5660

Čadek O, Souček O, Běhounková M, Choblet G, Tobie G, Hron J (2019) Long-term stability of Enceladus' uneven ice shell. Icarus 319:476–484

Canup RM (2010) Origin of Saturn's rings and inner moons by mass removal from a lost Titan-sized satellite. Nature 468:943–946

Canup RM, Ward WR (2006) A common mass scaling for satellite systems of gaseous planets. Nature 441:834

Canup RM, Ward WR (2009) Origin of Europa and the Galilean satellites. In: Pappalardo RT, McKinnon WB, Khurana KK (eds) Europa. vol 59. University of Arizona Press, Tucson

Carnahan E, Wolfenbarger NS, Jordan JS, Hesse MA (2021) New insights into temperature-dependent ice properties and their effect on ice shell convection for icy ocean worlds. Earth Planet Sci Lett 563:116886. https://doi.org/10.1016/j.epsl.2021.116886

Castillo-Rogez JC, Hemingway D, Rhoden AR, Tobie G, McKinnon WB (2018) Origin and evolution of Saturn's mid-sized moons. In: Schenk PM et al (eds) Enceladus and the icy moons of Saturn. University of Arizona Press, Tucson. https://doi.org/10.2458/azu_uapress_9780816537075-ch014

Caudal G (2017) The damping of forced librations of triaxial satellites with eccentric orbits: consequences on the dynamics of Mimas. Icarus 286:280–288

Chapman CR, McKinnon WB (1986) Cratering of planetary satellites. In: Burns JA, Matthews MS (eds) Satellites. University of Arizona Press, Tucson, pp 492–580

Charnoz S, Morbidelli A, Dones L, Salmon J (2009) Did Saturn's rings form during the Late Heavy Bombardment? Icarus 199:413–428

Charnoz S, Salmon J, Crida A (2010) The recent formation of Saturn's moonlets from viscous spreading of the main rings. Nature 465:752–754

Charnoz S, Crida A, Castillo-Rogez J, Lainey V, Dones L (2011) Accretion of Saturn's mid-sized moons during the viscous spreading of young massive rings: solving the paradox of silicate-poor rings versus silicate-rich moons. Icarus 210:635–643. https://doi.org/10.1016/j.icarus.2010.08.005

Chen EMA, Nimmo F (2008) Implications from Ithaca Chasma for the thermal and orbital history of Tethys. Geophys Res Lett 35(19):L19203

Chen EMA, Nimmo F, Glatzmaier GA (2014) Tidal heating in icy satellite oceans. Icarus 229:11–30

Choblet G, Tobie G, Sotin C, Behounkova M, Cadek O, Postberg F, Soucek O (2017) Powering prolonged hydrothermal activity inside Enceladus. Nat Astron. https://doi.org/10.1038/s41550-017-0289-8

Christou AA, Namouni F, Morais MHM (2007) The long term stability of coorbital moons of the satellites of Saturn: I. Conservative case. Icarus 192(1):106–116

Clark RN et al (2008) Compositional mapping of Saturn's satellite Dione withCassini VIMS and implications of dark material in the Saturn system. Icarus 193:372–386

Clark RN et al (2012) The surface composition on Iapetus: mapping results from Cassini VIMS. Icarus 218:831–860. https://doi.org/10.1016/j.icarus.2012.008

Collins GC, McKinnon WB, Moore JM, Nimmo F, Pappalardo RT, Prockter LM, Schenk PM (2009) Tectonics of the outer planet satellites. Planet Tecton 11:229

Collins GC, McKinnon WB, Moore JM, Nimmo F, Pappalardo RT, Prockter LM, Schenk PM (2010) Tectonics of the outer planet satellites. In: Watters T, Schultz R (eds) Planetary tectonics. Cambridge University Press, Cambridge, pp 264–350

Collins GS, Elbeshausen D, Davison TM, Robbins SJ, Hynek BM (2011) The size-frequency distribution of elliptical impact craters. Earth Planet Sci Lett 310:1–8. https://doi.org/10.1016/j.epsl.2011.07.023

Crida A, Charnoz S (2012) Formation of regular satellites from ancient massive rings in the Solar System. Science 338:1196–1212

Crida A, Charnoz S, Hsu H-W, Dones L (2019) Are Saturn's rings actually young? Nat Astron 3:967–970

Crow-Willard EN, Pappalardo RT (2015) Structural mapping of Enceladus and implications for formation of tectonized regions. J Geophys Res, Planets 120(5):928–950

Ćuk M, El Moutamid M (2022) Three-body resonances in the Saturnian system. Astrophys J Lett 926:L18

Ćuk M, Dones L, Nesvorný D (2016) Dynamical evidence for a late formation of Saturn's moons. Astrophys J 820:97

Cuzzi JN, Durisen RH (1990) Bombardment of planetary rings by meteroids: general formulation and effects of Oort cloud projectiles. Icarus 84:467–501

Cuzzi JN, Estrada PR (1998) Compositional evolution of Saturn's rings due to meteroid bombardment. Icarus 132:1–35

Denton CA, Rhoden AR (2022) Tracking the evolution of an ocean within Mimas using the Herschel impact basin. Geophys Res Lett 49:e2022GL100516

Denton CA, Johnson BC, Wakita S, Freed AM, Melosh HJ, Stern AS (2021) Pluto's antipodal terrains imply a thick subsurface ocean and hydrated core. Geophys Res Lett 48:e91596
Di Sisto RP, Brunini A (2007) The origin and distribution of the Centaur population. Icarus 190:224–235
Di Sisto RP, Brunini A (2011) Origin of craters on Phoebe: comparison with Cassini's data. Astron Astrophys 534:A68
Di Sisto RP, Rossignoli NL (2020) Centaur and giant planet crossing populations: origin and distribution. Celest Mech Dyn Astron 132:36
Di Sisto RP, Zanardi M (2013) The production of craters on the mid-sized Saturnian satellites by Centaur objects. Astron Astrophys 553:A79
Di Sisto RP, Zanardi M (2016) Surface ages of mid-size saturnian satellites. Icarus 264:90–101
Di Sisto RP, Fernandez JA, Brunini A (2009) On the population, physical decay, and orbital distribution of Jupiter family comets: numerical simulations. Icarus 203:140–154
Dombard AJ, McKinnon WB (2006) Elastoviscoplastic relaxation of impact crater topography with application to Ganymede and Callisto. J Geophys Res, Planets 111(E1):E01001
Dones L (1991) A recent cometary origin for Saturn's rings? Icarus 92:194–203
Dones L, Chapman CR, McKinnon WB, Melosh HJ, Kirchoff MR, Neukum G, Zahnle KJ (2009) Icy satellites of Saturn: impact cratering and age determination. In: Dougherty MK, Esposito LW, Krimigis SM (eds) Saturn from Cassini-Huygens. Springer, Dordrecht, pp 613–635. https://doi.org/10.1007/978-1-4020-9217-6_19
Dougherty MK, Khurana KK, Neubauer FM, Russell CT, Saur J, Leisner JS, Burton ME (2006) Identification of a dynamic atmosphere at Enceladus with the Cassini magnetometer. Science 311:1406–1409
Dubinski J (2019) A recent origin for Saturn's rings from the collisional disruption of an icy moon. Icarus 321:291–306
Durham WB, Stern LA (2001) Rheological properties of water ice: applications to satellites of the outer planets. Annu Rev Earth Planet Sci 29:295–330. https://doi.org/10.1146/annurev.earth.29.1.295
Durham WB, Kirby SH, Stern LA (1992) Effects of dispersed particulates on the rheology of water ice at planetary conditions. J Geophys Res 97:883–903
Durham WB, Prieto-Ballesteros O, Goldsby DL, Kargel JS (2010) Rheological and thermal properties of icy materials. Space Sci Rev 153(1–4):273–298. https://doi.org/10.1007/s11214-009-9619-1
Elbeshausen D, Wünnemann K, Collins GS (2009) Scaling of oblique impacts in frictional targets: implications for crater size and formation mechanisms. Icarus 204:716–731
Elbeshausen D, Wünnemann K, Collins GS (2013) The transition from circular to elliptical impact craters. J Geophys Res, Planets 118:2295–2309
Eluszkiewicz J (1990) Compaction and internal structure of Mimas. Icarus 4:215–225
Emery JP, Burr DM, Cruikshank DP, Brown RH, Dalton JB (2005) Near-infrared (0.8-4.0 μm) spectroscopy of Mimas, Enceladus, Tethys, and Rhea. Astron Astrophys 435:353–362. https://doi.org/10.1051/0004-6361:20042482
English NJ, Tse JS (2009) Mechanisms for thermal conduction in methane hydrate. Phys Rev Lett 103:015901. https://doi.org/10.1103/PhysRevLett.103.015901
Esposito LW (1986) Structure and evolution of Saturn's rings. Icarus 67:345–357
Ferguson SN, Rhoden AR, Kirchoff MR (2020) Small impact crater populations on Saturn's moon Tethys and implications for source impactors in the system. J Geophys Res, Planets 125:e2020JE006400
Ferguson SN, Rhoden AR, Kirchoff MR (2022a) Regional impact crater mapping and analysis on Saturn's moon Dione and the relation to source impactors. J Geophys Res, Planets 127:e2022JE007204
Ferguson SN, Rhoden AR, Kirchoff MR, Salmon JJ (2022b) A unique Saturnian impactor population from elliptical craters. Earth Planet Sci Lett 593:117652
Ferguson SN, Rhoden AR, Kirchoff M (2024) Mimas: a middle-aged mid-sized moon? EPSL, in press
Friedson AJ, Stevenson DJ (1983) Viscosity of rock-ice mixtures and applications to the evolution of icy satellites. Icarus 56:1–14
Fuller J, Luan J, Quataert E (2016) Resonance locking as the source of rapid tidal migration in the Jupiter and Saturn moon systems. Mon Not R Astron Soc 458:3867–3879. https://doi.org/10.1093/mnras/stw609
Gault DE (1970) Saturation and equilibrium conditions for impact cratering on the lunar surface: criteria and implications. Radio Sci 5(2):273–291
Gault DE, Wedekind JA (1978) Experimental studies of oblique impact. In: Lunar and Planetary Science Conference, 9th, Houston, Tex., March 13–17, 1978, proceedings. Volume 3. (A79-39253 16-91). Pergamon, New York, pp 3843–3875
Giese B, Wagner R, Neukum G, Helfenstein P, Thomas PC (2007) Tethys: lithospheric thickness and heat flux from flexurally supported topography at Ithaca Chasma. Geophys Res Lett 34(21):L21203
Giese B, Wagner R, Hussmann H, Neukum G, Perry J, Helfenstein P, Thomas PC (2008) Enceladus: an estimate of heat flux and lithospheric thickness from flexurally supported topography. Geophys Res Lett 35:L24204. https://doi.org/10.1029/2008GL036149

Giuppone CA, Roig F, Saad-Olivera X (2018) Modeling the evection resonance for Trojan satellites: application to the Saturn system. Astron Astrophys 620:A90
Goldreich PM, Mitchell JL (2010) Elastic ice shells of synchronous moons: implications for cracks on Europa and non-synchronous rotation of Titan. Icarus 209:631–638
Gomes R, Levison HF, Tsiganis K, Morbidelli A (2005) Origin of the cataclysmic late heavy bombardment period of the terrestrial planets. Nature 7041:466–469
Gough DO (1981) Solar interior structure and luminosity variations. Sol Phys 74:21–34
Greenberg R, Weidenschilling SJ (1984) How fast do Galilean satellites spin? Icarus 58:186–196
Greenberg R, Geissler P, Hoppa G, Tufts BR, Durda DD, Pappalardo R et al (1998) Tectonic processes on Europa: tidal stresses, mechanical response, and visible features. Icarus 135:64–78
Greenstreet S, Gladman B, McKinnon WB (2015) Impact and cratering rates onto Pluto. Icarus 258:267–288
Gyalay S, Nimmo F (2023) Estimates for Tethys' moment of inertia, heat flux distribution, and interior structure from its long-wavelength topography. J Geophys Res, Planets 128:e2022JE007550
Hammond NP, Phillips CB, Nimmo F, Kattenhorn SA (2013) Flexure on Dione: investigating subsurface structure and thermal history. Icarus 223:418–422. https://doi.org/10.1016/j.icarus.2012.12.021
Hansen CJ, Esposito EL, Stewart AIF, Colwell J, Hendrix A, Pryor W, Shemansky D, West R (2006) Enceladus' water vapor plume. Science 311:1422–1425
Hartmann WK (1984) Does crater "saturation equilibrium" occur in the Solar System? Icarus 60(1):56–74
Hay CFC, Matsuyama I (2019) Nonlinear tidal dissipation in the subsurface oceans of Enceladus and other icy satellites. Icarus 319:68–85
Hedman MM, Gosmeyer CM, Nicholson PD, Sotin C, Brown RH, Clark RN, Baines KH, Buratti BJ, Showalter MR (2013) An observed correlation between plume activity and tidal stresses on Enceladus. Nature 500:182–184
Helfenstein P, Parmentier EM (1985) Patterns of fracture and tidal stresses due to nonsynchronous rotation: implications for fracturing on Europa. Icarus 61:175–184
Hemingway DJ, Mittal T (2019) Enceladus's ice shell structure as a window on internal heat production. Icarus 332:111–131
Hemingway DJ, Zannoni M, Tortora P, Nimmo F, Asmar SW (2016) Dione's internal structure inferred from Cassini gravity and topography. 47th LPSC, abstract 1314
Hemingway D, Iess L, Tajeddine R, Tobie G (2018) The interior of Enceladus. In: Schenk PM et al (eds) Enceladus and the icy moons of Saturn. University of Arizona Press, Tucson, pp 57–77
Hemingway DJ, Rudolph ML, Manga M (2020) Cascading parallel fractures on Enceladus. Nat Astron 4:234–239
Hirata N (2016) Differential impact cratering of Saturn's satellites by heliocentric impactors. J Geophys Res, Planets 121:111–117
Holo SJ, Kite ES, Robbins SJ (2018) Mars obliquity history constrained by elliptic crater orientations. Earth Planet Sci Lett 496:206–214
Horedt GP, Neukum G (1984) Planetocentric versus heliocentric impacts in the Jovian and Saturnian satellite system. J Geophys Res 89:10405–10410
Howett CJA, Spencer JR, Pearl J, Segura M (2011) High heat flow from Enceladus' south polar region measured using 10–600 cm^{-1} Cassini/CIRS data. J Geophys Res, Planets 116(E3):E03003
Hsu HW, Postberg F, Sekine Y, Shibuya T, Kempf S, Horányi M et al (2015) Ongoing hydrothermal activities within Enceladus. Nature 519:207–210
Hurford TA, Helfenstein P, Hoppa GV, Greenberg R, Bills BG (2007) Eruptions arising from tidally controlled periodic openings of rifts on Enceladus. Nature 447(7142):292–294
Hussmann H, Sohl F, Spohn T (2006) Subsurface oceans and deep interiors of medium-sized outer planet satellites and large trans-Neptunian objects. Icarus 185:258–273
Hussmann H, Rodríguez A, Callegari N Jr, Shoji D (2019) Early resonances of Tethys and Dione: implications for Ithaca Chasma. Icarus 319:407–416
Hyodo R, Charnoz S (2017) Dynamical evolution of the debris disk after a satellite catastrophic disruption around Saturn. Astron J 154:34
Iess L, Rappaport NJ, Tortora P, Lunine J, Armstrong JW, Asmar SW, Somenzi L, Zingoni F (2007) Gravity field and interior of Rhea from Cassini data analysis. Icarus 190:585–593
Iess L, Stevenson DJ, Parisi M, Hemingway D, Jacobson RA, Lunine JI et al (2014) The gravity field and interior structure of Enceladus. Science 344(6179):78–80
Iess L, Militzer B, Kaspi Y, Nicholson P, Durante D, Racioppa P et al (2019) Measurement and implications of Saturn's gravity field and ring mass. Science 364:eaat2965
Johnson TV, McGetchin TR (1973) Topography on satellite surfaces and the shape of asteroids. Icarus 18:612–620. https://doi.org/10.1016/0019-1035(73)90064-X
Kamata S, Nimmo F (2015) Relaxation of long-wavelength topography and the thermal evolution of Rhea. 46th LPSC, abstract 1832

Karato S, Wu P (1993) Rheology of the upper mantle: a synthesis. Science 260:771–778. https://doi.org/10.1126/science.260.5109.771
Kattenhorn SA, Hurford T (2009) The tectonics of Europa. In: Pappalardo RT, McKinnon WB, Khurana KK (eds) Europa. University of Arizona Press, Tucson, pp 219–256
Kinczyk MJ, Byrne PK, Patterson GW (2022) Structure and morphology of Enceladus' cratered terrains. Lunar Planet Sci Conf 53:1144
Kirchoff MR, Schenk P (2009) Crater modification and geologic activity in Enceladus' heavily cratered plains: evidence from the impact crater distribution. Icarus 202:656–668
Kirchoff MR, Schenk P (2010) Impact cratering records of the mid-sized, icy saturnian satellites. Icarus 206:485–497
Kirchoff MR, Schenk P (2015) Dione's resurfacing history as determined from a global impact crater database. Icarus 256:78–89
Kirchoff MR, Bierhaus EB, Dones L, Robbins SJ, Singer KN, Wagner RJ, Zahnle KJ (2018) Cratering histories in the Saturnian system. In: Schenk PM et al (eds) Enceladus and the icy moons of Saturn. University of Arizona Press, Tucson, pp 267–284
Kirchoff MR, Dones L, Singer K, Schenk P (2022) Crater distributions of Uranus's mid-sized satellites and implications for outer Solar System bombardment. Planet Sci J 3:42
Kossacki KJ, Leliwa-Kopystyński J (1993) Medium-sized icy satellites: thermal and structural evolution during accretion. Planet Space Sci 41(10):729–741
Lainey V, Karatekin Ö, Desmars J, Charnoz S, Arlot JE, Emelyanov N et al (2012) Strong tidal dissipation in Saturn and constraints on Enceladus' thermal state from astrometry. Astrophys J 752:14
Lainey V, Jacobson RA, Tajeddine R, Cooper NJ, Murray C, Robert V et al (2017) New constraints on Saturn's interior from Cassini astrometric data. Icarus 281:286–296
Lainey V, Casajus LG, Fuller J, Zannoni M, Tortora P, Cooper N et al (2020) Resonance locking in giant planets indicated by the rapid orbital expansion of Titan. Nat Astron 4:1053–1058
Lainey V, Rambaux N, Tobie G, Cooper N, Zhang Q, Noyelles B, Baillie K (2024) A recently formed ocean inside Saturn's moon Mimas. Nature 626:280–282
Leliwa-Kopystyński J, Kossacki KJ (2000) Evolution of porosity in small icy bodies. Planet Space Sci 48:727–745
Levison HF, Duncan MJ, Zahnle K, Holman M, Dones L (2000) NOTE: planetary impact rates from ecliptic comets. Icarus 143:415–420
Lissauer JJ, Squyres SW, Hartmann WK (1988) Bombardment history of the Saturn system. J Geophys Res 93(B11):13776–13804
Lorenz RD, Shandera SE (2001) Physical properties of ammonian-rich ice: application to Titan. Geophys Res Lett 28(2):215–218. https://doi.org/10.1029/2000GL012199
Mackenzie RA, Iess L, Tortora P, Rappaport NJ (2008) A non-hydrostatic Rhea. Geophys Res Lett 35(5):L05204
Mah J, Brasser R (2019) The origin of the cratering asymmetry on Triton. Mon Not R Astron Soc 486:836–842
Mangold N, Allemand P, Duval P, Geraud Y, Thomas P (2002) Experimental and theoretical deformation of ice-rock mixtures: implications on rheology and ice content of Martian permafrost. Planet Space Sci 50:385–401
Martin ES (2016) The distribution and characterization of strike-slip faults on Enceladus. GRL 43:2456–2464
Martin B, Fyfe WS (1970) Some experimental and theoretical observations on the kinetics of hydration reactions with particular reference to serpentinization. Chem Geol 6:185–202
Martin ES, Patthoff DA, McDaniel J, Watters TR (2016) Exploring the formation of Dione's wispy terrains. 2016 GSA Annual Meeting Abstracts with Programs 48(7):Paper No. 48-12. https://doi.org/10.1130/abs/2016AM-285329
Martin ES, Kattenhorn SA, Collins GC, Michaud RL, Pappalardo RT, Wyrick DY (2017) Pit chains on Enceladus signal the recent tectonic dissection of the ancient cratered terrain. Icarus 294:209–217
Martin ES, Whitten JL, Kattenhorn SA, Collins GC, Southworth BS, Wiser LS, Prindle S (2023) Measurements of regolith thickness on Enceladus: uncovering the record of plume activity. Icarus 392:115369
Matson DL, Brown RH (1989) Solid-state greenhouses and their implications for icy satellites. Icarus 77:67–81
McEwen AS, Bierhaus EB (2006) The importance of secondary cratering to age constraints on planetary surfaces. Annu Rev Earth Planet Sci 34:535–567
McKinnon WB (2015) Effect of Enceladus's rapid synchronous spin on interpretation of Cassini gravity. Geophys Res Lett 42(7):2137–2143
Meyer J, Wisdom J (2008) Tidal evolution of Mimas, Enceladus, and Dione. Icarus 193:213–223
Moore JM (1984) The tectonic and volcanic history of Dione. Icarus 59:205–220
Moore JM, Ahern JL (1983) The geology of Tethys. 13th LPSC

Moore JM, Schenk PM (2007) Topography of endogenic features on Saturnian mid-sized satellites (abs.). In: Lunar planetary science conference. XXXVIII, p 2136
Moore JM, Hornder VM, Greeley R (1985) The geomorphology of Rhea: implications for geologic history and surface processes. J Geophys Res 90:C785–C795
Moore JM, Schenk PM, Bruesch LS, Asphaug E, McKinnon WB (2004) Large impact features on middle-sized icy satellites. Icarus 171:421–443
Morrison D et al (1984) Satellites of Saturn: geological perspective. In: Morrison D (ed) Saturn. University of Arizona Press, Tucson, pp 609–639
Mosqueira I, Estrada P, Turrini D (2010) Planetesimals and satellitesimals: formation of satellite systems. Space Sci Rev 153:431–446. https://doi.org/10.1007/s11214-009-9614-6
Movshovitz N, Nimmo F, Korycansky DG, Asphaug E, Owen JM (2015) Disruption and reaccretion of mid-sized moons during an outer Solar System late heavy bombardment. Geophys Res Lett 42:256–263
Nakajima A, Ida S, Kimura J, Brasser R (2019) Orbital evolution of Saturn's mid-sized moons and the tidal heating of Enceladus. Icarus 317:570–582
Nakamura T, Yoshikawa M (1995) Close encounters and collisions of short-period comets with Jupiter and its satellites. Icarus 116:113–130
Nesvorny D (2018) Dynamical evolution of the early Solar System. Annu Rev Astron Astrophys 56:137–174
Nesvorny D, Dones L, De Pra M, Womack M, Zahnle KJ (2023) Impact rates in the outer Solar System. Planet Sci J 4:139
Neukum G, Wagner R, Wolf U, Denk T (2006) The cratering record and cratering chronologies of the Saturnian satellites and the origin of impactors: Results from Cassini ISS data. EPSC, 610
Neveu M, Rhoden AR (2017) The origin and evolution of a differentiated Mimas. Icarus 296:183–196
Neveu M, Rhoden AR (2019) Evolution of Saturn's mid-sized moons. Nat Astron 3:543–552
Nimmo F (2004) Stresses generated in cooling viscoelastic ice shells: application to Europa. J Geophys Res, Planets 109(E12):E12001
Nimmo F (2018) Icy satellites: interior structure, dynamics, and evolution. Oxford research encyclopedia of planetary science. https://doi.org/10.1093/acrefore/9780190647926.013.29
Nimmo F, Korycansky DG (2012) Impact-driven ice loss in outer Solar System satellites: consequences for the late heavy bombardment. Icarus 219:508–510
Nimmo F, Bills BG, Thomas PC, Asmar S (2010) Geophysical implications of the long-wavelength topography of Rhea. J Geophys Res, Planets 115:E10008. https://doi.org/10.1029/2010JE003604
Nimmo F, Bills BG, Thomas PC (2011) Geophysical implications of the long-wavelength topography of the Saturnian satellites. J Geophys Res, Planets 116(E11):E11001
Nimmo F, Barr AC, Behounková M, McKinnon WB (2018) The thermal and orbital evolution of Enceladus: observational constraints and models. In: Schenk PM et al (eds) Enceladus and the icy moons of Saturn. University of Arizona Press, Tucson, pp 79–94
Nimmo F, Neveu M, Howett C (2023) Origin and evolution Enceladus' tidal dissipation. Space Sci Rev 219:57. https://doi.org/10.1007/s11214-023-01007-4
Noyelles B (2017) Interpreting the librations of a synchronous satellite – how their phase assesses Mimas' global ocean. Icarus 282:276–289
Noyelles B, Baillié K, Charnoz S, Lainey V, Tobie G (2019) Formation of the Cassini division – II. Possible histories of Mimas and Enceladus. Mon Not R Astron Soc 486:2947–2963. https://doi.org/10.1093/mnras/stz445
Park R, Mastrodemos N, Jacobson RA, Berne A, Vaughan AT, Hemingway DJ, Leonard EJ, Castillo-Rogez JC, Cockell CS, Keane JT, Konopliv AS, Nimmo F, Riedel JE, Simons M, Vance S (2024) The global shape, gravity field, and libration of Enceladus. J Geophys Res, Planets 129:e2023JE008054
Parmentier EM, Head JW (1981) Viscous relaxation of impact craters on icy satellite surfaces: determination of viscosity variation with depth. Icarus 47(100):111
Passey QR (1983) Viscosity of the lithosphere of Enceladus. Icarus 53:105–120
Passey QR, Shoemaker EM (1982) Craters and basins on Ganymede and Callisto: morphological indicators of crustal evolution. In: Morrison D (ed) Satellites of Jupiter. University of Arizona Press, Tucson
Patterson GW, Kattenhorn SA, Helfenstein P, Collins GC, Pappalardo RT (2018) The geology of Enceladus. In: Schenk PM et al (eds) Enceladus and the icy moons of Saturn. University of Arizona Press, Tucson, p 95
Patthoff DA, Kattenhorn SA (2011) A fracture history on Enceladus provides evidence for a global ocean. Geophys Res Lett 38(18):L18201
Patthoff DA, Pappalardo RT, Chilton HT, Thomas PT, Schenk P (2015) Diverse origins of Enceladus's ridge terrains including evidence for contraction. Lunar Planet Sci Conf 46:2870
Patthoff DA, Pappalardo RT, Golombek M, Chilton H, Crow-Willard E, Thomas PC (2022) Thrust faulting as the origin of dorsa in the trailing hemisphere of Enceladus. Icarus 375:114815. https://doi.org/10.1016/j.icarus.2021.114815

Peale SJ (1976) Orbital resonances in the Solar System. Annu Rev Astron Astrophys 14:214–246
Peale SJ, Greenberg R (2007) Coorbital collision as the source of Enceladus heat. In: AGU Fall Meeting abstracts, vol 2007, pp P12B-02
Peale SJ, Cassen P, Reynolds RT (1979) Melting of Io by tidal dissipation. Science 203:892–894
Petrenko VF, Whitworth RW (1999) Physics of ice. Oxford University Press, Oxford
Plescia JB (1983) The geology of Dione. Icarus 56:255–277
Plescia JB, Boyce JM (1982) Crater densities and geological histories of Rhea, Dione, Mimas and Tethys. Nature 295:285–290
Porco CC, Helfenstein P, Thomas PC, Ingersoll AP, Wisdom J, West R et al (2006) Cassini observes the active south pole of Enceladus. Science 311:1393–1401
Qi C, Stern LA, Pathare A, Durham WB, Goldsby DL (2018) Inhibition of grain boundary sliding in fine-grained ice by intergranular particles: implications for planetary ice masses. Geophys Res Lett 45(23):12757–12765. https://doi.org/10.1029/2018GL080228
Rambaux N (2014) Rotational and librational motion of Solar System bodies. In: Capitaine N (ed) Proceedings of the Journées 2013 "Systèmes de référence spatio-temporels": Scientific developments from highly accurate space-time reference systems, Observatoire de Paris, 16–18 September 2013, pp 235–240
Rambaux N, Castillo-Rogez JC, Williams JG, Karatekin O (2010) Librational response of Enceladus. Geophys Res Lett 37:L04202
Rappaport NJ, Iess L, Torora P, Anabtawi A, Asmar SW, Somenzi L, Zingoni F (2007) Mass and interior of Enceladus from Cassini data analysis. Icarus 190:175–178
Rhoden AR (2023) Mimas: frozen fragment, ring relic, or emerging ocean world? Annu Rev Earth Planet Sci 51:367–387
Rhoden AR, Walker ME (2022) The case for an ocean-bearing Mimas from tidal heating analysis. Icarus 376:114872
Rhoden AR, Militzer B, Huff EM, Hurford TA, Manga M, Richards MA (2010) Constraints on Europa's rotational dynamics from modeling of tidally-driven fractures. Icarus 210:770–784
Rhoden AR, Henning W, Hurford TA, Patthoff DA, Tajeddine R (2017) The implications of tides on the Mimas ocean hypothesis. J Geophys Res, Planets 122:400–410
Rhoden AR, Hurford TA, Spitale J, Henning W, Huff EM, Bland MT, Sajous S (2020) The formation of Enceladus' tiger stripe fractures from eccentricity tides. Earth Planet Sci Lett 544:116389
Rhoden AR, Rudolph ML, Manga M (2023) The challenges of driving Charon's cryovolcanism from a freezing ocean. Icarus 392:115391
Rhoden AR, Walker ME, Rudolph ML, Bland MT, Manga M (2024) The evolution of a young ocean within Mimas. 55th LPSC, abstract 1628
Richardson JE (2009) Cratering saturation and equilibrium: a new model looks at an old problem. Icarus 204(2):697–715. https://doi.org/10.1016/j.icarus.2009.07.029
Robbins SJ, Singer KN (2021) Pluto and Charon impact crater populations: reconciling different results. Planet Sci J 2:192
Robbins SJ, Bierhaus EB, Dones L (2015) Craters of the saturnian satellite system: I. Mimas. In: Planetary Crater Consortium 5. Abstract #1411
Robbins SJ, Bierhaus EB, Dones L (2024) Crater populations of the Saturnian satellites Mimas, Rhea, and Iapetus. J Geophys Res, Planets 129(1):e2023JE007941. https://doi.org/10.1029/2023JE007941
Roberts JH (2015) The fluffy core of Enceladus. Icarus 258:54–66. https://doi.org/10.1016/j.icarus.2015.05.033
Roberts JH, Stickle AM (2021) Breaking the symmetry by breaking the ice shell: an impact origin for the south polar terrain of Enceladus. Icarus 359:114302
Rossignoli NL, Di Sisto RP, Zanardi M, Dugaro A (2019) Cratering and age of the small Saturnian satellites. Astron Astrophys 627:A12
Rovira-Navarro M, Rieutord M, Gerkema T, Maas LRM, van der Wal W, Vermeersen B (2019) Do tidally-generated inertial waves heat the subsurface oceans of Europa and Enceladus? Icarus 321:126–140
Rovira-Navarro M, Katz RF, Liao Y, van der Wal W, Nimmo F (2022) The tides of Enceladus' porous core. J Geophys Res, Planets 127(5):e2021JE007117. https://doi.org/10.1029/2021JE007117
Rudolph ML, Manga M (2009) Fracture penetration in planetary ice shells. Icarus 199:536–541
Rudolph ML, Manga M, Walker M, Rhoden AR (2022) Cooling crusts create concomitant cryovolcanic cracks. Geophys Res Lett 49(5):e2021GL094421
Salmon J, Canup RM (2017) Accretion of Saturn's inner mid-sized moons from a massive primordial ice ring. Astrophys J 836:109
Salmon J, Charnoz S, Crida A, Brahic A (2010) Long-term and large-scale viscous evolution of dense planetary rings. Icarus 209:771–785

Schenk PM (1989) Crater formation and modification on the icy satellites of Uranus and Saturn: depth/diameter and central peak occurrence. J Geophys Res, Solid Earth 94:3813–3832
Schenk P (2011) Geology of Mimas. 42nd LPSC, abstract 1608
Schenk PM, Moore JM (2009) Eruptive volcanism on Saturn's icy moon Dione. Lunar Planet Sci 40. Abstract #2465
Schenk PM, Zahnle K (2007) On the negligible surface age of Triton. Icarus 192:135–149
Schenk P, White OL, Byrne PK, Moore JM (2018) Saturn's other icy moons: geologically complex worlds in their own right. In: Schenk PM et al (eds) Enceladus and the icy moons of Saturn. University of Arizona Press, Tucson, p 237
Schenk P, Kirchoff M, Hoogenboom T, Rivera-Valentín E (2020) The anatomy of fresh complex craters on the mid-sized icy moons of Saturn and self-secondary cratering at the rayed crater Inktomi (Rhea). Meteorit Planet Sci 55:2440–2460
Schmedemann N, Neukum G, Denk T, Wagner R (2009) Impact crater size-frequency distribution (SFD) on saturnian satellites and comparison with other solar-system bodies. In: 40th lunar planet. sci. conf., Abst. #1941
Schoenfeld A, Yin A (2024) Quantifying tidal versus non-tidal stresses in driving time-varying fluxes of Enceladus' plume eruptions. Icarus 415:116064
Scott RF (1967) Viscous flow of craters. Icarus 7:139–148. https://doi.org/10.1016/0019-1035(67)90058-9
Senft LE, Stewart ST (2011) Modeling the morphological diversity of impact craters on icy satellites. Icarus 214(1):67–81
Shoemaker EM, Wolfe RF (1981) Evolution of the Saturnian satellites: the role of impact. 12th LPSC, abstract 428
Shoshany Y, Prialnik D, Podolak M (2002) Monte Carlo modeling of the thermal conductivity of porous cometary ice. Icarus 157(1):219–227
Silber EA, Johnson BC (2017) Impact crater morphology and the structure of Europa's ice shell. J Geophys Res, Planets 122:2685–2701
Singer KN, McKinnon WB, Nowicki LT (2013) Secondary craters from large impacts on Europa and Ganymede: ejecta size-velocity distributions on icy worlds, and the scaling of ejected blocks. Icarus 226:865–884
Singer KN, Bland MT, Schenk PM, McKinnon WB (2018) Relaxed impact craters on Ganymede: regional variation and high heat flows. Icarus 306:214–224. https://doi.org/10.1016/j.icarus.2018.01.012
Singer KN, Mckinnon WB, Gladman B, Greenstreet S, Bierhaus EB, Stern SA, Parker AH, Robbins SJ, Schenk PM, Grundy WM, Bray VJ, Beyer RA et al (2019) Impact craters on Pluto and Charon indicate a deficit of small Kuiper Belt objects. Science 959:955–959. https://doi.org/10.1126/science.aap8628
Smith BA, Soderblom L, Beebe R, Boyce J, Bunker A, Collins SA, Hansen CJ, Johnson TV, Jim L, Terrile RJ, CarrM et al (1981) Encounter with Saturn: Voyager 1 imaging science results. Science 212:163–191
Soucek O, Behounkova M, Cadek O, Hron J, Tobie G, Choblet G (2019) Tidal dissipation in Enceladus' uneven, fractured ice shell. Icarus 328:218–231
Spencer JR, Pearl JC, Segura M, Flasar FM, Mamoutkine A, Romani P et al (2006) Cassini encounters Enceladus: background and the discovery of a south polar hot spot. Science 311:1401–1405
Spencer JR, Stern SA, Moore JM, Weaver HA, Singer KN, Olkin CB et al (2020) The geology and geophysics of Kuiper Belt object (486958) Arrokoth. Science 367:eaay3999
Spencer J, Beyer RA, Robbins SJ, Singer KN, Nimmo F (2021) The geology and geophysics of Charon. In: Stern SA et al (eds) The Pluto system after New Horizons. University of Arizona Press, Tucson, p 395–412
Squyres SW, Howell C, Liu MC, Lissauer JJ (1997) Investigation of crater "saturation" using spatial statistics. Icarus 125(1):67–82
Stephan K, Wagner R, Jaumann R, Clark RN, Cruikshank DP, Brown RH et al (2016) Cassini's geological and compositional view of Tethys. Icarus 274:1–22
Tajeddine R, Rambaux N, Lainey V, Charnoz S, Richard A, Rivoldini A, Noyelles B (2014) Constraints on Mimas' interior from Cassini ISS libration measurements. Science 346:322–324
Teodoro LFA, Kegerreis JA, Estrada PR, Cuk M, Eke VR, Cuzzi JN, Massey RJ, Sandnes TD (2023) A recent impact origin of Saturn's rings and mid-sized moons. Astrophys J 955:137
Thomas PC (2010) Sizes, shapes, and derived properties of the saturnian satellites after the Cassini nominal mission. Icarus 208:395–401
Thomas PC, Burns JA, Helfenstein P, Squyers S, Veverka J, Porco C, Turtle EP, McEwen A, Denk T, Giese B, Roatsch T, Johnson TV, Jacobson RA (2007) Shapes of the saturnian icy satellites and their significance. Icarus 190:573–584
Thomas PC, Tajeddine R, Tiscareno MS, Burns JA, Joseph J, Loredo TJ et al (2016) Enceladus's measured physical libration requires a global subsurface ocean. Icarus 264:37–47

Tian Z, Nimmo F (2020) Implications of second-order resonance for the thermal and orbital evolution of Mimas. Mon Not R Astron Soc 492:369–376

Tortora P, Zannoni M, Hemingway D, Nimmo F, Jacobson RA, Iess L, Parisi M (2016) Rhea gravity field and interior modeling from Cassini data analysis. Icarus 264:264–273. https://doi.org/10.1016/j.icarus.2015.09.022

Tricarico P (2014) Multi-layer hydrostatic equilibrium of planets and synchronous moons: theory and application to Ceres and Solar System moons. Astrophys J 782:99

Turcotte DL, Schubert D (2002) Geodynamics, 2nd edn. Cambridge University Press, Cambridge

Turtle EP, Pierazzo E (2001) Thickness of a Europan ice shell from impact crater simulations. Science 294:1326–1328

van Hoolst T, Baland R-M, Trihn A (2016) The diurnal libration and interior structure of Enceladus. Icarus 277:311–318

Verbiscer AJ, Peterson DE, Skrutskie MF, Cushing M, Helfenstein P, Nelson MJ, Smith JD, Wilson JC (2006) Near-infrared spectra of the leading and trailing hemispheres of Enceladus. Icarus 182:211–223. https://doi.org/10.1016/j.icarus.2005.12.008

Verbiscer AJ, Peterson DE, Skrutskie MF, Cushing M, Helfenstein P, Nelson MJ, Smith JD, Wilson JC (2008) Ammonia hydrate on Tethys trailing hemisphere. In: The science of Solar System ices: a cross-disciplinary workshop, abstract 9064. Lunar and Planetary Institute, Houston

Volk K, Malhotra R (2008) The scattered disk as the source of the Jupiter family comets. Astrophys J 687:714–725

Wagner R, Neukum G, Giese B, Roatsch T, Wolf U, Denk T, the Cassini ISS team (2006) Geology, ages and topography of Saturn's satellite Dione observed by the Cassini ISS camera. 37th LPSC, abstract 1805

Waite JH, Glein CR, Perryman RS, Teolis BD, Magee BA, Miller G et al (2017) Cassini finds molecular hydrogen in the Enceladus plume: evidence for hydrothermal processes. Science 356:155–159

White OL, Schenk PM (2011) Crater shapes on the saturnian satellites: new measurements using Cassini stereo images. Lunar Planet Sci 42:2136

White OL, Schenk PM, Dombard AJ (2013) Impact basin relaxation on Rhea and Iapetus and relation to past heat flow. Icarus 223(2):699–709

White OL, Schenk PM, Bellagamba AW, Grimm AM, Dombard AJ, Bray VJ (2017) Impact crater relaxation on Dione and Tethys and relation to past heat flow. Icarus 288:37–52

Whitten JS, Martin ES (2019) Icelandic pit chains as planetary analogs: using morphologic measurements of pit chains to determine regolith thickness. J Geophys Res 124:2983–2999. https://doi.org/10.1029/2019JE006099

Wisdom J, Dbouk R, Militzer B, Hubbard WB, Nimmo F, Downey BG, French RG (2022) Loss of a satellite could explain Saturn's obliquity and young rings. Science 377:1285–1289

Wong EW, Brasser R, Werner SC (2019) Impact bombardment on the regular satellites of Jupiter and Uranus during an episode of giant planet migration. Earth Planet Sci Lett 506:407–416

Wong EW, Brasser R, Werner SC (2021) Early impact chronology of the icy regular satellites of the outer Solar System. Icarus 358:article 114184

Wong E, Brasser R, Werner SC, Kirchoff MR (2023) Saturn's ancient regular satellites. Icarus 406:115763

Woronow A (1978) A general cratering-history model and its implications for the lunar highlands. Icarus 34(1):76–88

Woronow A (1985) Recovering crater production functions from densely cratered surfaces. In: Proc. 15th lunar planet. sci. conf., Part 2, 90, Suppl., pp. C825–C827

Yin A, Pappalardo RT (2015) Gravitational spreading, bookshelf faulting, and tectonic evolution of the South Polar Terrain of Saturn's moon Enceladus. Icarus 260:409–439

Zahnle K, Schenk P, Levison H, Dones L (2003) Cratering rates in the outer Solar System. Icarus 163:263–289

Zahnle K, Alvarellos JL, Dobrovolskis A, Hamill P (2008) Secondary and sesquinary craters on Europa. Icarus 194:660–674

Zannoni M, Hemingway D, Casajus LG, Tortora P (2020) The gravity field and interior structure of Dione. Icarus 345:113713

Zhang K, Nimmo F (2012) Late-stage impacts and the orbital and thermal evolution of Tethys. Icarus 218:348–355

Zhang Z, Hayes AG, Janssen MA, Nicholson PD, Cuzzi JN, de Pater I, Dunn DE (2017) Exposure age of Saturn's A and B rings, and the Cassini division as suggested by their non-icy material content. Icarus 294:14–42

Authors and Affiliations

Alyssa Rose Rhoden[1] · Sierra N. Ferguson[1] · William Bottke[1] · Julie C. Castillo-Rogez[2] · Emily Martin[3] · Michael Bland[4] · Michelle Kirchoff[1] · Marco Zannoni[5] · Nicolas Rambaux[6] · Julien Salmon[1]

✉ M. Zannoni
m.zannoni@unibo.it

1 Southwest Research Institute, 1050 Walnut St, Boulder, CO 80302, USA

2 Jet Propulsion Laboratory, California Institute of Technology, Pasadena, CA, USA

3 Center for Earth and Planetary Studies, National Air and Space Museum, Smithsonian Institution, Washington DC, USA

4 U.S. Geological Survey, Astrogeology Science Center, Flagstaff AZ, USA

5 Dipartimento di Ingegneria Industriale, Alma Mater Studiorum – Università di Bologna, Forlì, Italy

6 IMCCE, CNRS, Observatoire de Paris, PSL Université, Sorbonne Université, Université de Lille 1, UMR 8028 du CNRS, 77 Denfert-Rochereau, 75014 Paris, France

Space Science Reviews (2025) 221:56
https://doi.org/10.1007/s11214-025-01181-7

Constraining the Mass and Energy of Enceladus' Dissipation Systems

Carly J.A. Howett[1] · Georgina M. Miles[2] · Lynnae C. Quick[3]

Received: 9 October 2023 / Accepted: 7 June 2025 / Published online: 19 June 2025

Abstract
NASA's Cassini mission revealed endogenic activity at the south pole of Saturn's moon Enceladus. The activity is concentrated along four fractures in Enceladus' ice shell, which are much warmer than their surroundings and the source of Enceladus' plumes. This work provides a review of the current state of knowledge of the energy and mass lost by Enceladus through this activity. Specifically, we discuss the composition of the plumes, along with their spatial and temporal variation. The mass flux loss predicted for the three plume constituents (gas, dust and charged particles) is reviewed and a total mass flux of ejected material that subsequently escapes Enceladus is estimated to be 2.1×10^{11} kg over a Saturn year. Given that Enceladus' ocean is predicted to be 10^{19} kg this loss is sustainable in the very long term ($\sim$1.5 billion Earth years). However, unless a resupply mechanism (such as serpentinization) exists molecular hydrogen is expected to be depleted within $\sim$1 million Earth years. The difficulty in determining Enceladus' current heat flow is outlined, along with the advantages and disadvantages of the various techniques used to derive it. We find a robust lower limit for Enceladus' exogenic production is 7.3 GW. Tidal heating models show endogenic emission of this level is sustainable, and Enceladus may have long-term near-surface heating (a result supported by studies of Enceladus' geology). Finally, we offer suggestions for future observations, instrumentation, and missions. Enceladus remains a high-priority target for NASA, and as such it is highly likely that we will return to study this enigmatic world. Hopefully these missions will answer some of the questions that remain.

Keywords Satellites · Surface · Plume · Tides

1 Introduction

Along with Europa and Triton, Enceladus belongs to a small family of cryovolcanically active moons in our solar system (Fagents et al. 2022). The activity occurs inside Enceladus' South Polar Terrain (SPT) along four fractures, or sulci, commonly known as Tiger Stripes

✉ C.J.A. Howett
carly.howett@physics.ox.ac.uk

[1] Atmospheric, Oceanic and Planetary Physics, University of Oxford, Oxford, UK

[2] Southwest Research Institute, Boulder, CO, USA

[3] NASA Goddard Space Flight Center, Greenbelt, MD, USA

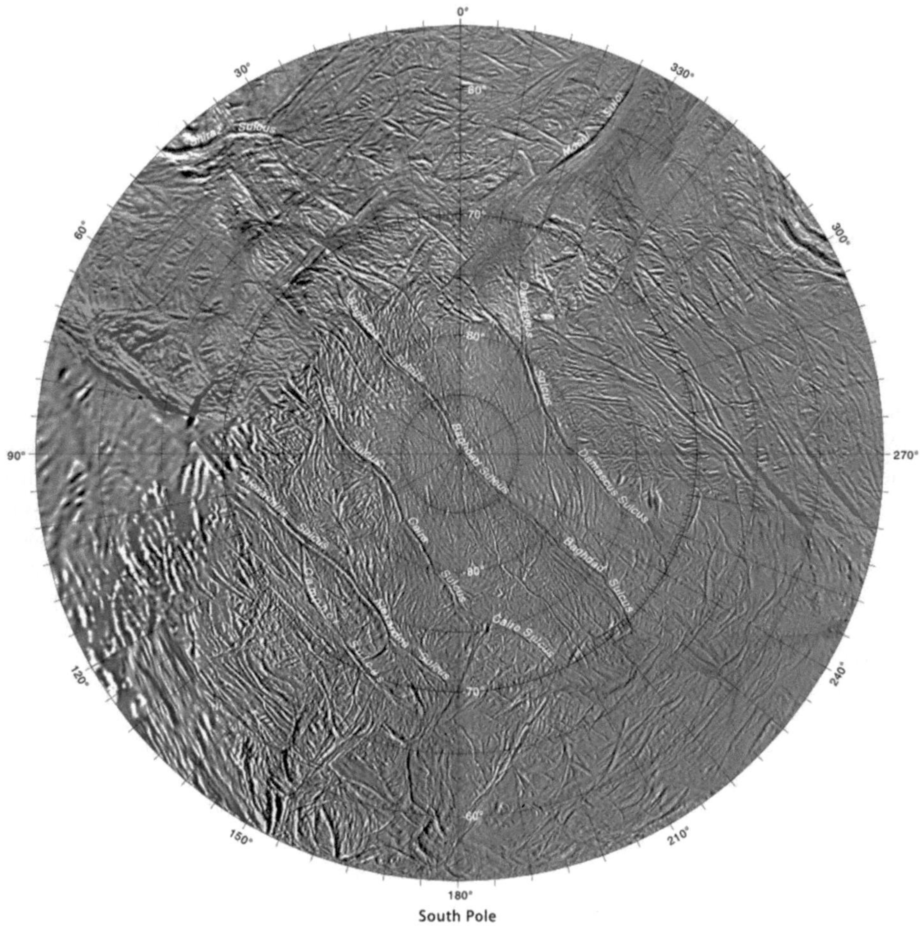

Fig. 1 Enceladus' South Pole, with the major sulci nomenclature included. Image from (USGS 2015)

(Fig. 1). The fractures are named after cities or countries referred to in The Arabian Nights, and the four largest fractures are Damascus, Baghdad, Cairo, and Alexandria Sulcus (see Fig. 1 for locations). These fractures are approximately 130 km long, with a central trough 2 km wide and 0.5 km deep. They are the source of both the plumes (Fig. 2) (Helfenstein and Porco 2015; Porco et al. 2014, 2006a; Spitale et al. 2015; Spitale and Porco 2007) and Enceladus' thermal emission (Fig. 3) (Spencer et al. 2006).

Enceladus' south polar plumes represent the only manifestation of its cryovolcanic activity. This contrasts with Europa and Triton, that display a variety of flow-like features and landforms that are indicative of effusive eruptions (Croft et al. 1995; Fagents 2003; Kattenhorn and Prockter 2014; Lesage et al. 2021; Miyamoto et al. 2005; Prockter et al. 2017; Quick et al. 2022, 2017). Triton's surface is also littered with features that may have been emplaced because of explosive cryovolcanism, including pitted cones, and ring and pit paterae and their deposits (Croft et al. 1995; Helfenstein et al. 1992).

Enceladus' plumes were seen to be continuously active throughout the Cassini mission (e.g., Ingersoll et al. 2020), and are expected to be at least decadal-long, starting well before Cassini's arrival (Jurac and Richardson 2001). By contrast Europa's putative plumes appear

Fig. 2 A mosaic of two Cassini ISS NAC images taken of Enceladus' plumes on 21st November 2009. Image credit: NASA PIA11688

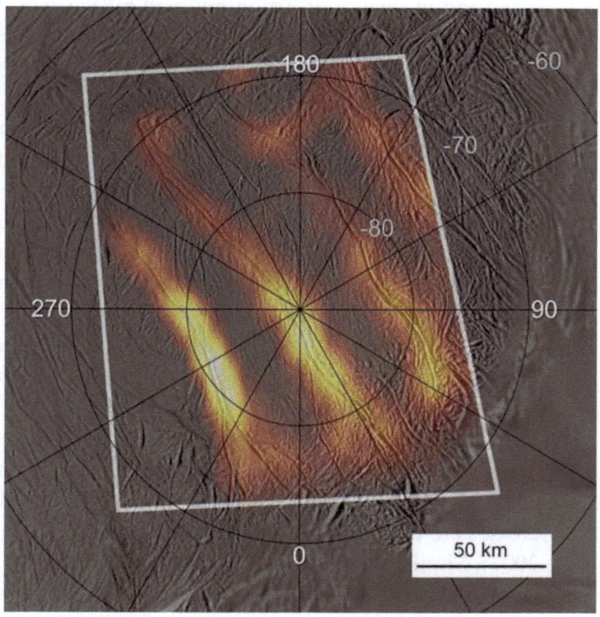

Fig. 3 Heat flow across Enceladus' South Polar Terrain. The warmest temperatures (bright yellow) are at ~180 K, and the background (uncolored inside the white box) is at ~72 K. Image credit: NASA PIA10361

to be someone elusive. While ground- and space-based telescopic observations have provided evidence for Europa's plumes, they may be sporadic with large, 200 km tall, plumes being infrequent outliers (Paganini et al. 2019; Rhoden et al. 2015; Roth et al. 2014a,b; Sparks et al. 2017, 2016). This is consistent with modeling (Fagents 2000; Quick et al. 2013) and past searches for plumes in the Galileo dataset (Phillips et al. 2000) which sug-

Table 1 A comparison of properties for Mars', Triton's, and Enceladus' plumes. Table adapted from Table 1 of Hansen et al. (2021) using values from Hansen et al. (2021) and references therein

	Mars	Triton	Enceladus
Erupting Volatile	CO_2	N_2 or H_2O	H_2O
Surface Gravity (m/s^2)	3.72	0.779	0.113
Plume Height (km)	0.08	8	1500
Plume Height Normalized to Triton Gravity (km)	2	8	60
Source Vent Diameter (m)	< 1	< 3000	~ 9
Exit Velocity (m/s)	20-300	20-40	450
Vapor Mass Flux	150 g/s	up to 400 kg/s	200 kg/s
Particle Mass Flux	30-150 g/s	< 10 kg/s	~ 50 kg/s
Eruption Temperature (K)	140	38-42	76-170
Eruption Duration	< 2 hours	1-3 years	continuous

gest that plume activity on Europa may be small-scale and transient in nature. We note that the community remains divided on whether Europa does indeed have plumes, and recent JWST observations found no evidence for strong active plumes (Villaneuva et al. 2023) at the time of their study. However, for the purposes of this review we assume Europa's plumes exist and use their constraints to better contextualize those of Enceladus.

Triton's plumes are quite different to both those on Europa and Enceladus. The activity was discovered in images of Triton returned by Voyager 2, and shown to be four N_2-rich plumes located between 50°S and 57°S. These plumes reached heights of up to 8 km above the surface (Soderblom et al. 1990). The initially favored eruption hypothesis for Triton's plumes was that they were solar driven (Kirk et al. 1995), similar to seasonal jets on Mars (Thomas et al. 2011). However, the discovery and characterization of Enceladus' plumes has led to a reevaluation of plausible hypotheses for eruptive venting on Triton (Hansen and Kirk 2015). Indeed, Hansen et al. (2021) note that the mass flux of Triton's plumes is more akin to the plumes at Enceladus than to the Martian jets (Table 1). In addition, new analyses by Hofgartner et al. (2022) suggest that neither the locations of the plumes, or of their fan-like deposits, strongly support the solar-driven model. While Kirk et al. (1995) invoked a cryovolcanic model for the formation of Triton's plumes, a separate endogenic eruption model was also proposed, in which the base of a polar cap composed of N_2 ice is melted by internal heat, and plumes form when the melt is exposed to the surface (Brown and Kirk 1994).

Venting on Enceladus, Europa, and Triton may be driven, at least partially, by tides and tidal stresses (e.g., Nimmo et al. 2014; Nimmo and Spencer 2015; Rhoden et al. 2015). Enceladus' plumes appear brightest at apoapsis— a time during its orbit when extensional tidal stresses are the greatest, allowing the fractures to open (Hedman et al. 2009; Hurford et al. 2007). Similarly, cryovolcanic activity may occur on Europa during times in its orbit when extensional stresses facilitate the opening of fractures and the exposure of subsurface water (Fagents et al. 2022). Internal energy from obliquity tides could drive cryovolcanism on Triton (Nimmo and Spencer 2015). Nevertheless, ocean-derived materials that participate in cryovolcanic eruptions on Europa and Triton may experience a more complex path to the surface compared to cryovolcanic materials on Enceladus. Owing to its small size, low surface gravity and thin south polar ice shell, ocean-based fluids erupt directly through south polar fissures (Čadek et al. 2016; Manga and Wang 2007; Nakajima and Ingersoll 2016; Postberg et al. 2011). Indeed, the presence of molecular hydrogen, salts and SiO_2 in

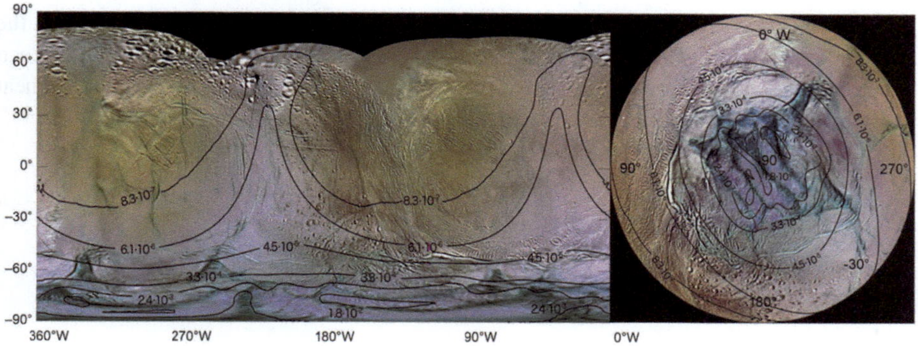

Fig. 4 Color map of Enceladus using ISS GRN/UV3, IR3/GRN and IR3/UV3 for RGB color (Schenk et al. 2011). Overlaid are the plume fallout predictions from Southworth et al. (2019). Image credit: Postberg et al. (2018)

Enceladus' plumes indicate that plume material emanates from an internal ocean that is in direct contact with a rocky mantle (Hsu et al. 2015; Postberg et al. 2011). Conversely, owing to their larger masses, thicker ice shells and higher surface gravities, cryovolcanic materials on Europa and Triton are likely to be sourced from discrete liquid reservoirs or melt lenses within their icy shells (Fagents 2003; Hofgartner et al. 2022; Schmidt et al. 2011).

Models of the predicted distribution of Enceladus plume deposits (Kempf et al. 2010; Southworth et al. 2019) have a strong spatial correlation with Enceladus' surface color change (Schenk et al. 2011) and changes in the water ice distribution (Postberg et al. 2018; Scipioni et al. 2017). From these works it is believed that the plume material that doesn't escape Enceladus' gravity falls back on its surface in two "wedge" shapes centered approximately along 45°W and 225°W (Fig. 4). Recent work has revealed no large-scale surface changes due to the emplacement of plume fall-out on Europa (Schenk 2020), modeling has shown that deposits emplaced by putative large Europan plumes may be difficult to identify (Quick and Hedman 2020). However, one exception may be the low-albedo deposits along ridges and other lineated features on Europa, which may have been emplaced during episodes of explosive cryovolcanism (Fagents 2000). On Triton, surface alteration by plume fallout is easy to spot, with over 120 fan-shaped dark deposits associated with plume activity being seen against the surrounding brighter surface. So many deposits suggest that plume activity could be a common occurrence on Neptune's largest moon (Hansen et al. 1990; Kirk et al. 1995).

Regardless of where their cryovolcanic materials originate, the mass- and energy-loss created by plumes have shaped these worlds. In the case of Enceladus the plume material has covered not only its surface, but sustains the E-ring, and coats and brightens the surface of its neighboring satellites (Verbiscer et al. 2007). The rest of this article delves into both the mass and energy loss Enceladus experiences from its activity. There is a particular emphasis on constraining the magnitude of these losses, their implications for Enceladus' interior, and the insights they offer for models of geological and internal processes on the icy moons.

While the energy loss in the SPT associated with the plumes and cryovolcanic activity via latent heat is significant, by far the principal mechanism of heat loss from the energy created by tidal heating is conduction of heat from the ocean through the ice shell. At a significantly lower intensity, summed over the surface of Enceladus it is estimated to be 5-8 times the energy associated with the plumes (Hemingway and Mittal 2019; Čadek et al. 2016; Spencer et al. 2018). At present we can only predict this heat loss using models of ice shell thickness (from gravity, topographic and libration data) and assumptions about the structure and properties of the ice shell to conduct heat. The longevity of Enceladus' ocean is currently unknown but will be governed by the balance of heat generated within and heat lost via global conduction and the more acute cryovolcanic activity in the SPT.

2 Observations of Enceladus' Mass Flux

2.1 Introduction

The predominant mass flux loss mechanism on Enceladus is the ejection of plume material that erupts from fissures in the South Polar Terrain (SPT). The discovery of Enceladus' plumes was arguably one of the most groundbreaking moments of the Cassini mission, and constraining the plumes' properties quickly became one of the mission's highest priorities. This is because it is hypothesized that the plume composition can shed light on the composition of and conditions of Enceladus' potentially life-supporting sub-surface ocean. The plume ejection has two parts: diffusive emission from along the fractures, sometimes referred to as curtains (Spitale et al. 2015), and from the collimated emission from the $\sim$100 discrete jets (Porco et al. 2014). We use the term plume to describe both these parts.

Enceladus' plume is comprised of three components: solid (dust) particles, gas, and ions. While water is the major constituent of each of these components their individual composition and size varies. Variation is also seen both spatially and temporally. Solid particles are the heaviest components, and as such are most difficult to accelerate to speeds above Enceladus' escape velocity with only 5-10% (by mass) escaping (Postberg et al. 2018). Most of the particles fall back down to coat Enceladus' surface in clearly distinguishable patterns (Fig. 4). However, the lighter neutral gas and charged particles are ejected at such high velocities that they almost all entirely escape Enceladus (Hansen et al. 2008). The escaping particles go on to populate the E-ring (Ingersoll and Ewald 2011; Kempf et al. 2010, 2008; Porco et al. 2006a) and the charged particles are rapidly coupled into Saturn's magnetosphere. So, plume material is found not just in the plume, but also in the E-ring and on Enceladus' surface.

2.2 Instruments Able to Measure Mass Flux and Composition

Cassini was well equipped to study Enceladus' activity with both in situ and remote sensing instrumentation. Both instrumentation types were used to determine the composition of the plume's vapor, solid and charged particles. In this section we briefly describe the instrumentation that was used to make these detections, along with their advantages and limitations.

2.2.1 Solid Particles (CDA/CAPS/VIMS)

The Cosmic Dust Analyzer (CDA) (Srama et al. 2004), Cassini Plasma Spectrometer (CAPS) (Young et al. 2004), the Visible and Infrared Mapping Spectrometer (VIMS) (Brown et al. 2004) and the Imaging Science Subsystem (ISS) (Porco et al. 2006b) all constrained the composition of the plume's solid particles. CDA and CAPS are in situ instruments, while VIMS and ISS remotely sensed the plumes, measuring their composition along its line of sight.

CDA is an impact ionization detector that works by ionizing dust and ice grains that impact its detector's metal target plate at high speed (>1 km/s). A time-of-flight mass spectrometer (TOF-MS) then produces a mass spectrum of the cations present. The mass resolution ($^m/_{\Delta m}$) varies from ~10 at 1 u (where u is the unified atomic mass unit) to ~50 at 200 u (Postberg et al. 2006). CDA recorded a few hundred spectra of Enceladus' plume, and many thousands of Saturn's E-ring (Postberg et al. 2018).

CAPS had three sensors: Electron Spectrometer (ELS), Ion Mass Spectrometer (IMS), and the Ion Beam Spectrometer (IBS) (Young et al. 2004). ELS and IMS were primary used to observe the charged nanograins in Enceladus' plume. ELS is a hemispherical top-hat electrostatic analyzer, which was employed during the Enceladus plume flybys was used to detect the direction and energy per charge of negatively charged nanograins. ELS could detect grains with a mass to charge ratio (m/q) of $10^3 - 10^4$ u/q, which made it sensitive to grains sizes between the small neutral gas grains and charged particles, and the larger (>0.4 μm diameter) macroscopic solid ice particles. Postberg et al. (2018) provides full operational details for ELS in the plume. CAPS' IMS provided electrostatic E/q analysis of positive ions between 1 and 34,700 eV/e and included a time-of-flight system for composition (m/q) analysis.

VIMS had both visible (VIS) and infrared (IR) channel which provided sensitivity over its 0.25 to 5.1 μm spectral range, with an effective field of view (iFOV) of 0.5 by 0.5 mrad. The two channels were illuminated via independent telescopes, that were typically operated in conjunction with one another to provide a single spectral cube. As Enceladus' plume is strongly forward scattering, with a low optical depth, it was challenging for VIMS to remotely observe. However, VIMS was able to constrain the plume characteristics by using occultation measurements.

ISS was Cassini's imaging device, made of two framing cameras: a narrow (NAC) and wide (WAC) angle cameras with FOVs of 0.35° and 3.5° across respectively. In both cameras the light was detected by a 1024 by 1024 Charge-Couple Device (CCD). The NAC had a pixel scale of 6 μrad/pixel, while the WAC's was ~60 μrad/pixel. Filter wheels allowed the camera to span from 200 to 1100 nm.

2.2.2 Gas Phase (UVIS/INMS)

The Ultraviolet Imaging Spectrograph (UVIS) (Esposito et al. 2004) was composed of an Extreme UltraViolet (EUV) and Far Ultraviolet (FUV) spectrograph, a High-Speed Photometer (HSP) and a Hydrogen Deuterium Absorption Cell (HDAC). EUV and FUV covered 563 to 1182 Å and 1115 to 1912 Å respectively. UVIS is a remote sensing instrument, that detected neutral gases in the plume by looking for absorption features in stellar and solar occultations in order to provide the composition integrated through the line of sight between UVIS and the star.

The Ion and Neutral Mass Spectrometer (INMS) (Waite et al. 2004) was a dual-ion source quadrupole mass spectrometer. It covered a mass-to-charge range of 0.5 to 8.5 and 11.5 to 99.5 u. It could operate in a closed- and open-ionization configuration, meaning both inert and reactive species and ions could be detected.

2.2.3 Charged Particles (CAPS and MIMI)

As described above, CAPS used ELS and IMS to measure the negatively and positively charged particles in the plume respectively. The Magnetospheric Imaging Instrument (MIMI) (Krimigis et al. 2004) had three sensors: Low-Energy Magnetospheric Measurement System (LEMMS), Charge-Energy Mass Spectrometer (CHEMS) and Ion and Neutral

Camera (INCA). LEMMS measured the velocity, direction and number of electrons, ions and protons that struck its detector. CHEMS measured both the charge and composition of particles that struck its detector, and INCA was able to image high velocity atoms.

2.3 The Composition and Variability of Enceladus' Plume

2.3.1 Solid Particles

As already mentioned, most of the plume particles fall back onto Enceladus' surface (Postberg et al. 2018), but 5-10% are able to escape Enceladus and sustain the E-ring (Ingersoll and Ewald 2011; Postberg et al. 2018). So, this means that the plume's solid particle composition can be either directly measured or inferred from E-ring studies.

Solid particles have been detected by CDA, CAPS and VIMS, in both the plume and the E-Ring.

However, since the E-ring is much larger than the plumes the in-situ instruments had more opportunity to sample it than the plumes. Most of the compositional information on plume/E-ring ice grains comes from CDA (Postberg et al. 2018), which showed the grains can be broadly separated into three types:

- Type 1 particles ($\sim$65% of E-ring particles) are almost pure water ice, with very low concentrations of alkali salts (Na/H_2O ratio of 10^{-7}) (Postberg et al. 2009).
- Type 2 particles ($\sim$25% of E-ring particles) are larger than Type 1 but roughly the same composition with some additional organic species (which manifest as 27 u to 31 u and/or 39 u to 45 u species) (Postberg et al. 2008). The ratio of organics to non-organics, and the concentration of silicate and salt varies between the particles.
- Type 3 particles ($\sim$10% of E-ring particles) are also much larger than Type 1 particles but are compositionally very different to the other types having a pronounced sodium signature (Na/H_2O $>10^{-3}$).

VIMS provided additional insight into the structure of the water of Enceladus' plume, showing that the water ice emerging from Cairo, Baghdad and Damascus was all crystalline (Dhingra et al. 2017). This implies that the water plume particles were formed at temperatures $>$130 K.

The concentration of these particles within the plume varies on short (diurnal/orbital), medium (monthly) and long (yearly) timescales. Since Enceladus is tidally locked to Saturn it takes the same time to orbit Saturn as it does to rotate (i.e., a day is the same length of time as its orbit). So, the terms orbital and diurnal can be used interchangeably. However, short-term variation in plume concentration varies as a function of its position (Hedman et al. 2013; Ingersoll et al. 2020; Ingersoll and Ewald 2017; Nimmo et al. 2014) (and not time of day) so we default to using the term orbital.

Hedman et al. (2013) used VIMS data to study changes in Enceladus' plume over its orbit. A position in an orbit is defined by mean anomaly (MA), which is the time from pericenter (when Enceladus is closest to Saturn) multiplied by 360° and divided by the orbit period. So pericenter is at 0° and apocenter (when Enceladus' is furthest from Saturn) has a mean anomaly of 180°. Hedman et al. (2013) also showed that the horizontal integrated plume brightness is three times greater at apocenter than pericenter, and that the orbital position appears to affect particle velocity (with lowest velocities at apocenter) but not the particle size distribution.

These initial findings were confirmed using ISS observations, which found the plume's peak brightness to occur at apoapse, notably with a scale height that didn't vary with orbital

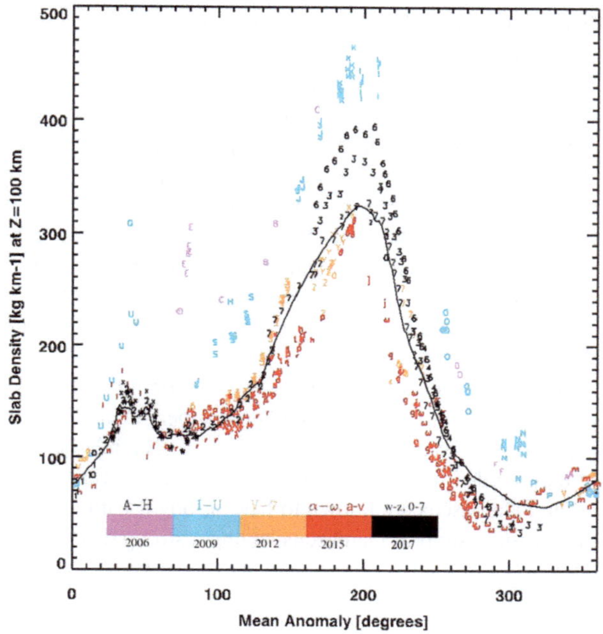

Fig. 5 Slab density verses mean anomaly (orbital position) on Enceladus. From Ingersoll et al. (2020)

position or plume brightness (Nimmo et al. 2014). This implies that (in contrast to Hedman et al. 2013) the particle velocities are constant. These results supported the theory that tidal stresses control Enceladus' plume emission (Nimmo et al. 2014). Further modelling showed that the location of Enceladus' jets, the location of warm surface regions and diurnally varying tidal stresses is coincident, and they could be sufficient to open Enceladus' fractures all the way from the surface to the sub-surface ocean (Porco et al. 2014).

A follow up ISS study observed a four-fold increase in plume brightness at apocenter than pericenter (Ingersoll and Ewald 2017). This was followed by the most comprehensive study of ISS images to date (Ingersoll et al. 2020), which analyzed 2415 ISS observations to find that the plumes change in brightness with orbital position is indeed very complex:

- The large maximum occurs slightly after apocenter (the maximum is at MA∼200°)
- There is a secondary maximum in the plume brightness, made up of two peaks with an MA of 25° and 60°.

This same study also found evidence of the plumes changing on monthly timescales. ISS observations of jets taken at the same mean anomaly were used to track their output and showed about half of the ten jets observed had changed over the course of a month (Ingersoll et al. 2020). Some of the jets turned on, while others disappear. Porco et al. (2014) also showed jet activity varied with time, but couldn't correlate the observed activity an individual jet to Enceladus' stress field. Instead, they suggest individual geysers turn off when they become clogged with ice from condensed vapor.

Between 2007 and 2017 there was a factor of two decrease in the plume brightness (Fig. 5) (Ingersoll et al. 2020; Ingersoll and Ewald 2017). This long-term brightness change is best fitted by a pair of sinusoids with a period of ∼4 years and ∼11 years (Porco et al. 2018), which agrees with the known tides on Enceladus from its resonance with Dione. Ingersoll et al. (2020) see evidence for variation in the launch velocity of plume particles, from their stable 2009-2015 value to an elevated one in 2017. This means their results are consistent with Hedman et al. (2013) but not with Nimmo et al. (2014).

While the total brightness of the plumes varies temporally, the composition of the plumes varies spatially. Generally, Type 3 grains are more concentrated close to the surface while Type 1 particles were less so, and Type 2 show a more complex variation pattern (Postberg et al. 2011). One explanation is that the ejection speeds of the smaller Type 1 grains are faster than the larger Type 3 ones (Postberg et al. 2011), perhaps due to a radius-dependent frictional force in the vent (Schmidt et al. 2008). Thus, the larger Type 3 particles are less able to escape Enceladus' gravity than the smaller Type 1 particles and are more highly concentrated close to the surface (Hedman et al. 2009; Kempf et al. 2008). This could also explain why Type 1 particles are more dominant in the E-ring (Khawaja et al. 2017), because they are more able to escape.

Type 2 particles appear to vary with the ground-track of the observations, being most pronounced over Damascus Sulcus (Postberg et al. 2011). Combined UVIS and VIMS analysis showed the dust/gas mass ratio above Baghdad and Damascus Sulci is about an order of magnitude higher than that of Alexandria and Cairo Sulci (Hedman et al. 2018). Surfaces around Damascus are known to have strong organic-ice features (Brown et al. 2006). Thus, these results support Damascus having a strong dust signal that is notably rich in organics.

2.3.2 Vapor

During Cassini's mission UVIS and INMS made numerous observations of Enceladus' plume vapor. Early observations established the gas was predominately water (Hansen et al. 2006; Waite et al. 2006) with a column density of 1.5×10^{16} H_2O molecules cm^{-2} over an altitude of 7 to 30 km (Hansen et al. 2006). INMS initially constrained the remainder of the gas plume to be ~3% CO_2, 4% unknown species with mass 28 u (CO, N_2 and C_2H_4 being some of the candidates), <0.5% NH_3, and trace amounts of acetylene and propane (Waite et al. 2006).

Analysis of UVIS data was able to provide some insight onto possible mass 28 u species. The data provided an upper limit of CO (<0.9%), N_2 (0.5%) and found evidence of C_2H_4 but couldn't define a mixing ratio (Hansen et al. 2006, 2017). However, later observations made by INMS in their open-source mode gave no detection at 28 u. In open source mode the particles are ionized without interaction with the instrument's walls, which leads to less fractionation of heavier particles. So perhaps the source of 28 u is actually a fragmentation product (e.g., CO or C_2H4) of a heavier organic molecules, rather than a specific 28 u mass one. This open-source data also enabled the mixing ratio of H_2 to be determined (0.4 to 1.4%) (Waite et al. 2017).

During the Cassini mission INMS refined the process of taking their data, finding a sweet spot between the fast speeds required for high signal-to-noise and the effect those high speeds had on producing fragmented species from high-speed wall impacts. Three encounters (E14, E17 and E18) in 2011 and 2012 particularly benefited from this work. They were taken at the sweet spot, with a relative speed of ~7.5 km/s (Postberg et al. 2018) and an altitude of 75 to 100 km above the fractures. Figure 6 shows the plume gas composition derived from data taken during the three flybys. The results are strongly consistent with one another, providing a confident determination of the gas species. The final volume mixing ratios are: H_2O (96-99%), CO_2 (0.3-0.8%), CH_4 (0.1-0.3%), NH_3 (0.4-1.3%) and H_2 (0.4-1.4%) (Waite et al. 2017). Compositional variations in the gas plume were observed, but are not currently linked to individual sources (Magee and Waite 2020, 2017; Waite et al. 2017)

The plume gas mass flux appears to not change notably with time. Analysis of five occultation observations made by UVIS showed the bulk water vapor coming from Enceladus

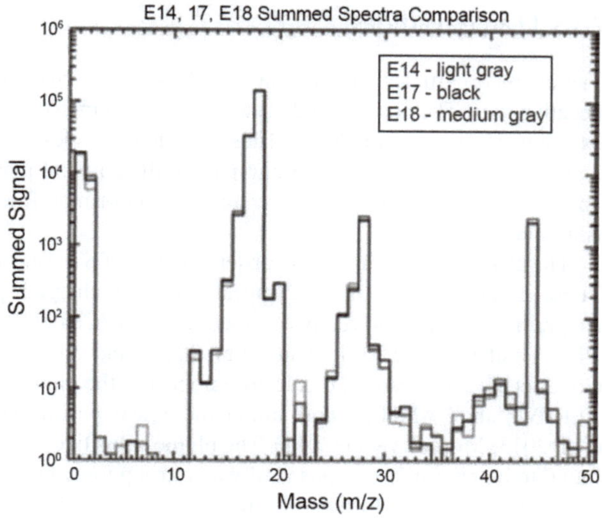

Fig. 6 Mass spectrum from INMS taken during E14 (light grey), E17 (black) and E18 (medium grey). From Postberg et al. (2018)

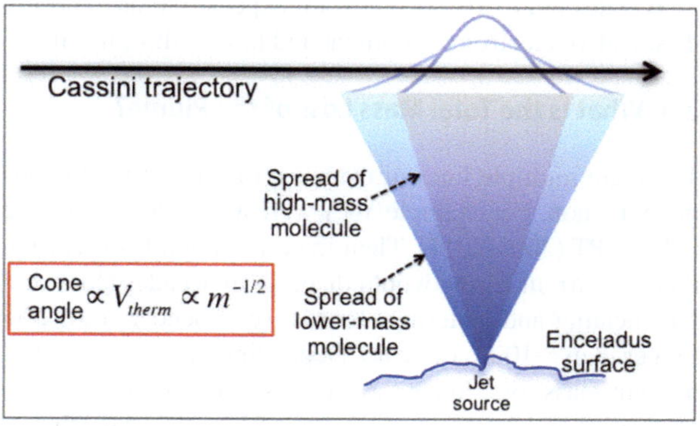

Fig. 7 The mass-dependent behavior of high-velocity molecules emitted by Enceladus' jets. It is assumed that all molecules are emitted at the same supersonic velocities. The figure shows that for species in thermal equilibrium the amount of spreading depends on their mass. From Perry et al. (2015)

was approximately constant with a surface density varying between 0.9×10^{16} cm^{-2} and 1.5×10^{16} H_2O molecules cm^{-2} (i.e. ±15%) and did not display correlation with orbital position (Hansen et al. 2017). However, interestingly the mass flux from individual gas jets was shown to significantly vary (Hansen et al. 2017; Teolis et al. 2017).

Spatial variations of the concentration of neutral gas within the plume are substantial. The gas is emitted from jets at supersonic velocities and at slower speeds from the diffuse emission (Hansen et al. 2008; Teolis et al. 2017). The fast-moving gas molecules leaving a jet will spread as a function of their molecular mass with low-mass molecules spreading more than high-mass ones (assuming they have the same eruption velocity and are in thermal equilibrium) (Fig. 7). Analysis of INMS data showed that this results in a spatial variation in the plume's composition (Perry et al. 2015), with heavy species (e.g., CO_2) spreading less than lighter ones (e.g., H_2 and H_2O) (Yeoh et al. 2015). INMS typically sampled the plume at altitudes between 50 and 100 km altitude. Through these encounters the spreading angle was shown to vary from 10° to 45°, depending on the temperature, velocity and mass (Hurley et al. 2015; Perry et al. 2015).

2.3.3 Charged Particles

The charged particles can be split into three types: charged nanograins, positive ions, and negative ions. The positive and negative ions are largely formed within the plume by charge exchange (Cravens et al. 2011). These charged particles, once formed, are eventually lost to Saturn's magnetosphere, becoming part of the ambient plasma torus (Johnson et al. 2006; Tokar et al. 2006). Therefore, their ability to inform on Enceladus' surface and sub-surface is limited.

The charged nanograins were observed by CAPS in Enceladus' plume (Jones et al. 2009). These grains are larger than the neutrals and other charged species, but smaller than the solid ice grains. They are indirectly assumed to be icy (Coates et al. 2010a,b; Hansen et al. 2011, 2008; Postberg et al. 2011; Waite et al. 2017), and carry both positive and negative charges (Jones et al. 2009) caused by electron impact in the plume itself (Hill et al. 2012).

RPWS and CAPS ELS both saw an increase in plasma density close to Enceladus (Coates et al. 2013; Morooka et al. 2011). The plume's positive ions were studied by CAPS IMS and found to be primarily composed of water group ions (O^+, OH^+, H_2O^+ and H_3O^+) (Young et al. 2005). H_3O^+ is formed via charge exchange, when magnetospheric ions interact with plume H_2O to form H_2O^+ and H_3O^+ (Cravens et al. 2011). Negative water ions (OH^-, O^-, H^-) are also observed in the plume

(Coates et al. 2010a). They are expected to have been formed inside the plume itself by dissociative electron attachment and have a short lifetime.

2.4 What Is the Total Mass Loss of the Plume?

There are multiple lines of evidence to support the hypothesis that plumes are a long-lived phenomenon. For example, they were active during every encounter Cassini had with Enceladus' SPT (2005-2017). Their material supports the E-ring, and without it then it is expected that the current E-ring would disperse in decades (Jurac and Richardson 2001). Finally, surface heating and cratering records (see Spencer et al. 2006) point to activity lasting much longer than ~1000s of years. Depending on the plume flux such activity could result in significant mass loss. Since the vapor, solid and ions are detected separately so are their mass loss rates, and in this section we explore the mass flux of these different plume particles separately.

UVIS data initially constrained the H_2O vapor from eruptions to be occurring at a rate of 150 to 350 kg/s (Hansen et al. 2006). Subsequent occultations supported these results, finding rates between 170 and 250 kg/s (Hansen et al. 2017, 2011). These initial estimates assumed ice temperatures which gave the gas a thermal velocity much higher than Enceladus' escape velocity (240 m/s), implying all the vapor escapes Enceladus. So, these rates imply that over each Saturn year (29.4 Earth years) Enceladus is losing $1.4 \times 10^{11} - 3.2 \times 10^{11}$ kg of H_2O through the gas plumes alone.

There is large uncertainty in the flux of solid particles emitted from Enceladus, with estimates varying from 3 to 65 kg/s (Dong et al. 2015; Ingersoll and Ewald 2011; Kempf et al. 2018, 2010; Schmidt et al. 2008). Assuming a conservative 20 kg/s (Kempf et al. 2018) this means 1.9×10^{10} kg of solid material is ejected from Enceladus' over the course of a Saturn year (6.3×10^8 kg over an Earth year). If 10% escapes Enceladus (Postberg et al. 2018) then the particle mass loss from Enceladus is 1.9×10^9 kg per Saturn year (6.3×10^7 kg over an Earth year). This is a sizeable amount but notably lower than the vapor loss.

Charged particles are expected to form in the plume by charge exchange, and it is difficult to determine their specific contribution to the total mass loss. Despite this some estimates do

Table 2 Summary of the mass flux estimates for the total vapor, solid and charge particles in Enceladus' plume in the literature (unless otherwise indicated)

Plume type	Mass flux (kg/s)	Reference
Vapor		
	150 - 350	(Hansen et al. 2006)
	100	(Schmidt et al. 2008)
	200	(Hansen et al. 2011)
	180 - 250	(Hansen et al. 2017)
H_2 Only	0.02 – 0.01	(Waite et al. 2017)
Solid		
	0.04	(Porco et al. 2006a)
	0.2	(Spahn et al. 2006)
	5	(Schmidt et al. 2008)
	50	(Ingersoll and Ewald 2011)
	20	(Postberg et al. 2018)
Charged Particles		
Nanograins	15-65	(Dong et al. 2015)
Negative and Positive Charged Particles	5.6	(Hill et al. 2012; Kempf et al. 2018)
Total Plasma	100	(Tokar et al. 2006)

exist. For nanograins the mass loss is expected to be between 15 kg/s and 65 kg/s (Dong et al. 2015; Kempf et al. 2018). The loss of positive and negative ions is expected to be lower, at 5.6 kg/s (Hill et al. 2012; Kempf et al. 2018). These charged particles are expected to escape Enceladus along field lines bent by Saturn's electromagnetic field (Dong et al. 2015; Hill et al. 2012). Assuming a conservative estimate of 15 kg/s for nanograins and 5.6 kg/s for the ions then Enceladus loses 1.9×10^{10} kg of charged particle material over the course of a Saturn year (6.5×10^8 kg over an Earth year).

The mass loss estimates described in this section are summarized in Table 2, showing that by far the largest mass loss is from the vapor. If mass fluxes of 200 kg/s are assumed for the vapor (Hansen et al. 2011), 20 kg/s for the solid particles (Kempf et al. 2018), 15 kg/s for the nanograins (Dong et al. 2015), and 5.6 kg/s for the charged ions (Postberg et al. 2018) then the total mass ejected is 241 kg/s, or $\sim 2.2 \times 10^{11}$ kg over a Saturn year (7.6×10^9 kg per Earth year). If everything escapes Enceladus, except for 90% of the solid particles, then the total mass lost is 223 kg/s, or $\sim 2.1 \times 10^{11}$ kg per Saturn year (7.0×10^9 kg per Earth year). If Enceladus' ocean is $\sim 10^{19}$ kg (Glein and Shock 2010) such loss rates are sustainable for ~fifty million Saturn-years (~1.5 billion Earth years).

Another constraint on the longevity of Enceladus' activity comes from the loss of molecular hydrogen (H_2), which has been estimated to be between 1×10^9 to 5×10^9 mol per year, equivalent to 0.02 to 0.1 kg/s, or 2.0×10^6 to 1.0×10^7 kg/year (Waite et al. 2017). Models indicate that the H_2 concentration in the ocean sourced plume will be between 10^{-7} to 10^{-4} mols (kg H_2O)$^{-1}$ (Postberg et al. 2011, 2009). So, an ocean of $\sim 10^{19}$ kg (Glein and Shock 2010) will not be an adequate reservoir of molecular hydrogen in less than one million Earth years (Waite et al. 2017), unless a resupply mechanism exists (such as serpentinization).

2.5 What Does the Mass Loss Tell Us About Enceladus' Interior?

The salt in the Type 3 particles and the minor component of plume gases (CO_2, NH_3, CH_4, H_2, heavy hydrocarbons and silica) provide constraints on Enceladus' ocean composition and dynamics. The leading explanation for this composition is that their source is a liquid water ocean in direct contact with a porous core (Hsu et al. 2015; McKinnon 2015; Postberg et al. 2009). This is consistent with the presence of the minor species, which require hydrothermal cycling between the interior and ocean/ice (see summary chapters by Glein et al. 2018 and Postberg et al. 2018 to explain their presence).

The question then becomes how do these particles form? One option is that they are formed by bubbles bursting at the ocean surface (from evaporation or upwelling of gasses). The spray that results can then be transported upwards by the vapor into space (Postberg et al. 2009). If this hypothesis is correct then the Type 3 grains are direct samples of the subsurface ocean, providing a direct constraint of the composition and salinity of Enceladus' ocean (0.5%) (Postberg et al. 2018).

Conversely the small Type 1 grains, which are almost pure water-ice, are likely to be sublimated ice or evaporated ocean water. The particle's trace sodium levels support the latter explanation, but it could be a mixture of them both (Postberg et al. 2009). This initial vapor then condenses into solid ice grains, a process that requires the vapor to pass through regions of local supersaturation. Such supersaturation conditions could be met by the fractures narrowing into small passageways. This process results in smaller ice grains than the freezing of ocean spray (which can then be accelerated to higher speeds by the escaping gas than their larger counterparts) and can explain the compositional differences between the Type 1 and Type 3 grain sizes.

Enceladus' surface tidal stresses can be constrained by the observed orbital variability of Enceladus' plumes. When the fractures are under compression they will be pressed together (or perhaps even shut), which is likely to reduce the material able to escape. The reverse is true in tension (fractures are pulled apart probably making it easier for material to escape). Current models of Enceladus' surface tension (e.g., Hurford et al. 2007 and Nimmo et al. 2007) show that fissures are expected to be under tension at apocenter and compressed at pericenter, which is in keeping with the brightest plumes being seen at apocenter (Hedman et al. 2013; Nimmo et al. 2014). Berne et al. (2024) showed strike-slip motion on Enceladus could explain the double peak observed in diurnal variations of jet activity, and also matches the tiger stripe's observed heat signature.

The ~monthly variation in jet activity can constrain the fracture size and ice deposition rates. If some of the plume's particles freeze onto the fracture walls as it ascends through the fractures this would both seal up the fracture, and release latent heat. Models of this process, that are able to explain the surface temperature observed, suggest that a crack 5 to 7.5 cm wide could seal in a few months (Nakajima and Ingersoll 2016).

2.6 What Constraints Does This Mass Loss Have on Models?

The observed variation in plume particle output on a variety of timescales indicates the venting process is a complex one. We know very little of the ocean-to-surface plumbing system, which could be anything between distributed porous flow through the ice to direct channels. However, any such model would have to explain the observed temporal variations. The orbital variability alone provides a large source of questions, such as how do wider fractures lead to more plume particles travelling at lower velocities, but with the same size distribution as those emitted by narrower fractures? Why does the particle density change over Enceladus' orbit, while the bulk gas density is (by comparison) constant? Why is the velocity a minimum when the flux is a maximum? How is the secondary peak in particle emission formed, is it due to emission from cross-cutting fractures as suggested by Ingersoll et al. (2020), or something else? Do cracks eventually get sealed by the freezing out of plume particles on the fracture walls? Do they ever reopen? Or more fundamentally, how do they start? Any proposed solution to these questions must also be able to explain longer term plume activity and variability. For example, it is important to determine whether Dione's induced 4- and 11-year tides are really able to cause the trends in plume variation observed. They also must address the long-lived nature of the activity, as all evidence points towards the plume activity being a long-term phenomenon.

The composition of the plumes sets some interesting constraint for interior ocean models. The sodium and potassium salts in the plume (Postberg et al. 2011, 2009) provides a strong constraint on how the salt-carrying plume particles are formed. For example, models that predict vapor is from the sublimation of ice have difficulty explaining their presence, since the ice is expected to be almost pure water-ice, salt free (e.g., Cooper et al. 2009; Kieffer et al. 2006; Loeffler and Baragiola 2009; Nimmo et al. 2007). Instead, the salts are better explained by liquid water models, which allow a subsurface ocean to be in direct contact with a rocky core. The water dissolved salts from the rocks, and eventually go onto be frozen at the surface into droplets that retain their salt content (Postberg et al. 2011). The timescale for this process is not well understood. Hsu et al. (2015) inferred it would take particles several months to transport through Enceladus' ocean. While Ames et al. (2025) use ocean models to show a stratified ocean cannot be made from seafloor to ice shell at Enceladus' polar regions, and thus any hydrothermal products will take upwards of 100s of years to reach the plume (unless the ice shell is not in quasi-equilibrium)

SiO_2 particles will grow to sizes larger than those observed over thousands of years (Spencer et al. 2018). So, the nano-sized silica particles point to a process that is ongoing, rather than one that's completed by reaching the maximum size of particles. If these models of silica growth are right then they themselves provide a constraint for ocean and habitability models, since the growth of such particles requires moderate salinity, alkaline pH and warm temperatures (>363 K) (Hsu et al. 2015; Sekine et al. 2015).

3 Enceladus' Energy Flux

3.1 Introduction

3.1.1 Enceladus' Inputs and Losses

Enceladus has two sources of energy: radiogenic heating from the radioactive decay of heavy elements in its core, and tidal heating due to its eccentric orbit of Saturn. The current day estimate of Enceladus' radiogenic heat is predicted to be $\leq$0.3 GW. Hemingway and Mittal (2019) estimate it to be <0.3 GW based on a chondritic radiogenic heating rate of $5\text{x}10^{-12}$ W kg^{-1} (Spohn and Schubert 2003) and a chondritic mass fraction of 0.5. Other studies estimate it to be $\sim$0.3 GW (Schubert et al. 2007; Travis and Schubert 2015), based on chondritic abundances of ^{235}U, ^{238}U, ^{232}Th, and ^{40}K. Either way, this is a lot lower than the 1.8 to 150 GW predicted by tidal heating (Lainey et al. 2020, 2017), and thus tidal heating is expected to be the dominant source of Enceladus' energy.

As previously mentioned, Enceladus and Dione are in a 2:1 mean motion resonance, which means Enceladus orbits Saturn twice, for every one of Dione's orbits. This resonance perturbs Enceladus' orbit, keeping it at a forced eccentricity of 0.0047 (Meyer 2008).

Furthermore, Enceladus is in a synchronous rotation of Saturn (i.e., its spin period and its orbital period are the same, and so the same hemisphere of Enceladus always faces Saturn). As such, Saturn's intense gravitational field deforms Enceladus along the line connecting the centers of the two bodies. This deformation creates a tidal bulge on Enceladus. Since the distance between Saturn and Enceladus is constantly changing (due to the non-zero eccentricity) the size of this tidal bulge is constantly varying too. Enceladus' own interior friction resists this change to its bulge, which results in heating. A more complete description of the complexities of tidal modeling are described in Nimmo et al. (2023, this collection).

In this section we consider how much of this energy Enceladus is able to dissipate as exogenic heat. Specifically, we describe the methods that have been used to constrain this heat flow, and their strengths and limitations.

3.1.2 Passive Thermal Models

In order to understand how much of Enceladus' heat flow is endogenic the exogenic component often needs to be well constrained. This is done through thermal models, which use the thermophysical properties of a surface to predict its passive surface temperature. This is usually achieved by solving the 1-dimensional heat equation:

$$\rho c_p \frac{\partial T(z,t)}{\partial t} = \frac{\partial}{\partial z}\left(\kappa \frac{\partial T(z,t)}{\partial z}\right) \tag{1}$$

Where ρ is the density, c_p is the specific heat capacity, κ is thermal conductivity, T is temperature, t is time and z is regolith depth. The 1-D heat equation can be simplified by writing it in terms of the surface's skin depth (l_s), which is defined as $l_s = \sqrt{\frac{\kappa}{\rho c_p \omega}}$, where ω is the angular rotation of the body. If the depth is now expressed in terms of the skin depth ($Z = \frac{z}{l_s}$) then the 1-D heat equation reduces to:

$$\frac{\partial T(Z,t)}{\partial t} = \omega \frac{\partial T(Z,t)}{\partial Z^2} \tag{2}$$

The upper boundary of these models balances the absorbed solar insolation against infrared emission to space:

$$\kappa \frac{\partial T(z,t)}{\partial z}\Big|_{z=0} = \varepsilon \sigma T(0,t)^4 - (1-A)\, F_\odot \tag{3}$$

Where $(1-A)\, F_\odot$ is the solar heating rate (i.e., the surface energy flux), A is the albedo, $F_\odot$ is the incident solar flux, ε is the infrared emissivity, σ is the Stefan–Boltzmann constant, and $T(0,t)$ is surface temperature. This boundary condition can also be rewritten in terms of skin depth:

$$\sqrt{\kappa \rho c_p \omega}\, \frac{\partial T(Z,t)}{\partial Z}\Big|_{z=0} = \varepsilon \sigma T(0,t)^4 - (1-A)\, F_\odot \tag{4}$$

Changing the boundary condition into these terms is useful, because all of the physical parameters are now combined into a single term, known as thermal inertia (Γ) which is defined as $\Gamma = \sqrt{\kappa \rho c_p}$. So, the boundary condition can be rewritten as:

$$\Gamma \sqrt{\omega}\, \frac{\partial T(Z,t)}{\partial Z}\Big|_{z=0} = \varepsilon \sigma T(0,t^4) - (1-A)\, F_\odot \tag{5}$$

In order to accurately set the upper boundary condition the albedo, thermal inertia, emissivity and rotation state of the surface must be known. Typically only the rotation state is well constrained, and the uncertainty of the other parameters translates into uncertainty in the surface temperatures predicted.

At the base of a passive model, it is assumed there is no heat flux through the boundary, that is:

$$\Gamma\sqrt{\omega}\frac{\partial T(Z,t)}{\partial Z}\,|_{z=d}=0 \tag{6}$$

This implies the temperature at the boundary is unchanged with both time and location. While we know this is not completely correct (heat is being transported from the relatively warm ocean to the cold outer ice shell), this heat flow is usually much lower than the solar forcing and thus this assumption holds. In order for these conditions to be met the boundary must be set much lower than the seasonal and diurnal skin depths.

Different models parametrize the required variables in different ways. The simplest models assume them to be constant (e.g., Spencer et al. 1989), while some assume specific heat, thermal conductivity and density vary with temperature (e.g., Hayne and Aharonson 2015). The choice of what to assume often depends upon how well the surface properties are known, and how accurately the problem needs to be solved. How such models can be used and constrained are explored in the next sections.

3.1.3 Deriving Enceladus' Surface Temperature

The Cassini spacecraft toured the Saturn system between 2004 and 2017 and provided a wealth of data from its suite of instruments on Enceladus. Cassini's Composite Infrared Spectrometer (CIRS) (Flasar et al. 2004) measured thermal infrared emission from Enceladus and as such provides some of the strongest constraint on Enceladus' heat flow. CIRS is a Fourier Transform Spectrometer (FTS), made of two interferometers that share a single scan mechanism and telescope. Wavenumbers between 10 and 600 cm^{-1} (16.7 to 1000 μm) are detected by two thermopile detectors known as focal plane 1 (FP1), which has a spatial resolution of 3.9 mrad. Higher wavenumbers, between 600 and 1100 cm^{-1} (9.1 to 16.7 μm), are detected by focal plane 3 (FP3). Focal plane 4 (FP4) is sensitive to wavenumbers between 1100 and 1400 cm^{-1} (7.1 to 9.1 μm). FP3 and FP4 both have a row of ten detectors, each of which has a 0.273 by 0.273 mrad field of view. This means FP1 has a spatial resolution $\sim$10 times lower than that of FP3 and FP4, but its wavelength coverage makes it more sensitive to lower temperatures as it covers the peak of the blackbody emission curve for all temperatures cooler than 320 K.

Figure 8 compares the wavelength coverage of CIRS' focal planes to blackbody emission at 30 K, 80 K and 150 K. As the figure shows FP1 is sensitive to the bulk of the emission from cooler temperatures, enabling it to reliably detect temperatures $>\sim$25 K (Howett et al. 2016). However, due to FP3 and FP4's shorter wavelength sensitivity they are only sensitive to warmer temperatures, with FP3 only being reliably sensitive to temperature greater than $\sim$80 K and FP4 to regions even warmer than that (Howett et al. 2019). In fact, on Enceladus FP4 was only sensitive to its warmest surface temperatures: its active tiger stripes. FP3 was sensitive to the tiger stripes and daytime emission, whereas FP1 was sensitive to the tiger stripes, day and nighttime emission. This means that CIRS is well suited to deriving Enceladus' surface temperatures, but a trade has to be made between its sensitivity to temperature and spatial resolution.

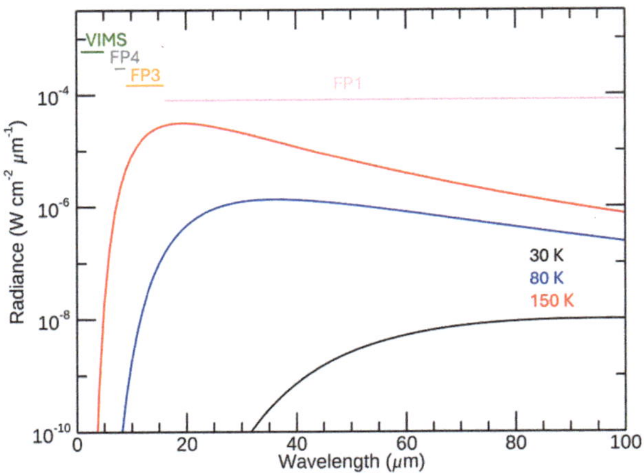

Fig. 8 A comparison of the radiance from blackbody emission at 30 K (black), 80 K (blue) and 150 K (red) to the wavelength range of the CIRS focal planes 1 (FP1, pink), 3 (FP3, orange) and 4 (FP4, grey), and the VIMS instrument (green)

The IR channel (0.85 to 5.1 μm) in Cassini's VIMS instrument enabled VIMS to constrain Enceladus' tiger stripe emission and surface properties (Filacchione et al. 2022; Goguen et al. 2013). Filacchione et al. (2022) used the position of the 3.6 μm continuum peak as a proxy for temperature, deriving daytime surface temperature maps for all of Saturn's large satellites, including Enceladus. VIMS is only sensitive to warm (>88 K) surface temperatures (Fig. 8), so the coverage of these maps is restricted to Enceladus' equatorial and mid-latitudes. Goguen et al. (2013) used a unique set of VIMS observations to constrain Enceladus' SPT surface temperature. During an unusually low (74 km) altitude pass of Enceladus VIMS took spectra every 0.025 seconds. The spatial resolution of each observation was 38 m, and 214 m along track (driven primarily by Cassini's 7.5 km/s velocity). Analysis of the 3 to 5 μm spectra provided a surface temperature constraint of Baghdad Sulcus of 197 $\pm$ 20 K over a width of 9 m.

3.2 Enceladus' Current Heat Flow

3.2.1 Thermal Measurements

The heat flow from Enceladus' South Polar Terrain (SPT) has been constrained using a variety of methods. Each one has its advantages and disadvantages, and none are without some large assumptions. The first constraint placed on the SPT's heat flow used an SPT FP3 observation along with a thermal inertia of 20 MKS graybody fit to constrain the emission to be 133 $\pm$ 12 K over 345^{+320}_{-160} km^2, to give a total estimate of Enceladus' endogenic emission of 5.8 $\pm$ 1.9 GW (Spencer et al. 2006). This study did not include any kinetic and latent heat of the plume, and (since they used FP3) were not sensitive to any low-temperature (<80 K) endogenic emission.

To convert this total endogenic power into a power per unit area an SPT bound needs to be assumed. To show the effect of the assumed area we use two regions, the first a great circle at 68°S, chosen because it encompasses the entire tiger stripe region (Fig. 9). However, this bound also includes a lot of non-stripe region, and so may lead to an underestimate of the emission. Thus, a second bound is also used, this one is formed by joining four corner points by great circles; the points are located at: 100°W/68°S, 170°W/°68S, 250°W/°70S and 345°W/°70S (Fig. 9). The circular bound provides an SPT estimate of 29,051 km^2; the rectangular bound an estimate of 16,655 km^2. Using these bounds the heat flow of Spencer et al. (2006) ranges from 200 $\pm$ 65 mW m^{-2} to 348 $\pm$ 114 mW m^{-2}.

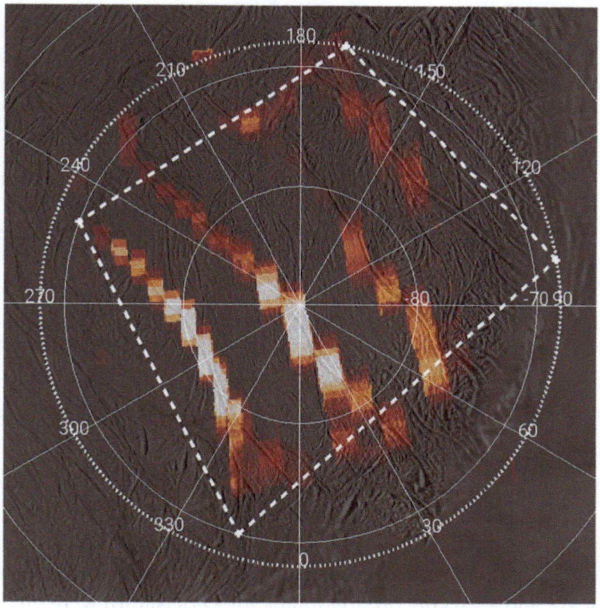

Fig. 9 Radiated 600-1100 cm^{-1} brightness of Enceladus' four tiger stripes, as determined from Cassini/CIRS FP3 using observations obtained on 12th March 2008 (Howett et al. 2011). Overlaid (dotted and dashed lines) are two potential SPT bounds. The dashed line is the most conservative estimate of SPT area only including the active tiger stripes. It assumes the SPT is bounded by great circles joining four points at: 100°W, 170°W, 250°W and 345°W/68°S, 68°S, 70°S and 70°S, providing an SPT area of 16,655 km^2. The dotted line defines the SPT as all terrain within 68°S, providing a larger SPT area of 29,051 km^2. Base map is PIA08417

To determine how much endogenic emission was coming from both low-temperature and high-temperature endogenic emission another study analyzed two FP1 stares of Enceladus' SPT (Howett et al. 2011). The stares, both taken in 2008, covered all of the SPT and a little of the surrounding terrain. However, since these observations were sensitive to cooler temperatures they were sensitive to both the endogenic and passive emission. So, in order to determine the endogenic component of the stares the passive surface emission had to be predicted. In order for the passive emission to be accurately predicted Enceladus' surface properties (albedo, thermal inertia) have to be well constrained (see Sect. 3.1.2). Since they are not well constrained, a range of reasonable values were adopted to test the sensitivity of the predicted power to them. An albedo (0.80) and surface thermal inertia (27 MKS) were assumed because they are the thermophysical properties derived closest to the SPT (in a band between 60°S to 50°S by Howett et al. 2010). Three thermal inertia scenarios were tested: 27 MKS at all depths, 27 MKS from the surface to 1 cm and then increasing to 100 MKS or 1000 MKS.

The FP1 response is a 2D Gaussian with a FWHM of 2.42 mrads (Flasar et al. 2004). Thus, FP1 is also sensitive to the spatial distribution of the endogenic emission across the SPT. To test this sensitivity to the distribution of endogenic emission three scenarios were tested: assuming the tiger stripes are the sole emission source and that the emission is uniform (scenario #1) or varying along the stripe according to FP3 results (as shown in Fig. 9) (scenario #2). The final scenario (#3) assumes all of the emission comes within 10% of the field of view (FOV). Scenario #3 was not expected to be realistic, but rather provided a lower limit to the endogenic emission.

These three thermal inertia and emission scenarios, along with the two FP1 observations, provided the 18 estimates of the SPT's endogenic emission shown in Table 3. As the Table shows for the most likely emission scenarios (i.e., #1 and #2) the predicted endogenic emission varied between 12.9 and 19.3 GW, which led to the final estimate of 15.8 ± 3.1 GW. This estimate did not include latent heat, which could add ~0.7 GW more to the total (Inger-

Table 3 Variation in derived endogenic emission assuming an albedo of 0.80 and three different subsurface conditions. See text for more details. From Howett et al. (2011)

Assumed conditions for thermal model	Endogenic power for Rev 61 uniform spatial distribution (GW)	Endogenic power for Rev 61 FP3-like spatial distribution (GW)	Endogenic power for Rev 61 central spatial distribution (GW)	Endogenic power for Rev 91 uniform spatial distribution (GW)	Endogenic power for Rev 91 FP3-like spatial distribution (GW)	Endogenic power for Rev 91 central spatial distribution (GW)
Scenario 1	16.6	13.8	8.2	19.3	16.9	11.9
Scenario 2	15.5	12.9	7.6	17.2	15.0	10.6
Scenario 3	16.6	13.8	8.1	16.6	14.5	10.2

soll and Pankine 2010). This translates to a heat flow between 543 ± 107 to 949 ± 186 mW^{-2} depending upon the SPT area assumed.

This ~16 GW total heat flow estimate was ~10 GW greater than the previous estimate, and at the time posed a problem to the scientific community. How could Enceladus be producing this much emission when the orbital evolution models at the time predict equilibrium rates about an order of magnitude lower than this (Meyer and Wisdom 2007)? Periodic release and a varying dissipation factor Q of Saturn were put forward as possible explanations. Since then, estimates of Q have evolved, and current estimates predict the equilibrium heating rate of Enceladus to be 50 to 150 GW (Lainey et al. 2020, 2017; also see Lainey et al. 2025 and Nimmo et al. 2023, this collection), well above the 15.8 GW measured in this study.

The results of Howett et al. (2011) are heavily model dependent, and do not resolve the heat flow from individual stripes themselves. So, a follow-on study was developed to combine FP1, FP3 and FP4 results to better understand the spatial variation of the endogenic emission (Howett et al. 2013; Spencer et al. 2018). It was assumed the emission is constant with time, and so observations taken at different epochs could be used to constrain it. It was also assumed that the emission had to be able to explain the observations at all wavelengths, so a single model could be used to explain emission across FP1, FP3 and FP4 wavelength ranges.

Two sets of observations were used: an FP3 and FP4 scan of the entire SPT taken in March 2008, and an FP1 scan across Damascus, Baghdad and Cairo taken in August 2010 (Fig. 10a and 10b). The first step was to divide the CIRS emission into wavelength bins, to ensure that different parts of the surface's blackbody emission were well characterized. The FP1 radiances were integrated over bins from 25-83 μm, and 83-200 μm. This was repeated for three wavelength regions in FP3: 9.5 to 11.1 μm, 11.1 to 13.3 μm and 13.3 to 16.7 μm, and between 7.7 and 9.1 μm for FP4.

A model of the width and magnitude of the low- and high-temperature endogenic emission along each of the stripes was then developed (Fig. 10e). Interestingly the initial model was unable to fit the emission from one side of Baghdad sulci and required an additional fracture to be added to provide an adequate fit (Figs. 10c and 10d). There was no previous indication from optical maps or other sources that this fracture was active. The lack of activity detection in optical maps is not unique. The warm subsurface regions observed by synthetic aperture radar (SAR) observations (discussed further below) also did not have other indications of activity (Le Gall et al. 2017).

The endogenic modelling results showed a high-temperature component of 3.1 GW, a low temperature component of 1.3 GW, which provided a total of 4.9 GW (including 0.5 GW

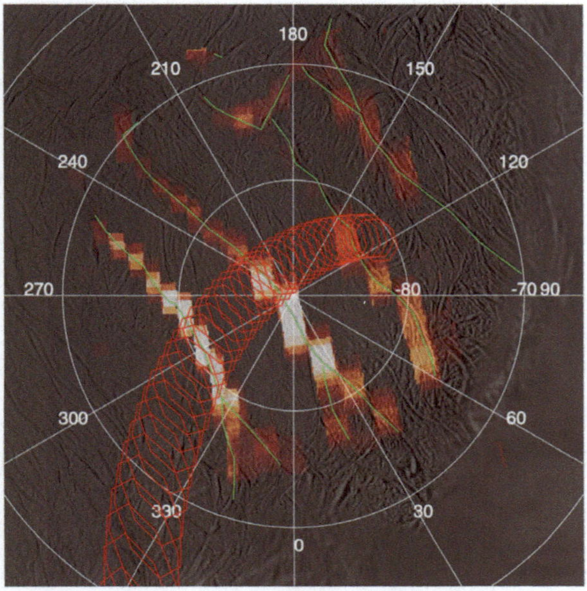

(a) - August 2010 FP1 scan across Damascus, Baghdad and Cairo is shown in red. The locations of the tiger stripes are shown in green. Overlaid on the March 2008 FP3 base map (Figure 4).

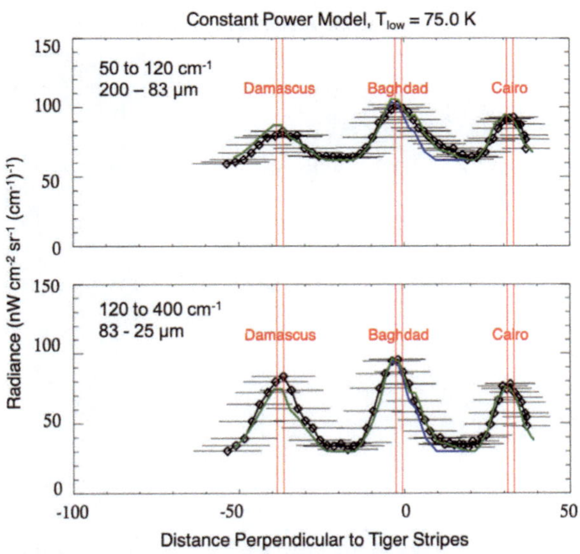

(b) Integrated FP1 radiance over two wavelength bands, 200-83 μm (top) and 83-25 μm (bottom) are shown by the black points. The black horizontal lines show the extent of each observation's FOV. The red vertical lines show the location of each of the three tiger stripes in the scan and are labeled accordingly. The model fits to these data are shown without (blue) and with (green) the additional Baghdad branch (position shown in (c)) included.

Fig. 10 The data (a and b) used to fit the temperature and emission width across the various tiger stripe fractures (c and d). The results from the fit are shown in (e). From Howett et al. (2013) and Spencer et al. (2018)

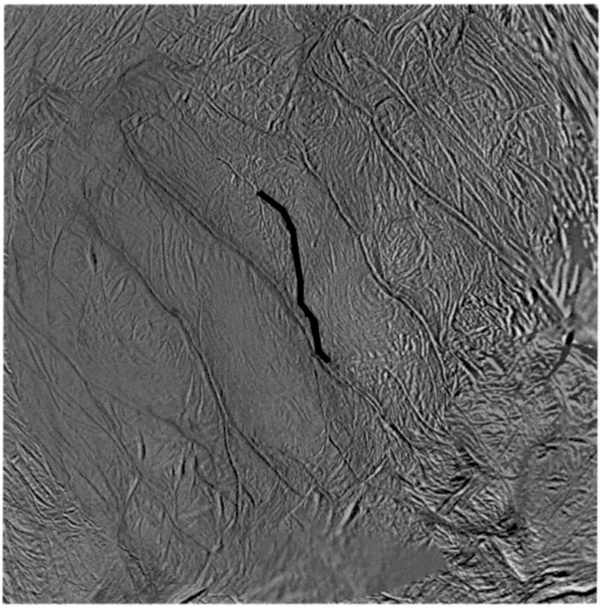

(c) Location of the additional warm stripe required to produce the model fit shown in (a).

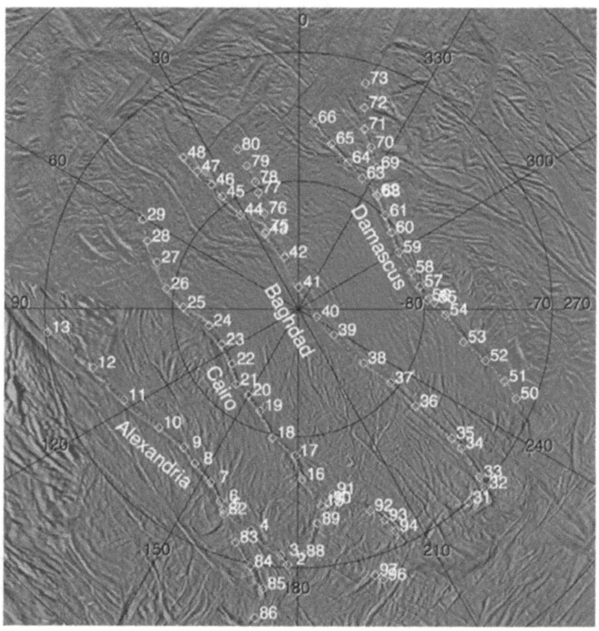

(d) The positions of the modeled emission temperatures and widths.

Fig. 10 (*Continued*)

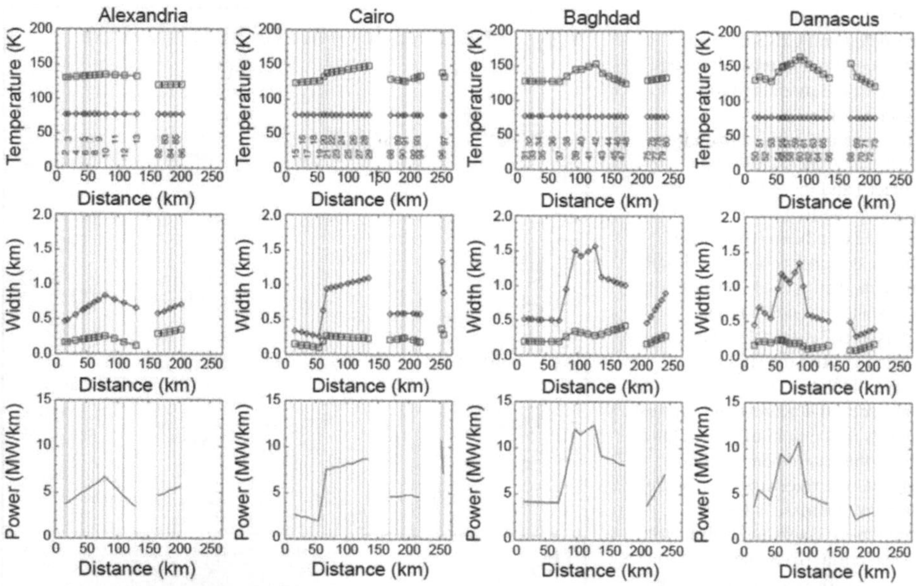

(e) Model emission widths and temperatures, and their corresponding power per unit length at each position indicated in the top figure (the same numbering scheme is used in the temperature plots as shown in the top figure). The points at the line were calculated, and the intermediate points were interpolated.

Fig. 10 *(Continued)*

of latent heat), or 169 to 294 mW m^{-2}. This method does not include any emission from between the stripes. Emission from any other small fractures (if more are active) were also not included. So, this method provided a robust solution for the lower bound of Enceladus' endogenic emission but not its total emission.

New work on modeling the interstitial (i.e., between stripe) regions of Enceladus is ongoing. Howett et al. (2022) (Fig. 11) showed the results of modeling a high-spatial resolution FP1 scan that cut across Damascus, Baghdad and Cairo sulci. These observations put the FP1 FOV in between the stripes, and so could measure their temperature directly. Since these are FP1 observations the passive emission component needed to be removed, but this time it was constrained using observations from the same scan that were taken close to the tiger stripes but outside the active regions. This requires the feasible assumption that the thermophysical properties inside and outside the stripes were similar, and if the model could fit the surface temperatures close to the stripes, then it would follow that it was modeling the passive temperatures inside the stripes. Depending on the passive model assumed, this work predicts the interstitial heat flow alone to be between 2.9 ± 0.5 GW and 4.0 ± 0.4 GW. This translates to an interstitial heating rate of (99 ± 17 to 174 ± 30 mW m^{-2}) and (138 ± 14 to 240 ± 24 mW m^{-2}).

Table 4 compares the results of the various studies that have modeled Enceladus' endogenic emission. It highlights the advantages and disadvantages of each study and shows that obtaining the "true" value of Enceladus' endogenic emission remains elusive. Arguably the most robust measurement of the stripe emission comes from Spencer et al. (2018), which sets a lower bound of 4.9 GW. If this is combined with the lowest bound of the between-stripe emission (2.4 GW), then the current best estimate of Enceladus' endogenic emission is

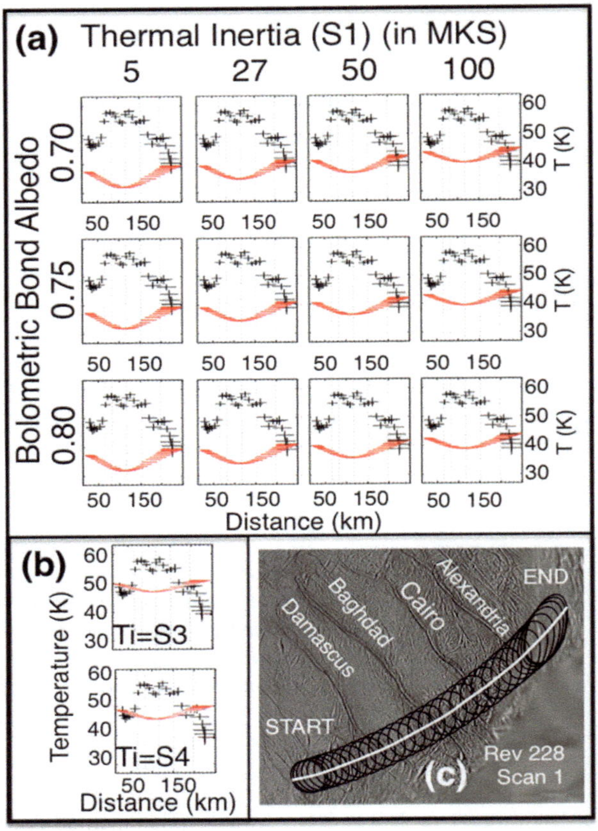

Fig. 11 A comparison of passive modeled and CIRS derived surface temperatures, showing none of the models fit the interstitial temperatures. (a) Temperature derived from Rev 228 (Dec 2015) CIRS data (black) and those predicted by passive thermal models (red) for a range of albedo (rows) and thermal inertia (columns). In all models the thermal inertia is constant with depth (S1). "Distance" is calculated as the distance from the start of the white line in (c) to where the line intersects the nearest and furthest points of the field of view. Dotted lines indicate where the line crosses the tiger stripes, from left to right: Damascus (both branches), Baghdad, Cairo and Alexandria. (b) Same as the top plot, but for seasonal models with an albedo 0.80 and a thermal inertia profile that changes with depth. (c) Footprints of the CIRS FP1 field of view (black). Tiger stripe names are indicated. Figure from Howett et al. (2022)

>7.3 GW (>251 mW m^{-2}). Current estimates of non-SPT emission are lacking, but there is preliminary evidence from CIRS of endogenic emission from Enceladus' north polar region (Miles et al. 2022).

3.2.2 Inference from Geology

A finite-element model was used to predict the heat flow required to form the periodic low-slope features found across Sarandib and Diyar Planitia, close to Enceladus' equator (Bland et al. 2007). The study used photoclinometry to show these regions have a periodic wavelength of 3 to 4 km and amplitudes between 100 and 400 m. Comparisons of the finite-element model to the photoclinometry results to predict a heat flow between 110 and 220 mW m^{-2} at the time of their formation. This suggests a strong localized heating source was present at the time of their formation, which is estimated to be 50 Ma to >3 Ga depending on the adopted model of impactor flux (Kirchoff and Schenk 2010; Porco et al. 2006a). Similar finite-element modeling work was also conducted on relaxed craters in Enceladus' equatorial and mid-latitude regions (North and East of Hamah Sulci). The results showed that heat flows of ∼150 mW m^{-2} were required to explain their relaxation, and their shape was best explained by short period of intense heating (Bland et al. 2012).

Models of flexural deformation at Enceladus's Harran Sulci (which covers equatorial to mid-northern latitudes) were used to constrain the likely heat flux at its formation (Giese et al. 2008). Depending upon the assumed surface porosity heat flows between 45 mW m^{-2}

Table 4 A comparison of the heat-flows predicted for Enceladus, both inside and outside its SPT. The heat flow in $mW\,m^{-2}$ is given assuming the SPT is the region south of 68°S. The lower limit given by Howett et al. (2022) plus Spencer et al. (2018) combines the lowest bound of Howett et al. (2022) (i.e., 2.9-0.5) to that of Spencer et al. (2018). Light grey indicates desired answered, while dark grey indicates solutions that require additional consideration

SPT studies

Source	Region	Power ($mW\,m^{-2}$) (Assuming SPT is inside 68°S)	Power (GW)	Include all SPT emission?	Need to remove passive?	Include high- and low-temperature?
Spencer et al. (2006)	SPT	200 ± 65	5.8 ± 1.9	Yes	No	No
Howett et al. (2011)	SPT	543 ± 107	15.8 ± 3.1	Yes	Yes	Yes
Spencer et al. (2018)	SPT	>169	>4.9	No (just tiger stripes)	No	Yes
Howett et al. (2022)	SPT	99 ± 17 to 138 ± 14	2.9 ± 0.5 GW to 4.0 ± 0.4 GW	No (just interstitial regions)	Yes	Yes
Spencer et al. (2018) + Howett et al. (2022)	SPT	>251	> 7.3	Yes	Yes	Yes

Non-SPT studies

Source	Region	Power ($mW\,m^{-2}$)	Notes
Le Gall et al. (2017)	~10 km north of the tiger stripes	500	Radar thermal anomaly 10 s of cm below surface (nothing in thermal emission)
Bland et al. (2007)	Sarandib and Diyar Planitia (equatorial)	110 to 220	Heat flow required at formation (50 Ma to >3 Ga)
Bland et al. (2012)	North and East of Hamah Sulci (equatorial and mid-northern latitudes)	150	Heat flow required at formation (<2 Ga)
Giese et al. (2008)	Harran Sulci (equatorial and mid-northern latitudes)	45 to 270	Heat flow required during formation ($0.45^{+1.1}_{-0.3}$ to $3.5^{+0.1}_{-0.4}$ Ga)

to 270 $mW\,m^{-2}$ are required to explain the observed topography, which is believed to be formed between $0.45^{+1.1}_{-0.3}$ and $3.5^{+0.1}_{-0.4}$ Ga (depending on whether a comet-like impactor chronology, or an asteroid-like one is assumed).

3.2.3 Inference from Radar

On 6th November 2011 Cassini's RADAR instrument observed Enceladus' south polar region from 500 km above the surface. The microwave radiometry (2.2 cm) results, which are sensitive to ~10 s of centimeters beneath Enceladus' surface, showed a subsurface with a warm, thermally anomalous region (Le Gall et al. 2017). The location of this thermal anomaly was ~10 km north of the tiger stripes and was not spatially related to any sulci similar to the tiger stripes. The temperature varied from $32^{+0.6}_{-2.6}$ K to $60^{+1.2}_{-4.8}$ K, while these are cold they are above the expected passive surface temperature in this region. The total difference was estimated to require a heat flow of $\sim$500 mW m^{-2}. Warmer surface temperatures were not recorded in this region by CIRS. One explanation is that the deeper radar observations sensed a dormant or "dying" sulcus (i.e., one that is no longer active and is slowly losing its excess heat). This sulcus is too cold to be distinguishable from the passive background at the surface (or is being masked from the surface by an insulating layer). Thus, it is feasible that there are other previously active sulci across Enceladus, but they remain undiscovered.

Further analysis of 2.2 cm disk-integrated radar observations of Enceladus shows the lower limit of its emissivity to be larger than expected (Le Gall et al. 2023). One possible explanation for this is that their initial assumption of no endogenic emission is incorrect. This would support the preliminary conclusion from CIRS (Miles et al. 2022) that actually Enceladus' endogenic emission is not confined to the SPT.

3.2.4 Conductive Heat Loss Estimates

Heat from Enceladus' relatively warm subsurface ocean is expected to conduct through its ice shell, eventually being radiated to space. The amount of heat lost depends upon the thickness of Enceladus' ice shell, which has been estimated using geodetic observations from Cassini data (Hemingway and Mittal 2019; Iess et al. 2015; Čadek et al. 2016; Schenk and McKinnon 2009; Thomas et al. 2016; Park et al. 2024). Differences in the ice shell thickness assumed lead to variations in the predicted conductive heat loss estimate for Enceladus. Hemingway and Mittal (2019) use a shell thickness model to predict losses between 20 and 35 GW, while Čadek et al. (2016) and Park et al. (2024) estimate values between 25 and 30 GW, and 18 and 28 GW respectively. These may be compare with equilibrium tidal heating rates between 1.8 and 150 GW depending on the assumptions made (Rhoden et al. 2015; Lainey et al. 2017; Nimmo et al. 2023).

4 What Further Observations Are Needed?

Cassini's amazing tour taught us a great deal about Enceladus' mass loss, but of course many questions remain. Due to Enceladus' small size, proximity to Saturn, and activity located at the poles (which are out of sight from Earth) future observations to solve these questions mostly require a spacecraft to return to the Saturn system. In fact, many of the outstanding questions either need high-spatial resolution coverage, or observations taken at regular time intervals. Both of these types of observation would be well served by a mission that was in orbit around Enceladus (and not Saturn). The following suggestions are not given in a priority order. Rather we defer to NASA's decadal survey (National Academies 2022) as a far more comprehensive study of the priories of Enceladus' exploration. We simply note that with Enceladus listed as a suggested target for New Frontiers 5 and 6, and an Enceladus Orbilander being the second-highest priority Flagship mission, future exploration of Enceladus is a very real possibility.

Flying a high-resolution mass spectrometer that is sensitive to heavy (>50 u) molecules through Enceladus' plumes would enable many of the outstanding composition questions to be resolved. For example, on the abundance of high-mass (e.g., C_2 and C_3 carbon) species. It would also enable the composition of lower-mass species to be differentiated (e.g., what are the species contributing to the 28 u feature)?

A high-spatial (<100 m) resolution thermal instrument could accurately determine how the thermal emission varies along and between the tiger stripes. This would enable the connection between plume emission and latent heat release to be studied and might determine whether the surface heating from latent heat release is higher over the jets. Since jets have denser plume emission, there is more material available to freeze out onto crack walls to release latent heat.

Thermal observations of the SPT by an instrument capable of detecting polar winter temperatures (~30 K) and resolving the regions between the stripes would enable the conduction of heat from the fissures to be studied. This would address questions on the heat flow through the ice shell, such as how efficient is the surface conduction, and how far away from the stripes is heated by them? If such an instrument were able to measure the global surface temperature, then small regions of undiscovered activity could be searched for, and a constraint put on Enceladus' global heat flow.

Observations that determine both the spatial distribution and temporal variability of the solid and vapor plume sources are required to constrain Enceladus sub-surface plumbing. The constraint on their spatial variation would enable questions about the source of the solid/vapor to be determined. For example, it could answer whether solids are ejected by curtains or jets (or both)? Do some sources only emit one (or the other)? These observations would be even more powerful if they were taken in addition to the thermal measurements because then the connection between the heat, solid and vapor sources could be determined.

A study of the solid and vapor particle's temporal variability with short- and long-term coverage would enable questions on their forcings to be determined. For example, are vapor and solids modulated by tides in the same way? Fully understanding their short and long-term variability (rather than patchy coverage) would better constrain the time lag between the tidal forcings and the plume's response, perhaps even addressing the question of why the plumes do not cease emission during periapsis (when the fractures should be closed)? Such observations made over ~decadal timescales could constrain the length and magnitude of global changes in plume brightness (like those seen at the end of the Cassini mission).

5 Conclusions

The Cassini mission revealed not only the presence of Enceladus' activity but allowed it to be studied in detail. Observations from the mission reveled surface warming around four fractures within Enceladus' SPT, and a plume that erupts from them. The energy and mass loss from both the surface warming and plumes were constrained by both in situ and remote observations.

The greatest source (~80%) of the mass loss via the plumes is from vapor, with between 100 and 250 kg/s being lost (see references in Table 2). The vapor is mostly water ice, with some carbon dioxide, ammonia, molecular hydrogen, and trace species, expected to come from evaporating ocean water. The loss of molecular hydrogen was specifically studied, as it provides support that Enceladus has active hydrothermal processes (Waite et al. 2017). Its current loss rate is estimated to be between 0.002 and 0.01 kg/s, so a very small fraction of the total vapor – but an important one!

This vapor almost always escapes Enceladus into space, since its low mass allows it to be more easily accelerated to speeds above Enceladus' escape velocity (Hansen et al. 2006). The bulk vapor mass loss rate seems to be approximately constant with time (albeit with individual jets increasing and reducing their specific vapor loss (Hansen et al. 2017; Teolis et al. 2017). However, the vapor has notable spatial variation, with heavier components (CO_2) having a smaller lateral spread than the lighter ones (H_2, H_2O) (Perry et al. 2015; Yeoh et al. 2015). So, all of this combines to Enceladus consistently losing a lot (100 to 250 kg/s) of (predominantly water) mass into space.

Solid particles provide the next highest mass flux (10%), with between 3 and 65 kg/s being ejected into space. However, these heavier particles are less able to reach Enceladus' escape velocity, and so between 90 and 95% will fall back onto Enceladus' surface (Schmidt et al. 2008). Unlike the vapor the mass-flux of these particles varies both temporally and spatially. The plumes are brightest (highest mass-flux) just after apoapsis (MA=180°) at MA=200°. There is also a secondary double-peaked maximum at MA of 25° and 60° (Hedman et al. 2013; Nimmo et al. 2014; Ingersoll et al. 2020). If this wasn't enough the total plume brightness also changes with the years (Ingersoll et al. 2020). Tidal stresses on Enceladus' surface are believed to be the cause of the orbital-length variation, with the plumes decreasing in output when the fractures are squeezed closed when the surface is under compression at apoapsis, and increasing when the fractures are pulled apart when the surface is under tension at periapsis (e.g., (Nimmo et al. 2014)). Tides are also thought to explain the multi-year changes, with the surface changing in response to Enceladus and Dione's 4- and 11-year tidal cycle (Porco et al. 2018).

The spatial variation of plume particles follows a more complex pattern than that of the vapor, linked to its composition. Broadly the plume particles can be separated into three types (Postberg et al. 2011). Type 1 particles make up the bulk of the plume (65%) and are small particles mostly made of pure water ice. Type 2 is the next most abundant (25% by mass) and are approximately speaking a larger version of Type 1, but with some additional silicates, salts and varying organics. Arguably the most interesting is Type 3 particles, which have the lowest abundance (~10% by mass) but have a very different composition from their counterparts, most notably containing silicon. They are believed to be frozen droplets of Enceladus' ocean, and the presence of salts and silicon imply that the ocean must be in direct contact with the core. They allow the composition of the ocean to be sampled, without disturbing it. This means that while the solid particles might not be the major source of mass flux, what is being lost provides unparalleled insight into Enceladus' interior (Postberg et al. 2011). Finally, nanograins and ions are estimated to have a mass flux of 15-65 kg/s and 5.6 kg/s respectively, roughly equivalent to that of the plume particles (Dong et al. 2015; Kempf et al. 2018). However, unlike the particles these charges species are small and so likely to escape - eventually ending up in Saturn's magnetosphere.

Our understanding of Enceladus' mass flux has evolved with observations and models. These current estimates provide a likely total mass flux of ~240 kg/s, meaning Enceladus could lose 7 billion kilograms of material every Earth year. Using the best estimate of the mass and composition of Enceladus' ocean, means activity at its current level could be sustained up to ~1.5 billon Earth years. The loss of molecular hydrogen constrains this further, limiting the activity to less than one million Earth years unless resupplied (e.g. by serpentinization).

A range of studies, using different instruments and/or models, show Enceladus to be a place of high surface heat flows, both now and in the past. Each of the studies tells a different part of the story, some setting bounds, and many investigating different epochs and regions of Enceladus. Despite this work it is difficult to know what Enceladus' current heat flow really is. The highest heat flows are observed in its South Polar Terrain, and the different methods used to constrain the SPT heat flow have provided a lower limit of tiger stripe emission of 169 $mW\,m^{-2}$ (~4.9 GW, Spencer et al. 2018). This can be reasonably extended to >251 $mW\,m^{-2}$ (>7.3 GW) by including the lowest bound of our best estimate of between-stripe emission (Howett et al. 2022) and the least conservative bound on the SPT area (i.e., all the regions south of 68°S). However, Enceladus SPT is only a small fraction of its total area, and so despite its high surface heat flows it alone only tells part of the story. The challenge remains in separating out the lower endogenic heat flows from the passive emission across the rest of Enceladus. This is particularly difficult in regions where it small relative to the passive emission - like the warmer equatorial and mid-latitude regions. Tentative evidence shows that this could be done by global radar studies (Le Gall et al. 2023) and in the considerably colder thermal regime of polar winter (Miles et al. 2022).

Studies of Enceladus' geology show that high heat flows are not restricted to our current epoch, but rather have existed across Enceladus for billions of years. These studies consider the conditions required either for formation, or to enable changes post-formation. The results imply that heat flows between ∼45 to 270 $mW\,m^{-2}$ were required for multiple regions across Enceladus' equatorial and mid-northern latitudes to form (Bland et al. 2007; Giese et al. 2008), which are comparable with the current SPT heat flow estimates. Age estimates using cratering fluxes put these formation events between ∼0.42 Ga and >3 Ga ago. These heat flows are consistent with the those required to form the relaxed craters seen across Enceladus' equatorial regions (110 to 200 $mW\,m^{-2}$), which are likely to have occurred within the last 2 Ga (Bland et al. 2012). Evidence of these ancient heat flows may still be present deep in Enceladus' surface, still detectable to long (>cm) wavelength instruments that probe deeper into Enceladus' surface than thermal ones. One such region was discovered ∼10 km north of the SPT by 2.2 cm radar, a thermally anomalous region with an inferred heat flow of 500 $mW\,m^{-2}$ (but nothing detectable in surface thermal emission) (Le Gall et al. 2017).

Enceladus is likely to have experienced high heat flows throughout its evolution. However, a final constraint on its current heat flow, both inside and outside of the SPT is lacking. We do not even know how common deep thermal anomalies are! While investigations into the wealth of data returned by the Cassini mission may provide further answers it is likely that additional missions are required to truly answer these questions. A mission that provides long-baseline thermal coverage of Enceladus at consistently high spatial resolutions, throughout both Enceladus' diurnal and seasonal cycles, would constrain its passive emission. This would enable Enceladus' non-SPT endogenic heat flows to be constrained, especially in regions of cold winter darkness where the passive component is minimized. Radar and sub-mm measurements would enable Enceladus' deeper sub-surface to be probed, enabling a picture to be constructed of the global heat flow from deep ancient thermally anomalous regions. These could then be used to finally constrain how Enceladus, and the Saturn-system as a whole, has evolved and what the future has in store.

Funding Information This work was partially funded by ISSI, NASA CDAP 80NSSC20K0477 and NNN12AA01C.

Declarations

Competing Interests The authors have no competing interests to declare that are relevant to the content of this article. CH is a Guest Editor of the collection but was not involved in the peer review process of this article.

References

Ames F, Ferreira D, Czaja A, Masters A (2025) Ocean stratification impedes particulate transport to the plumes of Enceladus. Commun Earth Environ 6:63. https://doi.org/10.1038/s43247-025-02036-3

Berne A, Simons M, Keane JT, Leonard EJ, Park RS (2024) Jet activity on Enceladus linked to tidally driven strike-slip motion along tiger stripes. Nat Geosci 17:385–391. https://doi.org/10.1038/s41561-024-01418-0

Bland MT, Beyer RA, Showman AP (2007) Unstable extension of Enceladus' lithosphere. Icarus 192:92–105. https://doi.org/10.1016/j.icarus.2007.06.011

Bland MT, Singer KN, McKinnon WB, Schenk PM (2012) Enceladus' extreme heat flux as revealed by its relaxed craters. Geophys Res Lett 39:L17204. https://doi.org/10.1029/2012GL052736

Brown RH, Kirk RL (1994) Coupling of volatile transport and internal heat flow on Triton. J Geophys Res Planets 99:1965–1981. https://doi.org/10.1029/93JE02618

Brown RH, Baines KH, Bellucci G, Bibring J-P, Buratti BJ, Capaccioni F, Cerroni P, Clark RN, Coradini A, Cruikshank DP, Drossart P, Formisano V, Jaumann R, Langevin Y, Matson DL, Mccord TB, Mennella V, Miller E, Nelson RM, Nicholson PD, Sotin C (2004) The Cassini Visual and Infrared Mapping Spectrometer (VIMS) investigation. Space Sci Rev 115:111–168. https://doi.org/10.1007/s11214-004-1453-x

Brown RH, Clark RN, Buratti BJ, Cruikshank DP, Barnes JW, Mastrapa RME, Bauer J, Newman S, Momary T, Baines KH, Bellucci G, Capaccioni F, Cerroni P, Combes M, Coradini A, Drossart P, Formisano V, Jaumann R, Langavin Y, Matson DL, McCord TB, Nelson RM, Nicholson PD, Sicardy B, Sotin C (2006) Composition and physical properties of Enceladus' surface. Science 1979(311):1425–1428. https://doi.org/10.1126/science.1121031

Čadek O, Tobie G, Van Hoolst T, Massé M, Choblet G, Lefèvre A, Mitri G, Baland R-M, Běhounková M, Bourgeois O, Trinh A (2016) Enceladus's internal ocean and ice shell constrained from Cassini gravity, shape, and libration data. Geophys Res Lett 43:5653–5660. https://doi.org/10.1002/2016GL068634

Coates WA, Lewis GR, Jones GH, Young DT, Crary FJ, Waite JH, Johnson RE, Hill TW, Sittler EC Jr (2010b) Negative ions at Titan and Enceladus: recent results. Faraday Discuss 147:293. https://doi.org/10.1039/c004700g

Coates JGH, Lewis GR, Wellbrock A, Young DT, Crary FJ, Johnson RE, Cassidy TA, Hill TW (2010a) Negative ions in the Enceladus plume. Icarus 206:618–622. https://doi.org/10.1016/j.icarus.2009.07.013

Coates AJ, Wellbrock A, Jones GH, Waite JH, Schippers P, Thomsen MF, Arridge CS, Tokar RL (2013) Photoelectrons in the Enceladus plume. J Geophys Res Space Phys 118:5099–5108. https://doi.org/10.1002/jgra.50495

Cooper JF, Cooper PD, Sittler EC, Sturner SJ, Rymer AM (2009) Old faithful model for radiolytic gas-driven cryovolcanism at Enceladus. Planet Space Sci 57:1607–1620. https://doi.org/10.1016/j.pss.2009.08.002

Cravens TE, Ozak N, Richard MS, Campbell ME, Robertson IP, Perry M, Rymer AM (2011) Electron energetics in the Enceladus torus. J Geophys Res Space Phys 116:A09205. https://doi.org/10.1029/2011JA016498

Croft S, Kargel J, Kirk RL, Moore J, Schenk P, Strom R (1995) The geology of Triton. In: Cruikshank DP (ed) Neptune and Triton. The University of Arizona Press, Tucson, pp 879–947

Dhingra D, Hedman MM, Clark RN, Nicholson PD (2017) Spatially resolved near infrared observations of Enceladus' tiger stripe eruptions from Cassini VIMS. Icarus 292:1–12. https://doi.org/10.1016/j.icarus.2017.03.002

Dong Y, Hill TW, Ye S-Y (2015) Characteristics of ice grains in the Enceladus plume from Cassini observations. J Geophys Res Space Phys 120:915–937. https://doi.org/10.1002/2014JA020288

Esposito LW, Barth CA, Colwell JE, Lawrence GM, McClintock WE, Stewart AIF, Keller HU, Korth A, Lauche H, Festou MC, Lane AL, Hansen CJ, Maki JN, West RA, Jahn H, Reulke R, Warlich K, Shemansky DE, Yung YL (2004) The Cassini ultraviolet imaging spectrograph investigation. Space Sci Rev 115:299–361. https://doi.org/10.1007/s11214-004-1455-8

Fagents S (2000) Cryomagmatic mechanisms for the formation of rhadamanthys linea, triple band margins, and other low-albedo features on Europa. Icarus 144:54–88. https://doi.org/10.1006/icar.1999.6254

Fagents SA (2003) Considerations for effusive cryovolcanism on Europa: the post-Galileo perspective. J Geophys Res Planets 108:5139. https://doi.org/10.1029/2003JE002128

Fagents SA, Lopes RMC, Quick LC, Gregg TKP (2022) Cryovolcanism. In: Gregg TKP, Lopes RMC, Fagents SA (eds) Planetary volcanism across the Solar System. Elsevier, Amsterdam, pp 161–234. https://doi.org/10.1016/B978-0-12-813987-5.00005-5

Filacchione G, Ciarniello M, D'Aversa E, Capaccioni F, Clark RN, Buratti BJ, Helfenstein P, Stephan K, Plainaki C (2022) Saturn's icy satellites investigated by Cassini - VIMS. V. Spectrophotometry. Icarus 375:114803. https://doi.org/10.1016/j.icarus.2021.114803

Flasar FM, Kunde VG, Abbas MM, Achterberg RK, Ade P, Barucci A, Bjoraker GL, Brasunas JC, Calcutt S, Carlson R, Esarsky CJC, Conrath BJ, Coradini A, Courtin R, Coustenis A, Edberg S, Edgington S, Ferrari C, Fouchet T, Gautier D, Gierasch PJ, Grossman K, Irwin P, Jennings DE, Lellouch E, Mamoutkine AA, Marten A, Meyer JP, Nixon CA, Orton GS, Owen TC, Pearl JC, Prangé R, Prangé P, Raulin F, Read PL, Romani PN, Samuelson RE, Segura ME, Showalter MR, Simon-Miller AA, Smith MD, Spencer JR, Spilker LJ, Taylor FW (2004) Exploring the Saturn system in the thermal infrared: the composite infrared spectrometer. Space Sci Rev 115:169–297. https://doi.org/10.1007/s11214-004-1454-9

Giese B, Wagner R, Hussmann H, Neukum G, Perry J, Helfenstein P, Thomas PC (2008) Enceladus: an estimate of heat flux and lithospheric thickness from flexurally supported topography. Geophys Res Lett 35:L24204. https://doi.org/10.1029/2008GL036149

Glein CR, Shock EL (2010) Sodium chloride as a geophysical probe of a subsurface ocean on Enceladus. Geophys Res Lett 37:L09204. https://doi.org/10.1029/2010GL042446

Glein CR, Postberg F, Vance SD (2018) The geochemistry of Enceladus: composition and controls. In: Schenk PM et al (eds) Enceladus and the icy moons of Saturn. The University of Arizona Press, Tucson. https://doi.org/10.2458/azu_uapress_9780816537075-ch003

Goguen JD, Buratti BJ, Brown RH, Clark RN, Nicholson PD, Hedman MM, Howell RR, Sotin C, Cruikshank DP, Baines KH, Lawrence KJ, Spencer JR, Blackburn DG (2013) The temperature and width of an active fissure on Enceladus measured with Cassini VIMS during the 14 April 2012 South Pole flyover. Icarus 226:1128–1137. https://doi.org/10.1016/j.icarus.2013.07.012

Hansen CJ, Kirk RL (2015) Triton's plumes–solar driven like Mars or endogenic like Enceladus? In: 46th Lunar and Planetary Science Conference, p 1832

Hansen CJ, McEwen AS, Ingersoll AP, Terrile RJ (1990) Surface and airborne evidence for plumes and winds on Triton. Science 1979(250):421–424. https://doi.org/10.1126/science.250.4979.421

Hansen CJ, Esposito L, Stewart AIF, Colwell J, Hendrix A, Pryor W, Shemansky D, Wast R (2006) Enceladus' water vapor plume. Science 1979(311):1422–1425. https://doi.org/10.1126/science.1121254

Hansen CJ, Esposito LW, Stewart AIF, Meinke B, Wallis B, Colwell JE, Hendrix AR, Larsen K, Pryor W, Tian F (2008) Water vapour jets inside the plume of gas leaving Enceladus. Nature 456:477–479. https://doi.org/10.1038/nature07542

Hansen CJ, Shemansky DE, Esposito LW, Stewart AIF, Lewis BR, Colwell JE, Hendrix AR, West RA, Waite JH, Teolis B, Magee BA (2011) The composition and structure of the Enceladus plume. Geophys Res Lett 38:L11202. https://doi.org/10.1029/2011GL047415

Hansen CJ, Esposito LW, Aye KM, Colwell JE, Hendrix AR, Portyankina G, Shemansky D (2017) Investigation of diurnal variability of water vapor in Enceladus' plume by the Cassini ultraviolet imaging spectrograph. Geophys Res Lett 44:672–677. https://doi.org/10.1002/2016GL071853

Hansen CJ, Castillo-Rogez J, Grundy W, Hofgartner JD, Martin ES, Mitchell K, Nimmo F, Nordheim TA, Paty C, Quick LC, Roberts JH, Runyon K, Schenk P, Stern A, Umurhan O (2021) Triton: fascinating moon, likely ocean world, compelling destination! Planet Sci J 2:137. https://doi.org/10.3847/PSJ/abffd2

Hayne PO, Aharonson O (2015) Thermal stability of ice on Ceres with rough topography. J Geophys Res Planets 120:1567–1584. https://doi.org/10.1002/2015JE004887

Hedman MM, Nicholson PD, Showalter MR, Brown RH, Buratti BJ, Clark RN (2009) Spectral observations of the Enceladus plume with Cassini-vims. Astrophys J 693:1749–1762. https://doi.org/10.1088/0004-637X/693/2/1749

Hedman MM, Gosmeyer CM, Nicholson PD, Sotin C, Brown RH, Clark RN, Baines KH, Buratti BJ, Showalter MR (2013) An observed correlation between plume activity and tidal stresses on Enceladus. Nature 500:182–184. https://doi.org/10.1038/nature12371

Hedman MM, Dhingra D, Nicholson PD, Hansen CJ, Portyankina G, Ye S, Dong Y (2018) Spatial variations in the dust-to-gas ratio of Enceladus' plume. Icarus 305:123–138. https://doi.org/10.1016/j.icarus.2018.01.006

Helfenstein P, Porco CC (2015) Enceladus' geysers: relation to geological features. Astron J 150:96. https://doi.org/10.1088/0004-6256/150/3/96

Helfenstein P, Veverka J, McCarthy D, Lee P, Hillier J (1992) Large quasi-circular features beneath frost on Triton. Science 1979(255):824–826. https://doi.org/10.1126/science.255.5046.824

Hemingway DJ, Mittal T (2019) Enceladus's ice shell structure as a window on internal heat production. Icarus 332:111–131. https://doi.org/10.1016/j.icarus.2019.03.011

Hill TW, Thomsen MF, Tokar RL, Coates AJ, Lewis GR, Young DT, Crary FJ, Baragiola RA, Johnson RE, Dong Y, Wilson RJ, Jones GH, Wahlund J-E, Mitchell DG, Horányi M (2012) Charged nanograins in the Enceladus plume. J Geophys Res Space Phys. https://doi.org/10.1029/2011JA017218

Hofgartner JD, Birch SPD, Castillo J, Grundy WM, Hansen CJ, Hayes AG, Howett CJA, Hurford TA, Martin ES, Mitchell KL, Nordheim TA, Poston MJ, Prockter LM, Quick LC, Schenk P, Schindhelm RN, Umurhan OM (2022) Hypotheses for Triton's plumes: new analyses and future remote sensing tests. Icarus 375:114835. https://doi.org/10.1016/j.icarus.2021.114835

Howett CJA, Spencer JR, Pearl J, Segura M (2010) Thermal inertia and bolometric Bond albedo values for Mimas, Enceladus, Tethys, Dione, Rhea and Iapetus as derived from Cassini/CIRS measurements. Icarus 206:573–593. https://doi.org/10.1016/j.icarus.2009.07.016
Howett CJA, Spencer JR, Pearl J, Segura M (2011) High heat flow from Enceladus' south polar region measured using 10-600 cm-1 Cassini/CIRS data. J Geophys Res E, Planets 116:E03003. https://doi.org/10.1029/2010JE003718
Howett C, Spencer J, Spencer D, Verbiscer A, Hurford T, Segura M (2013) Enceladus' enigmatic heat flow. In: American geophysical union, fall meeting. Abstract Id.P53E-06
Howett C.J.A, Spencer J.R., Hurford T, Verbiscer A, Segura M (2016) Thermal properties of Rhea's poles: evidence for a meter-deep unconsolidated subsurface layer. Icarus 272:140–148. https://doi.org/10.1016/j.icarus.2016.02.033
Howett CJA, Spencer JR, Hurford T, Verbiscer A, Segura M (2019) Maps of Tethys' thermophysical properties. Icarus 321:705–714. https://doi.org/10.1016/j.icarus.2018.12.018
Howett C, Nimmo F, Spencer J (2022) Constraining Enceladus' heat flow between its tiger stripes. In: Europlanet Science Congress, EPSC2022-219. https://doi.org/10.5194/epsc2022-219
Hsu H-W, Postberg F, Sekine Y, Shibuya T, Kempf S, Horányi M, Juhász A, Altobelli N, Suzuki K, Masaki Y, Kuwatani T, Tachibana S, Sirono S, Moragas-Klostermeyer G, Srama R (2015) Ongoing hydrothermal activities within Enceladus. Nature 519:207–210. https://doi.org/10.1038/nature14262
Hurford TA, Helfenstein P, Hoppa GV, Greenberg R, Bills BG (2007) Eruptions arising from tidally controlled periodic openings of rifts on Enceladus. Nature 447:292–294. https://doi.org/10.1038/nature05821
Hurley DM, Perry ME, Waite JH (2015) Modeling insights into the locations of density enhancements from the Enceladus water vapor jets. J Geophys Res Planets 120:1763–1773. https://doi.org/10.1002/2015JE004872
Iess L, Stevenson DJ, Parisi M, Hemingway D, Jacobson RA, Lunine JI, Nimmo F, Armstrong JW, Asmar SW, Ducci M, Tortora P (2015) The gravity field and interior structure of Enceladus. Science 1979(344):78–80. https://doi.org/10.1126/science.1250551
Ingersoll AP, Ewald SP (2011) Total particulate mass in Enceladus plumes and mass of Saturn's E ring inferred from Cassini ISS images. Icarus 216:492–506. https://doi.org/10.1016/j.icarus.2011.09.018
Ingersoll AP, Ewald SP (2017) Decadal timescale variability of the Enceladus plumes inferred from Cassini images. Icarus 282:260–275. https://doi.org/10.1016/j.icarus.2016.09.018
Ingersoll AP, Pankine AA (2010) Subsurface heat transfer on Enceladus: conditions under which melting occurs. Icarus 206:594–607. https://doi.org/10.1016/j.icarus.2009.09.015
Ingersoll AP, Ewald SP, Trumbo SK (2020) Time variability of the Enceladus plumes: orbital periods, decadal periods, and aperiodic change. Icarus 344:113345. https://doi.org/10.1016/j.icarus.2019.06.006
Johnson RE, Smith HT, Tucker OJ, Liu M, Burger MH, Sittler EC, Tokar RL (2006) The Enceladus and OH Tori at Saturn. Astrophys J 644:L137–L139. https://doi.org/10.1086/505750
Jones GH, Arridge CS, Coates AJ, Lewis GR, Kanani S, Wellbrock A, Young DT, Crary FJ, Tokar RL, Wilson RJ, Hill TW, Johnson RE, Mitchell DG, Schmidt J, Kempf S, Beckmann U, Russell CT, Jia YD, Dougherty MK, Waite JH, Magee BA (2009) Fine jet structure of electrically charged grains in Enceladus' plume. Geophys Res Lett 36:L16204. https://doi.org/10.1029/2009GL038284
Jurac S, Richardson JD (2001) The dependence of plasma and magnetic field correlations in the solar wind on geomagnetic activity. J Geophys Res Space Phys 106:29195–29205. https://doi.org/10.1029/2000JA000180
Kattenhorn SA, Prockter LM (2014) Evidence for subduction in the ice shell of Europa. Nat Geosci 7:762–767. https://doi.org/10.1038/ngeo2245
Kempf S, Beckmann U, Moragas-Klostermeyer G, Postberg F, Srama R, Economou T, Schmidt J, Spahn F, Grün E (2008) The E ring in the vicinity of Enceladus. I. Spatial distribution and properties of the ring particles. Icarus 193:420–437. https://doi.org/10.1016/j.icarus.2007.06.027
Kempf S, Beckmann U, Schmidt J (2010) How the Enceladus dust plume feeds Saturn's E ring. Icarus 206:446–457. https://doi.org/10.1016/j.icarus.2009.09.016
Kempf S, Horányi M, Hsu H-W, Hill TW, Juhász A, Smith HT (2018) Saturn's diffuse E ring and its connection with Enceladus. In: Schenk PM et al (eds) Enceladus and the icy moons of Saturn. The University of Arizona Press, Tucson. https://doi.org/10.2458/azu_uapress_9780816537075-ch010
Khawaja N, Postberg F, Schmidt J (2017) The compositional profile of the enceladian ice plume from the latest Cassini flybys. In: Lunar and Planetary Science Conference. Lunar and Planetary Science Institute, Houston, p 2005
Kieffer SW, Lu X, Bethke CM, Spencer JR, Marshak S, Navrotsky A (2006) A clathrate reservoir hypothesis for Enceladus' South Polar plume. Science 1979(314):1764–1766. https://doi.org/10.1126/science.1133519
Kirchoff MR, Schenk P (2010) Global impact cratering record of Saturn's moon dione: constraining the geological history. In: 41st Lunar and Planetary Science Conference

Kirk RL, Soderblom LA, Brown RH, Kieffer S, Kargel W (1995) Triton's plumes: discovery, characteristics and models. In: Cruikshank DP (ed) Neptune and Triton. The University of Arizona Press, Tucson, p 949

Krimigis SM, Mitchell DG, Hamilton DC, Livi S, Dandouras J, Jaskulek S, Armstrong TP, Boldt JD, Cheng AF, Gloeckler G, Hayes JR, Hsieh KC, Ip W-H, Keath EP, Kirsch E, Krupp N, Lanzerotti LJ, Lundgren R, Mauk BH, McEntire RW, Roelof EC, Schlemm CE, Tossman BE, Wilken B, Williams DJ (2004) Magnetosphere Imaging Instrument (MIMI) on the Cassini mission to Saturn/Titan. Space Sci Rev 114:233–329. https://doi.org/10.1007/s11214-004-1410-8

Lainey V, Jacobson RA, Tajeddine R, Cooper NJ, Murray C, Robert V, Tobie G, Guillot T, Mathis S, Remus F, Desmars J, Arlot JE, De Cuyper JP, Dehant V, Pascu D, Thuillot W, Le Poncin-Lafitte C, Zahn JP (2017) New constraints on Saturn's interior from Cassini astrometric data. Icarus 281:286–296. https://doi.org/10.1016/j.icarus.2016.07.014

Lainey V, Casajus LG, Fuller J, Zannoni M, Tortora P, Cooper N, Murray C, Modenini D, Park RS, Robert V, Zhang Q (2020) Resonance locking in giant planets indicated by the rapid orbital expansion of Titan. Nat Astron 4:1053–1058. https://doi.org/10.1038/s41550-020-1120-5

Lainey V, et al (2025) Space Sci Rev 221

Le Gall A, Leyrat C, Janssen MA, Choblet G, Tobie G, Bourgeois O, Lucas A, Sotin C, Howett C, Kirk R, Lorenz RD, West RD, Stolzenbach A, Massé M, Hayes AH, Bonnefoy L, Veyssière G, Paganelli F (2017) Thermally anomalous features in the subsurface of Enceladus's South Polar Terrain. Nat Astron 1:0063. https://doi.org/10.1038/s41550-017-0063

Le Gall A, Bonnefoy LE, Sultana R, Leyrat C, Janssen MA, Wall S, Lellouch E (2023) Microwaving Mimas, Enceladus, Tethys, Dione, Rhea, Iapetus and Phoebe: insights into the regolith properties and geological history of Saturn's icy satellites. Icarus 394:115446. https://doi.org/10.1016/j.icarus.2023.115446

Lesage E, Schmidt F, Andrieu F, Massol H (2021) Constraints on effusive cryovolcanic eruptions on Europa using topography obtained from Galileo images. Icarus 361:114373. https://doi.org/10.1016/j.icarus.2021.114373

Loeffler MJ, Baragiola RA (2009) Physical and chemical effects on crystalline H_2O_2 induced by 20 keV protons. J Chem Phys 130:114504. https://doi.org/10.1063/1.3079612

Magee BA, Waite H (2017) Neutral gas composition of Enceladus' plume – model parameter insights from Cassini-INMS. In: Lunar and planetary science XLVIII, p 2974

Magee B, Waite H (2020) Analysis of minor and trace organic species in Enceladus' plume gas. In: American geophysical union, fall meeting, p #P003-0003

Manga M, Wang C-Y (2007) Pressurized oceans and the eruption of liquid water on Europa and Enceladus. Geophys Res Lett 34:L07202. https://doi.org/10.1029/2007GL029297

McKinnon WB (2015) Effect of Enceladus's rapid synchronous spin on interpretation of Cassini gravity. Geophys Res Lett 42:2137–2143. https://doi.org/10.1002/2015GL063384

Meyer J (2008) Tidal evolution of Mimas, Enceladus and Dione. Icarus 193:213–223. https://doi.org/10.1016/j.icarus.2007.09.008s

Meyer J, Wisdom J (2007) Tidal heating in Enceladus. Icarus 188:535–539. https://doi.org/10.1016/j.icarus.2007.03.001

Miles GM, Howett CJA, Spencer JR (2022) Constraining Enceladus' energy emission outside the South Polar Terrain. In: Europlanet Science Congress, EPSC2022-1190. https://doi.org/10.5194/epsc2022-1190

Miyamoto H, Mitri G, Showman AP, Dohm JM (2005) Putative ice flows on Europa: geometric patterns and relation to topography collectively constrain material properties and effusion rates. Icarus 177:413–424. https://doi.org/10.1016/j.icarus.2005.03.014

Morooka MW, Wahlund J-E, Eriksson AI, Farrell WM, Gurnett DA, Kurth WS, Persoon AM, Shafiq M, André M, Holmberg MKG (2011) Dusty plasma in the vicinity of Enceladus. J Geophys Res Space Phys 116:A12221. https://doi.org/10.1029/2011JA017038

Nakajima M, Ingersoll AP (2016) Controlled boiling on Enceladus. 1. Model of the vapor-driven jets. Icarus 272:309–318. https://doi.org/10.1016/j.icarus.2016.02.027

National Academies of Sciences, Engineering, and Medicine (2022) Origins, worlds, and life: a decadal strategy for planetary science and astrobiology 2023–2032. National Academies Press, Washington DC. https://doi.org/10.17226/26522

Nimmo F, Spencer JR (2015) Powering Triton's recent geological activity by obliquity tides: implications for Pluto geology. Icarus 246:2–10. https://doi.org/10.1016/j.icarus.2014.01.044

Nimmo F, Spencer JR, Pappalardo RT, Mullen ME (2007) Shear heating as the origin of the plumes and heat flux on Enceladus. Nature 447:289–291. https://doi.org/10.1038/nature05783

Nimmo F, Porco C, Mitchell C (2014) Tidally modulated eruptions on Enceladus: Cassini ISS observations and models. Astron J 148:46. https://doi.org/10.1088/0004-6256/148/3/46

Nimmo F, Neveu M, Howett C.J.A (2023) Origin and evolution of Enceladus's tidal dissipation. Space Sci Rev 219:7. https://doi.org/10.1007/s11214-023-01007-4

Paganini L, Villanueva GL, Roth L, Mandell AM, Hurford TA, Retherford KD, Mumma MJ (2019) A measurement of water vapour amid a largely quiescent environment on Europa. Nat Astron 4:266–272. https://doi.org/10.1038/s41550-019-0933-6

Park RS, Mastrodemos N, Jacobson RA, Berne A, Vaughan AT, Hemingway DJ, Leonard EJ, Castillo-Rogez JC, Cockell CS, Keane JT, Konopliv AS, Nimmo F, Riedel JE, Simons M, Vance S (2024) The global shape, gravity field, and libration of Enceladus. J Geophys Res Planets 129:e2023JE008054. https://doi.org/10.1029/2023JE008054

Perry ME, Teolis BD, Hurley DM, Magee BA, Waite JH, Brockwell TG, Perryman RS, McNutt RL (2015) Cassini INMS measurements of Enceladus plume density. Icarus 257:139–162. https://doi.org/10.1016/j.icarus.2015.04.037

Phillips CB, McEwen AS, Hoppa GV, Fagents SA, Greeley R, Klemaszewski JE, Pappalardo RT, Klaasen KP, Breneman HH (2000) The search for current geologic activity on Europa. J Geophys Res Planets 105:22579–22597. https://doi.org/10.1029/1999JE001139

Porco CC, Helfenstein P, Thomas PC, Ingersoll AP, Wisdom J, West R, Neukum G, Denk T, Wagner R, Roatsch T, Kieffer S, Turtle E, Mcewen A, Johnson TV, Rathbun J, Veverka J, Wilson D, Perry J, Spitale J, Brahic A, Burns JA, Delgenio AD, Dones L, Murray CD, Squyres S (2006a) Cassini observes the active South Pole of Enceladus. Science 311:1393–1401. https://doi.org/10.1126/science.1123013

Porco C, West RA, Squyres S, Mcewen A, Thomas P, Murray CD, Delgenio A, Ingersoll AP, Johnson TV, Neukum G, Veverka J, Dones L, Brahic A, Burns JA, Haemmerle V, Knowles B, Dawson D, Roatsch T, Beurle K, Owen W (2006b) Cassini imaging science: instrument characteristics and anticipated scientific investigations at Saturn. Space Sci Rev 115:363–497. https://doi.org/10.1007/s11214-004-1456-7

Porco C, Dinino D, Nimmo F (2014) How the geysers, tidal stresses, and thermal emission across the South Polar Terrain of Enceladus are related. Astron J 148:45. https://doi.org/10.1088/0004-6256/148/3/45

Porco C, Mitchell C, Nimmo F, Tiscareno M (2018) Enceladus' plume temporal variability from analysis of Cassini ISS images. In: 49th Lunar and Planetary Science Conference

Postberg F, Kempf S, Srama R, Green SF, Hillier JK, McBride N, Grün E (2006) Composition of Jovian dust stream particles. Icarus 183:122–134. https://doi.org/10.1016/j.icarus.2006.02.001

Postberg F, Kempf S, Hillier JK, Srama R, Green SF, McBride N, Grün E (2008) The E-ring in the vicinity of Enceladus. II. Probing the moon's interior-the composition of E-ring particles. Icarus 193:438–454. https://doi.org/10.1016/j.icarus.2007.09.001

Postberg F, Kempf S, Schmidt J, Brilliantov N, Beinsen A, Abel B, Buck U, Srama R (2009) Sodium salts in E-ring ice grains from an ocean below the surface of Enceladus. Nature 459:1098–1101. https://doi.org/10.1038/nature08046

Postberg F, Schmidt J, Hillier J, Kempf S, Srama R (2011) A salt-water reservoir as the source of a compositionally stratified plume on Enceladus. Nature 474:620–622. https://doi.org/10.1038/nature10175

Postberg F, Clark RN, Hansen CJ, Coates AJ, Dalle Ore CM, Scipioni F, Hedman MM, Waite JH (2018) Plume and surface composition of Enceladus. In: Schenk PM et al (eds) Enceladus and the icy moons of Saturn. The University of Arizona Press, Tucson. https://doi.org/10.2458/azu_uapress_9780816537075-ch007

Prockter LM, Shirley JH, Dalton JB, Kamp L (2017) Surface composition of pull-apart bands in Argadnel Regio, Europa: evidence of localized cryovolcanic resurfacing during basin formation. Icarus 285:27–42. https://doi.org/10.1016/j.icarus.2016.11.024

Quick LC, Hedman MM (2020) Characterizing deposits emplaced by cryovolcanic plumes on Europa. Icarus 343:113667. https://doi.org/10.1016/j.icarus.2020.113667

Quick LC, Barnouin OS, Prockter LM, Patterson GW (2013) Constraints on the detection of cryovolcanic plumes on Europa. Planet Space Sci 86:1–9. https://doi.org/10.1016/j.pss.2013.06.028

Quick LC, Glaze LS, Baloga SM (2017) Cryovolcanic emplacement of domes on Europa. Icarus 284:477–488. https://doi.org/10.1016/j.icarus.2016.06.029

Quick LC, Fagents SA, Núñez KA, Wilk KA, Beyer RA, Beddingfield CB, Martin ES, Prockter LM, Hurford TA (2022) Cryolava dome growth resulting from active eruptions on Jupiter's moon Europa. Icarus 387:115185. https://doi.org/10.1016/j.icarus.2022.115185

Rhoden AR, Hurford TA, Roth L, Retherford K (2015) Linking Europa's plume activity to tides, tectonics, and liquid water. Icarus 253:169–178. https://doi.org/10.1016/j.icarus.2015.02.023

Roth L, Retherford KD, Saur J, Strobel DF, Feldman PD, McGrath MA, Nimmo F (2014a) Orbital apocenter is not a sufficient condition for HST/STIS detection of Europa's water vapor aurora. Proc Natl Acad Sci USA 111:E5123. https://doi.org/10.1073/pnas.1416671111

Roth L, Saur J, Retherford KD, Strobel DF, Feldman PD, McGrath MA, Nimmo F (2014b) Transient water vapor at Europa's South Pole. Science 1979(343):171–174. https://doi.org/10.1126/science.1247051

Schenk PM (2020) The search for Europa's plumes: no surface patterns or changes 1979–2007? Astrophys J 892:L12. https://doi.org/10.3847/2041-8213/ab6f78

Schenk PM, McKinnon WB (2009) One-hundred-km-scale basins on Enceladus: evidence for an active ice shell. Geophys Res Lett 36:L16202. https://doi.org/10.1029/2009GL039916

Schenk P, Hamilton DP, Johnson RE, McKinnon WB, Paranicas C, Schmidt J, Showalter MR (2011) Plasma, plumes and rings: Saturn system dynamics as recorded in global color patterns on its midsize icy satellites. Icarus 211:740–757. https://doi.org/10.1016/j.icarus.2010.08.016

Schmidt J, Brilliantov N, Spahn F, Kempf S (2008) Slow dust in Enceladus' plume from condensation and wall collisions in tiger stripe fractures. Nature 451:685–688. https://doi.org/10.1038/nature06491

Schmidt BE, Blankenship DD, Patterson GW, Schenk PM (2011) Active formation of 'chaos terrain' over shallow subsurface water on Europa. Nature 479:502–505. https://doi.org/10.1038/nature10608

Schubert G, Anderson JD, Travis BJ, Palguta J (2007) Enceladus: present internal structure and differentiation by early and long-term radiogenic heating. Icarus 188:345–355. https://doi.org/10.1016/j.icarus.2006.12.012

Scipioni F, Schenk P, Tosi F, D'Aversa E, Clark R, Combe J-P, Ore CMD (2017) Deciphering sub-micron ice particles on Enceladus surface. Icarus 290:183–200. https://doi.org/10.1016/j.icarus.2017.02.012

Sekine Y, Shibuya T, Postberg F, Hsu H-W, Suzuki K, Masaki Y, Kuwatani T, Mori M, Hong PK, Yoshizaki M, Tachibana S, Sirono S (2015) High-temperature water–rock interactions and hydrothermal environments in the chondrite-like core of Enceladus. Nat Commun 6:8604. https://doi.org/10.1038/ncomms9604

Soderblom LA, Kieffer SW, Becker TL, Brown RH, Cook AF, Hansen CJ, Johnson TV, Kirk RL, Shoemaker EM (1990) Triton's geyser-like plumes: discovery and basic characterization. Science 1979(250):410–415. https://doi.org/10.1126/science.250.4979.410

Southworth BS, Kempf S, Spitale J (2019) Surface deposition of the Enceladus plume and the zenith angle of emissions. Icarus 319:33–42. https://doi.org/10.1016/j.icarus.2018.08.024

Spahn F, Schmidt J, Albers N, Hörning M, Makuch M, Seiß M, Kempf S, Srama R, Dikarev V, Helfert S, Moragas-Klostermeyer G, Krivov AV, Sremčević M, Tuzzolino AJ, Economou T, Grün E (2006) Cassini dust measurement at Enceladus and implications for the origin of the E ring. Science 1979(311):1416–1418. https://doi.org/10.1126/science.1121375

Sparks WB, Hand KP, McGrath MA, Bergeron E, Cracraft M, Deustua SE (2016) Probing for evidence of plumes on Europa with HST/STIS. Astrophys J 829:121. https://doi.org/10.3847/0004-637X/829/2/121

Sparks WB, Schmidt BE, McGrath MA, Hand KP, Spencer JR, Cracraft M, Deustua SE (2017) Active cryovolcanism on Europa? Astrophys J 839:L18. https://doi.org/10.3847/2041-8213/aa67f8

Spencer JR, Lebofskyf LA, Sykes MV (1989) Systematic biases in radiometric diameter determinations. Icarus 78:337–354. https://doi.org/10.1016/0019-1035(89)90182-6

Spencer JR, Pearl JC, Segura M, Flasar FM, Mamoutkine A, Romani P, Buratti BJ, Hendrix AR, Spilker LJ, Lopes RMC (2006) Cassini encounters Enceladus: background and the discovery of a south polar hot spot. Science 1979(311):1401–1405. https://doi.org/10.1126/science.1121661

Spencer J, Nimmo F, Ingersoll AP, Hurford TA, Kite ES, Rhoden AR, Schmidt J, Howett CJA (2018) Plume origins and plumbing (ocean to surface). In: Schenk PM et al (eds) Enceladus and the icy moons of Saturn. The University of Arizona Press, Tucson. https://doi.org/10.2458/azu_uapress_9780816537075-ch008

Spitale JN, Porco CC (2007) Association of the jets of Enceladus with the warmest regions on its south-polar fractures. Nature 449:695–697. https://doi.org/10.1038/nature06217

Spitale JN, Hurford TA, Rhoden AR, Berkson EE, Platts SS (2015) Curtain eruptions from Enceladus'south-polar terrain. Nature 521:57–60. https://doi.org/10.1038/nature14368

Spohn T, Schubert G (2003) Oceans in the icy Galilean satellites of Jupiter? Icarus. https://doi.org/10.1016/S0019-1035(02)00048-9

Srama R, Ahrens TJ, Altobelli N, Auer S, Bradley JG, Burton M, Dikarev VV, Economou T, Fechtig H, Görlich M, Grande M, Graps A, Grün E, Havnes O, Helfert S, Horanyi M, Igenbergs E, Jessberger EK, Johnson TV, Kempf S, Krivov AV, Krüger H, Mocker-Ahlreep A, Moragas-Klostermeyer G, Lamy P, Landgraf M, Linkert D, Linkert G, Lura F, McDonnell JAM, Möhlmann D, Morfill GE, Müller M, Roy M, Schäfer G, Schlotzhauer G, Schwehm GH, Spahn F, Stübig M, Svestka J, Tschernjawski V, Tuzzolino AJ, Wäsch R, Zook HA (2004) The Cassini cosmic dust analyzer. Space Sci Rev 114:465–518. https://doi.org/10.1007/s11214-004-1435-z

Teolis BD, Perry ME, Hansen CJ, Waite JH, Porco CC, Spencer JR, Howett CJA (2017) Enceladus plume structure and time variability: comparison of Cassini observations. Astrobiology 17:926–940. https://doi.org/10.1089/ast.2017.1647

Thomas N, Portyankina G, Hansen CJ, Pommerol A (2011) HiRISE observations of gas sublimation-driven activity in Mars' southern polar regions: IV. Fluid dynamics models of CO_2 jets. Icarus 212:66–85. https://doi.org/10.1016/j.icarus.2010.12.016

Thomas PC, Tajeddine R, Tiscareno MS, Burns JA, Joseph J, Loredo TJ, Helfenstein P, Porco C (2016) Enceladus's measured physical libration requires a global subsurface ocean. Icarus 263:37–47. https://doi.org/10.1016/j.icarus.2015.08.037

Tokar RL, Johnson RE, Hill TW, Pontius DH, Kurth WS, Crary FJ, Young DT, Thomsen MF, Reisenfeld DB, Coates AJ, Lewis GR, Sittler EC, Gurnett DA (2006) The interaction of the atmosphere of Enceladus with Saturn's plasma. Science 1979(311):1409–1412. https://doi.org/10.1126/science.1121061

Travis BJ, Schubert G (2015) Keeping Enceladus warm. Icarus 250:32–42. https://doi.org/10.1016/j.icarus.2014.11.017

USGS (2015) Enceladus Nomenclature [WWW Document]. https://planetarynames.wr.usgs.gov/images/enceladus_comp.pdf

Verbiscer A, French R, Showalter M, Helfenstein P (2007) Enceladus: cosmic graffiti artist caught in the act. Science. https://doi.org/10.1126/science.1134681

Villaneuva GL, Hammel HB, Milam SN, Faggi S, Kofman V, Roth L, Hand KP, Paganini L, Stansberry J, Spencer J, Protopapa S, Strazzulla G, Cruz-Mermy G, Glein CR, Cartwright R, Liuzzi G (2023) Endogenous CO_2 ice mixture on the surface of Europa and no detection of plume activity. Science 381:1305–1308. https://doi.org/10.1126/science.adg4270

Waite JH, Lewis WS, Kasprzak WT, Anicich VG, Block BP, Cravens TE, Fletcher GG, Ip W-H, Luhmann JG, Mcnutt RL, Niemann HB, Parejko JK, Richards JE, Thorpe RL, Walter EM, Yelle RV (2004) The Cassini Ion and Neutral Mass Spectrometer (INMS) investigation. Space Sci Rev 114:113–231. https://doi.org/10.1007/s11214-004-1408-2

Waite JH, Combi MR, Ip WH, Cravens TE, McNutt RL, Kasprzak W, Yelle R, Luhmann J, Niemann H, Gell D, Magee B, Fletcher G, Lunine J, Tseng WL (2006) Cassini ion and Neutral Mass Spectrometer: Enceladus plume composition and structure. Science 1979(311):1419–1422. https://doi.org/10.1126/science.1121290

Waite JH, Glein CR, Perryman RS, Teolis BD, Magee BA, Miller G, Grimes J, Perry ME, Miller KE, Bouquet A, Lunine JI, Brockwell T, Bolton SJ (2017) Cassini finds molecular hydrogen in the Enceladus plume: evidence for hydrothermal processes. Science 1979(356):155–159. https://doi.org/10.1126/science.aai8703

Yeoh SK, Chapman TA, Goldstein DB, Varghese PL, Trafton LM (2015) On understanding the physics of the Enceladus south polar plume via numerical simulation. Icarus 253:205–222. https://doi.org/10.1016/j.icarus.2015.02.020

Young DT, Berthelier JJ, Blanc M, Burch JL, Coates AJ, Goldstein R, Grande M, Hill TW, Johnson RE, Kelha V, Mccomas DJ, Sittler EC, Svenes KR, Szegö K, Tanskanen P, Ahola K, Anderson D, Bakshi S, Baragiola RA, Barraclough BL, Black RK, Bolton S, Booker T, Bowman R, Casey P, Crary FJ, Delapp D, Dirks G, Eaker N, Funsten H, Furman JD, Gosling JT, Hannula H, Holmlund C, Huomo H, Illiano JM, Jensen P, Johnson MA, Linder DR, Luntama T, Maurice S, Mccabe KP, Mursula K, Narheim BT, Nordholt JE, Preece A, Rudzki J, Ruitberg A, Smith K, Szalai S, Thomsen MF, Viherkanto K, Vilppola J, Vollmer T, Wahl TE, Wüest M, Ylikorpi T, Zinsmeyer C (2004) Cassini plasma spectrometer investigation. Space Sci Rev 114:1–112. https://doi.org/10.1007/s11214-004-1406-4

Young DT, Berthelier J-J, Blanc M, Burch JL, Bolton S, Coates AJ, Crary FJ, Goldstein R, Grande M, Hill TW, Johnson RE, Baragiola RA, Kelha V, McComas DJ, Mursula K, Sittler EC, Svenes KR, Szegö K, Tanskanen P, Thomsen MF, Bakshi S, Barraclough BL, Bebesi Z, Delapp D, Dunlop MW, Gosling JT, Furman JD, Gilbert LK, Glenn D, Holmlund C, Illiano J-M, Lewis GR, Linder DR, Maurice S, McAndrews HJ, Narheim BT, Pallier E, Reisenfeld D, Rymer AM, Smith HT, Tokar RL, Vilppola J, Zinsmeyer C (2005) Composition and dynamics of plasma in Saturn's magnetosphere. Science 307:1262–1266. https://doi.org/10.1126/science.1106151

Space Science Reviews (2023) 219:57
https://doi.org/10.1007/s11214-023-01007-4

Origin and Evolution of Enceladus's Tidal Dissipation

Francis Nimmo[1] · Marc Neveu[2,3] · Carly Howett[4]

Received: 5 April 2023 / Accepted: 19 September 2023 / Published online: 6 October 2023

Abstract
Enceladus possesses a subsurface ocean beneath a conductive ice shell. Based on shell thickness models, the estimated total conductive heat loss from Enceladus is 25–40 GW; the measured heat output from the South Polar Terrain (SPT) is 4–19 GW. The present-day SPT heat flux is of order 100 mW m^{-2}, comparable to estimated paleo-heat fluxes for other regions of Enceladus. These regions have nominal ages of about 2 Ga, but the estimates are uncertain because the impactor flux in the Saturnian system may not resemble that elsewhere. Enceladus's measured rate of orbital expansion implies a low dissipation factor Q_p for Saturn, with $Q_p \approx 3 \times 10^3$ (neglecting the role of Dione). This value implies that Enceladus's present-day equilibrium tidal heat production (roughly 50 GW, but with large uncertainties) is in approximate balance with its heat loss. If Q_p is constant, Enceladus cannot be older than 1.5 Gyr (because otherwise it would have migrated more than is permissible). However, Saturn's dissipation may be better described by the "resonance-locking" theory, in which case Enceladus's orbit may have only evolved outwards by about 35% over the age of the Solar System. In the constant-Q_p scenario, any ancient tidal heating events would have been too energetic to be consistent with the observations. Because resonance-locking makes capture into earlier mean-motion orbital resonances less likely, the inferred ancient heating episodes probably took place when the current orbital resonance was already established. In the resonance-locking scenario, tidal heating did not change significantly over time, allowing for a long-lived ocean and a relatively stable ice shell. If so, Enceladus is an attractive target for future exploration from a habitability standpoint.

Keywords Satellites of Saturn · Tidal heating · Astrometry · Resonance locking · Orbital evolution · Impact flux

1 Introduction

One of the biggest surprises of the *Cassini* mission was the discovery of active venting and thermal anomalies at the South Pole of Enceladus (Porco et al. 2006; Spencer et al.

✉ F. Nimmo
fnimmo@ucsc.edu

1 Dept. Earth and Planetary Sciences, University of California Santa Cruz, CA 95064, USA

2 Dept. Astronomy, University of Maryland, College Park, MD 20742, USA

3 Planetary Environments Laboratory, NASA Goddard Space Flight Center, Greenbelt, MD 20771, USA

4 Dept. Physics, Oxford University, Oxford, OX1 3PU, UK

2006). Although it was immediately recognized that this activity was almost certainly due to tidal heating, the available tidal energy was initially thought to be insufficient (Meyer and Wisdom 2007). Later measurements of the orbital evolution of the Saturnian moons demonstrated that Saturn could in fact supply energy and angular momentum to these moons at a much higher rate than previously thought (Lainey et al. 2017, 2020; Jacobson 2022), thus potentially solving the Enceladus energy crisis. At about the same time, a combination of gravity (Iess et al. 2014), libration (Thomas et al. 2016) and chemical (Postberg et al. 2009) measurements revealed that Enceladus possessed a global ocean sandwiched between an ice shell and a low-density silicate core.

This paper will focus on the origin and evolution of tidal heating on Enceladus. There are two main reasons for doing so. First, apart from Io (Matsuyama et al. 2022), Enceladus is the best place in the solar system to understand the details of how tidal heating actually works, as well as its ultimate causes. Second, we need to understand how the balance between heat production and heat loss has evolved over time to understand the history of Enceladus's ocean. A long-lived ocean is more likely to be a habitable environment than one that has only appeared recently.

Section 2 will summarize the pertinent observations, while Sect. 3 will discuss the theoretical aspects of the present-day tidal heating. Section 4 will make use of both observations and theory to discuss how tidal heating on Enceladus may have evolved with time, and Sect. 5 will summarize our conclusions.

This article overlaps with several other in this collection, notably those by Rhoden et al., Howett et al., Tobie et al., Ćuk et al. and Fuller et al. It also provides an update on Nimmo et al. (2018), which covered a similar set of topics.

2 Observations

To understand the history of tidal heating on Enceladus, it is important to understand the general geologic setting and timing of events, as well as observational constraints on both its present-day and its ancient heat flux. Less obviously, it is also important to understand Enceladus's orbital evolution, because this is driven by the tides in Saturn which are also the ultimate heat source for the moons. A brief discussion of Dione is also included, because in some cases all we can determine is the energy budget of the Dione-Enceladus pair (see Sect. 3.1).

2.1 Global Geology and Ages

Broadly speaking, Enceladus can be divided into four regions (Crow-Willard and Pappalardo 2015; Patterson et al. 2018): the young, tectonized and geologically active south polar terrain (SPT); two tectonized regions of intermediate age, one centered on the trailing hemisphere (THT) and the other on the leading hemisphere (LHT); and the heavily cratered, ancient northern plains. The symmetry of this arrangement about the tidal axis is not understood, but the similarity of deformation at the LHT and the THT to that at the south pole suggests there were different episodes of heating at different locations on the satellite (Crow-Willard and Pappalardo 2015).

In principle, knowing the absolute ages of different terrains would provide a powerful constraint on Enceladus's evolution. Unfortunately, the only current way of deriving such ages is by combining crater counts with a model impact flux. In the outer solar system, model impact fluxes for heliocentric impactors are poorly-calibrated (Zahnle et al. 2003).

Furthermore, it is not entirely clear that the Saturn system has experienced the same impact history as the rest of the outer solar system (Ferguson et al. 2022), and indeed the system dynamics may have been perturbed only ~100 Myr ago (Wisdom et al. 2022). Bearing these caveats in mind, crater counts (Kirchoff and Schenk 2009) show that the cratered plains on Enceladus have a median age around 4 Gyr, while the tectonized ridged plains are somewhat younger, with median ages of about 2 Gyr.[1] The SPT region is very much younger, probably <1 Myr (Porco et al. 2006).

Although not as geologically active as Enceladus, Dione displays abundant tectonic features, both ridges and extensional troughs (Schenk et al. 2018). Based on cross-cutting relationships, some of the troughs are quite young, as is the large impact basin Evander (likely <2.5 Gyr and possibly <1 Gyr, (Kirchoff et al. 2018)). Although Dione's heavily cratered plains are probably >4 Gyr old (Dones et al. 2009), its smooth plains are roughly 2 Gyr old (Kirchoff et al. 2018) and may have experienced cryovolcanism (Schenk et al. 2018). The overall impression is of a body that has experienced moderate geological activity over an extended period, or several discrete intervals.

2.2 Direct Measurements of Present-Day Heat-Flux

Enceladus's endogenic heat flux results in significantly increased surface temperatures near its South Pole, with the warmest temperatures occurring along the tiger stripes (Howett et al. 2011). The magnitude of these elevated temperatures has been determined using 4–5 μm data from Cassini's Visible and Infrared Mapping Spectrometer (VIMS), and 7–1000 μm data from Cassini's Composite Infrared Spectrometer (CIRS). Due to the non-linearity of the Planck function, CIRS is sensitive to a wider range of surface temperatures than VIMS, and thus has been the preferred dataset for such study.

To convert the measured surface temperatures into the desired estimate of heat flow, two questions arise: (1) what fraction of the observed emission is endogenic versus exogenic, and (2) how much of the surface is emitting at this level? A full description of how different studies have approached these questions, along with their relative advantages and disadvantages, may be found in the article by Howett et al. (this collection). However, a brief overview is provided here for context.

Initial 9–16 μm observations from CIRS were used to infer a heat flow of 5.8 $\pm$ 1.9 GW (Spencer et al. 2006), but these wavelengths are insensitive to any low-temperature (<65 K) endogenic emission. So a follow up study using 16–1000 μm observations that covered the entire SPT was conducted. These longer wavelengths are sensitive to lower temperatures (>30 K), but require the removal of the exogenic emission. This study inferred a much higher total heat flow of 15.9 $\pm$ 3.1 GW (Howett et al. 2011). Further work showed that the SPT's four ≈150-km long tiger stripe fractures alone appear to be radiating 4.2 GW (Spencer et al. 2018), in which case the 35-km wide inter-tiger-stripe regions could be responsible for a few to roughly 10 GW of heat flow. Energy transport by latent heat from the tiger stripe plumes is small, roughly 0.5 GW (Ingersoll and Pankine 2010).

Thus, the total heat emitted by the south polar terrain is likely in the range 3.9–19 GW. The area south of 65°S constitutes 4.7% of the surface of Enceladus, so the corresponding average heat flux would be roughly 100-500 mW m^{-2}. Even the low end of this range is comparable to terrestrial heat fluxes. Furthermore, it roughly matches the heat fluxes inferred from attempting to explain the "ropy" tectonic texture observed between the tiger stripes by

[1] Here we are using model A of (Zahnle et al. 2003), because this model is heavily favored based on crater counts in the Pluto/Charon system (Singer et al. 2019).

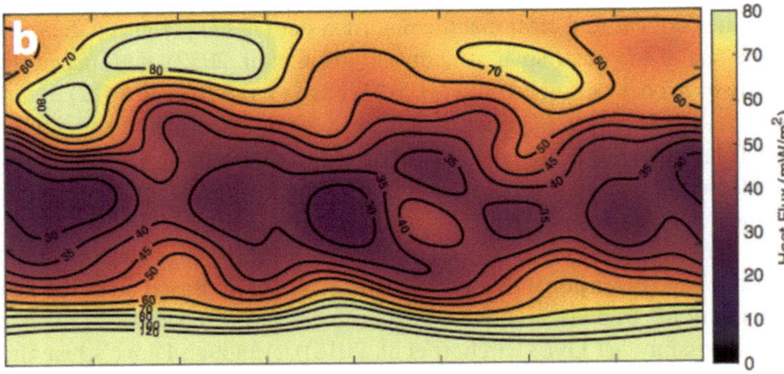

Fig. 1 Spatial variation in conductive heat flux due to ice shell thickness variations, from Hemingway and Mittal (2019)

compression (Bland et al. 2015). Even higher heat flux values ($\sim$500 mW m^{-2}) have been inferred for a region adjacent to the tiger stripes on the basis of microwave radiometry (Le Gall et al. 2017). However, this region does not show evidence of activity from either imaging or CIRS data, so the significance of this latter result is unclear. Geophysical constraints on ancient, as opposed to present-day, heat fluxes are discussed in Sect. 2.4 below.

2.3 Estimates of Global Present-Day Heat Loss

As discussed above, the CIRS instrument is most sensitive to heat fluxes emitted by the hottest parts of Enceladus. However, we expect a low level of heat flow across the rest of the body, which will be undetectable by CIRS but nonetheless constitute a major fraction of the overall heat budget. Thus, the *measured* heat flow must be less than the total, global heat flow by an amount that is uncertain but could be large. To estimate this latter heat flow, we turn to models of global variations in shell thickness.

Enceladus's ice shell is almost certainly conductive, because of the large shell thickness variations inferred primarily from gravity/topography measurements (see below). If the shell were convecting, these variations would have been smoothed out very rapidly by lateral flow of ice (Kamata and Nimmo 2017). Most recent works have also concluded that the amount of tidal heating in the ice shell is probably small compared to the global heat budget (e.g. Souček et al. 2019). If we assume a Cartesian, conductive shell and neglect internal heat production, then we can relate the local heat flux F_{cond} directly to the local shell thickness d:

$$F_{\text{cond}} = \frac{c}{d} \ln\left(\frac{T_m}{T_s}\right) \tag{1}$$

Here the thermal conductivity is taken to vary as c/T, where $c = 651$ W/m is a constant (Petrenko and Whitworth 1999), T_s is the surface temperature and T_m the temperature at the base of the ice shell. Any tidal heating in the shell would be concentrated at its base and effectively reduce T_m. Given a map of global shell thickness variations, equation (1) can be integrated to determine the total conductive heat loss across the ice shell.

The ice shell thickness and its variations are determined by a combination of libration (Thomas et al. 2016) and gravity/topography (Iess et al. 2014) measurements, where for the latter Airy isostasy is assumed. A good summary of these approaches, which also calculates the global heat loss, is found in Hemingway and Mittal (2019). Figure 1 is taken from their paper and shows the inferred spatial variation in conductive heat loss. The high heat flow ($\approx$160 mW m^{-2}) at the South Pole is a consequence of the low shell thickness ($\approx$5 km) there. Strikingly, the northern polar regions also exhibit high inferred heat fluxes, while in the equatorial regions heat fluxes are lower. This pattern is somewhat reminiscent of the heat flux variations expected for tidal heating in an ice shell (Ojakangas and Stevenson 1989).

Taking into account uncertainties in background ice shell thickness, density and so on, Hemingway and Mittal (2019) find a global conductive heat loss of 20–35 GW. This does not include the localized heat flow of $\approx$5 GW associated with the tiger stripes themselves, or the small contribution (0.5 GW) from advection of latent heat in the plumes. Thus, the global present-day heat loss at Enceladus is in the range 25–40 GW; Park et al. (submitted) find slightly smaller values (18–28 GW). In any event, these inferred heat flows provide one of our strongest observational constraints on the system.

At Dione, much less information is available. The available gravity and topography data are consistent with an ice shell of about 120 km thickness overlying an ocean, but non-ocean models are also permissible (Zannoni et al. 2020). In the ocean case, the total conductive heat flow would be roughly 25 GW, comparable to that estimated for Enceladus and an order of magnitude greater than the expected $\approx$2 GW of radiogenic heat production. Thus, the total heat loss in the Enceladus-Dione system might be as much as 65 GW.

2.4 Estimates of Ancient Heat Fluxes

Even if an area is not being heated at the present day, it is sometimes possible to make estimates of ancient heat fluxes using geophysical techniques. These estimates, although uncertain, are useful because they allow us to piece together a history of tidal heating.

There are three principal techniques which have been used at Enceladus. The first is to infer the thickness of the elastic lithosphere at the time of deformation, and then convert that thickness to a heat flux. This technique was originally developed for the terrestrial oceanic lithosphere (Watts et al. 2013). The second is to model the viscoelastic relaxation of impact craters, and to infer the minimum heat flux required to have caused the observed relaxation, given some assumed heating duration. The third is to calculate the heat flux required to cause an instability of the correct (observed) wavelength to develop when a section of ice shell is subject to extension.

An example of this first approach is by Giese et al. (2008). These authors modeled the flexural topography of an ancient equatorial rift zone on the edge of the trailing hemisphere region and obtained an effective elastic thickness of 0.3 km. They used a yield-strength envelope approach (McNutt 1984) to convert this thickness to a heat flux in the range 200-270 mW m^{-2}. The age of the deformation is uncertain, and model ages based on crater counts can differ by an order of magnitude depending on the assumptions made. Nonetheless, this heat flux probably records an event happening at least 1 Gyr ago (Giese et al. 2008).

Bland et al. (2012) looked at relaxed craters in two regions, one in the northern plains and one adjacent to the area investigated by Giese et al. (2008). Because of the small diameter of some of the relaxed craters, these authors found it was difficult to find conditions under which relaxation occurred. Even after appealing to a layer of insulating regolith to increase the effective surface temperature, they concluded that heat fluxes in excess of 150 mW m^{-2} would be required. Although the cratered terrains themselves are old, the heating event could have been significantly younger. The authors point out that heating can in some places apparently occur without accompanying deformation (unlike at the tiger stripes), and that areas of heating can be quite widespread.

Bland et al. (2007) examined two equatorial regions, including the one also analyzed by Giese et al. (2008). They argued that the ridges and troughs observed might be the result of an instability which can develop when the lithosphere undergoes extension. Although this constitutes a less direct estimate than the flexural approach, it produced comparable results: a heat flux of 110-220 $\mathrm{mW\,m^{-2}}$ and elastic thicknesses at the time of deformation of 0.4–1.4 km.

At Dione, a flexural study yielded an elastic thickness of 3.5 ± 1 km and a corresponding heat flux of 25-60 $\mathrm{mW\,m^{-2}}$ (Hammond et al. 2013), while impact crater relaxation studies yielded a lower bound on heat flux of 60 $\mathrm{mW\,m^{-2}}$ (White et al. 2017). These fluxes are an order of magnitude higher than the $\approx$6 $\mathrm{mW\,m^{-2}}$ expected if Dione currently possesses a deep ocean (see Sect. 2.3).

2.5 Astrometry

The ultimate source of the energy being tidally dissipated in Enceladus is the rotational kinetic energy of Saturn. Some of this rotational energy and angular momentum gets transferred to the satellites via the tides that these bodies raise on Saturn. The rate at which the energy and angular momentum transfer happens depends on the so-called dissipation factor of Saturn, Q_p, which in general is frequency-dependent.

A combination of long-term telescopic observations and radio tracking of the *Cassini* spacecraft have allowed Q_p to be determined at the frequencies of various satellites. For reasons explained below, it is more convenient to provide the results in terms of the outwards migration timescale of the satellite, t_{tide}, rather than Q_p, where

$$t_{\mathrm{tide}} = \left(\frac{1}{a}\frac{da}{dt}\right)^{-1} \tag{2}$$

where a is the semi-major axis and da/dt is the quantity that is measured by astrometry. An initial study which did not use spacecraft tracking yielded t_{tide} in the range 3–130 Gyr for Enceladus (Lainey et al. 2017), too uncertain to be very useful. A more recent study which incorporated tracking gave $t_{\mathrm{tide}} \approx 10$ Gyr, (Lainey et al. 2020). The equivalent value of Q_p at Enceladus's frequency for the latter study is 2030^{+3150}_{-1330}, neglecting the influence of Dione (see equation (6)). An independent evaluation using both astrometry and tracking yielded a Q_p at Enceladus's frequency of 3136 ± 1317 (Jacobson 2022) and thus a t_{tide} of about 9 Gyr, compatible within error with the Lainey et al. (2020) result.[2] The significance of these measurements will be discussed in much more detail below.

2.6 Summary

Enceladus is emitting observable heat at a current rate of about 4–19 GW. The total conductive heat loss across the ice shell is about 25–40 GW, while Dione's rate of heat loss, though uncertain, could be up to 25 GW. Present-day conductive heat losses at the SPT are around 160 $\mathrm{mW\,m^{-2}}$, which are compatible with estimates of ancient heat fluxes in both deformed equatorial and northern cratered regions. The age of these heating events is uncertain, but if current crater flux models are correct, the events are probably a few Gyr old. Astrometry

[2]One important difference in the results is that Lainey et al. (2020) find that Enceladus is currently moving outwards, while Jacobson (pers. comm.) finds that it is moving inwards. Because dissipation in Saturn and Enceladus respectively increase and decrease the semi-major axis, the *net* motion of Enceladus is hard to determine and increases uncertainty in Q_p.

has constrained the rate at which the inner satellites are moving outwards, and thus the rate at which Saturn is supplying energy and angular momentum to the satellites.

3 Origin of Present-Day Dissipation

3.1 Tidal and Equilibrium Heating

Tidal heating arises because, for a satellite in an eccentric orbit, the distance and direction to the primary is constantly changing and so too is the size of the tidal bulge (e.g. Murray and Dermott 1999). Unless the satellite is perfectly elastic, this continuous deformation will generate heat. The general expression for the rate of heat production if the eccentricity is small and the satellite's spin and orbital period are synchronous is

$$\dot{E} = \frac{21}{2}\frac{n^5 R_s^5}{G} e^2 \frac{k_{2s}}{Q_s} = 15\ \text{GW}\ \frac{k_{2s}/Q_s}{0.01} \tag{3}$$

Here n is the mean motion, R_s the satellite radius, G the gravitational constant, e the eccentricity and k_{2s} the so-called tidal Love number of the satellite, a measure of the response to the tidally-induced gravitational potential which indicates how deformable the satellite is. The quantity Q_s is the quality factor of the satellite, a measure of how dissipative it is. Note that here we are neglecting any contribution from obliquity tides or forced librations.

Although we do not have a direct measurement of k_{2s}/Q_s for Enceladus, we can estimate it as follows. If the tidal heat production is comparable to the heat loss (see Sect. 4.4), say 30 GW, then equation (3) yields a k_{2s}/Q_s of 0.02. The actual value of this quantity will depend on the viscosity and rigidity structure of the body (see Tobie et al. this collection).

Equation (3) therefore shows that, in general, we need information about the internal structure of the satellite, as parameterized by k_{2s}/Q_s, to calculate the rate of tidal heating. However, for a satellite in a mean motion resonance with another satellite, this requirement can be avoided. This is extremely useful, as it makes our conclusions insensitive to poorly-known details about individual satellites' structures.

A consequence of dissipative processes (Eq. (3)) is that the satellite's eccentricity will damp over time, thus reducing the tidal heating rate (Murray and Dermott 1999). Thus, an isolated satellite is expected to evolve to a circular orbit over time, as is the case with Triton. However, for a satellite in a mean-motion resonance (MMR), the eccentricity is simultaneously being excited, as a result of the torques being applied on the satellites by the central planet (Murray and Dermott 1999). In the specific case of the Enceladus-Dione 2:1 resonance, it is Enceladus's eccentricity that is getting excited (Meyer and Wisdom 2008b). There are thus competing eccentricity-exciting and eccentricity-damping mechanisms. As a result, the situation tends towards an equilibrium at which the eccentricity is constant. Furthermore, if this equilibrium is achieved, then the total tidal dissipation rate within the pair of satellites can be calculated without needing to know the satellite k_2/Q values.

If the eccentricities of the satellites in resonance are small, this equilibrium tidal heating rate may be written (Meyer and Wisdom 2007)

$$\dot{E}_{eq} = \frac{n_0 T_0}{\sqrt{1-e_0^2}} + \frac{n_1 T_1}{\sqrt{1-e_1^2}} - \frac{T_0+T_1}{L_0+L_1}\left(\frac{GM_p m_0}{a_0} + \frac{GM_p m_1}{a_1}\right) \tag{4}$$

where here subscripts refer to the inner (0) and outer (1) satellite, n is the mean motion, T is the torque on the satellite from Saturn, L is the orbital angular momentum

($= m[GM_p a(1-e^2)]^{1/2}$), a is the semi-major axis and m the satellite mass, assumed much smaller than Saturn's mass M_p. An equivalent way of calculating the equilibrium heating rate was introduced by Fuller et al. (2016), but we use equation (4) in what follows because it makes the dependence on Dione torques more obvious. Below we take $e_0 = 0.0047$ and $e_1 = 0.0022$.

A disadvantage of this equation is that it does not specify in which body the tidal heating is happening. Because the current Enceladus-Dione resonance is exciting Enceladus's eccentricity (Meyer and Wisdom 2008a), it has generally been assumed that the majority of heat production takes place in Enceladus. However, Dione also has a non-zero eccentricity and could thus (according to equation (3)) be dissipating energy too. It should also be noted that in some situations the instantaneous heating rate can oscillate about the equilibrium value (see Sect. 3.2 below).

The crucial quantity is the torque, which depends on the Q_p of Saturn at the relevant frequency:

$$T = \frac{3}{2}\frac{Gm^2 R_p^5 k_{2p}}{a^6 Q_p} \tag{5}$$

Here R_p is Saturn's radius (taken to be 58,232 km) and k_{2p} its tidal Love number (taken here to be 0.34 (Gavrilov and Zharkov 1977) and independent of frequency). As will be seen below, whether there is a significant torque acting on Dione (T_1) has an important bearing on our conclusions. The importance of equation (4) is that it allows the total tidal heating in Enceladus and Dione to be estimated without knowing their internal structures, as long as Q_p is known for both bodies.

The last piece of the puzzle is therefore to relate Q_p to the astrometrically-derived orbital evolution timescale $t_{\rm tide}$. This is done using (Fuller et al. 2016):

$$t_{\rm tide} = \frac{Q_p}{F}\frac{M_p}{m}\left(\frac{a}{R_p}\right)^5 \frac{1}{3k_{2p}} n^{-1} \approx 3\ {\rm Myr}\frac{Q_p}{F} \tag{6}$$

There is thus a linear relationship between Q_p and $t_{\rm tide}$. However, there is one complication, embedded in the factor F. For an isolated moon, $F = 1$ and no further discussion is required. For a moon in a mean-motion resonance, it is *a priori* unclear whether the motion of the outer moon is as a direct result of torques being exerted on it by Saturn, or whether its motion is due to transfer of torques applied to the inner moon. If the latter case, then the outwards motion of the two moons will be reduced by some factor, because the torque on the inner moon is being applied to the combined angular momentum of both bodies. Conversely, in order to match a specified rate of outwards motion, the torque in the second case must be larger than in the first case, and thus the Q_p must be smaller i.e. Saturn needs to be more dissipative. This additional factor is $F = [1 + L_1/L_0]^{-1}$. Unfortunately, for the Enceladus-Dione case, $F \approx 0.07$ and so whether direct tidal torques on Dione (T_1) are important matters a great deal, both for deriving Q_p and for determining the equilibrium heat flow (equation (4)). These issues are also discussed in Ćuk and El Moutamid (2023).

As explained in Sect. 4.1 below, Saturn torques most likely arise because there are peaks in Saturn's dissipation spectrum occurring at the satellite frequency. For torques to be operating on both bodies simultaneously, there would have to be two resonant peaks in Saturn, at the synodic periods of Enceladus and Dione. Furthermore, the period ratio of these two peaks would have had to be maintained at 2:1 as Saturn evolved. Although theory suggests that neither eventuality is very likely, astrometry finds non-zero Saturn torques (i.e. a finite

Table 1 Summary of Enceladus equilibrium heat flow calculations and Saturn dissipation factor at Enceladus frequency Q_p. Values for $t_{\rm tide}$ are based on (Lainey et al. 2020) and (Jacobson 2022); error bars in Q_p are estimated based on the latter. Q_p and $t_{\rm tide}$ are related via equation (6) and heat flow $\dot{E}$ is from equation (4). Note the large difference in whether direct torques on Dione (T_1) are included or not. The actual equilibrium heating rates are probably intermediate between these values (see text), and Q_p at the frequencies of Enceladus and Dione may not be the same

$t_{\rm tide}$ (Gyr)	Q_p	$\dot{E}$ (GW)	Notes
9	190 ± 80	190^{+150}_{-60}	No direct torques on Dione ($F = 0.07$, $T_1 = 0$)
9	2600 ± 1100	7^{+5}_{-2}	Direct torques on Dione ($F = 1$, $T_1 \neq 0$)

Q_p) on Dione (Lainey et al. 2020). As a result, the real situation is probably somewhere intermediate between the two cases presented in Table 1.

Table 1 summarizes the ranges of possible equilibrium tidal heating rates for a value of $t_{\rm tide} = 9$ Gyr, consistent with both Lainey et al. (2020) and Jacobson (2022). A smaller $t_{\rm tide}$ would mean more rapid evolution, a lower Q_p and higher heating rates. This Table illustrates the factor of ≈30 difference between neglecting and including the effects of torques on Dione. If the torques on Dione are zero, the equilibrium heating rate is in the range 130–350 GW. However, it is more likely that the torques on Dione contribute at some level. For instance, the lowest Dione torque case from Table 1 of Lainey et al. (2020) results in Q_p values of roughly 700 and 4000 at Enceladus and Dione, respectively, and an equilibrium heating rate of roughly 50 GW. Most of this energy is presumably going to heat Enceladus rather than Dione.

3.2 Periodic Heating

As explained above, for the Enceladus-Dione system there is a competition between eccentricity-forcing and eccentricity-damping which tends to produce a constant equilibrium eccentricity. However, this is not the only possible solution; as first pointed out by Ojakangas and Stevenson (1986), another possibility is that the eccentricity and heat production undergo periodic oscillations around their equilibrium values. This behaviour arises because of the feedbacks between the heating rate, the internal temperature (and thus viscosity) of the body, and the rate at which it dissipates energy. Periodic solutions are favored if the timescales of eccentricity damping and internal structure evolution are comparable.

Meyer and Wisdom (2008a) showed that the specific mechanism advocated by Ojakangas and Stevenson (1986) did not in fact produce periodic behaviour at Enceladus. However, Shoji et al. (2014) adopted a slightly different model set-up and found that for some parameter values periodic behavior could be generated. Oscillation periods were typically ≈100 Myr, but these authors assumed a constant Q_p of 18,000, which results in low mean heating rates (1.1 or 2.4 GW, see (Meyer and Wisdom 2007)). A lower Q_p would presumably have resulted in more rapid oscillations.

If it does experience oscillations, Enceladus could currently be in a state such that its instantaneous eccentricity and heating rate are below their equilibrium values, eccentricity damping is suppressed and the moon is moving deeper into the resonance. In principle, sufficiently good astrometric measurements should be able to determine whether the moon is moving into or out of the resonance; this would require a measurement of radial drift with a precision of better than 1 cm/yr. For Enceladus the competing effects of dissipation in Saturn and dissipation in the satellite are hard to disentangle (Lainey et al. 2020; Jacobson 2022).

3.3 Location of Heating

Just as equation (4) does not specify how tidal heating is distributed between Enceladus or Dione, nor does it specify where within a particular body the heat is being produced. Most recent authors have concluded that heating in either the ice shell (e.g. Souček et al. 2019) or the ocean (e.g. Hay and Matsuyama 2019; Rovira-Navarro et al. 2019) are insufficient, although one recent work on internal waves in a stratified ocean argued that this heating mechanism could be important (Rovira-Navarro et al. 2023). Nonetheless, it appears likely that the bulk of heating occurs within the low-density core (Roberts 2015; Choblet et al. 2017). How exactly the core dissipates heat is an open question, although the specific mechanism of tidal fluid pumping (Rovira-Navarro et al. 2022; Kamata 2023) appears to be insufficient. The location of tidal heating is explored in more detail in the article by Tobie et al. (this collection).

3.4 Comparison with Observations

Table 1 leaves us in a rather unsatisfactory state, as the high and low end of the tidal heat production estimates (5–350 GW) bracket the inferred heat loss rate of 25–65 GW (Sect. 2.3). The problem is that the direct Saturn torques on Dione are not known. The results of Lainey et al. (2020) suggest that a non-zero Dione torque does exist, but theoretical arguments suggest that it should not (see Sect. 3.1), and the uncertainties are large enough to also permit a zero-torque solution for Dione. Our current best guess is that the direct Saturn torques on Dione are small but non-zero, and that the heat production rate is intermediate, roughly 50 GW (Sect. 3.1). But the current uncertainties are large; reducing the uncertainties in these astrometric constraints would certainly improve our understanding of the problem.

At the low end of the heat production estimates shown in Table 1, Enceladus would be losing more heat than it is currently producing. The resulting ocean freezing timescale would be of order 10 Myr. Unless we are seeing Enceladus at a fortuitous time in its history, and the Saturn torques on Dione are large, this scenario seems unlikely.

At the high end, the heat production rate could be a factor of 6–16 times larger than the system conductive heat loss. An interesting consequence of this would be that Enceladus's ocean would be melting quite rapidly: 100 GW of excess heating would thin the ice shell to zero in about 1 Myr! This scenario also seems unlikely, not least because a rapidly-thinning ice shell would produce a network of global compressional features (Nimmo 2004) which are simply not seen on the surface. One could argue that Enceladus is fortuitously at a high heat production point of an oscillating cycle, but the oscillation period would have to be short compared with 1 Myr. Or perhaps Dione is currently experiencing the bulk of the heating right now, but has not yet had time for the effects to reach the surface. Neither possibility seems likely.

If there is some torque contribution to Dione, the heat production rate (and inferred Q_p) will be intermediate between the high and low-end values of Table 1. As described in Sect. 3.1, the lowest Dione torque case examined by Lainey et al. (2020) produces a heat production rate of about 50 GW. This intermediate value would roughly balance the conductive heat loss rate, suggesting that Enceladus's current structure (including its ocean) are potentially long-lived features. This result makes intuitive sense: if the heating rate exceeds the rate of heat loss, the ice shell will simply thin until thermal equilibrium is restored. Despite the uncertainties, it seems likely that for Enceladus, the present-day heat production and heat loss are roughly in balance. How that balance may have evolved over time is the subject of the next section.

4 Evolution of Dissipation

4.1 Semi-Major Axis Evolution

Recalling that $t_{\text{tide}} = a/\frac{da}{dt}$, equation (6) can be integrated to determine how the semi-major axis a varies with time, once Q_p is specified.

In the simplest case of a constant Q_p, a increases as roughly $t^{2/13}$ (Murray and Dermott 1999). More recently, it has been argued that Q_p is unlikely to be constant, and that the orbital evolution timescale t_{tide} is instead ultimately controlled by some aspect of the evolution timescale of the planet. This idea is known as the "resonance-locking" hypothesis. It was first introduced by (Fuller et al. 2016) and has subsequently been further elucidated elsewhere (Nimmo et al. 2018; Lainey et al. 2020; Ćuk and El Moutamid 2023), including the article by Fuller et al. (this collection). An alternative formulation for calculating gas giant dissipation has been presented by Terquem (2021).

The first flavor of the resonance-locking hypothesis (Fuller et al. 2016) focuses on the evolving frequency of gravity modes in Saturn's interior that are in resonance with tidal excitation by a satellite. This frequency changes on a timescale t_α. In this case we have

$$\frac{1}{a}\frac{da}{dt} = \frac{2}{3}\left[\frac{\Omega_p}{n}\left(\frac{1}{t_\alpha} - \frac{1}{t_p}\right) - \frac{1}{t_\alpha}\right] \tag{7}$$

Here Ω_p is Saturn's spin frequency and t_p is the spin evolution timescale. In this approach, if it is assumed that $t_p \to \infty$ then a simple analytical solution for $a(t)$ can be obtained (Nimmo et al. 2018). This solution typically produces satellites with diverging orbits, important for understanding their evolution through mean-motion resonances (see Sect. 4.3).

The second flavor (Lainey et al. 2020) is a special case of the first, in which the mode evolution timescale t_α is assumed proportional to the age of the planet t, such that

$$\frac{1}{a}\frac{da}{dt} \approx \frac{B}{t} \tag{8}$$

where B is a constant. If $t_{\text{tide}} \approx 9$ Gyr (Sect. 2.5), then $B \approx 1/2$, so we have a evolving as roughly $t^{1/2}$. This version is more likely to be associated with inertial waves and results in satellites maintaining near-equal period ratios as they migrate outwards.

Figure 2a plots an example of the semi-major axis evolution of Enceladus under these three different scenarios, ignoring the effects of any mean-motion resonances. The present-day evolution timescale t_{tide} is set to 9 Gyr (cf. Table 1) in all three cases. For the constant-Q_p model, Enceladus cannot be more than about 1.5 Gyr old. Conversely, using the Fuller et al. (2016) approach (equation (7)), Enceladus has only increased its semi-major axis by about 35% over 4.5 Gyr. Lastly, the Lainey et al. (2020) approach (equation (8)) yields an intermediate result with an Enceladus that could be as old as 3 Gyr. An important consequence of the resonance-locking hypotheses is that a young Enceladus is *not* required, despite its relatively rapid present-day outwards migration.

Figure 2b plots the corresponding evolution of Q_p. In both the Lainey et al. (2020) and Fuller et al. (2016) scenarios, Q_p increases with decreasing semi-major axis. This is important when considering past rates of tidal heating (see Sect. 4.2).

It should be emphasized that the results shown in Fig. 2 are subject to large uncertainties. Not only is the present-day outwards evolution rate of Enceladus quite uncertain, but the resonance-locking models make assumptions about the rate at which the internal structure of Saturn is changing which are likely to become progressively more inaccurate at earlier times.

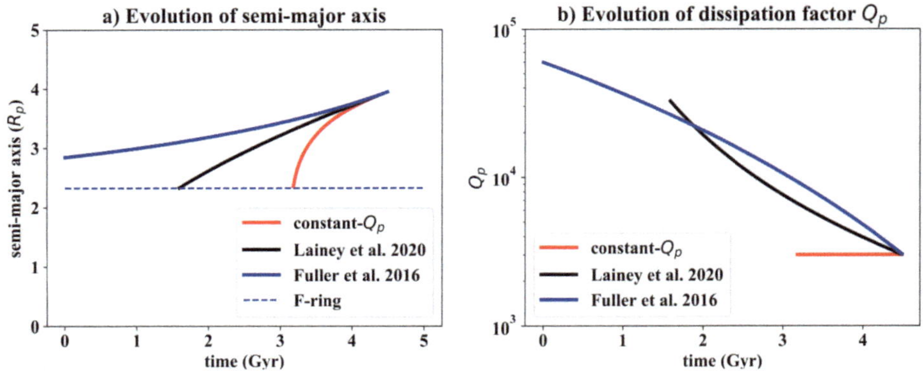

Fig. 2 a) Evolution of Enceladus's semi-major axis according to three different Saturn-Q scenarios. The present-day evolution timescale t_{tide} is set to 9 Gyr in all cases. In the constant-Q case, Q_p is set to 3100. In the Lainey et al. (2020) case (equation (8)), $B = 0.49$. In the Fuller et al. (2016) case (equation (7)), $t_\alpha = 13$ Gyr. The location of the F-ring denotes where a satellite will be torn apart by tides. Here we neglect any influence of the mean motion resonance with Dione. b) Corresponding values of Q_p as a function of time. Here we are taking $F = 1$ (equation (6)). Present day is at 4.5 Gyr

Furthermore, for Fig. 2b the Q_p values derived depend on whether torques from Dione are included or not (equation (6)), which adds an additional factor of 30 uncertainty. Finally, these calculations are only relevant while the Enceladus-Dione mean motion resonance is operating, and of course the age of that resonance is not known (see Sect. 4.3).

4.2 Tidal Heating Evolution

Figure 3 uses the same values as Fig. 2 to plot how equilibrium tidal heat production has evolved with time. For each of the three scenarios, lower (solid line) and upper (dashed line) values are plotted depending on whether torques on Dione are included in the calculations of Q_p (equation (6)) and $\dot{E}$ (equation (4)). As discussed above (Sect. 3.1) the true value will be somewhere between the two extremes. The green dashed lines represent the likely range of heat loss (25–65 GW) in the Enceladus-Dione system.

Even though the heat flow calculations only apply during the Enceladus-Dione MMR, it is clear that either constant-Q model would rapidly run into problems at earlier times, generating Io-like (1 W m^{-2}) heat fluxes. Unless the resonance was encountered recently, a constant-Q_p scenario can be ruled out. Conversely, the Lainey et al. (2020) heat flow increases more modestly with decreasing age, while the Fuller et al. (2016) model actually shows a small decrease at earlier times. The implications of these results will be discussed further in Sect. 4.4.

Although the values shown in Fig. 3 are specific to the Enceladus-Dione 2:1 resonance, equilibrium tidal heating rates for other resonances are not very different (Meyer and Wisdom 2007) and would show qualitatively the same behavior. Thus, if Enceladus encountered ancient resonances prior to its current one, we would expect heating rates similar to those in Table 1 and Fig. 3.

4.3 Constraints on Orbital Evolution

The considerable uncertainties in Enceladus's evolutionary path, which arise from the unknown nature and evolution of tidal dissipation inside Saturn and the uncertain present-day

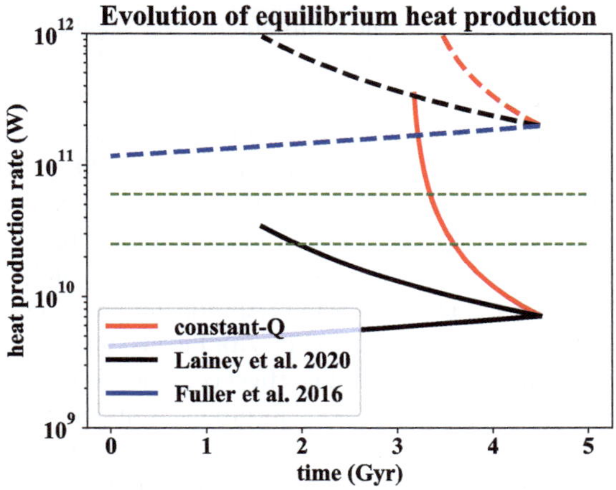

Fig. 3 Evolution of equilibrium heat production rate as a function of time, using the same parameters as for Fig. 2. Solid lines are calculated using equation (4) with $F = 1$, while dashed lines use $F = 0.07$, to reflect the effect of uncertainties in the role of Dione torques. The true values will lie between the two extremes. The horizontal dashed green lines denote the estimated heat loss rate of the Enceladus-Dione system (25–65 GW)

rates of orbital expansion of Enceladus and Dione, bring about a need for constraints on the past orbital evolution of these moons. These constraints fall into two main categories: orbital characteristics; and inferences of past heating episodes. As noted in Sect. 2.1, age constraints are based on impactor flux models that may not be relevant to the Saturn system, so their usefulness is limited. Further discussion on the orbital constraints may be found in the articles by Ćuk et al. (this collection) and the paper by Ćuk and El Moutamid (2023).

Orbital Characteristics There are three major obstacles in obtaining constraints based on orbital characteristics. First, due to the overall compactness of Saturn's mid-sized moon system, the orbital evolution of Enceladus is not solely tied to that of Dione, but also to those of the other moons in that system: Mimas, Tethys, and Rhea (Neveu and Rhoden 2019). All five moons are packed between 3 and 9 Saturn radii, which implies a high potential for encountering orbital resonances throughout their history. Titan's larger distance from Saturn makes it less likely that it has entered orbital resonances with the inner moons that played a major role in their orbital evolution.

Second, unlike secular (nonresonant) orbital evolution, the outcome of orbital (e.g., mean motion) resonances can be stochastic (e.g. Ćuk et al. 2020). This makes it difficult to constrain the orbital properties (semi-major axes, eccentricities, inclinations) of two satellites before a resonance is encountered even if these are known after the resonance is exited, and vice versa.

The third obstacle is a consequence of the first two: the age of the moons is unknown. In such a compact system where orbital semi-major axes and eccentricities are prone to large relative changes due to gravitational interactions between moons (Malhotra 1998), collision, ejection, and tidal disruption events (if a moon enters Saturn's Roche limit) are more likely (e.g. Nakajima et al. 2019). Saturn's ring-moons and moonlets, whose sizes increase with increasing distance from Saturn (Tiscareno et al. 2006) testify to ongoing accretion in the system (Charnoz et al. 2010). The rings, whose mass is 40% that of Mimas, have been argued to be much younger than the age of the solar system (Iess et al. 2019; Wisdom et al. 2022). If valid (see Crida et al. 2019, for counter-arguments), this inference suggests the geologically recent collision or disruption of one or more moons (e.g. Wisdom et al. 2022). More generally, material may have been cycled between forming mid-sized moons like Enceladus and Dione and forming rings several times since the formation of Saturn at the dawn of the solar system.

These obstacles notwithstanding, the secular orbital evolution of moons due to their tidal interaction with Saturn (Sect. 4.1) indicates that the orbits of moons expand with time. Superimposed on this interaction are moon-ring interactions, which result in comparatively fast outward migration of moons whose orbital semi-major axis is less than $R_{\text{F-ring}} \times 2^{2/3} = 222{,}000$ km (Charnoz et al. 2011), where $R_{\text{F-ring}}$ is the radius of the F ring (Fig. 2a). For example, the orbit of Mimas, which is currently within this zone, would have expanded from an orbit twice as close to the rings in just 0.7 Gyr from moon-ring interactions alone if the rings predate Mimas (Neveu and Rhoden 2019). This is comparable to a constant-Q_p rate of orbital expansion and much faster than that due to resonance locking at these orbital distances (Fig. 2a). Finally, a moon's orbit may contract if tidal dissipation inside the moon is high; this effect complicates interpretation of Enceladus's astrometry (see Sect. 3.1) as well as making its orbital evolution more uncertain.

The reliability of the above constraints then rests on the propensity of moons to enter orbital resonances. In the case of mean-motion resonances (MMRs), this propensity depends significantly on the mechanism of orbital expansion; only if satellites are on converging orbits is capture into an MMR possible (Murray and Dermott 1999). In the constant-Q_p picture, numerous converging resonances are encountered, making the moons' shorter orbital history unpredictable. At the other extreme, the special case of resonance locking suggested by the observations of Lainey et al. (2020) (equation (8)) preserves semi-major axis ratios between moons such that mean-motion resonances are never encountered unless two moons formed in resonance. Generally, the propensity for capture into mean-motion resonances is much lower under the resonance-locking hypothesis than if Q_p is constant. But, as explained above (Sect. 4.2), we regard the constant-Q_p case as unlikely.

The present-day existence of the Enceladus-Dione and Mimas-Tethys mean-motion resonances suggests that some evolution of orbits on converging tracks is likely to have occurred. For instance, Mimas's current eccentricity and Tethys's current inclination are both high, but are not being substantially excited at present; this suggests that these bodies experienced some other kind of resonance in the past (see Ćuk et al.). A past 3:2 Mimas:Enceladus resonance might have excited the eccentricity of either Mimas (Meyer and Wisdom 2008a; Tian and Nimmo 2020) or Enceladus (El Moutamid et al. 2019); the latter possibility might explain evidence for older heating episodes (Sect. 2.4). Other resonances that may have affected the inner Saturnian moons are less relevant to Enceladus, and are discussed in more detail in the article by Ćuk et al.

Past Heating Events The ongoing Enceladus-Dione and Mimas-Tethys mean-motion resonances, as well as evidence for past high heat flow on Enceladus (Sect. 2.4), Tethys (White et al. 2017) and Dione (Hammond et al. 2013), suggest that the non-converging orbits implied by equation (8) may not fully describe the moons' orbital evolution, despite being compatible with observed rates of orbital expansion within uncertainties. In the specific case of Enceladus, the inferred ancient heat fluxes (Sect. 2.4) are comparable to those measured at the present day, and are consistent with heating during previous resonances. For instance, with $Q_p = 2000$ the Mimas:Enceladus 3:2 resonance would have produced an equilibrium heating rate of 8.6 GW (Tian and Nimmo 2020).

In addition to improving the astrometry of the mid-sized moons, a path forward in obtaining better constraints may lie in meshing the relative chronologies of (1) resonances encountered for a given expansion mechanism (see Ćuk et al. this collection) and (2) episodes of past high heat flow from geological evidence at all five mid-sized moons, together with detailed N-body orbital evolution simulations of relevant resonances.

4.4 Comparison with Observations

Very little is known for sure about the orbital evolution of Enceladus and the other Saturnian moons. However, the constant-Q_p and resonance-locking hypotheses make predictions which are very different and can potentially be tested.

The most important difference between the two models is the predicted ancient heat fluxes (Fig. 3). Assuming an MMR is operating, the predicted heat fluxes in the constant-Q_p model rapidly reach values in excess of 1 $\mathrm{W\,m^{-2}}$ going back in time, implying a mean ice shell thickness of <1 km. Such a shell thickness is inconsistent with the existence of large, relatively unrelaxed craters, and is hard to reconcile with the apparently ancient nature of some parts of Enceladus. Unless the MMR was established recently - close to the present-day semi-major axis - the constant-Q_p mechanism does not agree with the observational constraints.

In contrast, in the resonance-locking case the predicted heating rates vary only weakly with time and the intermediate value of 50 GW derived earlier (Sect. 3.1) is roughly consistent with the present-day estimated heat loss and estimates of past heating. In this scenario, the ice shell could have experienced either modest thickening or thinning with time. Shell thinning would lead to compression, while shell thickening would lead to extension (e.g. Nimmo 2004). On Dione there is plentiful evidence for extension and essentially none for compression (Schenk et al. 2018), but on Enceladus the tectonic picture is much more complicated (Patterson et al. 2018). Both compressional and extensional features have been inferred, but the amount of global extension or contraction is unknown. This absence of a clear signal may simply be telling us that the ice shell thickness has stayed roughly constant with time, and that other sources of stress, such as diurnal tidal stresses, non-synchronous rotation or true polar wander, are dominating the tectonic record. A roughly constant ice shell thickness would be consistent with the resonance-locking theory.

The constant-Q_p model for Enceladus implies that Enceladus could be at most 1.5 Gyr old (Fig. 2). On the face of it, this is inconsistent with the crater-derived terrain ages (Sect. 2.1). However, these ages were derived using an assumed heliocentric impactor flux model which may not be appropriate to the Saturn system. The identification of a population of apparently planetocentric impactors (Ferguson et al. 2022) suggests that the heliocentric model is probably overestimating the age of Saturnian satellite terrains. Thus, it is not yet clear whether the "young Enceladus" paradigm can be ruled out on the basis of crater statistics.

The fact that some regions of Enceladus appear to have experienced ancient heating events could be interpreted in one of two ways. Either the spatial location of the current heating has shifted over time, perhaps due to true polar wander (Nimmo and Pappalardo 2006) or some kind of runaway feedback (Kang and Flierl 2020). Or the earlier events were caused by a different MMR, prior to capture into the current resonance. It is hard to distinguish between these two possibilities, because both the cratering timescale (see Sect. 2.1) and the orbital evolution timescale are very uncertain. The constant-Q_p model produces larger semi-major axis evolution and would thus imply more resonance crossings. However, the large orbital expansion would mean that earlier heating events would be much more energetic (Fig. 3), which is hard to reconcile with the fact that ancient estimated heat fluxes are similar to the present-day values (Sect. 2.4). In contrast, the resonance-locking model is less likely to result in resonant capture (Sect. 4.3) and would favour persistence of the current resonance and heating.

In principle, the composition of Enceladus's ocean might be used to place a constraint on how long it has been in existence. The problem is that the reaction rates of many silicates and organic compounds tend to be very fast compared to likely ocean lifetimes (e.g. Truong et al. 2019; Hao et al. 2022), so the ocean composition is controlled primarily not by reaction times but by the water-rock ratio, which depends on unknown details of the silicate core of Enceladus (e.g. how permeability varies with depth). The H_2 and CH_4 detected in Enceladus's plumes are plausibly the result of hydrothermal reactions with the core (Daval et al. 2022; Etiope and Sherwood Lollar 2013), but at least for H_2 such reactions are expected to be complete within a few hundred Myr. How to reconcile this inference with Enceladus's apparently protracted history of heating is not clear.

5 Summary & Future Work

Enceladus maintains a subsurface ocean beneath an icy shell, and Dione may have an ocean too. The estimated heat loss from the present-day Enceladus-Dione system based on conductive ice shell models is 25–65 GW (Sect. 2.3). This is a few to ten times larger than the measured heat output at the SPT of 4–19 GW (Sect. 2.2), because most of the heat is escaping at a low rate that makes it difficult to distinguish from the background insolation. Different regions of Enceladus appear to have experienced elevated heat fluxes (of order $100\ \mathrm{mW\,m^{-2}}$) in the past, comparable to those inferred for the present-day SPT (Sect. 2.4). These heating events are probably a few Gyr old, but the timing (which is based on crater counts) is uncertain. Dione also appears to have experienced heating episodes in the past, though perhaps of a lower amplitude.

Astrometry shows that Enceladus's orbit has expanded outwards at an unexpectedly rapid rate (Sect. 2.5). This observation implies high levels of dissipation within Saturn, and a low dissipation factor Q_p. If Q_p has stayed constant with time, Enceladus must be young, less than about 1.5 Gyr old. On the other hand, if the "resonance-locking" theory is correct (Sect. 4.1), Q_p was larger in the past and Enceladus's orbit may have expanded by only 35% over 4.5 Gyr. Because they are based on specific impactor flux models, crater-derived age estimates cannot distinguish between these two possibilities.

At the present day, Enceladus is in an MMR with Dione and there appears to be a rough balance between the amount of tidal heat being generated, and the amount of heat being lost (Sect. 4.2). If the constant-Q_p model is correct, this resonance can only have been established recently, as equilibrium heating rates increase very markedly with decreasing semi-major axis (Fig. 3). This fact also makes the occupation of earlier MMRs in the constant-Q_p case difficult to reconcile with the geophysical evidence (Sect. 2.4). In contrast, the resonance-lock model predicts roughly constant heat fluxes and a long-lived ocean, and is less likely to have caused capture into prior MMRs (Sect. 4.3).

Our understanding of Enceladus's thermal and orbital evolution is still very much in its infancy. The constant-Q_p model is hard to reconcile with our current knowledge of Enceladus's heating history (Sect. 4.4), but we are hampered by very uncertain dating techniques, and the considerable dynamical complexity of the Saturnian satellites. A putative, recent catastrophic event (Wisdom et al. 2022) further complicates matters.

One important step forwards would be to better understand the cratering history of the inner Saturnian satellites. The identification of a population of planetocentric impactors (Ferguson et al. 2022) is important and should be followed up, along with improved impactor flux models. The likely implications of the recent disruption of a satellite for the impactor population should likewise be explored. Thermal-orbital evolution models (e.g. Neveu and Rhoden 2019) constrained by geophysical inferences could be pursued, but our current ignorance of how to describe the Q_p of Saturn means that firm conclusions will be elusive.

Further advances may require observations of Enceladus by new spacecraft, as advocated by the 2023–2032 Planetary Science and Astrobiology Decadal Survey. In particular, such a spacecraft (e.g. Ermakov et al. 2021) would greatly improve our astrometric knowledge, determining Q_p with much better precision. It would also provide a direct measurement of the k_2/Q of Enceladus. This would in turn allow a calculation of the heat production in Enceladus (Eq. (3)), providing a direct test of the concept of equilibrium heating (Sect. 3.1). Improved global heat flow measurements (from infra-red observations) would allow a direct comparison between heat production and heat loss. Measurement of noble gas abundances in the plumes would help determine the history of outgassing and thermal activity (Truong et al. 2022).

Although there are considerable uncertainties, the resonance-locking model currently appears more consistent with observations than the constant-Q_p model. This implies that Enceladus is old, with a long-lived ocean, making it an attractive astrobiological target. While the next mission to Enceladus may be focused on habitability, we expect that it will also solve many of the puzzles outlined in this article, and we look forward to seeing the results.

Acknowledgements We thank Jim Fuller, Matija Cuk, and two anonymous reviewers for helpful comments. Support from NASA grants 80NSSC21K1802 (SSW) and 80NSSC19K0885 (CDAP), and the CRESST II agreement between NASA GSFC and Univ. Maryland, College Park (award number 80GSFC17M0002), is acknowledged.

Declarations

Competing Interests The authors declare that they have no competing interests.

References

Bland MT, Beyer RA, Showman AP (2007) Unstable extension of Enceladus' lithosphere. Icarus 192(1):92–105. https://doi.org/10.1016/j.icarus.2007.06.011

Bland MT, Singer KN, McKinnon WB, Schenk PM (2012) Enceladus' extreme heat flux as revealed by its relaxed craters. Geophys Res Lett 39(17):L17204. https://doi.org/10.1029/2012GL052736

Bland MT, McKinnon WB, Schenk PM (2015) Constraining the heat flux between Enceladus' tiger stripes: numerical modeling of funiscular plains formation. Icarus 260:232–245. https://doi.org/10.1016/j.icarus.2015.07.016

Charnoz S, Salmon J, Crida A (2010) The recent formation of Saturn's moonlets from viscous spreading of the main rings. Nature 465(7299):752–754

Charnoz S, Crida A, Castillo-Rogez JC, Lainey V, Dones L, Karatekin Ö, Tobie G, Mathis S, Le Poncin-Lafitte C, Salmon J (2011) Accretion of Saturn's mid-sized moons during the viscous spreading of young massive rings: solving the paradox of silicate-poor rings versus silicate-rich moons. Icarus 216(2):535–550

Choblet G, Tobie G, Sotin C, Běhounková M, Čadek O, Postberg F, Souček O (2017) Powering prolonged hydrothermal activity inside Enceladus. Nat Astron 1(12):841–847

Crida A, Charnoz S, Hsu H-W, Dones L (2019) Are Saturn's rings actually young? Nat Astron 3(11):967–970

Crow-Willard EN, Pappalardo RT (2015) Structural mapping of Enceladus and implications for formation of tectonized regions. J Geophys Res, Planets 120(5):928–950. https://doi.org/10.1002/2015JE004818
Ćuk M, El Moutamid M (2023) A past episode of rapid tidal evolution of Enceladus? Planet Sci J 4(7):119
Ćuk M, El Moutamid M, Tiscareno MS (2020) Dynamical history of the Uranian system. Planet Sci J 1(1):22
Ćuk M, El Moutamid M, Lari G, Neveu M, Nimmo F, Noyelles B, Rhoden A, Saillenfest M Long-Term Evolution. Submitted to SSR
Daval D, Choblet G, Sotin C, Guyot F (2022) Theoretical considerations on the characteristic timescales of hydrogen generation by serpentinization reactions on Enceladus. J Geophys Res, Planets 127(2):e2021JE006995
Dones L, Chapman CR, McKinnon WB, Melosh HJ, Kirchoff MR, Neukum G, Zahnle KJ (2009) Icy satellites of Saturn: impact cratering and age determination. In: Dougherty MK Esposito LW Krimigis SM (eds) Saturn from Cassini-Huygens. Springer, Dordrecht, pp 613–635. https://doi.org/10.1007/978-1-4020-9217-6_19
El Moutamid M, Ćuk M, Tiscareno M (2019) The orbital connection between Mimas and Enceladus. In: EPSC-DPS joint meeting 2019. EPSC Abstracts, vol 13. EPSC-DPS2019-149-1
Ermakov AI, Park RS, Roa J, Castillo-Rogez JC, Keane JT, Nimmo F, Kite ES, Sotin C, Lazio TJW, Steinbrügge G et al (2021) A recipe for the geophysical exploration of Enceladus. Planet Sci J 2(4):157
Etiope G, Sherwood Lollar B (2013) Abiotic methane on Earth. Rev Geophys 51(2):276–299
Ferguson SN, Rhoden AR, Kirchoff MR, Salmon JJ (2022) A unique Saturnian impactor population from elliptical craters. Earth Planet Sci Lett 593:117652
Fuller J, Luan J, Quataert E (2016) Resonance locking as the source of rapid tidal migration in the Jupiter and Saturn moon systems. Mon Not R Astron Soc 458(4):3867–3879. https://doi.org/10.1093/mnras/stw609
Fuller J, Guillot T, Mathis S Tidal Dissipation in Giant Planets. Submitted to SSR
Gavrilov SV, Zharkov VN (1977) Love numbers of the giant planets. Icarus 32(4):443–449. https://doi.org/10.1016/0019-1035(77)90015-X
Giese B, Wagner R, Hussmann H, Neukum G, Perry J, Helfenstein P, Thomas PC (2008) Enceladus: an estimate of heat flux and lithospheric thickness from flexurally supported topography. Geophys Res Lett 35(24):L24204. https://doi.org/10.1029/2008GL036149
Hammond NP, Phillips CB, Nimmo F, Kattenhorn SA (2013) Flexure on Dione: investigating subsurface structure and thermal history. Icarus 223(1):418–422. https://doi.org/10.1016/j.icarus.2012.12.021
Hao J, Glein CR, Huang F, Yee N, Catling DC, Postberg F, Hillier JK, Hazen RM (2022) Abundant phosphorus expected for possible life in Enceladus's ocean. Proc Natl Acad Sci 119(39):e2201388119
Hay HC, Matsuyama I (2019) Nonlinear tidal dissipation in the subsurface oceans of Enceladus and other icy satellites. Icarus 319:68–85
Hemingway DJ, Mittal T (2019) Enceladus's ice shell structure as a window on internal heat production. Icarus 332:111–131. https://doi.org/10.1016/j.icarus.2019.03.011
Howett CJA, Spencer JR, Pearl J, Segura M (2011) High heat flow from Enceladus' south polar region measured using 10-600 cm^{-1} Cassini/CIRS data. J Geophys Res, Planets 116:E03003. https://doi.org/10.1029/2010JE003718
Howett C, Miles G, Quick LC Constraining the mass and energy of dissipation systems. In preparation to SSR
Iess L, Stevenson DJ, Parisi M, Hemingway D, Jacobson RA, Lunine JI, Nimmo F, Armstrong JW, Asmar SW, Ducci M, Tortora P (2014) The gravity field and interior structure of Enceladus. Science 344(6179):78–80. https://doi.org/10.1126/science.1250551
Iess L, Militzer B, Kaspi Y, Nicholson P, Durante D, Racioppa P, Anabtawi A, Galanti E, Hubbard W, Mariani M et al (2019) Measurement and implications of Saturn's gravity field and ring mass. Science 364(6445):eaat2965
Ingersoll AP, Pankine AA (2010) Subsurface heat transfer on Enceladus: conditions under which melting occurs. Icarus 206(2):594–607
Jacobson RA (2022) The orbits of the main Saturnian satellites, the Saturnian system gravity field, and the orientation of Saturn's pole. Astron J 164(5):199
Kamata S (2023) Poroviscoelastic gravitational dynamics. J Geophys Res, Planets, 128:e2022JE007700
Kamata S, Nimmo F (2017) Interior thermal state of Enceladus inferred from the viscoelastic state of the ice shell. Icarus 284:387–393
Kang W, Flierl G (2020) Spontaneous formation of geysers at only one pole on Enceladus's ice shell. Proc Natl Acad Sci 117(26):14764–14768
Kirchoff MR, Schenk P (2009) Crater modification and geologic activity in Enceladus' heavily cratered plains: evidence from the impact crater distribution. Icarus 202(2):656–668. https://doi.org/10.1016/j.icarus.2009.03.034

Kirchoff MR, Bierhaus EB, Dones L, Robbins SJ, Singer KN, Wagner RJ, Zahnle KJ (2018) Cratering histories in the Saturnian system. In: Schenk PM, Clark RN, Howett CJA, Verbiscer AJ, Waite JH (eds) Enceladus and the icy moons of Saturn. University of Arizona Press, Tucson, pp 267–284
Lainey V, Jacobson RA, Tajeddine R, Cooper NJ, Murray C, Robert V, Tobie G, Guillot T, Mathis S, Remus F, Desmars J, Arlot J-E, De Cuyper J-P, Dehant V, Pascu D, Thuillot W, Le Poncin-Lafitte C, Zahn J-P (2017) New constraints on Saturn's interior from Cassini astrometric data. Icarus 281:286–296. https://doi.org/10.1016/j.icarus.2016.07.014
Lainey V, Casajus LG, Fuller J, Zannoni M, Tortora P, Cooper N, Murray C, Modenini D, Park RS, Robert V, Zhang Q (2020) Resonance locking in giant planets indicated by the rapid orbital expansion of Titan. Nat Astron 4:1053–1058. https://doi.org/10.1038/s41550-020-1120-5
Le Gall A, Leyrat C, Janssen MA, Choblet G, Tobie G, Bourgeois O, Lucas A, Sotin C, Howett C, Kirk R et al (2017) Thermally anomalous features in the subsurface of Enceladus's south polar terrain. Nat Astron 1(4):0063
Malhotra R (1998) Orbital resonances and chaos in the solar system. In: Lazzaro D et al (eds) Solar system formation and evolution, ASP Conference Series, vol 149, p 37
Matsuyama IN, Steinke T, Nimmo F (2022) Tidal heating in Io. Elements 18(6):374–378
McNutt MK (1984) Lithospheric flexure and thermal anomalies. J Geophys Res 89:11180–11194. https://doi.org/10.1029/JB089iB13p11180
Meyer J, Wisdom J (2007) Tidal heating in Enceladus. Icarus 188(2):535–539. https://doi.org/10.1016/j.icarus.2007.03.001
Meyer J, Wisdom J (2008a) Episodic volcanism on Enceladus: application of the Ojakangas Stevenson model. Icarus 198(1):178–180. https://doi.org/10.1016/j.icarus.2008.06.012
Meyer J, Wisdom J (2008b) Tidal evolution of Mimas, Enceladus, and Dione. Icarus 193(1):213–223
Murray CD, Dermott SF (1999) Solar system dynamics
Nakajima A, Ida S, Kimura J, Brasser R (2019) Orbital evolution of Saturn's mid-sized moons and the tidal heating of Enceladus. Icarus 317:570–582
Neveu M, Rhoden AR (2019) Evolution of Saturn's mid-sized moons. Nat Astron 3(6):543–552
Nimmo F (2004) Stresses generated in cooling viscoelastic ice shells: application to Europa. J Geophys Res, Planets 109:E12001
Nimmo F, Pappalardo RT (2006) Diapir-induced reorientation of Saturn's moon Enceladus. Nature 441(7093):614–616
Nimmo F, Barr AC, Běhounková M, McKinnon WB (2018) The thermal and orbital evolution of Enceladus: observational constraints and models. In: Schenk PM, Clark RN, Howett CJA, Verbiscer AJ, Waite JH (eds) Enceladus and the icy moons of Saturn. University of Arizona Press, Tucson, p 79. https://doi.org/10.2458/azu_uapress_9780816537075-ch005
Ojakangas GW, Stevenson DJ (1986) Episodic volcanism of tidally heated satellites with application to Io. Icarus 66(2):341–358. https://doi.org/10.1016/0019-1035(86)90163-6
Ojakangas GW, Stevenson DJ (1989) Thermal state of an ice shell on Europa. Icarus 81(2):220–241. https://doi.org/10.1016/0019-1035(89)90052-3
Park R, Mastrodemos N, Jacobson R et al (submitted) The global shape, gravity field and libration of Enceladus. J Geophys Res, Planets
Patterson GW, Kattenhorn SA, Helfenstein P, Collins GC, Pappalardo RT (2018) The geology of Enceladus. In: Schenk PM, Clark RN, Howett CJA, Verbiscer AJ, Waite JH (eds) Enceladus and the icy moons of Saturn. University of Arizona Press, Tucson, p 95
Petrenko VF, Whitworth RW (1999) The physics of ice. Oxford University Press
Porco CC, Helfenstein P, Thomas PC, Ingersoll AP, Wisdom J, West R, Neukum G, Denk T, Wagner R, Roatsch T, Kieffer S, Turtle E, McEwen A, Johnson TV, Rathbun J, Veverka J, Wilson D, Perry J, Spitale J, Brahic A, Burns JA, Del Genio AD, Dones L, Murray CD, Squyres S (2006) Cassini observes the active South pole of Enceladus. Science 311(5766):1393–1401. https://doi.org/10.1126/science.1123013
Postberg F, Kempf S, Schmidt J, Brilliantov N, Beinsen A, Abel B, Buck U, Srama R (2009) Sodium salts in E-ring ice grains from an ocean below the surface of Enceladus. Nature 459(7250):1098–1101. https://doi.org/10.1038/nature08046
Rhoden AR, Ferguson SN, Castillo-Rogez JC, Martin E, Bland M, Zannoni M, Rambaux N, Salmon J Geologic constraints on the formation and evolution of Saturn's mid-sized moons. Submitted to SSR
Roberts JH (2015) The fluffy core of Enceladus. Icarus 258:54–66
Rovira-Navarro M, Rieutord M, Gerkema T, Maas LR, van der Wal W, Vermeersen B (2019) Do tidally-generated inertial waves heat the subsurface oceans of Europa and Enceladus? Icarus 321:126–140
Rovira-Navarro M, Katz RF, Liao Y, van der Wal W, Nimmo F (2022) The tides of Enceladus' porous core. J Geophys Res, Planets, 127:e2021JE007117
Rovira-Navarro M, Matsuyama I, Hay HCC (2023) Thin-shell tidal dynamics of ocean worlds. Planet Sci J 4(2):23

Schenk P, White OL, Byrne PK, Moore JM (2018) Saturn's other icy moons: geologically complex worlds in their own right. In: Schenk PM, Clark RN, Howett CJA, Verbiscer AJ, Waite JH (eds) Enceladus and the icy moons of Saturn. University of Arizona Press, Tucson, p 237

Shoji D, Hussmann H, Sohl F, Kurita K (2014) Non-steady state tidal heating of Enceladus. Icarus 235:75–85. https://doi.org/10.1016/j.icarus.2014.03.006

Singer KN, McKinnon WB, Gladman B, Greenstreet S, Bierhaus EB, Stern SA, Parker AH, Robbins SJ, Schenk PM, Grundy WM, Bray VJ, Beyer RA, Binzel RP, Weaver HA, Young LA, Spencer JR, Kavelaars JJ, Moore JM, Zangari AM, Olkin CB, Lauer TR, Lisse CM, Ennico K, New Horizons Geology, Geophysics and Imaging Science Theme Team, New Horizons Surface Composition Science Theme Team, New Horizons Ralph and LORRI Teams (2019) Impact craters on Pluto and Charon indicate a deficit of small Kuiper belt objects. Science 363(6430):955–959. https://doi.org/10.1126/science.aap8628.

Souček O, Běhounková M, Čadek O, Hron J, Tobie G, Choblet G (2019) Tidal dissipation in Enceladus' uneven, fractured ice shell. Icarus 328:218–231

Spencer JR, Pearl JC, Segura M, Flasar FM, Mamoutkine A, Romani P, Buratti BJ, Hendrix AR, Spilker LJ, Lopes RMC (2006) Cassini encounters Enceladus: background and the discovery of a south polar hot spot. Science 311(5766):1401–1405. https://doi.org/10.1126/science.1121661

Spencer J, Nimmo F, Ingersoll AP, Hurford T, Kite E, Rhoden A, Schmidt J, Howett C (2018) Plume origins and plumbing: from ocean to surface. In: Schenk PM, Clark RN, Howett CJA, Verbiscer AJ, Waite JH (eds) Enceladus and the icy moons of Saturn. University of Arizona Press, Tucson, pp 163–174

Terquem C (2021) On a new formulation for energy transfer between convection and fast tides with application to giant planets and solar type stars. Mon Not R Astron Soc 503(4):5789–5806

Thomas PC, Tajeddine R, Tiscareno MS, Burns JA, Joseph J, Loredo TJ, Helfenstein P, Porco C (2016) Enceladus's measured physical libration requires a global subsurface ocean. Icarus 264:37–47. https://doi.org/10.1016/j.icarus.2015.08.037

Tian Z, Nimmo F (2020) Implications of second-order resonance for the thermal and orbital evolution of Mimas. Mon Not R Astron Soc 492(1):369–376

Tiscareno MS, Burns JA, Hedman MM, Porco CC, Weiss JW, Dones L, Richardson DC, Murray CD (2006) 100-metre-diameter moonlets in Saturn's A ring from observations of 'propeller' structures. Nature 440(7084):648–650

Tobie G, Auclair-Desrotour P, Behounkova M, Kervazo M, Soucek O Tidal dissipation processes in icy worlds. In preparation to SSR

Truong N, Monroe AA, Glein CR, Anbar AD, Lunine JI (2019) Decomposition of amino acids in water with application to in-situ measurements of Enceladus, Europa and other hydrothermally active icy ocean worlds. Icarus 329:140–147

Truong N, Glein C, Lunine J (2022) Constraining the outgassing history of Enceladus's ocean with noble gases. In: AAS/Division for planetary sciences meeting abstracts, vol 54, pp 206–02

Watts AB, Zhong SJ, Hunter J (2013) The behavior of the lithosphere on seismic to geologic timescales. Annu Rev Earth Planet Sci 41:443–468. https://doi.org/10.1146/annurev-earth-042711-105457

White OL, Schenk PM, Bellagamba AW, Grimm AM, Dombard AJ, Bray VJ (2017) Impact crater relaxation on Dione and Tethys and relation to past heat flow. Icarus 288:37–52. https://doi.org/10.1016/j.icarus.2017.01.025

Wisdom J, Dbouk R, Militzer B, Hubbard WB, Nimmo F, Downey BG, French RG (2022) Loss of a satellite could explain Saturn's obliquity and young rings. Science 377(6612):1285–1289

Zahnle K, Schenk P, Levison H, Dones L (2003) Cratering rates in the outer Solar system. Icarus 163(2):263–289. https://doi.org/10.1016/S0019-1035(03)00048-4

Zannoni M, Hemingway D, Gomez Casajus L, Tortora P (2020) The gravity field and interior structure of Dione. Icarus 345:113713. https://doi.org/10.1016/j.icarus.2020.113713

Space Science Reviews (2025) 221:6
https://doi.org/10.1007/s11214-025-01136-y

Tidal Deformation and Dissipation Processes in Icy Worlds

G. Tobie[1] · P. Auclair-Desrotour[2] · M. Běhounková[3] · M. Kervazo[1] · O. Souček[4] · K. Kalousová[3]

Received: 28 December 2023 / Accepted: 30 December 2024 / Published online: 16 January 2025

Abstract
Tidal interactions play a key role in the dynamics and evolution of icy worlds. The intense tectonic activity of Europa and the eruption activity on Enceladus are clear examples of the manifestation of tidal deformation and associated dissipation. While tidal heating has long been recognized as a major driver in the activity of these icy worlds, the mechanism controlling how tidal forces deform the different internal layers and produce heat by tidal friction still remains poorly constrained. As tidal forcing varies with orbital characteristics (distance to the central planet, eccentricity, obliquity), the contribution of tidal heating to the internal heat budget can strongly change over geological timescales. In some circumstances, the tidally-produced heat can result in internal melting and surface activity taking various forms. Even in the absence of significant heat production, tidal deformation can be used to probe the interior structure, the tidal response of icy moons being strongly sensitive to their hydrosphere structure. In the present paper, we review the methods to compute tidal deformation and dissipation in the different layers composing icy worlds. After summarizing the main principle of tidal deformation and the different rheological models used to model visco-elastic tidal response, we describe the dissipation processes expected in rock-dominated cores, subsurface oceans and icy shells and highlight the potential effects of tidal heating in terms of thermal evolution and activity. We finally anticipate how data collected by future missions to Jupiter's and Saturn's moons could be used to constrain their tidal response and the consequences for past and present activities.

✉ M. Běhounková
marie.behounkova@matfyz.cuni.cz

G. Tobie
gabriel.tobie@univ-nantes.fr

[1] Laboratoire de Planétologie et Géosciences, UMR 6112, CNRS Nantes Université, Université d'Angers, Le Mans Université, Nantes, France

[2] IMCCE, CNRS, Observatoire de Paris, PSL University, Sorbonne Université, Paris, France

[3] Charles University, Faculty of Mathematics and Physics, Department of Geophysics, V Holešovičkách 2, Praha, Praha 8, 180 00, Czech Republic

[4] Charles University, Faculty of Mathematics and Physics, Mathematical Institute, Sokolovská 83, Praha, Praha 8, 186 75, Czech Republic

1 Introduction

The discovery of huge volcanic eruptions on Io by Voyager 1 (Morabito et al. 1979), theoretically predicted by Peale et al. (1979) a few weeks prior to this historical flyby, confirmed that tidal dissipation can be a major driver for the evolution of the moons of the giant planets. The first observations of complex tectonic structures on Europa's surface (Smith et al. 1979) also highlighted the role of tidal forces, potentially maintaining a subsurface ocean at shallow depths (Cassen et al. 1979). The encounter with the Saturn system also revealed an abnormal geological activity on the small moon Enceladus, for which the role of tidal heating was suspected (Squyres et al. 1983), even though there was a lot of doubt at the time about the feasibility of generating sufficient power to melt such a small satellite (Peale et al. 1980; Poirier et al. 1983). On Titan, no direct observation of its surface was possible, but its high orbital eccentricity, which cannot be forced by any other satellite, suggests that tidal friction was limited and has little influence on its evolution (Sagan and Dermott 1982; Sears 1995).

The first observations by Voyager 1 and 2 hinted at the possibility of subsurface oceans in icy moons and at the role of tidal processes in explaining any sign of surface activity. The confirmations were provided decades later by the more detailed exploration by Galileo (1995-2003) at Jupiter and Cassini-Huygens (2004-2017) at Saturn. Magnetic measurements performed by Galileo confirmed the presence of a subsurface salty ocean in Europa from magnetic induction (Khurana et al. 1998). On Europa, the presence of a subsurface ocean at a relatively shallow depth (< 30 km) beneath a tectonically active ice shell is attributed to tidal forces which may fluctuate on geological timescales due to mutual gravitational interactions with Io and Ganymede, through an orbital resonance, known as the Laplace resonance (e.g. Hussmann and Spohn 2004). The ice shell thickness and thermal state of Europa's interior are thus expected to vary at a rhythm imposed by the Laplace resonance. The evolution of Europa is consequently intimately linked to that of Io and Ganymede.

Even more surprisingly, similar magnetic signatures were observed at Callisto and Ganymede (Kivelson et al. 2000; Zimmer et al. 2000), even though their interpretation as a signature of an electrically conductive subsurface ocean is more ambiguous due to the presence of a magnetic field generated in the iron core at Ganymede (Kivelson et al. 2002) and due to ionospheric effects altering the interpretation of magnetic induction of a subsurface ocean at Callisto (Hartkorn and Saur 2017). For Ganymede, the presence of a salty ocean has been confirmed from the observations of its oscillating auroral ovals by Hubble Space Telescope observations, whose limited oscillation amplitude implies an electrically conductive layer (Saur et al. 2015). On Callisto, we should wait for the exploration by the JUpiter ICy moon Explorer (Juice) to get confirmation of such an ocean (Van Hoolst et al. 2024). In addition to magnetic induction techniques, Juice will perform detailed measurements of tidal Love numbers k_2 and h_2 at Ganymede (De Marchi et al. 2022; Steinbrügge et al. 2015) and will determine with less precision the tidal Love number k_2 at Callisto (Cappuccio et al. 2022). Such measurements will provide key constraints on the hydrosphere structure and on the dissipation processes occurring in the different layers. The interpretation of these future data requires a detailed modeling of the tidal deformation and associated dissipation in these icy moons.

Thanks to Cassini gravity data, Titan is the first icy moon for which the tidal Love number has been estimated (Iess et al. 2012; Durante et al. 2019). The estimate of 0.616 ± 0.067 published by Durante et al. (2019) can be interpreted as the signature of a global ocean at a depth between 50 and 100 km for an average ocean density between 1150 $\mathrm{kg\,m^{-3}}$ (weakly salty water) to 1270 $\mathrm{kg\,m^{-3}}$ (~ 15 wt% salted ocean) (Mitri et al. 2014; Sotin et al.

2021). Idini and Nimmo (2024) proposed that the high value of the Love number reported by Durante et al. (2019) could be explained by a resonant, stably stratified ocean. More recently, by re-analysing the Cassini radio tracking data, Goossens et al. (2024) inferred a much smaller Love number, 0.375 ± 0.06, implying a less dense ocean below a thicker ice shell. It is still unclear which estimate is the most reliable. Moreover, the elevated orbital eccentricity of Titan (0.03) would imply limited dissipation in its interior during most of its history (Tobie et al. 2005a, 2006). However, Cassini data analysis suggests rapid orbital expansion of Titan implying much larger dissipation in Saturn than initially thought (Lainey et al. 2020). If such a rapid expansion is confirmed (which is still debated by Jacobson 2022), this would imply that Titan's eccentricity would decay more slowly, thus allowing more dissipation inside Titan than what was previously predicted. It is also possible that Titan's eccentricity was excited by interacting with another large moon, has disappeared since then, as proposed by Wisdom et al. (2022) to explain Saturn's obliquity and the ring youth. A detailed analysis of dissipation processes inside Titan and the consequences for its orbital evolution is needed to better understand the recent evolution of the Saturnian system.

Probably the biggest surprise in space exploration in the last two decades were the discoveries of active plumes of water vapour and icy grains emitted from the south pole of Enceladus by the Cassini spacecraft in 2005 (e.g. Dougherty et al. 2006; Hansen et al. 2006; Porco et al. 2006; Spencer et al. 2006; Waite et al. 2006). Two decades after the first suspicion of eruption activity on Enceladus (e.g. Squyres et al. 1983), the observations made by Cassini demonstrated that tides were a key driver of the activity of Enceladus, leading to tidal heating dominating its heat budget in a way similar to Io. The concentration of activity at the south pole suggested the presence of an internal ocean decoupling the ice shell from the silicate interior (Nimmo et al. 2007), at least in the southern hemisphere (Tobie et al. 2008; Běhounková et al. 2012). The ocean was then confirmed to be global from the detection of a physical libration indicating an average ice shell thickness initially estimated between 20 and 25 km (Thomas et al. 2016; Van Hoolst et al. 2016) and recently re-evaluated between 27 and 33 km from a lower estimate of the libration amplitude (Park et al. 2024). The interpretation of long-wavelength topography and low-degree gravity field (Iess et al. 2014; Čadek et al. 2016; Beuthe 2016; Tajeddine et al. 2017; Čadek et al. 2019) further indicates that the ice shell is strongly thinned at the south pole, possibly thinner than 5 km. The intense activity observed on Enceladus is clearly related to tidal forces, but the processes at the origin of tidal heat production, and how the energy is concentrated to the south polar terrain and evolves through geological time is still strongly debated (e.g. Nimmo et al. 2007; Tyler 2009; Kite and Rubin 2016; Hemingway and Mittal 2019; Souček et al. 2019; Neveu and Rhoden 2019; Kang et al. 2022a; Nimmo et al. 2023).

In this article, we review the main physical processes controlling the tidal deformation and dissipation in icy worlds (see Fig. 1) and the key controlling factors. We primarily focus on Enceladus and Europa, for which the most compelling evidence of tidally-induced processes has been observed. However, most of the processes discussed here are universal and may be relevant for many other icy worlds at present or during specific periods of their evolution. The content of this article is complementary to the article by Howett et al. (2025, this collection) dedicated to Enceladus heat budget and by Nimmo et al. (2023, this collection) dedicated to the tidally-controlled evolution of Enceladus. We also invite interested readers to consult previous reviews on similar topics, notably the synthesis of Hussmann et al. (2010) providing a broad overview on the energy sources and heat transport in icy worlds, and that of Bagheri et al. (2022a) providing complementary information on the tidal response of rocky and icy interiors. In Sect. 2, we present the general principle and computation technique to model tidal deformation and dissipation in planetary interiors. Section 3

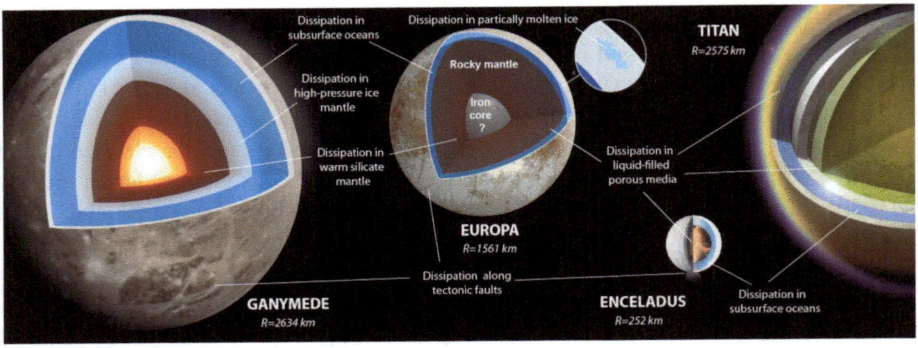

Fig. 1 Schematic illustration of interior structure and possible dissipation processes in a selection of icy worlds

describes the rheological properties of planetary materials composing icy world interiors, viscoelastic solids and partially molten rocks and ices. Section 4 is devoted to dissipation processes in rock-dominated layers and to the implications for hydrothermal and volcanic activities. Section 5 deals with tidal processes in subsurface oceans and coupling with the solid layers. Section 6 addresses the tidal response of icy shells, including convective ice shells, tidally-induced melting and porous flow in ices, and the tidal response of tectonic faults and water-filled cracks. Finally, Sect. 7 summarizes how future exploration missions may constrain tidal processes from geophysical measurements.

2 Principle of Tidal Deformation and Dissipation

2.1 Tidal Potential for a Satellite in 1:1 Spin-Orbit Resonance

The tidal forces exerted on a satellite in orbit around its central planet are the consequence of the difference in gravitational attraction created by the central planet throughout its satellite interior. As the gravitational force scales with $1/d^2$, d being the distance between the planet and the satellite, it is consequently slightly weaker at the far side than at the near side of the satellite. For a satellite in a stable orbit around its planet, the average gravitational force exerted at the center of mass is compensated by the centrifugal force due to the orbital motions around the center of mass of the satellite-planet system. If we subtract the average gravitational force exerted on the center of mass of the moon, we obtain a residual force field, which tends to stretch the satellite in the central planet direction and to flatten the poles.

For a satellite in 1:1 spin-orbit resonance on a circular orbit, the gravitational force and hence tidal force would remain constant, as both satellite-planet direction and distance remain fixed. In reality, the satellite orbit is never perfectly circular, so that any small eccentricity of the orbit results in time variations of the tidal force due to modulation of planet-satellite distance and direction. The satellite can also have a non-zero obliquity relative to the orbital plane and any long-term variation in the rotation rate (long-period libration or secular drift), which also add to modulation of the tidal force.

The tidal force is conveniently described as the gradient of a scalar field V_T, called the tidal potential, which can be developed in several components as a function of radius, r, co-latitude, θ, longitude, λ, and time, t (e.g. Wahr et al. 2009; Jara-Orué and Vermeersen 2016):

$$V_T(r,\theta,\lambda,t) = \frac{3GM_P R_S^2}{2a^3}\left(\frac{r}{R_S}\right)^2 [T_* + T_0 + T_1 + T_2 + T_3] \tag{1}$$

where

$$T_* = \frac{1}{6}\left(1 - 3\cos^2\theta\right), \tag{2}$$

$$T_0 = \frac{1}{2}\sin^2\theta\cos(2\lambda + 2bt), \tag{3}$$

$$T_1 = \frac{e}{2}\left(1 - 3\cos^2\theta\right)\cos(nt), \tag{4}$$

$$T_2 = \frac{e}{2}\sin^2\theta\,[3\cos(2\lambda)\cos(nt) + 4\sin(2\lambda)\sin(nt)], \tag{5}$$

$$T_3 = 2\sin(\theta)\cos(\theta)\sin(\epsilon)\cos(\lambda)\sin(\omega + nt). \tag{6}$$

with a, n, and e the semimajor axis, the mean motion and eccentricity of the satellite orbit, G the gravitational constant, M_P the planet mass, R_S the satellite radius, b the angular rate of non-synchronous rotation, ϵ the obliquity of satellite's spin axis, ω the argument of the pericenter measured with respect to the ascending node where the satellite's orbital plane crosses its equatorial plane, and t the time relative to the pericenter.

The term, T_*, describes the static flattening of the satellite, and T_0 describes the quasi-static stretching in the direction of the planet, which can slowly vary in the case of non-synchronous rotation. These two terms have no consequences for the heat budget of the satellite as they do not vary with time (or only very slowly). However, they control the shape of the satellite and the principal degree-two coefficients of the gravity field, which strongly depends on the density profile in the satellite interior and hence they are related to the moment of inertia of the satellite.

The terms T_1, T_2 and T_3, represent the diurnal tidal potential associated with orbital eccentricity and satellite obliquity, and are, therefore, responsible for the diurnal modulation of the satellite distortion. This modulation results in surface motions, internal mass redistribution and heat production by tidal friction.

2.2 Methods for Computing the Tidal Response of Ice-Covered Ocean Worlds

Different modeling strategies have been proposed in the literature to compute the displacements and stresses associated with tidal perturbation, as well as the resulting tidal dissipation. The standard approach used in most published studies consists of solving the equations of motion and Poisson's equation by assuming a spherically symmetric isotropic interior model. By neglecting any lateral variations in mechanical properties, the solutions can be separated into radial functions, classically denoted y_i, and spherical harmonics (e.g. Alterman et al. 1959; Takeuchi and Saito 1972), an approach commonly used to compute the spheroidal oscillation of self-gravitating interior. The set of equations to be solved can then be reorganized in a system of ordinary differential equations relying on six radial functions (noted y_i following the notation of Takeuchi and Saito 1972), describing the radial and lateral displacement (y_1 and y_3), the radial and tangential stresses (y_2 and y_4), the potential (y_5), and a modified gravity function (y_6). This formulation initially developed for the elastic case can be extended to the visco-elastic problem by using the correspondence principle (Biot 1954), which allows for the computation of the equivalent elastic solution in the frequency domain with complex moduli. As long as an effective complex shear and bulk modulus (see Sect. 3) can be defined for each internal layer in the frequency domain, the equivalent solution can be computed. For more details on the equation derivation and their numerical resolution, see, for instance, appendix A in Roberts and Nimmo (2008) and Kervazo et al. (2021).

The advantage of this approach is that the equations can be solved numerically for any linear rheology, interior structure and complex rheology profiles. Once the radial functions are determined, the displacement, stress, and associated dissipation rate can be determined at any point inside the satellite interior. At the surface, the radial functions also give access to the so-called Love numbers, k_2, h_2, l_2, which are parameters quantifying the amplitude of potential perturbation, radial and lateral displacements, respectively, induced by tidal forces. These parameters can be directly related to the observed time variations of gravity field and surface displacements. The inconvenience of this efficient computation technique is that any lateral variation in layer thickness or mechanical properties cannot be taken into account self-consistently. Some approximations are, however, possible to mimic the effect of long-wavelength variations in ice shell thickness and viscosity structures.

Another approach consists of solving directly the governing equations for viscoelastic deformation in the time domain in a full 3D geometry. The problem can be solved either using a spectral approach or finite-element techniques. However, compressibility and self-gravity are usually neglected for the sake of simplicity. The solution is also usually limited to a single layer, and the interaction with other layers is considered by using the appropriate boundary conditions or by their parameterization. To take into account only lateral variations in viscosity, a spectral approach is probably the most appropriate. This has been proposed notably to study the potential role of a low viscosity zone above a regional sea at Enceladus' south pole (Tobie et al. 2008; Běhounková et al. 2012). Such a spectral approach is particularly relevant for computing in a consistent way tidal dissipation from the 3D field of viscosity and hence allows for the full coupling with thermal convection codes (Běhounková et al. 2010, 2012, 2021) (see Sect. 6.1). To take into account other variations in mechanical properties, such as variations in ice shell thickness and local reduction of shear modulus associated with faults, a finite-element approach is required (Zhong et al. 2012; A et al. 2014; Souček et al. 2016, 2019; Berne et al. 2023). Such an approach is much more consistent, but highly demanding in computing resources. For viscoelasticity models, even though in theory any kind of rheological models could be considered, determining the viscoelastic memory term is computationally demanding, so an approximation consists in defining an effective viscosity using the Maxwell formalism but reproducing the dissipation rate expected for Andrade rheology (see Sect. 3). Alternatively, variations in shell thickness considering any viscoelastic rheology (though without any faults or any lateral variations in viscosity) can be treated with the thin shell approach of Beuthe (2018, 2019).

Once the different components of the stress field, σ_{ij}, and the strain field, ϵ_{ij}, due to tidal perturbation are determined, the tidal dissipation rate, h_{tide}, averaged over one tidal cycle, can be estimated at any point inside the satellite:

$$h_{\text{tide}}(r,\theta,\phi) = \frac{1}{T_{\text{tide}}} \int_t^{t+T_{\text{tide}}} \sigma_{ij} : \dot{\epsilon}_{ij} \mathrm{d}t'. \tag{7}$$

In the frequency domain, this corresponds to:

$$\tilde{h}_{\text{tide}}(r,\theta,\phi) = -\frac{\omega}{2}\left[\mathfrak{Im}(\tilde{\sigma}_{ij})\mathfrak{Re}(\tilde{\epsilon}_{ij}) - \mathfrak{Re}(\tilde{\sigma}_{ij})\mathfrak{Im}(\tilde{\epsilon}_{ij})\right], \tag{8}$$

with $\omega = 2\pi/T_{\text{tide}}$, $\tilde{x}$ indicating Fourier transforms (Tobie et al. 2005b) and $\mathfrak{Re}()$ and $\mathfrak{Im}()$ corresponding to the real and imaginary parts of complex quantities, respectively. In this formulation, both shear and bulk dissipation can be included as long as both shear and bulk dissipative parts are defined (see Sect. 3.2 and Kervazo et al. 2021 for more details).

By integrating over longitude and latitude and taking into account both shear and bulk dissipations, and assuming a spherically symmetric structure, the radial distribution of the tidal dissipation rate can be determined by the radial sensitivity functions to shear and bulk moduli, H_μ and H_K, introduced by Tobie et al. (2005b) and determined from the radial functions y_i:

$$\tilde{h}_{\text{tide}}(r) = -\frac{21}{10}\frac{\omega^5 R_s^4 e^2}{r^2}\left[H_\mu \mathfrak{Im}(\tilde{\mu}) + H_K \mathfrak{Im}(\tilde{K})\right], \tag{9}$$

with μ and K the shear and bulk moduli, respectively. Note that the radial sensitivity functions of Tobie et al. (2005b) have been complemented by angular sensitivity functions in Beuthe (2013), which makes it easier to compute spatial patterns of dissipation. Once integrated over the whole interior, we can show that the global dissipated power, P_{tide}, is then directly related to the imaginary part of the Love number, $\mathfrak{Im}(k_2)$:

$$P_{\text{tide}} = -\frac{21}{2}\mathfrak{Im}(k_2)\frac{(\omega R_s)^5}{G}e^2. \tag{10}$$

Note that the factors $\frac{21}{10}e^2$ and $\frac{21}{2}e^2$ in the above equations are based on the fact that 1:1 spin-orbit configuration is considered with a small eccentricity and neglecting the obliquity term. For any other orbital configuration, the appropriate factor should be re-derived for the corresponding tidal potential (see, for instance, Wisdom 2008; Efroimsky and Makarov 2014).

While the tidal response of solid bodies is particularly well-established and benefits from detailed constraints in the case of the Earth, the Moon or Mars, a particular challenge remains in the modeling of the tidal response of ice-covered ocean and its coupling with the ice shell. Models of tidal deformation based on the standard viscoelastic gravitational theory include the effect of the liquid layer. However, the dynamical response of such a layer is either considered in a simplified manner by ignoring the Coriolis effect, which breaks spherical symmetry (Takeuchi and Saito 1972; Kamata et al. 2015; Beuthe 2015b; Kervazo et al. 2021) or (most often) totally ignored by modeling the liquid layer as a static medium (Saito 1974; Sabadini et al. 1982; Spada 2008) or as a solid layer with a very low viscosity value (e.g. Nimmo et al. 2007), which ignored both inertial and Coriolis effects in the liquid layer.

In order to overcome these limitations, several authors adopted a more sophisticated formalism by self-consistently including the ocean-ice interactions into Laplace's tidal equations (LTEs) (e.g. Beuthe 2016; Matsuyama et al. 2018) (see Sect. 5 for further details). The visco-elastic adjustment of the overlying shell and underlying solid interior can strongly affect the oceanic response by both attenuating and shifting the resonant peaks of the dynamical oceanic tide. These can cause orders of magnitude changes in the dissipated energy flux if the oceanic configuration is close to a resonance (e.g. Matsuyama 2014; Beuthe 2016). Recently, Aygün and Čadek (2023) showed that the dissipation rate obtained using the LTE approach corrected by the damping effect of the ice shell can be significantly different from that obtained by solving the three-dimensional Navier-Stokes equations taking into account viscoelastic coupling with the solid layers. This clearly indicates that the ocean response and the deformation of the solid part must be solved in a self-consistent way, in particular when the ocean configuration is close to a resonant state.

3 Rheological Properties of Icy World Interiors

3.1 Viscoelastic Rheological Models

At periods of tidal forcing (a few days for icy worlds), planetary interiors exhibit anelastic behaviour, which can be described as a combination of elastic and viscous responses. Deformation of the tidally perturbed body is delayed relative to the forcing, and energy is dissipated during the deformation cycle. Part of the deformation is purely elastic and is determined by the elastic shear and bulk moduli of the materials. The other part is anelastic, and is determined by the various attenuation mechanisms that can occur in planetary materials (ice, brines, organics, rock, partially molten rocks, iron, etc). Various rheological models have been proposed to describe the viscoelastic behaviour of planetary interiors. The simplest viscoelastic model, called the Maxwell model (Fig. 2), is classically represented by an elastic spring and a viscous piston (or damper) in series. From these simple elements, a large number of viscoelastic materials can be established by assembling them in series or in parallel (Fig. 2). While most studies devoted to modeling planetary tides were classically based on the Maxwell model, more complex models taking into account rheological behaviours observed in the laboratory on analogous materials are now being used to model the response of planetary interiors to tidal forcing (Henning et al. 2009; Castillo-Rogez et al. 2011; Behounková and Cadek 2014; Henning and Hurford 2014; Kuchta et al. 2015; Renaud and Henning 2018; Tobie et al. 2019; Bagheri et al. 2022a; Amorim and Gudkova 2024; Musseau et al. 2024; Bierson 2024).

Whatever the assumed rheological models, the relationship between the deviatoric/shear stress σ'_{kl} and the deviatoric/shear strain ϵ'_{kl} during cyclic loading is conveniently described using a compliance function, $D(t)$, in the time domain (e.g. Jackson 2000):

$$2\epsilon'_{kl}(t) = \int_{-\infty}^{t} \dot{\sigma'}_{kl}(t')D(t-t')dt', \tag{11}$$

which corresponds to $\tilde{D}(\omega)$ in the frequency domain:

$$2\tilde{\epsilon'}_{kl}(\omega) = \tilde{\sigma'}_{kl}(\omega)\tilde{D}(\omega). \tag{12}$$

The dynamic compliance in the frequency domain is a complex variable, $\tilde{D}(\omega) = D_1(\omega) + \mathrm{i}D_2(\omega)$, from which the effective complex shear modulus can be directly calculated:

$$\mathfrak{Re}(\mu_{\text{eff}}) = \frac{D_1}{D_1^2 + D_2^2} \quad , \quad \mathfrak{Im}(\mu_{\text{eff}}) = -\frac{D_2}{D_1^2 + D_2^2} \tag{13}$$

Here, we briefly review these different conceptual viscoelastic models and derive their dynamic compliance functions and effective complex moduli, required to compute the viscoelastic tidal deformation as described in Sect. 2, for each of these models.

Maxwell Model As mentioned above, the Maxwell model consists of an elastic spring and a viscous piston (or damper) in series. The spring, therefore, deforms instantaneously in response to applied stress while the piston slides gradually until the stress is relaxed. For a forcing period close to or larger than the Maxwell time, defined as the ratio of the shear viscosity η to the shear modulus μ, $\tau_M = \eta/\mu$, the Maxwell rheology is a good approximation. It fails, however, to reproduce correctly the dissipation function over a wide range of temperature (or viscosity), as it underestimates the dissipation rate with decreasing temperature (increasing viscosity) (e.g. Castillo-Rogez et al. 2011; Efroimsky 2012) (see Fig. 3 and 4).

Maxwell rheology

$$D(t-t') = \frac{1}{\mu} + \frac{(t-t')}{\eta}$$

$$\tilde{D}(\omega) = \frac{1}{\mu} - \frac{\mathrm{i}}{\eta\omega}$$

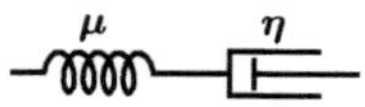

Kelvin-Voigt rheology

$$D(t-t') = \frac{1}{\mu_{\mathrm{KV}}}\left(1-\exp\left(-\frac{t-t'}{\eta_{\mathrm{KV}}/\mu_{\mathrm{KV}}}\right)\right)$$

$$\tilde{D}(\omega) = \frac{\mu_{\mathrm{KV}} - \mathrm{i}\omega\eta_{\mathrm{KV}}}{\mu_{\mathrm{KV}}^2 + \omega^2\eta_{\mathrm{KV}}^2}$$

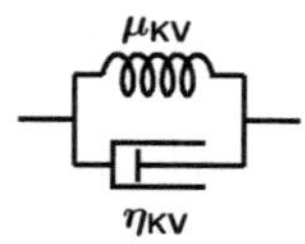

Burgers rheology

$$D(t-t') = \frac{1}{\mu} + \frac{1}{\mu_{\mathrm{B}}}\left(1-\exp\left(-\frac{t-t'}{\eta_{\mathrm{B}}/\mu_{\mathrm{B}}}\right)\right) + \frac{(t-t')}{\eta}$$

$$\tilde{D}(\omega) = \frac{1}{\mu} + \frac{\mu_{\mathrm{B}} - \mathrm{i}\omega\eta_{\mathrm{B}}}{\mu_{\mathrm{B}}^2 + \omega^2\eta_{\mathrm{B}}^2} - \frac{\mathrm{i}}{\eta\omega}$$

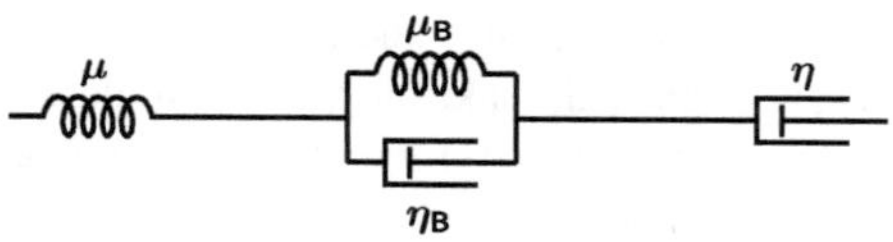

Andrade rheology

$$D(t-t') = \frac{1}{\mu} + \beta(t-t')^{\alpha} + \frac{(t-t')}{\eta}$$

$$D(\omega) = \frac{1}{\mu} + \beta(\mathrm{i}\omega)^{-\alpha}\Gamma(1+\alpha) - \frac{\mathrm{i}}{\eta\omega}$$

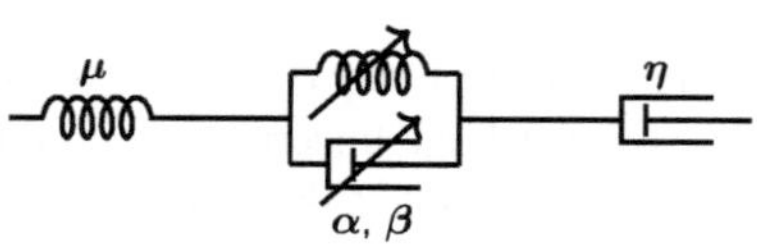

Sundberg-Cooper rheology

$$D(t-t') = \frac{1}{\mu} + \frac{1}{\mu_{\mathrm{B}}}\left(1-\exp\left(-\frac{t-t'}{\eta_{\mathrm{B}}/\mu_{\mathrm{B}}}\right)\right) + \beta(t-t')^{\alpha} + \frac{(t-t')}{\eta}$$

$$D(\omega) = \frac{1}{\mu} + \frac{\mu_{\mathrm{SC}} - \mathrm{i}\omega\eta_{\mathrm{SC}}}{\mu_{\mathrm{SC}}^2 + \omega^2\eta_{\mathrm{SC}}^2} + \beta(\mathrm{i}\omega)^{-\alpha}\Gamma(1+\alpha) - \frac{\mathrm{i}}{\eta\omega}$$

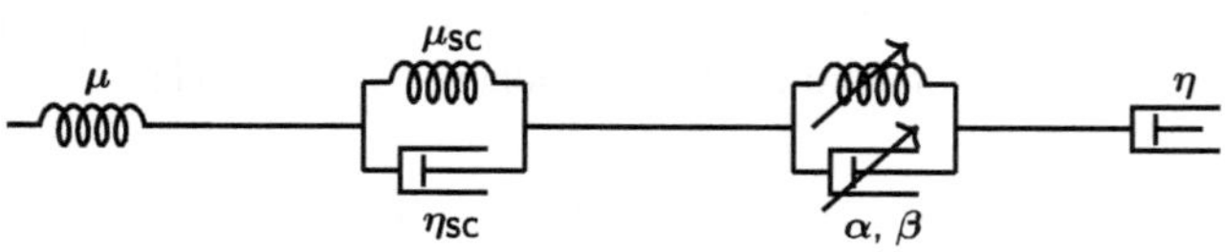

Fig. 2 Schematic representation of rheological models for the shear component and corresponding dynamical compliance formulation in time and frequency domain. See the main text for the definition of the different notations used for the mathematical description of each rheological conceptual model. In the Andrade model, Γ corresponds to the Gamma function defined as $\Gamma(z) = \int_0^\infty x^{z-1}e^{-x}dx$

Kelvin-Voigt Model This model consists of a parallel arrangement of viscous damper and elastic spring elements; it is characterized by the retardation time $\tau_{\mathrm{KV}} = \eta_{\mathrm{KV}}/\mu_{\mathrm{KV}}$ describing the retardation of strain upon the application of constant stress. The Kelvin-Voigt model has the peculiarity of not having instantaneous behaviour and of producing a delayed elasticity effect, which makes it not really relevant to describe visco-elastic response at typical tidal frequencies. A Kelvin-Voigt element, however, describes the recoverable deformation of the medium and is considered to describe more complex viscoelastic models mentioned below, so it is important to introduce it here.

Burgers Model This model corresponds to a Kelvin-Voigt element in series with the Maxwell element. The main advantage of this model is that it captures viscous creep and possible transient anelastic effects at forcing periods smaller than the Maxwell time, τ_M. This transient effect is mimicked by adding a second damper in parallel with a second spring, associated with two additional parameters, μ_B and η_B.

Andrade Model This rheological model is of particular interest for the tidally induced deformation of planetary bodies (e.g. Bierson 2024). This material can be seen as a Maxwell body to which an infinite number of springs and pistons are added in series. This extension, whose description is purely empirical, makes it possible to describe the body's transient response in a more accurate way than the Burgers model, by introducing only two additional parameters, α and β, which can be assessed experimentally. The Andrade model accounts for a large range of experimental data on various metals and minerals (e.g. Jackson 2000). The parameter α typically varies between 0.1 and 0.5 (e.g. $\alpha = 1/3$ in Jackson 2000) for olivine-rich rocks. It characterizes the duration of the transient phase of the first part (primary creep) of the material's response and represents diffusion along single grains (Jackson et al. 2002). The parameter β is related to the intensity of anelastic friction. Castillo-Rogez et al. (2011) have noted that, for various experimental data on olivine, β is proportional to $1/(\mu\tau_M^\alpha)$. Comparison with existing experimental data acquired at relatively high frequencies ($> 10^{-3}$ Hz) indicates that $\beta/(\mu\tau_M^\alpha)$ may typically vary between 0.1 and 10 (Walterová et al. 2023; Bierson 2024), however large uncertainties on this parameter still remain (e.g. Amorim and Gudkova 2024). The complex shear modulus displayed on Fig. 3 and 4 has been computed assuming $\alpha = 0.2 - 0.3$ and $\beta/(\mu\tau_M^\alpha) = 1$ for the Andrade model.

Sundberg-Cooper Mode Sundberg and Cooper (2010) discovered in their high-temperature olivine experiments that a Burgers-type attenuation peak connected to the elastically accommodated grain boundary sliding tended to appear in conjunction with an attenuation better characterized by the Andrade model. As neither the Burgers nor the Andrade formalism was able to account for this feature, they developed a composite rheological model combining the characteristics of both, known as the Sundberg-Cooper model.

Cole Model For the particular case of water ice, a specific visco-elastic model has been proposed by Cole and Durell (1995) based on cyclic loading tests. These mechanical tests revealed that viscoelastic response is primarily controlled by dislocation motions in ice crystals and, therefore, is related to the density of mobile dislocations in crystals. As the other viscoelastic models presented above, the mathematical formulation of this model remains empirical, but it has the merit of relying on microphysical processes through the dislocation density. Dynamic compliance of the Cole model is written as:

$$\tilde{D}^C(i\omega) = \frac{1}{\mu} + \delta D\left[1 - \frac{2}{\pi}\tan^{-1}\left(\exp(\alpha_g s)\right)\right] - i\delta D\frac{\alpha_g}{\exp(\alpha_g s) + \exp(-\alpha_g s)} - \frac{i}{\omega\eta}, \quad (14)$$

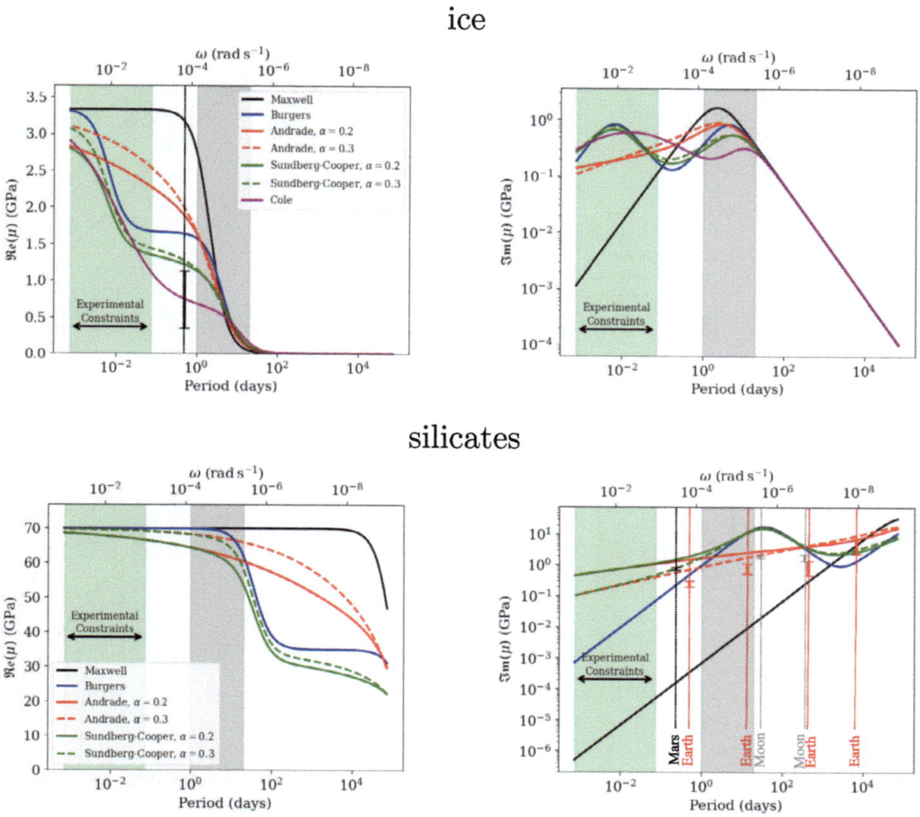

Fig. 3 Comparison of predicted real (left) and imaginary (right) parts of the complex shear modulus as a function of period/frequency for the different rheological models presented in the text, for ices (top) and silicates (bottom). The green bands indicate the typical frequency/period at which experimental data exists, while the grey band indicates the typical range for tidal deformation. For ices, the elastic shear modulus is set to 3.3 GPa and the shear viscosity to 10^{14} Pa s, and for silicates to 70 GPa and 10^{20} Pa s, respectively. The α parameter in the Andrade model is set to 0.2 or 0.3 and β to $1/\mu\tau_M^{\alpha}$. The parameters in the Cole model are set to: $\alpha_g = 0.53$, $\tau_d = 3300$ s, $\delta D = 1.4 \times 10^{-9}$ Pa^{-1}. Burgers and Sundberg-Cooper parameters are set to: $\mu_B = \mu_{SC} = 3.3$ GPa for ices, $\eta_B = \eta_{SC} = 60$ GPa s and $\mu_B = \mu_{SC} = 70$ GPa, $\eta_B = \eta_{SC} = 7 \cdot 10^{16}$ Pa s for silicates. As a comparison, the imaginary part of the shear modulus ($\sim 70/Q$ GPa) is provided using the Q factor estimates for Mars, the Earth and the Moon at different periods/angular frequencies (indicated by the vertical lines): Mars (black) at 5.5 hours, $Q = 95 \pm 10$ (Khan et al. 2018); the Moon (grey) at 28 days, $Q = 38 \pm 4$ (Williams and Boggs 2015), at 1 year, $Q = 41 \pm 9$ (Williams and Boggs 2015); the Earth (red) at 12.42 hours, $Q = 295 \pm 65$ (Ray et al. 2001), at 13.66 days, $Q = 115 \pm 30$ (Ding et al. 2021), at 27.5 days, $Q = 85 \pm 25$ (Zou et al. 2024), at 433 days, $Q = 55 \pm 15$ (Zou et al. 2024), at 18.6 years, $Q = 40 \pm 20$ (Benjamin et al. 2006)

where $s = \ln(\tau_d \omega)$, τ_d is the relaxation time, α_d is a parameter describing the width of the relaxation peak, and δD depends on the density of mobile dislocations, Δ, a restoring stress, σ_r, acting on the dislocation motions, an orientation factor, Ω, and the Burgers vector, b, such as $\delta D = \Delta \Omega b^2/\sigma_r$.

Comparison Between the Different Rheological Models and Existing Experimental and Geophysical Constraints The Maxwell rheology has been widely used to calculate tidal heating in planetary interiors of outer Solar System satellites, either in their rocky mantle or core

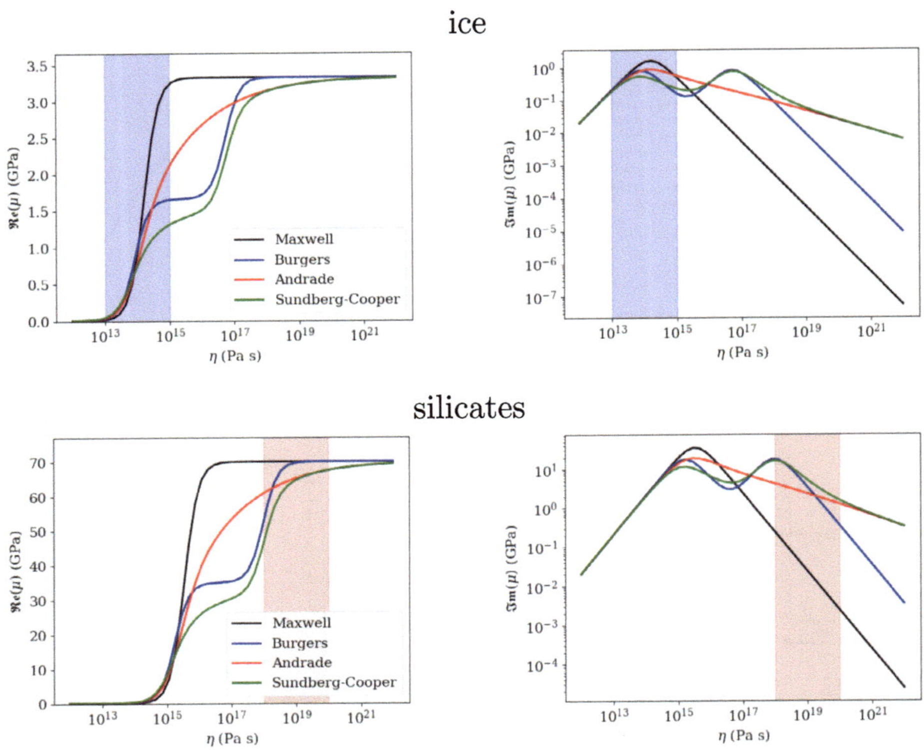

Fig. 4 Comparison of predicted real (left) and imaginary (right) parts of the complex shear modulus as a function of viscosity for the different rheological models presented in the text, for ices (top) and silicates (bottom). The blue and pink bands indicate the typical range of viscosity near the melting point for ices and silicates, respectively. The tidal period is set to 3.55 days, corresponding to the case of Europa. All the other rheological parameters are similar to those used in Fig. 3

(e.g Hussmann and Spohn 2004; Tobie et al. 2005b; Moore and Hussmann 2009), or in the ice shell (e.g. Ojakangas and Stevenson 1989; Tobie et al. 2003). This model is suitable for describing the relaxation properties of materials when the forcing period is close to or above their Maxwell time. However, it does not allow one to quantify the attenuation of materials over a wide range of frequencies and temperatures. On Earth, this model has been successfully used to describe the relaxation process associated with postglacial rebound, which occurs on time scales of a few thousand years (e.g. Peltier 1974), but extrapolation of Maxwell's model to tidal periods leads to dissipation values Q^{-1} several orders of magnitude lower than the global value derived from observations (e.g. Egbert and Ray 2001; Sotin et al. 2009). The Maxwell model also fails to reproduce the anelastic response observed in experimental studies performed on rocks at the seismic frequency range, i.e. at frequency several orders of magnitude higher than the Maxwell frequency (e.g. Takei et al. 2014; Faul and Jackson 2015). As shown on Fig. 3, the imaginary part of the shear modulus, $\Im m(\mu)$, computed for a reference viscosity of 10^{20} Pa s, is 4 to 5 orders of magnitude lower than the values derived from global Q^{-1} for the Earth (Ray et al. 2001) and Mars (Khan et al. 2018) when considering the Maxwell model. The Maxwell model leads to values comparable to observations only for the 18.6-yr lunar-nodal cycle. At shorter tidal periods, the Andrade and the Sundberg-Cooper models provide a dissipation rate more representative of the values inferred for Mars, the Earth and the Moon. This comparison is made assuming a single viscosity value of 10^{20} Pa s, while in reality more complex viscosity profiles should be considered for these three planetary bodies (Čížková et al. 2012; Karato 2010; Bagheri et al. 2019; Pou et al. 2022; Walterová et al. 2023). For each body, the viscosity profile and the rheological parameters should be adjusted to obtain a good fit. The main merit of the comparison displayed in Fig. 3 is to indicate which models between Andrade and Sundberg-Cooper provide a better representation of existing geophysical constraints. These models are also more suitable to reproduce the anelastic response of olivine and rock analogues at seismic frequencies (Jackson et al. 2002; Sundberg and Cooper 2010; McCarthy et al. 2011; Castillo-Rogez et al. 2011; McCarthy and Cooper 2016; Bierson 2024).

For ices, the most relevant geophysical constraints come from the tidal bending of Earth's ice shelf margins in Antarctica and Greenland (e.g. Vaughan 1995; Reeh et al. 2003; Hulbe et al. 2016; Rosier et al. 2017; Wild et al. 2017). The tidal deflection records of floating glaciers suggested an effective shear modulus at tidal frequencies 3 to 10 times smaller than the elastic modulus of natural polycrystalline ices (Sinha 1989). This reduction in apparent modulus is likely the consequences of significant viscous attenuation (e.g. Reeh et al. 2003; Wild et al. 2017) and of reduced stiffness due to the presence of bottom crevasse and a firn layer at the surface (e.g. Rosier et al. 2017). Direct comparisons cannot therefore be made between laboratory measurements of ice elastic modulus and effective elastic modulus inferred from fitting models to ice-shelf flexure profiles. However, the inferred effective modulus provides key information on how natural ices behaves at tidal frequencies, for which no laboratory constraints exist yet. A few experimental studies investigated the viscoelastic response of natural or synthetic ice samples to cyclic loading (Tatibouet et al. 1986, 1987; Cole and Durell 1995; Cole 1998; McCarthy and Cooper 2016), but tests were performed at frequencies higher than 10^{-3} rad s^{-1}. Mechanical tests revealed that the anelastic response of ice is related by the intracrystalline dislocation relaxation and to a lesser extent by grain boundary relaxation. Studies by Cole and Durell (1995) and McCarthy and Cooper (2016) indicated that the attenuation and shear modulus reduction is rather insensitive to grain size and appear mostly controlled by the density of mobile dislocations. Porosity, brine and crystal orientation also play a role by further enhancing attenuation and shear modulus reduction (Cole 1998). Mechanical tests by McCarthy and Cooper (2016) performed on fine-grained ice samples revealed a very dissipative behaviour, characterized by an abrupt shear modulus reduction and an strong increase of attenuation when frequency is reduced to 10^{-4} Hz. We can expect that shear modulus reduction and attenuation increase will be even more pronounced at tidal frequencies ($< 10^{-5}$ Hz).

Among the different rheological models, only the experimentally-based Cole model predicts a significant shear modulus reduction with increasing periods (or decreasing frequencies) (Fig. 4, top left), which seems also consistent with the behaviour observed by McCarthy and Cooper (2016). The Maxwell and Andrade models totally fail to reproduce the expected shear modulus reduction, while Burgers and Sundberg-cooper models can reproduce it partly for periods lower than 0.1 days. As shown by Cole and Durell (1995) and McCarthy and Cooper (2016), the viscoelastic response of ice depends on a variety of microstructure properties, notably on the density of mobile dislocations. Following the Cole model, the shear modulus reduction and dissipation enhancement with decreasing frequencies is related at first order to the dislocation density in ice crystals. Interestingly, for the examples displayed on Fig. 4 (top left), we can notice that the shear modulus predicted by the Cole model is typically in the same range as the apparent shear modulus inferred by tidal bending of Earth's floating glaciers. We should nevertheless keep in mind that the likely presence of crevasse should further reduce the apparent shear modulus (e.g. Nimmo 2004; Rosier et al. 2017) So more complex and realistic bending model should be used to infer the viscoelastic response of floating glaciers and to compare it to existing laboratory data on ice samples.

Extrapolation to icy moon conditions where tidal frequencies are $< 10^{-4}$ rad s^{-1} should be done with care. While the different rheological models converge to similar behaviour for the real part of the shear modulus at angular frequencies lower than 10^{-5} rad s^{-1} (Fig. 3, top left), they result in significant differences for the imaginary part, potentially up to one order of magnitude between the Cole and Maxwell models. Interestingly, the Cole model, which better reproduces the strong attenuation observed in laboratory data (Cole and Durell 1995; McCarthy and Cooper 2016), predicts the lowest dissipation at tidal frequencies.

In the absence of experimental data on the correct frequency range, it is difficult to conclude which model is the most appropriate to describe the viscoelastic response of ices and silicates in icy moon conditions. As shown by Bierson (2024), despite a wealth of experimental and seismic data, it is still difficult to conclude regarding the most appropriate rheological model and to determine the appropriate parameter values in order to correctly predict the tidal response of icy worlds. Due to its inability to reproduce experimental data and existing geophysical constraints, we can at least conclude that the Maxwell model is not appropriate. The Andrade model, which was originally developed to describe the deformation response of copper metal samples in the laboratory (Andrade 1910), shows promise for modelling transient creep in rock and ice (e.g. Faul and Jackson 2005; Sundberg and Cooper 2010; Faul and Jackson 2015; McCarthy and Cooper 2016; Bierson 2024). One of the features of Andrade's rheology is the softening of the strong frequency dependence of Maxwell's model, with a response characterized by a plateau, called an attenuation band in Andrade's case contrasting sharply with Maxwell's model, where the attenuation peak occurs at a mathematically exact frequency, with a sharp drop on either side. Such a reaction is different from a purely viscous response whose details are lost after the load is removed (irreversible), retaining some aspect of material"memory" (which can be reversible or irreversible, Efroimsky 2012). This memory depends not only on the static properties of the material (as is the case with the Kelvin-Voigt model), but also on how the microphysical properties of the material under consideration have evolved over time. On the other hand, one of the drawbacks of Andrade's model is that its two additional parameters are not directly associated with classical material property values such as viscosity or shear modulus, as is the case with Maxwell rheology, and which have to be assessed from laboratory data. By compiling a wide range of experimental data on rocks and analogous materials, Bierson (2024) showed that the Andrade parameters, α and β are poorly constrained. In particular, the β parameter may vary by several orders of magnitude, which may imply much more dissipative interiors that initially anticipated and affect the link between dissipation rate and viscosity (Amorim and Gudkova 2024).

Sundberg and Cooper (2010) proposed that the serial combination of an Andrade mechanism and Burgers rheology, now considered as the Sundberg-Cooper model, resulted in a better fit for laboratory data on olivine at high frequency ($> 10^{-3}$ Hz). The experiments of Sundberg and Cooper (2010) are of particular value to the planetary community, as they were conducted both with samples of mantle-like materials and at mantle-relevant temperatures. However, it is still unclear if the extrapolation of such models to lower frequency regimes predicts the correct dissipative function. The applicability of such a model to icy materials is also an open issue. Several experimental studies on rocks (e.g. Sundberg and Cooper 2010; Faul and Jackson 2015) and ices (e.g. Cole and Durell 1995; McCarthy and Cooper 2016) have revealed the complexity of the viscoelastic response of planetary materials by showing how sensitive it is to temperature, grain size, dislocation density, impurity concentration etc. There is still a lot of work to do to identify the correct rheological law at low-frequency regime.

Whatever the assumed rheological model, a peak of dissipation in ices is expected for viscosity values near the melting point ($10^{13}-10^{15}$ Pa s, Fig. 4). This has major consequences for the thermal state of the ice shell as further discussed in Sect. 6.1. Interestingly, a second dissipation peak predicted by Burgers and Sundberg-Cooper models for viscosity values of the order of 10^{17} Pa s which may provide an additional source of tidal heating closer to the surface, in the conductive lid and along active faults. Nevertheless, we should keep in mind that this secondary peak is a direct consequence of the mathematical formulation of the Burgers model and the Sundberg-Cooper model (which includes a Burgers element), it is not guaranteed that this formulation predicts the correct viscoelastic behaviour of ices at low frequency and low temperature. For silicates, these secondary peaks may also significantly increase the dissipation rate for viscosity values near the melting point ($\sim 10^{18}$ Pa s). This secondary peak, far from the Maxwell relaxation peak, corresponds to the peak observed at higher frequencies for silicate minerals at high temperatures and should be associated with similar relaxation mechanisms. As shown by Renaud and Henning (2018), this secondary peak, if really occurring, could significantly impact the tidal evolution of the rock interior of planetary bodies.

In summary, several empirical viscoelastic models have been proposed. Some of them reproduce, in a satisfactory manner, existing laboratory data obtained on cyclic loading tests, unfortunately at frequencies significantly higher than the typical tidal frequencies. Extrapolation of empirical functions to planetary conditions significantly different from those at which parameters have been derived in the laboratory should be considered with care when modeling the tidal response.

3.2 Bulk Dissipation

All the viscoelastic models presented above considered only shear deformation and stress as a source of mechanical dissipation. Global-scale seismic attenuation models of Earth indicate that bulk dissipation is small compared to shear dissipation (e.g. Widmer et al. 1991; Durek and Ekström 1996; Romanowicz and Mitchell 2015). As a consequence, bulk contribution has been classically ignored in studies dedicated to tidal deformation of planetary bodies. However, it might be non-negligible in some circumstances, in particular for partially molten interiors, such as Io (Beuthe 2013; Kervazo et al. 2021). Bulk dissipation can be considered by adding a term reflecting the bulk viscosity ζ associated with volumetric deformation, in addition to the standard shear viscosity associated with shear deformation. The constitutive equation relating stress σ_{kl} and strain ϵ_{kl}, including both the shear and bulk terms, of a Maxwell rheology is then written:

$$\dot{\sigma}_{kl} + \frac{\mu}{\eta}\left(\sigma_{kl} - \frac{1}{3}\sigma_{ii}\delta_{kl}\right) + \frac{K}{\zeta}\frac{1}{3}\sigma_{ii}\delta_{kl} = 2\mu\dot{\epsilon}_{kl} + \left(K - \frac{2}{3}\mu\right)\dot{\epsilon}_{ii}\delta_{kl}, \tag{15}$$

where δ_{kl} is the Kronecker symbol. The Fourier transform of Equation (15) leads to a relationship similar to Hooke's law for the purely elastic case:

$$\tilde{\sigma}_{kl} = \tilde{\lambda}(\omega)\tilde{\epsilon}_{ii}\delta_{kl} + 2\tilde{\mu}(\omega)\tilde{\epsilon}_{kl} \tag{16}$$

with

$$\tilde{\lambda}(\omega) = \tilde{K}(\omega) - \frac{2}{3}\tilde{\mu}(\omega), \tag{17}$$

$$\tilde{\mu}(\omega) = \frac{\mu\omega^2\eta^2}{\mu^2+\omega^2\eta^2} + i\frac{\mu^2\omega\eta}{\mu^2+\omega^2\eta^2}, \tag{18}$$

$$\tilde{K}(\omega) = \frac{K\omega^2\zeta^2}{K^2+\omega^2\zeta^2} + i\frac{K^2\omega\zeta}{K^2+\omega^2\zeta^2}, \tag{19}$$

where ω is the frequency, η the viscosity, μ the shear modulus, K the bulk modulus and λ the Lamé parameter. The tilde indicates the Fourier transform. These choices of $\tilde{\lambda}$, $\tilde{\mu}$ and $\tilde{K}$ assume that viscoelastic relaxation occurs for both shear deviatoric σ'_{ij} and bulk σ_{nn} stresses. In this case, unlike the pure shear case, the bulk modulus K is also affected and differs from the elastic case.

The dissipated power per unit volume averaged over one orbital is then determined by both $\mathfrak{Im}(\tilde{\mu})$ and $\mathfrak{Im}(\tilde{K})$ as (e.g. Beuthe 2013):

$$\tilde{h}_{tide}(r,\theta,\phi) = \omega\mathfrak{Im}(\tilde{\mu})\left(\tilde{\epsilon}_{kl}\tilde{\epsilon}^*_{kl} - \frac{1}{3}|\tilde{\epsilon}|^2\right) + \frac{\omega}{2}\mathfrak{Im}(\tilde{K})|\tilde{\epsilon}|^2, \tag{20}$$

with $*$ indicating a complex conjugate. Most of the time, $\mathfrak{Im}(\tilde{\mu}) \gg \mathfrak{Im}(\tilde{K})$, so that bulk dissipation can be neglected. However, for partially molten materials, as shown hereafter, $\mathfrak{Im}(\tilde{K})$ can become comparable to $\mathfrak{Im}(\tilde{\mu})$ as μ rapidly decreases with increasing melt fraction to an asymptotic zero value, while the bulk modulus K remains comparable to the solid phase even for a purely liquid phase (Fig. 5).

3.3 Role of Melts on the Rheology of Rocks and Ices

Melting is a first-order phase change that has drastic consequences on the rheological properties of the materials that build planetary interiors, as well as on their dynamics, on different spatial and time scales. Under certain temperature and pressure conditions, a heterogeneous material, such as a planetary mantle, can become multiphase as a result of partial melting. The presence of interstitial melts in rocks is known to affect both their elastic (e.g. Budiansky and O'Connell 1976; Mavko 1980; Takei 1998; Faul et al. 2004) and viscous (e.g. Hirth and Kohlstedt 1995a,b; Kohlstedt and Zimmerman 1996; Scott and Kohlstedt 2006) properties. At a given point in the partial melting process, the material changes from a solid matrix with fluid pores to a fluid solution with isolated floating crystal grains. When the grains lose contact with each other, the material loses most of its shear strength and is controlled by the viscous properties of the fluid (Fig. 5). The transition from solid-state creep rheology to melt-dominated creep rheology, i.e. from solid to fluid behaviour of the material, led to the concept of a rheological critical melt fraction (RCMF) associated with a sharp transition from the solid behaviour to the liquid behaviour (e.g. Renner et al. 2000). In practice, most theoretical and experimental studies devoted to the effect of partial melts on the rheology of rocks focus on viscosity as this parameter plays a prominent role in the dynamics. Following the pioneering work of Arzi (1978), suggesting a threshold value of about 25–30% for RCMF, a compilation of theoretical and experimental studies applied to a solid undergoing partial melting, gives a wide range of values for this critical fraction, from 26 to 62% (e.g. Vigneresse et al. 1996; Renner et al. 2000; Scott and Kohlstedt 2006; Costa et al. 2009). Another important aspect of partially molten rocks is the possible contribution of bulk dissipation. Once above the rheological critical melt fraction, small changes in melt fraction can have a large impact on the dissipation rate, and feedback between melt-induced tidal heating and melt production may be amplified by the bulk viscoelastic response (Kervazo et al. 2021).

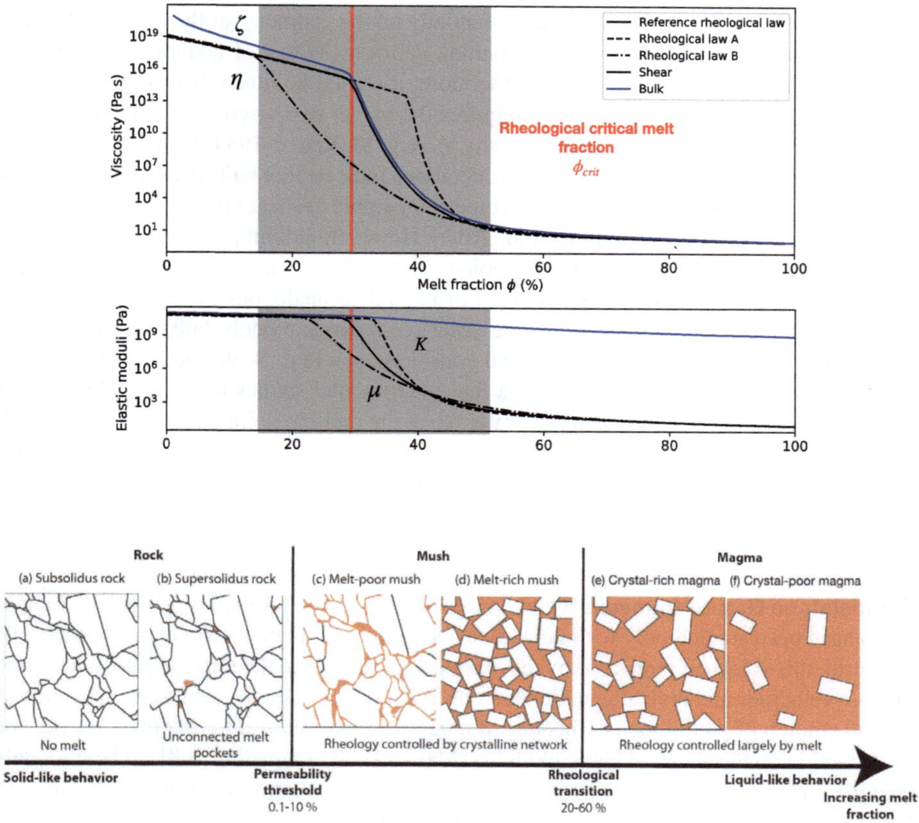

Fig. 5 *Top*. Effect of melt fraction ϕ on the viscoelastic parameters of a rocky mantle: shear viscosity η and modulus μ (in black); bulk viscosity ζ and modulus K (in blue). The three models correspond to three different rheological transitions where the transition between solid-state and liquid-state behaviors occurs. The red vertical line corresponds to the rheological critical melt fraction RCMF for the reference model (here $\phi_{\rm crit} = 30\%$). The "solid" shear viscosity (for $\phi = 0$), used as a reference value is that of silicate rocks close to their melting point, typically of the order of 10^{19} Pa s (Karato and Wu 1993). The liquid value (for $\phi = 1$) of the shear viscosity is of the order of 1 Pas, as observed for basaltic melts (e.g. Shaw 1969). For the mantle conditions typical of Io and Europa, the solid bulk modulus K is generally between 150 and 250 GPa. The solid shear modulus μ of a silicate mantle is typically 60 GPa. Concerning the values corresponding to the liquid case, ultrasonic velocity measurements indicate that for basaltic liquids, K is an order of magnitude lower than the value expected for mantle minerals, between 1 and 30 GPa (e.g. Murase and McBirney 1973). Figure adapted from (Kervazo et al. 2021). *Bottom*. Schematic illustration of possible physical configurations of melting in a partially molten silicate mantle. The ranges for permeability and rheological transition thresholds are uncertain and depend in particular on rock composition. Figure adapted from Sparks et al. (2019)

Constraints on Viscous Parameters The relationship between the (shear or bulk) viscosity of partially molten rock and the rate of partial melting is of critical importance in characterizing the rheological behaviour of silicate mantles. In partially molten materials undergoing diffusion creep, the presence of melt leads to a reduction in the material's effective shear viscosity compared to the situation without melt. One reason for this is that diffusion in the melt phase is generally much faster than in the solid phase (e.g. Takei and Holtzman 2009). Since the rate of diffusion creep depends on the rate at which material is transported at the grain scale, the presence of fast melting paths accelerates the overall rate of creep for a given stress. The extent of this effect depends essentially on the geometry of the grain-scale melting. The simplest model of grain-scale melting geometry is that of textural equilibrium, a state that minimizes surface energy. Furthermore, it should be noted that the complete description of a linear isotropic medium requires the use of two viscosities, implying a bulk viscosity, ζ, in addition to the shear viscosity, η, (e.g. McKenzie 1984; Scott and Stevenson 1986; Ricard et al. 2001). Theoretical considerations indicate that bulk viscosity ζ decreases with increasing melt fraction and may become comparable to shear viscosity η for melt fraction exceeding 10-20% (Schmeling et al. 2012). However, although viscosity has been the subject of more work than any other rheological property of natural silicate melts, there are no measurements of the bulk viscosity of natural or analogous systems applicable to the crust or mantle. Consequently, in large-scale geodynamic models, bulk viscosity is generally given by a law based on theoretical considerations (e.g. Scott and Stevenson 1986; Schmeling et al. 2012). For water ices, a few experimental studies have studied the effect of interstitial water on the viscosity of water ice at the melting point (De La Chapelle et al. 1999; Adams et al. 2021). Even a small amount of interstitial water can reduce significantly the effective viscosity of ice when approaching the melting point (Goldsby and Kohlstedt 2001; Adams et al. 2021). Based on the experimental work of (De La Chapelle et al. 1999), 5% of water content should result in a decrease of viscosity by one order of magnitude.

Constraints on Elastic Parameters Unlike viscosity, the elastic properties of partially molten rocks have been studied only for small values of the melt fraction (typically a few percent). On Earth, this is indeed motivated by the existence of regions in the upper mantle where low seismic velocities and high attenuation are observed, which is interpreted as the presence of a small amount of partial melts. Theoretical models were developed to describe the effect of melt fraction on the shear and bulk modulus (Mavko 1980; Schmeling 1985). These models quantify the dependence of seismic wave speeds and attenuation upon melt fraction and fluid-filled inclusions of specified shape (e.g., ellipsoids, grain boundary films, or grain-edge tubes). The viscoelastic behaviour of partially molten rock has also been investigated in the laboratory through forced torsional oscillation (Berckhemer et al. 1982; Bagdassarov and Dingwell 1993; Faul et al. 2004), on rock or glass; as well as ultrasonic studies of shear and compression waves (e.g. Dingwell and Webb 1989).

For water ice containing interstitial brines, a few studies have quantified the effect of water content on the elastic and attenuation properties of ice (e.g. Spetzler and Anderson 1968; Schwarz and Weeks 1977; McCarthy et al. 2019). As an example, according to Spetzler and Anderson (1968), for ice containing 2% NaCl, the shear modulus can drop by a factor of 2 even at a temperature as low as −20 °C which corresponds to brine fraction of about 6-7%.

3.4 Rheology of Unconsolidated Porous Materials

The rocky core of small icy moons, like Enceladus and other small moons, as well as the outer part of the rocky core of larger icy bodies likely exhibits a high degree of porosity due to the limited pressure (e.g. Vance et al. 2007; Roberts 2015). The low density of Enceladus's rock core (2450-2600 kg m^{-3}) as inferred from Cassini gravity data (Iess et al. 2014; Čadek et al. 2016; Beuthe et al. 2016), indicates a porosity of 20-30% (Choblet et al. 2017). Such a highly porous medium can be viewed as a highly fragmented/weakly cohesive material, where tidal deformation is likely associated with inter-granular friction during grain rearrangements or/and frictional sliding along microcracks (e.g. Seed et al. 1986; Ishibashi and Zhang 1993; Rollins et al. 1998; Wulff et al. 1999). Both mechanisms depend on the microstructure characteristics (grain distribution, microcrack density), which are unknown in the context of the icy moons' interiors. However, the mechanical behaviour of fragmented, unconsolidated media, has been intensively studied in laboratories in the context of geotechnical applications. Even though existing mechanical tests are usually performed at frequencies significantly higher ($> 10^{-2}$ Hz) than tidal frequencies, these measurements can be used to provide some mechanical constraints in planetary contexts.

The mechanical response of highly fragmented/granular media is classically parameterized using the effective shear modulus, μ_{eff}, and the dissipation function, Q_μ^{-1} (equal to twice the damping ratio, a quantity classically derived in laboratory mechanical tests), which control the response amplitude to cyclic tidal forcing and the fraction of mechanical energy that is converted into heat, respectively (e.g. Seed et al. 1986; Ishibashi and Zhang 1993; Rollins et al. 1998; Wulff et al. 1999; Brennan et al. 2005). An effective viscosity could also be defined assuming a Maxwell rheology as it is classically done when computing tidal deformation in viscoelastic bodies (e.g. Tobie et al. 2005b; Roberts 2015). However, it is more meaningful to directly represent the dissipation rate as a function of effective shear modulus and dissipation function (or damping ratio) as these are the quantities inferred directly from mechanical tests in the laboratory.

The real and imaginary parts of the complex shear modulus, $\tilde{\mu}$, used to compute the tidal deformation, are determined from the effective shear modulus, $\mu_{\text{eff}} = |\mu_c|$, and the local dissipation function, Q_μ^{-1}, as follows:

$$\mathfrak{Re}(\tilde{\mu}) = |\tilde{\mu}|\sqrt{1 - \left(Q_\mu^{-1}\right)^2}, \mathfrak{Im}(\tilde{\mu}) = |\tilde{\mu}|Q_\mu^{-1}. \tag{21}$$

The forcing frequency is expected to affect both shear modulus and dissipation function, especially in weakly cohesive unconsolidated materials, which may be representative of the water-filled porous rocky core of Enceladus as argued by Choblet et al. (2017). As explained by Shibuya et al. (1995), the damping ratio (dissipation function) in cohesive soils is expected to: (i) increase at frequencies below 0.1 Hz, due to creep effects of soil skeleton; (ii) remain more or less constant between 0.1 and 10 Hz; and (iii) increase at frequencies greater than 10 Hz because of the pore fluid viscosity. Cyclic loading tests performed on unconsolidated granular materials, i.e. mixture of sands and/or gravels (e.g. Seed et al. 1986; Rollins et al. 1998; Zhou et al. 2017) at low confining pressure, indicate that the effective shear modulus strongly decreases for cyclic strain exceeding 0.01-0.1%, for forcing frequency ranging between typically 0.01 and 1 Hz. These mechanical tests also demonstrate that the damping ratio increases with decreasing shear modulus, and can reach values of about 0.3-0.4, corresponding to a dissipation function of 0.6-0.8. While the occurrence of this reduction of shear modulus (and increase of dissipation) is observed in all samples in laboratory tests (e.g. Rollins et al. 1998; Brennan et al. 2005), the exact values of cyclic strain at which the change of mechanical behavior occurs may depend on the characteristics of the sample (grain size distribution, composition, porosity etc.), as well as on the frequency and strain history. The confining pressure also influences the dissipation rate, but this effect remains moderate (e.g. Zhou et al. 2017) at least for confining pressures up to 5 MPa, corresponding to the pressure at the surface of Enceladus' core. Unfortunately, no data exists at higher pressure to confirm this tendency. Even though the conditions inside Enceladus' core are different from those considered in laboratory tests (higher frequency, lower pressure), Choblet et al. (2017) assumed that a similar trend should occur when the core materials are subjected to cyclic strain. This assumption should be confirmed by future mechanical tests in the appropriate range.

In the absence of direct constraints, we can estimate the core mechanical properties from the total power required to explain the global ocean underneath a relatively thin ice shell. By computing the tidal dissipation in Enceladus' core, we show that the generation of 10-30 GW by tidal friction in Enceladus' unconsolidated core requires an effective shear modulus ranging between 10^7 and 10^8 Pa and a dissipation function between 0.2 and 0.8. For such a weak and dissipative core, the cyclic strain in the core would range between 0.005 and 0.015%. According to the data of Seed et al. (1986) and Rollins et al. (1998), the dissipation function typically ranges between 0.05 and 0.2 for gravelly soils subjected to a cyclic strain of 0.01% in amplitude at a frequency ranging between 0.1 and 1 Hz. No mechanical tests exist at very low frequencies ($< 10^{-5}$ Hz) relevant for tidal deformation, however, we anticipate the dissipation function to be higher and the reduction of shear modulus to be more pronounced than at 0.1-1 Hz. At very low frequency (< 0.01 Hz), dissipation is indeed expected to increase due to creep effects in the solid matrix (e.g. Shibuya et al. 1995). The likely presence of clays or organics in the porous core could favor this creep effect and increase dissipation. Although performed on materials of a different nature, such an increase of dissipation and shear modulus reduction are observed in mechanical tests performed at very low frequencies ($10^{-4} - 10^{-3}$ Hz) on polycrystalline aggregates (olivine, e.g. Faul and Jackson 2005 or ice, e.g. Cole and Durell 1995). Moreover, the cyclicity of tidal forcing may also result in strain accumulation and gradual build-up of pore pressure, favoring the weakness of the core material and maintaining it in a highly deformable state.

The tidally-induced motions of fluids in the porous unconsolidated rock matrix may also add additional dissipation as proposed by Liao et al. (2020), Rovira-Navarro et al. (2022) and Kamata (2023). Such dissipative processes, which can be estimated in the framework of the poroviscoelastic theory developed by Biot (1956), are controlled by the permeability and the viscosity of the fluids and are strongly frequency dependent. Efficient dissipation is expected to occur when the fluid phase oscillates in the solid matrix at frequencies approaching a characteristic frequency, f_c defined by Biot (1956) as:

$$f_c = \frac{\eta_{\mathrm{liq}} \phi^2}{2\pi \rho_{\mathrm{liq}} k}, \tag{22}$$

where η_{liq} is the viscosity of the liquid phase, ϕ the porosity, ρ_{liq} the density of the liquid phase and k Darcy's coefficient of permeability. The dissipation due to periodic fluid motions in the pores starts to get non-negligible for frequency larger than 0.01 f_c. For typical values of liquid viscosity ($10^{-3} - 10^{-2}$ Pa s), the porosity of $20 - 30\%$, permeability of $10^{-14} - 10^{-12}$ m^2 (Choblet et al. 2017), and water density of 1000 kg m^{-3}, dissipation would become non-negligible for frequencies larger than 300 Hz, so eight orders of magnitude larger than typical tidal frequencies ($\sim 10^{-5}$ s^{-1}). As shown by Rovira-Navarro et al. (2022), significant dissipation could occur at tidal frequency only if the porous core was highly permeable with permeability of 10^{-5} m^2 allowing high-amplitude Darcian flow and associated dissipation comparable to what is estimated on Enceladus (≥ 10 GW). But under such conditions, heat transfer out of the core would be extremely efficient at preventing any temperature increase in the core and any production of hot hydrothermal vents. For reasonable ranges of rock permeability, tidally-induced fluid motions are expected to be limited to the outer part of the porous core (depth < 25 km, Kamata 2023), leading to limited dissipation. The inner part of the core (depth > 25 km) should remain undrained, with almost no fluid motions in the pore and hence no fluid dissipation.

4 Tidally-Driven Hydrothermal and Volcanic Activities

4.1 Tidal Heating and Heat Transport in Water-Saturated Porous Core

As proposed by Roberts (2015) and further explored by Choblet et al. (2017), the silicate core of mid-sized moons, such as Enceladus, could be highly fragmented, consisting either of rock fragments embedded in an ice matrix as long as the internal temperature remains below the melting point of water ice, or of fragmented porous rock core filled with liquid water. Shortly after accretion, the interior of such small moons is expected to be frozen (Monteux et al. 2014) and to remain in this state until the interior reaches the melting point due to progressive warming produced by both radiogenic and tidal heating. As shown by Neveu and Rhoden (2019), depending on the size of the moon, its rock fraction and possible orbital resonance encounters, it may take from a few hundred million years to a few billion years before the ice melts and leading the formation of a water-filled porous rock core.

As shown by Roberts (2015) in the case of Enceladus, the maximum dissipation rate in a core composed of an ice-rock mixture is estimated to be about 10^{-7} W m^{-3}, while, for a water-filled porous core, the dissipation rate may exceed 10^{-6} W m^{-3} (Choblet et al. 2017). Tidal dissipation in an ice-rock interior is estimated to be at maximum of the order of 2 GW, which is not sufficient to explain the observed activity on Enceladus (Roberts 2015). In contrast, a water-filled fragmented porous core may generate more than 10 GW, for mechanical properties comparable to typical unconsolidated granular materials (Choblet et al. 2017, see also Sect. 3.4).

For an initially frozen interior, ice melting may be triggered when the moons pass through orbital resonances (e.g. Meyer and Wisdom 2008; Nakajima et al. 2019). During such periods of increased eccentricity, progressive warming until the melting point is expected to result in an enhancement of tidal heating, as viscosity and shear modulus get reduced, thus potentially leading to runaway melting until the whole icy interior melts and forms discrete fragmented porous core surrounded by a thick water ocean (Noyelles et al. 2019).

In the case of Enceladus, Choblet et al. (2017) showed that tidal heating in a weakly consolidated water-filled porous core should result in a warming of interstitial water powering efficient upward porous flow of hot water. The degree-two shape of the tidal potential results in a modulation of tidal heating as a function of latitude and longitude, with maximal tidal heating at the poles and along two specific meridians at 90° and −90° relative to the sub-/anti-Saturnian meridian (Fig. 6a). This moderate change in tidal heating results in a strong concentration of hot narrow upwellings in polar regions and along these meridians (Fig. 6c). These upwellings can sustain powerful (1–5 GW) hotspots at the seafloor, thus providing conditions for hydrothermal activities in the south polar region with water as hot as 363 K for core permeability between 10^{-14} and 10^{-13} m^{-2} (Fig. 6b, d). Such a concentration of heat release can thus explain the ice shell thinning observed at the south pole (Čadek et al. 2016; Beuthe et al. 2016; Čadek et al. 2019) and the associated jet activity (Porco et al. 2006, 2014). As long as the core remains weakly consolidated and permeable, such endogenic activity may be sustained, and could be operating for tens of millions to billions of years.

Similar tidally-heated water flow processes may also operate in other mid-sized moons, at least in the past (e.g. Castillo-Rogez et al. 2018). For instance, on Mimas, there is evidence for an ocean at present (Tajeddine et al. 2014; Lainey et al. 2024), which, based on the observation at Mimas's orbital pericenter drift, would be at relatively shallow depths (20-30 km). If Mimas has a weakly consolidated porous core with mechanical properties similar to Enceladus, strong tidal dissipation is expected, potentially of the order of 10-20 GW

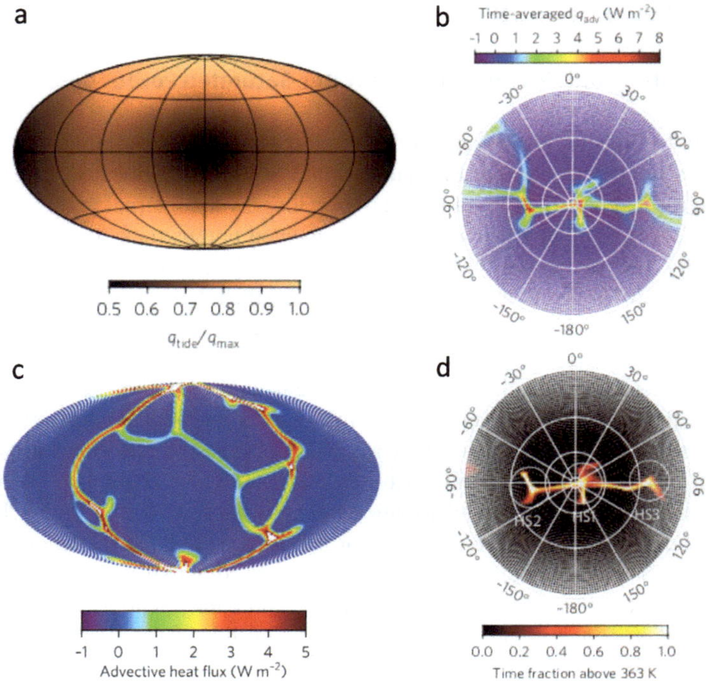

Fig. 6 a) Spatial distribution of tidal heat flux in the porous rock core of Enceladus; b) polar view distribution and c) global distribution of the advective heat flux through the surface of the porous core due to tidally-driven hot water flow, for a total power of 30 GW and a core permeability of 10^{-13} m^{-2}; d) Time fraction during which a given location at the rocky core/ocean interface remains at a temperature above 363 K (stereographic projection) considering a total simulation duration of 10 Myr (adapted from Choblet et al. 2017)

(Lainey et al. 2024). The absence of any sign of endogenic activity at its surface suggests that such a highly dissipative state is geologically recent (Rhoden 2023; Rhoden et al. 2024; Lainey et al. 2024). Lainey et al. (2024) showed that the ocean, if present, must have started forming between 5 and 15 Myr ago when the eccentricity was about 2.5 times higher than at present, which may have been the result of resonance crossings with other Saturn's moons (e.g. Ćuk and El Moutamid 2022; Ćuk et al. 2024, this collection). This result implies that Mimas might be an emerging ocean world, for which the timescale of ocean formation is directly controlled by the timescale of eccentricity damping (Lainey et al. 2024; Rhoden et al. 2024). The Mimas example highlights how crucial the understanding of orbital evolution is to better evaluate the past evolution of the icy moons. We refer interested readers to the articles by Nimmo et al. (2023) and Ćuk et al. (2024) in this collection for details on the long-term orbital evolution of the Saturnian system and the coupling with tidal dissipation.

Other moons, such as Saturn's moon Dione or Uranus's moon Ariel and Miranda, may also pass by a very dissipative state involving large heat production in a porous core and associated hydrothermal activity, as suggested from evidence of enhanced heat flux in the past based on crater relaxation (White et al. 2017; Hussmann et al. 2019; Beddingfield et al. 2022a,b). For larger bodies, Neptune's moon Triton (e.g. Gaeman et al. 2012), Pluto (e.g. Bagheri et al. 2022b) and Eris (e.g. Nimmo and Brown 2023), tidal dissipation and porous flow may also have played a role, but most likely during their early stage just after differentiation, before core consolidation and compaction which should be much faster due to higher pressure compared to much smaller moons like Enceladus and Mimas. Future work is needed to evaluate how such a dissipative state of their porous rocky core, even if brief, may have influenced the physico-chemical evolution of these bodies and the habitability of their internal oceans.

4.2 Coupling Between Tidal Heating and Convection: Implications for Magmatic Activities

Unlike smaller satellites where a significant fraction (if not all) of the rocky core can remain porous and unconsolidated during most of the satellite evolution, larger satellites such as Europa, Ganymede, Titan and Triton are expected to have a compact interior due to higher pressure and high temperature in the deep interior. Due to their large size, more energy accumulates during accretion and radiogenic heat is less efficiently removed from the deep interior than in smaller satellites where water transport in porous media is extremely efficient to limit internal temperature increase. For large moons like Europa and Triton, some porosity may be preserved in the silicate crust (Vance et al. 2016), but the depth at which water may circulate in non-compacted porous media probably does not extend 25-50 km. For even larger moons like Ganymede and Titan, the pressure at the base of the hydrosphere is probably too high to maintain any significant porosity. Below the porous upper crust, heat transfer will be dominated by heat conduction and advection similar to heat transfer in the Earth's mantle.

The thermal evolution of the rock mantle can be described by the equations characterizing heat transfer by convection and/or conduction or two-phase flow if melt is present. Tidal dissipation combined with radiogenic heating enters the energy conservation equation and represents the primary heat source for long-term evolution. While the contribution of radiogenic heating inevitably decays over time, tidal dissipation remains a dynamic energy source influenced by orbital parameters. In the case of multi-body systems, where mean-motion resonances can come into play, tidal dissipation may represent a highly variable and significant source of energy (e.g. Hussmann and Spohn 2004).

The precise ratio of the two contributions remains uncertain. For Io, the observed heat budget (65 - 125 TW; e.g., Lainey et al. 2009) significantly exceeds estimates for radiogenic heating (Hussmann et al. 2010). In contrast, for Europa's silicate mantle, radiogenic heating prevails, especially during the early phases of evolution. Only in the later stages of evolution and during periods of increased eccentricity is tidal dissipation speculated to become dominant (Běhounková et al. 2021). Tidal dissipation is expected to be even lower for Ganymedes's mantle due to its distance from Jupiter (Bland et al. 2009; Steinbrügge et al. 2023). However, even in this case, the tidal dissipation in the rock mantle could have played a key role in the past when Ganymede entered the Laplace resonance. For eccentricity two times larger in the past, tidal heating in Ganymede could have been comparable to present-day tidal heating in Europa's mantle, which can reach a level similar to radiogenic power (Steinbrügge et al. 2023).

Tidal energy is influenced not only by orbital parameters but also by the total mechanical responses of the moons. As explained in Sect. 2, the amplitude of tidal deformation is linked to the internal structure, while energy dissipation is sensitive to thermal conditions. In cases of extreme tidal dissipation, thermal runaways can occur, potentially initiating large-scale silicate melting causing an extreme volcanic activity as observed on Io (e.g., Lopes et al. 2004; Mura et al. 2020). This sensitivity of material properties to temperature and the presence of melt (see Sect. 3.3), gives rise to feedback between thermal evolution and tidal dissipation.

The interplay between thermal evolution and tidal heating becomes evident when we consider the sensitivity of the local tidal dissipation rate to temperature and viscosity (Fig. 4). Generally, the dissipation rate at tidal frequencies increases with increasing temperature and, consequently, with decreasing viscosity. This is a key characteristic that leads to positive feedback between the temperature and the magnitude of tidal dissipation.

Once the melting point is reached, the subsequent evolution depends on how efficiently the melt is extracted. In the extreme scenario, where the melt is instantaneously extracted to the mantle surface, the thermal runaway stops and any produced heat is balanced by melt extraction (Běhounková et al. 2021). Moore (2001) have proposed for Io the heat pipe mechanism, an equilibrium scenario in which heat produced by tidal dissipation is brought to the surface by the ascent of magma (see also Bierson and Nimmo 2016 and Steinke et al. 2020). In general, the melt ascent is described by the two-phase flow approach (see 1D two-phase approach for Io, Spencer et al. 2020, 2021), and can rapidly vary in time resulting in a disequilibrium between heat production and heat release.

Depending on the efficiency of melt extraction, significant melt can stay embedded in the solid matrix, affecting the mechanical properties of the molten region. The material begins to melt and gradually builds a system of interconnected channels. Once the local melt fraction reaches the disaggregation point (typically for 20-40% of melt, see Fig. 5), the rigidity is strongly affected and sharply decreases. Consequently, the tidal dissipation drops and the response to tidal loading changes (Moore 2001; Kervazo et al. 2021). Combined with strongly decreased viscosity and more effective heat transport, a new equilibrium near the disaggregation point could be theoretically reached.

The retention of melt, however, introduces an additional level of complexity (see also Sect. 3.3). A model of melt percolation within a solid-melt mixture (Miyazaki and Stevenson 2022) applied to Io suggests that the partial-melt layer ($\phi > 0.2$) rapidly separates into two phases, creating a subsurface magma ocean. In cases where significant melt is retained, resulting in the formation of a magmatic ocean or the presence of a substantial amount of melt approaching critical porosity, mechanical separation of layers may occur. This separation can induce alterations in global deformations, which may be detectable. Moreover, analogous effects to those considered in oceans can also come into play (see Sect. 5). For instance, in the case of the emergence of a thin magma ocean, dissipation within such a fluid layer can pass through resonance phases connected to an increase in tidal dissipation, potentially accelerating the melting process (Tyler et al. 2015). However, it is important to note that the feedback with orbital evolution remains a subject of ongoing research and is not yet fully understood.

The distribution of melt and the location of its production remains uncertain, even in the case of Io where, in the absence of a hydrosphere, significant volcanic activity gives direct insight into the melting of its mantle (e.g. Hamilton et al. 2013; de Kleer et al. 2019; Davies et al. 2024). Magnetic induction also indicated a subsurface molten layer containing a melt fraction of 20% or more (Khurana et al. 2011). The spatial distribution of the melt production is linked to the temperature profile and the mechanisms governing melt transport. A significant factor influencing melt distribution is the presence or absence of convective heat transfer. In situations where convection is absent, melt originates in the deepest regions of the mantle, just above a possible core where the temperature is the highest. Conversely, in the presence of convection, the subadiabatic temperature gradient is reached (the temperature is slightly higher under the stagnant lid than above the lower boundary layer) consistent with convection with significant internal heating characteristics. Moreover, the melt production beneath the lithosphere due to the decompressive melting is expected (e.g., Steinke et al. 2020; Běhounková et al. 2021).

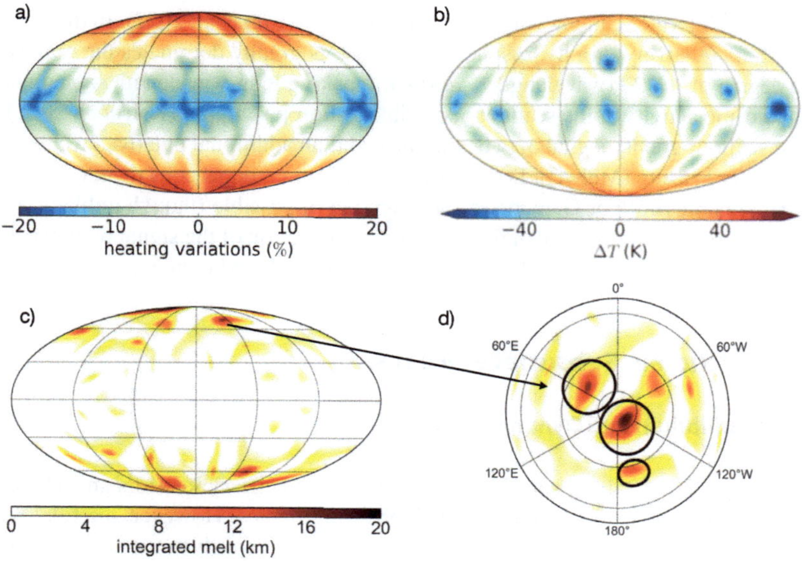

Fig. 7 Example of 3D simulations of tidally-heated thermal convection in Europa's mantle showing enhancement of tidal heating at high latitudes (a), resulting in a concentration of hot thermal plumes (b) and melt generation (c, d) towards the poles (adapted from Běhounková et al. 2021)

Due to tidally-induced changes in polar flattening and stretching in the planet direction, the maximum local heating rate is expected in the polar areas (provided the liquid layer lies beneath the deforming solid layer), and the minima are located in the equatorial area at sub- and anti-planet points in the case of a spin-orbit resonance configuration. Again, the presence of a fully liquid magma layer (Tyler et al. 2015) or a subsurface partially molten layer (e.g. Segatz et al. 1988; Tackley et al. 2001; Beuthe 2013; Bierson and Nimmo 2016; Steinke et al. 2020) maybe associated with bulk dissipation (Kervazo et al. 2021) can change this pattern substantially. At shorter wavelengths, dissipation is modulated by the temperature pattern originating in convection, and it is typically higher in the hot upwellings and lower in the cold downwellings. Conversely, tidal dissipation can break the relatively uniform distribution of convective planforms, can help focus the hot upwellings in the areas with the highest dissipation, and create sheet-like hot structures near the bottom boundary. For instance, in the case of Europa's silicate mantle, it has been shown that enhancement of tidal heating toward the poles tends to maintain hot upwellings at high latitudes, leading to higher melt production towards the poles (Fig. 7). For eccentricity twice higher than the present-day value (conditions potentially reached during the last 100 million years (Hussmann and Spohn 2004)), the volume of generated melts during magmatic episodes could be comparable to those involved in the Large Igneous Provinces observed on Earth (Běhounková et al. 2021). Similar tidally-induced magmatic events may occur on the seafloor of other icy moons, especially during periods of enhanced eccentricity, for instance on Titan, Ganymede or Triton.

The evidence of past or ongoing magmatic activity at Europa's seafloor relies on indirect observations as it is concealed under the icy shell. Gravity field measurements offer a potential tool to assess topographic gravity anomalies originating from volcanic activity (Pauer et al. 2010; Dombard and Sessa 2019; Koh et al. 2022; Mazarico et al. 2023). The anticipated primary contributors to the gravity field include the topography of the silicate/ocean and ice/ocean interfaces, along with surface topography (Koh et al. 2022). In the case of Europa, with its relatively thin hydrosphere, seafloor topography dominates the gravity signal up to a degree of $l = 22$ (wavelength > 200 km), making long-wavelength signatures of potential volcanic activity at the seafloor accessible to Europa Clipper (Koh et al. 2022). However, the seafloor gravity signal may be diminished if an additional putative layer containing materials like gypsum (Melwani Daswani et al. 2021) or carbonaceous matter (Reynard and Sotin 2023) is present. For Ganymede, the detection of possible (though unlikely) volcanic or tectonic activity is more challenging due to the attenuation of the signal with distance, the rock/ice interface being about 800 km below the surface (De Marchi et al. 2021).

5 Tidal Friction in Subsurface Oceans

5.1 The Equilibrium and Dynamical Tides

As indicated in Sect. 2, both solid and liquid layers in planetary interiors are subject to the gravitational tidal forcing generated by neighbouring bodies. However, the tidal response of the latter strongly differs from that of the former. The tidal dynamics of solid bodies is associated with long viscoelastic relaxation timescales (of the order of days for temperate ice layers and of the order of $\sim 10^3 - 10^4$ yr for rock mantles typically; see e.g. Karato and Spetzler 1990; Tobie et al. 2003; Castillo-Rogez et al. 2011; Sabadini et al. 2016; Bolmont et al. 2020). As a consequence, one can neglect inertial effects, which refer to the dynamical effects induced by the motion of particles (advection, Coriolis effects). The solid tidal response thus takes the form of a slightly delayed quasi-static viscoelastic adjustment, as described by the viscoelastic-gravitational theory (Takeuchi and Saito 1972; Tobie et al. 2005b, 2019) (see Sect. 2.2). This behaviour is equivalent to the large scale hydrostatic adjustment of fluid layers characterising the low-frequency limit $\sigma \to 0$, with σ being the forcing tidal frequency, which corresponds to the equilibrium tide (e.g. Ogilvie 2014). In the general case, however, the tidal response of fluid layers is frequency-dependent due to the excitation of waves, as detailed further. These waves form the so-called dynamical tide (Zahn 1975). As they are resonant for specific frequencies (or equivalent thicknesses of the fluid layer), they strongly affect the amplitude of the tidal fields and the resulting tidal dissipation.

In the equilibrium tide limit, a subsurface ocean can be treated as a non-elastic solid in the framework of the viscoelastic-gravitational theory (Beuthe 2015a,b). Under this approximation, the dynamical tide is filtered owing to the absence of inertial effects. As discussed by Kamata et al. (2015), although this static ocean formulation appears to be very convenient to compute the tidal response of multi-layered bodies as a whole, it delivers significant inaccuracies if the ocean thickness is not much larger than the resonant thicknesses associated with the excited tidal modes. One should then take into account the full dynamical oceanic response and the associated wave families. The only exception is the regime of 'creeping' flow where dissipative mechanisms are so strong that all inertial effects vanish. We note that Tyler et al. (2015) provide an analytical solution to the oceanic tidal equations in this creeping flow limit to model the tidal response of Io's suspected global magma ocean (Khurana et al. 2011).

The simplest framework that allows the dynamical tide to be examined is the thin-shell approximation, which consists in ignoring the vertical structure of the fluid layer and the dependence of the tidal fields upon the radial coordinate, thus leading to the shallow water equations (e.g. Vallis 2017). In this approach, the ocean is considered as a thin spherical shell of incompressible fluid with two-dimensional tidal fields reflecting the integrals over the layer, which is valid as long as the ocean depth H is much smaller than its distance to the centre of the body. The radial velocity is therefore assumed to be much smaller than the horizontal velocity, which makes it possible to ignore all the physical terms involving the vertical component of the velocity field in the momentum and continuity equations. We note that horizontal flows produce a vertical Coriolis acceleration that is not necessarily negligible with respect to pressure forces. However, this Coriolis acceleration induces substantial mathematical complications by coupling the vertical and horizontal structures of tidal flows together. It is therefore usually ignored by means of the traditional approximation, the limitations of which are thoroughly discussed by Gerkema and Shrira (2005) and Gerkema et al. (2008). This formalism has been widely used to study the tidal response of subsurface oceans in icy moons (e.g. Tyler 2011, 2014; Chen et al. 2014; Matsuyama 2014; Beuthe 2016; Hay and Matsuyama 2017; Matsuyama et al. 2018; Rovira-Navarro et al. 2023) and ocean planets (e.g. Webb 1980; Auclair-Desrotour et al. 2018; Motoyama et al. 2020; Tyler 2021; Farhat et al. 2022).

In the linear theory of tides, the tidal fields are considered as small perturbations with respect to the background fields of the body at rest. In particular, the surface elevation oscillations generated by the tidal forcing are assumed to be much smaller than the ocean depth, while their wavelengths are comparable to the radius of the body. Therefore, the non-linear advection of tidal flows is neglected. This yields the classical Laplace's tidal equations (LTE) (e.g. Longuet-Higgins 1968; Lamb 1993). The LTEs are the mass conservation and horizontal momentum equations given by (e.g. Tyler 2011; Matsuyama 2014)

$$\partial_t \xi + \nabla \cdot (H\boldsymbol{u}) = 0, \tag{23}$$

$$\partial_t \boldsymbol{u} + 2\boldsymbol{\Omega} \times \boldsymbol{u} = -g\nabla\xi + \nabla U - \boldsymbol{F}_{\mathrm{drag}}, \tag{24}$$

where ξ designates the tidal variation of the ocean's thickness, $\boldsymbol{u}$ the associated horizontal velocity field, g the gravity, U the forcing tidal gravitational potential, $\boldsymbol{\Omega}$ the spin vector, and $\boldsymbol{F}_{\mathrm{drag}}$ the drag force exerted on the layer per unit mass. The notations ∂_t, ∇, and $\nabla\cdot$, refer to the partial derivative with respect to time, to the horizontal gradient operator, and to the horizontal divergence operator, respectively. In the case of spherical or hemispherical ocean basins of uniform depth, the LTEs admit closed form solutions where the tidal fields are expanded in terms of the oceanic normal modes, also called free oscillations (Longuet-Higgins 1966, 1968; Longuet-Higgins and Pond 1970). These modes are associated with specific eigenfunctions and the corresponding eigenfrequencies.

In the simplest configuration where the ocean is global and inviscid (no drag forces), the eigenfunctions are called the Hough functions (Hough 1898) and commonly denoted by Θ (e.g. Lindzen and Chapman 1969; Lee and Saio 1997; Wang et al. 2016). The Hough functions sharing the same order[1] m are obtained by solving, for (Θ, Λ), the eigenvalues-eigenvectors problem formulated as (e.g. Lee and Saio 1997)

$$\left\{ \frac{d}{d\mu} \left[\frac{1-\mu^2}{1-\nu^2\mu^2} \frac{d}{d\mu} \right] - \frac{1}{1-\nu^2\mu^2} \left(\frac{m^2}{1-\mu^2} + m\nu \frac{1+\nu^2\mu^2}{1-\nu^2\mu^2} \right) \right\} \Theta = -\Lambda\Theta, \tag{25}$$

assuming regularity boundary conditions at the poles. In the above equation, $\mu = \cos\theta$, where θ refers to the colatitude, and the real parameter Λ is the eigenvalue associated with Θ. Besides, $\nu = 2\Omega/\sigma$ designates the spin parameter, which depends on the spin angular

[1] Analogous to the spherical harmonics Y_l^m, the order of the eigenfunction, m, refers to its integer wavenumber (see e.g. Varshalovich et al. 1988).

velocity, Ω, and the forcing frequency, σ, and quantifies the impact of rotation on the forced modes (e.g. Lee and Saio 1997). In the static limit ($\nu = 0$), the Hough functions converge towards the associated Legendre functions of the spherical harmonics, while rotational effects become stronger as ν increases in absolute value.

Several authors use this simplified theory to obtain analytical tidal solutions for the torque exerted on global or hemispherical oceans with free-surface conditions (e.g. Auclair-Desrotour et al. 2018, 2019; Motoyama et al. 2020; Farhat et al. 2022). However, more realistic configurations including non-linearities and more complex geometries require to switch to numerical modelling. In those configurations, the equations governing the tidal dynamics are efficiently solved using spectral methods, which consist in expanding the tidal fields in series of spherical harmonics as a first step, and in calculating the spherical harmonic coefficients by inverting a set of algebraic equations as a second step (Longuet-Higgins 1968; Tyler 2011, 2014; Chen et al. 2014; Matsuyama 2014; Matsuyama et al. 2018; Beuthe 2016). Such methods are, for instance, implemented in the Tidal Response of Planetary Fluids (`TROPF`) software package (Tyler 2019). Other approaches seem, nevertheless, more suitable in the case of complex boundary conditions (spatially-varying ocean depth, topography), such as the finite element method employed in the Ocean Dissipation in Icy Satellites (`ODIS`) model (Hay and Matsuyama 2017, 2019) or in most of state-of-the-art Earth oceanic tidal models (e.g. Egbert et al. 1994, 2004; Green et al. 2017; Schindelegger et al. 2018; Daher et al. 2021).

5.2 Several Families of Resonantly Excited Tidal Waves

In the tidal flows described by the shallow water approach, the fluid wave modes that carry the energy are divided into rotational-gravity and Rossby waves. The rotational-gravity and Rossby waves are sometimes referred to as the 'class I' and 'class II' oscillations, respectively (e.g. Longuet-Higgins 1968; Tyler 2021). The rotational-gravity waves refer to long-wavelength surface gravity waves modified by rotation. They have as a restoring force the gravity force exerted on the mass surplus resulting from an increase of the ocean surface elevation, and their typical phase speed is $c = (gH)^{1/2}$ (Vallis 2017). Each mode is associated with an eigenfrequency σ_{w}, and an equivalent depth $h = H\,(\sigma/\sigma_{\mathrm{w}})^2$, the latter referring to the ocean depth for which the mode would be resonant at the considered tidal forcing frequency (Taylor 1936; Hendershott 1981). The rotational-gravity waves exist for all values of the spin parameter ν introduced in Eq. (25) (Lee and Saio 1997). As $\nu \to 0$ (static limit), their spatial structure converges towards the spherical harmonics, which corresponds to pure gravity modes. Conversely, if $|\nu| \gg 1$ (low-frequency tidal forcing), the rotational-gravity waves are confined within an equatorial band. Their spatial structure is thus strongly altered by rotation in this configuration. The role played by rotational-gravity waves in the tidal energy dissipation has been examined both in the context of the subsurface oceans harboured by icy satellites (e.g. Tyler 2011, 2014; Chen et al. 2014; Matsuyama 2014; Matsuyama et al. 2018; Beuthe 2016; Rovira-Navarro et al. 2023) and that of the Earth's surface ocean (e.g. Webb 1980, 1982; Auclair-Desrotour et al. 2018; Tyler 2021; Farhat et al. 2022). As discussed further, these waves may be responsible for resonant amplifications of tidal heating in icy satellites, each mode being associated with a resonance occurring for $\sigma = \sigma_{\mathrm{w}}$ or, equivalently, for $h = H$.

Rossby waves, also called 'Rossby-Haurwitz' waves or 'planetary waves' (e.g. Longuet-Higgins 1968; Tyler 2021), are induced by rotation, their restoring force being the variation in Coriolis acceleration across the latitude. These waves can be excited by the tidal gravitational forcing if $|\nu| > 1$ (Lee and Saio 1997). While they are confined near the poles if the effect of rotation is weak, they tend to span large regions of the globe as ν increases. As Tyler (2008) indicates it in the case of Europa, large-amplitude Rossby waves can be resonantly excited by tidal forces due to obliquity – although these forces are subdominant – because their eigenfrequencies are close to obliquity forcing frequencies. This statement equally applies to other moons with suspected subsurface oceans sharing the same tidal regime, such as Titan and Enceladus (Hay and Matsuyama 2017; Matsuyama et al. 2018; Rovira-Navarro et al. 2020, 2023).

The rotational-gravity and Rossby waves described in the framework of the thin-shell approximation form the barotropic flow of the tidal response, namely the component for which the density variation is a monotonic function of the pressure variation (Vallis 2017). These surface waves are thus associated with uniform motions of fluid particles over the water column. The vertical structure of the fluid layer ignored in the thin-shell approximation is characterised by the stability of the stratification with respect to convection (Gerkema and Zimmerman 2008). In stably stratified regions, fluid particles moving across the vertical direction tend to be driven back to their initial positions by the Archimedean force, whereas they are free to go away in neutrally stratified (or unstratified) regions. The Archimedean force thus acts on particles as a spring whose stiffness is quantified by the squared Brunt-Väisälä (or buoyancy) frequency, $N^2 = g^2 \left(\partial \rho_0 / \partial p_0 - c_s^{-2}\right)$, with g, ρ_0, p_0, and c_s designating the local gravity, density, pressure, and sound speed, respectively (e.g. Gerkema and Zimmerman 2008). In Earth's ocean, N varies from $\sim 10^{-4}\ \mathrm{s}^{-1}$ in the deepest layers to $\sim 10^{-2}\ \mathrm{s}^{-1}$ in the 200 m-region beneath the surface (e.g. Gerkema et al. 2008).

The upward-travelling waves that have the Archimedean force as a restoring force are called internal gravity waves. They develop in stably stratified water columns and form the baroclinic flow of the tidal response, where the tidal density and pressure fields varies non-monotonically across the vertical direction (Vallis 2017). Similarly to surface waves, internal waves are associated with equivalent depths and eigenfrequencies characterising their resonances. It is noteworthy that this baroclinic response completely vanishes if the ocean is unstratified ($N = 0$), which yields a pure barotropic response. Nevertheless, both baroclinic and barotropic flows should exist in reality since an unstratified ocean is only a theoretical idealisation. For example, tidally forced internal gravity waves may be directly excited by eccentricity tidal forces in Enceladus, as proposed by Tyler (2011) in his stratified ocean scenario, which results in an increased tidal heating. They also contribute to enhance the tidal response of global ocean planets (Auclair-Desrotour et al. 2018).

The inertial (or gyroscopic) waves (e.g. Gerkema and Zimmerman 2008) are another – though secondary – type of internal waves that may be tidally excited in the oceans of icy moons (e.g. Rekier et al. 2019; Rovira-Navarro et al. 2019). These waves are restored by the Coriolis force, and thus develop in unstratified fluid layers. In the case where the ocean is stably stratified, inertial waves are mixed with gravity waves, which yields the so-called internal inertial-gravity waves (e.g. Gerkema and Zimmerman 2008). These waves are suspected to play an important role for energy dissipation in giant planets and stars (e.g. Ogilvie and Lin 2004; Ogilvie 2009, 2013; Rieutord and Valdettaro 2010; Lainey et al. 2017). For subsurface oceans, there are still a lot of debates and unknowns regarding their dynamical regimes (Soderlund 2019; Amit et al. 2020; Ashkenazy and Tziperman 2021; Bire et al. 2022; Kvorka and Čadek 2022; Soderlund et al. 2024) and the degree of oceanic stratification (e.g. Lobo et al. 2021; Kang et al. 2022b). Although the hydrostatic and traditional approximations used in the shallow water equations properly account for the predominating forced surface modes, they preclude the propagation of inertial modes. The latter are therefore studied in specific frameworks that relax these approximations by including, for instance, the full Coriolis force (non-traditional approximation; see e.g. Tort et al. 2014).

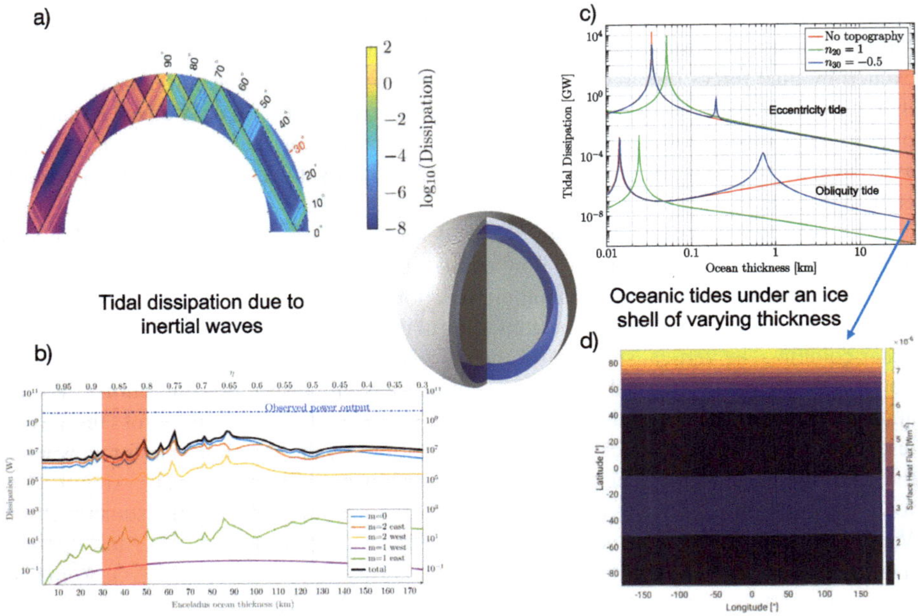

Fig. 8 a) Viscous dissipation amplitude due to tidally-excited inertial waves obtained for Ekman number of 10^{-9} and inner to outer ocean radius ratio of 0.73 (considered as roughly representative of Enceladus' ocean) and b) corresponding total dissipated as a function of ocean thickness (adapted from Rovira-Navarro et al. 2019). c) Corresponding tidal dissipation as a function of ocean thickness between models with an ocean of constant thickness and oceans with degree two and three thickness topography, assuming either eccentricity and obliquity tides, and d) corresponding time-averaged tidal dissipation patterns for a case with the obliquity tide case and an ocean with degree three topography (adapted from Rovira-Navarro et al. 2020). The pink band on panel b and c represents the estimated range of ocean thickness

By considering subsurface oceans of homogeneous densities for Europa and Enceladus, Rovira-Navarro et al. (2019) show that the inertial waves that could be indirectly forced by the equilibrium tide would lead to internal tidal flows of significant amplitudes (Fig. 8 a, c). Nevertheless, they note that the resulting tidally dissipated energy still remains several orders of magnitude smaller than Europa's radiogenic heating and Enceladus' observed heat flux (Fig. 8 c). Similar conclusions are drawn by Rekier et al. (2019) from the study of libration-induced inertial waves in Enceladus' ocean. More recently, Idini and Nimmo (2024) studied inertial waves in a stably-stratifed ocean in the context of Titan, and showed that resonant stratification may significantly amplify the tidal response.

5.2.1 Tidal Dissipation Mechanisms

Analogous to solid layers, the tidal response of fluid layers is accompanied with energy dissipation. In oceans, this energy dissipation is mainly caused by the friction of the barotropic tidal flows against the oceanic floor and the overlying ice shell. However the term 'friction' should be taken here in a general sense, as it refers to several dissipative mechanisms. Even though the configuration of subsurface oceans is significantly different the Earth's ocean, the Earth knowledge can provide useful physical constraints on the dissipative processes. On the Earth, tidal flows are primarily damped by the bottom drag due to turbulent friction in shallow seas. The work done by bottom drag is proportional to cubed velocity. As a consequence, it gets negligible in deep oceans where the tidal flows remain small compared to shallow seas/thin oceans.

Energy dissipation in the deep ocean is actually caused by another kind of friction that is directly related to the stable stratification of the water column. Through their interactions with the small scale topographic features of the oceanic floor, the long-wavelength barotropic tidal flows, excited by the forcing gravitational potential, are partly converted into gravity waves. These waves act as an energy sink for the oceanic tide since they remove kinetic energy from the tidal dynamics, this energy being eventually dissipated by turbulent and viscous frictions (e.g. Gerkema and Zimmerman 2008). This mechanism enters into the momentum equation as an additional linear drag, called 'Rayleigh drag' (e.g. Jayne and St. Laurent 2001; Tyler 2011; Matsuyama 2014; Auclair-Desrotour et al. 2018; Rovira-Navarro et al. 2023).

It is noteworthy that, in the linear theory, bottom drag and the drag of internal waves are both described by an effective Rayleigh drag parametrised by the so-called Rayleigh drag frequency (or coefficient), σ_R, and formulated as $\boldsymbol{f}_R = H\sigma_R \boldsymbol{u}$. While this effective Rayleigh drag is estimated to be around $\sigma_R \sim 10^{-5}\ s^{-1}$ for Earth (e.g. Webb 1980; Egbert and Ray 2001; Jayne and St. Laurent 2001; Farhat et al. 2022), no constraints exist for subsurface satellites, which allows a large range of values to be used (typically, $\sigma_R \sim 10^{-11}-10^{-5}\ \mathrm{s}^{-1}$; see e.g. Matsuyama et al. 2018). Note that, as discussed by Matsuyama et al. (2018), the Rayleigh drag, $\boldsymbol{f}_R$, can be used to mimic the dissipation generated by bottom friction, $\boldsymbol{f}_B$. These authors established a relationship between the dimensionless drag coefficient of the non-linear term describing bottom friction, C_d, and the Rayleigh drag coefficient, σ_R, by assuming that the two drags generate the same energy flux for the same horizontal velocity field. Finally, one should pay attention to the fact that the bottom and Rayleigh drag, $\boldsymbol{f}_B$ and $\boldsymbol{f}_R$, are both divided by the ocean depth when included in the shallow-water momentum equation given by Eq. (24), given that the drag is averaged over the water column in this equation (see e.g. Matsuyama et al. 2018; Motoyama et al. 2020).

5.3 Dynamical and Thermal Implications

Combined with the wave resonances detailed in Sect. 5.2, the above dissipative mechanisms have strong repercussions on the thermal-orbital evolution of planet-satellite systems, as they usually result in increased tidal heating and energy dissipation. The Earth's oceanic tides are typically responsible for nearly $\sim$90–95% of the tidally dissipated energy (e.g. Egbert and Ray 2001), meaning that they primarily drive the long-term evolution of the Earth-Moon distance and Earth length-of-day (e.g. Webb 1982; Daher et al. 2021; Tyler 2021; Farhat et al. 2022). Besides, their dependence upon the forcing frequency is consistent with the past variations of energy dissipation that can be inferred from geological data, which show that the gravitational tidal torque exerted on Earth during the Precambrian ($\sim 1.13 \times 10^{16}$ J) was roughly a quarter of the present-day torque ($\sim 4.51 \times 10^{16}$ J) in magnitude (e.g. Zahnle and Walker 1987; Farhat et al. 2022). In particular, the resonances of the predominant gravity modes markedly alter the climato-orbital history of the Earth-Moon system reconstructed in the recent solutions (e.g. Daher et al. 2021; Tyler 2021; Farhat et al. 2022).

Such resonant amplifications are likely in the icy satellites of the outer Solar system. Considering Europa, Tyler found that resonantly excited Rossby waves due to obliquity forcing could lead to tidal flows with a kinetic energy 2000 times larger than that of the flow generated by the dominant tidal forces (Tyler 2008). This first analysis by Tyler (2008), however, neglected the effect of dissipation which should damp the resonant flow and assumed a relatively high obliquity (0.05°). He also suggested that Enceladus' extreme heat flux could be explained by strong obliquity-forced tidal flows in its subsurface ocean if Enceladus' spin axis is tilted with respect to its orbital plane by at least 0.05° (Tyler 2009, 2011). The latter scenario, however, is discounted by Chen and Nimmo (2011) and Chen et al. (2014), who argue that Enceladus' obliquity should be actually less than 0.0015°, and that important ocean tidal heating is unlikely for most satellites of the outer Solar system. Several authors invoke the possibility of strong resonantly excited eccentricity tides, but only if the subsurface ocean is $\sim$1 km thick or thinner (e.g. Matsuyama 2014; Tyler 2014; Kamata et al. 2015). Even though such a strong dissipation may have occurred during some specific periods, a thin ($\leq$ 1 km) ocean on Enceladus is clearly inconsistent with the existing topography, gravity and rotation data collected by Cassini, which rather indicate a thick ocean, potentially as thick as 50-60 km in the south polar region (Čadek et al. 2019; Hemingway and Mittal 2019). Moreover, the strong damping effect of the overlying ice crust may further reduce the resonant response and associated dissipation (Beuthe 2016; Aygün and Čadek 2023).

Beyond these considerations on the possible present states of liquid water reservoirs within icy moons, it is worth bearing in mind the full dynamical process leading to such states. In particular, a liquid ocean attempting to freeze is likely to experience strong tidal heating during its evolution since it must pass through resonant configurations, which may even lead to an equilibrium state if the produced heat is sufficient to melt the lower bound of the ice crust (e.g. Tyler 2011). Also, as the above studies mostly rely on the shallow water approach, alternate scenarios take into account the heating induced by internal waves. The studies exploring these scenarios, which mainly focus on Enceladus, show that the energy dissipated by tidally forced gravity waves in stably stratified oceans may exceed that resulting from surface waves (e.g. Tyler 2011; Rovira-Navarro et al. 2023), while the resonant excitation of inertial waves can significantly increase the heat production (e.g. Rekier et al. 2019; Rovira-Navarro et al. 2019).

Subsurface water oceans are not the only fluid layers playing a role in the evolution of rocky bodies. Magma oceans, namely liquid layers of melted rocks, are expected to affect this evolution as well. Typically, the strong tidal heating occurring in the primitive Earth and lunar surface magma oceans of the Hadean had repercussions on the early dynamical and thermal evolution of the Earth-Moon system (e.g. Chen and Nimmo 2016). This tidal heating was, for instance, capable of prolonging the duration of the lunar magma ocean for $\sim$200−300 Myr (e.g. Meyer et al. 2010), and it may have partly controlled the crystallisation process in the Earth's solidifying crust through thermal convection (e.g. Monteux et al. 2023). Similarly, Tyler et al. (2015) show that the region of the parameter space where the tidal heat released in Io's magma ocean reaches the observed heat flux is wider when the ocean is treated as a fluid than when it is treated as a solid.

Furthermore, the increased energy dissipation induced by the dynamical tide allows subdominant tidal forces to generate heat flows exceeding those associated with the predominant forces for specific resonant configurations. With this effect in mind, Hay et al. (2020) investigate the possibility that Moon-Moon tides – namely the tides raised on moons by other moons – can lead to substantial energy dissipation. Although they could be in theory detectable on Ganymede by Juice's 3GM experiment (De Marchi et al. 2022), Moon-Moon tides are much weaker than the tides raised by the planet because of the great difference in mass of the perturbers. Therefore, their contribution to the tidal energy budget is negligible in the general case. Hay et al. (2020) argue, however, that this contribution can affect the total energy budget for very specific ocean thicknesses if the ocean is inviscid enough to build a strongly resonant behaviour, which corresponds to $\sigma_{\mathrm{R}} < 10^{-6}\ \mathrm{s}^{-1}$, typically. While this value of the drag frequency is less than the Earth value by one order of magnitude, Matsuyama et al. (2018) suggest that it is not unlikely regarding the icy moons of the outer Solar system, as σ_R can reach values as low as $\sigma_R \sim 10^{-11}$ s^{-1} in their estimates. Nevertheless, one should bear in mind that there are no constraints for the bottom drag coefficients in icy satellites. Such extreme values should thus be considered as theoretical lower limits rather than nominal values.

5.4 Coupling with the Overlying Ice Shell and the Solid Part

Contrary to present-day Earth's oceans, liquid water layers in icy satellites are overlaid by ice crusts whose thickness depends on the thermal-orbital history of the bodies. By constraining the thickness of Europa's ice shell from numerical experiments of thermal convection, Tobie et al. (2003) found it to be about 20–25 km, with lateral variations of about 5 km. Such constraints can also be obtained from the measurements of tidal gravity fields and vertical displacements, although uncertainties in the shear modulus feed directly into uncertainties in the ice shell thickness (e.g. Wahr et al. 2006; De Marchi et al. 2022). The coupling of the ocean with the overlying shell was first examined in the framework of the viscoelastic-gravitational theory, where all inertial effects are ignored (e.g. Tobie et al. 2003; Beuthe 2015a). This approach shows that the tidal response of the crust is controlled by its membrane spring constant and its effective Poisson's ratio, which quantify the elasticity and compressibility of the material, respectively (e.g. Beuthe 2015a). However, it does not capture the coupling of the oceanic tidal modes with the shell.

In order to remedy these limitations, several authors adopted a more sophisticated formalism by self-consistently including the ocean-crust interactions into the LTEs (e.g. Beuthe 2016, 2018, 2019; Matsuyama et al. 2018). These models are based upon the membrane theory, where the shell is assumed to be thin compared with the radius of the body (e.g. Beuthe 2008, 2018). They predict that the crustal constraint strongly attenuates the resonances of barotropic modes, thus reducing the associated oceanic energy dissipation (see e.g. Beuthe 2016). This damping effect is expected to be more pronounced on Enceladus than on Europa as it is straightforwardly related to the effective elasticity of the overlying shell, which increases as the size of the body decreases (Matsuyama et al. 2018).

Furthermore, it is noteworthy that the thickness variations of the shell play an important role as well, as the crustal thinning of Enceladus at the south pole can amplify the local tidal stresses by one order of magnitude (Beuthe 2018). Conversely, the horizontal thickness variations of the ice shell are themselves dependent upon the oceanic tidal heat flow, which is not homogeneous over the sphere (e.g. Matsuyama et al. 2018). Rovira-Navarro et al. (2020) investigated how degree-two and degree-three ice/ocean interface topography could affect the response of Enceladus's ocean to eccentricity and obliquity tides (Fig. 8 c and d). These results show that the ice shell thickness variations shift the resonant peak to different values of ocean thickness. However, for ocean thicknesses representative of present-day Enceladus, the produced total power remains unchanged for eccentricity tides and is reduced for obliquity tides when varying thickness is taken into account. Even though ice shell thinning should result in a higher dissipation rate at the south pole for obliquity tides, the produced heat flow from the ocean remains several orders of magnitude below the observed values. The predicted heat flow remains below 10^{-2} $mW\,m^{-2}$ while the observed heat flow exceeds 100 $mW\,m^{-2}$ (Spencer et al. 2013, 2018).

Last but not least, the solid part of an icy moon is not rigid, as described in simplified idealisations (e.g. Longuet-Higgins 1968). Instead, it responds visco-elastically to the surface loading and self-attraction variations of the ocean (e.g. Peltier 1974). To refine the Earth tidal theory, Hendershott (1972) introduced the deformations of solid regions in the LTEs that describe the global ocean tides, thus showing that the solid-ocean coupling leads to more excited harmonics. This effect is due to the differences between the spherical harmonics and the Hough functions featuring the horizontal structures of the tidally induced barotropic modes, the two sets of eigenfunctions being equivalent in the static limit solely (e.g. Lee and Saio 1997). In the context of icy moons, models including the deformations of solid regions and self-attraction variations show that the visco-elastic adjustment of the solid part slightly attenuates the resonant peaks of the dynamical tide. These peaks are, moreover, shifted by the solid response, which can cause orders of magnitude changes in the dissipated energy flux if the oceanic configuration is close to a resonance (e.g. Matsuyama 2014). By solving the three-dimensional Navier-Stokes equations taking into account viscoelastic coupling with the overlying ice shell and underlying rock layer, Aygün and Čadek (2023) show that the resonant modes both in terms of peak amplitude and frequency can be significantly modified, resulting in a dissipation power much smaller than what is predicted using the standard LTE approach.

6 Tidal Response of Icy Shells: Implications for Surface Activity

6.1 Tidally-Induced Melting in Convective Ice Shells

Depending mainly on the ice shell thickness and ice viscosity, the heat released from the moon's interior can be transferred through an ice shell by either conduction or convection. As tidal heating is strongly dependent on the viscosity with optimal dissipation for the viscosity of the order of $10^{13} - 10^{15}$ Pa s (see Sect. 3, Fig. 4), heat transfer and heat production by viscous friction are strongly coupled, in particular for convective ice shells. In a conductive ice shell, the heat transfer is very inefficient and the thermal profile is characterized by an almost linear increase of temperature with depth, resulting in only a small warm portion of the ice shell just above the ice/ocean interface, where viscosity is close to the optimal value for viscous dissipation. On the contrary, in a convective shell, the heat transfer is much more efficient and the thermal structure is characterized by a warm, nearly isothermal sublayer below a cold, highly-viscous conductive lid. The relative thickness of the convective and conductive sublayers depends on various factors, such as ice viscosity, amount of tidal heating, mobility/deformation of the lid etc. Consequently, tidal heating can vary by orders of magnitude between conductive and convective cases, as the thickness of the dissipative ice can vary from less than 1 km for a thin conductive layer to several tens of kilometers for a convective ice layer. Moreover, as viscosity varies laterally in a convective layer, tidal heating can be locally increased in regions where the local viscosity is equal to the optimal viscosity value for dissipation (Tobie et al. 2003). For optimal viscosity equal to the viscosity at the melting point, tidal heating is maximized at the hot thermal boundary layer which can favor the initiation of warm upwelling plumes (Běhounková et al. 2013).

Focusing tidal heating in warm upwelling plumes can lead to strong melting (Tobie et al. 2003), potentially resulting in surface disruptions due to melt drainage and collapse, proposed as a possible origin of chaotic terrains on Europa (Sotin et al. 2002) and for initiating the activity on Enceladus during periods of enhanced eccentricity (Běhounková et al. 2012). Accumulation of melts in tidally-heated plumes is expected to affect their dynamics owing to the density difference between the generated meltwater and the ice matrix. The produced meltwater represents a new source of buoyancy that can act to strengthen or weaken the thermal convection and should therefore be considered in the convection model. Solving the full problem of two-phase convection is numerically demanding and thus various approximations have been considered. At first order, one can neglect the differential motion between

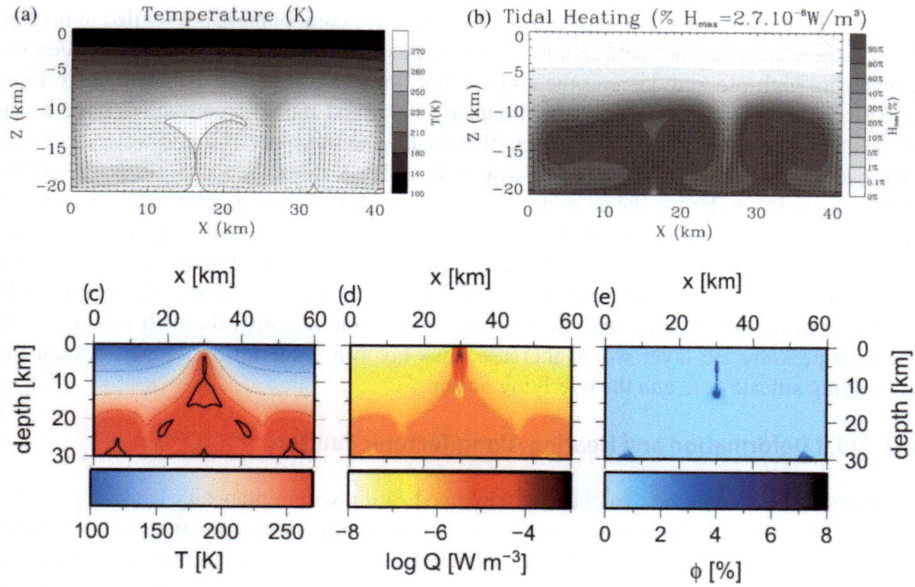

Fig. 9 *Top*: Temperature (a) and volumetric tidal heating rate (b) in a convecting ice shell of Europa with viscosity-dependent volumetric heating (assuming a Maxwell model) (Tobie et al. 2003). The black line in panel (a) marks the area where temperature equals the melting temperature of pure water ice. *Bottom*: Temperature (c), heating rate (d), and porosity (e) in Europa's ice shell with fixed frictional heating at the fault and viscosity-dependent volumetric heating (assuming an Andrade model) (Kalousová et al. 2016). Dashed contours in panel (c) mark temperatures of 150, 200, and 250 K, the thick contour denotes the melting temperature (270 K)

the two phases and assume that water is simply advected by the solid ice matrix. As illustrated in Fig. 9, melt produced by enhanced tidal heating in convective plumes (top) or at the base of active strike-slip faults (bottom) in the context of Europa results in a downward motion of the partially molten ice due to negative buoyancy. The melt volume and associated dissipation rate tend to increase during this downward migration which amplifies this effect. As shown by Tobie et al. (2003), the coupling between tidal heating and melt production leads to strong time variations in convective dynamics with intermittent periods of thermal plume growth followed by periods of melt-induced plume collapses. This results in significant variations in heat flux variations at the ice/ocean interface, with important consequences for mass and chemical exchanges between the ocean and overlying ice shell. Such downward motions of partially molten ice could entrain any salt and non-ice constituent present in the ice shell to the ocean and such cold salty water may significantly affect the dynamics of the underlying ocean.

More recent studies take into account the differential motion between the two phases (Kalousová et al. 2018; Kalousová and Sotin 2018, 2020). In such cases, the water velocity can be up to a few orders of magnitude larger than the ice velocity, depending mostly on the amount of available water, the ice permeability, the liquid water viscosity, and the convection regime (stationary or chaotic). This approach was used for the deep ice layers that are predicted in large moons such as Ganymede or Titan. These layers are made of high-pressure ice polymorphs, which are denser than liquid water at the corresponding pressures (hundreds of MPa to units of GPa). In that setting, the phase-change-induced buoyancy acts in the same direction as the thermal buoyancy and thus enhances the convective motions. In a pure ice setting (i.e. without the addition of salts or other non-ice compounds), a so-called temperate layer (where ice is at the melting temperature) is established at the boundary between the convecting high-pressure ice and the above-lying liquid water ocean. This temperate layer effectively eliminates the top thermal boundary layer and thus the convective pattern can be described as active upwelling plumes (that enable the transport of liquid water) and passive background downwelling (Kalousová et al. 2018; Kalousová and Sotin 2018, 2020). This process is similar to silicate melts ascending through silicate mantles (e.g. Ogawa and Yanagisawa 2011; Ogawa 2014). Even though the tidal dissipation is expected to be two orders of magnitude smaller in the high-pressure ice layer compared to the ice I layer above the liquid ocean, tidal heating during periods of enhanced eccentricity could significantly affect the heat budget in the high-pressure ice layer. Consequently, the melt production in the high-pressure ice layer would increase, thus favoring material exchange between the underlying silicate core and the overlying ocean.

6.2 Tidal Deformation and Heating Along Tectonic Faults

Tidal deformation and the associated dissipative heating within planetary ice shells are expected to be intensified by the presence of active faults or fault systems within these shells. Currently, the system consisting of four prominent faults located in the southern polar region of Enceladus, commonly referred to as "tiger stripes", stands as a preeminent exemplar of such structures among the known ocean worlds. Instruments aboard the Cassini spacecraft observed remarkable activity in the form of jets of vapor and ice particles emanating from these south polar fissures (Porco et al. 2006; Postberg et al. 2018). The modulation of brightness, reflecting the solid particles content, on the tidal timescale (e.g., Hedman et al. 2013; Nimmo et al. 2014; Ingersoll et al. 2020) implies tidal control of the process (Hurford et al. 2007). However, the exact interpretation of this activity signal has not been straightforward due to the more than five-hour observed lag between the timing of the predicted extensional stress maximum and the main activity peak (Nimmo et al. 2014). Several attempted explanations, including viscoelasticity (Nimmo et al. 2014; Běhounková et al. 2015), libration (Nimmo et al. 2014), or local hydraulic processes (Kite and Rubin 2016), have either been disproved, doubted, or raised further questions. Additionally, the measurements seem to indicate either weaker or absent modulation of the vapor mass flux on the tidal time scale (Hansen et al. 2020) with potential changes on shorter (hour) time scale (Denny et al. 2024).

Pleiner Sládková et al. (2021) and Berne et al. (2024) observed that the two maxima of jets' activity can be correlated with modeled strike-slip motions rather than normal stresses or displacements at the faults. Souček et al. (2024) proposed a model providing a physical explanation for this correlation, combining a global shell deformation model with detailed modeling of water and vapor transport processes within the partially flooded fissures, in line with models by Kite and Rubin (2016), and Nakajima and Ingersoll (2016). These interpretations thus support the view of tiger stripes as strike-slip faults penetrating the entire shell thickness, whose activity is maintained by incessant tidal forcing.

The effect of the presence of faults on the global tidal deformation of Enceladus has been investigated by means of numerical models of the tidal deformation in the studies by Souček et al. (2016), Běhounková et al. (2017), Souček et al. (2019), Pleiner Sládková et al. (2021), and its validity was corroborated by subsequent research efforts by Berne et al. (2023). It has been demonstrated that the thinning of the southern polar shell, in conjunction with the existence of these faults, synergistically culminates in a substantial amplification of tidal deformations, particularly when these faults are treated as frictionless slots. Consequently, this leads to a localized augmentation of dissipation in the proximity of the fault zones, particularly at the tips of these faults (Fig. 10). The cumulative impact of the presence of faults

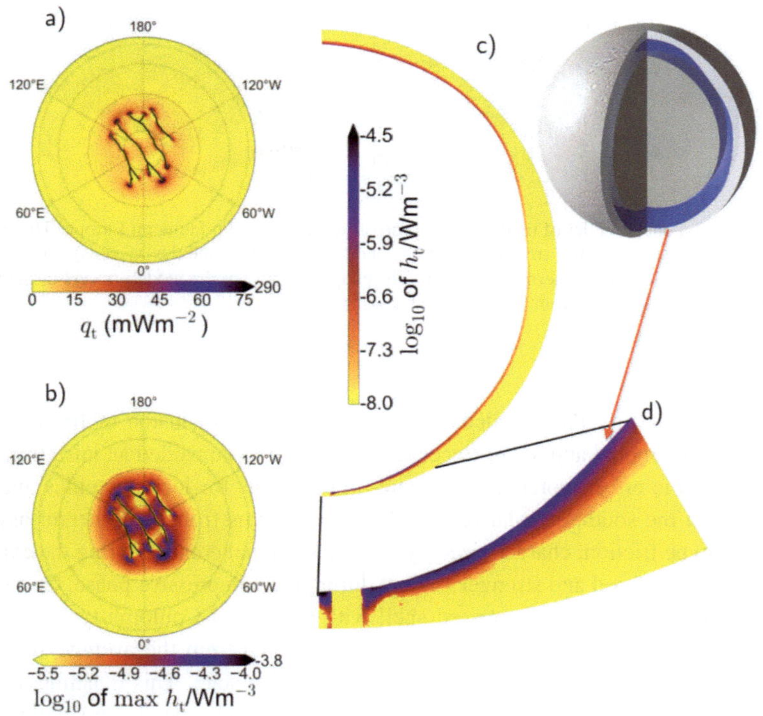

Fig. 10 Tidal dissipation in the SPR of Enceladus (computed for pseudo-Andrade rheological model and reference viscosity at the melting point $\eta_0 = 3 \cdot 10^{14}$ Pa s). Panels show a) Depth-integrated heating power expressed as equivalent tidal heat flux in the southern hemisphere. b) Maximum tidal heating rate with respect to depth. c) Tidal heating rate plotted at meridional cross-section ϕ=0°. d) Tidal heating rate near the south pole. The thickness of the ice shell in panel d is exaggerated by a factor of 3. The faults in this study are considered to be frictionless, consequently, the tidal heating is zero inside fault zones. The SPR (panels a and b) is shown in stereographic projection for latitudes 60°–90° S. Adapted from Souček et al. (2019)

on the dissipative heat budget of the southern polar region of Enceladus has been, under the above assumptions, estimated to be of the order of units of GW (Souček et al. 2019). Owing to the local effects, especially at the fault tips, it is conceivable that these effects could be responsible for generating local thermal heat flux or temperature anomalies such as the ones inferred by Le Gall et al. (2017). Even if tidal dissipation can be locally enhanced, the total power generated along such active faults estimated to be less than 2.1 GW (Souček et al. 2019) remains too small to explain the total heat budget of Enceladus estimated to 25-30 GW (Čadek et al. 2016; Choblet et al. 2017).

The phenomenon of friction occurring at the faults (neglected in Souček et al. 2016; Běhounková et al. 2017; Souček et al. 2019) introduces an additional source of dissipative heating (Pleiner Sládková et al. 2021). Simultaneously, it results in the damping of deformation (Pleiner Sládková et al. 2021; Berne et al. 2023), and consequently, a decrease of volumetric dissipation in the vicinity of the faults. The combined impact of these processes on the dissipative heat budget within the fault's immediate surroundings varies depending on the nature of the frictional interaction and can either amplify or reduce the overall dissipative heat budget in the fault's neighborhood. In the case of a simplified Coulomb-type friction model, the frictional contribution to the comprehensive dissipative heat budget of Enceladus's South Polar region has been estimated to be less than 1 GW.

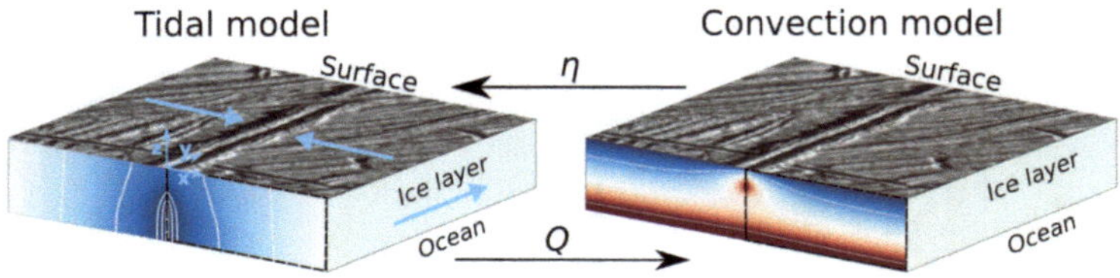

Fig. 11 Two-component model of tidal walking process at the strike-slip faults on Europa. The tidal model (left) solves for the (viscoelastic) strike-slip motions due to diurnal tides and the (viscous) convection model (right) computes the long-term thermal evolution. The coupling between the models occurs through the tidal heating rate and the viscosity. From Sládková et al. (2020)

As demonstrated by Pleiner Sládková et al. (2021), the inclusion of friction within a tidal deformation model applied to Enceladus's icy shell has unveiled an intriguing and unexpected dynamic effect, potentially carrying implications for local tectonics and surface morphology in the southern polar region. The effect stems from the inherent asymmetry of Coulomb-type friction, characterized by weaker frictional contact during the extensional phase of the tidal period and stronger contact during the compressive phase. Consequently, an initially hydrostatically pre-stressed shell with embedded Coulomb-type faults would reach after some time a novel dynamic "equilibrium" wherein the average stress conditions at the faults transition towards a more compressive state. This dynamic adjustment results in the development of predominantly compressive background stress within the SPR of Enceladus, reaching a magnitude comparable to that of the dynamic tidal stresses itself ($\sim$100 kPa). The possible implications of this background stress field have not been studied yet.

In studies conducted by Preblich et al. (2007) and Sládková et al. (2020), a similar mechanism involving Coulomb-type friction acting on tidally loaded faults in a viscoelastic shell was demonstrated to offer a self-consistent explanation for the concept of "tidal walking" (Fig. 11). This conceptual idea has been proposed by Hoppa et al. (1999) to account for observed lateral offsets on numerous lineaments interpreted as strike-slip faults on the surface of Europa (Schenk and McKinnon 1989). Through direct numerical modeling, it was confirmed that the asymmetry in frictional forces, in conjunction with viscoelastic relaxation occurring within the bulk material surrounding these faults, has the potential, under specific favorable conditions, to result in the gradual accumulation of irreversible slip along the faults. According to Sládková et al. (2020), the favorable conditions in question necessitate the fault's penetration from the surface either all the way to the internal ocean or to a sufficiently low-viscosity zone within the moon's icy shell. It was found that such conditions are unlikely under the present-day conditions on Europa and the present magnitude of tidal forcing. However, it cannot be ruled out that these conditions may have been realized during Europa's past history, particularly during periods of higher orbital eccentricity and, consequently, greater tidal forcing.

In summary, considering both the local and regional effects associated with the mechanical and thermal responses of faults within the outer shells of icy moons is undeniably crucial for obtaining a comprehensive understanding of their tidal behavior. Unfortunately, due to the substantial computational challenges involved in this type of modeling, there are currently only a limited number of studies that have addressed this complex phenomenon.

6.3 Tidal Dynamics in Water-Filled Cracks

Another notable instance of a tidally induced and highly localized process with significant potential implications for local thermal conditions is the phenomenon of tidal water plumbing within faults that traverse the entire shell, from the surface down to the internal ocean. This process involves periodic transitions between compressional and extensional conditions within a water-filled fault system, resulting in dynamic changes in the elevation of the water column. The movement of liquid water under idealized conditions, akin to those found in the Enceladean tiger stripe system, was investigated by Kite and Rubin (2016). Their research demonstrated that mechanical dissipation due to turbulent flow occurring within the water channel system itself can, under specific favorable circumstances, account for a significant portion (units of GW) of the endogenic heat flux $\sim$5 GW from the tiger stripes (Spencer et al. 2013). It is important to note that the authors considered a relatively thick shell in the SPR (30 $\pm$ 10 km), while more recent estimates (Beuthe et al. 2016; Čadek et al. 2016, 2019; Hemingway and Mittal 2019) suggest thinner shells (only a few km in the SPR). This would accordingly reduce the estimated heating contribution. Nevertheless, this mechanism remains a plausible component of the puzzle concerning the heat budget of the Enceladean SPR.

The water plumbing system proposed by Kite and Rubin (2016) assumed a direct communication with the underlying ocean and idealized geometry. In reality, the water system may be more tortuous and characterized by local enrichment in meltwater. A favorable "resonant" state, with peaking hydraulic dissipation, is more difficult to reach in the case of more complex geometry. However, even if a resonant state is not reached, the presence of meltwater, potentially produced locally under the action of enhanced tidal heating (Nimmo et al. 2007) or injected from the underlying ocean, may influence the mechanical properties of the crack system and thus further amplify tidal motions.

7 Geophysical Constraints on Tidal Processes by Future Exploration Missions

Tidal deformation and associated dissipation can be assessed using accurate gravimetric and altimetric measurements, as well as thermal emission in case of very strong and localized dissipation. So far, Titan is the only moon for which tidal response has been directly measured from the accurate determination of the time-varying gravity field (Iess et al. 2012; Durante et al. 2019). The inferred potential Love number, $k_2 = 0.616 \pm 0.067$, implies that Titan possesses an internal ocean under an ice shell whose thickness remains poorly constrained. As the tidal response depends on both the ice shell thickness and ocean density (Mitri et al. 2014) as well as possibly on the degree of ocean stratification (Idini and Nimmo 2024), it is only possible to provide a broad range of possible ice shell thickness ($\sim 50 - 100$ km) and ocean density (1150-1300 $\mathrm{kg\,m^{-3}}$). More recently, Goossens et al. (2024) re-evaluated the Love number to about 0.375, suggesting an ocean with a low density and under a thick ice shell. Future measurements are needed to be conclusive. Notably Dragonfly may provide additional constraints by monitoring very accurately the atmospheric pressure variations due to the gravitational tides (Charnay et al. 2022).

For the Galilean moons, in particular Ganymede and Europa, the precision of the observations to be performed by the ESA JUpiter ICy moons Explorer (Juice) mission (Grasset et al. 2013) and the NASA Europa Clipper mission (Howell and Pappalardo 2020) will be much better. The two missions will measure both the time-varying gravity field and surface deflection using radio tracking (Cappuccio et al. 2020; Mazarico et al. 2023) and altimetry, with laser technique for Juice (Steinbrügge et al. 2015) and radar technique for Europa Clipper (Steinbrügge et al. 2018). Europa Clipper will measure the tidal distortion by performing about 50 close flybys as low as 25 km (Roberts et al. 2023), while Juice will monitor the tidal response during a dedicated geophysical campaign on a low-altitude circular orbit (of at least 130 days at 500 km and possibly at 200 km during the extended mission) (Van Hoolst et al. 2024). The expected accuracy for the gravitational Love number, k_2, is ≤ 0.02 for Europa Clipper (Mazarico et al. 2023) and ≤ 0.0001 for Juice (Cappuccio et al. 2020). In the case of Ganymede, the imaginary part of k_2 will also be determined with a comparable accuracy by Juice, while for Europa it is still unclear if the precision will allow a quantification of the imaginary part of k_2 (Mazarico et al. 2023). Juice will also provide constraints on the gravitational Love number of Callisto with an accuracy comparable to Cassini at Titan (Cappuccio et al. 2022), which will permit the confirmation of the existence of a subsurface ocean.

The tidal response in terms of gravitational perturbations and surface deflections is at first order controlled by the ice shell thickness and the ocean density. The thermal structure of the ice shell which affects the viscosity and effective rigidity can also significantly affect the tidal response, as demonstrated in the case of Titan (Mitri et al. 2014). Even if the tidal response is measured very precisely, it is difficult to separate the effects of ice shell thickness, ocean density and thermal structure of the ice shell. By combining the Love numbers, k_2 and h_2, the effect of ocean density can be suppressed or at least significantly reduced, but still, the estimate of ice shell thickness will depend on the viscoelastic model assumption and on the assumed thermal profile in the ice shell. Additional geophysical measurements such as magnetic induction, radar sounding and gravity-topography ratio will be essential to constrain the ice shell thickness and the ocean properties (Van Hoolst et al. 2024; Roberts et al. 2023).

Owing to the very high precision of tidal monitoring by Juice at Ganymede, any subtle variation relative to the standard degree-two tidal response could be detected. For instance, Juice could detect even small changes in the pattern of tidal surface displacement and gravitational perturbations related to the non-zero obliquity of the satellite (Jara-Orué and Vermeersen 2016; Steinbrügge et al. 2023). Given the 1:1 spin-orbit resonance, the obliquity tides have the same period as the eccentricity tides and have to be considered in the determination of the Love numbers h_2 and k_2 from Juice data. Since both the orbital eccentricity and the obliquity can vary significantly (by 10% to almost 100%, Steinbrügge et al. 2023) over a few months, the amplitudes of the eccentricity and obliquity tides as well as their ratio will depend on time, and should be carefully monitored to correctly interpret the interior response. In addition to the main tidal interaction with Jupiter, the Galilean moons are subject to lower amplitude, time-dependent, tidal interactions with the other moons (Hay et al. 2020), resulting in multiple frequencies related to the Io-Europa-Ganymede-Callisto synodic frequencies. Even though much smaller than the main tidal response, these moon-moon tidal responses could be potentially detected at Ganymede using the 3GM radio science experiment onboard Juice (De Marchi et al. 2022), which may provide additional and independent constraints on the thickness of the ocean to within $\sim$10 km if one of the tides is resonant with a surface gravity wave of the ocean (De Marchi et al. 2022; Hay et al. 2022). Accurate tidal monitoring may also reveal lateral variations in ice shell thickness as well as the dynamical response of subsurface ocean and its coupling with the overlying ice shell (e.g. A et al. 2014; Beuthe 2016, 2018; Rovira-Navarro et al. 2023). Future modeling work dedicated to Ganymede taking into account the complexity of ocean-ice tidal coupling is required to anticipate the interpretation of the future very accurate data.

For Enceladus, the strong ice shell thickness variations and the fault system at the south pole make the tidal response much more complex (Běhounková et al. 2017; Souček et al. 2019; Marusiak et al. 2021; Berne et al. 2023). Detailed monitoring of tidal response would require an orbiter mission, similar to the GRAIL mission, performing dedicated gravity and altimetry mapping investigations (e.g. Ermakov et al. 2021; Genova et al. 2024). This would allow us to determine the time variations of the gravity field and surface deflection at degree two but also at higher degrees due to the complex response of Enceladus' uneven ice shell. The effect of the unconsolidated weak core may be also revealed by an amplification of the degree-two pattern in the equatorial region (Marusiak et al. 2021), which would provide quantification of the amount of energy involved in hydrothermal activity. The global tidal monitoring would also be complemented by regional mapping of thermal emission and fault dynamics using InSar and radar-penetrating techniques in the south polar terrain (e.g. Choblet et al. 2021). Monitoring of plume activity during the tidal cycle and the correlation of individual jet activity with the fault dynamics would also provide insights into the eruption dynamics and the efficiency and mode of heat and material transport from the ocean to the surface, essential for selecting landing sites for future astrobiological investigations (e.g. Choblet et al. 2021; MacKenzie et al. 2021).

8 Conclusion

In the present article, we have reviewed the different principles of tidal deformation and the mechanisms controlling the dissipation in icy worlds. With the exception of Io and Enceladus, we have no observational constraints on how much tidal heating is produced inside the interior of these moons. On Io, both thermal emission and astrometric data indicate a total dissipated power ranging between 65 and 125 TW (e.g. Spencer et al. 2000; Lainey et al. 2009). Such a strong dissipation and associated volcanic activity indicate a molten interior, however the depth and extent of the molten region still remains poorly constrained and debated (e.g. Hamilton et al. 2013; Tyler et al. 2015; Steinke et al. 2020; Kervazo et al. 2022; Davies et al. 2024). Detailed monitoring of Io's tidal response by a future dedicated mission will be required to provide more constraining insights on the dissipation processes.

For Enceladus, the total heat emitted from the active south polar region is estimated between 4 and 19 GW (Spencer et al. 2006; Howett et al. 2011; Spencer et al. 2018; Nimmo et al. 2023), but the total heat budget should also account for heat lost by conductive cooling through the ice shell outside the south polar region, estimated to 25-30 GW (e.g. Čadek et al. 2016). This leads to a total power ranging between 30 and 50 GW, more than 100 times larger than the expected radiogenic power. Modeling of tidal dissipation in the ice shell including highly deformable faults (Souček et al. 2019) clearly indicates that ice dissipation in the active south polar region cannot account for more than a few Gigawatts. Turbulent dissipation in the water-filled cracks could potentially account for a few more gigawatts if a resonant mode could be sustained (Kite and Rubin 2016), but it still remains insufficient to counterbalance the huge heat loss (> 25 GW) required to sustain a global ocean. Strong viscous dissipation in the ice shell, as proposed by Hemingway and Mittal (2019), will require a low viscosity ($< 10^{13} - 10^{14}$ Pa s) throughout the ice shell, which is clearly incompatible with the observed topography, requiring large variations in ice shell thickness and weak ice flow only possible for a high-viscosity conductive ice shell (Čadek et al. 2019).

For the present-day ocean/ice configuration as inferred from Cassini data, no oceanic dissipation model can explain the required heat production. Significant dissipation may occur only for thin ocean underneath a thick ice shell (e.g. Matsuyama 2014; Tyler 2014; Kamata et al. 2015), which may have occurred in the past, but which is far from the present-day configuration. The only identified dissipative mechanism that could produce at least 10 GW of power is dissipation in a highly deformable, weakly consolidated, water-filled porous rocky core (Choblet et al. 2017; Rovira-Navarro et al. 2022). The total power that could be produced is however highly dependent on the mechanical properties of such water-filled porous core, which remain poorly constrained. A variety of organic compounds and secondary minerals, such as clays, could significantly influence the dissipative properties of the core. It is likely that the chemical and mechanical properties of such a porous core evolve relatively rapidly on timescales of tens of million years owing to efficient water-rock alteration (e.g. Zandanel et al. 2021). So, even if strong dissipation in Saturn could maintain Enceladus in a dissipative state during long periods of time (Nimmo et al. 2023), it is unlikely that its interior properties remain constant over geological timescales.

The orbital configuration (eccentricity and semi-major axis) may have also significantly varied due to tidal expansion and orbital resonance encounters (e.g. Neveu and Rhoden 2019; Nimmo et al. 2023). Changes in orbital characteristics, interior structure and mechanical properties are strongly linked through tidal heating and tidally-induced water circulation. As shown recently by the evidence of a recently-formed ocean in Mimas (Lainey et al. 2024), the interior structure and total heat production can change on relatively brief periods of time (< 10 Myr), making the coupled thermo-orbital modeling of such mid-sized moons very challenging. Once the orbital eccentricity reaches a critical value, such mid-sized icy moons can go in a few million years from a frozen state to a fully molten state with a thick global ocean circulating in a porous rock core. Oceanic dissipation may also play a role at the beginning of the ocean growth, accelerating the melting of the overlying ice shell. Once the eccentricity is damped, tidal heating ceases and the interior progressively freezes to the initial state. If the analysis of Lainey et al. (2024) is correct, Mimas should have experienced such a melting event a few million years ago and should go back to a more dormant state in the coming few million years. Several other moons, such as Saturn's moon Dione or Uranus' moons Ariel and Miranda may have experienced similar melting/freezing events, which could explain evidence for past high heat flows (White et al. 2017; Nimmo 2023). Whether Enceladus experienced several cycles of melting/freezing is still an open issue. The recent activation of Mimas' internal dissipation suggests that, in the recent past, Enceladus may also have experienced a period with higher eccentricity and hence even higher heat production than at present (Noyelles et al. 2019). Future measurements including monitoring of degree-two tidal distortion as well as tidally-modulated deformation, eruption rate and thermal emission along the SPT faults will provide key information on the source of dissipation and how heat is released to the surface. Such measurements will require a dedicated polar orbiter. Future experimental works are required to quantify the dissipative function of weakly consolidated rocky materials and to better understand the role of low viscosity/fluid interstitial phase on the dissipation processes. A better understanding of the role of water in the mechanical properties of the fault zone is also required.

Europa has likely also experienced significant variations in ice shell thickness modulated by changes in eccentricity variations and hence in heat production (e.g. Hussmann and Spohn 2004). The geological record suggests that Europa should be now in a period of ice thickening (e.g. Figueredo and Greeley 2004; Doggett et al. 2009), even though the thickness and timing of the ice shell thickening is still uncertain. Depending on the ice shell thickness and eccentricity, different dissipation mechanisms may be predominant. During periods of a thin shell ($< 5 - 10$ km) and high eccentricity (at least twice the present-day value), dissipation along tectonic faults may be a dominant process, as the condition for propagating faults throughout the whole ice shell may be met, thus amplifying strike-slip motions (Sládková et al. 2020). Oceanic water injected in the cracks from the bottom of the ice shell may also add additional sources of dissipation if a resonant configuration similar to what has been proposed for Enceladus (Kite and Rubin 2016) is met. During periods of elevated eccentricity, tidal dissipation in the mantle may exceed the radiogenic power, thus resulting in internal melting and enhanced heat flux at high latitudes (Běhounková et al. 2021). This additional heat source favors the thinning of the ice shell, which should be even more pronounced in high latitudes under the joint contribution of dissipation in the mantle and in the ice shell. Ocean dissipation is not expected to play a key role during periods of thin shell/thick ocean.

During periods of thick shell and reduced eccentricity, dissipation is expected to be dominated by viscous dissipation in the convective part of the ice shell. For present-day eccentricity, such a dissipative process is expected to be sufficient to stop the ocean freezing when the ice shell thickness reaches a value of 20-40 km (e.g. Tobie et al. 2003; Howell 2021). During periods of lower eccentricity, much thicker ice shell may be reached, resulting in thin oceans potentially highly concentrated in salts. When such configuration is met, dissipation in the remaining oceanic layer could become a dominant process (e.g. Matsuyama 2014; Tyler 2014), preventing the ocean from full freezing even during periods of very low eccentricity. The amplitude and periodicity of eccentricity change on Europa are still poorly constrained. Modeling such eccentricity changes and the consequences on the ice shell state and thickness requires a better understanding of the coupling with Io and Ganymede via the Laplace resonance. Future measurements by Europa Clipper and Juice missions will provide some direct constraints on the strength of this coupling at present through astrometric measurements (e.g. Dirkx et al. 2017) as well as constraints on the tidal response of Europa (Mazarico et al. 2023; Steinbrügge et al. 2018), essential to determine its hydrosphere structure and to give insights on the dissipation processes. The interpretation of these measurements requires, however, experimental data on the viscoelastic response of ice and ice-salt mixture in the appropriate frequency regime.

For Ganymede and Callisto, tidal dissipation has a negligible contribution to their heat budget at present. However, tidal monitoring by the Juice mission will provide key constraints on their hydrosphere structure (Van Hoolst et al. 2024). For Callisto, the expected accuracy (Cappuccio et al. 2022) will not be sufficient to provide precise constraints on the ice shell thickness and ocean density, but it will be comparable to what has been achieved at Titan with Cassini (Durante et al. 2019). For Ganymede, very subtle time variations in the gravity field (Cappuccio et al. 2020) and surface deflections (Steinbrügge et al. 2015, 2023) will be measured, offering unprecedented insights on the tidal response of an ice-covered ocean world. Such measurements will potentially reveal resonant modes in the ocean excited by moon-moon interactions (De Marchi et al. 2022) or related to ocean stratification (e.g. Idini and Nimmo 2024). Future model developments are needed to carefully predict the oceanic response and its coupling with the overlying ice shell and deep structure. Even if tidal heating is negligible at present, it might have significantly contributed in the past during the initial tidal despinning to their tidally-locked spin-orbit resonances (Tobie et al. 2014; Journaux et al. 2020) and during periods of elevated eccentricities (Bland et al. 2009). The tidal despinning which should occur on brief periods of time shortly after accretion could potentially rise the internal temperature by 25-50 K (Tobie et al. 2014), thus favoring the initiation of ice melting in the undifferentiated proto-core. Depending on how Ganymede enters the Laplace resonance, it may have experienced prolonged periods of elevated eccentricity (up to 10 times the current value), leading to tidally dissipated power of the order of 1 TW comparable to the radiogenic power (Bland et al. 2009). Geological and compositional mapping by Juice will provide insights into the change in tectonic regime and crater relaxation, thus constraining any past heat flux and changes in ice shell thickness associated with enhanced tidal heating.

On Titan, the contribution of tidal heating at present is more uncertain. It has long been assumed that tidal dissipation should be small at present in order to explain its elevated eccentricity (Sagan and Dermott 1982; Sears 1995; Sohl et al. 1995; Tobie et al. 2005b). The indication of rapid orbital expansion of Titan due to strong dissipation in Saturn (Lainey et al. 2020) suggests that its eccentricity may decay more slowly that previously estimated. In this case, tidal heating may still be a significant, if not dominant, source of internal heating. Even though Titan is the first icy moon for which the tidal Love number, k_2, has been inferred from gravity measurements (Iess et al. 2012; Durante et al. 2019), little is known on its tidal response as the measurement uncertainty still remains too large to be conclusive. The apparently elevated Love number could be explained either a moderately to very salty ocean (Mitri et al. 2014), or even by a low-density ocean if oceanic resonance amplifies the tidal response (Idini and Nimmo 2024). Future measurements by Dragonfly may shed new light on the tidal deformation and dissipation process. Charnay et al. (2022) showed that the DragMet pressure sensor may be accurate enough to detect gravitational atmospheric tides, thus providing new insights on the interior response, independent from Cassini measurements. Interpretation of the detailed morphology of the crater Selk and of related tectonic features could provide information on the flexural and relaxation processes, providing indirect constraints on the heat flux history and hence on internal heating. Again, further modeling and experimental efforts are needed to anticipate the interpretation of these future measurements.

Acknowledgements Thoughtful comments from Francis Nimmo and an anonymous reviewer helped improve the manuscript.

Funding Open access publishing supported by the institutions participating in the CzechELib Transformative Agreement. G.T. and M. K. acknowledge support from the ANR OASIS project and from CNES for the preparation to the ESA Juice and NASA Europa Clipper mission. G.T. also acknowledges support from the ANR COLOSSe project. M.B., K.K and O.S. acknowledge support by the Czech Science Foundation grant no. 22-20388S. This work has been co-funded by the European Union (ERC, PROMISES, project #101054470). Views and opinions expressed are however those of the authors only and do not necessarily reflect those of the European Union or the European Research Council. Neither the European Union nor the granting authority can be held responsible for them.

Declarations

Competing Interests The authors declare that they have no competing financial or non-financial interests to declare that are relevant to the content of this article.

References

A G, Wahr J, Zhong S (2014) The effects of laterally varying icy shell structure on the tidal response of Ganymede and Europa. J Geophys Res, Planets 119(3):659–678

Adams CJ, Iverson NR, Helanow C, Zoet LK, Bate CE (2021) Softening of temperate ice by interstitial water. Front Earth Sci 9:702761

Alterman Z, Jarosch H, Pekeris C (1959) Oscillations of the Earth. Proc R Soc Lond Ser A, Math Phys Sci 252(1268):80–95

Amit H, Choblet G, Tobie G, Terra-Nova F, Čadek O, Bouffard M (2020) Cooling patterns in rotating thin spherical shells—application to Titan's subsurface ocean. Icarus 338:113509

Amorim DO, Gudkova T (2024) Constraining Earth's mantle rheology with Love and Shida numbers at the M_2 tidal frequency. Phys Earth Planet Inter 347:107144

Andrade ENDC (1910) On the viscous flow in metals, and allied phenomena. Proc R Soc Lond Ser A, Contain Pap Math Phys Character 84(567):1–12

Arzi AA (1978) Critical phenomena in the rheology of partially melted rocks. Tectonophysics 44(1–4):173–184

Ashkenazy Y, Tziperman E (2021) Dynamic Europa ocean shows transient Taylor columns and convection driven by ice melting and salinity. Nat Commun 12(1):6376

Auclair-Desrotour P, Mathis S, Laskar J, Leconte J (2018) Oceanic tides from Earth-like to ocean planets. Astron Astrophys 615:A23. https://doi.org/10.1051/0004-6361/201732249

Auclair-Desrotour P, Leconte J, Bolmont E, Mathis S (2019) Final spin states of eccentric ocean planets. Astron Astrophys 629:A132. https://doi.org/10.1051/0004-6361/201935905

Aygün B, Čadek O (2023) Impact of the core deformation on the tidal heating and flow in Enceladus' subsurface ocean. Authorea Preprints

Bagdassarov N, Dingwell DB (1993) Frequency dependent rheology of vesicular rhyolite. J Geophys Res, Solid Earth 98(B4):6477–6487

Bagheri A, Khan A, Al-Attar D, Crawford O, Giardini D (2019) Tidal response of Mars constrained from laboratory-based viscoelastic dissipation models and geophysical data. J Geophys Res, Planets 124(11):2703–2727

Bagheri A, Efroimsky M, Castillo-Rogez J, Goossens S, Plesa A-C, Rambaux N, Rhoden A, Walterová M, Khan A, Giardini D (2022a) Tidal insights into rocky and icy bodies: an introduction and overview. In: Schmelzbach C, Stahler SC (eds) Geophysical exploration of the Solar System. Advances in geophysics, vol 63. Elsevier, Amsterdam, pp 231–320

Bagheri A, Khan A, Deschamps F, Samuel H, Kruglyakov M, Giardini D (2022b) The tidal–thermal evolution of the Pluto-Charon system. Icarus 376:114871

Beddingfield CB, Cartwright RJ, Leonard E, Nordheim T, Scipioni F (2022a) Ariel's elastic thicknesses and heat fluxes. Planet Sci J 3(5):106

Beddingfield CB, Leonard E, Cartwright RJ, Elder C, Nordheim TA (2022b) High heat flux near Miranda's inverness corona consistent with a geologically recent heating event. Planet Sci J 3(7):174

Behounková M, Cadek O (2014) Tidal dissipation in heterogeneous bodies: Maxwell vs Andrade rheology. In: European Planetary Science Congress, vol 9, EPSC2014–409

Běhounková M, Tobie G, Choblet G, Čadek O (2010) Coupling mantle convection and tidal dissipation: applications to Enceladus and Earth-like planets. J Geophys Res, Planets 115(E9):E09011

Běhounková M, Tobie G, Choblet G, Čadek O (2012) Tidally-induced melting events as the origin of south-pole activity on Enceladus. Icarus 219(2):655–664. https://doi.org/10.1016/j.icarus.2012.03.024

Běhounková M, Tobie G, Choblet G, Čadek O (2013) Impact of tidal heating on the onset of convection in Enceladus's ice shell. Icarus 226(1):898–904

Běhounková M, Tobie G, Čadek O, Choblet G, Porco C, Nimmo F (2015) Timing of water plume eruptions on Enceladus explained by interior viscosity structure. Nat Geosci 8:601–604. https://doi.org/10.1038/ngeo2475

Běhounková M, Souček O, Hron J, Čadek O (2017) Plume activity and tidal deformation on Enceladus influenced by faults and variable ice shell thickness. Astrobiology 17:941–954. https://doi.org/10.1089/ast.2016.1629

Běhounková M, Tobie G, Choblet G, Kervazo M, Melwani Daswani M, Dumoulin C, Vance SD (2021) Tidally induced magmatic pulses on the Oceanic floor of Jupiter's moon Europa. Geophys Res Lett 48(3):e2020GL090077. https://doi.org/10.1029/2020GL090077

Benjamin D, Wahr J, Ray RD, Egbert GD, Desai SD (2006) Constraints on mantle anelasticity from geodetic observations, and implications for the J_2 anomaly. Geophys J Int 165(1):3–16

Berckhemer H, Kampfmann W, Aulbach E, Schmeling H (1982) Shear modulus and Q of forsterite and dunite near partial melting from forced-oscillation experiments. Phys Earth Planet Inter 29(1):30–41

Berne A, Simons M, Keane JT, Park RS (2023) Inferring the mean thickness of the outer ice shell of Enceladus from diurnal crustal deformation. J Geophys Res, Planets 128(6):e2022JE007712. https://doi.org/10.1029/2022JE007712

Berne A, Simons M, Keane JT, et al (2024) Jet activity on Enceladus linked to tidally driven strike-slip motion along tiger stripes. Nat Geosci 17:385–391. https://doi.org/10.1038/s41561-024-01418-0

Beuthe M (2008) Thin elastic shells with variable thickness for lithospheric flexure of one-plate planets. Geophys J Int 172(2):817–841. https://doi.org/10.1111/j.1365-246X.2007.03671.x
Beuthe M (2013) Spatial patterns of tidal heating. Icarus 223(1):308–329. https://doi.org/10.1016/j.icarus.2012.11.020
Beuthe M (2015a) Tides on Europa: the membrane paradigm. Icarus 248:109–134. https://doi.org/10.1016/j.icarus.2014.10.027
Beuthe M (2015b) Tidal Love numbers of membrane worlds: Europa, Titan, and co. Icarus 258:239–266. https://doi.org/10.1016/j.icarus.2015.06.008
Beuthe M (2016) Crustal control of dissipative ocean tides in Enceladus and other icy moons. Icarus 280:278–299. https://doi.org/10.1016/j.icarus.2016.08.009
Beuthe M (2018) Enceladus's crust as a non-uniform thin shell: I tidal deformations. Icarus 302:145–174
Beuthe M (2019) Enceladus's crust as a non-uniform thin shell: II tidal dissipation. Icarus 332:66–91
Beuthe M, Rivoldini A, Trinh A (2016) Enceladus's and Dione's floating ice shells supported by minimum stress isostasy. Geophys Res Lett 43(19):10088–10096. https://doi.org/10.1002/2016GL070650
Bierson CJ (2024) The impact of rheology model choices on tidal heating studies. Icarus 414:116026
Bierson C, Nimmo F (2016) A test for Io's magma ocean: modeling tidal dissipation with a partially molten mantle. J Geophys Res, Planets 121:2211–2224
Biot MA (1954) Theory of stress-strain relations in anisotropic viscoelasticity and relaxation phenomena. J Appl Phys 25(11):1385–1391
Biot MA (1956) Theory of propagation of elastic waves in a fluid-saturated porous solid. II. Higher frequency range. J Acoust Soc Am 28(2):179–191
Bire S, Kang W, Ramadhan A, Campin J-M, Marshall J (2022) Exploring ocean circulation on icy moons heated from below. J Geophys Res, Planets 127(3):e2021JE007025
Bland MT, Showman AP, Tobie G (2009) The orbital–thermal evolution and global expansion of Ganymede. Icarus 200(1):207–221. https://doi.org/10.1016/j.icarus.2008.11.016
Bolmont E, Breton SN, Tobie G, Dumoulin C, Mathis S, Grasset O (2020) Solid tidal friction in multi-layer planets: application to Earth, Venus, a super Earth and the TRAPPIST-1 planets. Potential approximation of a multi-layer planet as a homogeneous body. Astron Astrophys 644:A165. https://doi.org/10.1051/0004-6361/202038204
Brennan A, Thusyanthan N, Madabhushi S (2005) Evaluation of shear modulus and damping in dynamic centrifuge tests. J Geotech Geoenviron Eng 131(12):1488–1497
Budiansky B, O'Connell RJ (1976) Elastic moduli of a cracked solid. Int J Solids Struct 12(2):81–97
Čadek O, Tobie G, van Hoolst T, Masse M, Choblet G, Lefevre A, Mitri G, Baland R-M, Běhounková M, Bourgeois O, Trinh A (2016) Enceladus's internal ocean and ice shell constrained from Cassini gravity, shape, and libration data. Geophys Res Lett 46:5653–5660. https://doi.org/10.1002/2016GL068634
Čadek O, Souček O, Běhounková M, Choblet G, Tobie G, Hron J (2019) Long-term stability of Enceladus' uneven ice shell. Icarus 319(2018):476–484. https://doi.org/10.1016/j.icarus.2018.10.003
Cappuccio P, Hickey A, Durante D, Di Benedetto M, Iess L, De Marchi F, Plainaki C, Milillo A, Mura A (2020) Ganymede's gravity, tides and rotational state from JUICE's 3GM experiment simulation. Planet Space Sci 187:104902
Cappuccio P, Di Benedetto M, Durante D, Iess L (2022) Callisto and Europa gravity measurements from JUICE 3GM experiment simulation. Planet Sci J 3(8):199
Cassen P, Reynolds RT, Peale SJ (1979) Is there liquid water on Europa? Geophys Res Lett 6(9):731–734. https://doi.org/10.1029/GL006i009p00731
Castillo-Rogez JC, Efroimsky M, Lainey V (2011) The tidal history of iapetus: spin dynamics in the light of a refined dissipation model. J Geophys Res, Planets 116(E9):E09008
Castillo-Rogez J, Hemingway D, Rhoden A, Tobie G, McKinnon W (2018) Origin and evolution of Saturn's mid-sized moons. In: Schenk PM (eds) Enceladus and the icy moons of Saturn. University of Arizona Press, Tucson, pp 285–306
Charnay B, Tobie G, Lebonnois S, Lorenz RD (2022) Gravitational atmospheric tides as a probe of Titan's interior: application to dragonfly. Astron Astrophys 658:A108
Chen EMA, Nimmo F (2011) Obliquity tides do not significantly heat Enceladus. Icarus 214(2):779–781. https://doi.org/10.1016/j.icarus.2011.06.007
Chen EMA, Nimmo F (2016) Tidal dissipation in the lunar magma ocean and its effect on the early evolution of the Earth-Moon system. Icarus 275:132–142. https://doi.org/10.1016/j.icarus.2016.04.012
Chen EMA, Nimmo F, Glatzmaier GA (2014) Tidal heating in icy satellite oceans. Icarus 229:11–30. https://doi.org/10.1016/j.icarus.2013.10.024
Choblet G, Tobie G, Sotin C, Běhounková M, Čadek O, Postberg F, Souček O (2017) Powering prolonged hydrothermal activity inside Enceladus. Nat Astron 1(12):841–847
Choblet G, Tobie G, Buch A, Čadek O, Barge LM, Běhounková M, Camprubi E, Freissinet C, Hedman M, Jones G, et al (2021) Enceladus as a potential oasis for life: Science goals and investigations for future explorations. Exp Astron, 1–39

Čížková H, van den Berg AP, Spakman W, Matyska C (2012) The viscosity of Earth's lower mantle inferred from sinking speed of subducted lithosphere. Phys Earth Planet Inter 200:56–62

Cole DM (1998) Modeling the cyclic loading response of sea ice. Int J Solids Struct 35(31–32):4067–4075

Cole DM, Durell GD (1995) The cyclic loading of saline ice. Philos Mag A 72(1):209–229

Costa A, Caricchi L, Bagdassarov N (2009) A model for the rheology of particle-bearing suspensions and partially Molten rocks. Geochem Geophys Geosyst 10(3)

Ćuk M, El Moutamid M (2022) Three-body resonances in the Saturnian system. Astrophys J Lett 926(2):L18

Ćuk M, El Moutamid M, Lari G, Neveu M, Nimmo F, Noyelles B, Rhoden A, Saillenfest M (2024) Long-term evolution of the Saturnian system. Space Sci Rev 220:20. https://doi.org/10.1007/s11214-024-01049-2

Daher H, Arbic BK, Williams JG, Ansong JK, Boggs DH, Müller M, Schindelegger M, Austermann J, Cornuelle BD, Crawford EB, Fringer OB, Lau HCP, Lock SJ, Maloof AC, Menemenlis D, Mitrovica JX, Green JAM, Huber M (2021) Long-term Earth-Moon evolution with high-level orbit and ocean tide models. J Geophys Res, Planets 126(12):e06875. https://doi.org/10.1029/2021JE006875

Davies AG, Perry JE, Williams DA, Nelson DM (2024) Io's polar volcanic thermal emission indicative of magma ocean and shallow tidal heating models. Nat Astron 8(1):94–100

de Kleer K, Nimmo F, Kite E (2019) Variability in Io's volcanism on timescales of periodic orbital changes. Geophys Res Lett 46(12):6327–6332

De La Chapelle S, Milsch H, Castelnau O, Duval P (1999) Compressive creep of ice containing a liquid intergranular phase: rate-controlling processes in the dislocation creep regime. Geophys Res Lett 26(2):251–254

De Marchi F, Di Achille G, Mitri G, Cappuccio P, Di Stefano I, Di Benedetto M, Iess L (2021) Observability of Ganymede's gravity anomalies related to surface features by the 3GM experiment onboard ESA's JUpiter ICy moons Explorer (JUICE) mission. Icarus 354:114003. https://doi.org/10.1016/j.icarus.2020.114003

De Marchi F, Cappuccio P, Mitri G, Iess L (2022) Frequency-dependent Ganymede's tidal Love number k_2 detection by JUICE's 3GM experiment and implications for the subsurface ocean thickness. Icarus 386:115150. https://doi.org/10.1016/j.icarus.2022.115150

Denny KE, Hedman MM, Bockelée-Morvan D, Filacchione G, Capaccioni F (2024) Constraining time variations in Enceladus's water-vapor plume with near-infrared spectra from Cassini's visual and infrared mapping spectrometer. Planet Sci J 5(6):144. https://doi.org/10.3847/PSJ/ad4c69

Ding H, Chen Z, Pan Y, Zou C (2021) The complex Love numbers of long-period zonal tides retrieved from global GPS displacements: applications for determining mantle anelasticity. J Geophys Res, Solid Earth 126(9):e2021JB022380

Dingwell DB, Webb SL (1989) Structural relaxation in silicate melts and non-Newtonian melt rheology in geologic processes. Phys Chem Miner 16:508–516

Dirkx D, Gurvits L, Lainey V, Lari G, Milani A, Cimò G, Bocanegra-Bahamon T, Visser P (2017) On the contribution of PRIDE-JUICE to Jovian system ephemerides. Planet Space Sci 147:14–27. https://doi.org/10.1016/j.pss.2017.09.004

Doggett T, Greeley R, Figueredo P, Tanaka K (2009) Geologic stratigraphy and evolution of Europa's surface. In: Pappalardo RT, McKinnon WB, Khurana KK (eds) Europa. University of Arizona Press, Tucson, pp 137–160

Dombard AJ, Sessa AM (2019) Gravity measurements are key in addressing the habitability of a subsurface ocean in Jupiter's moon Europa. Icarus 325:31–38. https://doi.org/10.1016/j.icarus.2019.02.025

Dougherty M, Khurana K, Neubauer F, Russell C, Saur J, Leisner J, Burton M (2006) Identification of a dynamic atmosphere at Enceladus with the Cassini magnetometer. Science 311(5766):1406–1409

Durante D, Hemingway D, Racioppa P, Iess L, Stevenson D (2019) Titan's gravity field and interior structure after Cassini. Icarus 326:123–132

Durek JJ, Ekström G (1996) A radial model of anelasticity consistent with long-period surface-wave attenuation. Bull Seismol Soc Am 86(1A):144–158

Efroimsky M (2012) Tidal dissipation compared to seismic dissipation: in small bodies, Earths, and super-Earths. Astrophys J 746(2):150

Efroimsky M, Makarov VV (2014) Tidal dissipation in a homogeneous spherical body. I. Methods. Astrophys J 795(1):6

Egbert GD, Ray RD (2001) Estimates of M_2 tidal energy dissipation from TOPEX/Poseidon altimeter data. J Geophys Res 106(C10):22475–22502. https://doi.org/10.1029/2000JC000699

Egbert GD, Bennett AF, Foreman MGG (1994) TOPEX/POSEIDON tides estimated using a global inverse model. J Geophys Res 99(C12):24821–24852. https://doi.org/10.1029/94JC01894

Egbert GD, Ray RD, Bills BG (2004) Numerical modeling of the global semidiurnal tide in the present day and in the last glacial maximum. J Geophys Res, Oceans 109(C3):C03003. https://doi.org/10.1029/2003JC001973

Ermakov AI, Park RS, Roa J, Castillo-Rogez JC, Keane JT, Nimmo F, Kite ES, Sotin C, Lazio TJW, Steinbrügge G, et al (2021) A recipe for the geophysical exploration of Enceladus. Planet Sci J 2(4):157
Farhat M, Auclair-Desrotour P, Boué G, Laskar J (2022) The resonant tidal evolution of the Earth-Moon distance. Astron Astrophys 665:L1. https://doi.org/10.1051/0004-6361/202243445
Faul UH, Jackson I (2005) The seismological signature of temperature and grain size variations in the upper mantle. Earth Planet Sci Lett 234(1–2):119–134
Faul U, Jackson I (2015) Transient creep and strain energy dissipation: an experimental perspective. Annu Rev Earth Planet Sci 43:541–569
Faul UH, Fitz Gerald JD, Jackson I (2004) Shear wave attenuation and dispersion in melt-bearing olivine polycrystals: 2. Microstructural interpretation and seismological implications. J Geophys Res, Solid Earth 109(B6):B06202
Figueredo PH, Greeley R (2004) Resurfacing history of Europa from pole-to-pole geological mapping Icarus 167:287–312. https://doi.org/10.1016/j.icarus.2003.09.016
Gaeman J, Hier-Majumder S, Roberts JH (2012) Sustainability of a subsurface ocean within Triton's interior. Icarus 220(2):339–347
Genova A, Parisi M, Gargiulo AM, Petricca F, Andolfo S, Torrini T, Del Vecchio E, Glein CR, Cable ML, Phillips CB, et al (2024) Gravity investigation to characterize Enceladus's ocean and interior. Planet Sci J 5(2):40
Gerkema T, Shrira VI (2005) Near-inertial waves in the ocean: beyond the "traditional approximation". J Fluid Mech 529:195–219. https://doi.org/10.1017/S0022112005003411
Gerkema T, Zimmerman J (2008) An introduction to internal waves. Lecture notes. Royal NIOZ, Texel
Gerkema T, Zimmerman JTF, Maas LRM, van Haren H (2008) Geophysical and astrophysical fluid dynamics beyond the traditional approximation. Rev Geophys 46(2):RG2004. https://doi.org/10.1029/2006RG000220
Goldsby D, Kohlstedt DL (2001) Superplastic deformation of ice: experimental observations. J Geophys Res, Solid Earth 106(B6):11017–11030
Goossens S, van Noort B, Mateo A, Mazarico E, van der Wal W (2024) A low-density ocean inside Titan inferred from Cassini data. Nat Astron, 1–10
Grasset O, Dougherty M, Coustenis A, Bunce E, Erd C, Titov D, Blanc M, Coates A, Drossart P, Fletcher L, et al (2013) JUpiter ICy moons Explorer (JUICE): an ESA mission to orbit Ganymede and to characterise the Jupiter system. Planet Space Sci 78:1–21
Green JAM, Huber M, Waltham D, Buzan J, Wells M (2017) Explicitly modelled deep-time tidal dissipation and its implication for lunar history. Earth Planet Sci Lett 461:46–53. https://doi.org/10.1016/j.epsl.2016.12.038
Hamilton CW, Beggan CD, Still S, Beuthe M, Lopes RM, Williams DA, Radebaugh J, Wright W (2013) Spatial distribution of volcanoes on Io: implications for tidal heating and magma ascent. Earth Planet Sci Lett 361:272–286
Hansen CJ, Esposito L, Stewart A, Colwell J, Hendrix A, Pryor W, Shemansky D, West R (2006) Enceladus' water vapor plume. Science 311(5766):1422–1425
Hansen C, Esposito L, Colwell J, Hendrix A, Portyankina G, Stewart A, West R (2020) The composition and structure of Enceladus' plume from the complete set of Cassini UVIS occultation observations. Icarus 344:113461
Hartkorn O, Saur J (2017) Induction signals from Callisto's ionosphere and their implications on a possible subsurface ocean. J Geophys Res Space Phys 122(11):11677–11697. https://doi.org/10.1002/2017JA024269
Hay HCFC, Matsuyama I (2017) Numerically modelling tidal dissipation with bottom drag in the oceans of Titan and Enceladus. Icarus 281:342–356. https://doi.org/10.1016/j.icarus.2016.09.022
Hay HCFC, Matsuyama I (2019) Nonlinear tidal dissipation in the subsurface oceans of Enceladus and other icy satellites. Icarus 319:68–85. https://doi.org/10.1016/j.icarus.2018.09.019
Hay HCFC, Trinh A, Matsuyama I (2020) Powering the Galilean satellites with moon-moon tides. Geophys Res Lett 47(15):e88317. https://doi.org/10.1029/2020GL088317
Hay HCFC, Matsuyama I, Pappalardo RT (2022) The high-frequency tidal response of ocean worlds: application to Europa and Ganymede. J Geophys Res, Planets 127(5):e07064. https://doi.org/10.1029/2021JE007064
Hedman MM, Gosmeyer CM, Nicholson PD, Sotin C, Brown RH, Clark RN, Baines KH, Buratti BJ, Showalter MR (2013) An observed correlation between plume activity and tidal stresses on Enceladus. Nature 500(7461):182–184. https://doi.org/10.1038/nature12371
Hemingway DJ, Mittal T (2019) Enceladus's ice shell structure as a window on internal heat production. Icarus 332:111–131. https://doi.org/10.1016/j.icarus.2019.03.011
Hendershott MC (1972) The effects of solid Earth deformation on global ocean tides. Geophys J 29(4):389–402. https://doi.org/10.1111/j.1365-246X.1972.tb06167.x

Hendershott MC (1981) Long waves and ocean tides. In: Warren BA, Wunsch C (eds) Evolution of physical oceanography. MIT Press, Cambridge, pp 292–341

Henning WG, Hurford T (2014) Tidal heating in multilayered terrestrial exoplanets. Astrophys J 789(1):30

Henning WG, O'Connell RJ, Sasselov DD (2009) Tidally heated terrestrial exoplanets: viscoelastic response models. Astrophys J 707(2):1000

Hirth G, Kohlstedt DL (1995a) Experimental constraints on the dynamics of the partially molten upper mantle: deformation in the diffusion creep regime. J Geophys Res, Solid Earth 100(B2):1981–2001

Hirth G, Kohlstedt DL (1995b) Experimental constraints on the dynamics of the partially molten upper mantle: 2. Deformation in the dislocation creep regime. J Geophys Res, Solid Earth 100(B8):15441–15449

Hoppa G, Tufts BR, Greenberg R, Geissler P (1999) Strike-slip faults on Europa: global shear patterns driven by tidal stress. Icarus 141(2):287–298. https://doi.org/10.1006/icar.1999.6185

Hough SS (1898) On the application of harmonic analysis to the dynamical theory of the tides. Part II: on the general integration of Laplace's dynamical equations. Philos Trans R Soc Lond Ser A 191:139–185 https://doi.org/10.1098/rsta.1898.0005

Howell SM (2021) The likely thickness of Europa's icy shell. Planet Sci J 2(4):129

Howell SM, Pappalardo RT (2020) Nasa's Europa Clipper—a mission to a potentially habitable ocean world. Nat Commun 11(1):1311

Howett CJA, Spencer JR, Pearl J, Segura M (2011) High heat flow from Enceladus' south polar region measured using 10-600 cm^{-1} Cassini/CIRS data. J Geophys Res, Planets 116(E3):E03003. https://doi.org/10.1029/2010JE003718

Howett C, Miles G, Quick LC (2025) Constraining the mass and energy of dissipation systems. Space Sci Rev 221

Hulbe CL, Klinger M, Masterson M, Catania G, Cruikshank K, Bugni A (2016) Tidal bending and strand cracks at the kamb ice stream grounding line, West Antarctica. J Glaciol 62(235):816–824

Hurford TA, Helfenstein P, Hoppa GV, Greenberg R, Bills BG (2007) Eruptions arising from tidally controlled periodic openings of rifts on Enceladus. Nature 447(7142):292–294. https://doi.org/10.1038/nature05821

Hussmann H, Spohn T (2004) Thermal-orbital evolution of Io and Europa. Icarus 171:391–410. https://doi.org/10.1016/j.icarus.2004.05.020

Hussmann H, Choblet G, Lainey V, Matson DL, Sotin C, Tobie G, van Hoolst T (2010) Implications of rotation, orbital states, energy sources, and heat transport for internal processes in icy satellites. Space Sci Rev 153:317–348. https://doi.org/10.1007/s11214-010-9636-0

Hussmann H, Rodríguez A, Callegari N Jr, Shoji D (2019) Early resonances of Tethys and Dione: implications for Ithaca Chasma. Icarus 319:407–416

Idini B, Nimmo F (2024) Resonant stratification in Titan's global ocean. Planet Sci J 5:15. https://doi.org/10.3847/PSJ/ad11ef

Iess L, Jacobson RA, Ducci M, Stevenson DJ, Lunine JI, Armstrong JW, Asmar SW, Racioppa P, Rappaport NJ, Tortora P (2012) The tides of Titan. Science 337(6093):457. https://doi.org/10.1126/science.1219631

Iess L, Stevenson D, Parisi M, Hemingway D, Jacobson R, Lunine J, Nimmo F, Armstrong J, Asmar S, Ducci M, et al (2014) The gravity field and interior structure of Enceladus. Science 344(6179):78–80

Ingersoll AP, Ewald SP, Trumbo SK (2020) Time variability of the Enceladus plumes: orbital periods, decadal periods, and aperiodic change. Icarus 344:113345. https://doi.org/10.1016/j.icarus.2019.06.006

Ishibashi I, Zhang X (1993) Unified dynamic shear moduli and damping ratios of sand and clay. Soil Found 33(1):182–191

Jackson I (2000) Laboratory measurement of seismic wave dispersion and attenuation: recent progress. In: Karato S-I et al (eds) Earth's deep interior: mineral physics and tomography from the atomic to the global scale. Geophysical Monograph, vol 117. AGU, Washington DC, pp 265–289. https://doi.org/10.1029/GM117p0265

Jackson I, Fitz Gerald JD, Faul UH, Tan BH (2002) Grain-size-sensitive seismic wave attenuation in polycrystalline olivine. J Geophys Res, Solid Earth 107(B12):ECV-5

Jacobson RA (2022) The orbits of the main Saturnian satellites, the Saturnian system gravity field, and the orientation of Saturn's pole. Astron J 164(5):199

Jara-Orué H, Vermeersen B (2016) Tides on Jupiter's moon Ganymede and their relation to its internal structure. Geol Mijnb 95(2):191–201

Jayne SR, St. Laurent LC (2001) Parameterizing tidal dissipation over rough topography. Geophys Res Lett 28(5):811–814. https://doi.org/10.1029/2000GL012044

Journaux B, Kalousová K, Sotin C, Tobie G, Vance S, Saur J, Bollengier O, Noack L, Rückriemen-Bez T, Van Hoolst T, et al (2020) Large ocean worlds with high-pressure ices. Space Sci Rev 216:7. https://doi.org/10.1007/s11214-019-0633-7

Kalousová K, Sotin C (2018) Melting in high-pressure ice layers of large ocean worlds—implications for volatiles transport. Geophys Res Lett 45(16):8096–8103
Kalousová K, Sotin C (2020) Dynamics of Titan's high-pressure ice layer. Earth Planet Sci Lett 545:116416
Kalousová K, Souček O, Tobie G, Choblet G, Čadek O (2016) Water generation and transport below Europa's strike-slip faults. J Geophys Res, Planets 121(12):2444–2462
Kalousová K, Sotin C, Choblet G, Tobie G, Grasset O (2018) Two-phase convection in Ganymede's high-pressure ice layer—implications for its geological evolution. Icarus 299:133–147
Kamata S (2023) Poroviscoelastic gravitational dynamics. J Geophys Res, Planets 128(7):e2022JE007700
Kamata S, Matsuyama I, Nimmo F (2015) Tidal resonance in icy satellites with subsurface oceans. J Geophys Res, Planets 120(9):1528–1542. https://doi.org/10.1002/2015JE004821
Kang W, Bire S, Marshall J (2022a) The role of ocean circulation in driving hemispheric symmetry breaking of the ice shell of Enceladus. Earth Planet Sci Lett 599:117845
Kang W, Mittal T, Bire S, Campin J-M, Marshall J (2022b) How does salinity shape ocean circulation and ice geometry on Enceladus and other icy satellites? Sci Adv 8(29):eabm4665
Karato S-i (2010) Rheology of the Earth's mantle: a historical review. Gondwana Res 18(1):17–45
Karato S, Spetzler HA (1990) Defect microdynamics in minerals and solid state mechanisms of seismic wave attenuation and velocity dispersion in the mantle. Rev Geophys 28(4):399–421. https://doi.org/10.1029/RG028i004p00399
Karato S-i, Wu P (1993) Rheology of the upper mantle: a synthesis. Science 260(5109):771–778
Kervazo M, Tobie G, Choblet G, Dumoulin C, Běhounková M (2021) Solid tides in Io's partially molten interior-contribution of bulk dissipation. Astron Astrophys 650:A72
Kervazo M, Tobie G, Choblet G, Dumoulin C, Běhounková M (2022) Inferring Io's interior from tidal monitoring. Icarus 373:114737
Khan A, Liebske C, Rozel A, Rivoldini A, Nimmo F, Connolly J, Plesa A-C, Giardini D (2018) A geophysical perspective on the bulk composition of Mars. J Geophys Res, Planets 123(2):575–611
Khurana K, Kivelson M, Stevenson D, Schubert G, Russell C, Walker R, Polanskey C (1998) Induced magnetic fields as evidence for subsurface oceans in Europa and Callisto. Nature 395(6704):777–780
Khurana KK, Jia X, Kivelson MG, Nimmo F, Schubert G, Russell CT (2011) Evidence of a global magma ocean in Io's interior. Science 332(6034):1186–1189
Kite ES, Rubin AM (2016) Sustained eruptions on Enceladus explained by turbulent dissipation in tiger stripes. Proc Natl Acad Sci 113(15):201520507. https://doi.org/10.1073/pnas.1520507113
Kivelson MG, Khurana KK, Russell CT, Volwerk M, Walker RJ, Zimmer C (2000) Galileo magnetometer measurements: a stronger case for a subsurface ocean at Europa. Science 289(5483):1340–1343. https://doi.org/10.1126/science.289.5483.1340
Kivelson MG, Khurana KK, Volwerk M (2002) The permanent and inductive magnetic moments of Ganymede. Icarus 157(2):507–522. https://doi.org/10.1006/icar.2002.6834
Koh Z-W, Nimmo F, Lunine JI, Mazarico E, Dombard AJ (2022) Assessing the detectability of Europa's seafloor topography from Europa Clipper's gravity data. Planet Sci J 3(8):197. https://doi.org/10.3847/PSJ/ac82aa
Kohlstedt DL, Zimmerman ME (1996) Rheology of partially molten mantle rocks. Annu Rev Earth Planet Sci 24(1):41–62
Kuchta M, Tobie G, Miljković K, Běhounková M, Souček O, Choblet G, Čadek O (2015) Despinning and shape evolution of Saturn's moon iapetus triggered by a giant impact. Icarus 252:454–465
Kvorka J, Čadek O (2022) A numerical model of convective heat transfer in Titan's subsurface ocean. Icarus 376:114853
Lainey V, Arlot J-E, Karatekin Ö, Van Hoolst T (2009) Strong tidal dissipation in Io and Jupiter from astrometric observations. Nature 459:957–959
Lainey V, Jacobson RA, Tajeddine R, Cooper NJ, Murray C, Robert V, Tobie G, Guillot T, Mathis S, Remus F, Desmars J, Arlot J-E, De Cuyper J-P, Dehant V, Pascu D, Thuillot W, Le Poncin-Lafitte C, Zahn J-P (2017) New constraints on Saturn's interior from Cassini astrometric data. Icarus 281:286–296. https://doi.org/10.1016/j.icarus.2016.07.014
Lainey V, Casajus LG, Fuller J, Zannoni M, Tortora P, Cooper N, Murray C, Modenini D, Park RS, Robert V, et al (2020) Resonance locking in giant planets indicated by the rapid orbital expansion of Titan. Nat Astron 4(11):1053–1058
Lainey V, Rambaux N, Tobie G, Cooper N, Zhang Q, Noyelles B, Baillié K (2024) A recently formed ocean inside Saturn's moon Mimas. Nature 626(7998):280–282. https://doi.org/10.1038/s41586-023-06975-9
Lamb H (1993) Hydrodynamics, 6th edn. Cambridge University Press
Le Gall A, Leyrat C, Janssen MA, Choblet G, Tobie G, Bourgeois O, Lucas A, Sotin C, Howett C, Kirk R, Lorenz RD, West RD, Stolzenbach A, Massé M, Hayes AH, Bonnefoy L, Veyssière G, Paganelli F (2017) Thermally anomalous features in the subsurface of Enceladus's south polar terrain. Nat Astron 1:0063. https://doi.org/10.1038/s41550-017-0063

Lee U, Saio H (1997) Low-frequency nonradial oscillations in rotating stars. I. Angular dependence. Astrophys J 491(2):839–845. https://doi.org/10.1086/304980

Liao Y, Nimmo F, Neufeld JA (2020) Heat production and tidally driven fluid flow in the permeable core of Enceladus. J Geophys Res, Planets 125(9):e2019JE006209

Lindzen RS, Chapman S (1969) Atmospheric tides. Space Sci Rev 10(1):3–188. https://doi.org/10.1007/BF00171584

Lobo AH, Thompson AF, Vance SD, Tharimena S (2021) A pole-to-equator ocean overturning circulation on Enceladus. Nat Geosci 14(4):185–189

Longuet-Higgins MS (1966) Planetary waves on a hemisphere bounded by meridians of longitude. Philos Trans R Soc Lond Ser A 260(1111):317–350. https://doi.org/10.1098/rsta.1966.0053

Longuet-Higgins MS (1968) The eigenfunctions of Laplace's tidal equations over a sphere. Philos Trans R Soc Lond Ser A 262(1132):511–607. https://doi.org/10.1098/rsta.1968.0003

Longuet-Higgins MS, Pond GS (1970) The free oscillations of fluid on a hemisphere bounded by meridians of longitude. Philos Trans R Soc Lond Ser A 266(1174):193–223. https://doi.org/10.1098/rsta.1970.0006

Lopes RM, Kamp LW, Smythe WD, Mouginis-Mark P, Kargel J, Radebaugh J, Turtle EP, Perry J, Williams DA, Carlson R, et al (2004) Lava lakes on Io: observations of Io's volcanic activity from Galileo nims during the 2001 fly-bys. Icarus 169:140–174

MacKenzie SM, Neveu M, Davila AF, Lunine JI, Craft KL, Cable ML, Phillips-Lander CM, Hofgartner JD, Eigenbrode JL, Waite JH, et al (2021) The Enceladus orbilander mission concept: balancing return and resources in the search for life. Planet Sci J 2(2):77

Marusiak AG, Vance S, Panning MP, Běhounková M, Byrne PK, Choblet G, Daswani MM, Hughson K, Journaux B, Lobo AH, et al (2021) Exploration of icy ocean worlds using geophysical approaches. Planet Sci J 2(4):150

Matsuyama I (2014) Tidal dissipation in the oceans of icy satellites. Icarus 242:11–18. https://doi.org/10.1016/j.icarus.2014.07.005

Matsuyama I, Beuthe M, Hay HCFC, Nimmo F, Kamata S (2018) Ocean tidal heating in icy satellites with solid shells. Icarus 312:208–230. https://doi.org/10.1016/j.icarus.2018.04.013

Mavko GM (1980) Velocity and attenuation in partially molten rocks. J Geophys Res, Solid Earth 85(B10):5173–5189

Mazarico E, Buccino D, Castillo-Rogez J, Dombard AJ, Genova A, Hussmann H, Kiefer WS, Lunine JI, McKinnon WB, Nimmo F, et al (2023) The Europa Clipper gravity and radio science investigation. Space Sci Rev 219:30. https://doi.org/10.1007/s11214-023-00972-0

McCarthy C, Cooper RF (2016) Tidal dissipation in creeping ice and the thermal evolution of Europa. Earth Planet Sci Lett 443:185–194

McCarthy C, Takei Y, Hiraga T (2011) Experimental study of attenuation and dispersion over a broad frequency range: 2. The universal scaling of polycrystalline materials. J Geophys Res, Solid Earth 116(B9):B09207

McCarthy C, Nielson MA, Coonin AN, Minker JS, Domingos AA (2019) Acoustic and microstructural properties of partially molten samples in the ice–ammonia system. Geosciences 9(8):327

McKenzie D (1984) The generation and compaction of partially molten rock. J Petrol 25(3):713–765

Melwani Daswani M, Vance SD, Mayne MJ, Glein CR (2021) A metamorphic origin for Europa's ocean. Geophys Res Lett 48(18):e2021GL094143

Meyer J, Wisdom J (2008) Tidal evolution of Mimas, Enceladus, and Dione. Icarus 193(1):213–223

Meyer J, Elkins-Tanton L, Wisdom J (2010) Coupled thermal-orbital evolution of the early Moon. Icarus 208(1):1–10. https://doi.org/10.1016/j.icarus.2010.01.029

Mitri G, Meriggiola R, Hayes A, Lefevre A, Tobie G, Genova A, Lunine JI, Zebker H (2014) Shape, topography, gravity anomalies and tidal deformation of Titan. Icarus 236:169–177

Miyazaki Y, Stevenson DJ (2022) A subsurface magma ocean on Io: exploring the steady state of partially molten planetary bodies. Planet Sci J 3(11):256. https://doi.org/10.3847/PSJ/ac9cd1

Monteux J, Tobie G, Choblet G, Le Feuvre M (2014) Can large icy moons accrete undifferentiated? Icarus 237:377–387

Monteux J, Qaddah B, Andrault D (2023) Conditions for segregation of a crystal-rich layer within a convective magma ocean. J Geophys Res, Planets 128(5):e2023JE007805. https://doi.org/10.1029/2023JE007805

Moore WB (2001) The thermal state of Io. Icarus 154(2):548–550

Moore WB, Hussmann H (2009) Thermal evolution of Europa's silicate interior. In: Pappalardo RT, McKinnon WB, Khurana K (eds) Europa. University of Arizona Press, Tuscon, pp 369–380. https://doi.org/10.2307/j.ctt1xp3wdw.21

Morabito L, Synnott S, Kupferman P, Collins SA (1979) Discovery of currently active extraterrestrial volcanism. Science 204(4396):972–972

Motoyama M, Tsunakawa H, Takahashi F (2020) Tidal resonance of eigenmode oscillation in the early Earth's ocean and its acceleration effect on the Moon's orbital evolution. Icarus 335:113382. https://doi.org/10.1016/j.icarus.2019.07.016
Mura A, Adriani A, Tosi F, Lopes R, Sindoni G, Filacchione G, Williams D, Davies A, Plainaki C, Bolton S, et al (2020) Infrared observations of Io from Juno. Icarus 341:113607
Murase T, McBirney AR (1973) Properties of some common igneous rocks and their melts at high temperatures. Geol Soc Am Bull 84(11):3563–3592
Musseau Y, Tobie G, Dumoulin C, Gillmann C, Revol A, Bolmont E (2024) The viscosity of Venus' mantle inferred from its rotational state. Icarus 422:116245
Nakajima M, Ingersoll AP (2016) Controlled boiling on Enceladus. 1. Model of the vapor-driven jets. Icarus 272:309–318. https://doi.org/10.1016/j.icarus.2016.02.027
Nakajima A, Ida S, Kimura J, Brasser R (2019) Orbital evolution of Saturn's mid-sized moons and the tidal heating of Enceladus. Icarus 317:570–582
Neveu M, Rhoden AR (2019) Evolution of Saturn's mid-sized moons. Nat Astron 3:543–552. https://doi.org/10.1038/s41550-019-0726-y
Nimmo F (2004) What is the young's modulus of ice? Europa's icy shell meeting. Europa's Icy Shell, LPI Contribution 1195, 7005
Nimmo F (2023) Strong tidal dissipation at Uranus? Planet Sci J 4(12):241
Nimmo F, Brown ME (2023) The internal structure of Eris inferred from its spin and orbit evolution. Sci Adv 9(46):eadi9201
Nimmo F, Spencer JR, Pappalardo RT, Mullen ME (2007) Shear heating as the origin of the plumes and heat flux on Enceladus. Nature 447(7142):289–291. https://doi.org/10.1038/nature05783
Nimmo F, Porco C, Mitchell C (2014) Tidally modulated eruptions on Enceladus: Cassini ISS observations and models. Astron J 148(3):46. https://doi.org/10.1088/0004-6256/148/3/46
Nimmo F, Neveu M, Howett C (2023) Origin and evolution of Enceladus's tidal dissipation. Space Sci Rev 219:57. https://doi.org/10.1007/s11214-023-01007-4
Noyelles B, Baillie K, Charnoz S, Lainey V, Tobie G (2019) Formation of the Cassini division - II. Possible histories of Mimas and Enceladus. Mon Not R Astron Soc 486(2):2947–2963
Ogawa M (2014) A positive feedback between magmatism and mantle upwelling in terrestrial planets: implications for the Moon. J Geophys Res, Planets 119(11):2317–2330
Ogawa M, Yanagisawa T (2011) Numerical models of Martian mantle evolution induced by magmatism and solid-state convection beneath stagnant lithosphere. J Geophys Res, Planets 116(E8):E08008
Ogilvie GI (2009) Tidal dissipation in rotating fluid bodies: a simplified model. Mon Not R Astron Soc 396(2):794–806. https://doi.org/10.1111/j.1365-2966.2009.14814.x
Ogilvie GI (2013) Tides in rotating barotropic fluid bodies: the contribution of inertial waves and the role of internal structure. Mon Not R Astron Soc 429(1):613–632. https://doi.org/10.1093/mnras/sts362
Ogilvie GI (2014) Tidal dissipation in stars and giant planets. Annu Rev Astron Astrophys 52:171–210. https://doi.org/10.1146/annurev-astro-081913-035941
Ogilvie GI, Lin DNC (2004) Tidal dissipation in rotating giant planets. Astrophys J 610(1):477–509. https://doi.org/10.1086/421454
Ojakangas GW, Stevenson DJ (1989) Thermal state of an ice shell on Europa. Icarus 81(2):220–241. https://doi.org/10.1016/0019-1035(89)90052-3
Park R, Mastrodemos N, Jacobson R, Berne A, Vaughan A, Hemingway D, Leonard E, Castillo-Rogez J, Cockell C, Keane J, et al (2024) The global shape, gravity field, and libration of Enceladus. J Geophys Res, Planets 129(1):e2023JE008054
Pauer M, Musiol S, Breuer D (2010) Gravity signals on Europa from silicate shell density variations. J Geophys Res 115:E12005. https://doi.org/10.1029/2010JE003595
Peale SJ, Cassen P, Reynolds RT (1979) Melting of Io by tidal dissipation. Science 203(4383):892–894
Peale S, Cassen P, Reynolds R (1980) Tidal dissipation, orbital evolution, and the nature of Saturn's inner satellites. Icarus 43(1):65–72
Peltier WR (1974) The impulse response of a Maxwell Earth. Rev Geophys Space Phys 12:649–669. https://doi.org/10.1029/RG012i004p00649
Pleiner Sládková K, Souček O, Běhounková M (2021) Enceladus' tiger stripes as frictional faults: effect on stress and heat production. Geophys Res Lett 48(19):e2021GL094849. https://doi.org/10.1029/2021GL094849
Poirier JP, Boloh L, Chambon P (1983) Tidal dissipation in small viscoelastic ice moons: the case of Enceladus. Icarus 55(2):218–230
Porco CC, Helfenstein P, Thomas P, Ingersoll A, Wisdom J, West R, Neukum G, Denk T, Wagner R, Roatsch T, et al (2006) Cassini observes the active south pole of Enceladus. Science 311(5766):1393–1401
Porco C, DiNino D, Nimmo F (2014) How the geysers, tidal stresses, and thermal emission across the south polar terrain of Enceladus are related. Astron J 148(3):45. https://doi.org/10.1088/0004-6256/148/3/45

Postberg F, Clark RN, Hansen CJ, Coates AJ, Ore CMD, Scipioni F, Hedman MM, Waite JH, Dotson R (2018) Plume and surface composition of Enceladus University of Arizona Press, Tucson, pp 129–162. http://www.jstor.org/stable/j.ctv65sw2b.16

Pou L, Nimmo F, Rivoldini A, Khan A, Bagheri A, Gray T, Samuel H, Lognonné P, Plesa A-C, Gudkova T, et al (2022) Tidal constraints on the Martian interior. J Geophys Res, Planets 127(11):e2022JE007291

Preblich B, Greenberg R, Riley J, O'Brien D (2007) Tidally driven strike–slip displacement on Europa: viscoelastic modeling. Planet Space Sci 55(10):1225–1245. https://doi.org/10.1016/j.pss.2007.01.018

Ray RD, Eanes RJ, Lemoine FG (2001) Constraints on energy dissipation in the Earth's body tide from satellite tracking and altimetry. Geophys J Int 144(2):471–480

Reeh N, Christensen EL, Mayer C, Olesen OB (2003) Tidal bending of glaciers: a linear viscoelastic approach. Ann Glaciol 37:83–89

Rekier J, Trinh A, Triana SA, Dehant V (2019) Internal energy dissipation in Enceladus's subsurface ocean from tides and libration and the role of inertial waves. J Geophys Res, Planets 124(8):2198–2212. https://doi.org/10.1029/2019JE005988

Renaud JP, Henning WG (2018) Increased tidal dissipation using advanced rheological models: implications for Io and tidally active exoplanets. Astrophys J 857(2):98

Renner J, Evans B, Hirth G (2000) On the rheologically critical melt fraction. Earth Planet Sci Lett 181(4):585–594

Reynard B, Sotin C (2023) Carbon-rich icy moons and dwarf planets. Earth Planet Sci Lett 612:118172

Rhoden AR (2023) Mimas: frozen fragment, ring relic, or emerging ocean world? Annu Rev Earth Planet Sci 51:367–387

Rhoden AR, Walker ME, Rudolph ML, Bland MT, Manga M (2024) The evolution of a young ocean within Mimas. Earth Planet Sci Lett 635:118689

Ricard Y, Bercovici D, Schubert G (2001) A two-phase model for compaction and damage: 2. Applications to compaction, deformation, and the role of interfacial surface tension. J Geophys Res, Solid Earth 106(B5):8907–8924

Rieutord M, Valdettaro L (2010) Viscous dissipation by tidally forced inertial modes in a rotating spherical shell. J Fluid Mech 643:363–394. https://doi.org/10.1017/S002211200999214X

Roberts JH (2015) The fluffy core of Enceladus. Icarus 258:54–66

Roberts JH, Nimmo F (2008) Tidal heating and the long-term stability of a subsurface ocean on Enceladus. Icarus 194(2):675–689

Roberts JH, McKinnon WB, Elder CM, Tobie G, Biersteker JB, Young D, Park RS, Steinbrügge G, Nimmo F, Howell SM, et al (2023) Exploring the interior of Europa with the Europa Clipper. Space Sci Rev 219:46. https://doi.org/10.1007/s11214-023-00990-y

Rollins KM, Evans MD, Diehl NB, Daily III WD (1998) Shear modulus and damping relationships for gravels. J Geotech Geoenviron Eng 124(5):396–405

Romanowicz B, Mitchell B (2015) 1.25—deep Earth structure: Q of the Earth from crust to core. Treatise Geophys 1:789–827

Rosier SH, Marsh OJ, Rack W, Gudmundsson GH, Wild CT, Ryan M (2017) On the interpretation of ice-shelf flexure measurements. J Glaciol 63(241):783–791

Rovira-Navarro M, Rieutord M, Gerkema T, Maas LRM, van der Wal W, Vermeersen B (2019) Do tidally-generated inertial waves heat the subsurface oceans of Europa and Enceladus? Icarus 321:126–140. https://doi.org/10.1016/j.icarus.2018.11.010

Rovira-Navarro M, Gerkema T, Maas LRM, van der Wal W, van Ostayen R, Vermeersen B (2020) Tides in subsurface oceans with meridional varying thickness. Icarus 343:113711. https://doi.org/10.1016/j.icarus.2020.113711

Rovira-Navarro M, Katz RF, Liao Y, van der Wal W, Nimmo F (2022) The tides of Enceladus' porous core. J Geophys Res, Planets 127(5):e2021JE007117

Rovira-Navarro M, Matsuyama I, Hay HCFC (2023) Thin-shell tidal dynamics of ocean worlds. Planet Sci J 4(2):23. https://doi.org/10.3847/PSJ/acae9a

Sabadini R, Yuen DA, Boschi E (1982) Polar wandering and the forced responses of a rotating, multilayered, viscoelastic planet. J Geophys Res, Solid Earth 87(B4):2885–2903

Sabadini R, Vermeersen B, Cambiotti G (2016) Global Dynamics of the Earth: Applications of Viscoelastic Relaxation Theory to Solid-Earth and Planetary Geophysics. https://doi.org/10.1007/978-94-017-7552-6

Sagan C, Dermott SF (1982) The tide in the seas of Titan. Nature 300(5894):731–733. https://doi.org/10.1038/300731a0

Saito M (1974) Some problems of static deformation of the Earth. J Phys Earth 22(1):123–140

Saur J, Duling S, Roth L, Jia X, Strobel DF, Feldman PD, Christensen UR, Retherford KD, McGrath MA, Musacchio F, Wennmacher A, Neubauer FM, Simon S, Hartkorn O (2015) The search for a subsurface ocean in Ganymede with Hubble Space Telescope observations of its auroral ovals. J Geophys Res Space Phys 120(3):1715–1737. https://doi.org/10.1002/2014JA020778

Schenk PM, McKinnon WB (1989) Fault offsets and lateral crustal movement on Europa: evidence for a mobile ice shell. Icarus 79(1):75–100. https://doi.org/10.1016/0019-1035(89)90109-7
Schindelegger M, Green JAM, Wilmes SB, Haigh ID (2018) Can we model the effect of observed sea level rise on tides? J Geophys Res, Oceans 123(7):4593–4609. https://doi.org/10.1029/2018JC013959
Schmeling H (1985) Numerical models on the influence of partial melt on elastic, anelastic and electric properties of rocks. Part I: elasticity and anelasticity. Phys Earth Planet Inter 41(1):34–57
Schmeling H, Kruse JP, Richard G (2012) Effective shear and bulk viscosity of partially molten rock based on elastic moduli theory of a fluid filled poroelastic medium. Geophys J Int 190(3):1571–1578
Schwarz J, Weeks W (1977) Engineering properties of sea ice. J Glaciol 19(81):499–531
Scott T, Kohlstedt D (2006) The effect of large melt fraction on the deformation behavior of peridotite. Earth Planet Sci Lett 246(3–4):177–187
Scott DR, Stevenson DJ (1986) Magma ascent by porous flow. J Geophys Res, Solid Earth 91(B9):9283–9296
Sears WD (1995) Tidal dissipation in oceans on Titan. Icarus 113(1):39–56
Seed HB, Wong RT, Idriss I, Tokimatsu K (1986) Moduli and damping factors for dynamic analyses of cohesionless soils. J Geotech Eng 112(11):1016–1032
Segatz M, Spohn T, Ross M, Schubert G (1988) Tidal dissipation, surface heat flow, and figure of viscoelastic models of Io. Icarus 75(2):187–206
Shaw HR (1969) Rheology of basalt in the melting range. J Petrol 10(3):510–535
Shibuya S, Mitachi T, Fukuda F, Degoshi T (1995) Strain rate effects on shear modulus and damping of normally consolidated clay. Geotech Test J 18(3):365–375
Sinha NK (1989) Elasticity of natural types of polycrystalline ice. Cold Reg Sci Technol 17(2):127–135. https://doi.org/10.1016/S0165-232X(89)80003-5
Sládková K, Souček O, Kalousová K, Běhounková M (2020) Tidal walking on Europa's strike slip faults - insight from numerical modeling. J Geophys Res, Planets. https://doi.org/10.1029/2019je006327
Smith BA, Soderblom LA, Beebe R, Boyce J, Briggs G, Carr M, Collins SA, Cook AF, Danielson GE, Davies ME, et al (1979) The Galilean satellites and Jupiter: Voyager 2 imaging science results. Science 206(4421):927–950
Soderlund KM (2019) Ocean dynamics of outer Solar System satellites. Geophys Res Lett 46(15):8700–8710
Soderlund KM, Rovira-Navarro M, Le Bars M, Schmidt BE, Gerkema T (2024) The physical oceanography of ice-covered moons. Annu Rev Mar Sci 16(1):25–53
Sohl F, Sears WD, Lorenz RD (1995) Tidal dissipation on Titan. Icarus 115(2):278–294
Sotin C, Head JW, Tobie G (2002) Europa: tidal heating of upwelling thermal plumes and the origin of lenticulae and chaos melting. Geophys Res Lett 29(8):1233. https://doi.org/10.1029/2001GL013844
Sotin C, Tobie G, Wahr J, McKinnon WB, McKinnon W, Khurana K (2009) Tides and tidal heating on Europa. In: Pappalardo RT, McKinnon WB, Khurana KK (eds) Europa. University of Arizona Press, Tucson, pp 85–118. https://doi.org/10.2307/j.ctt1xp3wdw.10
Sotin C, Kalousová K, Tobie G (2021) Titan's interior structure and dynamics after the Cassini-Huygens mission. Annu Rev Earth Planet Sci 49:579–607
Souček O, Hron J, Běhounková M, Čadek O (2016) Effect of the tiger stripes on the deformation of Saturn's moon Enceladus. Geophys Res Lett 43(14):7417–7423. https://doi.org/10.1002/2016GL069415
Souček O, Běhounková M, Čadek O, Hron J, Tobie G, Choblet G (2019) Tidal dissipation in Enceladus' uneven, fractured ice shell. Icarus 328:218–231. https://doi.org/10.1016/j.icarus.2019.02.012. ISSN 0019-1035
Souček O, Běhounková M, Lanzendörfer M, Tobie G, Choblet G (2024) Variations in plume activity reveal the dynamics of water-filled faults on Enceladus. Nat Commun 15(1):7405
Spada G (2008) Alma, a fortran program for computing the viscoelastic Love numbers of a spherically symmetric planet. Comput Geosci 34(6):667–687
Sparks R, Annen C, Blundy J, Cashman K, Rust A, Jackson M (2019) Formation and dynamics of magma reservoirs. Philos Trans R Soc A 377(2139):20180019
Spencer JR, Rathbun JA, Travis LD, Tamppari LK, Barnard L, Martin TZ, McEwen AS (2000) Io's thermal emission from the Galileo photopolarimeter-radiometer. Science 288(5469):1198–1201
Spencer JR, Pearl JC, Segura M, Flasar FM, Mamoutkine A, Romani P, Buratti BJ, Hendrix AR, Spilker LJ, Lopes RMC (2006) Cassini encounters Enceladus: background and the discovery of a south polar hot spot. Science 311(5766):1401–1405. https://doi.org/10.1126/science.1121661
Spencer JR, Howett CJA, Verbiscer A, Hurford TA, Segura M, Spencer DC (2013) Enceladus heat flow from high spatial resolution thermal emission observations. Eur Planet Sci Congr 8:1–2
Spencer J, Nimmo F, Ingersoll AP, Hurford T, Kite E, Rhoden A, Schmidt J, Howett C (2018) Plume origins and plumbing: from ocean to surface. In: Schenk PM et al (eds) Enceladus and the icy moons of Saturn. University of Arizona Press, Tucson, pp 163–174
Spencer DC, Katz RF, Hewitt IJ (2020) Magmatic intrusions control Io's crustal thickness. J Geophys Res, Planets 125(6):e2020JE006443

Spencer DC, Katz RF, Hewitt IJ (2021) Tidal controls on the lithospheric thickness and topography of Io from magmatic segregation and volcanism modelling. Icarus 359:114352
Spetzler H, Anderson DL (1968) The effect of temperature and partial melting on velocity and attenuation in a simple binary system. J Geophys Res 73(18):6051–6060
Squyres SW, Reynolds RT, Cassen PM, Peale SJ (1983) The evolution of Enceladus. Icarus 53(2):319–331
Steinbrügge G, Stark A, Hussmann H, Sohl F, Oberst J (2015) Measuring tidal deformations by laser altimetry. A performance model for the Ganymede laser altimeter. Planet Space Sci 117:184–191
Steinbrügge G, Schroeder DM, Haynes MS, Hussmann H, Grima C, Blankenship DD (2018) Assessing the potential for measuring Europa's tidal Love number h2 using radar sounder and topographic imager data. Earth Planet Sci Lett 482:334–341
Steinbrügge G, Hussmann H, Tobie G, Castillo-Rogez J, Lainey V, Stark A (2023) The tidal deformation and tidal evolution of Ganymede. Planet Space Sci 28:46
Steinke T, van Sliedregt D, Vilella K, Van der Wal W, Vermeersen B (2020) Can a combination of convective and magmatic heat transport in the mantle explain Io's volcanic pattern? J Geophys Res, Planets 125(12):e2020JE006521
Sundberg M, Cooper RF (2010) A composite viscoelastic model for incorporating grain boundary sliding and transient diffusion creep; correlating creep and attenuation responses for materials with a fine grain size. Philos Mag 90(20):2817–2840
Tackley PJ, Schubert G, Glatzmaier GA, Schenk P, Ratcliff JT, Matas J-P (2001) Three-dimensional simulations of mantle convection in Io. Icarus 149(1):79–93
Tajeddine R, Rambaux N, Lainey V, Charnoz S, Richard A, Rivoldini A, Noyelles B (2014) Constraints on Mimas' interior from Cassini ISS libration measurements. Science 346(6207):322–324
Tajeddine R, Soderlund KM, Thomas PC, Helfenstein P, Hedman MM, Burns JA, Schenk PM (2017) True polar wander of Enceladus from topographic data. Icarus 295:46–60
Takei Y (1998) Constitutive mechanical relations of solid-liquid composites in terms of grain-boundary contiguity. J Geophys Res, Solid Earth 103(B8):18183–18203
Takei Y, Holtzman BK (2009) Viscous constitutive relations of solid-liquid composites in terms of grain boundary contiguity: 1. Grain boundary diffusion control model. J Geophys Res, Solid Earth 114(B6):B06205
Takei Y, Karasawa F, Yamauchi H (2014) Temperature, grain size, and chemical controls on polycrystal anelasticity over a broad frequency range extending into the seismic range. J Geophys Res, Solid Earth 119(7):5414–5443
Takeuchi H, Saito M (1972) Seismic surface waves. Methods Comput Phys, Adv Res Appl 11:217–295. https://doi.org/10.1016/B978-0-12-460811-5.50010-6
Tatibouet J, Perez J, Vassoille R (1986) High-temperature internal friction and dislocations in ice ih. J Phys 47(1):51–60
Tatibouet J, Perez J, Vassoille R (1987) Study of grain boundaries in ice by internal friction measurement. J Phys, Colloq 48(C1):C1-197
Taylor GI (1936) The oscillations of the atmosphere. Proc R Soc Lond Ser A 156(888):318–326. https://doi.org/10.1098/rspa.1936.0150
Thomas PC, Tajeddine R, Tiscareno MS, Burns JA, Joseph J, Loredo TJ, Helfenstein P, Porco C (2016) Enceladus's measured physical libration requires a global subsurface ocean. Icarus 264:37–47. https://doi.org/10.1016/j.icarus.2015.08.037
Tobie G, Choblet G, Sotin C (2003) Tidally heated convection: constraints on Europa's ice shell thickness. J Geophys Res, Planets 108(E11):5124. https://doi.org/10.1029/2003JE002099
Tobie G, Grasset O, Lunine JI, Mocquet A, Sotin C (2005a) Titan's internal structure inferred from a coupled thermal-orbital model. Icarus 175(2):496–502
Tobie G, Mocquet A, Sotin C (2005b) Tidal dissipation within large icy satellites: applications to Europa and Titan. Icarus 177(2):534–549. https://doi.org/10.1016/j.icarus.2005.04.006
Tobie G, Lunine JI, Sotin C (2006) Episodic outgassing as the origin of atmospheric methane on Titan. Nature 440(7080):61–64
Tobie G, Čadek O, Sotin C (2008) Solid tidal friction above a liquid water reservoir as the origin of the south pole hotspot on Enceladus. Icarus 196(2):642–652
Tobie G, Lunine J, Monteux J, Mousis O, Nimmo F (2014) The origin and evolution of Titan. In: Müller-Wodarg I et al (eds) Titan: interior, surface, atmosphere, and space environment. Cambridge University Press, pp 29–62
Tobie G, Grasset O, Dumoulin C, Mocquet A (2019) Tidal response of rocky and ice-rich exoplanets. Astron Astrophys 630:A70. https://doi.org/10.1051/0004-6361/201935297
Tort M, Dubos T, Bouchut F, Zeitlin V (2014) Consistent shallow-water equations on the rotating sphere with complete Coriolis force and topography. J Fluid Mech 748:789–821. https://doi.org/10.1017/jfm.2014.172

Tyler RH (2008) Strong ocean tidal flow and heating on moons of the outer planets. Nature 456(7223):770–772. https://doi.org/10.1038/nature07571

Tyler RH (2009) Ocean tides heat Enceladus. Geophys Res Lett 36(15):L15205. https://doi.org/10.1029/2009GL038300

Tyler R (2011) Tidal dynamical considerations constrain the state of an ocean on Enceladus. Icarus 211(1):770–779. https://doi.org/10.1016/j.icarus.2010.10.007

Tyler R (2014) Comparative estimates of the heat generated by ocean tides on icy satellites in the outer Solar System. Icarus 243:358–385. https://doi.org/10.1016/j.icarus.2014.08.037

Tyler R (2019) TROPF (v1.0). Zenodo. https://doi.org/10.5281/zenodo.3523274

Tyler RH (2021) On the tidal history and future of the Earth-Moon orbital system. Planet Sci J 2(2):70. https://doi.org/10.3847/PSJ/abe53f

Tyler RH, Henning WG, Hamilton CW (2015) Tidal heating in a magma ocean within Jupiter's moon Io. Astrophys J Suppl Ser 218(2):22. https://doi.org/10.1088/0067-0049/218/2/22

Vallis GK (2017) Atmospheric and Oceanic fluid dynamics: fundamentals and large-scale circulation. https://doi.org/10.1017/9781107588417

Van Hoolst T, Baland R-M, Trinh A (2016) The diurnal libration and interior structure of Enceladus. Icarus 277:311–318

Van Hoolst T, Tobie G, Vallat C, et al (2024) Geophysical characterization of the interiors of Ganymede, Callisto and Europa by ESA's JUpiter ICy moons Explorer. Space Sci Rev 220:54. https://doi.org/10.1007/s11214-024-01085-y

Vance S, Harnmeijer J, Kimura J, Hussmann H, DeMartin B, Brown JM (2007) Hydrothermal systems in small ocean planets. Astrobiology 7(6):987–1005

Vance SD, Hand KP, Pappalardo RT (2016) Geophysical controls of chemical disequilibria in Europa. Geophys Res Lett 43(10):4871–4879. https://doi.org/10.1002/2016GL068547

Varshalovich DA, Moskalev AN, Khersonskii VK (1988) Quantum theory of angular momentum. https://doi.org/10.1142/0270

Vaughan DG (1995) Tidal flexure at ice shelf margins. J Geophys Res, Solid Earth 100(B4):6213–6224

Vigneresse JL, Barbey P, Cuney M (1996) Rheological transitions during partial melting and crystallization with application to felsic magma segregation and transfer. J Petrol 37(6):1579–1600

Wahr JM, Zuber MT, Smith DE, Lunine JI (2006) Tides on Europa, and the thickness of Europa's icy shell. J Geophys Res, Planets 111(E12):E12005. https://doi.org/10.1029/2006JE002729

Wahr J, Selvans ZA, Mullen ME, Barr AC, Collins GC, Selvans MM, Pappalardo RT (2009) Modeling stresses on satellites due to nonsynchronous rotation and orbital eccentricity using gravitational potential theory. Icarus 200(1):188–206

Waite JH Jr, Combi MR, Ip W-H, Cravens TE, McNutt RL Jr, Kasprzak W, Yelle R, Luhmann J, Niemann H, Gell D, et al (2006) Cassini ion and Neutral Mass Spectrometer: Enceladus plume composition and structure. Science 311(5766):1419–1422

Walterová M, Běhounková M, Efroimsky M (2023) Is there a semi-Molten layer at the base of the lunar mantle? J Geophys Res, Planets 128(7):e2022JE007652

Wang H, Boyd JP, Akmaev RA (2016) On computation of Hough functions. Geosci Model Dev 9(4):1477–1488. https://doi.org/10.5194/gmd-9-1477-2016

Webb DJ (1980) Tides and tidal friction in a hemispherical ocean centred at the equator. Geophys J 61(3):573–600. https://doi.org/10.1111/j.1365-246X.1980.tb04833.x

Webb DJ (1982) Tides and the evolution of the Earth-Moon system. Geophys J 70:261–271. https://doi.org/10.1111/j.1365-246X.1982.tb06404.x

White OL, Schenk PM, Bellagamba AW, Grimm AM, Dombard AJ, Bray VJ (2017) Impact crater relaxation on Dione and Tethys and relation to past heat flow. Icarus 288:37–52

Widmer R, Masters G, Gilbert F (1991) Spherically symmetric attenuation within the Earth from normal mode data. Geophys J Int 104(3):541–553

Wild CT, Marsh OJ, Rack W (2017) Viscosity and elasticity: a model intercomparison of ice-shelf bending in an Antarctic grounding zone. J Glaciol 63(240):573–580

Williams JG, Boggs DH (2015) Tides on the Moon: theory and determination of dissipation. J Geophys Res, Planets 120(4):689–724

Wisdom J (2008) Tidal dissipation at arbitrary eccentricity and obliquity. Icarus 193(2):637–640

Wisdom J, Dbouk R, Militzer B, Hubbard WB, Nimmo F, Downey BG, French RG (2022) Loss of a satellite could explain Saturn's obliquity and young rings. Science 377(6612):1285–1289

Wulff A-M, Hashida T, Watanabe K, Takahashi H (1999) Attenuation behaviour of tuffaceous sandstone and granite during microfracturing. Geophys J Int 139(2):395–409

Zahn JP (1975) The dynamical tide in close binaries. Astron Astrophys 41:329–344

Zahnle K, Walker JCG (1987) A constant daylength during the Precambrian era? Precambrian Res 37(2):95–105. https://doi.org/10.1016/0301-9268(87)90073-8

Zandanel A, Truche L, Hellmann R, Myagkiy A, Choblet G, Tobie G (2021) Short lifespans of serpentinization in the rocky core of Enceladus: implications for hydrogen production. Icarus 364:114461
Zhong S, Qin C, Wahr J (2012) Can tidal tomography be used to unravel the long-wavelength structure of the lunar interior? Geophys Res Lett 39(15)
Zhou W, Chen Y, Ma G, Yang L, Chang X (2017) A modified dynamic shear modulus model for rockfill materials under a wide range of shear strain amplitudes. Soil Dyn Earthq Eng 92:229–238
Zimmer C, Khurana KK, Kivelson MG (2000) Subsurface oceans on Europa and Callisto: constraints from Galileo magnetometer observations. Icarus 147(2):329–347. https://doi.org/10.1006/icar.2000.6456
Zou C, Ding H, Luan W (2024) Anelasticity of the lower mantle inferred from the pole and lunar monthly tides using global doris coordinate time series. Glob Planet Change 236:104415

Space Science Reviews (2024) 220:70
https://doi.org/10.1007/s11214-024-01104-y

The Composition of Saturn's Rings

Kelly E. Miller[1] · Gianrico Filacchione[2] · Jeffrey N. Cuzzi[3] · Philip D. Nicholson[4] · Matthew M. Hedman[5] · Kevin Baillié[6] · Robert E. Johnson[7] · Wei-Ling Tseng[8] · Paul R. Estrada[3] · Jack Hunter Waite[9] · Mauro Ciarniello[2] · Cécile Ferrari[10] · Zhimeng Zhang[11] · Amanda Hendrix[12] · Julianne I. Moses[13] · Hsiang-Wen Hsu[14]

Received: 5 October 2023 / Accepted: 20 August 2024 / Published online: 9 September 2024

Abstract
The origin and evolution of Saturn's rings is critical to understanding the Saturnian system as a whole. Here, we discuss the physical and chemical composition of the rings, as a foundation for evolutionary models described in subsequent chapters. We review the physical characteristics of the main rings, and summarize current constraints on their chemical composition. Radial trends are observed in temperature and to a limited extent in particle size distribution, with the C ring exhibiting higher temperatures and a larger population of small particles. The C ring also shows evidence for the greatest abundance of silicate material, perhaps indicative of formation from a rocky body. The C ring and Cassini Division have lower optical depths than the A and B rings, which contributes to the higher abundance of the exogenous neutral absorber in these regions. Overall, the main ring composition is strongly dominated by water ice, with minor silicate, UV absorber, and neutral absorber components. Sampling of the innermost D ring during Cassini's Grand Finale provides a new set of in situ constraints on the ring composition, and we explore ongoing work to understand the linkages between the main rings and the D ring. The D ring material is organic- and silicate-rich and water-poor relative to the main rings, with a large population of small grains. This composition may be explained in part by volatile losses in the D ring, and current constraints suggest some degree of fractionation rather than sampling of the bulk D ring material.

Keywords Ring particle composition, mixing and particle size distribution · Ring radial and vertical structure · Ring atmosphere

1 Introduction

The rings are a prominent and unique feature in the Saturnian system, and they serve both as a witness plate and a driver for the evolution of the system via interactions with both the moons and Saturn itself. In this chapter we review current knowledge of the physical and chemical composition of Saturn's rings, with a focus on the main rings and the innermost D ring. We begin in Sect. 1 with an overview of the ring structure and main evolutionary processes affecting the rings. In Sect. 2, we describe the physical properties of the main rings, including their brightness, particle size distribution and temperature. Remote sensing constraints on the main rings' chemical composition are then reviewed in Sects. 3, 4 and 5.

Extended author information available on the last page of the article

Fig. 1 An optical mosaic of the dark side of Saturn's rings, obtained by the narrow-angle camera of the Cassini ISS instrument. From left to right, we see the broad C, B and A rings at increasing radii, flanked by the narrow F ring at the extreme right. Note the numerous narrow gaps, appearing here as black bands in the C ring, the Cassini Division and in the outer part of the A ring. Courtesy: NASA/JPL-Caltech/Space Science Institute

Section 6 reviews our current knowledge of the D ring, which is thought to be the immediate source for the inflowing ring material that was measured in situ by Cassini and is described in Sect. 7. Current constraints suggest this material includes a gas-phase component, and we review our knowledge of the ring atmosphere and possible connections to the inflowing material in Sect. 8. Finally, we briefly explore the relationship between the main ring composition and the ring moons in Sect. 9, before summarizing our conclusions in Sect. 10.

1.1 Background on Ring Structure

1.1.1 Terminology and Dimensions

Saturn's dense rings (or "main rings") are traditionally divided into three major components known as the A, B and C rings (see Fig. 1). The outermost or A ring extends from a radius of 136,770 km relative to Saturn's center inward to $\sim 122,000$ km; the outer edge at ~ 2.25 Saturn radii (R_S, where Saturn's equatorial radius is 60,268 km) essentially marks the Roche Limit for self-gravitating icy particles (Tiscareno et al. 2013a). The middle or B ring extends from 117,570 km inward to $\sim 92,000$ km, and is both the brightest and most opaque of the main rings, containing approximately 2/3 of the system's total mass. The inner or C ring extends from the B ring inward to 74,490 km, or ~ 1.25 R_S. Located between the A and B rings is the 4500 km wide Cassini Division ("CD"), which Voyager data showed to be comparable in brightness and optical depth to the C ring and which is now considered to be a distinct fourth component of the main rings.

In comparison to their radial width of $\sim 62,000$ km, or almost 10 Earth radii, Saturn's rings are extremely thin. Theoretical estimates based on collisional equilibrium suggest that their centrifugal scale height H is likely to be of order 10 m, a value that is supported by a limited number of experimental estimates ranging from < 30 m in the C ring and Cassini Division to ~ 6 m in the A ring (Hedman et al. 2007b; Jerousek et al. 2011).

Although it is less opaque and almost certainly contains much less material than the main rings, we include in this chapter the nearby D ring, primarily because of the *in situ* sampling by Cassini on its final orbits that observed material hypothesized to originate from this region. The D ring extends inwards from the C ring almost to the outer reaches of the planet's atmosphere at a radius of $< 68,000$ km. Invisible in Earth-based images and in most optical depth profiles, the D ring is dominated by a series of narrow, dusty ringlets and is most readily seen in images taken in forward-scattering geometries (see Sect. 6).

Limited *in situ* sampling near the F ring region outside of the A ring was also performed by Cassini. This ring, which is both shepherded and perturbed by the ring moons Pandora and Prometheus, is discussed further in Ciarniello et al. (2024, this collection), along with the dusty G ring that is associated with the satellite Aegaeon.

Not included in this volume is the very tenuous and isolated E ring, which is associated with the satellite Enceladus, or the even more distant "Phoebe ring" discovered by the Spitzer space telescope (Verbiscer Astron et al. 2009). For information on these rings the interested reader is directed to the reviews by Hedman et al. (2018) and Murray and French (2018) in the book "Planetary Ring Systems" (Tiscareno and Murray 2018).

1.1.2 Optical Depths and Ring Structure

The density of a planetary ring is usually expressed in terms of its normal optical depth τ, which in general is a function of the wavelength of observation λ and the particle size distribution $n(a)$:

$$\tau(\lambda) = \pi \int_{a_{\min}}^{a_{\max}} Q_{\text{ext}}(a, \lambda)\, a^2\, n(a)\, da \tag{1}$$

where $n(a)da$ is the number of ring particles per unit ring area with particle radii between a and $a + da$ and $Q_{\text{ext}}(a, \lambda)$ is their extinction efficiency at wavelength λ. For quasi-spherical particles, Q_{ext} approaches 0 when the size parameter $x = 2\pi a/\lambda$ is much less than 1, but approaches an asymptotic value between 1 and 2 for $x \gg 1$ (Cuzzi 1985; French and Nicholson 2000).

Detailed radial profiles of optical depth for Saturn's rings are available from stellar occultations observed by the Ultraviolet Imaging Spectrometer (UVIS; Esposito et al. 2004) at $\lambda \simeq 0.15$ μm and by the Visual and Infrared Mapping Spectrometer (VIMS; Brown et al. 2004) at $\lambda = 2.9$ μm, for both of which $Q_{\text{ext}} \simeq 1$, as well as radio occultations observed by the Radio Science Subsystem (RSS; Kliore Astron et al. 2004) at Ka-band ($\lambda = 0.94$ cm), X-band (3.4 cm) and S-band (13 cm) (Cuzzi et al. 2009). [1]

A second useful quantity is the surface mass density of the ring for a given distance (*i.e.*, the mass per unit ring area), expressed as:

$$\Sigma = \frac{4\pi}{3} \int_{a_{\min}}^{a_{\max}} \rho\, a^3\, n(a)\, da \tag{2}$$

where ρ is the average internal density of the ring particles. For a ring composed of spherical particles with a narrow range of sizes and $Q_{\text{ext}} \simeq 1$, the ratio Σ/τ is approximately equal to $\frac{4}{3}\rho\bar{a}$, where $\bar{a}$ is the mean particle radius. The inverse quantity τ/Σ is known as the opacity of the ring, and is a measure of its optical cross section per unit mass .[2]

1.2 Relationship of Composition to Ring Age

Evolutionary processes alter the physical and chemical composition of the rings, and an understanding of the types and timescales of these forms of alteration is an essential element

[1]For most of the regions in the main rings, $a \gg \lambda$ for the UVIS and VIMS occultation data and $Q_{\text{ext}} \simeq 1$, but for the RSS occultation data, $a \simeq \lambda$ and $Q_{\text{ext}} \simeq 2$ due to the particles' broad diffraction lobes at cm-wavelengths, as well as the coherent nature of the spacecraft's radio signal.

[2]The quantity κ is generally used in radiative transfer to indicate the opacity (cm^2/g) of a medium, or its cross section (for absorption, scattering, or both) per unit surface mass density. Then the incremental optical depth $d\tau$ of a thin layer dz is $d\tau = \kappa\rho dz$. Integrating this expression over the thickness of a ring layer, the normal optical depth of the ring is $\tau = \kappa\Sigma$, where Σ is the surface mass density in g/cm^2. A related quantity is the extinction coefficient k whereby $d\tau = kdz$ and thus $k = nQ\pi a^2$ (cm^{-1}), for particles with volume density n (cm^{-3}) and a single particle size a, having (absorption, scattering, or both) efficiency Q (dimensionless)..

of the determination of the ring age. Here, we summarize the primary processes known to drive evolution of the main rings. For an in-depth discussion of the ring age, see Crida et al. (2024, this collection).

1.2.1 Physical Evolution

The physical evolution of the rings is controlled by interparticle collisions and the rings' self-gravity, both of which determine the rings' viscosity and thickness as described below, as well as by gravitational interactions with distant and embedded satellites. In order to estimate the dynamical age of the rings, the two critical parameters are the surface mass density Σ and the effective kinematic viscosity of the rings ν, which control the rate at which keplerian shear within the rings leads to their radial spreading over time. The latter rate is determined by the outward viscous angular momentum flux across the rings

$$F_\upsilon(r) = \frac{3}{2}\nu\Sigma\Omega r, \tag{3}$$

where $\Omega \approx \sqrt{GM_S/r^3}$ is the local keplerian angular velocity at a radius r from Saturn in the rings (Lynden-Bell and Pringle 1974; Longaretti 2018).

Theoretical models suggest that there are at least three important sources of viscosity: the normal "local viscosity" due to random velocities and the resulting collisions between individual ring particles,[3] a "nonlocal viscosity" due to transmission of momentum across large particles in densely-packed rings, and the transfer of momentum by gravitational interactions associated with self-gravity wakes (Daisaka et al. 2001; Longaretti 2018). These self-gravity wakes are 100-m- to km-scale temporary aggregations of particles that form within the Roche limit but are subsequently sheared out and disrupted by tidal forces and the keplerian angular velocity gradient within the rings. The relative importance of these processes in different parts of the rings is not well-known, but gravitational interactions are expected to be dominant in the A ring, where self-gravity wakes are ubiquitous (Colwell et al. 2006; Hedman et al. 2007a).

Empirical estimates of ν come from two sources: the damping of density and bending waves, and the balancing of the viscous torque $2\pi r F_\upsilon$ with gravitational torques exerted at satellite resonances. For the A ring, estimates of the former lead to $\nu = 40$ to $200\ \mathrm{cm^2\ s^{-1}}$ (Tiscareno et al. 2007), comparable to the self-gravity-wake viscosity of $\sim 100\ \mathrm{cm^2\ s^{-1}}$. Balancing the viscous and satellite torques, on the other hand, Tajeddine et al. (2017) find somewhat lower values for the A ring of $\nu = 10 - 50\ \mathrm{cm^2\ s^{-1}}$ and a similar value for the B ring of $\sim 30\ \mathrm{cm^2\ s^{-1}}$. There are few reliable estimates in the C ring or Cassini Division.

Under the (probably erroneous) assumption that the dominant contribution to viscosity is local, the measured value of ν can be used to estimate an upper limit on the root mean square (rms) random velocity of particles in the ring plane c via the expression (Goldreich and Tremaine 1978):

$$c^2 = 2\nu\Omega\,\frac{1+\tau^2}{\tau}. \tag{4}$$

If the velocity ellipsoid is isotropic, then the ring's centrifugal scale height is simply $H \simeq c/\Omega$. The lower values quoted above for ν imply that $c \leq 0.05 - 0.25\ \mathrm{cm\ s^{-1}}$ and $H \leq 4 - 20$ m in the A ring and that $c \leq 0.2\ \mathrm{cm\ s^{-1}}$ and $H \leq 10$ m in the B ring, consistent with

[3] Akin to molecular viscosity in a fluid medium.

the more direct estimates given above. In the Cassini Division (and perhaps also the C ring), where self-gravity wakes are absent, the same argument leads to $H \simeq 3-5$ m (Colwell et al. 2009). In all cases, the ring thickness appears to be comparable to the size of the largest ring particles (see Sect. 2.3 below).

A lower limit on the velocity dispersion comes from the requirement that the rings be stable against axisymmetric and/or wake-type instabilities due to their self-gravity. Known as the Toomre criterion (Toomre 1964), this specifies a minimum value for the dimensionless ratio:

$$Q = \frac{c\Omega}{\pi G \Sigma} \tag{5}$$

of about 1. Numerical simulations show that stable rings with self-gravity wakes typically have $Q \simeq 2$ (see Fig. 14.7 of Schmidt et al. 2009), leading to estimates of $c \simeq 0.10$ and 0.15 cm s^{-1} in the A and B rings, respectively.[4] For further discussion of the dynamical and structural evolution timescales of the rings, see Crida et al. (2024, this collection).

1.2.2 Compositional Evolution

Apart from their dynamical evolution, ring particles also undergo compositional evolution because they are continuously exposed to exogenous processes such as photolysis and meteoritic particle bombardment, which are capable of altering their original state and chemical makeup. The composition of the rings is one of the key measurements that is useful in constraining their age. This latter topic is discussed in detail in Crida et al. (2024, this collection), and is briefly discussed here as context for the ring composition as a whole.

While the majority of compositional data for the rings comes from remote sensing at UV to infrared wavelengths, and therefore represents the ring particles' surface materials, these data are still useful constraints for the bulk composition of the rings, with the exception of the middle of C ring. The continuous low-velocity collisions among nearby ring particles, and mixing of their subsurfaces due to extrinsic bombardment by meteoroids (known as "impact gardening") is commonly thought to result in a very efficient mixing of the exogenous and original ring material over timescales of 1 Myr or so, even at depths comparable to the size of the largest ring particles (Elliott and Esposito 2011). Similarly, the ring layer itself is well-mixed, as ring particles move vertically on their inclined orbits, colliding and bouncing around, fragmenting and re-forming, spending time at the faces as well as in the midplane and getting equally exposed on average (Morishima et al. 2010). Comparison of results from visible and near-IR studies with those obtained from microwave observations that sample the bulk of the ring material (Zhang et al. 2017a,b, 2019, see also Sect. 5) generally supports this mixing framework. Interestingly though, microwave data identify a unique part of the mid-C Ring where there is evidence favoring particles having icy mantles and rocky cores.

Owing to their huge surface-area-to-mass ratio (see Table 1), the rings are highly susceptible to the effects of micrometeoroid bombardment and the resulting ballistic transport of their impact ejecta. In particular, ring composition (which is predominantly icy, $\gtrsim 95\%$ by mass) is sensitive to the continuous infall of non-icy meteoroidal material that may accumulate over time. The CDA team's final analysis of an extensive set of *in situ* observations collected over the more than a dozen years of the Cassini tour provides the total mass flux of

[4]Note that the keplerian angular velocity $\Omega \simeq 1.29 \times 10^{-4}$ s^{-1} in the outer A ring and $\sim 1.66 \times 10^{-4}$ s^{-1} in the middle B ring.

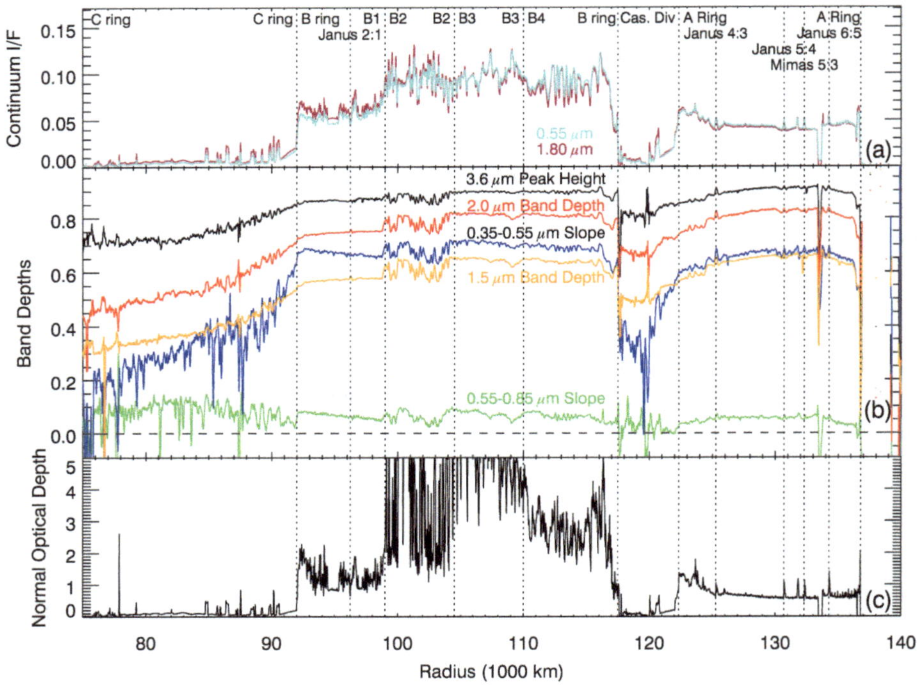

Fig. 2 Radial profiles of the main rings of Saturn derived from Cassini-VIMS observations. The upper panel shows the reflectivity of the rings at two continuum wavelengths of 0.55 and 1.8 μm, with the locations of the major ring boundaries and the strongest density waves identified by vertical dotted lines. The middle panel shows the fractional depths of the water ice bands at 1.55 and 2.0 μm, as well as the peak height at 3.6 μm and the blue and red slopes, rebinned to a uniform sampling resolution of 20 km and smoothed to 100 km. The lower panel shows an optical depth profile of the rings obtained from an occultation of the star γ Crucis, binned to 10 km resolution. (From Tiscareno et al. 2019)

extrinsic, non-icy material impinging on the rings (Kempf et al. 2023; see also discussion in Crida et al. 2024). The main conclusions from this study indicate that the magnitude of the measured flux entering the Saturnian system, between 0.69 and 2.7×10^{-16} kg m^{-2} s^{-1}, is roughly a factor of two smaller than previously derived estimates. Moreover, the dynamical origin of these extrinsic micrometeoroids is not from the Oort Cloud as previously assumed, but predominantly from the Edgeworth-Kuiper Belt (Kempf et al. 2023). The latter dynamical population is characterized by much lower entry speeds upon crossing Saturn's Hill sphere, and therefore is more strongly affected by gravitational focusing by the planet. As a result, the rate at which the rings are polluted is also faster. This polluting material is linked to the neutral absorbers, discussed further in Sect. 3.2.2.

2 Physical Properties of the Main Rings

The physical properties of the rings include characteristics such as their brightness and optical depth profiles, particle size distribution, surface mass density, temperature and thermal behavior.

Table 1 Radii, surface areas, surface mass densities, and estimated masses for the principal Saturnian rings

Ring region	R_{in} (km)	R_{out} (km)	Ring surface area (km^2)	Σ (kg m^{-2})	Mass (kg)	Mass(M_{Mimas})	Ref.
A ring	122,340	136,770	1.18×10^{10}	150-400	3.7×10^{18}	0.10	a
Cass. Div.	117,930	122,340	3.3×10^{9}	15–190	1.7×10^{17}	0.004	b
B ring	92,000	117,570	1.68×10^{10}	600	1.01×10^{19}	0.27	c
C ring	74,490	92,000	9.2×10^{9}	10–60	3.4×10^{17}	0.009	d
Total	74,490	136,770	4.11×10^{10}		1.43×10^{19}	0.38	

[a] 3-component model (Tiscareno et al. 2007; Tiscareno and Harris 2018)

[b] 2-component model (Colwell et al. 2009; Tiscareno et al. 2013b)

[c] average value used (Hedman and Nicholson 2016)

[d] 3-component model (Baillié et al. 2011; Hedman and Nicholson 2014)

2.1 Brightness and Optical Depth Profiles

Figure 2 compares two radial brightness profiles of the main rings in reflected light (I/F) as measured in the visible and near-infared (top), as well as a profile of optical depth (bottom) measured by the VIMS instrument on the Cassini spacecraft. The spectral profiles in the middle panel of Fig. 2 are discussed further in Sect. 3 below. In the A ring, both the optical depth τ and I/F profiles are relatively flat, with an average $\tau \simeq 0.5$ in both the UV (Colwell et al. 2010) and near-IR (Hedman et al. 2013). There are two relatively narrow gaps in its outer part, known as the Encke and Keeler gaps.

The B ring is somewhat brighter than the A ring and much more structurally complex, with τ_{UV} and τ_{IR} ranging from a minimum of ~ 0.9 in the inner parts (known as the B1 region; see labels at top of Fig. 2) to > 5 in parts of the central (or B3) region. The microwave optical depths exceed the radio science detection threshold in the densest parts of the B Ring, partly because $Q_{ext} \simeq 2$.

The C ring, on the other hand, is relatively transparent. It is dominated by an undulating profile with an average $\tau \simeq 0.05 - 0.10$, punctuated by a series of narrow, more opaque features known as "plateaux" where $\tau \simeq 0.5$, as well as several narrow gaps, some of which contain optically-thick ringlets.

Located between the A and B rings, the Cassini Division shows many similarities to the C ring, with an average $\tau \simeq 0.10 - 0.20$, comprised of eight narrow gaps and three isolated ringlets (Nicholson et al. 2018).

The sharp outer edges of the A and B rings are controlled by strong satellite resonances (Nicholson et al. 2018), but their more gradual inner edges take the form of "ramps", where τ decreases more-or-less linearly towards the much less opaque Cassini Division and C ring. These inner edges, and the inner edge of the D Ring, appear to be unconfined. In fact, many physical parameters change gradually across these ramps, which are believed to be generated by the process of ballistic transport (Durisen et al. 1989, 1996; Estrada et al. 2015). They are thus important sites for assessing the degree of long-term evolution of the rings, at least on a regional scale (for further discussion see Crida et al. 2024).

2.2 Surface Mass Density

The surface mass densities of the main ring regions (Table 1) have been estimated from the radial wavelengths of the many density and bending waves in the rings that are driven by external satellites and planetary internal oscillations (Tiscareno et al. 2007; Colwell et al. 2009; Baillié et al. 2011; Hedman and Nicholson 2014, 2016). Numerous estimates for the A ring show that $150 < \Sigma < 400$ kg m^{-2} (Tiscareno et al. 2007; Tiscareno and Harris 2018), with an average value in the central A ring of 350 kg m^{-2}. In the Cassini Division Σ is much lower but varies substantially, ranging from ~ 15 in the inner part to almost 200 kg m^{-2} in the outer ramp (Colwell et al. 2009; Tiscareno et al. 2013b). Reliable data are much scarcer in the B ring, but one study finds that $\Sigma \simeq 600$ kg m^{-2} over much of this region (Hedman and Nicholson 2016), with surprisingly little radial variation. For the C ring Σ ranges from ~ 10 to 60 kg m^{-2}, with the highest values in the central undulating region (Baillié et al. 2011; Hedman and Nicholson 2014).

Based on the above surface mass densities, we can estimate the mass of the main ring components, as listed in Table 1. In units of the mass of the satellite Mimas ($M_{\rm Mimas} = 3.75 \times 10^{19}$ kg), we find that the overall mass of the rings is $\sim 0.38\ M_{\rm Mimas}$. This value was confirmed when direct gravity measurements were made by Cassini on its final orbits, yielding $M_{\rm rings} = (0.41 \pm 0.13)\ M_{\rm Mimas}$ (Iess et al. 2019).

Indirect estimates of the average particle size in different ring regions can be made by combining the optical depth with the local surface mass density. As noted in Sect. 1.1.2, $\Sigma/\tau \simeq \frac{4}{3}\rho\bar{a}$, where ρ is the internal mass density of a particle and $\bar{a}$ is the mean particle radius. Assuming an average particle density of 500 kg m^{-3}, suitable for porous water ice,[5] we find that $\bar{a} \simeq 100$ cm in the A ring, $25 - 60$ cm in the B ring and $10 - 20$ cm in the C ring.

An unexpected result of such comparisons is that within both the B and C rings there appears to be little local correlation between Σ and τ, with τ varying much more than does Σ. This lack of correlation suggests that variations in the particle size distribution may be the cause of much of the observed variability in τ, rather than variations in the total amount of ring material per unit area. This is not currently well-understood. In the B ring, moreover, some local brightness variations may reflect variations in the albedo or phase function of the ring particles, rather than variations in either τ or Σ (Estrada and Cuzzi 1996).

2.3 Particle Size Distribution

The most direct constraints on the particle size distribution come primarily from the multi-wavelength RSS occultation data from Voyager (Zebker et al. 1985), estimates of the near-forward scattering cross-section of the rings at optical and near-IR wavelengths from Earth-based stellar occultation (French and Nicholson 2000), and comparisons of the measured values of optical depth τ in the Cassini UVIS (UV) and VIMS (near-IR) data with those obtained from microwave (RSS) data (Jerousek 2018; Jerousek et al. 2020).

The areal density $n(a)$ of particles of radius a can be modeled as a power law, $n(a) \propto a^{-q}$ for $a_{\rm min} < a < a_{\rm max}$. The power-law index q and the upper size limit for particles $a_{\rm max}$ are both found to vary across the rings, as estimated using various methods. Representative estimates by ring location are provided in Table 2. In the A and C rings and in the Cassini Division, Zebker et al. (1985) and Cuzzi et al. (2009) report a differential index $q = 2.7 - 3.2$ and $a_{\rm max} \simeq 2 - 10$ m. Because of its large optical depth, estimates of $a_{\rm max}$ and q have a much greater uncertainty for the B ring.

The value of $a_{\rm min}$ is more poorly defined, but occultations in the near-IR suggest that it varies from a few mm in the outer A ring and C ring to as much as 30 cm in the B

[5] "Ice" or "icy" are used interchangeably with "water ice" throughout the chapter, unless a different compound (e.g. methane) is noted.

Table 2 Particle size distributions in the main rings. Values are averaged by region for differential indices for model power-law particle size distributions in Saturn's rings. Fits are derived from Voyager RSS data (Zebker et al. 1985), Earth-based 28 Sgr occultation in 1989 (French and Nicholson 2000) and Cassini UVIS, VIMS & RSS data (Jerousek 2018; Jerousek et al. 2020)

Ring Region	Radial location (km)	Avg. q	Min. a_{min} (cm)	Max. a_{max} (m)
C Ring background	74500 - 91980	3.16[a]	0.42[a]	11[a]
C Ring Plateaux		3.08[a]	0.36[a]	13[a]
C Ring ramp	90620 - 91980	3.16[a]	0.38[a]	13[a]
B Ring[b]	91980 - 117500	~ 2.8[c,d]	0.5[d]	7[e]
Cassini Division background	117500 - 122100	2.85[f]	0.41[a]	8.9[a]
Cassini Division Triple Band	120550 - 120800	2.94[a]	0.61[a]	7.6[a]
Cassini Division ramp	120900 - 122100	3.07[a]	0.33[a]	20[a]
A Ring inside Encke Gap	122100 - 133410	2.9 ± 0.1[d,g]	~ 2.5[d]	$5 - 10$[d,g]
A Ring outside Encke Gap	133410 - 136800	3.05 ± 0.15[d,g]	0.4[d]	$5 - 10$[d,g]

[a]Jerousek et al. (2020)

[b]Constraints are limited to regions of the B ring with lower optical depth

[c]French and Nicholson (2000)

[d]Jerousek (2018)

[e]inner B ring, Jerousek (2018)

[f]mean of values from Zebker et al. (1985) and Jerousek et al. (2020)

[g]Zebker et al. (1985)

and inner A rings (French and Nicholson 2000; Cuzzi et al. 2009; Harbison et al. 2013). Figure 3 summarizes the published estimates of power-law indices across the rings, while Fig. 4 shows the corresponding values of a_{min} and a_{max}.

Inverting the near-forward scattered signal from RSS, Zebker et al. (1985) derived a power-law index between $q = 2.7$ in the A ring and $q = 3.11$ in the C ring, with particles ranging from 1 mm to ~ 10 m. They also estimated that the Cassini Division has a slightly shallower (*i.e.,* smaller q) particle size distribution than the C ring, with $q = 2.79$, and maximum sizes of $4.5 - 7.5$ m. Following the method of Zebker et al. (1985), Jerousek et al. (2020) determined separate power-law indices in the background C ring, in the C ring plateaux, in the background Cassini Division, and in the Triple Band that connects the inner Cassini Division outward to the ramp to the inner edge of the A ring.

Zebker et al. (1985) also found that the size distribution steepens from the inner A ring ($q = 2.7$) to the outer A ring ($q = 3.03$), with the largest values of q being found outside the Encke Gap at a radius of 133,500 km. This result was confirmed by subsequent studies (see Fig. 3). An abrupt jump in q has also been observed by Becker et al. (2016) in diffraction signatures at the edges of the Encke gap: they measured $q = 2.9$ and $a_{min} = 15$ mm at the inner edge and $q = 3.1$ and $a_{min} = 9.3$ mm at its outer edge. This trend continues to the outer edge of the A ring, where the minimum size is 4.4 mm with a power-law index of 3.2. This change in the size distribution may be related to an abrupt change in the A ring's photometric properties (color and phase function) at the Encke gap (Tiscareno et al. 2019). The relative lack of small particles in the inner A ring is consistent with the analyses by Marouf et al. (1986) and Jerousek et al. (2016). Becker (2016) suggests that this depletion in sub-mm particles may indicate that the collisions in this region are too gentle to release

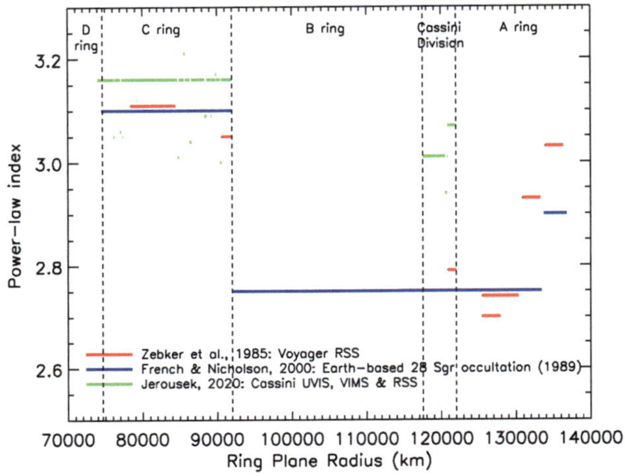

Fig. 3 Particle size distributions across Saturn's main rings: power-law indices inferred from Voyager RSS data (Zebker et al. 1985), 28 Sgr Earth-based occultation (French and Nicholson 2000), and Cassini UVIS, VIMS and RSS (Jerousek et al. 2020)

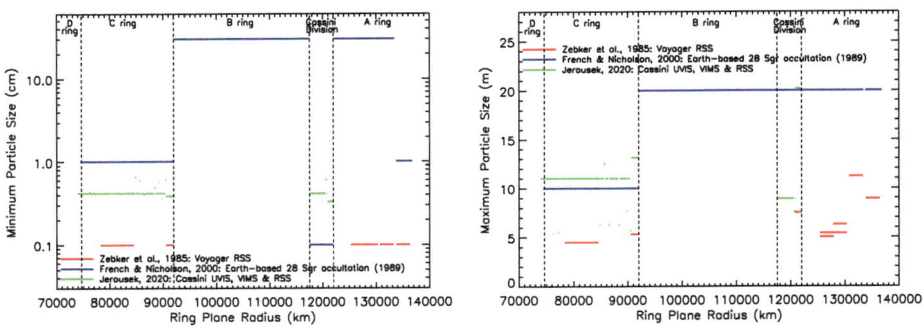

Fig. 4 Particle size distributions across Saturn's main rings: minimal (left) and maximal (right) limits for power-law distributions inferred from Voyager RSS data (Zebker et al. 1985), 28 Sgr Earth-based occultation (French and Nicholson 2000), and Cassini UVIS, VIMS and RSS (Jerousek et al. 2020)

enough energy to remove micron-sized dust from the surfaces of larger aggregates (Albers and Spahn 2006).

From the analysis of Cassini UVIS stellar occultation scans crossing the edges of the Encke and Keeler gaps, the R6, Huygens and Titan ringlets, as well as the outer edges of the A and B rings, Eckert et al. (2021) found minimum particle sizes ranging from 4.5 mm to 66 mm and average power law indices ranging from 3.0 to 3.2. Consistent with the results of Becker et al. (2016) and the RSS multi-wavelength profiles, this indicates that edges perturbed by satellite resonances tend to host more numerous sub-cm particles than the sharp edges of other ringlets. This result appears consistent with the predator-prey model detailed in Esposito et al. (2012), based on the competition between aggregation and fragmentation mechanisms in regions perturbed by satellite resonances. The kinetic model of Brilliantov et al. (2015) confirmed that the observed distributions could be consistent with the steady-state of such an aggregation/fragmentation mechanism.

Particles are much smaller in the D and F rings, with photometric and spectrophotometric data suggesting average sizes of a few tens of microns (Hedman et al. 2011; Vahidinia et al. 2011; Hedman and Stark 2015), although some regions of the F ring may contain a substantial population of cm-sized particles (Cuzzi et al. 2014b). See Sect. 6.1 for further discussion of particle sizes in the D ring.

2.4 Embedded Moonlets

There is also evidence for a small population of much larger ring particles that are up to a few kilometers in size, distributed across the main rings. Daphnis (3.8 km in radius) and Pan (14.1 km in radius) define the upper limit of this population, followed by a population of $\sim$ 100- to 1000-m- size objects, mainly observed by Cassini-ISS in the A and B rings (Spahn and Sremčević 2000; Tiscareno et al. 2006, 2008, 2010). These 100- to 1000-m objects are known as "propellers" because of the shape of the denser regions they produce in the rings surrounding depletion zones located ahead of the object (which is itself unseen) on its interior and behind the object on its exterior. Propellers form when the orbiting objects are too small to clear a full azimuthal gap but are large enough to clear small azimuthal regions of the rings in their immediate vicinity (Spahn et al. 2018). More massive moons result in perturbations that can open a gap all the way around the rings, as with Pan and Daphnis, which clear the Encke and Keeler gaps in the outer A ring (Nicholson et al. 2018, see Sect. 9 for further discussion). Propellers in the A ring are primarily detected in three narrow belts between radii of 126,750 and 132,000 km (Tiscareno et al. 2008), and their differential size distribution may be modeled as a steep power-law of index $q \simeq 6$ (Tiscareno et al. 2010).

Finally, from the analysis of Cassini UVIS stellar occultations, Baillié et al. (2013) reported the detection of a population of "ghosts" (small regions of low local optical depth, or "holes") in the C ring and in the Cassini Division, which they believe are consistent with the depletion zones surrounding propellers. Their estimate of the upper limit of the particle size distribution is consistent with the existence of a population of icy particles of up to 12 meters in the C ring and $\sim$ 50 m in the Cassini Division. These objects do not contribute significantly to the measured normal optical depth but may result in the "streaky texture" identified in the C ring by Tiscareno et al. (2019) in Cassini ISS images.

From the distribution of the ghost widths, Baillié et al. (2013) were able to model the size distribution of these boulders as a power-law with a differential index q = 1.6 in the C ring and q = 1.8 in the Cassini Division assuming the density of water ice. These power-law indices for objects between a few tens of meters and a few hundreds of meters are thus quite different from the estimated values for propellers in the A ring or for smaller ring particles. It is uncertain at present how the assumption of pure water ice for the ghost objects affects the retrieved size distribution, and further work in this area is needed. Ghost objects in the C ring and Cassini Division may represent the coherent shards of one or more consolidated objects that broke up to form the rings, as they exist well inside the Roche limit for solid water ice and are therefore unlikely to have accreted *in situ*. The increased abundance of non-icy material in the central C ring has been cited as additional evidence for the break up of a Centaur or other solid body (Zhang et al. 2017b, ; see also Sect. 5).

2.5 Ring Brightness Temperatures Measured by CIRS

The numerous radial scans of the main rings made with the Composite Infrared Spectrometer (CIRS; Flasar et al. 2004) FP1 detector have provided radial profiles of brightness temperatures, both on their lit and unlit sides (for a review see Spilker et al. 2018). The brightness temperature, which describes the temperature of a blackbody emitting the measured amount of radiation at a given wavelength, is a complex function of the ring's 3-dimensional structure, particle properties such as spin state, Saturn's season, the radius from Saturn (which is a supplemental heating source), and viewing geometry. On the lit side of the rings, the brightness temperature generally decreases from the C ring to the A ring except for the Cassini Division, reflecting the decreasing flux from the planet. Saturn remains a significant source of heating for particles within the close, thin C ring (Spilker et al. 2013). The higher temperatures seen in the C ring and Cassini Division are additionally attributed to the albedos of the particles being lower in these regions, as well as to the lower optical depths in these regions that results in reduced particle mutual shadowing.

On the unlit side, the optically-thin C ring and Cassini Division rings appear hotter than the more opaque A ring, while the optically thick B ring is the coldest of all, reflecting the effects of mutual shadowing between the particles as well as the difficulty with which heat may be transferred across the ring plane (Spilker et al. 2006; Morishima et al. 2010). Any deconvolution of the properties of the ring particles and regolith grains from the brightness temperature profiles is therefore ring-model dependent.

Morishima et al. (2010) have derived the ring particle bolometric Bond albedo A_V as a function of Saturn distance, assuming a multi-particle-layer model that mixes slow and fast rotating particles. The albedo for the A and B rings is roughly constant, with $A_V \sim 0.55$, while the C ring albedo seems to decrease towards the planet from 0.4 to 0.1. In this study, the infrared emissivity was fixed to unity. Were the true infrared emissivity to be a bit lower, the fitted temperatures would favor lower albedos to reproduce a given observed temperature. Other thermal models for the C ring provide average values of about $A_V \sim 0.2$ (see Spilker et al. 2018), similar to but slightly higher than the value of ~ 0.15 derived from visible and NIR observations (Smith et al. 1981; Cooke 1991; Cuzzi et al. 2009). In comparison, the empirical slab model of Altobelli et al. (2014) yields an albedo $A_V \sim 0.45$ that is approximately constant with distance to the planet, albeit with large fluctuations.

If the apparently increasing albedo from the C ring to the A and B rings is real, a correlation between local optical depth and Bond albedo should exist. Such a correlation would be compatible with a darkening pollutant, such as one delivered by meteoroid bombardment, comprising a larger portion of the optically thin C ring (Cuzzi et al. 2009). At this stage, no thermal model has taken into account this variation of albedo in terms of pollutant fraction.

Radial variations in ring brightness temperatures may also originate from changes in emissivity that can stem from compositional or structural causes, such as a change in the regolith grain sizes covering the ring particles. Morishima et al. (2012) have investigated radial variations in regolith properties of particles, assuming grains are made of pure water ice. Any variation of ring emissivity versus distance has then been interpreted as due to variations in the size distribution of grains with distance (see Sect. 3.1).

2.6 Comparison to Ring Temperatures Measured by VIMS

The differences in albedo, optical depth, proximity to Saturn and regolith grain size discussed in the previous sections are the main cause of the temperature variation observed across the rings. While Cassini CIRS data directly measure the rings' brightness temperature, as discussed in Sect. 2.5, VIMS data permit an indirect measurement based on the spectral properties of water ice. Specifically, the position of the 3.6 μm continuum peak exhibited by crystalline water ice is temperature-dependent and can be determined from optical constants measured in the laboratory. As Mastrapa et al. (2009) have demonstrated, the imaginary part of the refractive index of crystalline water ice shows a change in its minimum around 3.6 μm for a range of temperatures between 20 and 150 K.

A similar behaviour has been observed by Clark et al. (2012) in reflectance measurements of small grains of pure water ice at standard illumination conditions (phase = 30°)

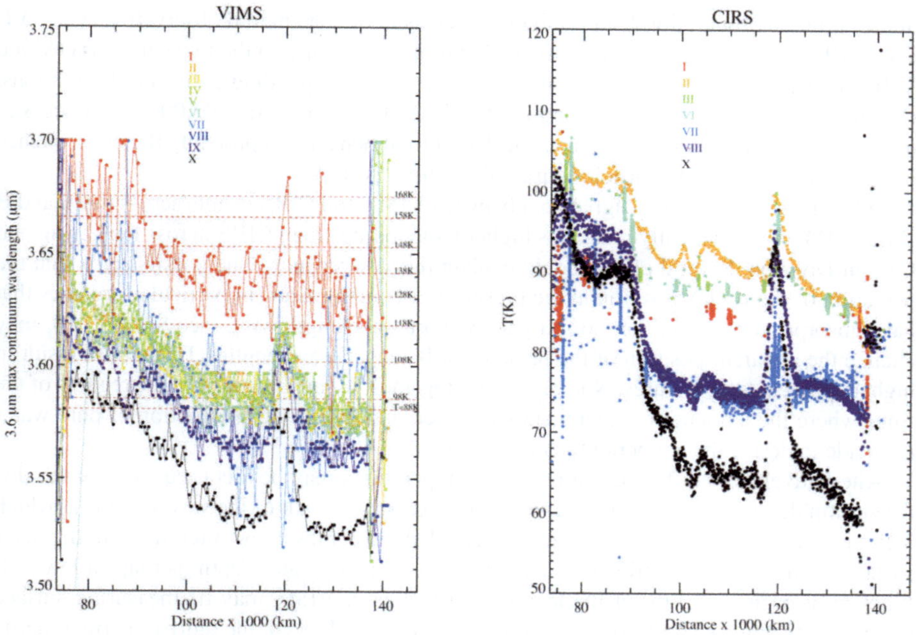

Fig. 5 Left panel: VIMS-retrieved 3.6 μm continuum peak wavelength and corresponding water ice temperature scale (horizontal dashed lines) for the 10 observations considered in Filacchione et al. (2014). From observation (I) to (X) the sub-solar latitude, or solar elevation angle, decreases from -23.5° to 2.6°. When the sub-solar latitude is negative VIMS observes the south face of the ring plane, (I) to (IX); when sub-solar latitude is positive, VIMS observes the north face, (X). Right panel: CIRS-retrieved ring's temperature during 7 of the 10 observations. For some CIRS observations, like in (VIII), the data shown are taken across two consecutive radial scans; this explains the differences observed across the C ring

for sample temperatures varying between 88 K and 172 K. By determining the wavelength of the 3.6 μm peak in reflectance, it is therefore possible to derive the temperature of the ice grains that are the main constituent of the ring particles. The 3.6 μm peak shifts towards shorter wavelengths when the ice is cooled, moving from about 3.675 μm at T = 172 K to about 3.581 μm at T = 88 K (Filacchione et al. 2016). By applying this method to ten VIMS radial mosaics of the rings sampled at 400 km/bin in the radial direction, Filacchione et al. (2014) derived radial profiles of the rings' temperature on the sunlit side as shown in Fig. 5. The observations were taken between 2004 near the southern solstice when the solar elevation angle B_0 was -23.5°, and 2010 near equinox ($B_0 = 2.6°$), allowing tracking of the diurnal temperature changes over one half of the seasonal cycle.

Before equinox the south side of the ring plane was illuminated, while after equinox the sun shone on the north face. The maximum temperature was measured in 2004 when the solar elevation angle, B_0, was at a maximum (observation I, red curve in Fig. 5). Ring particles cool down moving from southern summer to equinox. At equinox the solar rays are parallel to the ring plane and the particles receive the minimum solar illumination. As a consequence of this effect the peak's position shifts to shorter wavelengths and the minimum temperatures are observed across the rings (observation X, black curve in Fig. 5). When both solar and antisolar ansae observations[6] are available, such as in the two paired observations (V)–(VI) and (VIII)–(IX), it is possible to distinguish diurnal variations: on the two solar ansa (V and VIII), higher temperatures are measured than for the corresponding antisolar observations, (VI and IX). This effect is particularly evident in the C ring inward of 92,000 km. Since Saturn's thermal emission is isotropic, a similar thermal behavior is probably the consequence of the cooling of the ring particles during the eclipse period.

While there is substantial agreement between VIMS and CIRS results across the A and B rings, VIMS systematically measures higher temperatures than CIRS across the C ring and Cassini Division. Several effects could explain this discrepancy. One possibility is that the deviation of the peak's position in the presence of contaminants, in particular organics that have absorption bands in the 3.6 μm peak's short-wavelength wing (see Sect. 3.2.3), may change the apparent position of the peak towards longer wavelengths. This would result in higher apparent temperatures. Notably, the C ring and Cassini Division are the regions of the rings where the amount of contaminants is larger, and where the assumption of pure water ice made to derive the temperature is less robust.

Alternatively, the difference in reported temperatures may be a consequence of the different skin depths l_λ sampled by the two instruments as a result of the wavelengths at which they operate. While VIMS is sensitive to very shallow depths a few microns from the outer surface of the particles, CIRS measures temperature at a greater depth, perhaps a few millimeters or even a centimeter in clean ice. In this case, VIMS may be measuring surface temperatures while CIRS retrieves volume temperatures. Indeed, the diurnal thermal inertia for the main rings is estimated from eclipse cooling data to be $\sim 10\,\mathrm{J\,m^{-2}\,K^{-1}\,s^{-0.5}}$, with the lowest values in the Cassini Division. The corresponding thermal skin depth is on the order of 1 mm, greater than the characteristic VIMS penetration but less than that for CIRS (Ferrari et al. 2005; Morishima et al. 2011).

Overall, the comparison between VIMS' "indirect" temperature measurements and CIRS brightness temperatures reveals that they show the same radial trends, with temperatures on the sunlit side increasing towards Saturn.

3 UV/Optical/IR Observations of the Main Ring Composition

Cassini UV and near-IR spectral data show that Saturn's ring particles are predominantly made of crystalline water ice with minor contamination from non-icy materials (Cuzzi et al. 2009; Cuzzi, Filacchione and Marouf 2018a). While water ice is original to the rings, part of the non-icy materials that is responsible for the darkening and red-colored appearance at visible wavelengths is exogenous in origin. In this respect, when comparing the current composition of the rings with their original composition, one has to formulate some hypothesis about the ab-initio conditions proposed by the different formation scenarios, whether the rings are remnants of the Saturnian protoplanetary nebula or fragments of a destroyed moon or comet (Harris 1984; Esposito and De Stefano 2018). For example, rings originating in the protoplanetary nebula might be expected to contain some silicates and some volatile ices such as NH_3 or CO_2, while material tidally stripped from a differentiated icy satellite (Canup 2010) might be almost pure water ice. A disrupted comet might have a very different composition from a satellite's mantle. The different ring formation scenarios are discussed in Sect. 5 of Crida et al. (2024, this collection).

[6] Here "solar ansa" and "antisolar ansa" refer to the ansa closer to the subsolar (or noon) and antisolar (or midnight) points respectively. The antisolar ansa is therefore closer to Saturn's shadow, but both acquisitions are from the illuminated portion of the rings.

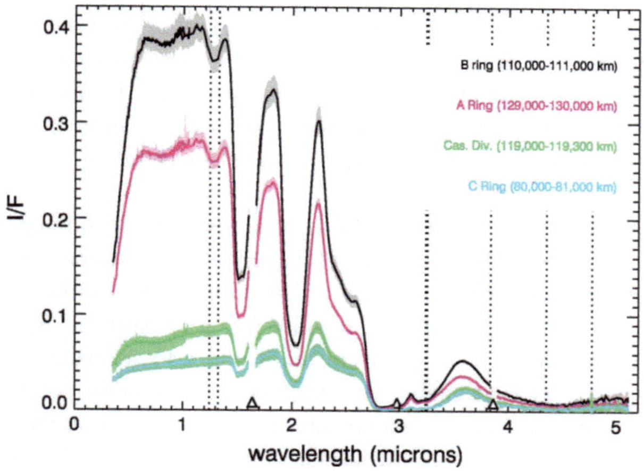

Fig. 6 Reflectance spectra of the sunlit side of the main rings of Saturn derived from Cassini-VIMS observations in 2005 at phase angles of $13 - 41°$. Vertical dotted lines mark the locations of "hot pixels" in the detector with higher than average dark current, while triangles mark the locations of gaps in the order-sorting filter, where the spectral data are less secure and thus not plotted. The solid curves show the mean I/F spectrum for each region, while the shaded bands indicate the range of values observed in each of these locations. These ranges primarily reflect variations in the absolute brightness of the ring, rather than statistical uncertainties in the spectra themselves. Note the strong water–ice absorption bands at 1.5, 2.0 and 3.0 μm, as well as weaker ice bands in the A and B ring spectra at 1.25 and 1.05 μm, and the steep slope at wavelengths below 0.6 μm in the A and B rings. Also note the variations in the spectral slope at continuum wavelengths between 0.6 and 2.3 μm among the different spectra. (From Hedman et al. 2013)

In Sect. 3.1 we discuss evidence for crystalline water ice as the dominant ring component, before turning in Sect. 3.2 to a description of endmember materials added to water ice (silicates and rocky material, carbon-bearing material and other ices) and their distribution across ring regions as derived from synergistic analysis and modeling of Cassini data. In Sects. 3.2.1 to 3.2.4 below, we describe spectral features that are observed in the data, as well as how these features have been interpreted in the context of radial trends and other associations with regions of the main rings. Modeling of the data is discussed separately in Sects. 4.1 and 4.2. The data and interpretations described in Sects. 3.2 to 4.2 are associated with spectra at nm to μm wavelengths, and constrain the surface composition of the ring particles. Regolith gardening processes may homogenize the particles on timescales of 1 Myr (Elliott and Esposito 2011). Data and interpretation related to the ring particle interiors are discussed in Sect. 5.

3.1 Water

The reflectance spectrum of the main rings from the near-UV ($\lambda = 0.30$ μm) to the near-IR ($\lambda = 5.2$ μm) is dominated by crystalline H_2O frost (Fig. 6), with typical grain sizes of a few tens to 100 μm (Clark and McCord 1980; Poulet et al. 2003; Cuzzi et al. 2009, 2018b; Filacchione et al. 2012, 2014). The comparative purity of the ice is supported by the very high albedo of the rings in the visible region, where I/F $\simeq 0.5$ at low phase angles, as well as an extremely low albedo in the $2.7 - 3.1$ μm region where water ice is strongly absorbing, and with the very strong absorption edge in the rings' far-UV spectrum at ~ 0.17 μm that is characteristic of fine-grained H_2O ice (Bradley et al. 2010, 2013).

Spilker et al. (2005) derived emissivity spectra for the A, B and C rings from CIRS data. Particles much larger than the observing wavelengths (16.7 to 1000 μm) were found to dominate the spectrum for most of the FP1 spectral range, since the observed emissivity spectra appeared free of features within the error bars across this range. Of most interest, however, they found a clear roll-off in the averaged ring particle emissivity at frequencies below 50 cm^{-1} ($\lambda > 200$ μm) that they attributed to a combination of particle size effects along with the low absorption of pure water ice in this spectral region (see their Fig. 5 in particular).

The shape of the roll-off is sensitive to the size distribution and upper and lower size limits of ring particles, as well as the nature of the regolith that covers them. With a radiative transfer model making use of a four-stream calculation along with the delta-M method on a single layer with randomly-distributed regolith-free ring particles, Spilker et al. (2005) found that a size distribution with power-law index $q = 3.4$ reasonably fits the spectrum of the B ring. However, no robust analysis has been performed on the possible range of these parameters, nor on the fraction of contaminants in the water ice. Near the wavelength of the roll-off, the transparency of water ice makes the influence of sub-millimeter-sized water-ice particles apparent, but seems to preclude the presence of grains smaller than a few tens of micrometers. The fact that the CIRS results show the influence of sub-cm size particles so clearly, and that Cassini radio occultation results that sense only free-floating particles have larger minimum-particle-size cutoffs (Sect. 5), suggests that the spectral signature of sub-cm size particles seen in the CIRS data might actually be a manifestation of regolith emissivity, which Spilker et al. (2005) did not model.

Morishima et al. (2012) derived the spectral emissivity from a large number of spectra accumulated by the CIRS FP1 channel at various seasons and viewing geometries before the end of 2010. Good fits were obtained using pure water-ice and a power-law distribution of grain sizes ranging from ~ 1 μm to 1 cm. For the main rings, they found that emissivity varies with ring temperature range (i.e. radial distance, season or phase angle). This observed emissivity is interpreted to be that of the regolith covering ring particles.

Despite many searches, there appear to be no unambiguous spectral indicators of other materials in the UV to near-IR spectral range (Cuzzi et al. 2009, 2018b), or in the mid-IR spectrum measured by the CIRS instrument (Edgington et al. 2008). However, the steep decrease in the main rings' albedo shortward of ~ 0.6 μm — which is especially apparent in the A and B rings in Fig. 6, and leads to their distinctly reddish color at visible wavelengths (Estrada and Cuzzi 1996; Cuzzi et al. 2002) — is *not* characteristic of pure water ice. This is discussed further in Sects. 3.2.1 and 3.2.2.

Data from the Cassini-VIMS instrument have been used by several investigators to spatially map the variations in the purity, regolith grain size and even the temperature of the water ice. This is done using the relative depths of the water-ice bands at 1.55 and 2.0 μm, the height and exact wavelength of the spectral peak at ~ 3.6 μm, and the average spectral slope between 0.35 and 0.55 μm (referred to below as the "UV slope").[7] Examples of such measurements are shown in the middle panel of Fig. 2, where it is apparent that the strengths of the water ice bands closely track the steepness of the UV slope. This has led to the interpretation that the material responsible for the UV absorption is embedded within the icy grains (*i.e.,* it is an "intra-mix" in spectral parlance; see Sect. 4). However, the observed variations in band depth could be due predominantly to variations in the mixing ratio of the non-icy material or to systematic variations in the water ice grain size in the regolith (Cuzzi et al. 2009).

[7] The band depths are defined as $B_\lambda = (I_0 - I_\lambda)/I_0$, where I_0 is the average continuum intensity level outside the band, while the spectral slope is defined by $S = (I_{\lambda 2} - I_{\lambda 1})/I_0$. Here I_0 is the average intensity level over the interval $[\lambda 1, \lambda 2]$, and I_λ is the calibrated I/F at wavelength λ.

Overall, the strongest water-ice bands are found in the outer A and middle B rings, and the weakest in the C ring and Cassini Division. However, the tranisitions between these regions are much more gradual than might be expected, given the sharpness of the underlying transitions in optical depth. Across the C ring and in the part of the A ring beyond $\sim 136,000$ km there is a steady weakening of the ice bands and of the UV slope (Nicholson et al. 2008; Tiscareno et al. 2019).

3.2 Non-water Components

The primary spectral features that have been ascribed to non-water components are referred to as the "ultraviolet absorber" (Sect. 3.2.1) and the "neutral absorber" (Sect. 3.2.2). Since organic material is a significant component in the Cassini Grand Finale *in situ* ring measurements (Sect. 7.2.1) and different carbon-based materials have been hypothesized as the cause of these two sets of spectral characteristics, we discuss spectral evidence for organic material in the main rings further in Sect. 3.2.3. Finally, in Sect. 3.2.4 we briefly address the lack of evidence in the main rings for volatile ices such as CH_4 or CO_2, which were however both observed in the in situ data in the D ring region (Sect. 7.2.1).

3.2.1 Ultraviolet Absorber

Two competing explanations have been offered for the steep decrease in the main rings' albedo shortward of ~ 0.5 μm (Fig. 6): the presence of small amounts of carbon or carbon-bearing material such as tholins within the ice grains (Cuzzi and Estrada 1998; Poulet et al. 2003), or the presence of nano-phase inclusions of metallic iron or iron minerals such as hematite (Fe_2O_3) or troilite (FeS) (Clark et al. 2012). These two models have very different cosmogonic implications (see further discussion in Sect. 3.2.3 below).

Iron metal or hematite, perhaps in the form of extremely tiny nanograins, has been proposed as a darkening and/or reddening agent (Clark et al. 2012). This suggestion has really only been made for the C Ring, which is far less red than the A and B rings (Sect. 4.2). Meteoroid bombardment and ballistic transport would predict that the C Ring is indeed more polluted by extrinsic meteoroids. However, radiative transfer models tend to find that carbon-tholin-silicate mixtures provide better fits to C ring spectra than iron metal or hematite (Sect. 4.2, Cuzzi et al. 2018b; Ciarniello et al. 2024).

While carbon-based materials therefore provide the better spectral fits, the Grand Finale data provide some new evidence for the presence of Fe-based materials in the rings. In analyses of CDA data from Cassini's Grand Finale (see Sect. 7), Fischer et al. (2018) reported two distinct compositional populations: iron-rich grains (93% of them sulfides) providing more than half of the detections; and iron-poor grains (most of them silicates) providing less than half. A dynamical analysis of these detected grains, expanded upon by Trieloff et al. (2023), found that the silicate component plausibly derives from ejecta off the surfaces of Saturn's retrograde irregular moons, while the iron sulfide-rich component is plausibly exogenic to the Saturn system. Iron sulfides and silicates have not been found to provide optimal fits to spectra of the C ring, and less so of the A and B rings. However, any measurement of the composition of non-icy material is of interest.

More recently, Linti et al. (2024) performed compositional analysis of silicate grains detected by CDA during the Grand Finale that were traced to the B, C, and D rings. They found that the compositions were homogeneous, independent of the source region, and that the grains were Fe-poor in comparison to chondritic materials, as well as in comparison to the Fischer et al. (2018) analysis. They suggest these results may point to chemical fractionation occurring in the rings that results in the small, Fe-poor silicate grains that were sampled, and larger, Fe-enriched grains that may not have been measured by *in situ* techniques.

3.2.2 Neutral Absorber

As noted above, the water-ice band depths in the C ring and Cassini Division are low in comparison with those in the A and B rings, as seen in Fig. 2. At the same time, the UV slope is shallower, which makes the rings' visible colors less red. These changes in the UV, visible and near-IR spectrum are accompanied by a decrease in the visual albedo in these same spectral regions that has generally been attributed to the addition of a spectrally-neutral absorber from an external source (Cuzzi and Estrada 1998). The higher concentrations of this material in the C ring and Cassini Division can then readily be explained in terms of these regions' lower mean surface mass densities and thus greater susceptibility to "contamination" by non-icy extrinsic material (e.g. Filacchione et al. 2014).

This neutral absorber has no specific spectral signature, but on cosmogonic grounds it has usually been identified either as carbonaceous material such as in carbonaceous meteorites or comets, amorphous elemental carbon from the *in situ* breakdown of accreted organic material, or perhaps silicates such as in stony meteorites. While most recent modellers assume this material is carbonaceous in nature, silicates remain a possibility. Future analyses of spectral data from the CIRS FP3 channel ($9-17$ μm) might permit detection of the characteristic emission or absorption features due to silicates in the $10-11$ μm region, but will be challenging due to the relatively low ring temperatures and poor signal-to-noise ratio of these data (Edgington et al. 2008).

3.2.3 Spectral Evidence for Organic Material

The question remains whether the neutral absorber in the rings (and probably also the UV absorber) is predominantly some sort of carbonaceous material, perhaps in the form of tholins, polycyclic aromatic hydrocarbons (PAHs) or amorphous carbon; or a combination of silicates and oxidized iron. It seems likely that this journal can only be finally settled by analysis of *in situ* data such as that discussed in Sect. 7 below, but there are some additional remote sensing observations that bear on this journal.

In their study of Cassini VIMS spectra of the rings, which covered a wide range of phase, incidence and emission angles, Filacchione et al. (2014) reported a detection of the fundamental — but very weak — absorption bands due to CH bond stretching. The analysis, performed on a low phase ($\approx 6°$), high signal-to-noise ring mosaic containing a statistically significant dataset (>8,000 individual spectra), allowed the detection of faint aliphatic CH_2 bands at 3.42 and 3.52 μm (Fig. 7). Conversely, the other main spectral feature associated with organic matter and particularly with C-C or aromatic bonds in polycyclic aromatic hydrocarbons (PAHs) at 3.29 μm,[8] which was detected by VIMS on the dark terrains of Iapetus, Hyperion and Phoebe (Coradini et al. 2008; Dalle Ore et al. 2012; Cruikshank et al. 2007, 2008), is not recognizable in VIMS rings spectra. These aromatic features are thought to be produced in warmer, higher-energy environments such as the protosolar or proto-Saturnian nebula. This feature seems to be absent in the rings, though its proximity to the very strong water ice absorption at ~ 3.0 μm makes any detection very difficult, if not impossible.

The detection of the aliphatic features across the rings and the absence of the aromatic band led Filacchione et al. (2014) to suggest an ultimate origin for this material in the diffuse interstellar medium where aliphatic bonds outnumber the aromatic ones (Pendleton and

[8] Dalle Ore et al. (2012) report the aromatic mode at 3.284 μm. The positions of the 4 aliphatic bands of CH_2-CH_3 symmetric-asymmetric are in fact not fixed, but have a small range of variability (~ 20 nm) depending on the specific molecule.

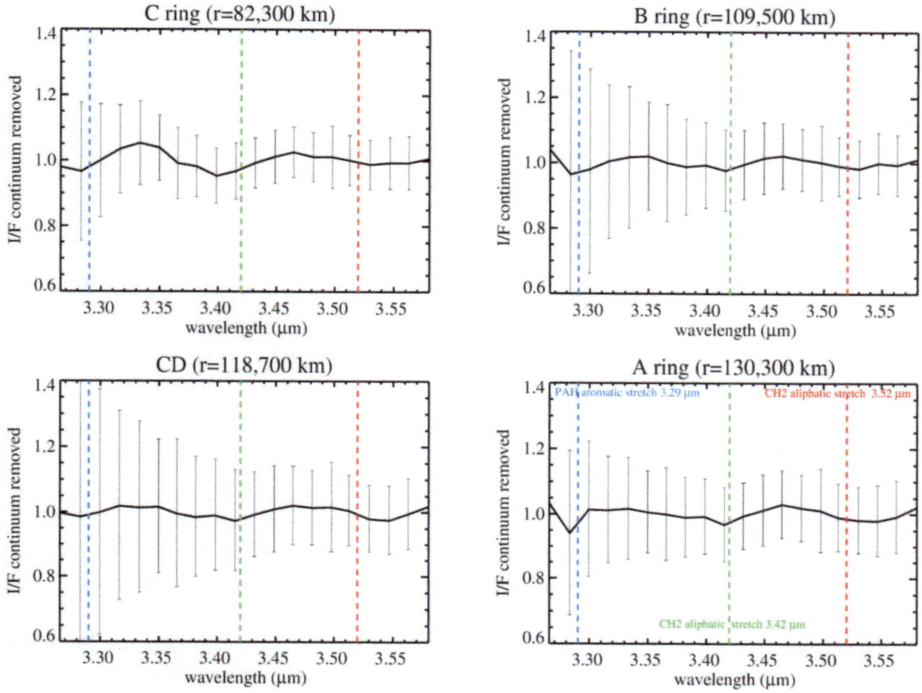

Fig. 7 Continuum-removed IR reflectance spectra by VIMS showing the two aliphatic hydrocarbon absorption features at 3.42–3.52 μm across the A, B, and C rings and CD. From (Filacchione et al. 2014)

Allamandola 2002; Raponi et al. 2020). Although seen in all major ring regions, the 3.42 μm band was stronger in the C ring and in the Cassini Division (Fig. 8), implying a likely association with accumulated exogenous, interplanetary debris that is now suspected to originate in the Kuiper Belt (Kempf et al. 2023). This spatial distribution matches very well with the quantitative spectral fits reported by Ciarniello et al. (2019), who found higher abundances of organics in the C ring and Cassini Division (see Tables 3 and 6).

A careful study of the B ring by Clark et al. (2019), on the other hand, failed to find any evidence for an aliphatic band at the 0.2% level, after the co-addition of almost 22,000 individual VIMS spectra from regions selected to be uncontaminated by saturnshine.[9] One way to reconcile these two observations, of course, is to attribute the CH band to the "neutral absorber" brought in by interplanetary bombardment. As noted above, this aliphatic material is more visible in parts of the rings with lower optical depths such as the C ring or Cassini Division, and much less abundant in the A and B rings.

3.2.4 Other Ices

Most common ices (*e.g.*, CO, CO_2, NH_3 and especially N_2 and CH_4) are too volatile to survive over geological timescales in Saturn's rings. Of course, merely being volatile is no guarantee of escape from Saturn's gravity field, but the gaseous species are likely to be photoionized and then effectively swept away into the magnetosphere on seasonal timescales

[9]Reflected light from the planet is dominated by strong methane bands, and thus carries an imprint of a similar C-H stretch feature at ~ 3.3 μm.

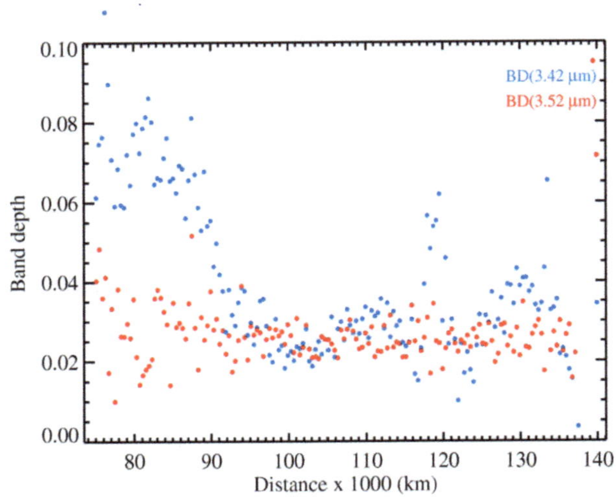

Fig. 8 Radial profiles showing the variability of the aliphatic hydrocarbon band depths at 3.42–3.52 μm. From (Filacchione et al. 2014)

(Cuzzi et al. 2009). The detection of other, less-volatile organic ices is again complicated by the presence of saturnshine, which imprints strong CH_4 bands in the rings' spectra except at large phase angles or at longitudes well beyond the terminator.

Nevertheless, CO_2 or NH_3 ice should be readily detectable, if they are present. Models of the near-UV spectrum suggest that there is very little or no NH_3 ice in the rings (Cuzzi et al. 2018b), while near-IR spectra of the rings show no sign of the strong CO_2 feature at 4.2 μm seen on Iapetus and several other mid-size icy satellites (Clark et al. 2012; Filacchione et al. 2014).

4 Regolith Radiative Transfer Modeling of the Main Rings

4.1 Overview of Regolith Radiative Transfer Modeling

Traditional regolith radiative transfer (RRT) modeling, following either Hapke (1993, 2013) or Shkuratov et al. (1999, 2005, 2012), treats regolith grains on the surface of a moon or an individual ring particle as independent, isolated scattering objects with some "single-scattering" albedo and phase function. The basic versions of these approaches presume the particles have no internal structure on scales of a wavelength; being internally uniform they are treated using some average real and imaginary refractive indices as a function of wavelength.

This approach has frequently been adapted to two similar-sounding regolith structures, which however have important differences from the standpoint of compositional interpretation (Fig. 9). One regolith structure is the so-called *intramixture* in which individual regolith grains are composed of some primary material with tiny embedded particles of different materials at sizes much smaller than the wavelength of interest. The refractive indices of these mixed grains are treated with Effective Medium Theory (for reviews and references see Bohren and Huffman 1983; Cuzzi et al. 2014a). On the other hand, a physical mixture of grains having multiple different sizes and/or compositions is called an *intimate mixture*, like salt and pepper or grains of sand on a beach. Traditional Hapke theory provides standard ways to treat this heterogeneous mix (Roush 1994; Hendrix and Hansen 2008), using an average grain albedo and phase function calculated and adopted for the entire mixture,

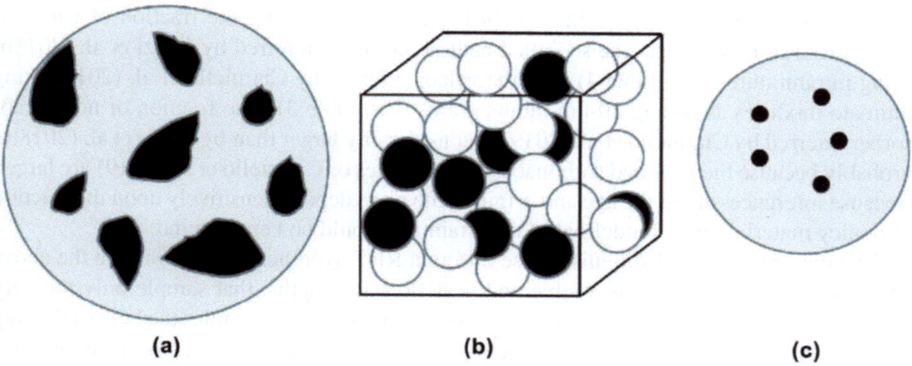

Fig. 9 Schematic diagram of different modalities for two-component mixtures. (a) areal mixture, blue circle represents instrument field of view; (b) intimate mixture; (c) intramixture, blue circle represents a single grain. From Ciarniello et al. (2011)

with different components weighted by their cross sections. RRT models can even include combinations of both kinds - an intimate mixture of grains containing different kinds of intramixtures (Cuzzi et al. 2018b). In the even more advanced adaptations these regolith grains can themselves have *internal scatterers*, distinct from any internal intramixed composition (Chap. 6 of Hapke 1993). The most basic assumption is that the regolith mixture has a single typical regolith grain size. However, multiple size distributions, different size distributions for different compositions, or other refinements are often incorporated, all at the cost of additional parameters (Poulet et al. 2003; Clark et al. 2012).

A complication of realistic RRT is that longer wavelengths may be insensitive to scattering by grains much smaller than the wavelength, instead selecting out grains (or clusters of grains) with size perhaps closer to the wavelength in question to interact most strongly with. This may be why RRT analyses often tend to infer "optimum" regolith grain sizes that increase with wavelength. For example, compare VIMS implications of particle size in Fig. 15.27 of Cuzzi et al. (2009) with those from UVIS analyses by Bradley et al. (2010). This effect is likely to be most important in the analysis of observations covering wide spectral ranges, from perhaps fractions of a micron well into the IR. Poulet et al. (2003) dealt with this problem by assuming a broad regolith grain size distribution, not unrealistic in general.

Intramixures tend to maximize the absorbing power per unit mass of nonicy material. Grains of even strongly colored material that is highly absorbing (large imaginary index) can become totally absorbing over wide ranges of wavelength at fairly small sizes, suppressing any wavelength dependence (i.e. becoming less strongly colored). Tholins are good examples of this. While their imaginary indices are strong functions of wavelength through the visible and NIR, grains of the pure material quickly become so opaque that their characteristic reddish intrinsic color caused by strong wavelength dependence is lost if they are more than several microns in size. This drove Cuzzi and Estrada (1998) to adopt a tholin-in-ice intramixing model to fit the very strong redness of the A and B rings in Voyager imaging data. The same approach was used by Poulet et al. (2003), and Cuzzi et al. (2018b) still find that this approach provides the best fits to HST-STIS ring spectra.

Even for material with strong but wavelength-independent intrinsic absorption, large grains become so opaque at even small sizes that their darkening power becomes proportional to their surface area, not their volume, weakening the absorption per unit mass as they get larger. An example of this might be seen in comparing the fraction of carbonaceous material in the rings (the so-called neutral absorber) inferred by Cuzzi et al. (2018b) using intramixtures (see Table 4) with the values inferred by Ciarniello et al. (2019) using intimate mixtures assuming 10-μm-radius grains (see Table 3). The fraction of nonicy absorber inferred by Ciarniello et al. (2019) is considerably larger than by Cuzzi et al. (2018b), probably because the assumed carbonaceous grain sizes of Ciarniello et al. (2019) are larger. Because inferences about the age and origin of the rings depend sensitively upon the fraction of nonicy material, these model-based uncertainties should be kept in mind.

One possible way to distinguish these different RRT hypotheses is to compare the above abundance inferences (both at visible and near-IR wavelengths, that sample only the very surfaces of ring particles) with cm-wavelength radiometry which samples the bulk of the ring particles. At the long wavelengths of microwave emission and scattering, intimate mixtures containing purely absorbing grains that are tens of microns or more in size behave like intramixtures. At present, the abundance of nonicy material over most of the rings inferred from 2.2 cm radiometry by Zhang et al. (2017a,b, 2019) seems to be closer to the results of Cuzzi et al. (2018b) than those of Ciarniello et al. (2019), possibly providing support for the lower fractions of nonicy material obtained from the intramixing model.

Finally, a recent wrinkle has emerged regarding ring RRT, that the planetary surfaces community has long been aware of: unresolved macroscopic shadows on/in the surfaces being observed make a surface look darker. If this effect is ignored, one infers an abundance of darkening material that is too large. It would not be unreasonable for ring particles, if they have a grape-cluster aggregate nature, to be even more strongly affected by this shadowing effect than other smoother surfaces. Hapke (1993, Chap. 12) has advanced a first-principles, geometrically-inspired, forward model along these lines, with several parameters. The role of shadowing in the rings context is also discussed by Cuzzi et al. (2017, see also references therein), where a simple one-parameter model is advanced to modify a basically Lambertian underlying smooth particle phase function. This simpler model turns out to be fairly similar in properties to the more complicated forward model of Hapke (1993).[10]

In the approach suggested by Cuzzi et al. (2017), the particle phase function is measured using observations over a wide range of viewing and illumination geometries. The shadowing behavior is manifested in the shape of the observed phase function, and this phase function defines the correction from the directly observed rough-surface, shadowed, effective ring particle albedo that rings analyses normally discuss, to a theoretical smooth-particle albedo with a Lambertian phase function and composition unbiased by shadowing. This is most easily done in low-optical-depth regions where multiple scattering is not a complication, but can be incorporated into a proper multiple-(ring particle) scattering model of arbitrary optical depth. The smooth particle albedo can then be modeled in terms of its underlying composition by the simplest forms of Hapke or Shkuratov theory. Of course, all of the above caveats apply as well to ring moon surfaces as to ring particle surfaces.

One prediction of this approach is that the shadowing parameter characterizing the observed particle phase function depends on the intrinsic, smooth-surface reflectance of the particle material - with particles appearing "smoother" at wavelengths where they are intrinsically brighter because shadows are filled in by nearby lit facets. A look at recent phase functions from icy moons provides some support for this hypothesis (Pitman et al. 2010), but more comparisons need to be done for the probably much craggier ring particles. [11]

[10] As the Hapke model is parameterized assuming shadows are formed by craters, one might not expect the quantitative results (slopes) to be meaningful for clumpy aggregate ring particles.

[11] An undergraduate class could test this conjecture in a large, empty, blacked out room with a flashlight and a light meter.

Along these lines, Cooke (1991) contains some carefully derived and highly relevant results, even if only in the Voyager clear filter. For example, the particle phase functions in all the outer C ring plateaux, and certain parts of the very inner C ring, are steeper (suggesting stronger shadowing, requiring a larger albedo correction) than the phase function in the mid-C Ring. The implications of this effect for the underlying intrinsic ring particle material is explored further in Cuzzi et al. (2024) using C ring regions intentionally chosen to be complementary to those of Cooke (1991) but covering Cassini UV3, BL1, and RED filters, as well as by Elliott and Esposito (2021) over all the rings using the basic Hapke approach.

4.2 Compositional Inferences from Modeling

Having described the approach, terminology, and caveats generally associated with spectral modeling in Sect. 4.1, we summarize here models of ring composition to date, as well as their primary conclusions. In one of the first attempts to apply a physics-based light-scattering model to the VIMS data, Hedman et al. (2013) modeled a selected subset of VIMS spectra using the 1D reflectance model of Shkuratov et al. (1999). Their principal results, shown in Fig. 10, indicate that while effective scattering lengths (a surrogate for grain size) are larger in the outer A and B rings than in the C ring or Cassini Division, the volume fraction of the UV absorber is in fact similar in the Cassini Division to that in the A and outermost B ring. The volume fraction of the UV absorber then increases monotonically inwards across the B ring, before decreasing again in the outer C ring. The above behavior supports the idea that the UV absorber is uniformly mixed with the water ice and is thus probably embedded within the icy grains, but also suggests that its volume mixing ratio generally increases with proximity to Saturn. In the Hedman models this component also appears to have a local maximum in the middle C ring, although some caution is appropriate here, as described below.

The model of Hedman et al. (2013) requires all the observed spectral properties to correspond to one of their three components: water ice, UV absorber, or neutral absorber. If the reality is more complex than three components, this approach may lead to analytical artifacts. For example, Fig. 14 of Cuzzi et al. (2002) shows very similarly increasing short-wavelength reddening inwards to the B ring edge, but nothing similar in the mid-C ring. Given the likelihood of excess non-icy material of *some kind* in the mid-C Ring (see Sect. 5), inclusion of an additional component in the model to account for rocky material may be more appropriate. Similarly, the Hedman et al. (2013) model ascribes a considerable amount of fine-scale photometric structure in the mid-B ring to grain size variations that affect spectral slopes. This approach of simplifying complex natural materials that may be present in the rings, such as chondritic grains, via modeling of a few key representative characteristics or pure components, such as amorphous carbon and silicates, is a common compromise that reproduces some features of the ring spectra within a more tractable parameter space. However, reality is more complex than any model, and there may well be other radial variations on those scales that became convolved into the grain size parameter.

The models shown in Fig. 10 confirm the much higher concentration of a neutral absorber within the Cassini Division and especially the C ring, where its volume fraction is $5-10$ times higher than in the A and B rings. It is noteworthy that the abundance of the neutral absorber seems to increase monotonically inwards across the C ring, whereas the microwave measurements show a peak in the abundance of the nonicy material near the mid-C ring (Sect. 5), much like that of the UV absorber shown here. These two different radial trends may further support different origins for the UV and neutral absorbers, and a link between the UV absorber and the non-icy material from microwave observations. The models of

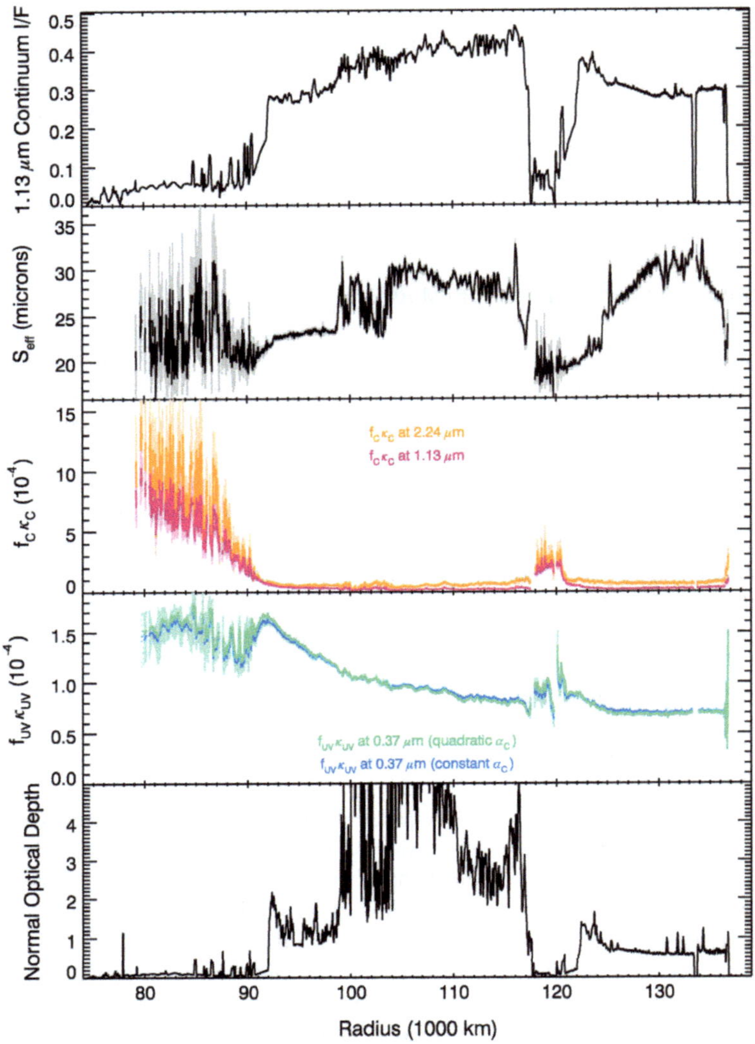

Fig. 10 Radial profiles of spectral parameters for the rings of Saturn derived from selected Cassini-VIMS observations and the 1D light-scattering model of Shkuratov et al. (1999). The upper panel shows the low-phase reflectivity of the rings at the continuum wavelength of 1.13 μm, The second panel shows the effective scattering length in the regolith, which may be loosely interpreted as an average "grain size". The third and fourth panels show the volume-weighted absorption coefficients of the continuum (or neutral) and UV absorbers, respectively. The lower panel shows an optical depth profile of the rings obtained from an occultation of the star γ Crucis, binned to 10 km resolution. (from Hedman et al. 2013)

Hedman et al. (2013) shown in Fig. 10 do not attempt to identify either the UV absorber or the neutral absorber, only their spatial variations.

Clark et al. (2012) pointed out the similiarity between the dark regions on Iapetus and other satellites with properties of the C ring and Cassini Division in terms of the shape and central wavelength of the strong water ice band at ~ 3.0 μm as well as the visible wavelength spectral slope. They suggested on the basis of this similarity that the best overall spectral fit, including the UV slope and infrared water bands, may be obtained with nano-scale inclusions (i.e. an intramixture) of iron and/or iron-oxides, rather than the so-called "ice-tholins". However, significant overall spectral and albedo differences between the satellites and the rings preclude any obvious similarities in the composition of even the C Ring, much less the A and B rings, with these dark regions (e.g. Figs. 15-34 and 15-37 in Cuzzi et al. 2009). In addition, recent spectral observations of the satellites in question actually find better fits from carbon-tholin mixtures than iron-hematite mixtures (Ciarniello et al. 2024). More recently, spectral fits to UV-visible STIS data from the Hubble Space Telescope that bridge the gap between the Cassini UVIS and VIMS datasets indicate that a good match to the steep UV slope seen in the A and B ring spectra can be obtained with tholins[12] embedded in fine-grained water ice (Cuzzi et al. 2018b). Attempts to fit this region of the rings' spectra using nanophase iron-bearing materials were found to be unsatisfactory. These models are discussed further below, and illustrated in Fig. 13.

Additional clues as to the composition of the non-icy component(s) of the main rings are provided by more sophisticated modeling of the VIMS spectra of Saturn's rings and moons based on Hapke's theory of light-scattering in planetary regoliths (Hapke 2012). Ciarniello et al. (2019) showed that the main ring spectra can be reproduced by intramixtures of water ice grains with inclusions of organic materials providing the UV absorption (Titan tholins; Imanaka et al. 2004, 2012; West et al. 2014), intimately mixed with sub-micron water ice grains and a variable abundance of other compounds contributing as neutral absorbers, either in pure form (amorphous carbon or amorphous silicates) or embedded in water ice (nanophase hydrated iron oxides, carbon, silicates, crystalline hematite, metallic iron, troilite). Corresponding optical constants are shown in Fig. 11.

Given the lack of diagnostic VIS-IR features, the nature of the neutral absorber and its corresponding absolute volumetric abundance cannot be unequivocally determined remotely. The latter in fact depends on the particular compound selected and on the corresponding grain size. In this respect, the model VIS-IR spectrum of opaque and spectrally-bland materials is virtually independent of the grain size, provided they are large enough to completely absorb light passing through them. This prevents a robust grain size characterization, thus potentially biasing the inferred abundance of the corresponding compound.

Nonetheless, assuming a common composition and similar properties of the neutral absorber across the rings, results from spectral modeling can still be interpreted in relative terms. Keeping this in mind, Ciarniello et al. (2019) determined the abundance of the non-icy materials across regions of the main rings characterized by different optical depths, under the assumption that the neutral absorber is represented by 10-μm-size amorphous carbon grains. For the A and B rings the best unmixing fits were achieved by assuming intimate mixing of three types of grains: (1) water ice grains intramixed with tholins, (2) submicron (0.2 μm) water ice grains, and (3) carbon grains that are 10 μm.

For the C ring it is necessary to add a further population of water ice grains containing troilite intramixture inclusions in order to match the higher broad-band absorption and to introduce a positive continuum spectral slope across the whole VIS-NIR spectral range. Their findings are summarized in Table 3 and in Fig. 12, and indicate that the content of tholin increases inwards across the main rings, while the fraction of darkening materials (amorphous carbon and troilite) is anti-correlated with optical depth. (See Sect. 4.1 however, for a cautionary note about a possible role for on-particle shadowing in lowering apparent particle albedo.)

[12] Physically, the decrease in reflectivity is attributed to a $\pi - \pi^*$ transition in the carbon rings common in tholins and other PAHs (Jaffe and Orchin 1962; Clar 1964; Birks 1970; Salama et al. 1996; Malloci et al. 2004).

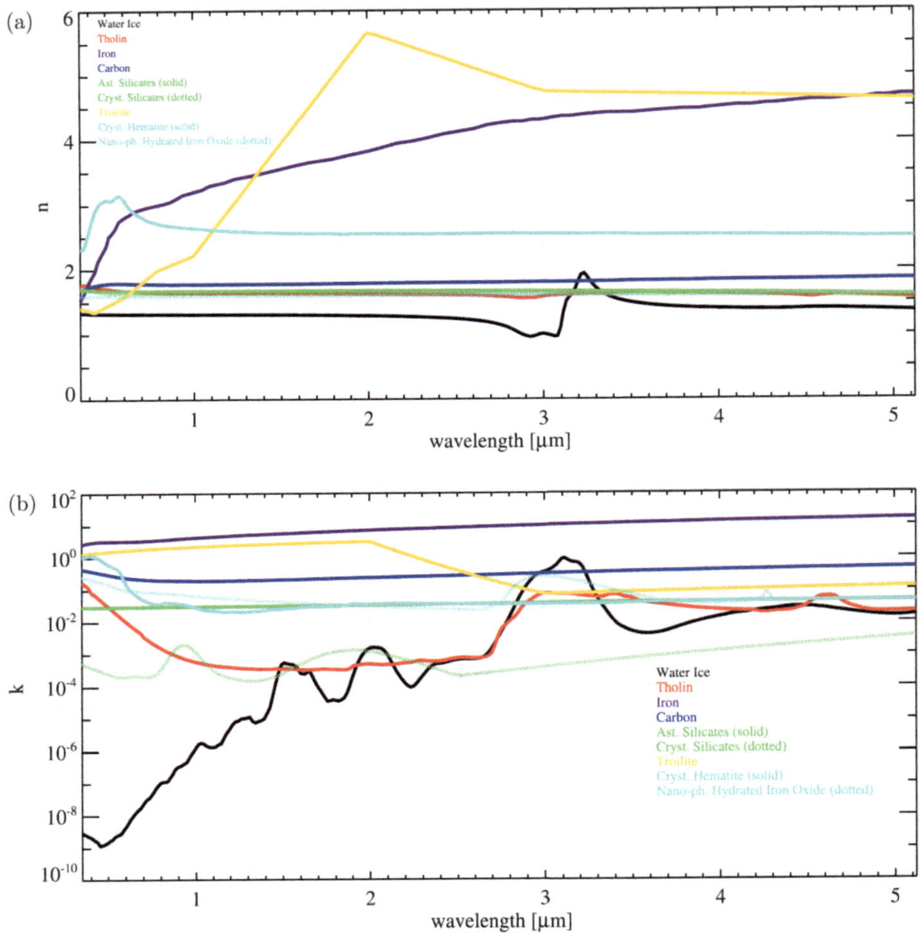

Fig. 11 Real (top) and imaginary (bottom) part of the refractive index of different materials used for ring spectral modeling. Note the decreasing absorption coefficient of water ice at shorter wavelengths, while those of tholins, hematite, and (to a lesser extent) carbon are rapidy increasing. From Ciarniello et al. (2019)

These results are consistent with the trends found by Hedman et al. (2013) for the UV and the neutral absorber, respectively (see Fig. 10). Also, the inferred size of the icy grains tends to be smaller in the C ring, in agreement with the scattering lengths or regolith grain sizes estimated by Hedman et al. (2013) and Filacchione et al. (2014).

The models discussed above were based entirely on VIS-IR data obtained by the Cassini-VIMS instrument that span the wavelength range $0.35 - 5.2$ μm. By extending such physical models for the ring particle reflectivity into the UV it is possible to derive further clues about the rings' composition, complementary to those obtained in the VIS-IR region. The near-UV spectral range, in fact, is very sensitive to the presence of chromophores, or UV-absorbers, including tholins, silicates, nanophase iron and amorphous carbon mixed with the dominant water ice.

Cuzzi et al. (2018b) fitted high-quality HST spectra of different ring regions using Hapke (1993) theory, with additional corrections made to take into account the shadowing effects caused by the particles' roughness (Cuzzi et al. 2017). Lacking such a correction, the shad-

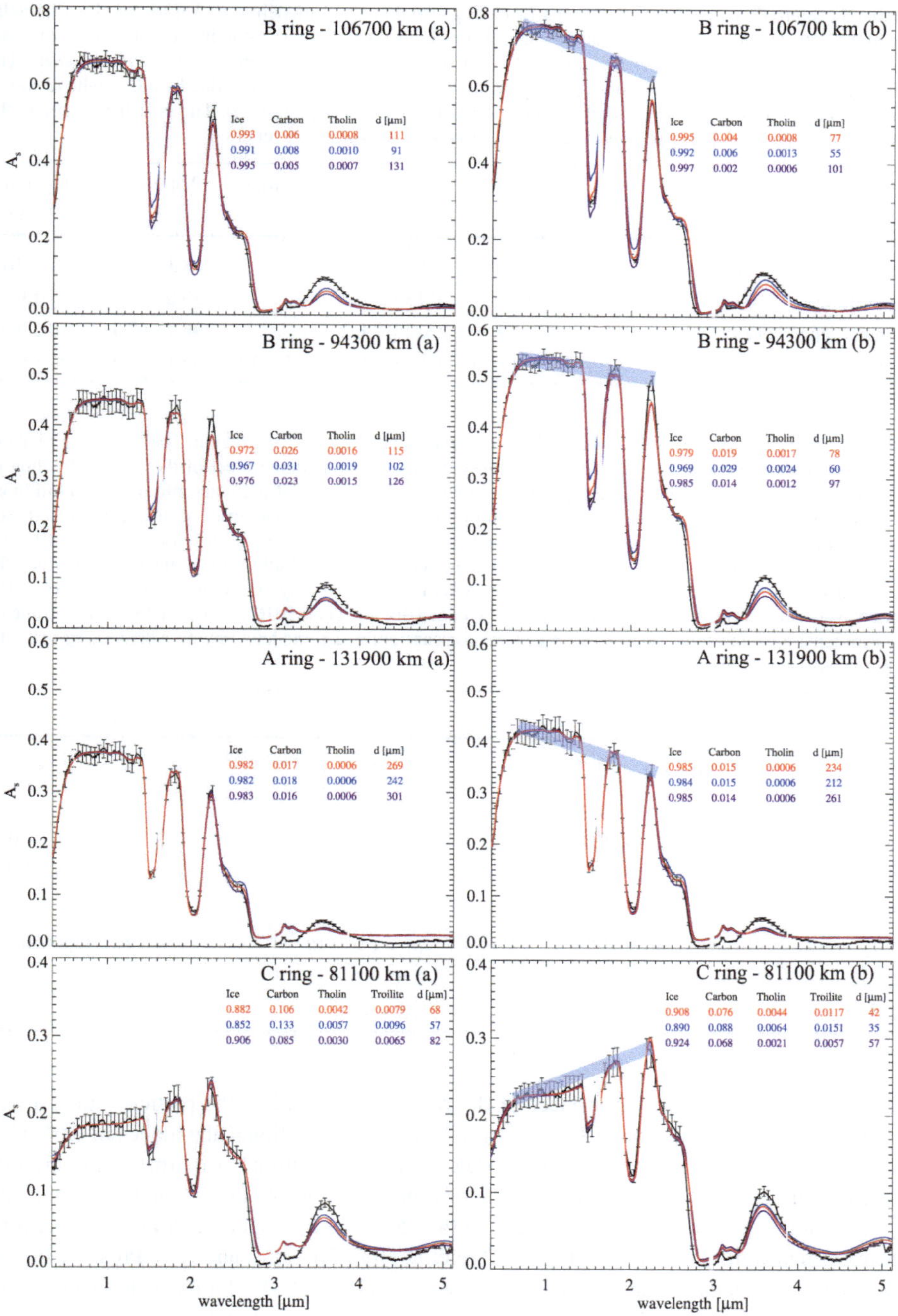

Fig. 12 Bond albedo spectra of Saturn's main rings derived from Cassini VIMS observations (black lines) and the best-fitting spectral models (red curve: best fit; blue and violet curves: models with accuracy comparable to the best fit). For each ring region two solutions for the Bond albedo spectra are shown (a, b), as resulting from ring photometric modeling performed in Ciarniello et al. (2019). Missing parts of the spectrum indicate instrumental order sorting filters. Note the positive continuum slope in the C ring between 0.6 and 2.3 μm, compared with the flat or blueish slope in the A and B rings, characteristic of mostly pure water ice. Adapted from Ciarniello et al. (2019)

Table 3 Inferred composition of Saturn's main rings from spectral modeling of VIMS data by Ciarniello et al. (2019). The models assume water ice grains with embedded tholin (ice-tholin intramixture) in intimate mixture with sub-micron water ice grains (0.2 μm size) and amorphous carbon grains (10 μm size). The C ring model requires an additional population of water ice grains with embedded troilite, whose size is assumed equal to that of the ice-tholin population (indicated as icy-grain size). The possible ranges in the total percentage of volumetric abundance of the different constituents are reported

Ring region	Radius [km]	τ	Carbon	Tholin	Troilite	Water ice	Icy-grain size [μm]
outer A ring	131900	0.6	1.39-1.78	0.061-0.063	-	98.15-98.54	212-301
mid B ring	106700	4.6	0.24-0.81	0.13-0.57	-	99.09-99.70	55-131
inner B ring	94300	1.4	1.42-3.08	0.12-0.24	-	96.73-98.46	60-126
mid C ring	81100	0.1	6.76 - 13.30	0.21-0.64	0.57- 1.5	85.18-92.45	35 - 82

Table 4 Total fractional abundances (F) of tholin, carbon and silicate by ring region. The total average abundance F_j for component j is given for different ring regions. F_j is the product of the volume fraction of grains containing component j , multiplied by the volume fraction f_j of pollutant intramixed within those grains. The range shown is inferred from best-fitting models across a wide array of different Hapke-based models that all fit comparably well. This systematic sensitivity testing provides a sense of the uncertainties in the spectral modeling results. Non-asterisked values come from the bulk of the simulations in which the grain size for each population is allowed to vary within a plausible range between 3–10 μm diameter. The grain size is constrained by the location of the UV edge (Bradley et al. 2010). Asterisked models assume all constituents have the same grain size; in general constant grain size models provide worse fits to the data. From Cuzzi et al. (2018b)

Ring	F_{tho}	F_{carb}	F_{sil}
A	$(4-10) \times 10^{-3}$	$(2-4) \times 10^{-4}$	$(0.5-7) \times 10^{-5}$
A*	9×10^{-3}	1.5×10^{-4}	1.6×10^{-3}
B	$(2.5-3) \times 10^{-3}$	$(0.5-3) \times 10^{-5}$	$(0.5-3) \times 10^{-5}$
B	1.2×10^{-2}	$(2-6) \times 10^{-6}$	$(0-1) \times 10^{-6}$
B*	8×10^{-3}	2.5×10^{-7}	1.9×10^{-3}
C	$(2-10) \times 10^{-3}$	$(5-15) \times 10^{-3}$	$(1-30) \times 10^{-4}$
C*	8×10^{-3}	2.6×10^{-2}	2.5×10^{-4}
C′*	7×10^{-3}	4.5×10^{-2}	2.0×10^{-4}

ows caused by the irregular surfaces of the ring particles darken the particle and can lead typical smooth particle models to overestimate the actual fractional abundance of darkening material. The main ring spectra were modeled with combinations of intimate and intramixture modalities of 10-μm-size grains made of water ice, amorphous carbon, tholins and silicates. Tholins appear only within a fraction ($4-50\%$) of the water ice grains, arguing for a heterogeneous mixture. Here, HP and LP refer to High Pressure- and Low-Pressure formation regimes (Imanaka et al. 2004, 2012; West et al. 2014). The LP tholins tend to have more aromatic bonds and the HP tholins tend to have more aliphatic bonds (Cuzzi et al. 2018b). In general neither iron nor hematite, in nanophase form or in larger grains, was found to provide satisfactory spectral fits to the HST data for the A or B rings. Examples of the best fits from Cuzzi et al. (2018b) are shown in Fig. 13, while the corresponding model parameters are listed in Table 4.

Two fits are given for the B Ring, where we see that the tholin abundance can exceed one percent or so only if the amorphous carbon and silicates are both reduced to extremely low

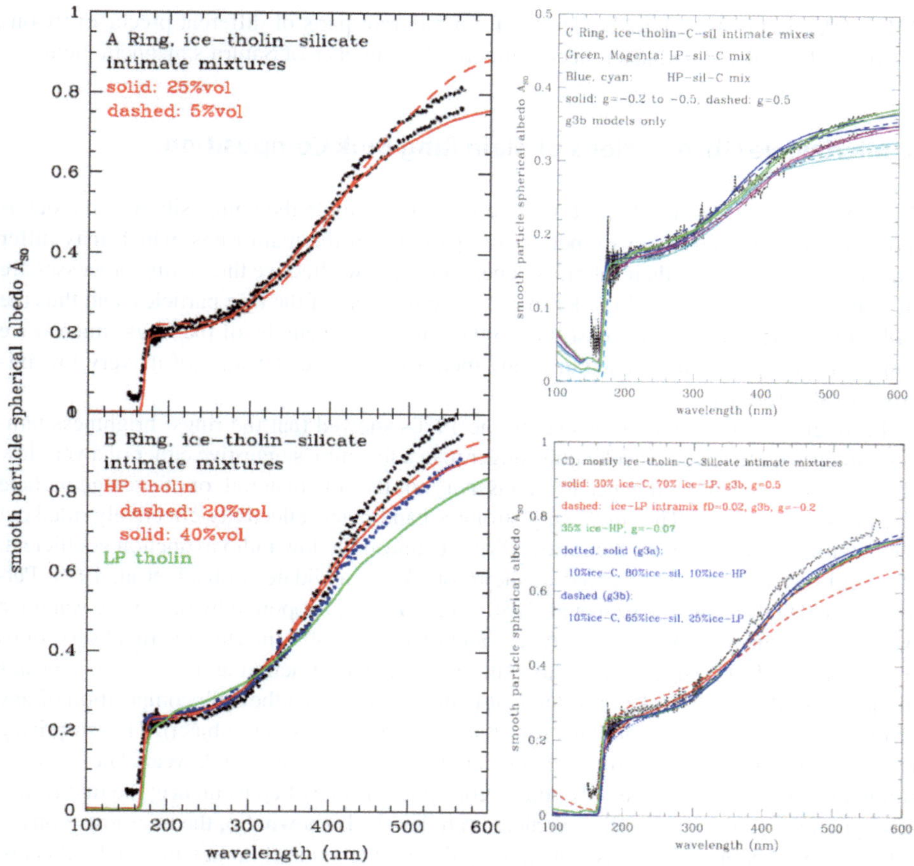

Fig. 13 "Smooth particle" Bond albedo of Saturn's main ring (A, B, C, CD) particles derived from HST observations (heavy and light dotted curves, representing two different particle shadowing parameters) along with the best-fitting spectral modeling results (colored curves).Several variants of Hapke models are shown, to provide a sense of potential model-based uncertainties. The parameter g is the assumed regolith grain scattering asymmetry, with negative values referring to backscattering grains and positive values to forward scattering grains. Models denoted g3a and g3b are slightly different ways Hapke theory can be implemented. For details of the models see Cuzzi et al. (2018b)

levels. For the A ring, the upper limit of about one percent tholin also occurs only in models with almost no amorphous carbon. The main difference between fits with variable (A, B) and constant (A*, B*) regolith grain sizes is that the latter have more silicates — but still only a fraction of a percent. For the C Ring, the variable grain size model (non-asterisked) has much more silicate than do the corresponding A and B Ring models. Underneath the nominal C ring fit, we show two alternate fits with constant grain sizes for comparison with the A* and B* models: one for the nominal rough surface (C*), and one where the particles are assumed to be smooth (C′*). The latter needs more amorphous carbon (*i.e.,* darkening agent) than does the nominal case.

As noted above, the radial distribution of the neutral absorber (*i.e.,* amorphous carbon or silicates) supports an exogenous origin by meteoroid or cometary bombardment (Cuzzi and Estrada 1998), which more effectively contaminates regions of low surface mass density, while the radial trend of the UV absorber (tholin) may reflect an intrinsic origin (Cuzzi et al.

2018b). Alternatively, it may be derived from different parts of different precursor moons (Crida et al. 2024), or perhaps reflect indirectly the influence of Saturn's magnetic field.

5 Microwave Observations of Main Ring Bulk Composition

The UV, visible and near-IR spectra, of course, only sample the composition of the outermost layers of the decimeter to meter-sized particles in the main rings, which may differ from the composition of their interiors depending on how effective the mixing processes are (Elliott and Esposito 2011). In order to probe the interiors of the ring particles, and thus the bulk of the ring material, we must turn to the longer wavelengths of the rings' microwave emission spectrum, which penetrate many meters of solid ice because of its very low microwave absorptivity.

Early ground-based measurements in the 1970s showed that the rings' brightness temperature was quite low at wavelengths longer than a few mm, suggestive either of a very low absorption coefficient (and hence low emissivity) of the ring material, or of average particle sizes less than ~ 1 cm. The rings' very strong S-band radar reflectivity effectively ruled out the latter possibility, leaving the former. With its unusually low radio extinction coefficient, pure, cold water ice soon emerged as the most likely candidate (Pollack et al. 1973; Pettengill and Hagfors 1974; Cuzzi et al. 1980), subsequently supported by the strong water ice bands seen in the near-IR spectrum (Clark and McCord 1980). Interferometric observations by Grossman (1990) suggested a maximum non-icy mass fraction assumed to be silicates in the A and B rings of $\sim 1\%$, a value that is much lower than the bulk composition of any of the mid-sized saturnian satellites (or of any known solar system object). This surprising result was not challenged or improved upon until the Cassini mission 25 years later. The assumption that silicates are the only nonicy constituent is one of convenience; the microwave absorption properties of all other candidates are less well known. So, the microwave observations should be interpreted as constraining the total amount of nonicy material (silicates, organics, etc).

The radiometer channel of the Cassini RADAR instrument (Elachi et al. 2004) observed Saturn's rings at a wavelength of 2.2 cm between 2004 and 2008 with resolutions of order 2000 km, providing an ideal window through which to study the non-icy material fraction within the bulk of the ring mass. At 2.2 cm, the absorptivity of water ice is negligible compared to that of most non-icy materials, and thus the intrinsic thermal radiation from the ring layer is dominated by its non-icy components.

The Cassini radiometer observations show an exceptionally high brightness when the C ring is in front of Saturn, at near-zero azimuthal angles. Besides directly-transmitted planetary thermal radiation, a large portion of the scattered radiation by the ring particles must be in the forward direction, suggesting a high porosity of 70-75% for the C ring particles. The ring particles could achieve such a quasi-equilibrium, highly porous regolith as a result of a balance between frequent fragmentation and excavation of debris due to micrometeoroid bombardment, ballistic transport to relocate the debris more gently on other particle surfaces, and inter-particle collisions to compact the particles. Most regions in the C ring contain about 1-2% silicates by volume (Fig. 14; Zhang et al. 2017b). This composition can be converted into an exposure age between 10^2 to 10^8 years if the ring system is assumed to be initially pure ice that has been continuously contaminated by in-falling micrometeoroids (Cuzzi and Estrada 1998; Kempf et al. 2023). Such a calculation assumes that the C ring optical depth and surface density has not changed significantly during that time, and also that the present-day micrometeoroid flux is representative. This absolute time scale is inversely

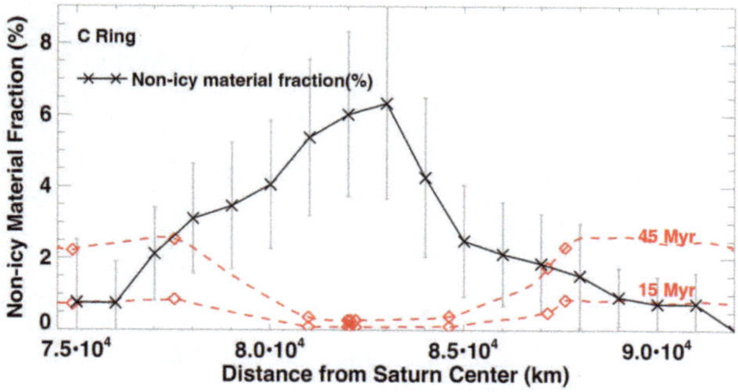

Fig. 14 The black curve shows the non-icy material fraction in the C ring derived from Cassini-RADAR observations. Red dashed curves show the predicted evolution of the non-icy material fraction over 15 and 45 Myr, assuming that meteoroid bombardment is the only source of contamination and that the meteoroid flux has remained constant over the past few tens of Myr. Red diamonds indicate the positions where opacity measurements have been made using density waves. The broad minimum in the predicted nonicy fraction is caused by the small opacities (high surface mass densities) observed in that part of the C Ring (Baillié et al. 2011; Hedman and Nicholson 2014). Figure from Zhang et al. (2017b)

proportional not only to the flux at infinity, but also to the amount of gravitational focusing by Saturn that the micrometeoroids experience before encountering the rings. See Sect. 2 of Crida et al. (2024, this collection) for a more extensive discussion of the exposure age, its relationship to the ring age, and associated assumptions.

An enhanced abundance of non-icy material is found in the middle C ring. Modeling of the middle C ring (Zhang et al. 2017b) reproduces the observed strong forward-directed scattering phase function as well as radially-variable thermal emission and low opacity (Hedman and Nicholson 2013; Zhang et al. 2017b) in the middle C ring when silicate material is included. Notably, the best fits are obtained when the silicate material is not well mixed with the water ice, but is instead present as large chunks in the cm to m size range, coated with porous icy mantles. In this case, the bulk composition may be on the order of 54 vol.% silicate. When the silicate material is instead modeled as volumetrically mixed with water, the non-ice component peaks at 6-11 vol.% at a radius of 83,000 km (Fig. 14; Zhang et al. 2017b). The significantly higher non-ice concentration in the middle C ring cannot be explained by direct meteoroid deposition alone. A possible scenario is that the entire C ring has been continuously polluted by meteoroid bombardment since it first formed, and the middle C ring includes additional contamination from a rocky body such as a Centaur that was broken apart by tidal forces. In this case, the spatial extent of the non-icy enhancement in the middle C ring suggests that the Centaur was likely to have been captured and integrated into the rings as recently as $\sim 10--20$ Myr ago (Zhang et al. 2017b).

Zhang et al. (2017a) found non-icy volume fractions of $\sim 0.3--0.5\%$ in the inner and outer B ring and as little as $\sim 0.1--0.2\%$ in the central region. For the A ring interior to the Encke gap, the derived non-icy volume fraction is $\sim 0.2--0.3\%$ everywhere. Finally, the Cassini Division has a non-icy fraction of $\sim 1--2\%$, similar to most regions in the C ring. The overall pollution exposure times calculated for the A and B rings and the Cassini Division from microwave constraints range from $\sim 30--150$ Myr, in line with the $\sim$ $15--90$ Myr derived for most regions in the C ring. For further discussion of the exposure age, see Sect. 2 of Crida et al. (2024, this collection).

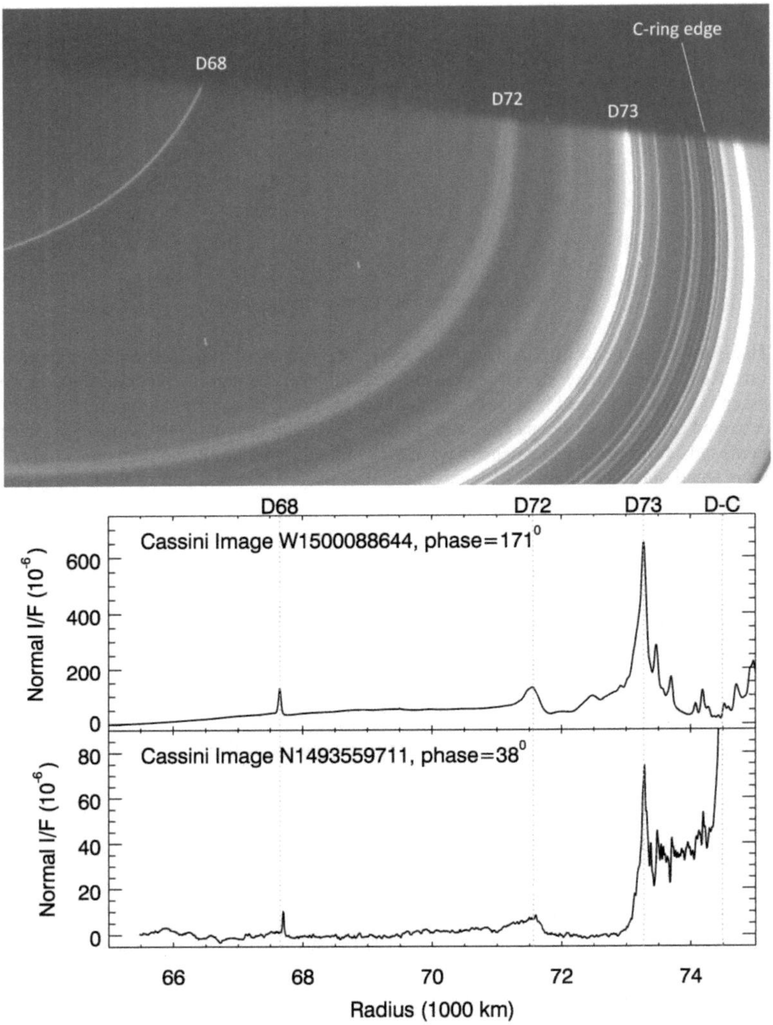

Fig. 15 Overview of the D-ring's structure. The top image shows a Cassini image of the D ring obtained at a phase angle of 154° (N1541397571), while the bottom image shows radial brightness profiles of the ring from selected images obtained at high and low phase angles (adapted from Hedman et al. 2007a). In all cases, the locations of the three named ringlets and the inner edge of the C ring are marked

6 D Ring

Having reviewed the physical and chemical composition of the main rings, we now turn to the innermost D ring. The D ring is a relatively tenuous but complex ring that extends from the main rings down almost to the planet's upper atmosphere. Any material coming from the main rings and entering the planet's equatorial atmosphere must pass through this region, so the structure and composition of this ring is relevant to connecting the composition of the main rings with the material flowing into the planet.

Figure 15 provides a general overview of the D ring's structure (Hedman et al. 2007a). The most prominent features in this ring are three ringlets designated D68, D72 and D73 after their distance from Saturn's center in thousands of km. Exterior to D73 there is a region that contains a collection of additional ringlets when viewed at high phase angles, but appears instead as a faint shelf of material when viewed at low phase angles. Meanwhile, D72 and D68 are surrounded by a broad, diffuse sheet of material that is only clearly visible at high phase angles. This diffuse sheet of material gradually fades interior to D68 with a scale length of order 1000 km and no distinct inner edge.

6.1 Constraints on Particle Size and Composition

Remote sensing data provide some constraints on the composition and typical particle size in the D ring, and are the only data source for the bulk D ring properties. The fact that all parts of the ring become brighter at higher phase angles implies that the majority of the visible particles are less than 100 microns across (Hedman et al. 2007a; Hedman and Stark 2015). Larger particles may exist in the region exterior to D73, which is both unusually bright at low phase angles and has a measurable optical depth in some stellar occultations (Hedman et al. 2007a). The sudden appearance of clumps in the D68 ringlet near the end of the Cassini mission provides indirect evidence that larger objects may reside within the ringlet (Hedman 2019; A'Hearn et al. 2021), but these objects have not yet been directly observed. Thus far, no evidence for larger particles has been uncovered elsewhere in the D ring. At the same time, sub-micron particles are very inefficient at scattering light at visible wavelengths, so there are at present no strong constraints on the distribution of such tiny particles within the D ring, a fact that should be kept in mind when assessing the *in situ* data from CDA, MIMI and INMS described in Sect. 7.2.

The small size of the visible particles in the ring limits the amount of compositional information that can be extracted from the remote sensing data. Fortunately, near-infrared spectra of the D ring obtained by VIMS do provide some useful constraints on the composition of this material. As shown in Fig. 16, many different parts of the D ring show a clear and deep absorption band around 3 μm due to water-ice. This indicates that water-ice is a major component of the D ring, as it is for the rest of the rings. The question then becomes how much non-icy material could be in this region.

It is reasonable to expect that the D ring contains a higher fraction of non-icy material than the main rings because the observed spectral trends across the main rings clearly indicates that the concentration of non-icy materials becomes higher closer to the planet (Hedman et al. 2013; Filacchione et al. 2014). We can place a rough constraint on the amount of non-icy material in the outer and middle D ring based on the observed depth of the very strong 3 μm ice band. This particular band is visible at high phase angles because the real index of water-ice approaches unity, thus inhibiting surface scattering. This dip becomes weaker if non-icy material is added to the particles because this tends to increase the real index away from unity.

Fig. 16 shows the predicted ratio of brightness between 2.9 μm and 2.5 μm for dielectric spheres computed from Mie theory[13] as a function of particle size and composition for various mixtures of water ice and amorphous carbon. Note that the optical constants used for these calculations are obtained using measurements for water ice and amorphous carbon by Mastrapa et al. (2009) and Rouleau and Martin (1991), respectively, and the effective optical constants for the mixtures are computed using formulae from Cuzzi et al. (2014a). These calculations indicate that adding even 20% by volume of amorphous carbon to ice is enough to significantly change the depth of the ice band; this limit scales roughly proportionally with

[13]Readers are referred to van de Hulst (1981) for an introduction to Mie theory.

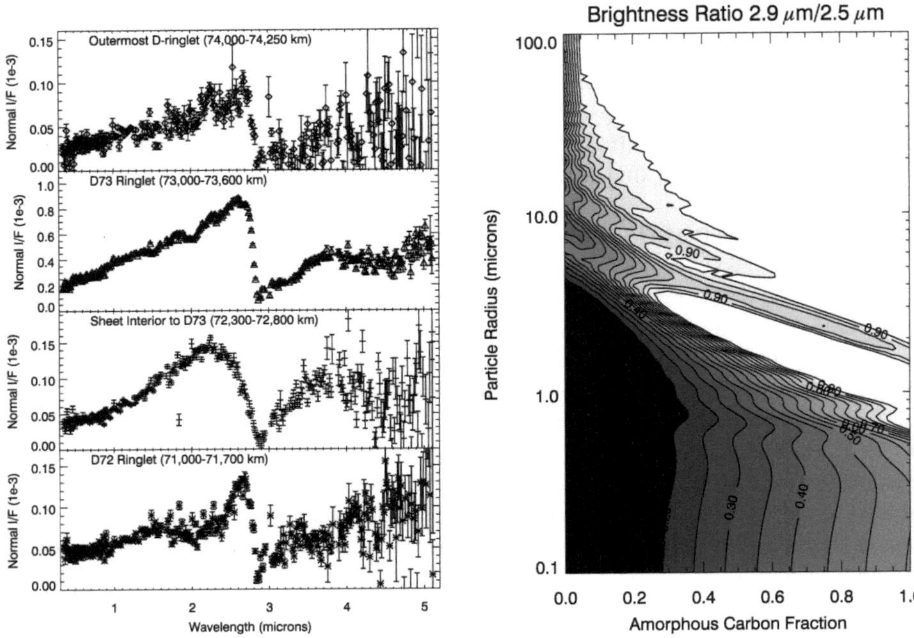

Fig. 16 Compositional information from near-infrared D-ring spectra. The left image shows spectra of various regions in the D ring derived from VIMS observations obtained at a phase angle of 167° (Hedman et al. 2007a). Note that all these spectra show a strong band at 3 μm due to the fundamental water-ice absorption band. These spectra have ratios for particle brightness at 2.5 μm and 2.9 μm that range between R = 0.04 and R = 0.24. The right panel shows as contours the predicted values for R as a function of particle size and composition assuming the particles are spheres with optical constants given by a mixture of water ice and amorphous carbon

the real index of the contaminant at 2.9 μm. Thus the strength of these ice bands suggests that the material in at least the outer and middle D ring is still predominantly ($\geq$ 80 vol.%) water ice. However, the ice-band strengths in the innermost D ring have not yet been analyzed in detail.

6.2 Evidence for Disturbances

Another important aspect of the D ring relevant for understanding the flow of material between the rings and the planet is that unlike the main rings, the D ring is time variable on time scales of years to decades (Hedman et al. 2007a). Between the Voyager flybys in 1980/81 and the arrival of Cassini in 2004 the structure of the middle D ring changed dramatically, with the brightest feature in the entire ring during the Voyager encounters (Showalter 1996) turning into a much more subtle feature 25 years later. During the Cassini mission, various parts of this ring were perturbed by sudden events, either impacts or temporal anomalies in the planet's magnetosphere (Hedman et al. 2007a, 2009, 2015; Hedman and Showalter 2016).

Finally, the innermost narrow feature in the ring, D68, developed a collection of bright clumps towards the end of the Cassini mission (Hedman 2019; A'Hearn et al. 2021). This localized brightening is most easily explained as the result of disruption of source bodies trapped in this region, releasing additional dust into the system. These sorts of changes could well have influenced the rate at which material flows from the rings into the planet.

7 Inflowing Material

During Cassini's Grand Finale orbits, the spacecraft passed between the inner edge of the rings and Saturn's upper atmosphere, to a depth of 1360 km above the 1 bar level. Initial expectations were that only Saturn's atmosphere would be detected by the suite of *in situ* instruments during these proximal orbits. As the spacecraft completed the proximal orbits, it became evident that an abundance of material was measured by the Ion Neutral Mass Spectrometer (INMS, Waite et al. 2004), the Magnetospheric Imaging Instrument (MIMI, Krimigis et al. 2004), and the Cassini Dust Analyzer (CDA, Srama et al. 2004) at or near the equatorial plane flowing into the atmosphere from the rings (Waite et al. 2018; Mitchell et al. 2018; Hsu et al. 2018). This chemically complex material included grains from nanometers to tens of nanometers in size, with mixing ratios that increased relative to the H_2 and He atmosphere with increasing altitude (Perry et al. 2018), supporting its origin exogenous to Saturn's atmosphere. Here, we review the current understanding of the material that has been observed flowing into Saturn's atmosphere from the rings, which breaks into two categories: ring rain at mid-latitudes coming in along magnetic field lines, and the equatorial inflow observed *in situ* by Cassini.

7.1 Charged Ring Rain at Mid-Latitudes

The first evidence for the flow of material from Saturn's rings into the atmosphere was measurement of ionospheric electron density profiles by Pioneer and Voyager (Kliore Astron et al. 1980; Tyler et al. 1981), which indicated a paucity of electrons with respect to the modeled densities (Waite et al. 1979; Atreya and Waite 1981). These previous chemical models required high electron densities to reproduce observations of long-lived H^+ ions. The addition of charge-exchange reactions with water as an additional sink for H^+ ions reduced the required electron densities and reconciled theory and data, leading to the identification of "ring rain" (Connerney and Waite 1984). Further analysis and modeling of this phenomenon identified that the main influx was due to movement of charged material along magnetic field lines, with deposition at mid-latitudes especially in the southern hemisphere (Northrop and Hill 1982; Connerney 1986; Ip 1983).

More recently, ground-based observations revealed that H_3^+ ion emission in Saturn's atmosphere is latitudinally variable, with a symmetric pattern offset from the equator that maps along magnetic field lines to areas of the rings with greater optical depth, which was interpreted as additional evidence of ring rain (O'Donoghue et al. 2013). Moore et al. (2015) reproduced the observed H_3^+ observations with an atmospheric chemistry model that includes a shifting balance between two loss pathways for H^+: reaction with water or reaction with vibrationally excited H_2. This leads to a complex relationship between H_3^+ and H_2O. Charge exchange with vibrationally excited H_2 produces H_3^+, and competes with the water pathway for removal of protons, subsequently decreasing the electron density. This in effect slows the rate of loss of H_3^+ by electron recombination. Combined with constraints on the electron density (Kliore et al. 2009) and haze latitudinal distribution (Connerney 1986), both of which are removed by water, O'Donoghue et al. (2013, 2019) utilized the H_3^+ distribution to infer the presence and abundance of ring rain.

Notably, Cravens (1987) identified quenching with atomic H as a major loss pathway for vibrationally excited H_2, and photoionization of atomic H is a major source for H^+ (Connerney and Waite 1984). Atomic H may play an important role in the inflow of ring material, including water, at the equator (Mitchell et al. 2018, see also Sect. 7.2 below). Observations of Saturn's Lyman-alpha brightness between 5°N and 35°N suggest that the mid-latitude abundance of atomic H may be higher than current model predictions by a factor of two to three (Ben-Jaffel et al. 2023). Better understanding of the latitudinal distribution and source of atomic hydrogen may therefore be an important step forward for understanding the Saturn-rings system.

The Cassini Grand Finale orbits provided a unique opportunity to directly measure the mass flux and composition of ring rain particles, which includes the small fraction of high-speed ejecta from micrometeoroid impacts across the rings that can be charged and travel along magnetic field lines (e.g., Northrop and Connerney 1987; Ip et al. 2016). Measuring the flux of these nanograins plays an important role in constraining the rings' age, origin, and their remaining lifetime (discussed in more detail in Crida et al. 2024), as well as for understanding certain observed properties of Saturn's atmosphere itself.

During Cassini's Grand Finale, the CDA instrument detected inflowing grains with peaks in distribution in the northern and especially southern mid-latitudes when the instrument was pointed in the plasma-ram direction, as well as at the equator when the instrument was pointed towards the Keplerian-ram direction (Hsu et al. 2018). Mapping of this distribution along magnetic field lines to the B and C rings with the observed north-south flux asymmetry was interpreted as evidence that this material constitutes direct observation of charged ring rain.

One of the interesting aspects of this observation is that in general, many of the detected particles are composed of silicates, which is in contrast to the water ice-rich composition of the main rings from which they originated, with between 8 and 33 weight% silicate grains reported in the CDA data (Hsu et al. 2018). There are a few ways in which this may be explained: (a) The detected grains could be biased towards non-icy material by the nature of the high-velocity initial impact into the rings that charged the grains, perhaps because the ice resident in the ring particles is more easily vaporized and recondenses back onto shadowed ring surfaces or is reduced to grains too tiny for CDA sampling; (b) impact ejecta grains retaining ice could be larger or have slower ejection speeds, and could be more closely restricted to low latitudes that were not as well sampled by CDA; (c) photosputtering of icy-silicate ring ejecta could readily convert the water ice into vapor as the grains flow into the planet's atmosphere at high latitudes; or (d) CDA deadtime effects (Kempf 2008) could lead to biasing of the grain composition sampling during flux peaks.

These possibilities may be assessed by a parallel analysis of the total amount of water that is entering the planet's midlatitudes. CDA measured a total flux of nanometer-sized grains arriving at Saturn consistent with a mass-deposition rate (MDR) of $\sim 300 - 1200$ kg s^{-1}, 70% of which fell near the midplane, and the remaining 30% ($\sim 100 - 370$ kg s^{-1}) at the so called ring rain latitudes (Hsu et al. 2018). This water-depleted inflow is insufficient to explain the amount of charged water products needed ($\sim 430 - 2800$ kg s^{-1}) to account for the observed latitudinal pattern of H_3^+ emission in Saturn's atmosphere (O'Donoghue et al. 2019). This may suggest that the water has been liberated from the observed grain population, and that this missing water is still somehow making it into the atmosphere, perhaps in vapor form. If one were to reconstitute the CDA grain population with sufficient water so that it were similar in composition to the inner edge of the C ring ($\sim$ 6% non-icy by mass, see Fig. 14), this would give a lower bound of an additional $\sim 470 - 1750$ kg s^{-1} of water (Durisen and Estrada 2023) which is more consistent with the O'Donoghue et al. (2019) result.

This water liberation may be explained by photosputtering of the icy material *en route* from the rings to the planet. This rate can be simply estimated from the far-UV flux at Saturn (between 10-1200 Å) of approximately 10^{-3} J m^{-2}s^{-1}, scaled from 1 AU (Colina et al. 1996; Chadney 2017). A 20-nm-radius grain, the best-fit size for CDA data (Hsu et al. 2018), then

Table 5 Composition of inflowing ring material in weight percent. Values presented here compare deconvolutions from Waite et al. (2018), Miller et al. (2020), and Serigano et al. (2022) and are scaled to account for the effects of diffusive fractionation in Saturn's atmosphere following methods from Waite et al. (2018). Organics ("Org") are held constant for comparison

Reference	CH_4	H_2O	NH_3	CO	N_2	CO_2	Org
Waite et al. (2018)	16%	24.2%	2.4%	20%	20%	0.5%	37%
Miller et al. (2020)	9.0%	17.9%	3.6%	15.7%	16.4%	0.4%	37%
Serigano et al. (2022)	12.6%	16.6%	2.2%	15.2%	13.7%	2.7%	37%

receives about 5×10^{-18} J s^{-1}. The latent heat of water is 2800 J g^{-1} so the energy needed to photosputter away a 20-nm-radius grain is about 0.9×10^{-13} J, giving an exposure time of about five hours, comparable to the time the grain takes to wander along field lines into the planet's upper atmosphere at ring rain latitudes (Hsu et al. 2018). The most likely effect of the CDA deadtime biasing the composition would be to increase the silicate/water ratio of the detected material. These numbers therefore may give an upper limit estimate to the amount of ring rain water that could be derived via this mechanism. Nevertheless, while additional work is needed it is possible that water ice is photosputtered off of icy ejecta grains, and is a significant source of Saturn's mid-latitude ring rain.

7.2 Equatorial Inflow

In addition to the flow of charged material inward from the rings along magnetic field lines, neutral material from the inner edge of Saturn's D ring can be decelerated by impacts with Saturn's extended atmosphere, and subsequently deorbited (Mitchell et al. 2018; Perry et al. 2018). The highest influx of material observed during the Grand Finale was centered around the equator and measured by INMS, which detected masses across its full range from 1 to 99 unified atomic mass units (u; Waite et al. 2018; Miller et al. 2020; Serigano et al. 2020, 2022). This dataset provides a new window into the composition of Saturn's rings, but there are some challenges to its interpretation. Cassini's relative velocity compared to the inflowing ring material was approximately 30 km s^{-1}. This high velocity may have led to impact fragmentation of material entering the instrument, such that the measured masses may represent impact products of the ring materials rather than direct measurement of unaltered ring materials.

Furthermore, data on neutrals from INMS in this time period were collected using the closed source, which has a higher sensitivity but can lead to artifacts caused by interactions of incoming material with the instrument. Such effects include a time delay in detection of materials that adsorb more strongly to the surface, such as water. This effect leads to compositional fractionation, which requires recreation of the original composition via integration of the data from each orbit into a single composite mass spectrum (Magee et al. 2009). Finally, while ionization fragmentation can yield unique signatures for many compounds, convolution of those signatures leads to degeneracies that cannot be resolved with the unit mass resolution of INMS. With these caveats in mind, some important conclusions can still be drawn about the composition of the inflowing ring material.

7.2.1 Composition

The INMS measurements included gas, both from the rings and from the nominal H_2 and He atmospheric gas component, and volatilized fragments generated by dust impacting at over 30 km s^{-1} into the instrument's titanium antechamber surface. Unexpectedly, there was a high abundance of material heavier than He with identifiable compounds (CH_4, CO_2, $CO/N_2/C_2H_4$, H_2O, NH_3), as well as a host of organic fragments. Together, these compounds comprised a measured mass density of 1.6×10^{-16} g cm^{-3}, as compared to the mean atmospheric mass density in this region of 4.7×10^{-15} g cm^{-3} (Waite et al. 2018). This material was measured at altitudes above the turbopause,[14] and the presence of these heavy masses with increasing mixing ratios at higher altitudes indicates that they originate from material falling into Saturn's atmosphere rather than being atmospheric constituents (Perry et al. 2018).

Overall, the inflowing equatorial material was found to be quite water-poor in comparison to remote sensing constraints. By mixing ratio, water is still the single most abundant compound, comprising approximately 25 molar% of the material measured by INMS (Miller et al. 2020; Serigano et al. 2020), or an estimated 20 vol.% assuming all components have approximately the same porosity (Table 5). This is notably lower than the main ring composition, which is greater than 90 vol.% water (see Sect. 3.1 and Table 6). However, the D ring composition is also water-poor compared to the main rings (Sect. 6.1) as is the mid C ring (Sect. 5), and the overall trend suggests enrichment of non-water material moving inward through the rings. Other abundant compounds in the equatorial inflow include organics with fragments dominated by compounds with 4 C atoms and comprising between 10 and 20 wt.% (g per 100 g), CH_4 comprising approximately 10 to 16 wt.% of the INMS sample, NH_3 at 2 to 3 wt.%, CO_2 at $<$ 3 wt.%, and CO, C_2H_4 and/or N_2 with a combined 20 to 30 wt.%.[15]

7.2.2 Evidence for a Gas-Grain Mixture

Based on comparisons of proximal orbits that crossed the ring plane (e.g. Rev290-292[16]) with the final plunge (Rev293) that did not pass through the ring plane, Miller et al. (2020) concluded that the inflowing material was latitudinally fractionated, with more refractory components depleted at higher latitudes (see their Fig. 3). They interpreted this as inflow of the more volatile components CH_4 and CO (or N_2) as gas, and the remaining, refractory components as grains, and estimated a gas to dust molar ratio of 0.7.

New analysis presented in this work of the scale heights of Saturn's atmospheric constituents and the major classes of inflowing material provides additional evidence for a gas-grain division (Fig. 17). Here, 15 u represents CH_4, 18 u represents H_2O, 28 u represents the combination of CO, N_2, and C_2H_4, 44 u represents CO_2, and 56 u is representative of organics. The scale height H and pressure P are described by

$$H = \frac{kT}{mg} \tag{6}$$

$$P = P_0 \exp\left(-\frac{z - z_0}{H}\right) \tag{7}$$

[14]The turbopause is defined as the vertical height in the atmosphere at which diffusion dominates over turbulent mixing.

[15]The INMS data cannot mass resolve CO, C_2H_4 and N_2. The confinement of the 28 u peak around the equator in high altitude proximals provides some evidence that more refractory C_2H_4 contributes to the 28 u counts (see Fig. 1b of Perry et al. 2018). Whether the remainder of 28 u material is dominated by CO or N_2 depends on the original source for the material. CO is much more abundant in cometary sources, for example.

[16]Here we refer to each different revolution around Saturn as a "Rev." Earlier literature on the Grand Finale dataset also uses the terms "Prox" for proximal orbit, or "Orbit"; these terms all refer to the same concept.

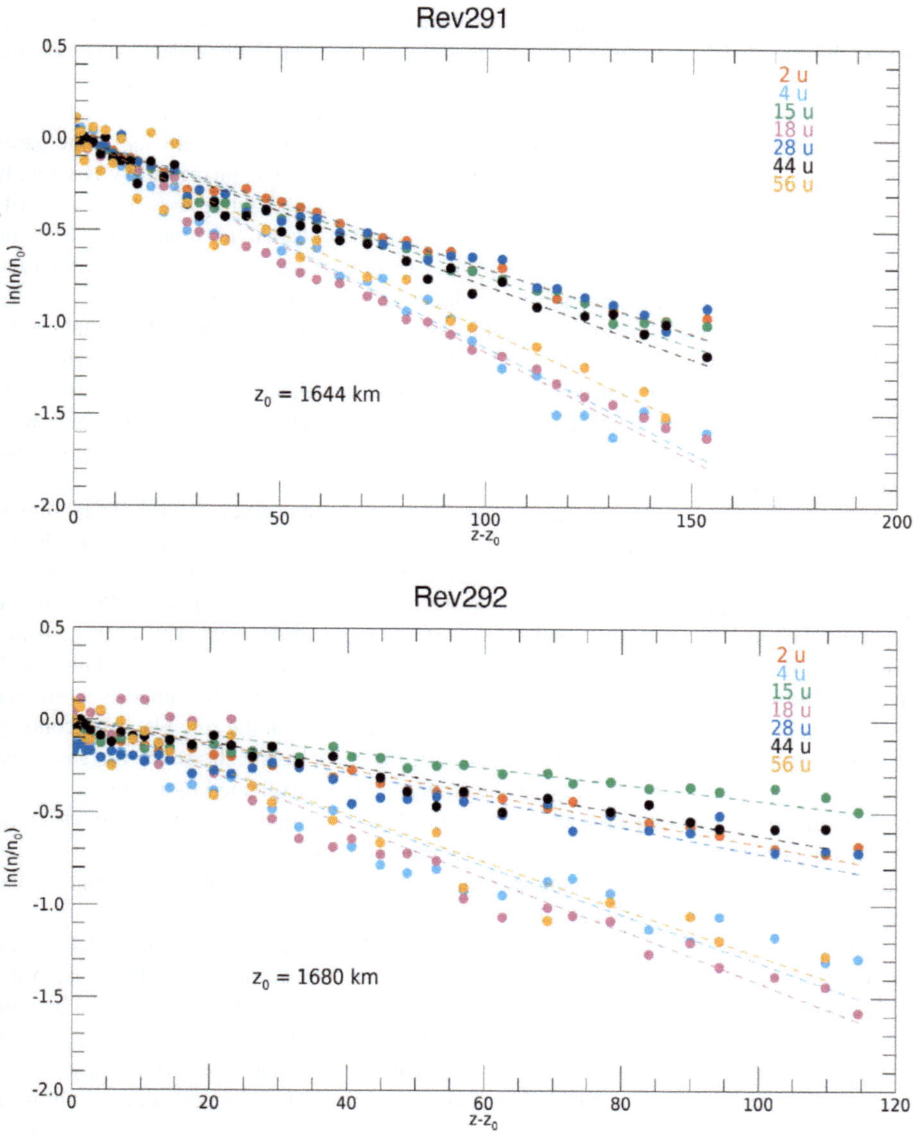

Fig. 17 Calculated scale heights for marker masses for atmospheric H_2 (2 u) and He (4 u), as well as for inflowing CH_4 (15 u), H2O (18 u), $CO/N_2/C_2H_4$ (28 u), CO_2 (44 u), and organics (56 u). Symbols show measured density values, and dashed lines show linear fits to the data. The different scale heights for H_2 and He are evidence that the measurements are made above the turbopause where diffusion velocity dominates. The different scale heights between volatile and refractory inflowing masses suggests that refractory materials are falling more rapidly because of contributions from grains

where k is the Boltzmann constant, T is the temperature, m is the compound mass, g is the gravitational acceleration, and z is the altitude. While scale heights calculated from the slope of the dashed fits in Fig. 17 match within formal uncertainty (which is not shown for visual clarity), two separate trends or groups are visible. The scale heights for H_2 and He differ by a factor of 2 ($\sim$ 140 km versus 70 km), matching the 2× mass difference between the compounds. This provides confirmation that the measurements are taken above the turbopause, where the scale height for upward-moving material becomes mass-dependent. Masses in the downward-flowing material that have been associated with the gas phase (Miller et al. 2020) have scale heights that are most similar to the bulk atmosphere (H_2). This suggests diffusive coupling of these gases with the atmosphere; that is, the downward velocity of the gases is controlled by the rate of diffusion through H_2 because the gas-phase inflow is strongly coupled with the atmosphere. Masses that have been associated with grains of less-volatile material (18 u and 56 u; Miller et al. 2020) have scale heights that are smaller than the bulk atmosphere. This may indicate that these masses are not yet coupled to the bulk atmosphere at the altitudes measured, and are falling at higher velocities in the form of grains.

The presence of inflowing grains is further supported by analyses of the Langmuir probe (LP) dataset. Johansson et al. (2022) modeled secondary emission produced by high velocity impacts of water vapor molecules on the Langmuir probe, and derived a median mixing ratio for gaseous water molecules of 10^{-5}. This value is approximately an order of magnitude lower than the water mixing ratio reported by Waite et al. (2018), suggesting that the majority of water measured by INMS may be present as grains.

Ionospheric chemistry also points towards the presence of grains in the equatorial inflow. Photochemical modeling identifies loss processes for H^+ and H_3^+ via reaction with a heavy compound M, which may be CH_4 or H_2O (Cravens et al. 2019). The modeled mixing ratio for M, f_M, is the same order of magnitude as the mixing ratio measured by INMS for water. At higher latitudes, f_M increases, while the equatorial region is dominated by a second heavy compound R that reacts with H_3^+ but not H^+. N_2 and CO are suggested as candidates for R. While the photochemical model of Cravens et al. (2019) therefore also supports latitudinal fractionation, the higher abundances of volatile CO and/or N_2 at the ring plane are contrary to the findings of Miller et al. (2020). However, the modeled reactions are for gas-phase compounds, and the depletion in f_M at the equator may be explained by retention of most water in grains at lower latitudes, while CO and/or N_2 may be in the gas phase at all latitudes.

This explanation is consistent with further modeling by Vigren et al. (2022), who consider the role of grain charging in this environment. Their model results require higher mixing ratios of M-type molecules and lower mixing ratios of R-type molecules than reported by Miller et al. (2020). Similar to Cravens et al. (2019), their photochemical model considers reactions of gas-phase water with H_3^+ and H^+, but does not include solid-phase water in icy grains. Vigren et al. (2022) report that the number density of grains required by their model is approximately equal to the difference between the number of positive ions, n_i, and the number of electrons, n_e. This difference is greater at lower latitudes, which may be suggestive of an increased grain population near the equator. They suggest an upper limit of ~ 40 ppm for M, and compare this to the 200 to 400 ppm abundances reported by Miller et al. (2020). This may suggest a molar gas to dust ratio for M molecules on the order of 40:200 (0.2) to 40:400 (0.1) if the grains are dominated by water ice, somewhat lower than the value of 0.7 estimated by Miller et al. (2020).

Modeling of Saturn's atmospheric chemistry by Moses et al. (2023) suggests that the physical state of the inflow as a gas-grain mixture rather than as gas only is significant for interpretation of the inflow timescales; see Sect. 7.2.4. Additional work to understand the distribution of gas and grains in the inflowing material is needed.

7.2.3 Comparison of Mass Deposition Rate Derivations

In situ measurements by Cassini captured instantaneous compositional snapshots of parcels of Saturn's atmosphere along each proximal orbit. To determine the mass deposition rate

Table 6 Comparison of ring compositions via in situ and remote sensing measurements. Remote sensing values summarize the upper and lower limits for non-water and water ice components respectively. In situ compositions include the average value from deconvolution results in Serigano et al. (2022) and Miller et al. (2020). Other ices includes NH_3 and CO_2. Gas includes CH_4, N_2, and CO. Rocky material includes silicates and troilite

	In Situ					Remote Sensing					
	INMS inflow			SUDA inflow		D ring[a]	C ring edges[b]	C ring mid[c]	B ring[d]	CD[e]	A ring[d]
	molar%[f]	weight%[g]	volume%	weight%[g]	volume%	volume%	volume%	volume%	volume%	volume%	volume%
water	24	16	31	67–92	87–97	>80	>92	>75[h]	>95	>92	>97
organics	34	50	55	–	–	<20	<6	<14	<4	<6	<2
other ices	11	8	14	–	–	–	–	–	–	–	–
gas	30	26	–	–	–	–	–	–	–	–	–
rocky	–	–	–	8–33	3–13	–	<2	<54[i]	<1	<2	<1

[a]see Sect. 6

[b](Cuzzi et al. 2018b; Zhang et al. 2017b); away from the central, anomalous peak (see Fig. 14)

[c](Zhang et al. 2017b; Ciarniello et al. 2019)

[d](Cuzzi et al. 2018b; Zhang et al. 2017b; Ciarniello et al. 2019)

[e]quantitative constraints from Zhang et al. (2017b), plus similarity to C ring from Hedman et al. (2013)

[f]effectively molecules per 100 molecules

[g]grams per 100 grams

[h]from measurement techniques that probe the surface

[i]from measurement techniques that probe the bulk volume

(MDR) required to produce these measured compositions, the downward velocity of ring material traversing through each parcel and the total surface area represented by each parcel must be constrained. To date, there are five reported calculations of a large equatorial mass influx of ring material based on data from Cassini's Ion Neutral Mass Spectrometer (INMS; Waite et al. 2004), including two different methods and three independent calculations (Waite et al. 2018; Perry et al. 2018; Yelle et al. 2018; Serigano et al. 2020, 2022). The results are summarized in Table 7. Calculations based on measurements by the MIMI Ion Neutral CAmera (INCA) of the equatorial influx (Mitchell et al. 2018), and by CDA of grains at the equator as well as at mid-latitudes are also included (Hsu et al. 2018). Ground-based measurements from the Keck telescope constraining the mass influx of charged water at mid-latitudes from modeling of H_3^+ densities are provided for comparison (O'Donoghue et al. 2019).

MIMI detected high energy impacts of large molecules in a tight band with a half width of +/- 1.4 degrees centered around the equator, when the spacecraft was moving through the upper atmosphere at a relative velocity of greater than 30 km s^{-1}. Particles were found to have a limited mass range greater than 1 $\times 10^4$ u. Assuming a mass density of 1 g cm^{-3} for the grains, the inferred particle sizes range from 1.6 to 2.5 nm. MIMI's INCA sensor detected a dust number density of 0.1 cm^{-3} (Mitchell et al. 2018). Utilizing terminal velocity from their model of collisional deceleration of dust grains (2.5 km s^{-1} at an altitude of 3000 km), and assuming a grain size of 15,000 u, Mitchell et al. (2018) scale the measurement for regions within 1° latitude of the equator to calculate a grain MDR of 5 kg s^{-1}.

The MIMI value is several orders of magnitude lower than the MDR calculated from INMS data, and the values do not appear to be reconcilable via reasonable adjustment of the model parameters. The MIMI influx calculation scales approximately linearly with grain mass. Use of the terminal velocity yields the upper limit for mass influx, and choice of a different altitude consistent with the same atmospheric model would lead to lower values for the MDR. Imposing a 10× increase on the atmospheric H density leads to a 3× increase in the mass influx. The most likely explanation may be measurement by the two instruments of different grain populations based on size (Perry et al. 2018), discussed further below.

INMS data are dominated by signal at the lowest altitudes, where ring material is diffusively coupled to the atmosphere, e.g. below 1700 km. Therefore, in the Waite et al. (2018) calculations of influx rate, diffusive velocities[17] are used rather than terminal velocity. Measurements are still made above the turbopause, where diffusion dominates over turbulent mixing; this is evidenced by the different scale heights measured for H_2 and He (Fig. 17; see also Fig. 3 of Waite et al. 2018).

Waite et al. (2018) base their velocities off of bounding limits for the downward velocity of methane, with the lower limit (45 m s^{-1}) given by the limiting flux approximation, and the upper limit (100 m s^{-1}) adapted from a simple hydrostatic model (Bell et al. 2014). Velocities for different compounds are then scaled using diffusion coefficients (Mason and Marrero 1970) for each of the major molecular components identified by INMS. These velocities are much lower than the terminal velocities utilized by Mitchell et al. (2018). Note that using higher velocities at the same altitudes would imply higher mass influxes to explain the observed mixing ratios for inflowing material. The Waite et al. (2018) calculation uses a wider spread of the material than Mitchell et al. (2018) at 4° north and south of the

[17] "Diffusive velocities" are defined from a simplified diffusion equation, which is easily seen to be just a rearrangement of terms in equation (9) to $F_i = D_i(X_i N_{H_2}/H_i)$ where the term in parentheses approximates a vertical gradient. The same equation can then be written as $F_i = (D_i/H_i)X_i N_{H_2} \equiv v_D N_i$ where $v_D = D_i/H_i$ is thereby defined as the diffusive velocity, multiplying the local number density.

Table 7 Mass deposition rates (MDR) calculated for inflowing material measured during Cassini's Grand Finale. Inflows are categorized as mid-latitude ring rain ("Mid") or as equatorial inflow ("Eq")

Reference	Dataset	Material included	MDR (kg/s)	Latitude	Notes
Waite et al. (2018)	INMS	Volatiles + nanograins	4800 - 45000	Eq.	
Perry et al. (2018)	INMS	Volatiles + nanograins	4000 - 30000	Eq.	Altitudes and velocities comparable to MIMI (Mitchell et al. 2018)
Perry et al. (2018)	INMS	Volatiles + nanograins	10000 - 200000	Eq.	Sedimentation velocities, and mid-altitudes comparable to Waite et al. (2018), Yelle et al. (2018)
Yelle et al. (2018)	INMS final plunge	CH4	1300 - 11000	Eq.	Here we assume 10^{10} km^2 surface area for comparison
Serigano et al. (2020)	INMS	CH_4, H_2O, NH_3	1390 - 9580	Eq.	Following Yelle et al. (2018)
Serigano et al. (2022)	INMS	All compounds	19700 - 80600	Eq.	
Mitchell et al. (2018)	MIMI	Particles 10,000 to 40,000 u	5.5	Eq.	Some evidence for flow of charged material on magnetic field lines
Hsu et al. (2018)	CDA	Grains	224 - 854	Eq.	
Hsu et al. (2018)	CDA	Grains	100 - 370	Mid	
O'Donoghue et al. (2019)	Keck telescope	Water	432 - 2870	Mid	Based on models of observed H_3^+ densities

equator, based off of the half-width for INMS counts. The mass influx numbers calculated over three separate orbits ranged from 4800 to 45,000 kg s^{-1} of infalling material at masses heavier than the H_2 and He atmospheric compounds (Waite et al. 2018). The mass influx variation from orbit to orbit differs by 2.5, while differences from the two bounding cases for velocity are a factor of 3.8; uncertainty in velocity contributes more strongly to the range of values reported.

Similar influx estimates (10,000 to 200,000 kg s^{-1} for the mid-altitude range) were determined by Perry et al. (2018), with velocity estimates calculated based on Epstein drag:

$$v_s = \frac{0.36\rho_p g(2a)}{m_{H_2} n_{H_2} v_{H_2}} \tag{8}$$

where v_s is the sedimentation velocity, ρ_p is the particle mass density, g is the gravitational acceleration, a is the particle radius, and v_{H_2}, n_{H_2}, and m_{H_2} are the mean thermal speed, number density, and mass for atmospheric H_2. They calculate velocities at 1700 km of between 3 and 30 m s^{-1}. Perry et al. (2018) also report that use of INMS data at altitudes (3000 km) and terminal velocities (1 to 3 km s^{-1}) matching the method and region used for the MIMI MDR result in a $\sim 50\%$ reduction in the calculated INMS MDR. However, this method still results in an exceptionally high inflow rate, and at the higher altitudes it is likely that material is undercounted in the INMS dataset because of the lower signal to noise ratio.

Perry et al. (2018) concluded that the MDR discrepancy between MIMI and INMS was best explained by the size distribution of the infalling material, with a peak near a mass of 500 u and INMS measuring the smallest grains. Such small grains may be more likely to avoid sublimation (Hamil et al. 2018), consistent with the minimal effects observed on atmospheric chemistry (Moses et al. 2023). These data imply an extremely steep size distribution if the INMS and MIMI grains are derived from the same population. However, the physical process of deorbiting ring material via collisions with atmospheric gases (Mitchell et al. 2018; Perry et al. 2018) is likely to fractionate the original grain population as a function of grain radius a. While grains with a larger radius are more likely to have collisions with atomic H (an a^2 dependence), the greater momentum of grains with higher mass may require more collisions to deorbit (an a^3 dependence). Further work is needed to understand the relationship between size distribution of the measured material, the size distribution of the source material in the D ring, and any connection to the main rings e.g. by ballistic transport (Durisen and Estrada 2023).

The Yelle et al. (2018) work draws on a different atmospheric model, based on Koskinen and Guerlet (2018), and utilizes the methane mixing ratio in the deep atmosphere reported from occultations (Koskinen and Guerlet 2018) as well as mixing ratios calculated from the INMS data to construct and solve a diffusion profile of the form:

$$X_i = (H_i F_i)/(N_{\mathrm{H_2}} D_i), \tag{9}$$

where X_i is the mixing ratio of compound i, H is the scale height, F is the flux, $N_{\mathrm{H_2}}$ is the density of H_2, the dominant atmospheric component, and D is the diffusion coefficient for compound i. They utilize the same source for coefficients for diffusion through H_2 gas as Waite et al. (2018): Mason and Marrero (1970). Their separate analysis confirms that the region sampled is dominated by diffusion, and calculates similar mass influxes independent of the underlying atmospheric model.

Serigano et al. (2020) and Serigano et al. (2022) build on this approach to calculate the MDR through the conversion

$$\mathrm{MDR} = F_i m_i \times 2\pi R_{\mathrm{S}}^2 \theta, \tag{10}$$

where m is the molecular mass, and θ is the latitudinal width over which the influx is spread. For θ, they utilize a latitudinal width of 16°, chosen to match the drop in signal for minor constituents at ±8° from the equator. They follow the same formulation for F_i as Yelle et al. (2018), adding H_2O and NH_3 (Serigano et al. 2020), as well as deconvolved compounds across the INMS mass range (Serigano et al. 2022). Their results suggest that the MDR may reach 74,000 kg s^{-1}.

In conclusion, the MDR from the rings to Saturn's atmosphere inferred from the INMS dataset is measured to be of the order of 10^4 kg s^{-1}, orders of magnitude higher than the 10^2 kg s^{-1} flux to the rings from meteoroid bombardment (Durisen and Estrada 2023). However, this calculation of the meteoroid flux assumes that only 10% of the material is retained as observable, absorbing material (Cuzzi and Estrada 1998; Estrada et al. 2015), in which case the total meteoroid influx becomes much more comparable to the INMS influx to Saturn's atmosphere. This inferred MDR is concentrated at the ring plane and is dependent on the measured number densities of the infalling material, as well as the downward velocities imposed. Analysis via two different atmospheric models plus separate consideration of Epstein drag suggests that velocities are on the order of < 100 m s^{-1}. The use of higher velocities, such as the thermal velocity of the bulk atmosphere, results in higher influxes. The MDR of fine-grained material therefore appears to be exceptionally high. Modeling of ring dynamics suggests that ballistic transport in the rings and, to a lesser degree, mass loading could lead to inward flow of the inner B ring and C ring with an MDR on the order of 10^4 kg s^{-1} (Durisen and Estrada 2023), comparable to the INMS MDR. These values therefore allow a possible explanation for the INMS observations, but do not exclude the possibility that the observed D ring inflow phenomenon may be transient, and Cassini may have observed an unusually high rate as discussed below in Sect. 7.2.4; see also discussion of changes in the D ring in Sect. 6.2.

7.2.4 Atmospheric Observational Inferences and Their Implications for Time Variability

Moses et al. (2023) place constraints on the duration and/or properties of the equatorial ring inflow based on comparisons between Saturn's observed atmospheric composition and the modeled impact of the inflow on atmospheric chemistry. They demonstrate that the non-detections of HCN and HC_3N in Saturn's stratosphere from infrared observations, and the low inferred stratospheric abundance of CO_2 and H_2O from infrared, ultraviolet, and sub-millimeter observations, are inconsistent with coupled ion-neutral photochemical models that adopt ring-inflow compositions from either Serigano et al. (2022) or Miller et al. (2020). Moses et al. (2023) conclude that the large inflow rates inferred from INMS measurements must be a recent phenomenon (i.e., $\lesssim 4$ months to 50 years since the onset of such large ring inflow rates), or the ring material must consist primarily of small grains that do not ablate in the atmosphere at the depths perceived by CIRS. Even a $\lesssim$50-year-old continuous ring-inflow event would provide observational signatures of excess HC_3N, HCN, and CO_2 in Saturn's stratosphere, unless the material flowing in from the rings is primarily present as unablated grains smaller than ~50 nm for entry velocities of 25 km s^{-1} (Hamil et al. 2018) ,[18] with only the most volatile compounds (CO, N_2, CH_4) being present as vapor; furthermore, this volatile vapor component must spread globally before reaching the top of

[18]For grains with a radius of 10 nm, slightly smaller than the CDA best-fit size of 20 nm (Hsu et al. 2018), and density of 917 kg/m^3, ablation occurs at velocities above 15 km/s (Hamil et al. 2018). These velocities are significantly higher than the velocities calculated by MIMI (Mitchell et al. 2018), which reach a maximum value of 2.5 km/s, and INMS (Perry et al. 2018; Yelle et al. 2018; Waite et al. 2018), which cite velocities on the order of 30 to 100 m/s for diffusively coupled material.

the stratosphere to prevent observational consequences in the stratosphere. Inflow of gas-phase CO and/or N_2 in particular cannot be reconciled with longer inflow timescales. The effects of assuming 28 u instead represents gas-phase C_2H_4 have not yet been examined.

Of the three atmospheric products that constrain the inflow timescale (HC_3N, HCN, and CO_2), HCN is the longest-lived, persisting for ~ 2000 years at 1 mbar and ~ 30 years at 1 μbar. Therefore, even if the inflow of material is a transient but repeating phenomenon, the times when this process is inactive may be more common than when it is active. The possible division of inflowing material into gas-phase and solid-phase components is consistent with compositional constraints discussed in Sect. 7.2.1. Saturn's atmospheric chemistry places time constraints on the inflow of gas-phase material, but does not directly constrain the inflow timescale for solid-phase material unless it ablates. The short timescales for the present gas inflow rate may be tied to recent changes in the D ring (Sect. 6.2).

8 Ring Atmosphere

The exposure of the ring particles to photons, charged particles, and meteoroids produces a toroidal gaseous envelope of neutrals that is referred to as the ring atmosphere. Atoms and molecules are subsequently scattered from this atmosphere, creating a system-wide neutral corona that extends into Saturn's atmosphere and out into the magnetosphere well beyond the A ring (Johnson et al. 2006; Tseng et al. 2010, 2013a; Cuzzi et al. 2009). This extended ring atmosphere is superimposed on the toroidal atmospheres produced by outgassing from Enceladus, atmospheric escape from Titan, and, to a lesser extent, by sputtering of the icy satellites and the grains in the tenuous E, F and G rings, as seen in modeling of their H_2 and H contributions to the atmosphere over the ring plane (Tseng et al. 2011, 2013b). The presence of an ambient gas with molecules precipitating into Saturn's atmosphere was predicted prior to Cassini's arrival (Carlson 1980), and Cassini's Saturn Orbital Insertion (SOI) maneuver where the spacecraft crossed the ring plane provided the only in situ measurement of the ring atmosphere region over the B ring during the outbound trajectory.[19] The simplicity of the ion mass spectrum, which consists of narrow peaks of O^+ and O_2^+ (Tokar et al. 2005), was surprising. It was the least contaminated mass spectrum obtained by the Cassini Plasma Spectrometer (CAPS; Young et al. 2004) during the mission. These SOI data are the basis of a model for the production, composition and spatial distribution of the ring atmosphere, dominated by O_2 (e.g. Johnson et al. 2006). That species other than oxygen must be present below the CAPS detection limits is suggested by reflectance observations of the rings (see Sects. 3.2 and 4.2) and, indirectly, by plasma observations outside the rings discussed below.

During Cassini's outbound SOI trajectory the CAPS instrument detected oxygen ions north of the B ring, through the Cassini division, out to the inner edge of the A ring (Tokar et al. 2005). Oxygen ions were again detected when CAPS was turned back on beyond the outer edge of the A ring as Cassini moved into the magnetosphere (Elrod et al. 2012, 2014). Although other water products were seen by INMS closer to the ring plane over the A ring (Waite et al. 2005), the dominance of oxygen ions was roughly consistent with our understanding of the radiation effects at the icy Jovian satellites (Johnson et al. 2004). That is, when ice is the dominant surface constituent, radiolysis and photolysis result in the ejection of water products, a significant fraction of which are O_2 and H_2. These volatiles do not recondense at the expected ring particle temperatures, resulting in residence times over the B-ring of the order of a half year for O_2 and an order of magnitude larger for H_2 (Johnson

[19] Readers are referred to Fig. 1 of Tokar et al. (2005) for a diagram of the SOI trajectory.

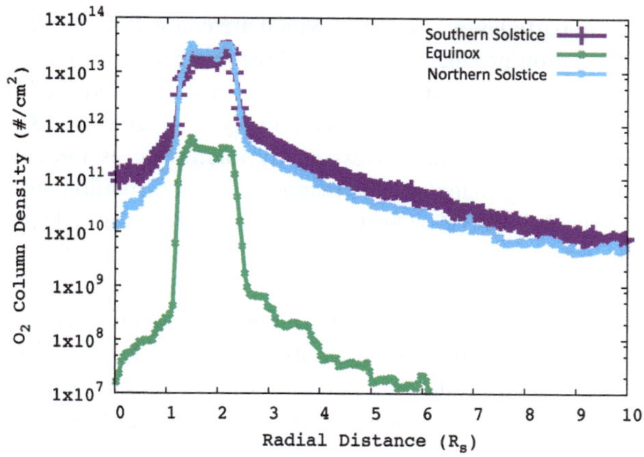

Fig. 18 Radial distribution of the modeled neutral O_2 column density for illumination of the A, B and C rings at three seasons with their minimum photodecomposition O_2 source rates: Southern Solstice (SOI, $2\text{x}10^{27}$ O_2 s^{-1}), Equinox ($2\text{x}10^{25}$ O_2 s^{-1}) and Northern Solstice (Grand Finale, $2\text{x}10^{27}$ O_2 s^{-1}). The respective injection rates into Saturn's atmosphere are $5\text{x}10^{26}$ O_2 s^{-1} (27 kg s^{-1}), $5\text{x}10^{23}$ O_2 s^{-1} (0.03 kg s^{-1}), and $2.5\text{x}10^{26}$ O_2 s^{-1} (27 kg s^{-1}). (Tseng et al. 2010; Tseng and Ip 2011; Tseng et al. 2013a)

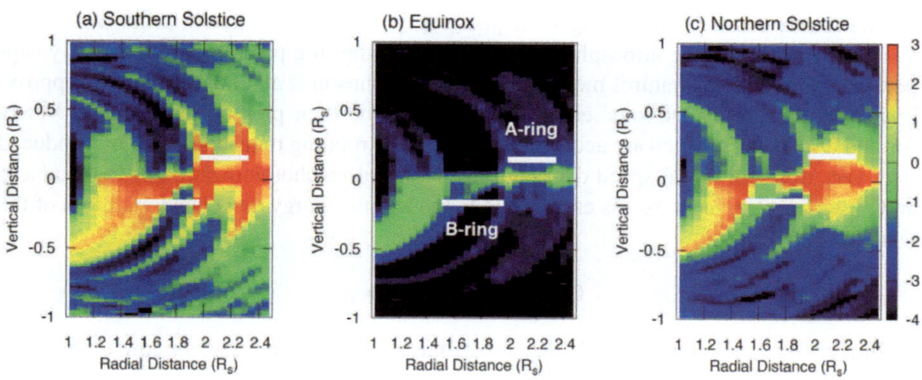

Fig. 19 O_2^+ ion density (cm^{-3}) in log10 scale as indicated by the color bar at right for the three seasons and neutral O_2 source rates in Fig. 18: (a) Southern Solstice (SOI), (b) Equinox and (c) Northern Solstice (Grand Finale). (Tseng et al. 2010; Tseng and Ip 2011; Tseng et al. 2013a)

et al. 2006). Because the inflowing magnetospheric plasma is suppressed at the edge of the A ring, photolysis is the dominant source of decomposition of the ice particles. Since the southern side of the ring plane was illuminated at SOI, the detection of a significant component of oxygen ions north of the ring plane indicates that a toroidal O_2, and concomitant H_2, atmosphere permeates and co-orbits with the ring particles with an O_2 centrifugal scale height $\sim$0.026 R_{ss} at 2 R_{ss}, varying as $r^{3/2}$. Their lifetimes are primarily determined by photoionization and photodissociation producing O, O^+, O_2^+ and H, H^+ and H_2^+. Although ionization predominantly occurred south of the ring plane at SOI, the CAPS detection of O^+ and O_2^+ north of the plane indicates that the B ring is not fully opaque to the UV and/or to the flow of ions along field lines through the plane (Johnson et al. 2006). Of course, the estimates of the source rates and the lifetimes needed to interpret the CAPS data depend strongly on the uncertain fate of the chemistry initiated by the interaction of the radiation products with the ring particles, a fraction of which are charged. In the following the fate of ionization and dissociation products is discussed using a rough estimate of their recycling on the ice grains (Johnson et al. 2006). Using this estimate, ~ 270 kg/s of O_2 is emitted from the A, B and C rings at solstice initiated by the photolysis of ice (Fig. 18).

The motion of the ionic products is especially interesting as the magnetic equator is $\sim 0.046\ R_S$ north of the ring plane and the magnetic field co-rotation speed falls below the neutral orbital speed at $R \lesssim 1.86\ R_S$. Due to this geometry, ions formed and then "picked up" (accelerated) by the rotating field south of the ring plane and inside of $\sim 1.86\ R_S$ precipitate into Saturn's southern hemisphere (Luhmann et al. 2006; Bouhram et al. 2006) as seen in Fig. 19. This accounts for a molecular ion contribution to ring rain (see more detail in Fig. 4 in Tseng et al. 2010). This contribution consists of on the order of 0.5 to 1 percent of the oxygen neutral source rate from the ring particles, up to ~ 0.3 kg s^{-1}, suggesting a minor contribution relative to the grain deposition rates of Hsu et al. (2018) and O'Donoghue et al. (2019).

Pick-up ions that do not precipitate into Saturn will oscillate along the field lines about the magnetic equator prior to impacting a ring particle. Such ions have a centrifugal scale height of O_2^+ varying from $\lesssim 0.1$-to-$0.16\ R_S$ across the region where CAPS ring atmosphere data were acquired ($\sim 1.8 - 2.08 R_S$). Therefore, although neutral molecules, including O_2, have not been directly measured, the observed population of ions oscillating about the magnetic equator with lifetimes limited by impact with ring particles indicates the presence of a neutral source that we call the ring atmosphere.

This picture of a ring atmosphere dominated near the ring plane by O_2 is strongly supported by the ion temperatures measured by CAPS (Johnson et al. 2006), which are approximated by the sum of the ions' energy at formation and their pick-up gyro energy. Newly formed O_2^+ and O^+, which are accelerated by Saturn's rotating magnetic field, are produced with very different initial speed distributions. Dissociative photoionization of O_2 produces an O^+ with a significant excess energy due to the kinetic energy associated with loss of the O atom:

$$O_2 + hv \rightarrow O + O^+ + e^-$$

In contrast, O_2^+ is formed with very little excess energy because of the low mass of the electron that is lost:

$$O_2 + hv \rightarrow O_2^+ + e^-$$

Since the pick-up contribution to the ion temperature depends on the relative speeds between the neutrals that orbit at the Keplerian orbital velocity, and ions confined by the rotating magnetic field orbiting at the magnetic field co-rotation velocity, the O_2^+ temperature was seen to go through a minimum at $\sim 1.86\ R_S$, where the Keplerian and co-rotation velocities are roughly equal (see Fig. 1b of Johnson et al. 2006). That is, because of the low excess energy of formation of O_2^+, its temperature is dominated by the pick-up energy, whereas the O^+ temperature is dominated by its excess energy and does not exhibit as strong of a radial influence from the pick-up energy. The radial temperature trends of O_2^+ and O^+ therefore support their relationship to neutral O_2.

Because the distance of the magnetic equator above the ring plane is of the order of the neutral scale height, those ions moving about the magnetic equator can collide with and charge exchange with neutrals, producing hot neutrals with energies determined primarily by the ion's speed. These scattered neutrals create a highly extended component of the ring atmosphere indicated by the radial column densities in Fig. 18. It is seen that a fraction of the scattered O_2, as well as H_2 not shown here, precipitates into Saturn's atmosphere as neutrals with estimated rates at three phases given in the caption. The absolute magnitude of the ring atmosphere density and, hence, these precipitation rates, depends on the still-uncertain recycling of hydrogen and oxygen on the ring particles.

Although the much lighter H_2 dominates O_2 in the extended ring atmosphere, it is not the dominant source of H_2 or H over the ring plane. Modelling of significant contributions to hydrogen in this region indicate it also comes from Saturn's extended atmosphere, escape from Titan's atmosphere, and the Enceladus torus (Tseng et al. 2011, 2013b) each with very different centrifugal scale heights. This toroidal atmosphere produces the Lyman-alpha flux in this region (e.g. Ben-Jaffel et al. 2023) as well as the hot proton plasma detected by Cassini at the inner (Kollmann et al. 2018) and outer (Cooper et al. 2018) edges of the main rings. However, the observed seasonal and longitudinal dependence of the Lyman-alpha flux (Ben-Jaffel et al. 2023) are suggestive of a ring source of hydrogen. These data have also been used to help constrain hydrogen recycling on the ring particles. Of particular interest is the role of the incident H and H^+ on the particles of the inner D-ring ringlets that has yet to be modeled.

The extension of the neutral ring atmosphere beyond the edge of the A ring ($R_S = 2.27$) seen in Fig. 18 contributes to the local plasma out to large distances. As Cassini did not fly over the rings again until the end of the mission, by which time the CAPS instrument was unfortunately shut off, the model of the O_2 ring atmosphere was further confirmed by analysis of the plasma data outside the A ring (e.g. Elrod et al. 2014, 2012; Christon et al. 2013). That is, since the oxygen plasma density roughly tracks the local neutral density, the plasma outside the A ring exhibited a variability consistent with a steep drop in the solar flux onto the ring plane as Saturn approached equinox, as indicated by the drop in the source rate in Fig. 18. Post-equinox, when the magnetic equator is above the sunlit ring plane, the same processes occur. Therefore, the tilt of the ring plane to the sun leads to an extended ring atmosphere that varies with inclination to the solar illumination (Tseng et al. 2010).

To explain the radiation processing by the magnetospheric plasma adsorbed at the outer edge of the A ring, Tseng and Ip (2011) suggested that the erosion of the icy ring particles in this region might be sustained by water from Enceladus. Refreshed ice grains in this region were also suggested to account for the observed delay in the post-equinox repopulation of O_2^+ plasma (Christon et al. 2013). Using a direct simulation Monte Carlo model and a plume outflow rate of 10^{28} molecules $H_2O\ s^{-1}$, Cassidy and Johnson (2010) find a mass deposition rate from Enceladus' neutral cloud onto the rings of approximately 100 kg s^{-1}, which is equivalent to a quarter of the mass of the main rings in approximately 1 Gyr if the plume rate were constant. (Note that the active lifetime of the Enceladus plume is at present poorly constrained.) However, using a similar plume rate Smith and Richardson (2021) find a much lower water product deposition rate. Further research into contributions from Enceladus to the rings is needed, including how such contributions would affect regional composition and albedo in the rings.

Even though INMS detected other water products closer to the ring plane during SOI (Waite et al. 2005) when CAPS was briefly off, the absence of non-ice products was striking. This is especially the case given the material from Enceladus that is deposited onto the ring particles (Jurac and Richardson 2007; Farrell et al. 2008; Cassidy and Johnson 2010). This material is likely processed in two ways. Due to collisions of ring particles, contaminants more refractory than H_2O can be dispersed into the ring particles, depleting surface concentrations. In addition, any desorbed volatiles, or surface species made volatile by reactions with the water products, can be scattered into the outer magnetosphere or into Saturn's atmosphere as discussed above for the photolytically produced O_2. That the latter process occurs is confirmed by the seasonal variation of ions of mass 28 u as well as O_2^+ observed by the MIMI CHEMS (Charge-Energy-Mass Spectrometer) sensor at suprathermal energies (Christon et al. 2014). Further analysis of CHEMS data presented at the Cassini Science Seminar in 2018 by Hamilton and colleagues showed that C^+ and CO^+ also exhibit a seasonal variation suggestive of a photolytic ring source that is seen to peak in fall of 2014 and persist through to the Grand Finale (Hamilton et al. 2018). This was interpreted to be due to CO_2 deposited on the rings from the Enceladus torus. The observed lifetimes of CO^+ and CO_2^+ seen in the plasma appears to correlate with their photoionization rates producing CO^+ and CO_2^+ over the rings, which are scattered into the outer magnetosphere as neutrals by charge exchange with the dominant O_2.

Although the D-ring particles are also dominated by ice so that photolysis can contribute water products to the ring atmosphere, they contain a higher concentration of more refractory materials. Whereas the plasma flux onto the other rings is highly suppressed, Kollmann et al. (2018) examined the hot plasma quenching going from the outer regions of Saturn's atmosphere onto the inner edge of the D-ring. In this way they were able to estimate the extended scale height of Saturn's atmosphere as well as sense the effect of the ambient gas-phase species over the inner ringlets, which are exposed to hot neutrals and energetic ions. Surprisingly, their estimated flux, 10^{16} H_2O/m^2 is remarkably similar to that estimated by Tseng et al. (2010) for the contribution to the column density in this region of O_2 formed from the A, B and C rings. Therefore, the D-ring contribution to its local atmosphere might indeed be small, in spite of any chemistry induced in the particles by the incident protons and hydrogen. These interactions should produce a not-yet modeled gas component originating from this region that would also contribute to Saturn's atmosphere. If this D ring atmosphere is related to the CH_4-rich inflowing gas measured by INMS during the Grand Finale, then its composition must be quite different than the ring atmosphere over the main rings.

8.1 Possible Ring Atmosphere Connection to the Inflowing Material

We have seen that the composition of the inflowing material differs substantially from the observed composition in the main rings and in the D ring, namely in how water-poor the inflowing material is (Table 6). Here, we consider whether this difference in composition is intrinsic to the material, or whether the initial source composition may be identical and the observed differences could be caused by different evolutionary processes occurring in Saturn's ring environment. In the latter case, more volatile compounds may be lost via such processes, and the material observed in situ may represent the remaining residue. This volatile component may contribute to the ring atmosphere, including in the D ring region. We consider this component here. Since the D ring bulk composition is poorly constrained (Sect. 6.1), we assume that it is related to the main ring composition with a greater abundance of non-icy material.

A large mass of material more volatile than water (e.g. CH_4, CO) was observed by INMS, and the latitudinal variation suggests it is present as a volatile component rather than as an impact fragment from more refractory material (Miller et al. 2020; Serigano et al. 2020). However, compounds more volatile than water are not observed in the main rings (e.g. Sect. 3.2.4), and should be lost more quickly than water ice in the absence of a lag deposit. Therefore, to calculate the missing water mass, we make the simplifying assumption here that all non-water masses detected by INMS can be attributed to impact fragments of refractory organic material related to the neutral and/or UV absorbers in the main rings (Sects. 3.2.2 and 3.2.1). In this simplified scenario, there are two observed components (water and organic products), and the missing mass of water is the mass required to reconcile the relative abundance of these two components with remote observations of the D ring or main rings.

The range in observed inflowing mass from INMS is bracketed by values in Serigano et al. (2020) (see Table 7), between 2080 and 74000 kg s^{-1}. The observed inflowing composition was approximately 16 to 24 wt.% water (Table 5). If we take 24 wt.% as representative, this implies that the inflowing non-water mass is 1,581 to 56,260 kg s^{-1}. For the source material, we take the lower limit of 80 vol.% water ice in the bulk D ring composition (Fig. 16), and assume a density ratio for water/organics of 0.9/1.4 (Reynard and Sotin 2023) i.e., equal porosity for the two compositions. This corresponds to 72 wt.% water ice in the bulk D ring. In order to reconcile the observed composition with the bulk D ring, between 3,566 and 126,857 kg s^{-1} of water must be volatilized or lost in the ring environment. Estimates of the ring atmosphere production rate for the main rings (Fig. 18) peak at 2×10^{27} O_2 molecules s^{-1}, which corresponds to approximately 120 kg s^{-1} of water ice, a value that is significantly lower than the required water mass loss ($3,566 - 126,857$ kg s^{-1}). The ring atmosphere in the D ring region has not yet been modeled, but as discussed above, estimates by Kollmann et al. (2018) of plasma quenching suggest that the D ring atmosphere flux may be relatively small. If true, then this estimate of 120 kg s^{-1} may be an upper limit. Steady ongoing production of the water-poor inflow composition is therefore not realistic, since the water loss rate required to keep pace with the mass deposition rate ($3,566 - 126,857$ kg s^{-1}) is much greater than the expected rate of water volatilization in the D ring region.

An alternative explanation is that ring "residue" may accumulate, since the observed inflow may be a transient process (Sect. 7.2.4). From Moses et al. (2023), the time constraints on the inflow are between 4 and 600 months. This time range, together with the water mass loss rate from above ($3,566 - 126,857$ kg s^{-1}) implies a total water mass loss between 4×10^{10} and 2×10^{14} kg. We can generate a rough estimate for the total mass of the ring atmosphere by taking an upper limit on the column density, 1×10^{14} cm^{-2} (Fig. 18), and the surface area of the rings, 4×10^{20} cm^2 (Table 2.2). This yields on the order of 4×10^{34} molecules of O_2, or 2×10^9 kg. For the extended ring atmosphere from 2.5 to 10 R_S, we calculate an additional 5×10^8 kg using an upper limit column density of 1×10^{12} cm^{-2} (Fig. 18). We therefore estimate a total ring atmosphere mass of $\sim 3 \times 10^9$ kg. The estimated water loss that would be needed to reconcile the inflowing composition with the bulk D ring composition (4×10^{10} and 2×10^{14} kg) is likely orders of magnitude greater than the mass of the ring atmosphere. We would therefore expect that this magnitude of water loss may be significant, and have a noticeable effect on the ring system.

While the total required water mass loss appears large in comparison to the estimated ring atmosphere mass, it may be possible that the water loss has occurred over a sufficiently long timescale that it can be reconciled with the expected ring atmosphere production and loss rates. As calculated above, the production rate for the ring atmosphere is on the order of 120 kg of water s^{-1}. To remove 4×10^{10} to 2×10^{14} kg of water into the ring atmosphere would then require between 10 and 50,000 years. This is rapid on geologic timescales, and further suggests that the observed inflow may be transient if the material is derived from the bulk D ring. This additionally assumes concentration of residue material at the inner edge of the D ring. The overall contribution of the D ring region to the ring atmosphere has not yet been modeled but is expected to be small in comparison to the main rings. Loss rates for the ring atmosphere into Saturn's atmosphere are < 30 kg s^{-1} (Fig. 18), and transfer of water vapor products outward to the magnetosphere may be required instead if the INMS material does represent D ring residue. While we have assumed here that all inflowing material may be represented by a combination of water and complex organic material, this simplified scenario does not explain the presence of CH_4 and CO or N_2 in the vapor phase (Sect. 7.2.1).

A related scenario may involve water loss from a subset of the D ring material, and selective deorbiting of that residue into Saturn's atmosphere. The "tholin" material used for spectral fitting in Cuzzi et al. (2018b) produces the best fits when it is concentrated into a subset of grains that comprises 2-40% of the overall population of grains. We suggest that these grains may be volatilized more quickly due to increased absorption of sunlight by the darker organics, and subsequently higher temperatures for organic-bearing grains. Water loss may reduce the size of these organic-rich grains, which in turn may increase the likelihood that they fall into Saturn's atmosphere since small grains may be decelerated more rapidly by impacts with Saturn's extended atmosphere (Mitchell et al. 2018). The INMS material is hypothesized to be very small in size (e.g. Perry et al. 2018), and if the inflowing material selectively draws from the organic-rich material in the D ring, such a scenario may reduce the water loss mass required to reconcile INMS observations with the bulk D ring composition.

9 Relationship Between the Rings and Ring Moons

While inwardly connected with Saturn through the observed mass inflow discussed above, from a compositional standpoint the main rings also appear to be outwardly linked with Saturn's ring moons, and in particular with the innermost ones. These include small satellites orbiting within (Pan, Daphnis) and just outside the A ring (Atlas), shepherding the F ring (Prometheus and Pandora), and the co-orbitals Janus and Epimetheus (orbiting at $\sim$151,000 km).

As discussed in greater detail in Ciarniello et al. (2024, this collection), Cassini remote sensing observations at UV-VIS-IR wavelengths show significant spectral similarities between the main rings and these ring moons. Both are characterized by the dominance of water ice features with absorption bands at 1.5, 2 and 3 μm, and by UV absorption at wavelengths shorter than $\sim$ 0.55 μm, indicative of the presence of non-icy material. Although such spectral features can also be observed on the rest of Saturn's icy moons, spectral indicator analysis (see Sect. 6 in Ciarniello et al. 2024) reveals similar trends between the main rings and inner ring moons, for which the intensity of the UV absorption is in general positively correlated with the albedo and the water ice band depths. This is not the case for the mid-sized icy satellites, thus suggesting a similar compositional paradigm for the rings and the inner ring moons in which the UV absorber is embedded in water ice grains, with an additional endmember as a broad-band absorber (see Sect. 4).

Further evidence of a compositional link between the main rings and ring moons is provided by the radial trend of the ring moon spectral shapes, which become progressively similar to that of the A ring moving inward from Janus and Epimetheus, which appear the most polluted, to Pan. In fact, this radial trend may result from the accumulation of nearby A ring particles on the ring moon surfaces. This accumulation is more effective for the inner moons, and counteracts the contamination from meteoritic bombardment that modifies the surface composition inherited from the rings. An additional contribution may also come from a seasonal flow of neutral species from the ring atmosphere that are scattered beyond the edge of the rings by molecules and ions generated from the rings by photoionization and subsequently accelerated by the magnetic field (Tseng et al. 2010).

Ring moons are thought to originate within or in the vicinity of the A ring, accumulating ring material on their external layers in the process, and then moving outward due to the gravitational torque of the rings (Charnoz et al. 2010; see also Sect. 5 of Ciarniello et al. 2024 and Sect. 3.2.2 of Blanc et al. 2024). As a consequence, Janus and Epimetheus, orbiting at larger distance from Saturn with respect to the other ring moons, are likely older and may have endured a longer exposure to meteoritic bombardment. In this respect, it is interesting to note that these two moons appear spectrally similar to the C ring, the most contaminated of the main rings, thus pointing to a similar role for the pollution mechanism. However, we note that contrary to the main rings, ring moons are affected by additional exogenous processes such as contamination from E-ring grains and the effects of charged particle fluxes (see Sect. 8 of Ciarniello et al. 2024), thus requiring particular care in comparisons with contamination/alteration processes in the main rings.

10 Conclusions

From this retrospective look at the Cassini dataset and at previous remote observations of the Saturnian ring system, a few themes for the ring composition emerge. Overall, the bulk composition of Saturn's rings is overwhelmingly dominated by water ice. This is especially true in the optically thick B ring, which dominates the ring mass. However, remote sensing constraints suggest that even the "water-poor" D ring may be as much as 80 vol.% water. This composition is consistent with fairly minimal signs of pollution or contamination of the rings by external impactors. While microwave observations suggest that some silicate material may remain "hidden" from surficial spectroscopy in the C ring, in general the dynamic evolution of the rings suggests they are well-mixed and this composition is representative of the bulk. Thus, the effective exposure age of the rings to micrometeoroid flux may be relatively young, on the order of 10^8 years.

While the composition is dominated by water throughout, compositional trends are observed with both radial distance (Table 6) and optical depth. The abundance of non-water components tends to increase going from the A ring towards the D ring. This trend mirrors the inward increase in temperature seen by VIMS and CIRS data on the sunlit side. Similarly, the abundance of small particles increases from the inner A ring to the D ring. The Cassini Division deviates from these trends, and instead points to the influence of optical depth on the ring composition. Namely, optical depths are lower in the C ring and Cassini Division, and as a result these regions are more susceptible to pollution from external impactors. Microwave data show that the composition of the middle C ring may also be silicate-rich, with icy mantles concealing interiors that may point towards the recent break up of a small rocky body.

Data collected during Cassini's Grand Finale of material from the rings support this trend of enhanced abundance of non-water components moving inwards towards Saturn. However, the degree of enhancement from INMS measurements on the smallest size fraction is extreme. This material is comprised of small grains as well as gas that includes a 28 u component (likely dominated by CO with some C_2H_4 and possibly some N_2) and methane. Such highly volatile material has not previously been detected in the main rings from remote sensing observations. While inflow of charged "ring rain" material to the mid-latitudes is likely a steady process, constraints from Saturn's atmospheric chemistry suggest that the more massive equatorial inflow detected during the Grand Finale may represent a recent event. This may be related to the appearance of bright clumps in the D68 ringlet in 2014, just a few years before the measurements.

The composition of this material may be partially reconciled with the bulk D ring composition via photolysis and sublimation processes, which are known to generate the ring atmosphere. While these processes in the D ring region have not been specifically modeled, estimates for the local gas density from plasma flux onto the rings suggest that D ring contributions to the ring atmosphere may be relatively small. If the equatorial inflow does represent evolved material from the main rings, the timescale for devolatilization must be much longer than the inflow timescale ($\sim 10^5\times$) and the ring "residue" must accumulate and be effectively fractionated by the inflow process. The simplified scenario presented here may broadly allow reconciliation of the inflow composition with the bulk composition, but it does not explain the presence of the vapor component in the INMS inflow. This processing of the ring material may be related to the ring "cleaning" suggested by Crida et al. (2019).

Ballistic transport indicates mass inflow of material from across the rings with total inward fluxes consistent with the equatorial inflow abundances. However, additional work is needed to explore the relationship of these processes to the compositionally- and size-fractionated material in the D ring region. Similarly, future work should examine how the Grand Finale inflow fits within the broader context of ring evolution, and the degree to which these processes are stochastic or smooth in time.

Acknowledgements We are grateful to M. Tiscareno and to an anonymous reviewer for their constructive feedback. We also thank V. Lainey for organizing the discussions that led to this work, and to A. Crida for editing. This research was supported by the International Space Science Institute (ISSI) in Bern, through ISSI Workshop "New Vision of the Saturnian System in the Context of a Highly Dissipative Saturn" (Bern, 9-13 May 2022). This research has made use of NASA's Astrophysics Data System. KEM acknowledges support from SwRI. GF and MC acknowledge support from INAF-IAPS.

Author Contribution The idea for this review article came from GF and JNC. All authors contributed to the literature search. New data analysis was performed by KB, CF, MMH, JC, KEM and WLT. All authors contributed to drafting and revisions.

Funding Open access funding provided by Istituto Nazionale di Astrofisica within the CRUI-CARE Agreement. Nothing to declare.

Data Availability Cassini data used in this work are archived in NASA's Planetary Data System (PDS) and publicly available.

Declarations

Competing Interests The authors have no competing interests to declare.

References

A'Hearn JA, Hedman MM, Hamilton DP (2021) Modeling Saturn's D68 clumps as a co-orbital satellite system. Planet Sci J 2(2):74. https://doi.org/10.3847/PSJ/abed57. arXiv:2103.05119 [astro-ph.EP]

Albers N, Spahn F (2006) The influence of particle adhesion on the stability of agglomerates in Saturn's rings. Icarus 181(1):292–301. https://doi.org/10.1016/j.icarus.2005.10.011

Altobelli N, Lopez-Paz D, Pilorz S, et al (2014) Two numerical models designed to reproduce Saturn ring temperatures as measured by Cassini-CIRS. Icarus 238:205–220. https://doi.org/10.1016/j.icarus.2014.04.031

Atreya SK, Waite JH (1981) Saturn ionosphere: theoretical interpretation. Nature 292(5825):682–683. https://doi.org/10.1038/292682a0

Baillié K, Colwell JE, Lissauer JJ, et al (2011) Waves in Cassini UVIS stellar occultations. 2. The C ring. Icarus 216:292–308. https://doi.org/10.1016/j.icarus.2011.05.019

Baillié K, Colwell JE, Esposito LW, et al (2013) Meter-sized moonlet population in Saturn's C ring and Cassini division. Astron J 145(6):171. https://doi.org/10.1088/0004-6256/145/6/171

Becker TM (2016) Saturn's rings: measuring particle size distributions using Cassini UVIS occultation data. PhD thesis, University of Central Florida

Becker TM, Colwell JE, Esposito LW, et al (2016) Characterizing the particle size distribution of Saturn's A ring with Cassini UVIS occultation data. Icarus 279:20–35. https://doi.org/10.1016/j.icarus.2015.11.001

Bell JM, Hunter Waite J, Westlake JH, et al (2014) Developing a self-consistent description of Titan's upper atmosphere without hydrodynamic escape. J Geophys Res Space Phys 119(6):4957–4972. https://doi.org/10.1002/2014JA019781

Ben-Jaffel L, Moses JI, West RA, et al (2023) The enigmatic abundance of atomic hydrogen in Saturn's upper atmosphere. Planet Sci J 4(3):54. https://doi.org/10.3847/PSJ/acaf78

Birks JB (1970) Photophysics of aromatic molecules. Wiley, London

Blanc M, Crida A, Shibaike Y (2024) Space Sci Rev 220

Bohren CF, Huffman DR (1983). Absorption and scattering of light by small particles

Bouhram M, Johnson RE, Berthelier JJ, et al (2006) A test-particle model of the atmosphere/ionosphere system of Saturn's main rings. Geophys Res Lett 33(5):L05106. https://doi.org/10.1029/2005GL025011

Bradley ET, Colwell JE, Esposito LW, et al (2010) Far ultraviolet spectral properties of Saturn's rings from Cassini UVIS. Icarus 206(2):458–466. https://doi.org/10.1016/j.icarus.2009.12.021

Bradley ET, Colwell JE, Esposito LW (2013) Scattering properties of Saturn's rings in the far ultraviolet from Cassini UVIS spectra. Icarus 225(1):726–739. https://doi.org/10.1016/j.icarus.2013.04.008

Brilliantov N, Krapivsky PL, Bodrova A, et al (2015) Size distribution of particles in Saturn's rings from aggregation and fragmentation. Proc Natl Acad Sci 112(31):9536–9541. https://doi.org/10.1073/pnas.1503957112

Brown RH, Baines KH, Bellucci G, et al (2004) The Cassini visual and infrared mapping spectrometer (vims) investigation. Space Sci Rev 115(1–4):111–168. https://doi.org/10.1007/s11214-004-1453-x

Canup RM (2010) Origin of Saturn's rings and inner moons by mass removal from a lost Titan-sized satellite. Nature 468:943–946. https://doi.org/10.1038/nature09661

Carlson RW (1980) Photo-sputtering of ice and hydrogen around Saturn's rings. Nature 283(5746):461. https://doi.org/10.1038/283461a0

Cassidy TA, Johnson RE (2010) Collisional spreading of Enceladus' neutral cloud. Icarus 209(2):696–703. https://doi.org/10.1016/j.icarus.2010.04.010

Chadney J (2017) Modelling the Upper Atmosphere of Gas-Giant Exoplanets Irradiated by Low-Mass Stars. https://doi.org/10.1007/978-3-319-63351-0

Charnoz S, Salmon J, Crida A (2010) The recent formation of Saturn's moonlets from viscous spreading of the main rings. Nature 465(7299):752–754. https://doi.org/10.1038/nature09096

Christon SP, Hamilton DC, Difabio RD, et al (2013) Saturn suprathermal O_2^+ and mass-28^+ molecular ions: long-term seasonal and solar variation. J Geophys Res Space Phys 118(6):3446–3463. https://doi.org/10.1002/jgra.50383

Christon SP, Hamilton DC, Mitchell DG, et al (2014) Suprathermal magnetospheric minor ions heavier than water at Saturn: discovery of $^{28}M^+$ seasonal variations. J Geophys Res Space Phys 119(7):5662–5673. https://doi.org/10.1002/2014JA020010

Ciarniello M, Capaccioni F, Filacchione G, et al (2011) Hapke modeling of Rhea surface properties through Cassini-VIMS spectra. Icarus 214(2):541–555. https://doi.org/10.1016/j.icarus.2011.05.010

Ciarniello M, Filacchione G, D'Aversa E, et al (2019) Cassini-VIMS observations of Saturn's main rings: II. A spectrophotometric study by means of Monte Carlo ray-tracing and Hapke's theory. Icarus 317:242–265. https://doi.org/10.1016/j.icarus.2018.07.010

Ciarniello M, Filacchione G, Nicholson PD, et al (2024) Space Sci Rev 220

Clar F (1964) Polycyclic hydrocarbons. Academic Press, NY

Clark RN, McCord TB (1980) The rings of Saturn: new near-infrared reflectance measurements and a 0.326-4.08 μm summary. Icarus 43(2):161–168. https://doi.org/10.1016/0019-1035(80)90116-5

Clark RN, Cruikshank DP, Jaumann R, et al (2012) The surface composition of Iapetus: mapping results from Cassini VIMS. Icarus 218(2):831–860. https://doi.org/10.1016/j.icarus.2012.01.008

Clark RN, Brown RH, Cruikshank DP, et al (2019) Isotopic ratios of Saturn's rings and satellites: implications for the origin of water and Phoebe. Icarus 321:791–802. https://doi.org/10.1016/j.icarus.2018.11.029

Colina L, Bohlin RC, Castelli F (1996) The 0.12-2.5 micron absolute flux distribution of the Sun for comparison with solar analog stars. Astron J 112:307. https://doi.org/10.1086/118016

Colwell JE, Esposito LW, Sremčević M (2006) Self-gravity wakes in Saturn's A ring measured by stellar occultations from Cassini. Geophys Res Lett 33(7):L07201. https://doi.org/10.1029/2005GL025163
Colwell JE, Cooney JH, Esposito LW, et al (2009) Density waves in Cassini UVIS stellar occultations. 1. The Cassini division. Icarus 200(2):574–580. https://doi.org/10.1016/j.icarus.2008.12.031
Colwell J, Esposito L, Jerousek R, et al (2010) Cassini uvis stellar occultation observations of Saturn's rings. Astron J 140(6):1569
Connerney JEP (1986) Magnetic connection for Saturn's rings and atmosphere. Geophys Res Lett 13(8):773–776. https://doi.org/10.1029/GL013i008p00773
Connerney JEP, Waite JH (1984) New model of Saturn's ionosphere with an influx of water from the rings. Nature 312(5990):136–138. https://doi.org/10.1038/312136a0
Cooke ML (1991) Saturn's rings: photometric studies of the C ring and radial variation in the Keeler gap. PhD thesis, Cornell University, New York
Cooper J, Johnson R, Kollmann P, et al (2018) Plasma, neutral atmosphere, and energetic radiation environments of planetary rings. Planet Ring Syst Prop Struct Evol, 363–398
Coradini A, Tosi F, Gavrishin AI, et al (2008) Identification of spectral units on Phoebe. Icarus 193(1):233–251. https://doi.org/10.1016/j.icarus.2007.07.023
Cravens TE (1987) Vibrationally excited molecular hydrogen in the upper atmosphere of Jupiter. J Geophys Res 92(A10):11,083–11,100. https://doi.org/10.1029/JA092iA10p11083
Cravens TE, Moore L, Waite Jr J, et al (2019) The ion composition of Saturn's equatorial ionosphere as observed by Cassini. Geophys Res Lett 46(12):6315–6321
Crida A, Charnoz S, Hsu HW, et al (2019) Are Saturn's rings actually young? Nat Astron 3:967–970. https://doi.org/10.1038/s41550-019-0876-y
Crida A, Estrada PR, Nicholson PD, et al (2024) Space Sci Rev 220
Cruikshank DP, Dalton JB, Dalle Ore CM, et al (2007) Surface composition of Hyperion. Nature 448(7149):54–56. https://doi.org/10.1038/nature05948
Cruikshank DP, Wegryn E, Dalle Ore CM, et al (2008) Hydrocarbons on Saturn's satellites Iapetus and Phoebe. Icarus 193(2):334–343. https://doi.org/10.1016/j.icarus.2007.04.036
Cuzzi JN (1985) Rings of Uranus: not so thick, not so black. Icarus 63(2):312–316. https://doi.org/10.1016/0019-1035(85)90014-4
Cuzzi JN, Estrada PR (1998) Compositional evolution of Saturn's rings due to meteoroid bombardment. Icarus 132(1):1–35. https://doi.org/10.1006/icar.1997.5863
Cuzzi JN, Pollack JB, Summers AL (1980) Saturn's rings: particle composition and size distribution as constrained by observations at microwave wavelengths II. Radio interferometric observations. Icarus 44(3):683–705. https://doi.org/10.1016/0019-1035(80)90137-2
Cuzzi JN, French RG, Dones L (2002) HST multicolor (255-1042 nm) photometry of Saturn's main rings. I: radial profiles, phase and opening angle variations, and regional spectra. Icarus 158(1):199–223. https://doi.org/10.1006/icar.2002.6851
Cuzzi J, Clark R, Filacchione G, et al (2009) Ring particle composition and size distribution. In: Dougherty MK, Esposito LW, Krimigis SM (eds) Saturn from Cassini-Huygens. Springer, Dordrecht, p 459. https://doi.org/10.1007/978-1-4020-9217-6_15
Cuzzi JN, Estrada PR, Davis SS (2014a) Utilitarian opacity model for aggregate particles in protoplanetary nebulae and exoplanet atmospheres. Astrophys J Suppl Ser 210(2):21. https://doi.org/10.1088/0067-0049/210/2/21. arXiv:1312.1798 [astro-ph.EP]
Cuzzi JN, Whizin AD, Hogan RC, et al (2014b) Saturn's F ring core: calm in the midst of chaos. Icarus 232:157–175. https://doi.org/10.1016/j.icarus.2013.12.027
Cuzzi JN, Chambers LB, Hendrix AR (2017) Rough surfaces: is the dark stuff just shadow?; Who knows what evil lurks in the hearts of men? The shadow knows!. Icarus 289:281–294. https://doi.org/10.1016/j.icarus.2016.10.018
Cuzzi JN, Filacchione G Marouf EA (2018a) The rings of Saturn. In: Tiscareno MS, Murray CD (eds) Planetary ring systems. Properties, structure, and evolution, pp 51–92. https://doi.org/10.1017/9781316286791.003
Cuzzi JN, French RG, Hendrix AR, et al (2018b) HST-STIS spectra and the redness of Saturn's rings. Icarus 309:363–388. https://doi.org/10.1016/j.icarus.2018.02.025
Cuzzi JN, Marouf EA, French RG, et al (2024) Saturn's F Ring is Intermittently Shepherded by Prometheus. Sci Adv. https://doi.org/10.1126/sciadv.adl6601 [astro-ph.EP]
Daisaka H, Tanaka H, Ida S (2001) Viscosity in a dense planetary ring with self-gravitating particles. Icarus 154(2):296–312. https://doi.org/10.1006/icar.2001.6716
Dalle Ore CM, Cruikshank DP, Clark RN (2012) Infrared spectroscopic characterization of the low-albedo materials on Iapetus. Icarus 221(2):735–743. https://doi.org/10.1016/j.icarus.2012.09.010
Durisen RH, Estrada PR (2023) Large mass inflow rates in Saturn's rings due to ballistic transport and mass loading. Icarus 400:115221. https://doi.org/10.1016/j.icarus.2022.115221. arXiv:2305.13578 [astro-ph.EP]

Durisen RH, Cramer NL, Murphy BW, et al (1989) Ballistic transport in planetary ring systems due to particle erosion mechanisms I. Theory, numerical methods, and illustrative examples. Icarus 80(1):136–166. https://doi.org/10.1016/0019-1035(89)90164-4

Durisen RH, Bode PW, Dyck SG, et al (1996) Ballistic transport in planetary ring systems due to particle erosion mechanisms. III. Torques and mass loading by meteoroid impacts. Icarus 124(1):220–236. https://doi.org/10.1006/icar.1996.0200

Eckert S, Colwell JE, Becker TM, et al (2021) Sizes of the smallest particles at Saturn's ring edges. Icarus 357:114224. https://doi.org/10.1016/j.icarus.2020.114224

Edgington SG, Spilker LJ, Leyrat C, et al (2008) Emissivity in the thermal IR: composition and polarization in Saturn's rings with Cassini/CIRS: part 2. In: AAS/Division for Planetary Sciences Meeting abstracts #40, pp 29.04

Elachi C, Allison MD, Borgarelli L, et al (2004) Radar: the Cassini Titan radar mapper. Space Sci Rev 115(1–4):71–110. https://doi.org/10.1007/s11214-004-1438-9

Elliott JP, Esposito LW (2011) Regolith depth growth on an icy body orbiting Saturn and evolution of bidirectional reflectance due to surface composition changes. Icarus 212(1):268–274. https://doi.org/10.1016/j.icarus.2010.10.031

Elliott J, Esposito L (2021) Macroscopic roughness, composition and age of saturns rings and FUV phase curve analysis. In: AGU Fall Meeting abstracts, pp P35E–2167

Elrod MK, Tseng WL, Wilson RJ, et al (2012) Seasonal variations in Saturn's plasma between the main rings and Enceladus. J Geophys Res Space Phys 117(A3):A03207. https://doi.org/10.1029/2011JA017332

Elrod MK, Tseng WL, Woodson AK, et al (2014) Seasonal and radial trends in Saturn's thermal plasma between the main rings and Enceladus. Icarus 242:130–137

Esposito LW, De Stefano M (2018) Space age studies of planetary rings. In: Tiscareno MS, Murray CD (eds) Planetary ring systems. Properties, structure, and evolution. Cambridge University Press, pp 3–29. https://doi.org/10.1017/9781316286791.001

Esposito LW, Barth CA, Colwell JE, et al (2004) The Cassini ultraviolet imaging spectrograph investigation. Space Sci Rev 115(1–4):299–361. https://doi.org/10.1007/s11214-004-1455-8

Esposito LW, Albers N, Meinke BK, et al (2012) A predator-prey model for moon-triggered clumping in Saturn's rings. Icarus 217(1):103–114. https://doi.org/10.1016/j.icarus.2011.09.029

Estrada PR, Cuzzi JN (1996) Voyager observations of the color of Saturn's rings. Icarus 122(2):251–272. https://doi.org/10.1006/icar.1996.0124

Estrada PR, Durisen RH, Cuzzi JN, et al (2015) Combined structural and compositional evolution of planetary rings due to micrometeoroid impacts and ballistic transport. Icarus 252:415–439. https://doi.org/10.1016/j.icarus.2015.02.005

Farrell WM, Kaiser ML, Gurnett DA, et al (2008) Mass unloading along the inner edge of the Enceladus plasma torus. Geophys Res Lett 35(2):L02203. https://doi.org/10.1029/2007GL032306

Ferrari C, Galdemard P, Lagage PO, et al (2005) Imaging Saturn's rings with CAMIRAS: thermal inertia of B and C rings. Astron Astrophys 441(1):379–389. https://doi.org/10.1051/0004-6361:20053100

Filacchione G, Capaccioni F, Ciarniello M, et al (2012) Saturn's icy satellites and rings investigated by Cassini-VIMS: III - radial compositional variability. Icarus 220(2):1064–1096. https://doi.org/10.1016/j.icarus.2012.06.040. arXiv:1203.6230 [astro-ph.EP]

Filacchione G, Ciarniello M, Capaccioni F, et al (2014) Cassini-VIMS observations of Saturn's main rings: I. Spectral properties and temperature radial profiles variability with phase angle and elevation. Icarus 241:45–65. https://doi.org/10.1016/j.icarus.2014.06.001

Filacchione G, D'Aversa E, Capaccioni F, et al (2016) Saturn's icy satellites investigated by Cassini-VIMS. IV. Daytime temperature maps. Icarus 271:292–313. https://doi.org/10.1016/j.icarus.2016.02.019. arXiv:1602.02890 [astro-ph.EP]

Fischer C, Postberg F, Altobelli N, et al (2018) Mineral dust in the Saturnian system. In: European Planetary Science Congress, EPSC2018–1279

Flasar FM, Kunde VG, Abbas MM, et al (2004) Exploring the Saturn system in the thermal infrared: the composite infrared spectrometer. Space Sci Rev 115(1–4):169–297. https://doi.org/10.1007/s11214-004-1454-9

French RG, Nicholson PD (2000) Saturn's rings II. Particle sizes inferred from stellar occultation data. Icarus 145(2):502–523. https://doi.org/10.1006/icar.2000.6357

Goldreich P, Tremaine SD (1978) The velocity dispersion in Saturn's rings. Icarus 34(2):227–239. https://doi.org/10.1016/0019-1035(78)90164-1

Grossman AW (1990) Microwave imaging of Saturn's deep atmosphere and rings. PhD thesis, California Institute of Technology

Hamil O, Cravens TE, Reedy NL, et al (2018) Fate of ice grains in Saturn's ionosphere. J Geophys Res Space Phys 123(2):1429–1440. https://doi.org/10.1002/2017JA024616

Hamilton DC, Christon SP, Mitchell DG, et al (2018) A large seasonal variation of energetic C^+ and CO^+ abundances in Saturn's magnetosphere probably resulting from changing ring illumination. In: Cassini science symposium, pp 29–30. https://lasp.colorado.edu/home/wp-content/uploads/2018/08/Cassini-Science-Symposium-Program-with-cover-web.pdf

Hapke B (1993) Theory of reflectance and emittance spectroscopy. Cambridge University Press. https://doi.org/10.1017/CBO9780511524998

Hapke B (2012) Theory of reflectance and emittance spectroscopy, 2nd edn. Cambridge University Press, Cambridge. https://doi.org/10.1017/CBO9781139025683

Hapke B (2013) Comment on "A critical assessment of the Hapke photometric model" by Y. Shkuratov et al. J Quant Spectrosc Radiat Transf 116:184–190. https://doi.org/10.1016/j.jqsrt.2012.11.002

Harbison RA, Nicholson PD, Hedman MM (2013) The smallest particles in Saturn's A and C rings. Icarus 226(2):1225–1240. https://doi.org/10.1016/j.icarus.2013.08.015. arXiv:1312.2927 [astro-ph.EP]

Harris AW (1984) The origin and evolution of planetary rings. In: Greenberg R, Brahic A (eds) IAU colloq. 75: Planetary rings. University of Arizona Press, Tucson, pp 641–659

Hedman MM (2019) Bright clumps in the D68 ringlet near the end of the Cassini mission. Icarus 323:62–75. https://doi.org/10.1016/j.icarus.2019.01.007. arXiv:1901.02043 [astro-ph.EP]

Hedman MM, Nicholson PD (2013) Kronoseismology: using density waves in Saturn's C ring to probe the planet's interior. Astron J 146(1):12. https://doi.org/10.1088/0004-6256/146/1/12. arXiv:1304.3735 [astro-ph.EP]

Hedman MM, Nicholson PD (2014) More Kronoseismology with Saturn's rings. Mon Not R Astron Soc 444(2):1369–1388. https://doi.org/10.1093/mnras/stu1503. arXiv:1407.6702 [astro-ph.EP]

Hedman MM, Nicholson PD (2016) The B-ring's surface mass density from hidden density waves: less than meets the eye?. Icarus 279:109–124. https://doi.org/10.1016/j.icarus.2016.01.007. arXiv:1601.07955 [astro-ph.EP]

Hedman MM, Showalter MR (2016) A new pattern in Saturn's D ring created in late 2011. Icarus 279:155–165. https://doi.org/10.1016/j.icarus.2015.09.017. arXiv:1509.04755 [astro-ph.EP]

Hedman MM, Stark CC (2015) Saturn's G and D rings provide nearly complete measured scattering phase functions of nearby debris disks. Astrophys J 811(1):67. https://doi.org/10.1088/0004-637X/811/1/67. arXiv:1508.00261 [astro-ph.EP]

Hedman MM, Burns JA, Showalter MR, et al (2007a) Saturn's dynamic D ring. Icarus 188(1):89–107. https://doi.org/10.1016/j.icarus.2006.11.017

Hedman MM, Nicholson PD, Salo H, et al (2007b) Self-gravity wake structures in Saturn's A ring revealed by Cassini VIMS. Astron J 133(6):2624–2629. https://doi.org/10.1086/516828

Hedman MM, Burns JA, Tiscareno MS, et al (2009) Organizing some very tenuous things: resonant structures in Saturn's faint rings. Icarus 202(1):260–279. https://doi.org/10.1016/j.icarus.2009.02.016

Hedman MM, Nicholson PD, Showalter MR, et al (2011) The christiansen effect in Saturn's narrow dusty rings and the spectral identification of clumps in the F ring. Icarus 215(2):695–711. https://doi.org/10.1016/j.icarus.2011.02.025. arXiv:1102.5116 [astro-ph.EP]

Hedman MM, Nicholson PD, Cuzzi JN, et al (2013) Connections between spectra and structure in Saturn's main rings based on Cassini VIMS data. Icarus 223(1):105–130. https://doi.org/10.1016/j.icarus.2012.10.014. arXiv:1210.4727 [astro-ph.EP]

Hedman MM, Burns JA, Showalter MR (2015) Corrugations and eccentric spirals in Saturn's D ring: new insights into what happened at Saturn in 1983. Icarus 248:137–161. https://doi.org/10.1016/j.icarus.2014.10.021. arXiv:1410.3761 [astro-ph.EP]

Hedman M, Postberg F, Hamilton D, et al (2018) Dusty rings. In: Tiscareno MS, Murray CD (eds) Planetary ring systems. Properties, structure, and evolution. Cambridge University Press, Cambridge, pp 308–337

Hendrix AR, Hansen CJ (2008) Ultraviolet observations of Phoebe from the Cassini UVIS. Icarus 193(2):323–333. https://doi.org/10.1016/j.icarus.2007.06.030

Hsu HW, Schmidt J, Kempf S, et al (2018) In situ collection of dust grains falling from Saturn's rings into its atmosphere. Science 362(6410):aat3185. https://doi.org/10.1126/science.aat3185

Iess L, Militzer B, Kaspi Y, et al (2019) Measurement and implications of Saturn's gravity field and ring mass. Science 364(6445):aat2965. https://doi.org/10.1126/science.aat2965

Imanaka H, Khare BN, Elsila JE, et al (2004) Laboratory experiments of Titan tholin formed in cold plasma at various pressures: implications for nitrogen-containing polycyclic aromatic compounds in Titan haze. Icarus 168(2):344–366. https://doi.org/10.1016/j.icarus.2003.12.014

Imanaka H, Cruikshank DP, Khare BN, et al (2012) Optical constants of Titan tholins at mid-infrared wavelengths (2.5-25 μm) and the possible chemical nature of Titan's haze particles. Icarus 218(1):247–261. https://doi.org/10.1016/j.icarus.2011.11.018

Ip WH (1983) On plasma transport in the vicinity of the rings of Saturn: a Siphon flow mechanism. J Geophys Res 88(A2):819–822. https://doi.org/10.1029/JA088iA02p00819

Ip WH, Liu CM, Pan KC (2016) Transport and electrodynamical coupling of nano-grains ejected from the Saturnian rings and their possible ionospheric signatures. Icarus 276:163–169. https://doi.org/10.1016/j.icarus.2016.04.004

Jaffe H, Orchin M (1962) Theory and applications of ultraviolet spectroscopy. Wiley, London

Jerousek R (2018) Determining the small-scale structure and particle properties in Saturn's rings from stellar and radio occultations. PhD thesis, University of Central Florida

Jerousek RG, Colwell JE, Esposito LW (2011) Morphology and variability of the Titan ringlet and Huygens ringlet edges. Icarus 216(1):280–291. https://doi.org/10.1016/j.icarus.2011.09.001

Jerousek RG, Colwell JE, Esposito LW, et al (2016) Small particles and self-gravity wakes in Saturn's rings from UVIS and VIMS stellar occultations. Icarus 279:36–50. https://doi.org/10.1016/j.icarus.2016.04.039

Jerousek RG, Colwell JE, Hedman MM, et al (2020) Saturn's C ring and Cassini division: particle sizes from Cassini UVIS, VIMS, and RSS occultations. Icarus 344:113565. https://doi.org/10.1016/j.icarus.2019.113565

Johansson FL, Vigren E, Waite JH, et al (2022) Implications from secondary emission from neutral impact on Cassini plasma and dust measurements. Mon Not R Astron Soc 515(2):2340–2350

Johnson RE, Carlson RW, Cooper JF, et al (2004) Radiation effects on the surfaces of the Galilean satellites. In: Bagenal F, Dowling TE, McKinnon WB (eds) Jupiter. The planet, satellites and magnetosphere, vol 1, pp 485–512

Johnson RE, Luhmann JG, Tokar RL, et al (2006) Production, ionization and redistribution of O_2 in Saturn's ring atmosphere. Icarus 180(2):393–402. https://doi.org/10.1016/j.icarus.2005.08.021

Jurac S, Richardson JD (2007) Neutral cloud interaction with Saturn's main rings. Geophys Res Lett 34(8):L08102. https://doi.org/10.1029/2007GL029567

Kempf S (2008) Interpretation of high rate dust measurements with the Cassini dust detector CDA. Planet Space Sci 56(3-4):378–385. https://doi.org/10.1016/j.pss.2007.11.022

Kempf S, Altobelli N, Schmidt J, et al (2023) Micrometeoroid infall onto Saturn's rings constrains their age to no more than a few hundred million years. Sci Adv 9(19):eadf8537 https://doi.org/10.1126/sciadv.adf8537

Kliore Astron J, Patel IR, Lindal GF, et al (1980) Structure of the ionosphere and atmosphere of Saturn from pioneer 11 Saturn radio occultation. J Geophys Res Space Phys 85(A11):5857–5870

Kliore Astron J, Anderson JD, Armstrong JW, et al (2004) Cassini Radio Sci Space Sci Rev 115(1–4):1–70. https://doi.org/10.1007/s11214-004-1436-y

Kliore A, Nagy A, Marouf E, et al (2009) Midlatitude and high-latitude electron density profiles in the ionosphere of Saturn obtained by Cassini radio occultation observations. J Geophys Res Space Phys 114(A4)

Kollmann P, Roussos E, Kotova A, et al (2018) Saturn's innermost radiation belt throughout and inward of the D-ring. Geophys Res Lett 45(20):10,912–10,920. https://doi.org/10.1029/2018GL077954

Koskinen TT, Guerlet S (2018) Atmospheric structure and helium abundance on Saturn from Cassini/UVIS and CIRS observations. Icarus 307:161–171. https://doi.org/10.1016/j.icarus.2018.02.020

Krimigis S, Mitchell D, Hamilton D, et al (2004) Magnetosphere imaging instrument (MIMI) on the Cassini mission to Saturn/Titan In: Russell CT (ed) The Cassini-Huygens Mission: orbiter in situ investigations, vol 2. Springer, Dordrecht, pp 233–329. https://doi.org/10.1007/978-1-4020-2774-1_3

Linti S, Postberg F, Hsu HW, et al (2024) Iron depletion in mineral dust grains from Saturn's main rings. Mon Not R Astron Soc 529(2):1207–1221. https://doi.org/10.1093/mnras/stae603

Longaretti PY (2018) Theory of narrow rings and sharp edges. In: Tiscareno MS, Murray CD (eds) Planetary ring systems. Properties, structure, and evolution. Cambridge University Press, pp 225–275. https://doi.org/10.1017/9781316286791.010

Luhmann JG, Johnson RE, Tokar RL, et al (2006) A model of the ionosphere of Saturn's rings and its implications. Icarus 181(2):465–474. https://doi.org/10.1016/j.icarus.2005.11.022

Lynden-Bell D, Pringle JE (1974) The evolution of viscous discs and the origin of the nebular variables. Mon Not R Astron Soc 168:603–637. https://doi.org/10.1093/mnras/168.3.603

Magee BA, Waite JH, Mandt KE, et al (2009) INMS-derived composition of Titan's upper atmosphere: analysis methods and model comparison. Planet Space Sci 57(14-15):1895–1916. https://doi.org/10.1016/j.pss.2009.06.016

Malloci G, Mulas G, Joblin C (2004) Electronic absorption spectra of PAHs up to vacuum UV. Towards a detailed model of interstellar PAH photophysics. Astron Astrophys 426:105–117. https://doi.org/10.1051/0004-6361:20040541

Marouf EA, Tyler GL, Rosen PA (1986) Profiling Saturn's rings by radio occultation. Icarus 68(1):120–166. https://doi.org/10.1016/0019-1035(86)90078-3

Mason E, Marrero T (1970) The diffusion of atoms and molecules. In: Advances in atomic and molecular physics, vol 6. Elsevier, Amsterdam, pp 155–232

Mastrapa RM, Sandford SA, Roush TL, et al (2009) Optical constants of amorphous and crystalline H_2O-ice: 2.5-22 μm (4000-455 cm^{-1}) optical constants of H_2O-ice. Astrophys J 701(2):1347–1356. https://doi.org/10.1088/0004-637X/701/2/1347

Miller KE, Waite JH, Perryman RS, et al (2020) Cassini INMS constraints on the composition and latitudinal fractionation of Saturn ring rain material. Icarus 339:113595. https://doi.org/10.1016/j.icarus.2019.113595

Mitchell DG, Perry ME, Hamilton DC, et al (2018) Dust grains fall from Saturn's D-ring into its equatorial upper atmosphere. Science 362(6410):aat2236. https://doi.org/10.1126/science.aat2236

Moore L, O'Donoghue J, Müller-Wodarg I, et al (2015) Saturn ring rain: model estimates of water influx into Saturn's atmosphere. Icarus 245:355–366. https://doi.org/10.1016/j.icarus.2014.08.041

Morishima R, Spilker L, Salo H, et al (2010) A multilayer model for thermal infrared emission of Saturn's rings II: albedo, spins, and vertical mixing of ring particles inferred from Cassini CIRS. Icarus 210(1):330–345. https://doi.org/10.1016/j.icarus.2010.06.032

Morishima R, Spilker L, Ohtsuki K (2011) A multilayer model for thermal infrared emission of Saturn's rings. III: thermal inertia inferred from Cassini CIRS. Icarus 215(1):107–127. https://doi.org/10.1016/j.icarus.2011.06.042. arXiv:1209.3797 [astro-ph.EP]

Morishima R, Edgington SG, Spilker L (2012) Regolith grain sizes of Saturn's rings inferred from Cassini-CIRS far-infrared spectra. Icarus 221(2):888–899. https://doi.org/10.1016/j.icarus.2012.09.012. arXiv:1209.3799 [astro-ph.EP]

Moses JI, Brown ZL, Koskinen TT, et al (2023) Saturn's atmospheric response to the large influx of ring material inferred from Cassini INMS measurements. Icarus 391:115328. https://doi.org/10.1016/j.icarus.2022.115328. arXiv:2211.05253 [astro-ph.EP]

Murray CD, French RS (2018) The F ring of Saturn. In: Tiscareno MS, Murray CD (eds) Planetary ring systems. Properties, structure, and evolution. Cambridge University Press, pp 338–362. https://doi.org/10.1017/9781316286791.013

Nicholson PD, Hedman MM, Clark RN, et al (2008) A close look at Saturn's rings with Cassini VIMS. Icarus 193(1):182–212. https://doi.org/10.1016/j.icarus.2007.08.036

Nicholson PD, French RG, Spitale JN (2018) Narrow rings, gaps, and sharp edges. In: Tiscareno MS, Murray CD (eds) Planetary ring systems. Properties, structure, and evolution. Cambridge University Press, pp 276–307 https://doi.org/10.1017/9781316286791.011

Northrop TG, Connerney JEP (1987) A micrometeorite erosion model and the age of Saturn's rings. Icarus 70(1):124–137. https://doi.org/10.1016/0019-1035(87)90079-0

Northrop T, Hill JR (1982) Stability of negatively charged dust grains in Saturn's ring plane. J Geophys Res Space Phys 87(A8):6045–6051

O'Donoghue J, Stallard TS, Melin H, et al (2013) The domination of Saturn's low-latitude ionosphere by ring 'rain'. Nature 496(7444):193–195. https://doi.org/10.1038/nature12049. arXiv:1310.7293 [astro-ph.EP]

O'Donoghue J, Moore L, Connerney J, et al (2019) Observations of the chemical and thermal response of 'ring rain' on Saturn's ionosphere. Icarus 322:251–260. https://doi.org/10.1016/j.icarus.2018.10.027

Pendleton YJ, Allamandola LJ (2002) The organic refractory material in the diffuse interstellar medium: mid-infrared spectroscopic constraints. Astrophys J Suppl Ser 138(1):75–98. https://doi.org/10.1086/322999

Perry ME, Waite JH, Mitchell DG, et al (2018) Material flux from the rings of Saturn into its atmosphere. Geophys Res Lett 45(19):10,093–10,100. https://doi.org/10.1029/2018GL078575

Pettengill GH, Hagfors T (1974) Comment on radar scattering from Saturn's rings. Icarus 21(2):188–190. https://doi.org/10.1016/0019-1035(74)90135-3

Pitman KM, Buratti BJ, Mosher JA (2010) Disk-integrated bolometric bond albedos and rotational light curves of saturnian satellites from Cassini visual and infrared mapping spectrometer. Icarus 206(2):537–560. https://doi.org/10.1016/j.icarus.2009.12.001

Pollack JB, Summers A, Baldwin B (1973) Estimates of the sizes of the particles in the rings of Saturn and their cosmogonic implications. Icarus 20(3):263–278. https://doi.org/10.1016/0019-1035(73)90003-1

Poulet F, Cruikshank DP, Cuzzi JN, et al (2003) Compositions of Saturn's rings A, B, and C from high resolution near-infrared spectroscopic observations. Astron Astrophys 412:305–316. https://doi.org/10.1051/0004-6361:20031123

Raponi A, Ciarniello M, Capaccioni F, et al (2020) Infrared detection of aliphatic organics on a cometary nucleus. Nat Astron 4:500–505. https://doi.org/10.1038/s41550-019-0992-8. arXiv:2009.14476 [astro-ph.EP]

Reynard B, Sotin C (2023) Carbon-rich icy moons and dwarf planets. Earth Planet Sci Lett 612:118172 https://doi.org/10.1016/j.epsl.2023.118172

Rouleau F, Martin PG (1991) Shape and clustering effects on the optical properties of amorphous carbon. Astrophys J 377:526. https://doi.org/10.1086/170382

Roush TL (1994) Charon: more than water ice? Icarus 108(2):243–254. https://doi.org/10.1006/icar.1994.1059
Salama F, Bakes ELO, Allamandola LJ, et al (1996) Assessment of the polycyclic aromatic hydrocarbon–diffuse interstellar band proposal. Astrophys J 458:621. https://doi.org/10.1086/176844
Schmidt J, Ohtsuki K, Rappaport N, et al (2009) Dynamics of Saturn's dense rings. In: Dougherty MK, Esposito LW, Krimigis SM (eds) Saturn from Cassini-Huygens. Springer, Dordrecht, pp 413–458. https://doi.org/10.1007/978-1-4020-9217-6_14
Serigano J, Hörst SM, He C, et al (2020) Compositional measurements of Saturn's upper atmosphere and rings from Cassini INMS. J Geophys Res, Planets 125(8):e06427. https://doi.org/10.1029/2020JE006427. arXiv:2005.06554 [astro-ph.EP]
Serigano J, Hörst SM, He C, et al (2022) Compositional measurements of Saturn's upper atmosphere and rings from Cassini INMS: an extended analysis of measurements from Cassini's grand finale orbits. J Geophys Res, Planets 127(6):e07238. https://doi.org/10.1029/2022JE007238
Shkuratov Y, Starukhina L, Hoffmann H, et al (1999) A model of spectral albedo of particulate surfaces: implications for optical properties of the moon. Icarus 137(2):235–246. https://doi.org/10.1006/icar.1998.6035
Shkuratov YG, Stankevich DG, Petrov DV, et al (2005) Interpreting photometry of regolith-like surfaces with different topographies: shadowing and multiple scattering. Icarus 173(1):3–15. https://doi.org/10.1016/j.icarus.2003.12.017
Shkuratov Y, Kaydash V, Korokhin V, et al (2012) A critical assessment of the Hapke photometric model. J Quant Spectrosc Radiat Transf 113(18):2431–2456. https://doi.org/10.1016/j.jqsrt.2012.04.010
Showalter MR (1996) Saturn's D ring in the Voyager images. Icarus 124(2):677–689. https://doi.org/10.1006/icar.1996.0241
Smith HT, Richardson JD (2021) The 3d structure of Saturn magnetospheric neutral tori produced by the Enceladus plumes. J Geophys Res Space Phys A028:775
Smith BA, Soderblom L, Beebe R, et al (1981) Encounter with Saturn: Voyager 1 imaging science results. Science 212(4491):163–191
Spahn F, Sremčević M (2000) Density patterns induced by small moonlets in Saturn's rings? Astron Astrophys 358:368–372
Spahn F, Hoffmann H, Rein H, et al (2018) Moonlets in dense planetary rings. In: Tiscareno MS, Murray CD (eds) Planetary ring systems. Properties, structure, and evolution. Cambridge University Press, pp 157–197. https://doi.org/10.1017/9781316286791.008
Spilker LJ, Pilorz SH, Edgington SG, et al (2005) Cassini CIRS observations of a roll-off in Saturn ring spectra at submillimeter wavelengths. Earth Moon Planets 96(3–4):149–163. https://doi.org/10.1007/s11038-005-9060-8
Spilker LJ, Pilorz SH, Wallis BD, et al (2006) Cassini thermal observations of Saturn's main rings: implications for particle rotation and vertical mixing. Planet Space Sci 54(12):1167–1176. https://doi.org/10.1016/j.pss.2006.05.033
Spilker L, Ferrari C, Morishima R (2013) Saturn's ring temperatures at equinox. Icarus 226(1):316–322. https://doi.org/10.1016/j.icarus.2013.06.002
Spilker LJ, Ferrari C, Altobelli N, et al (2018) Thermal properties of rings and ring particles. In: Tiscareno MS, Murray CD (eds) Planetary ring systems. Properties, structure, and evolution. Cambridge University Press, pp 399–433. https://doi.org/10.1017/9781316286791.015
Srama R, Ahrens TJ, Altobelli N, et al (2004) The Cassini cosmic dust analyzer. Space Sci Rev 114(1–4):465–518. https://doi.org/10.1007/s11214-004-1435-z
Tajeddine R, Nicholson PD, Longaretti PY, et al (2017) What confines the rings of Saturn? Astrophys J Suppl Ser 232(2):28. https://doi.org/10.3847/1538-4365/aa8c09. arXiv:1710.08462 [astro-ph.EP]
Tiscareno MS, Harris BE (2018) Mapping spiral waves and other radial features in Saturn's rings. Icarus 312:157–171. https://doi.org/10.1016/j.icarus.2018.04.023. arXiv:1708.03702 [astro-ph.EP]
Tiscareno MS, Murray CD (2018) Planetary ring systems: properties, structure, and evolution. Cambridge University Press, Cambridge. https://doi.org/10.1017/9781316286791
Tiscareno MS, Burns JA, Hedman MM, et al (2006) 100-metre-diameter moonlets in Saturn's A ring from observations of 'propeller' structures. Nature 440(7084):648–650. https://doi.org/10.1038/nature04581
Tiscareno MS, Burns JA, Nicholson PD, et al (2007) Cassini imaging of Saturn's rings. II. A wavelet technique for analysis of density waves and other radial structure in the rings. Icarus 189(1):14–34. https://doi.org/10.1016/j.icarus.2006.12.025. astro-ph. arXiv:astro-ph/0610242
Tiscareno MS, Burns JA, Hedman MM, et al (2008) The population of propellers in Saturn's A ring. Astron J 135(3):1083–1091. https://doi.org/10.1088/0004-6256/135/3/1083. astro-ph. arXiv:0710.4547
Tiscareno MS, Burns JA, Sremčević M, et al (2010) Physical characteristics and non-Keplerian orbital motion of "propeller" moons embedded in Saturn's rings. Astrophys J Lett 718(2):L92–L96. https://doi.org/10.1088/2041-8205/718/2/L92. arXiv:1007.1008 [astro-ph.EP]

Tiscareno MS, Hedman MM, Burns JA, et al (2013a) Compositions and origins of outer planet systems: insights from the Roche critical density. Astrophys J Lett 765(2):L28. https://doi.org/10.1088/2041-8205/765/2/L28. arXiv:1302.1253 [astro-ph.EP]

Tiscareno MS, Hedman MM, Burns JA, et al (2013b) Probing the inner boundaries of Saturn's A ring with the Iapetus -1:0 nodal bending wave. Icarus 224(1):201–208. https://doi.org/10.1016/j.icarus.2013.02.026. arXiv:1302.5712 [astro-ph.EP]

Tiscareno MS, Nicholson PD, Cuzzi JN, et al (2019) Close-range remote sensing of Saturn's rings during Cassini's ring-grazing orbits and grand finale. Science 364(6445):aau1017. https://doi.org/10.1126/science.aau1017

Tokar RL, Johnson RE, Thomsen MF, et al (2005) Cassini observations of the thermal plasma in the vicinity of Saturn's main rings and the F and G rings. Geophys Res Lett 32(14):L14S04. https://doi.org/10.1029/2005GL022690

Toomre A (1964) On the gravitational stability of a disk of stars. Astrophys J 139:1217–1238. https://doi.org/10.1086/147861

Trieloff M, Fischer C, Postberg F, et al (2023) Dynamical analysis of mineral dust in the Saturnian system. In: EGU General Assembly 2023. Vienna, Austria, 24–28 Apr 2023, EGU23-16469. https://doi.org/10.5194/egusphere-egu23-16469

Tseng WL, Ip WH (2011) An assessment and test of Enceladus as an important source of Saturn's ring atmosphere and ionosphere. Icarus 212(1):294–299. https://doi.org/10.1016/j.icarus.2010.12.003

Tseng WL, Ip WH, Johnson RE, et al (2010) The structure and time variability of the ring atmosphere and ionosphere. Icarus 206(2):382–389. https://doi.org/10.1016/j.icarus.2009.05.019

Tseng WL, Johnson RE, Thomsen MF, et al (2011) Neutral H_2 and H_2^+ ions in the Saturnian magnetosphere. J Geophys Res Space Phys 116(A3):A03209. https://doi.org/10.1029/2010JA016145

Tseng WL, Johnson RE, Elrod MK (2013a) Modeling the seasonal variability of the plasma environment in Saturn's magnetosphere between main rings and Mimas. Planet Space Sci 77:126–135. https://doi.org/10.1016/j.pss.2012.05.001. arXiv:1112.5511 [astro-ph.EP]

Tseng WL, Johnson RE, Ip WH (2013b) The atomic hydrogen cloud in the saturnian system. Planet Space Sci 85, 164–174. https://doi.org/10.1016/j.pss.2013.06.005. arXiv:1302.3270 [astro-ph.EP]

Tyler GL, Eshleman VR, Anderson JD, et al (1981) Radio science investigations of the Saturn system with Voyager 1: preliminary results. Science 212(4491):201–206. https://doi.org/10.1126/science.212.4491.201

Vahidinia S, Cuzzi JN, Hedman M, et al (2011) Saturn's F ring grains: aggregates made of crystalline water ice. Icarus 215(2):682–694. https://doi.org/10.1016/j.icarus.2011.04.011

van de Hulst HC (1981) Light scattering by small particles. Courier Corporation.

Verbiscer Astron J, Skrutskie MF, Hamilton DP (2009) Saturn's largest ring. Nature 461(7267):1098–1100. https://doi.org/10.1038/nature08515

Vigren E, Dreyer J, Eriksson AI, et al (2022) Empirical photochemical modeling of Saturn's ionization balance including grain charging. Planet Sci J 3(2):49

Waite JH, Atreya SK, Nagy AF (1979) The ionosphere of Saturn: predictions for pioneer 11. Geophys Res Lett 6(9):723–726. https://doi.org/10.1029/GL006i009p00723

Waite JH, Lewis W, Kasprzak W, et al (2004) The Cassini Ion and Neutral Mass Spectrometer (INMS) investigation. In: Russell CT (ed) The Cassini-Huygens Mission: orbiter in situ investigations, vol 2. Springer, Dordrecht, pp 113–231. https://doi.org/10.1007/978-1-4020-2774-1_2

Waite JH, Cravens TE, Ip WH, et al (2005) Oxygen ions observed near Saturn's A ring. Science 307(5713):1260–1262. https://doi.org/10.1126/science.1105734

Waite JH, Perryman RS, Perry ME, et al (2018) Chemical interactions between Saturn's atmosphere and its rings. Science 362(6410):aat2382. https://doi.org/10.1126/science.aat2382

West R, Lavvas P, Anderson C, et al (2014) Titan's haze. In: Titan, pp 285. https://doi.org/10.1017/CBO9780511667398.011

Yelle RV, Serigano J, Koskinen TT, et al (2018) Thermal structure and composition of Saturn's upper atmosphere from Cassini/Ion Neutral Mass Spectrometer measurements. Geophys Res Lett 45(20):10,951–10,958. https://doi.org/10.1029/2018GL078454

Young D, Berthelier J, Blanc M, et al (2004) Cassini Plasma Spectrometer investigation. In: Russell CT (ed) The Cassini-Huygens Mission: orbiter in situ investigations, vol 2. Springer, Dordrecht, pp 1–112. https://doi.org/10.1007/978-1-4020-2774-1_1

Zebker HA, Marouf EA, Tyler GL (1985) Saturn's rings: particle size distributions for thin layer models. Icarus 64(3):531–548. https://doi.org/10.1016/0019-1035(85)90074-0

Zhang Z, Hayes AG, Janssen MA, et al (2017a) Exposure age of Saturn's A and B rings, and the Cassini division as suggested by their non-icy material content. Icarus 294:14–42. https://doi.org/10.1016/j.icarus.2017.04.008

Zhang Z, Hayes AG, Janssen MA, et al (2017b) Cassini microwave observations provide clues to the origin of Saturn's C ring. Icarus 281:297–321. https://doi.org/10.1016/j.icarus.2016.07.020
Zhang Z, Hayes AG, de Pater I, et al (2019) VLA multi-wavelength microwave observations of Saturn's C and B rings. Icarus 317:518–548. https://doi.org/10.1016/j.icarus.2018.08.014

Authors and Affiliations

Kelly E. Miller[1] · Gianrico Filacchione[2] · Jeffrey N. Cuzzi[3] · Philip D. Nicholson[4] · Matthew M. Hedman[5] · Kevin Baillié[6] · Robert E. Johnson[7] · Wei-Ling Tseng[8] · Paul R. Estrada[3] · Jack Hunter Waite[9] · Mauro Ciarniello[2] · Cécile Ferrari[10] · Zhimeng Zhang[11] · Amanda Hendrix[12] · Julianne I. Moses[13] · Hsiang-Wen Hsu[14]

✉ G. Filacchione
gianrico.filacchione@inaf.it

K.E. Miller
kmiller@swri.edu

J.N. Cuzzi
jeffrey.cuzzi@nasa.gov

P.D. Nicholson
pdn2@cornell.edu

M.M. Hedman
mhedman@uidaho.edu

K. Baillié
kevin.baillie@obspm.fr

R.E. Johnson
rej@virginia.edu

W.-L. Tseng
wltseng@ntnu.edu.tw

P.R. Estrada
paul.r.estrada@nasa.gov

J.H. Waite
hunterwaite@gmail.com

M. Ciarniello
mauro.ciarniello@inaf.it

C. Ferrari
ferrari@ipgp.fr

Z. Zhang
zzm19881204@gmail.com

A. Hendrix
ahendrix@psi.edu

J.I. Moses
jmoses@spacescience.org

H.-W. Hsu
sean.hsu@lasp.colorado.edu

1 Space Science, Southwest Research Institute, 9503 W. Commerce, San Antonio, TX, 78227, USA

[2] INAF, IAPS Institute for Space Astrophysics and Planetology, via del Fosso del Cavaliere, 100, Rome, 00133, Italy

[3] Space Sciences Div., NASA Ames Research Center, MS 245-3, Moffett Field, CA, 94035, USA

[4] Department of Astronomy, Cornell University, Ithaca, NY, 14853, USA

[5] Department of Physics, University of Idaho, Moscow, ID, 83844, USA

[6] IMCCE, Observatoire de Paris, PSL Research University, CNRS, Sorbonne Universités, UPMC Univ Paris 06, Univ. Lille, 77 Av. Denfert-Rochereau, Paris, 75014, France

[7] Engineering Physics, University of Virgina, Charlottesville, VA, 22902, USA

[8] Department of Earth Sciences, National Taiwan Normal University, Taipei, 116, Taiwan

[9] Waite Science, LLC, 16284 Narwhal Drive, Pensacola, FL, 32507, USA

[10] Université de Paris Cité, Institut de physique du globe de Paris, CNRS, Paris, F-75005, France

[11] Division of Geological and Planetary Sciences, California Institute of Technology, Pasadena, CA, 91125, USA

[12] Planetary Science Institute, Tucson, AZ, 85719, USA

[13] Space Science Institute, Boulder, CO, 80301, USA

[14] Laboratory for Atmospheric and Space Physics, University of Colorado-Boulder, Boulder, CO, 80303, USA

Space Science Reviews (2024) 220:72
https://doi.org/10.1007/s11214-024-01103-z

The Origin and Composition of Saturn's Ring Moons

Mauro Ciarniello[1] · Gianrico Filacchione[1] · Philip D. Nicholson[2] · Matthew M. Hedman[3] · Sebastien Charnoz[4] · Jeffrey N. Cuzzi[5] · Maryame El Moutamid[6] · Amanda R. Hendrix[7] · Nicolas Rambaux[8] · Kelly E. Miller[9] · Olivier Mousis[10,11] · Kevin Baillié[8] · Paul R. Estrada[5] · J. Hunter Waite[9]

Received: 2 October 2023 / Accepted: 20 August 2024 / Published online: 17 September 2024

Abstract
Here we review the origin, evolution, and compositional properties of Saturn's ring moons. This class of eleven small satellites includes objects orbiting near the outer edge of the main rings (Pan, Daphnis, Atlas, Prometheus, Pandora, Janus, Epimetheus) and "ring-embedded" moons (Aegeon, Methone, Anthe, Pallene) orbiting inward of Enceladus and associated with either diffuse or partial rings. We discuss current formation scenarios, according to which ring moons could originate either in the main rings from accretion onto original seeds denser than the ring material, or outside the A ring from spontaneous accretion of ring particles, and then evolve outwards due to gravitational torque from the rings. Remote sensing observations of the ring moons from the Cassini mission are analyzed in the broader context of Saturn's icy moons and main rings observations. Spectroscopic data support a compositional paradigm similar to the main rings, dominated by water ice, and smaller amounts of two separate contaminants, in the form of a UV absorber and a spectrally neutral darkening material. Global radial trends in the spectral properties of the ring moons suggest that the surface composition is significantly affected by a complex interplay of exogenous processes, among which the contamination from nearby A ring particles, meteoritic bombardment, charged particle flux, and E ring particle accumulation, depending on the corresponding magnitude at the ring moon orbital distance and exposure time. These processes modify the original composition inherited by the rings and, coupled with the fact that the surface composition is likely representative only of the ring moon outer layers, make it difficult to trace back the present composition to a given ring moon formation scenario.

Keywords Ring moons · Composition · Origins · Surfaces · Ices · Saturn

1 Introduction

Among Saturn's 146 known satellites, as of 2023 (Sheppard et al. 2023), are eleven small inner satellites that are dynamically-related to the rings and which we refer to here as "ring moons". Ring moons play a crucial role in preserving the rings from viscous expansion, a result of the collisions among ring particles which causes their diffusion through viscous interactions (Charnoz et al. 2009). Orbiting near the outer edge of the rings, ring moons are at the right distances from Saturn to generate strong orbital resonances, which, in addition

Extended author information available on the last page of the article

to the gravitational influence of the larger Mimas, are capable of restraining the ring viscous radial spreading (Lissauer and Cuzzi 1982; Tiscareno et al. 2007; Tajeddine et al. 2017a). The 'dynamical age' of the rings is thus closely linked to the presence, number and size of its associated ring moons (as well as Mimas), while being fundamentally driven by viscous spreading due to internal collisional interactions (Salmon et al. 2010).

The histories of Saturn's ring moons and the ring system are strongly related among them since they probably share at least aspects of their origin. Numerical simulations show that collisions among ring particles can clump them together to form larger "dynamical ephemeral bodies", up to tens of meters in size, which then undergo tidal disruption (Weidenschilling et al. 1984; Davis et al. 1984). However, if the accretion of ring material occurs onto a seed with density larger than that of ring particles, the aggregate can survive disruption to form a small moon (Charnoz et al. 2007, 2009). As a consequence, many of the ring moons could be the result of a recent aggregation of ring particles, that successively evolved outside the main ring system. Alternatively, spontaneous accretion of ring material might lead to the in situ formation of moons, if occurring at distances where the moonlet self-gravity overcomes local Saturn's tidal forces. As a result of these formation processes, the ring moons preserve to some extent the composition of the ring particles from which were accreted, and can be considered as a "bridge" between the main rings and mid-sized icy satellites. Nonetheless, ring moon surfaces are also affected by a complex interplay of exogenous processes, such as contamination of nearby ring material (Buratti et al. 2019; Hedman et al. 2020), bombardment from magnetospheric particles (Roussos et al. 2008; Paranicas et al. 2018; Hedman et al. 2020), and contamination from meteoritic dust (Cuzzi et al. 2009), which modifies the physical and compositional properties of the material inherited from the main rings.

In this article we discuss the ring moon formation processes and investigate the surface composition of the ring moons in comparison to the main rings and the mid-sized icy satellites. This would help to improve our understanding of the processes determining the evolution of the Saturn's system in general, and, by consequence, of the main rings, in an effort of finally constraining their primordial composition, origin and age.

We proceed by first introducing the different ring moon dynamical families (Sect. 2), then we summarize their structural and morphological properties (Sects. 3 and 4) and review current ring moon formation and orbital evolution scenarios (Sect. 5); the current understanding of the ring moon surface composition from Cassini remote sensing data, in comparison to Saturn's ring and mid-sized icy satellites, is presented in Sect. 6 and it is complemented by a review of the composition of the F and G rings in Sect. 7; the exogenic processes affecting ring moon surface properties are addressed in Sect. 8; finally, in Sect. 9 we summarize the main findings and discuss the implications of the observed ring moon composition in the context of Saturn's ring moon origin and evolution.

2 Ring Moon Dynamical Families

Following Thomas et al. (2013, 2018b), we may conveniently divide the ring moons into three families, based on their distinct orbital or dynamical environments:

- The inner ring moons, Pan, Daphnis, Atlas, Prometheus and Pandora, occupy orbits either within gaps in the rings (Pan and Daphnis), in close proximity to the A ring (Atlas), or straddling the narrow F ring (Prometheus and Pandora).
- The coorbital satellites, Janus and Epimetheus, share a common orbit located between the F ring and Mimas, exchanging their relative positions every 4 yrs.

- Aegaeon, Methone, Anthe and Pallene are all embedded within either narrow diffuse rings or partial rings/arcs. Aegaeon orbits within the G ring, located between the coorbitals and Mimas, while the other three occupy orbits between those of Mimas and Enceladus. Along with tiny Polydeuces, they are the smallest of the inner Saturnian moons.

In addition, there are four objects, Telesto, Calypso, Helene and Polydeuces that are not strictly-speaking 'ring moons', but small bodies that librate around the L_4 and L_5 triangular Lagrange points of the mid-sized satellites Tethys and Dione, respectively. These objects are probably most like the objects in the last group because they orbit within the E ring like Anthe, Methone and Pallene. Furthermore, just as the orbits of these moons are strongly perturbed by Tethys and Dione, the orbits of Aegaeon, Anthe and Methone are heavily influenced by mean-motion resonances with Mimas. The dynamical state and history of all these moons are therefore strongly influenced by Saturn's mid-sized satellites.

The ring moons' principal orbital characteristics are summarized in Table 1, based largely on fits to Voyager and Cassini optical navigation data by Spitale et al. (2006), Jacobson et al. (2008), Hedman et al. (2009, 2010), and Cooper et al. (2008, 2015). Most of these objects suffer significant orbital perturbations due to interactions with other satellites and the values listed in Table 1 are decade-long averages based on numerical integrations.

2.1 Inner Ring Moons

Pan is responsible for clearing and maintaining the 322 km-wide Encke gap in the outer A ring, at whose edges and in the adjoining ring material it drives a complex pattern of km-scale oscillations collectively referred to as 'satellite wakes' (Showalter et al. 1986; Weiss et al. 2009), from which its mass has been estimated.

Daphnis plays a similar role in the 35 km-wide Keeler gap, but is significantly smaller than Pan. It too drives oscillatory perturbations on the gap edges, from which its mass has been estimated (Weiss et al. 2009). Cassini observations have revealed occasional and irregular changes in its semimajor axis and orbital period, of unknown origin (Jacobson 2014; Santana et al. 2019).

Orbiting about 900 km outside the edge of the A ring, Atlas is similar to Pan in size and shape. Their non-spherical shapes could be the result of either accretion of ring particles (Charnoz et al. 2007) or the result of low-velocity collisions (Leleu et al. 2018).

Atlas orbit is mildly chaotic, apparently due to overlapping 54:43 mean-motion resonances (henceforth MMRs) with Prometheus (Cooper et al. 2015; Renner et al. 2016). Despite the initial assumption that probably inspired its name, Atlas does not 'hold up' the edge of the A ring. This task is instead currently largely accomplished by the 7:6 mean-motion resonance with Janus (Spitale and Porco 2009; Tajeddine et al. 2017a; Nicholson et al. 2018).

About 3400 km beyond the outer edge of the A ring lies the narrow, multi-stranded F ring, on either side of which orbit the satellites Prometheus and Pandora. Initially referred to as 'shepherd satellites', due to their perceived role in confining the F ring, subsequent studies have shown that the true situation is more complex, although many of the unusual periodic structures seen in this ring have been shown to be due to perturbations by Prometheus (Murray and Giuliatti Winter 1996; Murray et al. 2005). For a more detailed discussion of the F ring's structure and dynamics, the reader is referred to the reviews by Colwell et al. (2009) and Murray and French (2018). Observations of Prometheus and Pandora with HST in the mid-1990s, and some models, suggested that their orbits are chaotic, with discontinuities in their mean motions associated with the times — every 6.2 yrs — when the apocenter of the former was aligned with the pericenter of the latter (Poulet and Sicardy 2001; French et al.

Table 1 Orbital data for Saturn's ring moons[1]. The JPL Horizons ID (Code), average semi-major axis (a_{mean}), eccentricity (Ecc.), inclination (Incl.) and period are reported for each target. For reference, also significant ring features are included, for which the corresponding radius (Rad.) is indicated. All moons outside Janus orbit within the diffuse E ring

Satellite/Ring	Code	a_{mean}/Rad. (km)	Ecc.	Incl. (deg)	Period (days)	Source
A ring inner edge		122,050				French et al. (2017)
Encke gap inner edge		133,423				French et al. (2017)
Pan[2]	618	133,584	0.00001	0.0	0.575051	Jacobson et al. (2008)
Encke gap outer edge		133,745				French et al. (2017)
Keeler gap inner edge		136,487				Tajeddine et al. (2017b)
Daphnis[2]	635	136,506	0.00003	0.004	0.594080	Jacobson et al. (2008)
Keeler gap outer edge		136,522				French et al. (2017)
A ring outer edge		136,770				El Moutamid et al. (2016)
Atlas	615	137,666	0.00116	0.003	0.601693	Cooper et al. (2015)
Prometheus[3]	616	139,378	0.00223	0.008	0.612990	Cooper et al. (2015)
F ring core		140,221				Cuzzi et al. (2018a)
Pandora[3]	617	141,712	0.00419	0.050	0.628503	Cooper et al. (2015)
Janus(a)[4]	610	151,462	0.0068	0.164	0.694661	Cooper et al. (2015)
Janus(b)[4]	610	151,441	0.0068	0.164	0.694517	Cooper et al. (2015)
Epimetheus(a)[4]	611	151,414	0.0097	0.352	0.694329	Cooper et al. (2015)
Epimetheus(b)[4]	611	151,489	0.0097	0.352	0.694850	Cooper et al. (2015)
Aegaeon[5]	653	167,494	0.00024	0.001	0.808111	Hedman et al. (2010)
Methone[5]	632	194,230	0.0000	0.013	1.009575	Jacobson et al. (2008)
Anthe[5]	649	197,655	0.0012	0.017	1.036517	Cooper et al. (2008)
Pallene[5]	633	212,280	0.0040	0.181	1.153746	Jacobson et al. (2008)
Telesto[6]	613	$\sim$ 294,670	0.001	1.18	1.88780	SAT441
Calypso[6]	614	$\sim$ 294,670	0.001	1.50	1.88780	SAT441
Helene[7]	612	$\sim$ 377,410	0.007	0.21	2.73691	SAT441
Polydeuces[7]	634	$\sim$ 377,410	0.019	0.18	2.73691	SAT441

[1]Geometric elements fitted to the numerically-integrated orbit, or average epicyclic elements

[2]Pan & Daphnis are modeled as precessing ellipses

[3]Prometheus and Pandora interact chaotically at intervals of $\sim$ 6.2 yrs when their pericenters are in anti-alignment

[4]Janus & Epimetheus share a common mean orbit, exchanging their relative positions every 4.0 yrs. Orbit (a) refers to the periods 2002-5 and 2010-3, while orbit (b) refers to the periods 2006-9 and 2014-7

[5]Aegaeon, Methone, Anthe, and Pallene orbit within faint dusty rings (see text)

[6]Telesto & Calypso librate about the L_4 and L_5 points of Tethys, respectively

[7]Helene & Polydeuces librate about the L_4 and L_5 points of Dione, respectively

2003). The chaos was later confirmed numerically but shown to be due to a series of overlapping 121:118 MMRs between the two moons (Goldreich and Rappaport 2003; Cooper et al. 2015). The dusty F ring itself lies in a dynamically chaotic region, but is anchored by a more massive large-particle core composed of multiple short arcs lying in locally-stable sites of a Corotation-Eccentricity resonance with Prometheus (Cuzzi 2024; Cuzzi et al. 2024). Nevertheless, important questions (such as its uniqueness) remain unresolved. Its mass is very uncertain, though probably much less than that of Prometheus or Pandora, and collisional events observed by Cassini suggest that it contains many small moonlets embedded within the icy dust that dominates its visual appearance (Charnoz et al. 2005; Beurle et al. 2010; Attree et al. 2012; Murray and French 2018).

2.2 Coorbital Moons

Approximately 11,000 km beyond the F ring we find the co-orbital satellites, Janus and Epimetheus, the only known example in the Solar System of a pair of similar-mass objects trapped in a horseshoe libration about their mutual L_4 and L_5 Lagrange points (Dermott and Murray 1981). Every 4.0 years, the two moons suffer a close approach and exchange angular momentum, one moving closer to Saturn while the other moves further away. Their mass ratio is $\sim$3:1, and modeling of the libration has yielded increasingly accurate masses for both objects (Yoder et al. 1989; Nicholson et al. 1992; Jacobson et al. 2008).

The present configuration is stable, but the outward migration of the satellites due to their gravitational interaction with the A ring[1] is likely to lead eventually to a 'tightening' of the co-orbital resonance, and its conversion from horseshoe motion to a Trojan-style libration of Epimetheus about the L_4 or L_5 point of Janus (Lissauer et al. 1985).

2.3 Ring-Embedded Moons

Between the orbits of the coorbital satellites and that of Enceladus, Cassini discovered four small moons, three of which are in MMRs with the mid-sized satellite Mimas and each of which is closely associated with either a diffuse ring or a partial ring (sometimes referred to as a 'ring arc'.)

The sub-km-sized satellite Aegaeon is actually embedded within the G ring (Hedman et al. 2010), a diffuse ring located midway between the orbits of Janus and Mimas and discovered by Pioneer 11. Cassini's discovery of Aegaeon resolved a long-standing mystery as to this ring's origin, which is now thought to arise from material ejected from Aegaeon as a result of impacts by interplanetary objects. Aegaeon is trapped in a 7:6 inner MMR with Mimas, as is the concentration of G ring particles that forms a distinct arc surrounding that moon.

The other three moons, Methone, Anthe and Pallene are located between the orbits of Mimas and Enceladus (Hedman et al. 2009). Anthe is in a 10:11 outer MMR with Mimas (Cooper et al. 2008), while Methone is in a 14:15 outer resonance (Spitale et al. 2006). Each is associated with a partial ring or arc that appears to be trapped in the same MMR, but lengthy libration periods of 1-2 yr make this difficult to prove observationally. Pallene, on the other hand, is associated with a narrow but azimuthally-complete dusty ring. It is very close to, but apparently not trapped in, a 19:16 inner MMR with Enceladus (Cooper et al. 2015).

These four satellites seem very likely to be the ultimate sources of the associated ring material, which is probably ejecta produced by high-velocity impacts of interplanetary dust particles, in the same manner that the jovian rings are believed to be derived from its embedded moons (Burns et al. 1999). The added wrinkle here is that, at least for Aegaeon, Methone and Anthe, some or all of this ejecta remains trapped in the same resonance as its parent moon, and thus forms an arc rather than a complete ring.

[1] More precisely, the reaction torque associated with the several strong density waves driven in the main rings by the satellites; see the discussion in Sect. 5.

3 Radii, Masses and Mean Densities

The radii, masses and mean densities of the inner small moons, where known, are summarized in Table 2, adapted from Thomas and Helfenstein (2020) and references therein. The masses listed here are based on analysis of their mutual gravitational perturbations or — in the case of Pan and Daphnis — their effects on the nearby rings. Helene's mass has recently been estimated from its perturbation of the orbit of Dione (Jacobson 2022). In the case of Aegaeon, Methone and Pallene, their near-ellipsoidal shapes have been used to estimate their densities directly, assuming hydrostatic equilibrium (Thomas et al. 2013).

The most striking feature here is the low densities; all but Anthe — whose mass and density are unknown — have mean densities appreciably less than that of pure water ice ($\sim$ 920 kg m^{-3}), and at least a factor of two less than that of the saturnian mid-sized satellites. The implication is that all of these objects are extremely porous, with a fraction of void space (or porosity) ranging from at least 30% for Janus and Epimetheus to at least 70% for Methone and Pallene, even assuming a negligible rocky component. The mean densities vary systematically with the objects' sizes: from a minimum of $\sim 250 - 300$ kg m^{-3} for Daphnis, Pallene, and Methone, to ~ 400 kg m^{-3} for Pan and Atlas, ~ 500 kg m^{-3} for Prometheus and Pandora, and 630 kg m^{-3} for Janus and Epimetheus. In this respect, the smallest ring moon Aegaeon is an exception, with a density of ~ 540 kg m^{-3}. Particularly noteworthy is the similarity in density shown by the pair of co-orbitals and by the two F ring shepherds. The trends seen in the ring-moon densities suggest an internal compression that increases with size, though even Janus and Epimetheus have very low central pressures, of order 3×10^5 Pa.

As is elaborated on in Sect. 5, the very high porosities are more consistent with the hypothesis that these objects accreted from ring material than with the idea that they are collisional shards resulting from the destruction of larger satellites, though we discuss below the possibility that all or many of these moons may harbour a denser core, which may be such a collisional shard.

Support for the accretional model is provided by the distinctive 'flying-saucer' shapes of Pan and Atlas, as revealed in high-resolution Cassini images (Porco et al. 2005, 2007; Charnoz et al. 2007; Buratti et al. 2019), whose prominent equatorial ridges might be due to ongoing accretion of icy material from the adjacent rings. Further discussions of the sizes, shapes and surface morphologies of the inner moons are to be found in Porco et al. (2007), Thomas et al. (2013, 2018b), Buratti et al. (2019), and Rambaux et al. (2022) as summarized in Sect. 4 below.

4 Ring Moon Shapes and Morphology

Ring moons range from a few kilometers to $\sim$ 100 km in radius (see Table 2), and many of these low-density, highly tidally constrained, rapidly rotating satellites exhibit irregular shapes. Figure 1, adapted from Thomas et al. (2018b), illustrates the diversity of shapes of these satellites in the context of Saturn's small moons. The images were acquired by the Cassini spacecraft during its journey through the saturnian system and especially at the end of the mission during the Grand Finale orbits. The shapes vary from ravioli-like structures with an equatorial ridge (Atlas, Pan and Daphnis), to elongated ones for Prometheus and Pandora, irregular covered with craters for Janus and Epimetheus, and the surprising smooth triaxial ellipsoids of Methone, Pallene and Aegaeon.

Table 2 Radii, masses, mean densities, surface gravities, albedos, and brightness coefficients for Saturn's small moons. For each satellite the best spatial resolution of ISS NAC images is reported

Satellite	R_{mean} (km)	Mass (kg)	ρ (kg m^{-3})	g (mm s^{-2})	Albedo[2] ($\alpha = 10°$)	Brightness Coefficient[3] ($\alpha = 30°$)	ISS NAC images best sp. res. (km/pixel)
Pan	13.7 ± 0.3	$4.30 \pm 0.22 \times 10^{15}$	400 ± 31	0.1-1.7	0.52 ± 0.04	$0.6492 \pm 0.0058 \pm 0.0142$	1.269
Daphnis	3.9 ± 0.5	$6.8 \pm 1.7 \times 10^{13}$	276 ± 144	0.0-0.3	0.56 ± 0.04	-	0.438
Atlas	14.9 ± 0.2	$5.753 \pm 0.015 \times 10^{15}$	412 ± 19	0.1-1.7	0.49 ± 0.08	$0.6254 \pm 0.0084 \pm 0.0084$	0.190
Prometheus	42.8 ± 0.7	$1.500 \pm 0.009 \times 10^{17}$	460 ± 21	0.7-5.6	0.67 ± 0.07	$0.7928 \pm 0.0127 \pm 0.0130$	0.201
Pandora	40.0 ± 0.3	$1.3684 \pm 0.013 \times 10^{17}$	509 ± 12	2.2-6.1	0.62 ± 0.08	$0.7600 \pm 0.0053 \pm 0.0057$	0.312
Epimetheus	58.6 ± 0.3	$5.2605 \pm 0.0013 \times 10^{17}$	625 ± 8	6.6-10-9	0.41 ± 0.04	$0.5305 \pm 0.0082 \pm 0.0027$	0.154
Janus	89.0 ± 0.5	$18.9548 \pm 0.0045 \times 10^{17}$	642 ± 10	11.1-16.9	0.37 ± 0.03	$0.4628 \pm 0.0060 \pm 0.0026$	0.178
Aegaeon[1]	0.33 ± 0.06	$7.82 \pm 3.00 \times 10^{10}$	539 ± 140	0.009-0.013	0.25 ± 0.23	$0.1793 \pm 0.0050 \pm 0.0326$	0.091
Methone[1]	1.45 ± 0.03	$3.92 \pm 1.00 \times 10^{12}$	307 ± 30	0.081-0.013	0.42 ± 0.05	$0.5558 \pm 0.0033 \pm 0.0115$	0.027
Anthe[4]	$< 1?$						5.290
Pallene[1]	2.23 ± 0.07	$1.15 \pm 0.40 \times 10^{13}$	251 ± 75	0.011-0.016	0.29 ± 0.11	$0.5228 \pm 0.0066 \pm 0.0164$	0.220
Telesto	12.3 ± 0.3				0.61 ± 0.04	$0.9028 \pm 0.0100 \pm 0.0220$	0.063
Calypso	9.5 ± 0.4				0.69 ± 0.14	$1.0999 \pm 0.0176 \pm 0.0463$	0.136
Helene	18.1 ± 0.2	7.1×10^{15}	291 ± 8		0.74 ± 0.15	$1.0251 \pm 0.0148 \pm 0.0113$	0.411
Polydeuces	1.3 ± 0.3				0.86 ± 0.06	$0.6075 \pm 0.0089 \pm 0.0794$	0.012

[1] Density & gravity inferred from shape

[2] Cassini ISS CL1/CL2, 610 nm

[3] The first error bars are statistical, and the second are systematic errors due to uncertainties in the moon's average size

[4] From Thomas et al. (2018b)

Fig. 1 Photomontage of the saturnian ring moons and Lagrangian satellites adapted from Thomas et al. (2018b). The images were all taken with Cassini's Narrow Angle Camera (NAC) and scale bars in kilometers indicate the relative sizes of each group of satellites

The Cassini-Huygens mission included several close flybys of these satellites, obtaining images with the Narrow-Angle Camera of the Imaging Science Subsystem (ISS) (Porco et al. 2004) at resolutions of $\sim$ 200 m/px (ranging from 5.3 km/px to 27 m/px; see Table 2 for the maximum resolution achieved at each body). The processing and analysis of these images is described in a series of papers by Thomas et al. (2013, 2018b) and Thomas and Helfenstein (2020), which provide each satellite's global topography as well as fundamental quantities such as its mean density, mean radius and best-fitting ellipsoid. These parameters were deduced using a combination of limb measurements and control point analyses (Thomas et al. 2013, and references therein).

Most notably, Thomas and Helfenstein (2020) discuss Digital Terrain Models (DTMs; shape data are from Thomas et al. 2018a, available in the Planetary Data System) that provide descriptions of the shapes of the bodies in terms of tessellated plate models whose estimated radial uncertainties range from tens to hundreds of meters. Recently, spherical

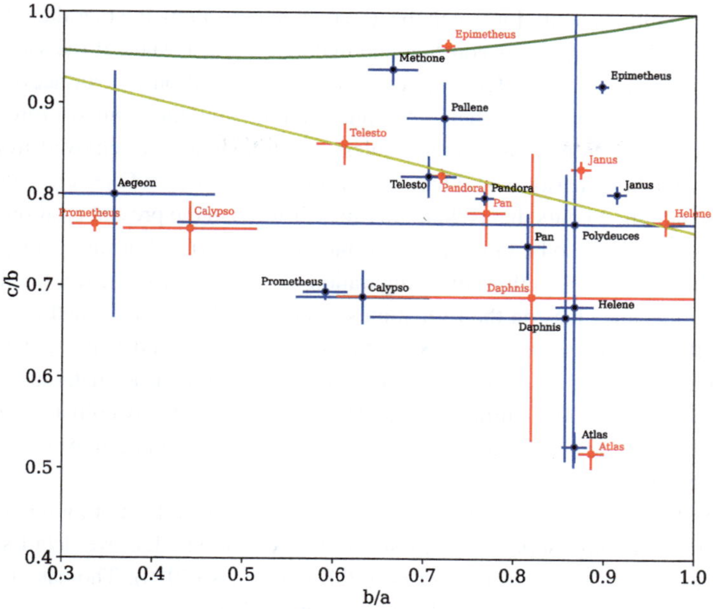

Fig. 2 Principal axis ratios b/a and c/b with their uncertainties for the small satellites of Saturn. The green curve represents the expected trend for a Roche ellipsoid configuration. The yellow curve is the "Omega sequence" of Dobrovolskis (2019). Blue points are from the best-fit ellipsoids of Thomas and Helfenstein (2020) and red points are from the degree-2 ellipsoid shapes of Rambaux et al. (2022) after spherical harmonic decomposition (figure adapted from Rambaux et al. 2022)

harmonic analysis of these DTMs has continued this descriptive work, to describe some geological features and to estimate fundamental quantities such as gravitational coefficients and librational amplitudes (Rambaux et al. 2022).

The hydrostatic equilibrium shape of a body in a strong tidal environment is specified by the Roche ellipsoid (Roche 1851; Chandrasekhar 1969). The first Cassini observations of ring moons and their analysis suggested that these bodies had accreted around an initial seed with shapes close to the appropriate Roche ellipsoid (computed using Roche lobe procedure in Porco et al. 2007). However, subsequent observations have led to more precise values of the semi-major axes of the best-fit ellipsoids (Thomas and Helfenstein 2020), whose ratios (c/b vs b/a) are reported in Fig. 2, indicating a departure from the values expected in Porco et al. (2007). In addition, the topography is disturbed by geological structures at various morphological wavelengths, such as ridges and depressions, or by impact processes that modify the final shapes for highly cratered bodies.

Rambaux et al. (2022) proposed to use the spherical harmonic decomposition to filter out the effect of craters (corresponding to short surface wavelengths) and to extract the global shape from the degree-2 coefficients (long wavelengths). The spherical harmonic decomposition is fitted to the DTM and computed until degree-14 for most satellites to obtain accurate low-degree harmonics, representing the shape underneath the cratered surface. The values of the semi-major axes ratio determined by this method are shown in Fig. 2. Among the ring moons only Epimetheus comes closer to the Roche ellipsoid curve. The ring satellites Pandora, Pan, Janus and the small bodies Telesto and Helene are closer to the so-called "Omega sequence" line defined by Dobrovolskis (2019). This curve represents the special case when the rotational velocity above which the force is repulsive at the ends of the a-axis (Ω_0) is equal to the rotation velocity for which the gravity potential along the meridian $x = 0$ (passing through the b and c axes) is constant and maximum (Ω_{bc}). Thus, if the satellite rotates at this rotation velocity $\Omega_0 = \Omega_{bc}$, any cohesionless material on the surface would accumulate at the ends of the a-axis (being potential minima) and the repulsive force will eject the material. In this configuration cohesionless regolith should be removed from the surface, however such scenarios would require a fast rotating body (at $\Omega_0 = \Omega_{bc}$) with rotation period of about 1 to 2 hours that is 20 to 60 times faster than the present rotational rate.

Some bodies appear to be more flattened (ratio $c/b < 0.8$) and/or more elongated (ratio $b/a < 0.8$) than the Roche ellipsoid configuration or the Omega sequence. Such variety of shapes provide some clues on the formation scenarios of these objects and is the result of possibly different histories (see also Sect. 5). Atlas, Daphnis, and Pan present equatorial ridges, which may have resulted from either the deposition of ring material on an initial object (Charnoz et al. 2007; Quillen et al. 2021) or from low-velocity collisions (Leleu et al. 2018), while the shape of elongated bodies such as Prometheus and Calypso might be more consistent with the latter mechanism.

Many of these satellites have distinctive geologic structures illustrating unique histories not yet included in any global scenario of satellite formation. Here we detail six of these features, based on studies by Porco et al. (2007), Thomas (2010), Thomas et al. (2013), Thomas and Helfenstein (2020) and Rambaux et al. (2022).

(i) The aforementioned equatorial ridges of Atlas, Pan and Daphnis. Such features are few kilometers above a presumed crestless body for Atlas and Pan, and tens of meters for Daphnis. The latitudinal spread of these bands is around $\pm 20°$ for Atlas, $\pm 15°$ for Pan, while Daphnis is characterized by two small off-center equatorial ridges centered at 2° S and 22° N (see Fig. 10 of Thomas and Helfenstein (2020) and Fig. 3 of Buratti et al. (2019)). It has been noted that these equatorial ridges have a polygonal longitudinal structure, as illustrated in the silhouettes shown in Fig. 5 of Thomas and Helfenstein (2020) or in the spherical harmonic mappings in the Appendix of Rambaux et al. (2022) (Figs. A.1 and A.7).

(ii) Spherical harmonic decomposition shows that the satellites Calypso, Pandora, and Prometheus have pronounced north-south longitudinal bands visible at degree-4 similar to those of Bennu and Ryugu (Roberts et al. 2021) despite their very different sizes.

(iii) It is noticeable that these three bodies are also quite elongated toward Saturn. The two satellites Daphnis and Telesto also show signs of elongation.

(iv) Helene and Telesto show large degree-3 spherical harmonics leading to a high-latitude offset of the topography of $+40°$ and $+15°$ respectively.

(v) Janus and Epimetheus do not show a clear pattern for the high-degree coefficients of their topography. This behavior could be the result of important processes of surface craterization as directly visible in the images of these two satellites.

(vi) Finally, several of these satellites show grooves, or long linear depressions in the regolith covering their surfaces (Morrison et al. 2009; Thomas and Helfenstein 2020). The origin of these grooves is not yet well understood and several scenarios have been proposed such as fracturing of the body below the regolith, secondary impacts, or even fissures in regolith (e.g. Asphaug 2020). Tidal stresses as for the grooves on Phobos seems unlikely because the satellites are beyond the synchronous orbit and are thus moving away from Saturn (Thomas et al. 2018b).

5 Ring Moon Origins and Evolution

The origin of moons within the ring system is not an easy thing to explain. Indeed forming icy satellites within the Roche Limit of Saturn for ice, is, by definition, prevented by the tidal forces from the planet. Accretion of ring material into moonlets is only possible beyond some distance where the self-gravity of the moonlet exceeds the planet's tidal force. But there are many definitions of the Roche Limit, depending on the details of the physics that is considered (pure gravity, frictional forces, etc.). The general form of the Roche Limit is given by:

$$R_L = k R_P \left(\frac{\rho_P}{\rho_r}\right)^{1/3} \tag{1}$$

where R_P and ρ_p are the planet's radius and average density, ρ_r is the density of the ring material and k is a prefactor that depends on the physics that is considered. For the case of non-porous ice $\rho_r \simeq 900$ kg m^{-3}, but studies of self-gravity wakes in Saturn's rings suggest that the ice may be very porous, with an average density of ~ 420 kg m^{-3} (Porco et al. 2008). For a cohesionless body $k = 2.4552$ (Chandrasekhar 1969).

Holsapple and Michel (2006) argue that we should consider two different versions of the Roche Limit. The "Fluid Roche Limit" (FRL) for a cohesionless body is the minimum distance at which material can accrete from the ring material into a self-gravitating object. For Saturn's rings the Fluid Roche Limit is $\sim 2.43\ R_P \simeq 132{,}000$ km for icy material with a density of 900 kg m^{-3}, but is much further from the planet for lower density material. Porco et al. (2007) reports that densities of Saturn's inner moons are all lower than 900 kg m^{-3} suggesting that they are all currently orbiting inward of their Fluid Roche Limit. Dermott et al. (1979) have shown that a satellite evolving inward its Fluid Roche Limit may lose material from its surface and feed the surrounding system, leading to a complex interplay between the satellite and the rings, implying also modifications of its shape. If it evolves inward, at some point it should break into objects small enough for their tensile stress to counter-balance tides. This is the "Rigid Roche Limit" (RRL, see Holsapple and Michel 2006, for a detailed calculation), that is the distance at which an already-constituted satellite with tensile stress is destroyed by tides. Whereas a body with no material resistance may be formed or destroyed at the FRL, the RRL may be well inside the FRL for the same density because (i) the body may have a non-hydrostatic shape, and (ii) internal cohesion within the satellite (as measured by the friction angle for example) acts against its destruction. So an already constituted object may survive against tides, inward the region where it formed, because of internal material resistance. For example a body with the same density (~ 640 kg m^{-3}) and shape as Janus has a fluid Roche Limit at $\sim 151{,}000$ km from Saturn, but would have a rigid Roche Limit of $\sim 108{,}000$ km for an assumed friction angle of 20° (Holsapple and Michel 2006).

The take-home message is that if ring-moons are only made of water ice with the same average density as the surrounding ring particles themselves, it does not seem easy that they could form spontaneously at less than $\sim 130{,}000$ km from Saturn for non-porous ice, and $\sim 168{,}000$ km for porous ice. However, once formed they might survive against tidal disruption down to about $\sim 110{,}000$ km.

Gravitational accretion of a swarm of particles in a disk under the tidal forces of the planet ("tidally modified accretion") was theorized by Canup and Esposito (1995, 1996). They study the conditions under which colliding particles may gravitationally stick under the effect of inelastic collisions. They found two necessary conditions: the final size of the body must be less than the Hill sphere of the equivalent mass body, and the final energy must be negative in the local rotating frame (Eqs. 28 to 30 in Canup and Esposito 1995). As a result the closer the bodies to the planet, the more dissipative the collision must be in order for them to stick. For equal size bodies, if the sum of their radii is larger than about 70% of their mutual Hill radius, then capture is impossible because the two-bodies would overflow their mutual Hill sphere. As the Hill Sphere radius is proportional to the distance to the planet ($R_H = a_s(m/(3M_P))^{1/3}$) where R_H, a_s, m and M_P are the Hill radius, the distance to the planet, the body mass and the planet's mass respectively), this condition becomes more and more restrictive closer to the planet. For particles accreting in Saturn's A ring, the mass-ratio of two colliding particles must be less than 0.01 to stay gravitationally bound (Table 1 of Canup and Esposito 1995). As a result, local simulations of accretion in such a tidally modified environment show that growth is strongly limited: larger bodies can only grow by accumulating much smaller objects, while small particles are gravitationally scattered. This results in bi-modal size distribution where "moonlets" co-exist with a swarm of high-velocity and small particles. However the largest objects have about the same size as the largest initial particles in the system, since they just accumulate a layer of small particles. The resulting size of the largest objects is then similar to that of the largest objects that were put at the start of the simulation. In these conditions it seems very difficult to grow spontaneously objects as big as Pan or Daphnis from a swarm of cm or meter sized particles. This may suggest that moonlets in the A ring may be formed by some other process, or that a big initial seed (or shard) must have been present. Canup and Esposito (1995) assumed that all particles have the same density, but allowing particles to have different densities (like porous and dense ice, or silicate-rich material) may offer interesting possibilities.

For these reasons, it has been proposed (Porco et al. 2007; Charnoz et al. 2007) that the inner ring moons may have formed from material with a density higher than that of typical ring particles, such as silicate-rich material or dense ice with low porosity (originating, maybe, from the formation of the rings via tidal destruction of a satellite or a passing object; Crida et al. 2024, this collection). The favored scenario would be a shard of dense material covered with much lower-density ring particles. If a silicate chunk is embedded within the rings it will accumulate ice-rich small particles up to a limit where the ringmoon's Hill sphere is completely filled with material, beyond which material cannot be accreted anymore. As shown in (Porco et al. 2007) there exists a critical moon density, ρ_c below which accretion is not possible anymore, $\rho_c = M_p/(\eta {a_s}^3)$ where M_p is the planet's mass, a_s the satellite's semi-major axis, and $\eta \sim 0.53$ (this factor arises from the volume of the Hill sphere; Charnoz et al. 2011). As long as the ringmoon density is $> \rho_c$ accretion continues indefinitely, but it stops when the moon's mean density is equal to ρ_c. At the distance of Pan, $\rho_c \sim 450$ kg m^{-3}, very close to Pan's estimated density of about 420 kg m^{-3} (Porco et al. 2007). This is why Charnoz et al. (2007) and (Porco et al. 2007) propose that a dense shard may be embedded in all ringmoons.

In summary, the processes described above suggest that two "formation" mechanisms may still be active in Saturn's rings. (i) Any chunks of dense material (or any material denser than the Roche Critical density, ρ_c) may rapidly capture an icy envelope made of ring material, and then be viscously transported outward. This scenario may be relevant for all ring-moons that are — or once were — embedded in the A ring. (ii) The concurrent process is the spontaneous accretion of icy ring material at or near the Fluid Roche Limit of Saturn (beyond the A ring, typically) that may form new ring-moons via gravitational instability. This second process may take place in Saturn's F ring also, giving birth to a population of transient moonlets detected inside this ring (Beurle et al. 2010).

In this context, two alternative processes have been proposed to explain the ring moons' observed shape described in Sect. 4. In the first case, it is the result of the accretional process of ring material onto this initial seed (Charnoz et al. 2007; Quillen et al. 2021). In this scenario, as the body migrates outwards, material from the rings falls onto its equator and forms the equatorial ridge seen around some moons. In the second case, the shape of the moon is produced during a low-impact-angle and slow-speed collision between two similar progenitors (Leleu et al. 2018). Such a process is in agreement with the pyramidal regime where bodies of similar sizes accrete to form larger bodies (Charnoz et al. 2010; Crida and Charnoz 2012), and could not only reproduce the ravioli-like shapes but also the highly elongated shape of Prometheus.

After their formation, the larger objects would have evolved rapidly outwards due to gravitational torques exerted by the nearby ring material.[2] Embedded ring-moons, if they are large enough to open a gap in the rings (like Pan or Daphnis), will evolve according to the so-called "Type 2" migration scenario (see, e.g. Crida et al. 2010) and remain "locked" in the rings, simply following the ring's viscous evolution. This process may be relevant for moonlets embedded in the A ring, but not for isolated moonlets beyond Atlas. Simulations of the rings' viscous evolution show that beyond $\sim 120{,}000$ km the rings should spread outward (neglecting the torques exerted by moons exterior to the rings) (Salmon et al. 2010).

The smallest of the present ring moons, Daphnis and Pan, are securely imprisoned within their respective gaps, but presumably will emerge from the rings in due course as the A ring edge moves outwards across the Roche limit for icy bodies and the ring material is converted into more moonlets (Charnoz et al. 2010; Tajeddine et al. 2017a). These moons may have accreted relatively recently in the overall history of Saturn's rings. If the ring moons do not open a full gap, then their dynamical evolution may be much more complex, and maybe even stochastic as in the case of "propellers" (see e.g. Tiscareno et al. 2008; Crida et al. 2010; Spahn et al. 2018)).

Once outside the rings, the radial evolution of the ring moons will accelerate due to what is now a one-sided torque from the rings. Since this torque scales as the square of the satellite's mass, in a manner analogous to tidal torques between planets and satellites, larger moons will drift outwards more rapidly than their smaller cousins, leading inevitably to collisions and subsequent mergers into a smaller number of progressively larger objects. This process naturally leads to a sorting of the resulting objects, with younger, smaller moons close to the rings and older, larger ones further away, similar to what is observed at Saturn. Indeed simulations of ring viscous evolution evolution coupled to moonlets formation at the Fluid Roche Limit show that masses and locations of moons inward of Mimas are reproduced, at least qualitatively (Charnoz et al. 2010). The end product is a few larger satellites such as Janus and Epimetheus.

The time for a moon to move from the edge of the A ring to the distance of the F ring depends on its mass, but is typically a few 10 s to ~ 100 Myr.

Although the outer edge of the A ring is currently defined by the 7:6 MMR with Janus (Spitale and Porco 2009; El Moutamid et al. 2016; Nicholson et al. 2018), this presumably was not always the case and will not be in the distant future, when Janus will have moved further away from the rings and this particular resonance will no longer coincide with the FRL for ice.

Tajeddine et al. (2017a) have estimated the tidal and ring torques on the current satellites and their resulting radial drift rates $\dot{a}$, as summarized in Table 3. The ring torque on each satellite was obtained by summing over the contributions from all of the particular satellite's Lindblad resonances that fall within the main rings, combined with the ring surface mass

[2]Note that for all moons interior to Mimas, the ring torques significantly exceed the torque due to tides raised in Saturn, at least for $Q_{\mathrm{Saturn}} > 10^3$; see Table 3 below.

Table 3 Radial evolution timescales for Saturn's inner moons due to saturnian tides[1] and ring torques. (Adapted from Tajeddine et al. (2017a))

Satellite	Δa (km)	$\dot{a}_{\rm Saturn}$ (km/Myr)	$\dot{a}_{\rm rings}$ (km/Myr)	$\tau_{\rm orb}$ (yr)	Notes
Pan	−3190	0.036	90.1	3.5×10^7	2
Atlas	900	0.047	233	3.9×10^6	3
Prometheus	2610	1.06	274	9.5×10^6	4
Pandora	4940	0.83	44.2	1.10×10^8	4
Janus	14,680	8.0	39.5	3.1×10^8	5
Epimetheus	14,680	2.2	10.8	1.13×10^9	5
coorbital res.	−32	—	−4.4	7.3×10^6	6
Mimas	48,770	46.4	4.7	9.5×10^8	7

[1]Assuming that $k_2/Q_{\rm Saturn} = 1.59 \times 10^{-4}$, or $Q_{\rm Saturn} = 1600$. (Lainey et al. 2012)

[2]Neglecting shepherding torques from the Encke Gap edges

[3]Neglecting retarding effects of Prometheus 54:53 resonance

[4]Neglecting Prometheus/Pandora/F ring interactions and transient Epimetheus resonances

[5]Neglecting capture into coorbital resonance

[6]From Lissauer et al. (1985), updated to current masses and ring surface density

[7]Neglecting retarding effects of Tethys 4:2 resonance

densities determined by Tiscareno et al. (2007). They found orbital evolution timescales $\tau_{\rm orb} = \Delta a/\dot{a}$, where Δa is the present distance of the moon from the edge of the A ring, that range from 4×10^6 yr for Atlas to 3×10^8 yr for Janus and ~ 1 Gyr for Epimetheus and Mimas. The shorter time scales for the inner satellites, despite their smaller masses, reflect the fact that Δa is much smaller, as well as the larger number of first-order resonances that fall within the A and B rings.

However, Tajeddine et al. (2017a) note that these evolutionary time scales are uncertain because (i) the drift rates would have been faster when the satellites were closer to the A ring, while (ii) the satellites would have been smaller in the past if the present bodies are the result of mergers, leading to slower drift rates. These calculations update earlier estimates by Borderies et al. (1984), who assumed higher ring surface mass densities and (probably) satellite masses corresponding to solid ice, leading generally to faster evolution rates.

Goldreich and Tremaine (1982) first recognized that the orbital expansion of the small-moons outward and interacting with the A ring is much shorter than the age of the Solar System (from 20 Myrs to 200 Myrs). They proposed that some resonance with other Saturn satellites may stabilize their expansion (without identifying them) or that gap-opening within the A ring may have lowered the ring-torque they experience, and thus increase the timescale of their orbital expansion.

The orbital history of the ring moons is indeed complicated by the likelihood of capture into, and possibly escape from, mean motion resonances between various pairs of satellites during their radial migration (Poulet and Sicardy 2001; Cooper et al. 2015).

Numerical integration of the current orbits of the inner satellites by Cooper et al. (2015) have identified several examples. In addition to the 54:53 MMR between Atlas and Prometheus, mentioned in Sect. 2 above, these authors found transient 17:15 MMRs between Prometheus and Epimetheus and 21:19 MMRs between Pandora and Epimetheus, both of which are dependent on the latter's coorbital phase. Earlier numerical integrations also showed temporary resonance trapping and orbital 'chaos' in the ring-moon system (Poulet and Sicardy 2001). The existence of such resonances, even temporary ones, means that the timescales in Table 3 are likely to be lower limits, inasmuch as the resonance will effectively act to transfer a portion of the ring torque on the small, inner moon to the more massive, outer moon, in such a way as to maintain a constant ratio of orbital periods.

An interesting special case is the coorbital resonance between Janus and Epimetheus, the largest ring moons interior to Mimas. Lissauer et al. (1985) studied the effect of ring torques on this resonance, finding that it is rapidly tightening over time (*i.e.,* the current 48 km separation between the two satellites' orbits is decreasing). Updating their numerical values, we find that the age of this resonance is only ~ 7 Myr, and that it will be converted into a Trojan-style configuration in another 9 Myr.

Also, Lissauer et al. (1984) noted that, for the current orbit of Janus, its strong 2:1 density wave falls only a few hundred km outwards of the unique 'flat spot' in the inner B ring, at which location Janus would probably have been trapped in the 2:1 MMR with Enceladus. They suggested that a disruptive collision some 10^8 or so years ago may have freed Janus from this resonance.

Although none of the existing ring moons is in resonance with any of the mid-sized satellites at the present time, Borderies et al. (1984) pointed out that Pandora is only $\sim$ 180 km interior to the 3:2 MMR with Mimas, and that it might be expected to be captured into this resonance in the fairly near future. Such a capture would almost 'freeze' the radial evolution of Pandora, as the resonance will act to transfer the ring torque to the more massive satellite. But Tajeddine et al. (2017a) find that the outward drift rates of Pandora and the Mimas 3:2 resonance are almost equal, raising doubts as to whether or not the capture will ever occur. Mimas, in turn, is in a stable 4:2 MMR with Tethys, which is likely to slow down its orbital expansion relative to that given in Table 3. For the above reasons, the times given here should be taken as indicative of the current orbital expansion rates, but not as realistic estimates of the ages of individual moons.

In conclusion, there are many still open questions concerning the origin and evolution of the small moons of Saturn. Do they form a long-lived or a transient population? When did they form? Are they as old as the main ring system or much younger? See also Crida et al. (2024, this collection) for more discussion.

6 Composition and Regolith Physical Properties

As of writing, the investigation of the physical and compositional properties of the regolith covering the surface of the ring moons was made possible through the extensive observation campaigns carried out by remote sensing instruments onboard the Cassini mission, namely the Imaging Science Subsystem (ISS; Porco et al. 2004), the Visible and Infrared Mapping Spectrometer (VIMS; Brown et al. 2004), and the Ultraviolet Imaging Spectrograph (UVIS; Esposito et al. 2004). These experiments characterized the spectrophotometric properties of the ring moons over a broad wavelength interval, from the far-UV to near-IR (~ 0.1 - 5.1 μm).

In this section we provide a summary of the main results from remote sensing observations of the ring moons, reporting on 1) their photometric properties at visible wavelengths (Sect. 6.1), 2) UV-VIS-IR spectroscopic investigations (Sect. 6.2), and 3) results of spectral modeling efforts (Sect. 6.3). The ring moon data are discussed in the broader context of the spectrophotometric properties of Saturn's mid-sized icy moons and main rings.

6.1 Visual Albedos and Photometric Properties

Thus far, there have not been many detailed photometric studies of Saturn's smaller moons simply because the available data are rather limited. For example, Cassini was unable to observe most of these moons at phase angles below 1°, so there is little information about the magnitude and extent of their opposition surges (Thomas et al. 2018b). Thus the only small moons with reasonably secure geometric albedos are Janus and Epimetheus, which are large enough to be observed with Earth-based telescopes such as HST. Interestingly, Verbiscer et al. (2007) found that these objects have geometric albedos of 0.71 ± 0.02 and 0.73 ± 0.03, respectively, significantly lower than those of the mid-sized satellites Mimas, Enceladus, Tethys, Dione and Rhea. They attributed this to the fact that the coorbital moons are in a region that probably has a much lower flux of the E-ring particles that are thought to coat the surfaces of the mid-sized satellites.

More recently, Thomas et al. (2018b) estimated whole-disk albedos for Saturn's satellites, including the small moons, at an effective phase angle (α) of 10° using extrapolations of the available Cassini photometric data (see Table 2). These numbers are somewhat lower than the geometric albedos (which are evaluated at 0° phase by definition), but generally follow similar trends with orbital location, confirming the importance of E-ring flux to the moons' brightnesses. Generally, among the ring moons, the albedos are highest for Prometheus and Pandora, perhaps due to a coating of icy material from the nearby F ring, somewhat lower for the inner ring moons and the coorbitals, and lowest for the ring-embedded moons, Aegaeon, Methone and Pallene. These three objects are darker than expected, suggesting that other processes might be influencing the smallest moons' surface brightness.

One limitation with the albedo estimates for the small moons derived by Thomas et al. (2018b) is that they have rather large error bars. These uncertainties are not primarily due to errors in the measured brightness of the moons, but instead arise because these moons are far from spherical and so the apparent brightness of the moon depends not only on its phase angle but also on the orientation of the moon relative to the observer and the Sun, which is not accounted for in standard albedo calculations.

To address these issues and provide tighter constraints on the moons' light-scattering properties, Hedman et al. (2020) used a technique that accounts for the varying lighting geometry across each moon's surface. Specifically, they estimated a phase-dependent ratio of the moon's observed brightness at a wavelength of 650 nm to its predicted brightness, assuming that its shape is a perfect ellipsoid and that its surface follows a perfect Lambertian or Minnaert scattering law. This ratio is referred to as the "brightness coefficient", and below we provide a brief description of how it is derived. The observed brightness can expressed in terms of the effective area of the target A_{eff},[3] given by

$$A_{eff} = \sum_x \sum_y I/F_{x,y} \Omega_{pixel} D^2 \tag{2}$$

where $I/F_{x,y}$ is the measured radiance factor of the pixel x, y in the target image, Ω_{pixel} is the solid angle subdued by the pixel, and D is the distance from the target. Assuming that the surface reflectance R follows a Minnaert scattering law, this is given by $R = B(\cos i)^k(\cos e)^{1-k}$ (notice that $k = 1$ reduces to the Lambertian case), where B is

[3] This is the equivalent area of a target with I/F = 1 required to provide the observed brightness

the "brightness coefficient" (a phase-depentent factor, constant for each image). Thus, the predicted brightness can be expressed in terms of the predicted effective area given by

$$A_{pred} = \int_{++} B(\cos i)^k (\cos e)^{1-k} (\cos e) r_{eff}^2 \cos\lambda d\lambda d\phi, \tag{3}$$

where the integral is performed over the portion of the target both illuminated and observed $(++)$, λ and ϕ are latitude and longitude on the object, and r_{eff} effective radius. For an ellipsoidal object with dimensions a, b, and c, the effective radius is given by

$$r_{eff}^2 = \left[b^2c^2\cos^2\lambda\cos^2\phi + a^2c^2\cos^2\lambda\sin^2\phi + a^2b^2\sin^2\phi\right]^{1/2} \tag{4}$$

Then, B can be straightforwardly derived by equating A_{eff} and A_{pred}, providing:

$$B = \frac{A_{eff}}{\int_{++}(\cos i)^k(\cos e)^{1-k}(\cos e) r_{eff}^2 \cos\lambda d\lambda d\phi}, \tag{5}$$

where A_{eff} is obtained from observations and the denominator can be computed numerically. As anticipated above, the"brightness coefficient" is a function of the phase angle (through A_{eff}), and as such it does not equal the moon's classical geometric or Bond albedo. Nonetheless, trends in the brightness coefficients across Saturn's larger moons match the trends in their geometric albedos, making it a useful parameter for comparing how efficiently the surfaces of different moons scatter light. Furthermore, the scatter in this quantity among images taken at similar phase angles is much smaller than that of the classical geometric albedo.

Figure 3 shows the brightness coefficients at a phase angle of 30° for all of Saturn's moons interior to Titan (see also Table 2 for small moons' values). The larger moons (Mimas, Enceladus, Tethys, Dione, Rhea) show the trend of brightening with increasing E-ring flux previously observed by Verbiscer et al. (2007) and Pang et al. (1984). Furthermore, Janus, Epimetheus, Mimas, Tethys, Dione and Rhea all show longitudinal brightness variations consistent with the expected asymmetries in the E ring flux. Note that the E-ring extends across the orbits of all these moons, primarily because the E-ring particles have a wide range of eccentricities while most have semi-major axes not far from that of Enceladus' orbit. This means that most of the E-ring particles encountering Janus, Epimetheus and Mimas are near their orbital pericenters, so they are moving faster than the moons and primarily strike their trailing sides. By contrast, the E-ring particles striking Tethys, Dione and Rhea are primarily near their orbital apocenters and so are moving slower than the moons, leading to a particle flux that is greater on the moon's leading sides. It turns out that Janus, Epimetheus and Mimas all have brighter trailing sides, while Tethys, Dione and Rhea are all brighter on their leading sides, consistent with the basic idea that impacting E ring particles brighten the surfaces of Saturn's moons.

However, the smaller moons show several interesting deviations from the trends observed in the larger moons, indicating that other processes are also influencing the surfaces of these moons. First of all, the smallest moons Aegaeon, Methone and Pallene are all much darker than one would predict based on the observed trend from Janus/Epimetheus to Mimas and Enceladus. All three of these moons reside within the inner flank of the E ring, which clearly influences the brightness of the larger moons. Currently, the most likely explanation for the low surface brightness of these tiny moons is that they are all exposed to much larger fluxes of high-energy charged particles, which are observed to be concentrated in regions between the rings and the orbits of the larger moons (Hedman et al. 2020).

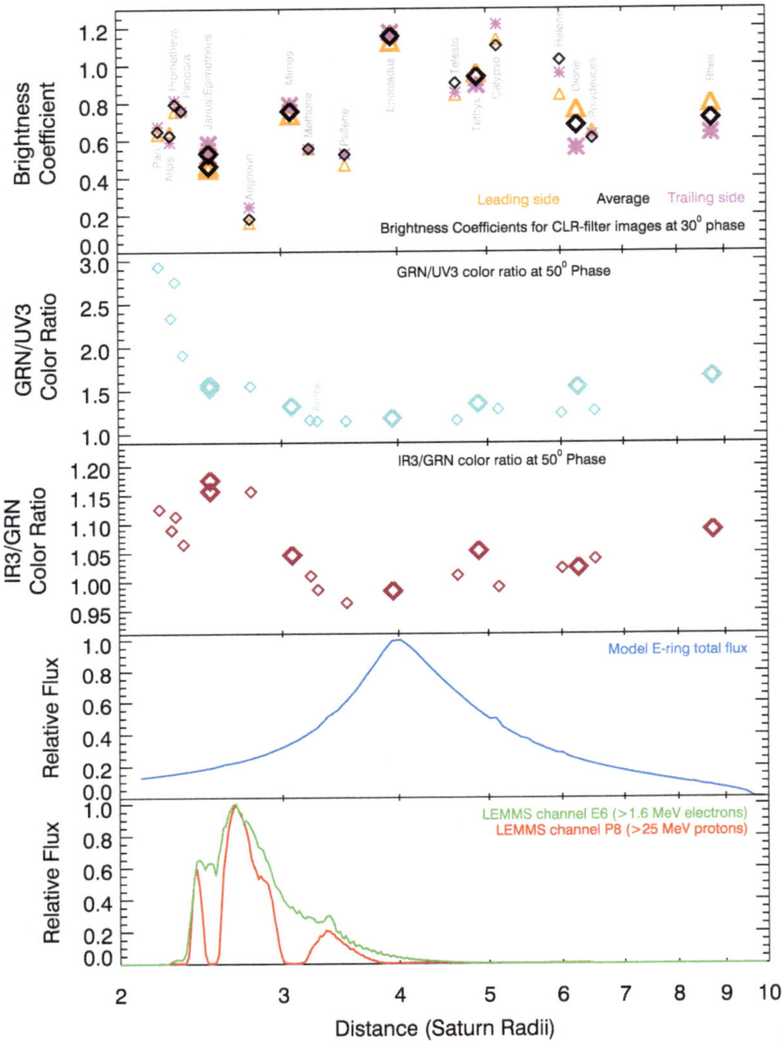

Fig. 3 Radial profiles of the brightness coefficients and GRN/UV3 and IR3/GRN color ratios of Saturn's icy moons, compared to the modeled flux of E-ring particles into the moons and the measured relative flux of high energy protons and electrons from the Cassini's Low-Energy Magnetospheric Measurement System (LEMMS) instrument (Krimigis et al. 2004). Diamonds, triangles, and asteriks refer to average, leading side, and trailing side values, respectively. The largest moons (Epimetheus, Janus, Mimas, Enceladus, Tethys, Dione, and Rhea) are indicated with larger symbols. Daphnis is not included in this plot because its signal was difficult to isolate from nearby ring material in the available low-phase images. Anthe is not included in the top panel because there were no suitable images of this moon at phase angles around 30°. Figure adapted from Hedman et al. (2020)

By contrast, several other small moons appear to be brighter than one would expect. Most dramatically, Calypso is much brighter than either Tethys or Telesto, while Helene is brighter than Dione or Polydeuces. Again, this suggests that something besides the local flux of E-ring particles is influencing the brightness of these particular bodies, although we note that Thomas et al. (2018b) report slight different albedo trend for these last objects. It is also worth noting that Prometheus and Pandora are brighter than any of the other inner moons (Pan, Atlas, Janus and Epimetheus). Since Prometheus and Pandora both orbit close to Saturn's F ring, this suggests that material from the water-ice rich F ring may be able to brighten both of these moons (Hedman et al. 2020).

Although the discussion above is based on the investigation of the brightness coefficient of the ring moons computed at a phase angle of 30°, future studies might benefit from replicating the exercise of Hedman et al. (2020) at a phase angle of $\sim 55°$. The full-disk brightness from a spherical object can be expressed as

$$I = J \frac{A_s P(\alpha)}{4}, \tag{6}$$

where is the solar irradiance at the target heliocentric distance, α is the phase angle and $P(\alpha)$ is full-disk phase function. It can be easily shown that, for a spherical object

$$\frac{A_s P(\alpha)}{4} = \frac{B(\alpha) P_L(\alpha)}{4} \tag{7}$$

where $P_L(\alpha)$ is the full-disk phase function of a sphere whose surface follows a lambertian scattering law. Also, it is found empirically that typical full-disk phase functions of spherical objects intersect at $\sim 55°$ (Cuzzi et al. 2017), thus $P(55°) \sim P_L(55°)$. This implies, from Eq. (7), that the brightness coefficient at 55° provides a good estimate of the target Bond albedo $B(55°) \sim A_s$, allowing us to compare systematically the intrinsic albedo properties of the different targets, independently from their particular scattering behavior.[4] We note that the above discussion is strictly valid for spherical objects, yet the described approach appears desirable also for the more or less irregular ring moons.

6.2 Colors and UV-VIS-IR Spectral Properties

Given the small size of the ring moons, and the resulting low signal levels, the VIMS experiment was only able to collect VIS-IR spectroscopic data for the brighter ones, namely Pan, Daphnis, Atlas, Prometheus, Pandora, Janus, Epimetheus, and Methone (Buratti et al. 2010; Filacchione et al. 2010, 2012, 2013; Buratti et al. 2019). The average spectra of these moons are shown in Fig. 4 and show definite similarities with the spectra of Saturn's main rings and the inner mid-sized moons. They are characterized by prominent water ice absorption features at 1.5, 2.0, and ~ 3 μm and a significant reflectance downturn shortward of ~ 0.55 μm.[5] The latter has been interpreted as resulting from contamination by UV-absorbing chromophores generally thought to be organic in nature (Ciarniello et al. 2011; Filacchione et al. 2013; Cuzzi et al. 2018b; Ciarniello et al. 2019, and Sect. 6.3), but for the C ring and several outer satellites an iron-rich composition has been suggested instead (Clark et al. 2012).

[4]There is some precedent for this approach. Normalized phase functions of plausible shapes tend to pass through a node in the vicinity of 50° phase angle (the so-called 'Russell's Law' of vintage planetary science lore). Thus the observed *brightness* of the object at this phase angle, which is the product of the (constant) spherical albedo and the normalized phase function, is nearly phase-function-independent. In this sense the relative values of the brightness coefficients are good proxies for the relative albedos of the objects. For more information see Veverka (1971). The shape of the phase function itself probably contains information as to the effects of shadowing due to surface roughness (Hapke 2012; Cuzzi et al. 2017).

[5]For Daphnis (where the VIS part of the spectrum is missing) and Methone (spectrum very noisy at visible wavelengths), the UV downturn cannot be characterized with VIMS data. Nonetheless, ISS multi-spectral observations of Methone indicate significant UV absorption for this target (Hedman et al. 2020).

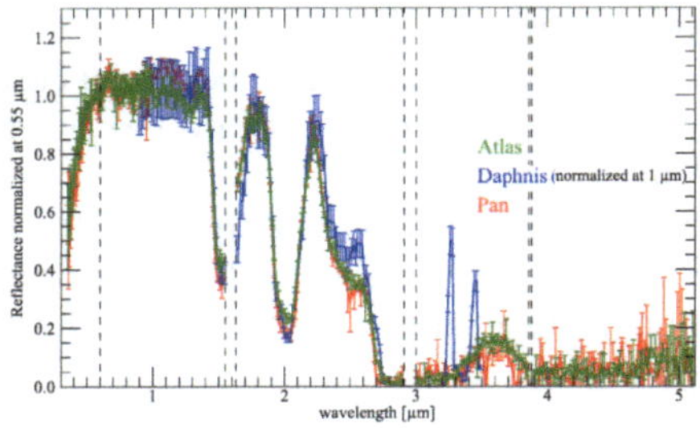

(a) Inner ring moons within/in proximity of the A ring: Pan, Daphins, and Atlas

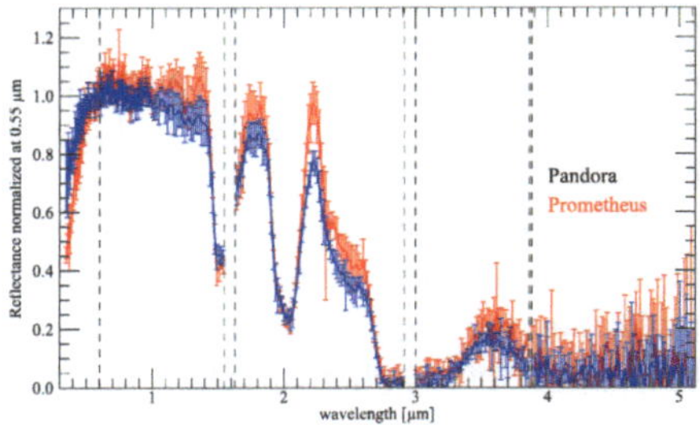

(b) Inner ring moons in proximity of the F ring: Prometheus and Pandora

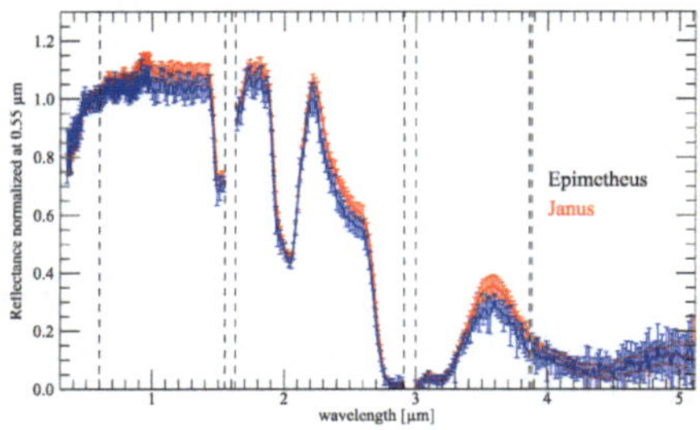

(c) Coorbital satellites: Janus and Epimetheus

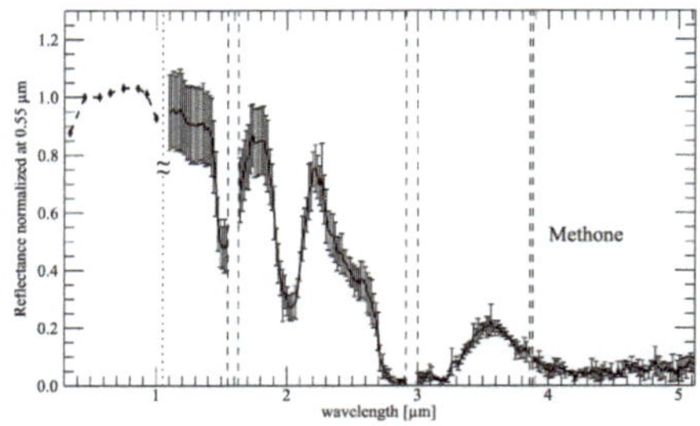

(d) Methone

Fig. 4 Average full-disk reflectance spectra of the ring moons, as computed from available VIMS data (dashed lines indicate instrument order sorting filters junctions). See Filacchione et al. (2010, 2012) for data processing details. Before averaging, each full-disk spectrum from a given observation of the selected target has been corrected to a reference phase angle of 50° (see Appendix A), in order to minimize spectral variability induced by varying observation geometry (Ciarniello et al. 2011; Filacchione et al. 2022). Methone reflectance at visible wavelengths is from ISS observations in different filters interpolated at 57.7° phase angle, and further corrected to 50° phase angle. The corresponding IR part of the spectrum, from VIMS observations, has been arbitrarily scaled around a putative water ice absorption at 1.05 µm (dotted line), so that the reflectance at 1.098 µm (right shoulder of the band) matches the interpolated value at 0.981 µm (left shoulder of the band)

Additional features, detectable for selected moons, are the 3.1 µm Fresnel peak of water ice seen on Janus and possibly Methone, and the faint 1.25 µm water ice absorption seen on Atlas, Pan, Prometheus and possibly also Pandora and Methone. Apart from water ice, no evidence of other volatiles is seen, including CO_2 with its diagnostic feature at 4.26 µm (e.g., Chaban et al. 2007), although this falls in a rather noisy spectral region (Buratti et al. 2019).

Interestingly, Pan, Daphnis, and Atlas (orbiting within and in close proximity to the A ring, respectively), appear to group separately in terms of spectral shape from the coorbital Janus and Epimetheus (orbiting at larger heliocentric distances between the F ring and Mimas), with the former being characterized by larger UV absorption and water ice band depths, and a smaller NIR spectral slope, suggesting a link with radial distance. Given this, one might expect that Prometheus and Pandora, orbiting just inside and outside the F ring, might share similar spectral shapes. However this is not the case, with Prometheus' spectrum being relatively similar to Atlas' and Pan's, while Pandora displays a distinctively smaller UV absorption.

This is better appreciated by comparing spectral indicators across the different ring moons, which we put here in a broader context with Saturn's main rings and mid-sized inner moons (plus their Lagrangian satellites). The suite of VIMS spectral indicators we consider are (see Filacchione et al. 2010, 2012, 2013, for further details):

- the spectral slope of the target's full-disk I/F spectrum normalized at 0.55 μm, from a linear best-fit in the 0.35-0.55 μm range: referred to as the UV-VIS slope;
- the spectral slope of the target spectrum normalized at 0.55 μm, from a linear best-fit in the 0.55-0.95 μm range: referred to as the VIS-NIR slope;
- the band depth of the 1.5- μm water ice absorption: $BD_{1.5\,\mu m}$;
- the band depth of the 2- μm water ice absorption: $BD_{2\,\mu m}$.

The UV-VIS and VIS-NIR slopes are proxies for the level of contaminants and chromophores mixed with the water ice (Clark et al. 2012; Ciarniello et al. 2021), while the $BD_{1.5\,\mu m}$ and $BD_{2\,\mu m}$ are sensitive to both the water ice grain size (Clark and Lucey 1984; Stephan et al. 2021) and the level of contaminants (Ciarniello et al. 2021). Furthermore, we integrate this information with indicators derived from ISS observations of Saturn's icy moons, namely the brightness coefficient at 650 nm (Hedman et al. 2020, and Sect. 6.1), a proxy for the amount of dark contaminants mixed with water ice, and the GRN/UV3 and IR3/GRN color filter ratios, analogues of the UV-VIS and VIS-NIR slopes from VIMS. These brightness ratios are derived from sets of calibrated images obtained through the relevant filters at nearly the same time and viewing geometry. We simply compute the integrated brightness of the moon (I/F) for each image in the set, and plot the ratios of the appropriate integrated brightness estimates. Scatter plots from different combinations of VIMS and ISS spectral indicators are shown in Figs. 5 and 6.

The $BD_{1.5\,\mu m}$ vs. $BD_{2\,\mu m}$ scatter plot (Fig. 5) shows that the water ice band depths of both main rings and inner moons are strongly correlated, with the corresponding values aligning along two partially overlapping diagonal branches, extending from small to high values of the two indicators. This is consistent with differences in water ice grain size and/or amount of contaminants (darkening materials) among the different targets, as these properties affect the $BD_{1.5\,\mu m}$ and $BD_{2\,\mu m}$ in a similar way. Along the icy moon branch, the ring moons (except Methone) populate the extremal positions, with Pan and Prometheus displaying the deepest water ice absorptions, and Janus and Epimetheus the lowest. The mid-sized icy moons and the lagrangian satellites populate the part of the branch at intermediate values. Along the main ring branch, the lowest values are registered for the C ring, falling below Janus and Epimetheus, while the largest band depths occur in the A and B rings, similar to Pan and Prometheus. In order to discern whether the $BD_{1.5\,\mu m}$ - $BD_{2\,\mu m}$ correlation is primarily driven by a variation of 1) the regolith grain size or of 2) the amount of darkening contaminants among the different targets, we can take into account also their albedo.

This is addressed in the brightness coefficient (B) vs. $BD_{2\,\mu m}$ scatter plot (Fig. 6). Here, the mid-sized icy moons with the Lagrangian satellites and the ring moons tend to distribute along two separate branches, where B and $BD_{2\,\mu m}$ are positively correlated. This would be consistent with variations primarily due to the amount of contaminants, whose increasing abundance simultaneously reduces the 2- μm band depth and the brightness coefficient along the two branches.[6] In this respect, the ring moons, being on average darker than the mid-sized icy moons and Lagrangian satellites, are likely to host larger amounts of contaminants

[6]Notice that a simultaneous reduction of the 2- μm band depth and visible reflectance can be also consistent with the effect of increasing grain size, in the case of coarse water ice regolith. This would require grains larger than a few hundreds of microns (Stephan et al. 2021), for which the 2- μm band starts saturating. However, the typical estimated water ice grain sizes of Saturn's icy moons and main rings are generally (much) smaller (Cuzzi et al. 2009; Bradley et al. 2010; Ciarniello et al. 2011; Filacchione et al. 2012; Clark et al. 2012; Bradley et al. 2013; Cuzzi et al. 2018b).

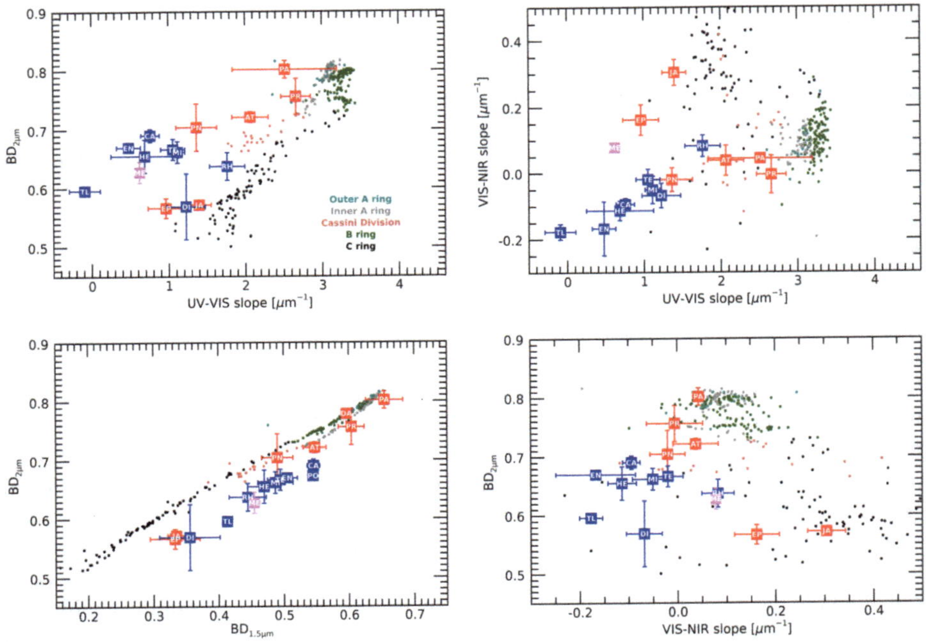

Fig. 5 Scatter plots of available VIMS spectral indicators for Saturn's ring moons (red symbols and pink for Methone), mid-sized icy moons with Lagrangian satellites (blue), and main rings (dots). Main ring spectral indicators are from a ring radial spectrogram[7] with 200-km resolution, built from the VIMS' radial mosaic observation sequence S36-SUBML001 acquired on 19 December 2007 (Filacchione et al. 2014). Spectral indicators of the moons are computed from VIMS full-disk reflectance data after photometric correction to 50° phase angle; a similar photometric correction has been applied also to main ring data (See Appendix A). UV-VIS and VIS-NIR slopes for Methone are from ISS data (see Fig. 4). Abbreviations: Pan (PA), Daphnis (DA), Atlas (AT), Prometheus (PR), Pandora (PN), Epimetheus (EP), Janus (JA), Mimas (MI), Methone (ME), Enceladus (EN), Tethys (TE), Telesto (TL), Calypso (CA), Dione (DI), Helene (HE), Polydeuces (PO)[8], and Rhea (RH)

on their surfaces. The existence of two branches, with the E-ring moons having a smaller band depth than ring moons with similar brightness coefficient, might also indicate that the former have a smaller characteristic regolith grain size, consistent with accumulation on their surfaces of fine water ice particles from the E ring. A similar argument, based on differences in the abundance of darkening contaminant(s), can be applied also for the main rings, where the A and B rings, intrinsically brighter than the C ring (Cuzzi et al. 2018b; Ciarniello et al. 2019), display larger water ice band depths.

In the UV-VIS slope vs. $BD_{2\,\mu m}$ scatter plot (Fig. 5), the different targets tend to globally align along a relatively broad diagonal distribution, with larger band depths corresponding to larger slopes. Nonetheless, different classes of object tends to cluster separately. The main rings populate a diagonal branch shifted to higher spectral slopes with respect to the inner

[7]Radial spectrograms are bi-dimensional arrays reporting the spectrum of the rings at different radial distances, sampled along a regular grid (Filacchione et al. 2012). The spectrum at a given position of the grid is obtained by averaging spectra of pixels falling in a given radial distance bin, acquired during a given VIMS radial mosaic observation.

[8]For Polydeuces only one VIMS IR sub-pixel observation was available at a relatively large phase angle of 120°, for which no photometric corrections has been applied.

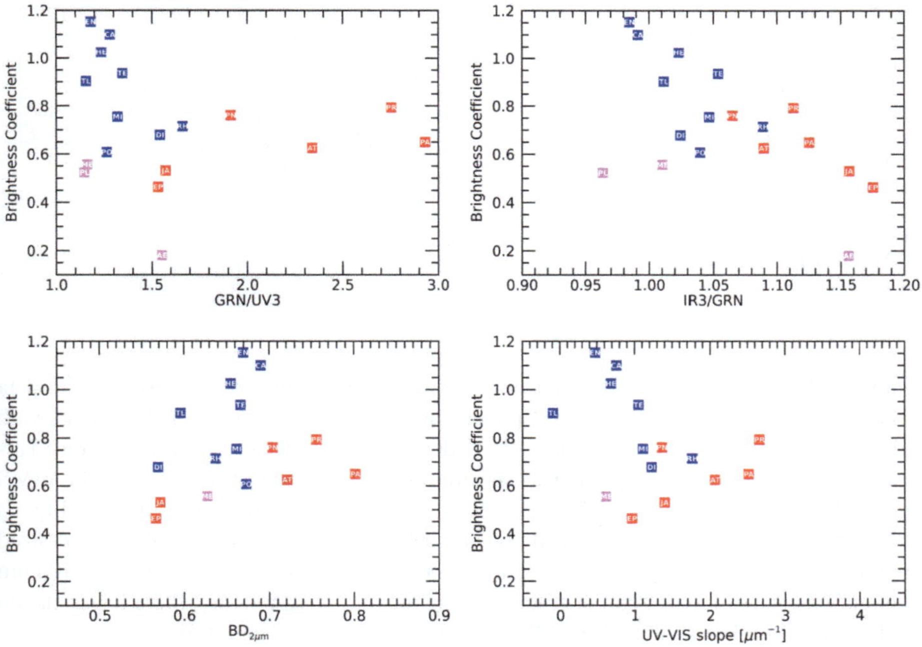

Fig. 6 Scatter plots of available VIMS and ISS spectral indicators for Saturn's ring moons (red symbols and pink for the ring-embedded moons Aegaeon, Methone, and Pallene) and mid-sized icy moons with Lagrangian satellites (blue) . The brightness coefficient is derived from ISS clear filter observations (650 nm) reported at 30° phase angle, while color ratios are computed at 50°. The UV-VIS slope and $BD_{2\ \mu m}$ are from VIMS full-disk reflectance observations after photometric correction to 50° phase angle (See Appendix A). The UV-VIS slope for Methone is from ISS data (see Fig. 4). Abbreviations: Pan (PA), Daphnis (DA), Atlas (AT), Prometheus (PR), Pandora (PN), Epimetheus (EP), Janus (JA), Mimas (MI), Aegaeon (AE), Methone (ME), Pallene (PA), Enceladus (EN), Tethys (TE), Telesto (TL), Calypso (CA), Dione (DI), Helene (HE), Polydeuces (PO), and Rhea (RH)

moons, with the A and B rings having the largest slopes and band depths. The ring moons (except Methone) span a similar region of the scatter-plot, but shifted at slightly smaller UV-VIS slope values for similar $BD_{2\ \mu m}$. As such, they appear as a transition population between the main rings and the inner mid-sized moons plus their Lagrangian satellites, which, with the exception of Rhea and Dione, and including Methone, are located in a portion of the scatter plot at even smaller spectral slope. The overall positive correlation of the UV-VIS slope and $BD_{2\ \mu m}$ for the main rings and ring moons would support the idea that the contaminant(s) providing the UV absorption for these targets is likely embedded in the water ice grains (Nicholson et al. 2008; Ciarniello et al. 2011; Hedman et al. 2013; Cuzzi et al. 2018b; Ciarniello et al. 2019). The UV-absorber would be different from the broad-band absorber which reduces the overall albedo and spectral contrast (see above).

This interpretation would be consistent with the distributions of the ring moons (excluding the ring-embedded moons) in the *B* vs GRN/UV3 scatter plot (Fig. 6), aligning along a branch over which the two indicators are positively correlated. This simple paradigm, however, might not apply to the E-ring moons, which do not display a clear correlation between the UV-VIS slope and $BD_{2\ \mu m}$, and are aligned along a different branch in the *B* vs GRN/UV3 plot, where these quantities are roughly anti-correlated. Notice that similar considerations are also supported from the *B* vs UV-VIS slope scatter plot (Fig. 6), consistent

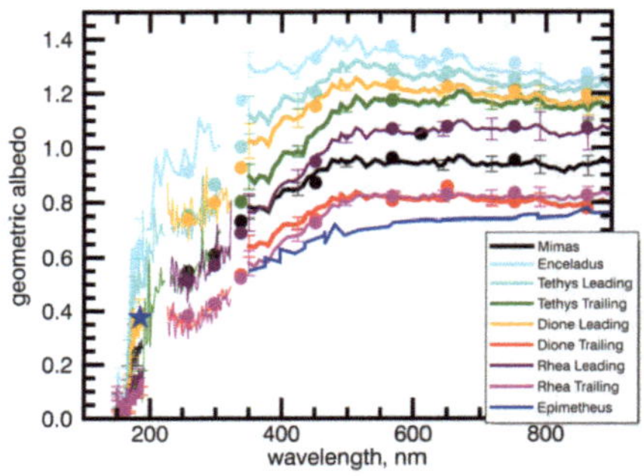

Fig. 7 UV-visible spectrum of Epimetheus compared with the inner icy moons of Saturn (Mimas, Enceladus, Tethys, Dione, Rhea). Epimetheus composite spectrum uses data from Cassini VIMS and UVIS (single point at 185 nm). Inner icy moons composite spectra use data from Cassini UVIS, ISS, VIMS and HST (after Hendrix et al. 2018)

with the *B* vs GRN/UV3 plot, as expected. This also appears in agreement with UVIS data (Buratti et al. 2019) of Epimetheus; Epimetheus was the only of the small satellites with adequate-enough signal-to-noise for UVIS to determine a reflectance value at 185 nm. This value is compared with the Epimetheus spectrum from VIMS and similar UVIS-ISS-VIMS and Hubble Space Telescope (HST) composite spectra for the mid-sized moons (Fig. 7) to demonstrate that Epimetheus, while having a lower albedo than most of the mid-sized moons at visible wavelengths, also displays a shallower UV absorption in the 300-400 nm region. This is a different trend than seen among the mid-sized moons, where lower albedo surfaces are correlated with a deeper UV absorption (Hendrix et al. 2018).

Ring moons (except Methone) separate the E-ring moons and main rings also in the UV-VIS slope vs. VIS-NIR slope scatter plot. The main rings distribute roughly along a diagonal branch where the VIS-NIR slope and UV-VIS slope are anticorrelated, and which is broadly aligned with a similar ring moon branch, although shifted at both smaller VIS-NIR slope and UV-VIS slope values. Instead, the E-ring moons, align along a diagonal branch where the VIS-NIR slope and UV-VIS slope are positively correlated, orthogonal to the ring moon and main ring distributions. This would be consistent with a single contaminant providing both the UV absorption and VIS-NIR coloring on the moons populating the E ring, while, as suggested above, for the ring moons and main rings two separate contaminants, a UV and a broad band absorber, can be envisaged (Cuzzi and Estrada 1998; Hedman et al. 2013; Cuzzi et al. 2018b; Ciarniello et al. 2019).

A similar grouping of the different targets, as in the VIS-NIR slope vs. UV-VIS slope and UV-VIS slope vs. $BD_{2\,\mu m}$ cases, can be observed in the VIS-NIR slope vs. $BD_{2\,\mu m}$ scatter plot (Fig. 5). The main rings and ring moons (except Methone) populate roughly parallel branches where VIS-NIR slope vs. $BD_{2\,\mu m}$ are anticorrelated, with the ring moons shifted at slightly smaller values of VIS-NIR slope. This population separates the main rings from the E-ring moons, which tend to cluster at smaller VIS-NIR slope and $BD_{2\,\mu m}$ values. For the main rings and ring moons (except Methone), this would be consistent with the effect of a broad-band absorber mixed with water ice, simultaneously reducing the spectral contrast of the water ice aborption features and reddening the otherwise neutral-to-blue VIS-NIR spectrum. This is also consistent with Fig. 6, where for the inner ring moons *B* increases at decreasing IR3/GRN. A similar trend is also observed for the largest E-ring moons, and roughly for the ring-embedded moons.

6.3 Clues from Spectral Modeling of the Rings and Mid-Sized Icy Moons

Extensive spectral modeling efforts of the UV-VIS-IR observations of the main rings (Cuzzi and Estrada 1998; Poulet et al. 2003; Bradley et al. 2010; Hedman et al. 2013; Cuzzi et al.

2018b; Ciarniello et al. 2019) and the mid-size icy moons (Ciarniello et al. 2011; Clark et al. 2012; Filacchione et al. 2012) have been made using regolith radiative transfer theories developed by Shkuratov et al. (1999, 2005, 2012) and Hapke (2012). The goal has been to test compositional paradigms for the surfaces of these targets, primarily to characterize the nature and amount of contaminants mixed with the water ice. For the ring moons, a similar approach has been attempted so far only for VIMS data on Epimetheus by Buratti et al. (2010), which suggested that its VIS-IR spectrum can be modeled with an intimate mixture of water ice and $\sim 5\%$ of Triton tholin,[9] the latter being responsible for the UV absorption and the reddish VIS-NIR spectral slope (Khare et al. 1994). Small amounts ($<1\%$) of Titan tholins (Imanaka et al. 2004, 2012; West et al. 2014) have been invoked in spectral modeling of the main rings as a UV absorber embedded in water ice grains (known as an intramixture), along with an additional contaminant to provide broad-band absorption, either in pure form (e.g. carbon, silicates), or embedded in water ice (carbon, silicates, iron-rich phases) (see Hedman et al. 2013; Cuzzi et al. 2018b; Ciarniello et al. 2019; Miller et al. 2024, this collection).

Simulations fitting the main ring spectra require increasing amounts of a broad-band absorber moving from the intrinsically brighter A and B rings to the darker C ring and Cassini Division (Cuzzi et al. 2018b; Ciarniello et al. 2019). This simultaneously reduces the overall albedo, mutes the spectral contrast of the water ice bands and UV absorption, and reddens the spectrum at VIS-NIR wavelengths. This is consistent with the trends of Fig. 5, where, for the main rings, the UV-VIS slope is (i) larger for the A and B rings than for the C ring and Cassini Division; (ii) increases with $BD_{2\,\mu m}$; and (iii) decreases with increasing VIS-NIR slope.

The ring moons display spectral shapes similar to the main rings and are distributed in the spectral indicator scatter plots along branches roughly parallel to those of the rings. Furthermore, the albedo of the ring moons is correlated with UV-VIS slope and $BD_{2\,\mu m}$ and anticorrelated with VIS-NIR slope (Fig. 6). This suggests that the ring moons share a similar surface composition with the main rings, and that the abundance of the broad-band absorber is a driver of the observed spectral variability, affecting ring moons and main rings in a similar way. In this respect, increasing amounts of such a spectrally bland contaminant, which would determine the spectral transition from the bright and reddish (in the UV-VIS region) A and B rings to the dark and less-red C ring (Cuzzi et al. 2018b), might also explain the spectral differences in moving from the inner ring moons (Pan, Daphnis, Atlas, Prometheus and Pandora) towards the co-orbital moons (Janus and Epimetheus).

Spectral models based on water-ice grains containing small ($\leq 0.4\%$) amounts of tholin (to provide the UV absorption), intimately mixed with a variable fraction of amorphous carbon (acting as a broad-band absorber), have been employed to fit the UV-NIR spectra of the mid-sized icy moons orbiting within the E-ring, Mimas, Enceladus, Tethys, Dione, Rhea (Ciarniello et al. 2011; Filacchione et al. 2012). However, the observed positive correlation of the UV-VIS slope and VIS-NIR slope slopes, both inversely correlated with the brightness coefficient B, cannot be explained solely in terms of variations in the abundance of a broad-band absorber. This likely points to a role of nearly pure water ice grains from the E ring (Hendrix et al. 2018; Hedman et al. 2020), limiting the usefulness of a comparison between the ring moons and icy satellites embedded in the E-ring.

An alternative compositional paradigm suggested by Clark et al. (2012) posits that the spectral signatures of the darkening materials in the Saturn system are dominated by iron-rich compounds (nano-phase iron and nano-phase iron oxide). Spectral models including

[9] Tholins are a wide class of macro-molecular organic compounds obtained from irradiation of simple carbon-bearing materials such as CH_4, C_2H_6, in presence of N_2 or H_2O (Khare et al. 1984; McDonald et al. 1994).

complex mixtures of variable amounts of these materials, water ice, and CO_2 provide good fits to the different terrains on Iapetus (bright, intermediate, and dark), Phoebe, Hyperion (dark and bright materials), Dione and the F ring. Clark et al. (2012) point out the similarity between the spectrum of Iapetus's bright material with the Cassini Division and C ring, and that of Hyperion with Epimetheus.

This might suggest that iron-rich compounds could also explain the VIS-NIR spectral variability among the main rings and ring moons. However, from a quantitative perspective, the A and B rings and the inner ring moons out to Prometheus are much redder (i.e., display a larger UV absorption; Filacchione et al. 2012) and brighter than are the Iapetus (Blackburn et al. 2011) and Hyperion (Thomas et al. 2007) bright materials (the reddest among the ones modeled in Clark et al. 2012). Thus it remains an open question whether models including iron-rich contaminants as darkening and UV-absorbing materials would be able to match the spectral properties of these targets.

Ongoing work (Hendrix et al. 2022) to simulate the UV-visible spectrum of the dark leading hemisphere of Iapetus shows that spectral models including carbon-rich tholins appear to better match the spectral shape of Iapetus, compared with models that consider irons only (with no tholins). This suggests that the tholin-based compositional paradigm might have the potential to explain the entire range of spectral variability from the brighter/redder to darker/less red rings and ringmoons.

A quantitative comparison between the physical and compositional properties of rings and ring moons would require conducting a systematic effort of regolith radiative transfer (RRT) modeling on available spectral observations of the latter. Ring moon RRT models would face similar difficulties as those encountered in ring modeling which we briefly address here (see a complete discussion of spectral modeling caveats in Sect. 4.1 of Miller et al. 2024, this collection). In this respect, whereas the mixing modality of the UV absorber, and tholin in particular, as emebedded in water ice grains (intramixture) appears as a fairly consistent result from ring models (Poulet et al. 2003; Cuzzi et al. 2018b; Ciarniello et al. 2019), this is not the same for the neutral absorber. Recent ring models of Cuzzi et al. (2018b) and Ciarniello et al. (2019) provide reasonable fits of the UV-VIS spectral properties of the main rings, by including the neutral absorbers using different mixing modalities, that is as embedded in water ice grains or in a "salt-and-pepper" fashion (intimate mixture). Intramixtures with water ice maximize the absorbing power of the non-icy materials, making it proportional to their mass fraction, and typically requiring smaller abundances of contaminants with respect to intimate mixtures. Furthermore, pure grains of highly absorbing materials, as neutral absorbers in intimate mixtures, might turn completely opaque if larger than a given grain size, thus suppressing any grain-size dependency on the corresponding spectral properties. In this regime, the absorbing power becomes proportional to the total cross-section of the contaminant, implying that larger grains have the same darkening power (same total cross-section) of smaller ones, but with larger total mass fraction. All this implies that the inferred fraction of contaminants from spectral modeling strongly depends on the chosen mixing modality. An additional complication is represented by the shadowing induced by macroscopic surface roughness, both affecting ring particles and ring moons. Rougher surfaces would have larger contributions from shadows, thus reducing the total amount of scattered light. This is reflected in a reduction of the surface Bond albedo of the ring particles and ring moons (Cuzzi et al. 2017, 2018b), which, if not accounted for, can be erroneously interpreted in terms of larger abundances of non-icy contaminants when RRT is performed. Such aspects become particularly relevant if the absolute values of the inferred amounts of non-icy materials are used to constrain the effectiveness of contamination processes among these. For example, following these arguments and adopting

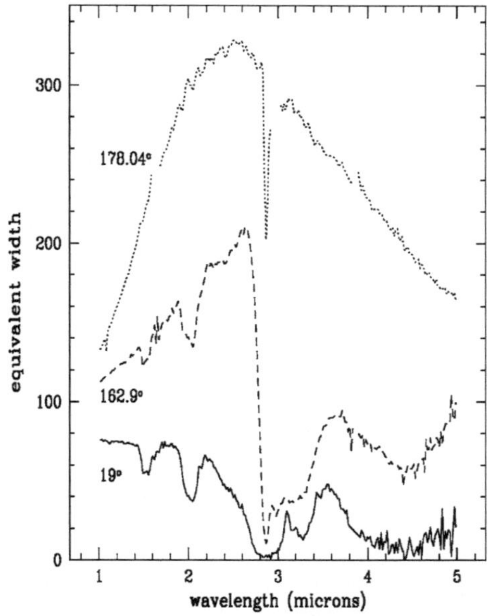

Fig. 8 Infrared spectra of the F-ring equivalent width at phase angles 178.04°, 162.9°, and 19° from VIMS observations. The equivalent width is computed as the radial integral of the ring brightness. The spectra at lower phase angles clearly display the 1.5-, 2- and 3- μm water ice bands. The high phase angle spectrum (178.04°) shows the Christiansen feature at 2.8 μm. The spectra at 178.04° and 9° have been scaled down by a factor of 9 and 30, respectively. From Vahidinia et al. (2011)

an intra-particle mixing paradigm as in recent RRT modeling of the main rings, we might expect that the only available estimation of the tholin fraction for a ring moon (Epimetheus; Buratti et al. 2010) might be revised towards smaller abundances, possibly closer to main ring values.

7 Composition of the F-Ring and G-Ring

Outside of the main rings are a variety of dusty rings, the three most prominent being the F, G and E rings. At the moment, we have the most compositional information about the E ring thanks to the extensive data obtained by Cassini's Cosmic Dust Analyzer (Srama et al. 2004), which directly sampled particles within this ring and Enceladus' plume. These data reveal that the majority of particles in the E ring are composed primarily of water ice, consistent with near-infrared spectral data. However, some of the particles in this ring were also found to contain organics, salts and silicates (Postberg et al. 2009, 2011; Hsu et al. 2015; Postberg et al. 2018).

By contrast, information about the compositions of the F and G rings is much more limited because these rings were considered too hazardous for Cassini to fly through and directly sample. Furthermore, since the visible particles in these rings are extremely small (around 1-100 microns), their spectral and photometric properties are more sensitive to the particle size distribution and structure than to composition, because diagnostic absorption bands are strongly suppressed.

Despite these limitations, the remote sensing data do provide some useful constraints on the composition of the F ring, which is a relatively bright ring and so multiple high signal-to-noise visible and near-infrared spectra were obtained by VIMS (Clark et al. 2008; Vahidinia et al. 2011). Infrared spectra obtained at low to moderate phase angles show all the standard signatures of water ice (Fig. 8). There is also a decrease in brightness shortwards of 0.5 μm that might be due to the same UV absorption seen in the main rings and moons, but

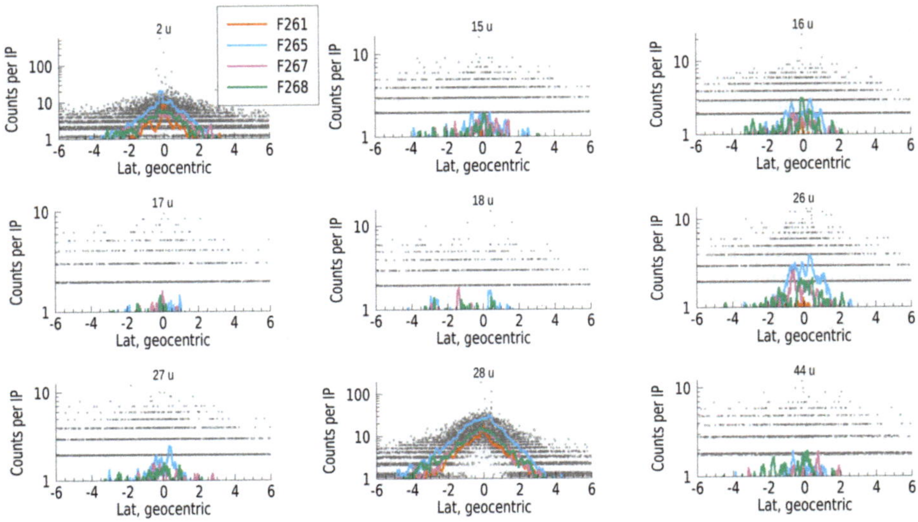

Fig. 9 In situ data from INMS during orbits near the F ring reveal the presence of icy grains. Grey points are individual datapoints from the four passes, and the overplotted lines show the smoothed data

the interpretation of this feature is complicated by a broad brightness peak centered around 0.5 μm ikely due to a relatively high fraction of sub-micron grains (Clark et al. 2012).

Meanwhile, observations at extremely high phase angles(Fig. 8) show a strong dip near 2.8 μm which corresponds to the Christiansen frequency of crystalline water ice (Vahidinia et al. 2011; Hedman et al. 2011). This spectral feature arises at wavelengths where the real refractive index of the material approaches 1, which greatly reduces the scattering efficiency in these geometries. Since only water-ice is known to have a real refractive index near 1 at these wavelengths, this means that the F-ring particles need to be composed primarily of crystalline water ice like the rest of the rings.

By contrast, the signal-to-noise ratio of the available Cassini spectra of the G ring is very low, making even the comparatively strong 3 μm water ice absorption band difficult to quantify (although it does appear to be present). Robust constraints on the G ring's composition will therefore likely have to await future studies that co-add data from the entire span of the Cassini mission to obtain sufficiently high signal-to-noise spectra of this ring.

Additional compositional information about the F ring can be obtained from in situ data collected by the Ion and Neutral Mass Spectrometer (INMS Waite et al. 2004) in the closed source neutral mode during four orbits that crossed the ring plane outside of the F ring, shown in Fig. 9. The data are dominated by 2 u and 28 u. Signal at 2 u may be a mixture of water ion fragments and H_2 sourced from the main ring atmosphere (Tseng et al. 2011), with possible contributions from Enceladus (Jurac and Richardson 2007; Cassidy and Johnson 2010). However, the similar distributions for 2 u and 28 u support a common origin such as icy grains rather than gaseous material, which should show different scaleheights for the different masses (Perry et al. 2017). While the molecular water peak at 18 u and fragment peak at 17 u are relatively low, this may be related to the speed of the spacecraft. High velocity impacts of ice grains have been found to increase the measured 2 u to 18 u ratio, with titanium vapor produced from the INMS antechamber by the impacts likely scavenging the resulting oxygen (Walker et al. 2015). The 28 u peak is likely related to CO and/or oxidation at the filament of organic carbon. Note that data shown in Fig. 9 show counts per

IP, which are related to environmental density by a mass-dependent ram factor. This ram factor leads to nearly a factor of four apparent increase for 28 u as compared to 2 u.

The pattern of peak heights with 28 u > 26 u > 27 u may be associated with the presence of unsaturated organics with an O or N heteroatom. However, multiple different compounds may contribute to the signal, and no single compound was found to match the pattern. Some of these compounds also produce mass fragments at 15 and 16 u, and a further subset also produce fragments at 44 u. Masses associated with atomic H (1 u) and O_2 (32 u) were not part of the "closed source neutral" measurement sequence, and it is possible that 16 u may be associated with O_2 instead. While 32 u was measured in "open source ion" mode, without the enhancement from the closed source the signal is below detection limits.

8 Radial Variability of the Spectral Indicators and Exogenic Processes Affecting Surface Properties

Icy satellites in the Saturn's system are subject to a series of different exogenic processes affecting their surfaces, including contamination from ring material in the nearby A, F and E rings (Hendrix et al. 2018; Buratti et al. 2019; Hedman et al. 2020), the pollution from meteoritic dust (Cuzzi et al. 2009), and bombardment by charged particles (Roussos et al. 2008; Paranicas et al. 2018; Hedman et al. 2020). Concerning ring moons, these processes modify the composition of the ring material from which they accreted, to determine their surface composition and the resulting spectral properties we observe today. The magnitude of the effects of such exogenic processes on the different objects orbiting Saturn, including the main rings, depend primarily on the object position, and can be put in relation with radial spectral/albedo gradients in the system.

In Fig. 10 the spectral indicators from VIMS data (see Sect. 6.2: UV-VIS and VIS-IR slopes, $BD_{1.5\,\mu m}$, and $BD_{2\,\mu m}$) for the main rings and icy moons are reported as a function of the radial distance from Saturn. This complements, for the icy moons, the information of Fig. 3 reporting radial trends of the brightness coefficient, GRN/UV3 and IR3/GRN color ratios from ISS observations, and the relative fluxes of E-ring grains and high energy protons and electrons. Looking at Fig. 10, it can be noted that VIMS spectral indicators of the ring moons inside Mimas' orbit progressively approach A-ring values moving inward from Janus/Epimetheus to Pan, with an increase of the water ice band depths ($BD_{1.5\,\mu m}$ and $BD_{2\,\mu m}$) and UV-VIS slope, and a reduction of the VIS-NIR slope, for which the trend is entirely driven by the particularly high values of Janus and Epimetheus. Such combined variation of the ring moon spectral indicators would appear consistent with a progressive enrichment with increasing radial distance of a broad-band-absorbing contaminant (see Sect. 6), resulting in a modification of the A-ring-like spectral properties of Pan/Atlas towards those of Janus/Epimetheus. It must be mentioned, however, that this picture is not fully consistent with the observed trend in the brightness coefficients (Fig. 3), with Prometheus and Pandora displaying the largest albedos among the ring moons. Furthermore, Prometheus sticks out from the expected radial trend also for what concerns the water ice band depths and the UV-VIS slope, displaying values larger than expected. This indicates that, at least for Prometheus and Pandora, an additional driver of the spectral properties should be considered.

Assuming that B and A rings preserve the least altered ring material (see Miller et al. 2024, this collection), and that their spectral properties (including the signature of the UV absorber, considered intrinsic to the ring system) are shared by the newly formed ring moons, the overall enrichment with radial distance of a broad-band absorber, can be naturally explained by contamination from bombardment of meteoritic dust (Cuzzi et al. 2009).

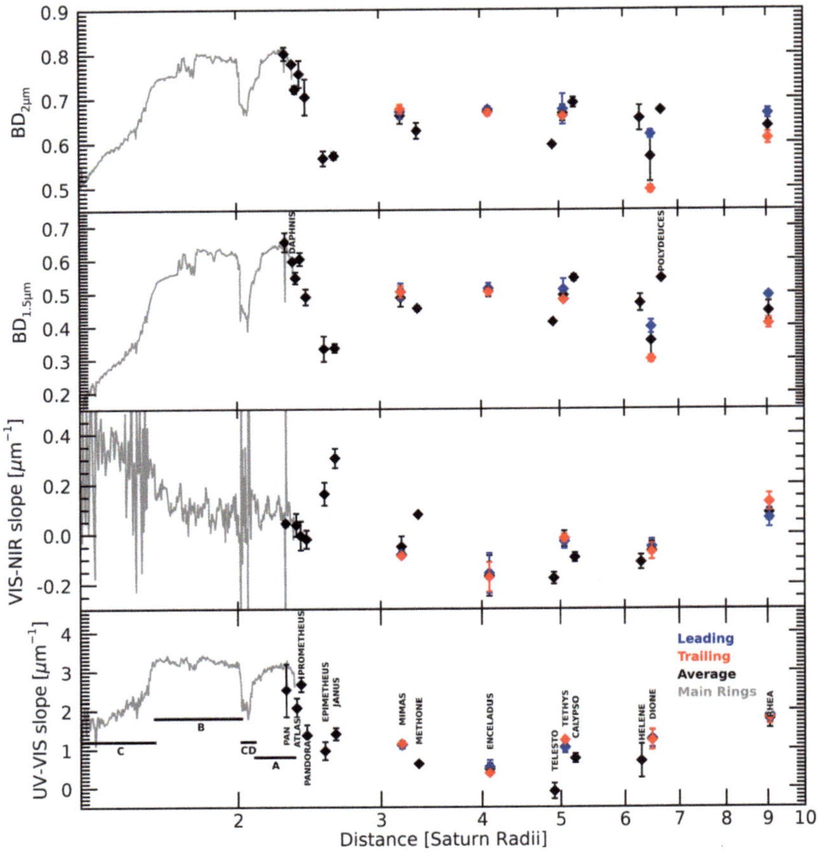

Fig. 10 Radial distribution of spectral indicators (UV-VIS and VIS-NIR slopes, $BD_{1.5\ \mu m}$, and $BD_{2\ \mu m}$) for Saturn's main ring (gray curves) and moons (diamonds) for which VIMS observations are available. Main ring values are derived from radial spectrograms with a 200-km radial sampling (see also Fig. 5). The respective radial extent of the different ring regions is indicated by horizontal bars (CD: Cassini Division). For Saturn's inner mid-sized moons (Mimas, Enceladus, Tethys, Dione, and Rhea) also leading and trailing hemisphere values are reported. UV-VIS slope and VIS-NIR slope slopes for Methone are from ISS data (see Fig. 4)

This would be consistent with the observed spectral properties of Janus and Epimetheus, possibly formed before the other inner ring moons (see Sect. 5), and then exposed for a longer time to such exogenous contamination. In this respect, it is interesting to note that their spectral indicators are similar to the C ring, where the contamination from the meteoritic flux is thought to be more effective than on the A and B rings (Cuzzi et al. 2018b; Ciarniello et al. 2019), because of the smaller optical depth (i. e. surface to mass ratio). Such similarity is confirmed by ultraviolet-visible composite spectra as shown in Fig. 11, with Epimetheus having a similar UV-visible spectral shape as the C ring. If so, the C ring on one side, and Janus/Epimetheus on the other, might represent the most contaminated endmembers for the main rings and inner ring moons, respectively.

An additional contribution affecting the surface composition of the ring moons inside Mimas' orbit and consistent with the observed spectral trends is the contamination from the nearby A ring particles. This would be more effective for ring moons close to the rings, and would contrast the effects of the meteoritic flux making the ring moon surface spectrally

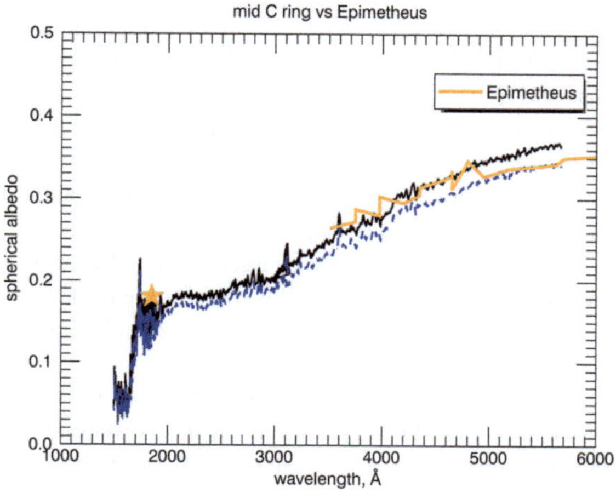

Fig. 11 UV-visible spectrum of Epimetheus compared with a) the C ring. Epimetheus composite spectrum uses data from Cassini VIMS and UVIS (single point at 185 nm). C ring composite spectrum (black and blue curves) uses data from Cassini UVIS and Hubble Space Telescope Space Telescope Imaging Spectrograph (HST STIS) (after Cuzzi et al. 2018b)

more similar to the A ring (Buratti et al. 2019). Ring-related material is also lost by photo-ionization both into the planet and outwards beyond the rings into the magnetosphere. Photo-ionized molecules and atoms in the neutral ring atmosphere are accelerated in the magnetic field, and collide with neutrals of all species; this sends many of these large, but not yet charged, molecules beyond the edge of the rings at 2-3R_S and off into calmer regions of the magnetosphere where they may affect the surface properties of the inner moons. The loss rate is far larger near solstice than near equinox, by this process - a seasonal flow. In particular, near solstice, the column density of H_2 due to the rings in the magnetosphere out to perhaps $12R_S$, is only about 3 times smaller than the contribution due to Enceladus (Tseng et al. 2011; Tseng and Ip 2011) - that is, there is outward as well as inward mixing. The abundance of O_2 scattered outwards from the rings is only about an order of magnitude smaller than that contributed by Enceladus (Tseng et al. 2010, and see Miller et al. 2024, this collection, for further details). A quantitative treatment of mass exchange and associated compositional implications for the rings and inner ringmoons remains to be done.

Moving outward from Epimetheus/Janus orbits, along with the meteoritic dust flux, the two major processes affecting the icy moon surfaces are the contamination from E-ring grains and the bombardment by charged particles (Hendrix et al. 2018; Hedman et al. 2020). The effect of the surface accumulation of E-ring icy grains, originating from Enceladus' plumes, is likely responsible for the observed spectral trends among the mid-sized icy moons up to Janus and Epimetheus, with spectral slopes and color ratios increasing away from Enceladus, accompanied by a progressive reduction of the moons' visual albedo (Verbiscer et al. 2007; Hendrix et al. 2018; Hedman et al. 2020), in agreement with the decreasing modeled E-ring particle total flux (Fig. 3). E ring particles preferentially accumulate on the leading side of moons outside Enceladus' orbit, and on the trailing side for the inner ones. This is consistent with the observed visual albedo longitudinal asymmetries discussed in Sect. 6.1, and appear also in agreement with the distribution of the water ice band depths on Dione and Rhea, with values larger on their leading sides than on the trailing. The influence of high-energy particle fluxes on the surfaces of Saturn's moons is best demonstrated by the low surface brightnesses of the small moons Aegaeon, Methone and Pallene shown in Fig. 3. These three moons are all located within the inner flank of the E ring, but are significantly darker than one would predict based on the brightness trends observed among the larger moons Janus, Epimetheus, Mimas and Enceladus. The most likely explanation for the anomalously low brightnesses of these small moons is that they are exposed to a much higher flux of high-energy protons. At suitably high energies, these protons are concentrated in a series of three belts that lie between the main rings and the orbits of Janus/Epimetheus, Mimas and Enceladus (Roussos et al. 2008; Paranicas et al. 2018). It turns out that those larger moons are efficient enough sinks for protons that they actually create localized depletions in the charged particle flux in their vicinity, while the smaller moons Aegaeon, Methone and Pallene are not such efficient proton sinks and so are exposed to much higher fluxes of charged particles. Furthermore, the brightnesses of all these moons are reasonably well correlated with the ratios of charged particle flux to E-ring particle flux experienced by each moon (Hedman et al. 2020). It is therefore reasonable to conclude that while E-ring particles can brighten the surfaces of Saturn's moons, high-energy charged particles like protons can somehow darken ice-rich surfaces.

At the moment, the physical mechanism by which the charged particles darken these moon's ice-rich surfaces is still unclear. In principle, the surfaces could become darker either due to changes in their chemistry or in the structure of their regoliths, but there are challenges with either of these interpretations. In particular, Methone has water ice band depths that are just slightly deeper than Mimas and Enceladus, so there is no indication that the radiation is causing significant changes in the structure of its regolith. At the same time, Methone has similar color ratios as Mimas and Enceladus, so there is also not any obvious compositional difference between these moons. The one potentially interesting spectral trend in this context are the strongly positive VIS-NIR slopes of Janus, Epimetheus, and Aegaeon. Such strong positive slopes are unusual for Saturn's moons, and while Janus/Epimetheus are not exposed to particularly high fluxes of high-energy protons, both Janus/Epimetheus and Aegaeon are exposed to high fluxes of moderately high-energy electrons (Roussos et al. 2008; Hedman et al. 2020).

As a final remark, we note that Hendrix et al. (2018) showed that connections could be made between the UV-visible spectral shapes of the mid-sized icy moons and the amount of E-ring grain and charged particle bombardment (namely the corotating plasma) experienced at the orbital locations of each moons. They suggested that the relatively low albedo of Rhea, and its relatively strong UV absorption, could be related to organics - emplaced by E-ring grain bombardment - on the surface, in combination with aging, via plasma bombardment-related weathering, of those E-ring organics. In this respect, Epimetheus' low albedo level and weak UV absorption (both smaller than Rhea's) do not appear fully consistent with the trend observed in the mid-sized icy moons (see Sect. 6.2), requiring further investigation.

9 Summary and Final Remarks

The surface composition of the inner ring moons, up to Janus and Epimetheus, as inferred from the UV-VIS-IR spectroscopic data of remote sensing instrument onboard the Cassini mission, appears linked to that of the main ring particles from which they accreted. This is supported by the overall observed spectral similarity between ring and ring moons and the similar trends of the spectral parameters (see Sect. 6 and Figs. 4 and 5), which suggests a common compositional paradigm. Nonetheless, the radial variability of the inner ring moon spectral properties can be also connected to the complex interplay of different exogenous processes affecting their surfaces. In this respect, the progressive departure from an A-ring like spectrum, moving from the innermost Pan to the outermost Janus/Epimetheus, can be interpreted in terms of a reduction with radial distance of the A-ring particle flux on the ring moon surfaces giving way to pollution from meteoritic dust. As mentioned above in Sect. 8, the similarity between Janus/Epimetheus surface composition and the C ring might be interpreted as a result of a more effective pollution on these targets compared to the other ring moons and main rings, respectively. However, the occurrence of leading/trailing asymmetries on Janus and Epimetheus is likely the result of the accumulation of E-ring particle on these targets, and additional contamination might come from the nearby G-ring, fed by the low-albedo Aegeon. Also, the flux of energetic electrons is relatively large at Janus and Epimetheus orbital distance. These sources of exogenous material and the potentially darkening effect of energetic particles are only relevant for Janus and Epimetheus and not for the C ring. As such, the interpretation of the Janus/Epimetheus vs. C ring spectral similarity may not be straightforward as it would appear. In this respect, it is worth mentioning that Prometheus and Pandora, the ring moons with the largest albedo, possibly witness the effects of another source of exogenous material, represented by the accumulation of icy F-ring particles brightening their surfaces. Yet, Prometheus, being marginally brighter than Pandora, has significantly larger UV-slope and water ice band depths, which appears difficult to explain with the sole effect of F-ring grain accumulation. Further insights into the investigation of the compositional link between the main rings and the inner ring moons and the (different) exogenous processes affecting their surface spectrophotometric properties could be provided by complementing the results from the spectral indicator analysis with a systematic and quantitative comparison of their intrinsic albedos. In this respect, a key quantity to be compared would be the Bond albedo of the ring particles with that of the ring moons: this would allow us to overcome the bias encountered when comparing the reflectance of the main rings (a many-particle-thick layer, but with finite optical depth) with the reflectance of a ring moon surface (a semi-infinite optical depth medium). The Bond albedo of the main ring particles is starting to appear in the literature (Cuzzi et al. 2018b; Ciarniello et al. 2019), even if particle-surface shadowing effects have generally not been accounted for, while for the ring moons these albedos could be approximated by the brightness coefficient at a phase angle of $\sim 55°$, computed following the approach of Hedman et al. (2020) (Sect. 6.1).

Ring-embedded moons (Aegeon, Methone, Anthe, and Pallene) appear as a separate class with respect to the rest of the ring moons and as such they would be less spectrally-linked to the main rings. Although, at the time of writing, there is limited hyperspectral information for this class of objects (namely only the IR spectrum of Methone), brightness coefficient vs color ratio scatter plots for Aegeon, Methone, and Pallene (Fig. 6) indicate they cluster separately from mid-sized icy moons and ring moons, displaying remarkably lower albedos with respect to satellites at comparable radial distance from Saturn. This is currently interpreted as a result of the relatively large fluxes of high-energy charged particles on their surfaces, dominating over the brightening effect of E-ring particle accumulation.

The comparison of ring moon spectral indicators with those of the midsized icy moons and their Lagrangian satellites reveal qualitatively different trends (Figs. 5 and 6). Notably, the UV absorption in ring moons is positively correlated with their surface brightness and water ice absorption band depths, and inversely correlated with the VIS-IR slope, supporting a compositional paradigm including at least two populations of grains, one of water ice particles with an embedded UV absorber (see Sect. 6.3 and references therein) and a second component providing a broad-band absorption. On the contrary, for the mid-sized icy moons and their Lagrangians the UV absorption is positively correlated with the VIS-IR slope while decreasing with the surface albedo, possibly pointing to a different compositional paradigm in which the same carrier might provide both the UV absorption and the surface darkening. This would be consistent with a scenario in which the average surface composition of these objects is primarily driven by the accumulated flux of E-ring particles (as evidenced by the albedo vs orbital distance trend), however assuming an aging of the E-ring particles due to exposure to radiolytic processes of their organic content, making them redder/darker.

As such, grains reaching midsized icy moons farther from Enceladus (as Rhea and Dione), would experience a longer radiolytic processing provide larger UV-reddening and darkening of their surfaces (Hendrix et al. 2018). However, Janus and Epimetheus, whose albedo leading/trailing hemisphere asymmetry suggests evidence of accumulation of E-ring grains on their surfaces and should receive larger fluxes than Rhea, do not seem to fit in this scenario, as they appear darker than Rhea while having a smaller UV-absorption. This points to a more complex picture, in which the resulting spectral properties of Janus and Epimetheus are the result of those inherited by the rings with superposition of the different exogenous mechanisms discussed above, including further processing of the organics in the E-ring particles accumulated on their surfaces by the relatively large fluxes of high energy electrons (Fig. 3), leading to carbonization which would progressively turn the grains from red to black.

In the current ring moon formation models these bodies could form 1) in the rings from original seeds, denser than the ring material, that may be shards from the destruction of a larger satellite or 2) at/near the Fluid Roche limit (outside the A ring) directly from ring material (Sect. 5). Disentangling between these two scenarios by means of compositional arguments based on spectroscopic observation does not appear as an easy task. The spectral similarity between main rings and ring moons, might just be representative of the ring moons outer layers, dominated by ring material, and hiding a potential inner seed not accessible by UV-VIS-IR spectroscopy. Furthermore, the evidence of a complex interplay of exogenous processes affecting the ring moon surfaces, as witnessed by the spectral radial trends, modifies the original composition inherited from the rings, to a degree which depends on the ring moon position and exposure time, i. e. its orbital evolution.

Once formed, embedded ring moons migrate outwards following the rings' viscous spreading, while isolated ring moons, emerged from the rings or formed just outside the FRL, are pushed by the one-sided ring torque. With the exception of Pan, orbiting inside the A ring, the ringmoon orbital evolution timescales, obtained by comparing the present distance from the A ring edge with the computed velocity of the outward radial drift, increase with the ringmoon radial distance, from $\sim 4 \times 10^6$ yr for Atlas, $\sim$ 1 Gyr for Epimetheus (Sect. 5). Although it might be tempting to interpret these values as actual ringmoon ages, this is hindered by the fact the drift rate changes over time due to 1) the increase of the distance from the A ring, 2) possible (temporary) trapping in mean motion resonances with other satellites and, 3) variations in the ringmoon mass resulting from merging processes. These arguments appear even more relevant in the light of recent estimations of the age of Saturn's rings from the degree of pollution delivered by micrometeoroid infall (Kempf et al. 2023), posing an upper limit to $\sim$ 400 Myr, which would be inconsistent with the above mentioned orbital evolution timescales (see also Crida et al. 2024, this collection). Nonetheless, these should still be meaningful in relative terms, and it appear broadly supported from a spectroscopic perspective, with Janus and Epimetheus, likely the oldest of the inner ring moons, being characterized by a more polluted appearance, possibly linked to a longer exposure to contamination processes.

The observed shapes of the inner ring moons may be either the result of ring material accretion onto the initial seed, which occurring preferentially at equatorial latitudes can also account for the equatorial ridge observed on Atlas, Pan, and Daphnis, or by a low-speed collision of two similar progenitors (Sect. 4). In the former case, one might hope to observe residual compositional differences between the ring moon central component and the equatorial ridge, if the internal seed is not completely hidden by the accumulated material. In this respect, Buratti et al. (2019) provided some information in terms of the disk-resolved ISS IR3/UV3 color ratio, as measured during the Cassini's ring-grazing orbits: for Pan and Atlas, the IR3/UV3 is 19% and 16% larger in the cratered terrains of their central components than for their equatorial ridges, respectively. If the equatorial ridge is more representative of the ring material than the central component, and given that inner moons are typically redder possibly because of the larger ring particle influence, we might expect the equatorial ridge to be redder than the rest of the moon. However, it appears to be the opposite for Atlas and Pan, a somewhat puzzling result. Similar information is available also for Pandora, with a IR3/UV3 color ratio 15% lower in the craters than in the smooth equatorial material. This seems in line with the argument above, yet Pandora is in general much less red in the UV-VIS than the A ring and might be significantly influenced by F ring grains (which may be polluted like the low-optical depth portions of the main rings), thus it is not straightforward to interpret such variability as an effect of preferential main ring material accumulation at equatorial/intermediate latitudes. One way to distinguish observationally between the "ring material accretion at equatorial latitudes" and the"low-speed collision" scenarios would be to determine the rotational libration amplitude of these bodies in order to estimate their moments of inertia (Rambaux et al. 2022). In the first scenario, the body is heterogeneous (the solid internal seed plus less-dense ring material at the equator) while in the second case, the mixing would probably lead to a rather homogeneous body. The libration of small moons may be deduced directly from high-resolution images (Tiscareno et al. 2009) or indirectly using astrometric measurements of the periapse variations (Lainey et al. 2019, 2023).

Despite the impressive wealth of data provided by the Cassini mission, tracing back the present compositional, morphological and orbital properties of the ring moons to the primordial phases of their formation in the rings endures as a difficult challenge. As summarized above, the series of processes modifying the surfaces of these bodies, and their dynamical evolution, are not fully unraveled yet, and will need to be quantitatively characterized by future theoretical and laboratory investigations. Hopefully, as recently achieved for the main rings, this will allow us to put independent and absolute constraints on the age of the ring moons and produce a more complete picture of how the Saturnian system evolved. Exploration of the Saturn system in the future might also allow us to sample the interior of the ring moons in the context of the nearby ring particles. This would pose crucial constraints on the formation processes and, by comparing surface and sub-surface properties, shed light on the exogenous mechanisms affecting these small satellites.

Appendix A: Photometric Correction of VIMS Data

VIMS full-disk spectra of Saturn's moons have been photometrically corrected to a reference phase angle of 50° before averaging procedures and computation of spectral indicators. This has been done by assuming that the icy moons spectrophotometric properties can be described by means of the Kaasalainen-Shkuratov model (Kaasalainen et al. 2001; Shkuratov et al. 2011), as disscussed in Filacchione et al. (2022). This implies that the full-disk spectral reflectance of a given target ($FDR_\lambda(\alpha)$) after normalization at 0.55 µm can be expressed as

$$\frac{FDR_\lambda(\alpha)}{FDR_{0.55\ \mu\mathrm{m}}(\alpha)} = \frac{A^{eq}_\lambda(\alpha)}{A^{eq}_{0.55\ \mu\mathrm{m}}(\alpha)} \tag{A.1}$$

where the subscript λ indicate different wavelengths, α is the phase angle and A^{eq} is the equigonal albedo. Given this, the normalized full-disk reflectance photometrically corrected at 50° can be expressed as

$$\frac{FDR_\lambda(50°)}{FDR_{0.55\ \mu\mathrm{m}}(50°)} = \left(\frac{FDR_\lambda(\alpha_0)}{FDR_{0.55\ \mu\mathrm{m}}(\alpha_0)}\right)_{VIMS} \frac{A^{eq}_{0.55\ \mu\mathrm{m}}(\alpha_0)}{A^{eq}_\lambda(\alpha_0)} \frac{A^{eq}_\lambda(\alpha)}{A^{eq}_{0.55\ \mu\mathrm{m}}(\alpha)} \tag{A.2}$$

with $\left(\frac{FDR_{\lambda}(\alpha_0)}{FDR_{0.55\ \mu m}(\alpha_0)}\right)_{VIMS}$ being the normalized full-disk reflectance spectrum from VIMS data. Filacchione et al. (2022) derived the equigonal albedos of Saturn's mid-sized inner moons (Mimas, Enceladus, Tethys, Dione, and Rhea) assuming a polynomial expression

$$A_{\lambda}^{eq}(g) = a_{0,\lambda} + a1_{1,\lambda}(\alpha) + a_{2,\lambda}(\alpha)^2 \tag{A.3}$$

with a set of different parameters ($a_{0,\lambda}$, $a_{1,\lambda}$, $a_{2,\lambda}$) for each target. Since spectrohotometric models for the rest of the Saturn's inner moons in the complete VIMS spectral range have not yet been derived at the time of writing, we perform the photometric correction of the observations of the different targets by means of Eq. (A.3), assuming for simplicity the set of parameters ($a_{0,\lambda}$, $a_{1,\lambda}$, $a_{2,\lambda}$) derived for the case of Rhea, with the exception of the mid-sized inner moons, for which the respective set of parameters were used.

With a similar approach, we also performed an empirical photometric correction of the main ring radial spectrogram presented in Sect. 6. This requires the simplifying assumpions that ring particles share spectrophotometric properties similar to the icy moons and that multiple scattering of light between ring particles is negligible. We judge this an acceptable approximation, considering the limited impact of the photometric correction for the main ring case, since the VIMS observations used to build the radial spectrogram were acquired with a phase angle of 38.8°, relatively close to the reference phase angle of 50°.

Acknowledgements This research was supported by the International Space Science Institute (ISSI) in Bern, through the ISSI Workshop "New Vision of the Saturnian System in the Context of a Highly Dissipative Saturn". MC and GF acknowledge internal support from INAF-IAPS. KEM acknowledges internal support from SwRI. This work has made use of NASA's Astrophysics Data System. We thank P. Thomas for valuable comments and suggestions and two anonymous referees for helpful reviews that improved this article.

Author Contribution The idea for this review article came from GF and JC. All authors contributed to the literature search. New data analysis was performed by MC, FG, MMH, KEM, and AH. All authors contributed to drafting and revisions.

Funding Open access funding provided by Istituto Nazionale di Astrofisica within the CRUI-CARE Agreement. The project leading to this publication has received funding from the Excellence Initiative of Aix-Marseille Université–A*Midex, a French "Investissements d'Avenir program" AMX-21-IET-018. This research holds as part of the project FACOM (ANR-22-CE49-0005-01_ACT) and has benefited from a funding provided by l'Agence Nationale de la Recherche (ANR) under the Generic Call for Proposals 2022.

Data Availability Cassini data used in this work are archived in NASA's Planetary Data System (PDS).

Declarations

Competing Interests The authors have no competing interests to declare.

References

Asphaug E (2020) Interiors of small bodies and moons. Nat Commun 11:1564. https://doi.org/10.1038/s41467-020-15458-8

Attree NO, Murray CD, Cooper NJ, et al (2012) Detection of low-velocity collisions in Saturn's F ring. Astrophys J Lett 755(2):L27. https://doi.org/10.1088/2041-8205/755/2/L27

Beurle K, Murray CD, Williams GA, et al (2010) Direct evidence for gravitational instability and moonlet formation in Saturn's rings. Astrophys J Lett 718(2):L176–L180. https://doi.org/10.1088/2041-8205/718/2/L176

Blackburn DG, Buratti BJ, Ulrich R (2011) A bolometric bond albedo map of iapetus: observations from Cassini vims and iss and Voyager iss. Icarus 212(1):329–338. https://doi.org/10.1016/j.icarus.2010.12.022

Borderies N, Goldreich P, Tremaine S (1984) Unsolved problems in planetary ring dynamics. In: Greenberg R, Brahic A (eds) Planetary rings. University of Arizona Press, Tucson, pp 713–734

Bradley ET, Colwell JE, Esposito LW, et al (2010) Far ultraviolet spectral properties of Saturn's rings from Cassini UVIS. Icarus 206(2):458–466. https://doi.org/10.1016/j.icarus.2009.12.021

Bradley ET, Colwell JE, Esposito LW (2013) Scattering properties of Saturn's rings in the far ultraviolet from Cassini UVIS spectra. Icarus 225(1):726–739. https://doi.org/10.1016/j.icarus.2013.04.008

Brown RH, Baines KH, Bellucci G, et al (2004) The Cassini Visual And Infrared Mapping Spectrometer (VIMS) investigation. Space Sci Rev 115(1–4):111–168. https://doi.org/10.1007/s11214-004-1453-x

Buratti BJ, Bauer JM, Hicks MD, et al (2010) Cassini spectra and photometry 0.25-5.1 μm of the small inner satellites of Saturn. Icarus 206(2):524–536. https://doi.org/10.1016/j.icarus.2009.08.015

Buratti BJ, Thomas PC, Roussos E, et al (2019) Close Cassini flybys of Saturn's ring moons Pan, Daphnis, Atlas, Pandora, and Epimetheus. Science 364:eaat2349. https://doi.org/10.1126/science.aat2349

Burns JA, Showalter MR, Hamilton DP, et al (1999) The formation of Jupiter's faint rings. Science 284:1146. https://doi.org/10.1126/science.284.5417.1146

Canup RM, Esposito LW (1995) Accretion in the Roche zone: coexistence of rings and ring moons. Icarus 113(2):331–352. https://doi.org/10.1006/icar.1995.1026

Canup RM, Esposito LW (1996) Accretion of the Moon from an impact-generated disk. Icarus 119(2):427–446. https://doi.org/10.1006/icar.1996.0028

Cassidy TA, Johnson RE (2010) Collisional spreading of Enceladus' neutral cloud. Icarus 209(2):696–703. https://doi.org/10.1016/j.icarus.2010.04.010

Chaban GM, Bernstein M, Cruikshank DP (2007) Carbon dioxide on planetary bodies: theoretical and experimental studies of molecular complexes. Icarus 187(2):592–599. https://doi.org/10.1016/j.icarus.2006.10.010

Chandrasekhar S (1969) Ellipsoidal figures of equilibrium. The Silliman Foundation lectures. Yale University Press, New Haven

Charnoz S, Porco CC, Déau E, et al (2005) Cassini discovers a kinematic spiral ring around Saturn. Science 310(5752):1300–1304. https://doi.org/10.1126/science.1119387

Charnoz S, Brahic A, Thomas PC, et al (2007) The equatorial ridges of pan and atlas: terminal accretionary ornaments? Science 318(5856):1622. https://doi.org/10.1126/science.1148631

Charnoz S, Dones L, Esposito LW, et al (2009) Origin and evolution of Saturn's ring system. In: Dougherty MK, Esposito LW, Krimigis SM (eds) Saturn from Cassini-Huygens, p 537. https://doi.org/10.1007/978-1-4020-9217-6_17

Charnoz S, Salmon J, Crida A (2010) The recent formation of Saturn's moonlets from viscous spreading of the main rings. Nature 465(7299):752–754. https://doi.org/10.1038/nature09096

Charnoz S, Crida A, Castillo-Rogez JC, et al (2011) Accretion of Saturn's mid-sized moons during the viscous spreading of young massive rings: solving the paradox of silicate-poor rings versus silicate-rich moons. Icarus 216(2):535–550. https://doi.org/10.1016/j.icarus.2011.09.017. arXiv:1109.3360 [astro-ph.EP]

Ciarniello M, Capaccioni F, Filacchione G, et al (2011) Hapke modeling of Rhea surface properties through Cassini-VIMS spectra. Icarus 214(2):541–555. https://doi.org/10.1016/j.icarus.2011.05.010

Ciarniello M, Filacchione G, D'Aversa E, et al (2019) Cassini-VIMS observations of Saturn's main rings: II. A spectrophotometric study by means of Monte Carlo ray-tracing and Hapke's theory. Icarus 317:242–265. https://doi.org/10.1016/j.icarus.2018.07.010

Ciarniello M, Moroz LV, Poch O, et al (2021) VIS-IR spectroscopy of mixtures of water ice, organic matter, and opaque mineral in support of small body remote sensing observations. Minerals 11(11):1222. https://doi.org/10.3390/min11111222

Clark RN, Lucey PG (1984) Spectral properties of ice-particulate mixtures and implications for remote sensing. 1. Intimate mixtures. J Geophys Res 89:6341–6348. https://doi.org/10.1029/JB089iB07p06341

Clark RN, Curchin JM, Jaumann R, et al (2008) Compositional mapping of Saturn's satellite Dione with Cassini VIMS and implications of dark material in the Saturn system. Icarus 193(2):372–386. https://doi.org/10.1016/j.icarus.2007.08.035

Clark RN, Cruikshank DP, Jaumann R, et al (2012) The surface composition of Iapetus: mapping results from Cassini VIMS. Icarus 218(2):831–860. https://doi.org/10.1016/j.icarus.2012.01.008
Colwell JE, Nicholson PD, Tiscareno MS, et al (2009) The structure of Saturn's rings. In: Dougherty MK, Esposito LW, Krimigis SM (eds) Saturn from Cassini-Huygens. Springer, Dordrecht, pp 375–412. https://doi.org/10.1007/978-1-4020-9217-6_13
Cooper NJ, Murray CD, Evans MW, et al (2008) Astrometry and dynamics of Anthe (S/2007 S 4), a new satellite of Saturn. Icarus 195(2):765–777. https://doi.org/10.1016/j.icarus.2008.01.006
Cooper NJ, Renner S, Murray CD, et al (2015) Saturn´s inner satellites: orbits, masses, and the chaotic motion of atlas from new Cassini imaging observations. Astron J 149(1):27. https://doi.org/10.1088/0004-6256/149/1/27. arXiv:1406.6492 [astro-ph.EP]
Crida A, Charnoz S (2012) Formation of regular satellites from ancient massive rings in the Solar System. Science 338(6111):1196. https://doi.org/10.1126/science.1226477. arXiv:1301.3808 [astro-ph.EP]
Crida A, Papaloizou JCB, Rein H, et al (2010) Migration of a moonlet in a ring of solid particles: theory and application to Saturn's propellers. Astron J 140(4):944–953. https://doi.org/10.1088/0004-6256/140/4/944. arXiv:1006.1573 [astro-ph.EP]
Crida A, Estrada PR, Nicholson PD, et al (2024) The age and origin of Saturn's rings. Space Sci Rev
Cuzzi JN (2024) Corrigendum to "Saturn's F ring core: calm in the midst of chaos" [Icarus 232 (April) (2014) 157–175]. Icarus 413:116011. https://doi.org/10.1016/j.icarus.2024.116011
Cuzzi JN, Estrada PR (1998) Compositional evolution of Saturn's rings due to meteoroid bombardment. Icarus 132(1):1–35. https://doi.org/10.1006/icar.1997.5863
Cuzzi J, Clark R, Filacchione G, et al (2009) Ring particle composition and size distribution. In: Dougherty MK, Esposito LW, Krimigis SM (eds) Saturn from Cassini-Huygens. Springer, Dordrecht, pp 459–509. https://doi.org/10.1007/978-1-4020-9217-6_15
Cuzzi JN, Chambers LB, Hendrix AR (2017) Rough surfaces: is the dark stuff just shadow?: who knows what evil lurks in the hearts of men? The shadow knows! Icarus 289:281–294. https://doi.org/10.1016/j.icarus.2016.10.018
Cuzzi JN, Filacchione G, Marouf EA (2018a) The rings of Saturn. In: Tiscareno MS, Murray CD (eds) Planetary ring systems. Properties, structure, and evolution. Cambridge University Press, Cambridge, pp 51–92. https://doi.org/10.1017/9781316286791.003
Cuzzi JN, French RG, Hendrix AR, et al (2018b) HST-STIS spectra and the redness of Saturn's rings. Icarus 309:363–388. https://doi.org/10.1016/j.icarus.2018.02.025
Cuzzi JN, Marouf EA, French RG, et al (2024) Saturn's F ring is intermittently shepherded by Prometheus. Sci Adv. https://doi.org/10.1126/sciadv.adl6601 [astro-ph.EP]
Davis DR, Weidenschilling SJ, Chapman CR, et al (1984) Saturn ring particles as dynamic ephemeral bodies. Science 224(4650):744–747. https://doi.org/10.1126/science.224.4650.744
Dermott SF, Murray CD (1981) The dynamics of tadpole and horseshoe orbits II. The coorbital satellites of Saturn. Icarus 48(1):12–22. https://doi.org/10.1016/0019-1035(81)90148-2
Dermott SF, Gold T, Sinclair AT (1979) The rings of Uranus: nature and origin. Astron J 84:1225–1232. https://doi.org/10.1086/112533
Dobrovolskis AR (2019) Classification of ellipsoids by shape and surface gravity. Icarus 321:891–928. https://doi.org/10.1016/j.icarus.2018.11.023
El Moutamid M, Nicholson PD, French RG, et al (2016) How Janus' orbital swap affects the edge of Saturn's A ring? Icarus 279:125–140. https://doi.org/10.1016/j.icarus.2015.10.025. arXiv:1510.00434 [astro-ph.EP]
Esposito LW, Barth CA, Colwell JE, et al (2004) The Cassini ultraviolet imaging spectrograph investigation. Space Sci Rev 115(1–4):299–361. https://doi.org/10.1007/s11214-004-1455-8
Filacchione G, Capaccioni F, Clark RN, et al (2010) Saturn's icy satellites investigated by Cassini-VIMS. II. Results at the end of nominal mission. Icarus 206(2):507–523. https://doi.org/10.1016/j.icarus.2009.11.006
Filacchione G, Capaccioni F, Ciarniello M, et al (2012) Saturn's icy satellites and rings investigated by Cassini-VIMS: III - radial compositional variability. Icarus 220(2):1064–1096. https://doi.org/10.1016/j.icarus.2012.06.040. arXiv:1203.6230 [astro-ph.EP]
Filacchione G, Capaccioni F, Clark RN, et al (2013) The radial distribution of water ice and chromophores across Saturn's system. Astrophys J 766(2):76. https://doi.org/10.1088/0004-637X/766/2/76. arXiv:1301.6611 [astro-ph.EP]
Filacchione G, Ciarniello M, Capaccioni F, et al (2014) Cassini-VIMS observations of Saturn's main rings: I. Spectral properties and temperature radial profiles variability with phase angle and elevation. Icarus 241:45–65. https://doi.org/10.1016/j.icarus.2014.06.001
Filacchione G, Ciarniello M, D'Aversa E, et al (2022) Saturn's icy satellites investigated by Cassini - VIMS. V. Spectrophotometry. Icarus 375:114803. https://doi.org/10.1016/j.icarus.2021.114803. arXiv:2111.15541 [astro-ph.EP]

French RG, McGhee CA, Dones L, et al (2003) Saturn's wayward shepherds: the peregrinations of Prometheus and Pandora. Icarus 162(1):143–170. https://doi.org/10.1016/S0019-1035(02)00050-7

French RG, McGhee-French CA, Lonergan K, et al (2017) Noncircular features in Saturn's rings IV: absolute radius scale and Saturn's pole direction. Icarus 290:14–45

Goldreich P, Rappaport N (2003) Chaotic motions of Prometheus and Pandora. Icarus 162(2):391–399. https://doi.org/10.1016/S0019-1035(02)00080-5

Hapke B (2012) Theory of reflectance and emittance spectroscopy, 2nd edn. Cambridge University Press, Cambridge. https://doi.org/10.1017/CBO9781139025683

Hedman MM, Murray CD, Cooper NJ, et al (2009) Three tenuous rings/arcs for three tiny moons. Icarus 199(2):378–386. https://doi.org/10.1016/j.icarus.2008.11.001

Hedman MM, Cooper NJ, Murray CD, et al (2010) Aegaeon (Saturn LIII), a G-ring object. Icarus 207(1):433–447. https://doi.org/10.1016/j.icarus.2009.10.024. arXiv:0911.0171 [astro-ph.EP]

Hedman MM, Nicholson PD, Showalter MR, et al (2011) The Christiansen effect in Saturn's narrow dusty rings and the spectral identification of clumps in the F ring. Icarus 215(2):695–711. https://doi.org/10.1016/j.icarus.2011.02.025. arXiv:1102.5116 [astro-ph.EP]

Hedman MM, Nicholson PD, Cuzzi JN, et al (2013) Connections between spectra and structure in Saturn's main rings based on Cassini VIMS data. Icarus 223(1):105–130. https://doi.org/10.1016/j.icarus.2012.10.014. arXiv:1210.4727 [astro-ph.EP]

Hedman MM, Helfenstein P, Chancia RO, et al (2020) Photometric analyses of Saturn's small moons: Aegaeon, Methone, and Pallene are dark; Helene and Calypso are bright. Astron J 159(4):129. https://doi.org/10.3847/1538-3881/ab659d. arXiv:1912.09192 [astro-ph.EP]

Hendrix AR, Filacchione G, Paranicas C, et al (2018) Icy Saturnian satellites: disk-integrated UV-IR characteristics and links to exogenic processes. Icarus 300:103–114. https://doi.org/10.1016/j.icarus.2017.08.037

Hendrix AR, French RG, Buratti BJ, et al (2022) The ultraviolet-visible spectra of Iapetus: unique insights into the composition of the dark material. In: AGU Fall Meeting abstracts, pp P22B–2102

Holsapple KA, Michel P (2006) Tidal disruptions: a continuum theory for solid bodies. Icarus 183(2):331–348. https://doi.org/10.1016/j.icarus.2006.03.013

Hsu HW, Postberg F, Sekine Y, et al (2015) Ongoing hydrothermal activities within Enceladus. Nature 519(7542):207–210. https://doi.org/10.1038/nature14262

Imanaka H, Khare BN, Elsila JE, et al (2004) Laboratory experiments of Titan tholin formed in cold plasma at various pressures: implications for nitrogen-containing polycyclic aromatic compounds in Titan haze. Icarus 168(2):344–366. https://doi.org/10.1016/j.icarus.2003.12.014

Imanaka H, Cruikshank DP, Khare BN, et al (2012) Optical constants of Titan tholins at mid-infrared wavelengths (2.5-25 μm) and the possible chemical nature of Titan's haze particles. Icarus 218(1):247–261. https://doi.org/10.1016/j.icarus.2011.11.018

Jacobson RA (2014) The small saturnian satellites – chaos and conundrum. In: AAS/Division of Dynamical Astronomy Meeting #45, p 304.05

Jacobson RA (2022) The orbits of the main saturnian satellites, the saturnian system gravity field, and the orientation of Saturn's pole. Astron J 164(5):199. https://doi.org/10.3847/1538-3881/ac90c9

Jacobson RA, Spitale J, Porco CC, et al (2008) Revised orbits of Saturn's small inner satellites. Astron J 135(1):261–263. https://doi.org/10.1088/0004-6256/135/1/261

Jurac S, Richardson JD (2007) Neutral cloud interaction with Saturn's main rings. Geophys Res Lett 34(8):L08102. https://doi.org/10.1029/2007GL029567

Kaasalainen M, Torppa J, Muinonen K (2001) Optimization methods for asteroid lightcurve inversion. II. The complete inverse problem. Icarus 153(1):37–51. https://doi.org/10.1006/icar.2001.6674

Kempf S, Altobelli N, Schmidt J, et al (2023) Micrometeoroid infall onto Saturn's rings constrains their age to no more than a few hundred million years. Sci Adv 9(19):eadf8537. https://doi.org/10.1126/sciadv.adf8537

Khare BN, Sagan C, Arakawa ET, et al (1984) Optical constants of organic tholins produced in a simulated Titanian atmosphere: from soft X-ray to microwave frequencies. Icarus 60(1):127–137. https://doi.org/10.1016/0019-1035(84)90142-8

Khare BN, Sagan C, Thompson WR, et al (1994) Optical properties of poly-HCN and their astronomical applications. Can J Chem 72:678–694

Krimigis SM, Mitchell DG, Hamilton DC, et al (2004) Magnetosphere Imaging Instrument (MIMI) on the Cassini Mission to Saturn/Titan. Space Sci Rev 114(1–4):233–329. https://doi.org/10.1007/s11214-004-1410-8

Lainey V, Karatekin Ö, Desmars J, et al (2012) Strong tidal dissipation in Saturn and constraints on Enceladus' thermal state from astrometry. Astrophys J 752(1):14. https://doi.org/10.1088/0004-637X/752/1/14. arXiv:1204.0895 [astro-ph.EP]

Lainey V, Noyelles B, Cooper N, et al (2019) Interior properties of the inner saturnian moons from space astrometry data. Icarus 326:48–62. https://doi.org/10.1016/j.icarus.2019.01.026. arXiv:1903.03603 [astro-ph.EP]

Lainey V, Rambaux N, Cooper N, et al (2023) Characterising the interior of five inner saturnian moons using Cassini ISS data. Astron Astrophys 670:L25. https://doi.org/10.1051/0004-6361/202244757

Leleu A, Jutzi M, Rubin M (2018) The peculiar shapes of Saturn's small inner moons as evidence of mergers of similar-sized moonlets. Nat Astron 2:555–561. https://doi.org/10.1038/s41550-018-0471-7. arXiv:1805.08682 [astro-ph.EP]

Lissauer JJ, Cuzzi JN (1982) Resonances in Saturn's rings. Astron J 87:1051–1058. https://doi.org/10.1086/113189

Lissauer JJ, Peale SJ, Cuzzi JN (1984) Ring torque on janus and the melting of Enceladus. Icarus 58(2):159–168. https://doi.org/10.1016/0019-1035(84)90034-4

Lissauer JJ, Goldreich P, Tremaine S (1985) Evolution of the Janus-Epimetheus coorbital resonance due to torques from Saturn's ring. Icarus 64(3):425–434. https://doi.org/10.1016/0019-1035(85)90066-1

McDonald GD, Thompson WR, Heinrich M, et al (1994) Chemical investigation of Titan and Triton tholins. Icarus 108(1):137–145. https://doi.org/10.1006/icar.1994.1046

Miller KE, Filacchione G, Cuzzi JN, et al (2024) The compostion of Saturn's Rings. Space Sci Rev 220:70. https://doi.org/10.1007/s11214-024-01104-y

Morrison SJ, Thomas PC, Tiscareno MS, et al (2009) Grooves on small saturnian satellites and other objects: characteristics and significance. Icarus 204(1):262–270. https://doi.org/10.1016/j.icarus.2009.06.003

Murray CD, French RS (2018) The F ring of Saturn. In: Tiscareno MS, Murray CD (eds) Planetary ring systems. Properties, structure, and evolution. Cambridge University Press, Cambridge, pp 338–362. https://doi.org/10.1017/9781316286791.013

Murray CD, Giuliatti Winter SM (1996) Periodic collisions between the moon Prometheus and Saturn's F ring. Nature 380(6570):139–141. https://doi.org/10.1038/380139a0

Murray CD, Chavez C, Beurle K, et al (2005) How Prometheus creates structure in Saturn's F ring. Nature 437(7063):1326–1329. https://doi.org/10.1038/nature04212

Nicholson PD, Hamilton DP, Matthews K, et al (1992) New observations of Saturn's coorbital satellites. Icarus 100(2):464–484. https://doi.org/10.1016/0019-1035(92)90111-J

Nicholson PD, Hedman MM, Clark RN, et al (2008) A close look at Saturn's rings with Cassini VIMS. Icarus 193(1):182–212. https://doi.org/10.1016/j.icarus.2007.08.036

Nicholson PD, French RG, Spitale JN (2018) Narrow rings, gaps, and sharp edges. In: Tiscareno MS, Murray CD (eds) Planetary ring systems. Properties, structure, and evolution. Cambridge University Press, Cambridge, pp 276–307. https://doi.org/10.1017/9781316286791.011

Pang KD, Voge CC, Rhoads JW, et al (1984) The E ring of Saturn and satellite Enceladus. J Geophys Res 89:9459–9470. https://doi.org/10.1029/JB089iB11p09459

Paranicas C, Hibbitts C, Kollmann P, et al (2018) Magnetospheric considerations for Solar System ice state. Icarus 302:560–564. https://doi.org/10.1016/j.icarus.2017.12.013

Perry M, Cravens T, Smith H, et al (2017) In-situ measurements of ions and neutrals near Saturn's f-ring. In: European Planetary Science Congress, pp EPSC2017–951

Porco CC, West RA, Squyres S, et al (2004) Cassini imaging science: instrument characteristics and anticipated scientific investigations at Saturn. Space Sci Rev 115(1–4):363–497. https://doi.org/10.1007/s11214-004-1456-7

Porco CC, Baker E, Barbara J, et al (2005) Cassini imaging science: initial results on Saturn's rings and small satellites. Science 307(5713):1226–1236. https://doi.org/10.1126/science.1108056

Porco CC, Thomas PC, Weiss JW, et al (2007) Saturn's small inner satellites: clues to their origins. Science 318(5856):1602. https://doi.org/10.1126/science.1143977

Porco CC, Weiss JW, Richardson DC, et al (2008) Simulations of the dynamical and light-scattering behavior of Saturn's rings and the derivation of ring particle and disk properties. Astron J 136(5):2172–2200. https://doi.org/10.1088/0004-6256/136/5/2172

Postberg F, Kempf S, Schmidt J, et al (2009) Sodium salts in E-ring ice grains from an ocean below the surface of Enceladus. Nature 459(7250):1098–1101. https://doi.org/10.1038/nature08046

Postberg F, Schmidt J, Hillier J, et al (2011) A salt-water reservoir as the source of a compositionally stratified plume on Enceladus. Nature 474(7353):620–622. https://doi.org/10.1038/nature10175

Postberg F, Clark RN, Hansen CJ, et al (2018) Plume and surface composition of Enceladus. In: Schenk PM, Clark RN, Howett CJA, et al (eds) Enceladus and the icy moons of Saturn. University of Arizona Press, Tucson, p 129. https://doi.org/10.2458/azu_uapress_9780816537075-ch007

Poulet F, Sicardy B (2001) Dynamical evolution of the Prometheus-Pandora system. Mon Not R Astron Soc 322(2):343–355. https://doi.org/10.1046/j.1365-8711.2001.04128.x

Poulet F, Cruikshank DP, Cuzzi JN, et al (2003) Compositions of Saturn's rings A, B, and C from high resolution near-infrared spectroscopic observations. Astron Astrophys 412:305–316. https://doi.org/10.1051/0004-6361:20031123

Quillen AC, Zaidouni F, Nakajima M, et al (2021) Accretion of ornamental equatorial ridges on Pan, Atlas and Daphnis. Icarus 357:114260. https://doi.org/10.1016/j.icarus.2020.114260. arXiv:2007.13227 [astro-ph.EP]

Rambaux N, Lainey V, Cooper N, et al (2022) Spherical harmonic decomposition and interpretation of the shapes of the small saturnian inner moons. Astron Astrophys 667:A78. https://doi.org/10.1051/0004-6361/202243355

Renner S, Cooper NJ, El Moutamid M, et al (2016) Origin of the chaotic motion of the saturnian satellite atlas. Astron J 151(5):122. https://doi.org/10.3847/0004-6256/151/5/122. arXiv:1602.01967 [astro-ph.EP]

Roberts JH, Barnouin OS, Daly MG, et al (2021) Rotational states and shapes of Ryugu and Bennu: implications for interior structure and strength. Planet Space Sci 204:105268. https://doi.org/10.1016/j.pss.2021.105268

Roche E (1851) La figure d'une masse fluide, soumise à l'attraction d'un point éloigné. Mem Acad Sci Montpellier 1:20

Roussos E, Krupp N, Armstrong TP, et al (2008) Discovery of a transient radiation belt at Saturn. Geophys Res Lett 35(22). https://doi.org/10.1029/2008GL035767

Salmon J, Charnoz S, Crida A, et al (2010) Long-term and large-scale viscous evolution of dense planetary rings. Icarus 209(2):771–785. https://doi.org/10.1016/j.icarus.2010.05.030. arXiv:1006.0633 [astro-ph.EP]

Santana T, Tajeddine R, El Moutamid M, et al (2019) Daphnis' chaotic orbit. In: EPSC-DPS joint meeting 2019, pp EPSC–DPS2019–1085

Sheppard SS, Tholen DJ, Alexandersen M, et al (2023) New Jupiter and Saturn satellites reveal new moon dynamical families. Res Notes AAS 7(5):100. https://doi.org/10.3847/2515-5172/acd766

Shkuratov Y, Starukhina L, Hoffmann H, et al (1999) A model of spectral albedo of particulate surfaces: implications for optical properties of the Moon. Icarus 137(2):235–246. https://doi.org/10.1006/icar.1998.6035

Shkuratov YG, Stankevich DG, Petrov DV, et al (2005) Interpreting photometry of regolith-like surfaces with different topographies: shadowing and multiple scattering. Icarus 173(1):3–15. https://doi.org/10.1016/j.icarus.2003.12.017

Shkuratov Y, Kaydash V, Korokhin V, et al (2011) Optical measurements of the Moon as a tool to study its surface. Planet Space Sci 59(13):1326–1371. https://doi.org/10.1016/j.pss.2011.06.011

Shkuratov Y, Kaydash V, Korokhin V, et al (2012) A critical assessment of the Hapke photometric model. J Quant Spectrosc Radiat Transf 113(18):2431–2456. https://doi.org/10.1016/j.jqsrt.2012.04.010

Showalter MR, Cuzzi JN, Marouf EA, et al (1986) Satellite "wakes" and the orbit of the Encke gap moonlet. Icarus 66(2):297–323. https://doi.org/10.1016/0019-1035(86)90160-0

Spahn F, Hoffmann H, Rein H, et al (2018) Moonlets in dense planetary rings. In: Tiscareno MS, Murray CD (eds) Planetary ring systems. Properties, structure, and evolution, pp 157–197. https://doi.org/10.1017/9781316286791.008

Spitale JN, Porco CC (2009) Time variability in the outer edge of Saturn's A-ring revealed by Cassini imaging. Astron J 138(5):1520–1528. https://doi.org/10.1088/0004-6256/138/5/1520

Spitale JN, Jacobson RA, Porco CC, et al (2006) The orbits of Saturn's small satellites derived from combined historic and Cassini imaging observations. Astron J 132(2):692–710. https://doi.org/10.1086/505206

Srama R, Ahrens TJ, Altobelli N, et al (2004) The Cassini cosmic dust analyzer. Space Sci Rev 114(1–4):465–518. https://doi.org/10.1007/s11214-004-1435-z

Stephan K, Ciarniello M, Poch O, et al (2021) VIS-NIR/SWIR spectral properties of H_2O ice depending on particle size and surface temperature. Minerals 11(12):1328. https://doi.org/10.3390/min11121328

Tajeddine R, Nicholson PD, Longaretti PY, et al (2017a) What confines the rings of Saturn? Astrophys J Suppl Ser 232(2):28. https://doi.org/10.3847/1538-4365/aa8c09. arXiv:1710.08462 [astro-ph.EP]

Tajeddine R, Nicholson PD, Tiscareno MS, et al (2017b) Dynamical phenomena at the inner edge of the Keeler gap. Icarus 289:80–93. https://doi.org/10.1016/j.icarus.2017.02.002 arXiv:1709.01974 [astro-ph.EP]

Thomas PC (2010) Sizes, shapes, and derived properties of the saturnian satellites after the Cassini nominal mission. Icarus 208(1):395–401. https://doi.org/10.1016/j.icarus.2010.01.025

Thomas P, Helfenstein P (2020) The small inner satellites of Saturn: shapes, structures and some implications. Icarus 344:113355. https://doi.org/10.1016/j.icarus.2019.06.016

Thomas PC, Armstrong JW, Asmar SW, et al (2007) Hyperion's sponge-like appearance. Nature 448(7149):50–56. https://doi.org/10.1038/nature05779

Thomas PC, Burns JA, Hedman M, et al (2013) The inner small satellites of Saturn: a variety of worlds. Icarus 226(1):999–1019. https://doi.org/10.1016/j.icarus.2013.07.022

Thomas P, Joseph J, Ansty T (2018a) Saturn Small Moon Shape Models Bundle V1.0. NASA Planetary Data System, p. 3. https://doi.org/10.26033/ewy3-jy61

Thomas PC, Tiscareno MS, Helfenstein P (2018b) The inner small satellites of Saturn, and hyperion. In: Schenk PM, Clark RN, Howett CJA, et al (eds) Enceladus and the icy moons of Saturn. University of Arizona Press, Tucson, p 387. https://doi.org/10.2458/azu_uapress_9780816537075-ch019

Tiscareno MS, Burns JA, Nicholson PD, et al (2007) Cassini imaging of Saturn's rings. II. A wavelet technique for analysis of density waves and other radial structure in the rings. Icarus 189(1):14–34. https://doi.org/10.1016/j.icarus.2006.12.025. arXiv:astro-ph/0610242 [astro-ph]

Tiscareno MS, Burns JA, Hedman MM, et al (2008) The population of propellers in Saturn's A ring. Astron J 135(3):1083–1091. https://doi.org/10.1088/0004-6256/135/3/1083. arXiv:0710.4547 [astro-ph]

Tiscareno MS, Thomas PC, Burns JA (2009) The rotation of Janus and Epimetheus. Icarus 204(1):254–261. https://doi.org/10.1016/j.icarus.2009.06.023. arXiv:0904.3515 [astro-ph.EP]

Tseng WL, Ip WH (2011) An assessment and test of Enceladus as an important source of Saturn's ring atmosphere and ionosphere. Icarus 212(1):294–299. https://doi.org/10.1016/j.icarus.2010.12.003

Tseng WL, Ip WH, Johnson RE, et al (2010) The structure and time variability of the ring atmosphere and ionosphere. Icarus 206(2):382–389. https://doi.org/10.1016/j.icarus.2009.05.019

Tseng WL, Johnson RE, Thomsen MF, et al (2011) Neutral H_2 and H_2^+ ions in the saturnian magnetosphere. J Geophys Res Space Phys 116(A3):A03209. https://doi.org/10.1029/2010JA016145

Vahidinia S, Cuzzi JN, Hedman M, et al (2011) Saturn's F ring grains: aggregates made of crystalline water ice. Icarus 215(2):682–694. https://doi.org/10.1016/j.icarus.2011.04.011

Verbiscer A, French R, Showalter M, et al (2007) Enceladus: cosmic graffiti artist caught in the act. Science 315(5813):815. https://doi.org/10.1126/science.1134681

Veverka J (1971) The meaning of Russell's law. Icarus 14(2):284–285. https://doi.org/10.1016/0019-1035(71)90066-2

Waite JH, Lewis WS, Kasprzak WT, et al (2004) The Cassini Ion and Neutral Mass Spectrometer (INMS) investigation. Space Sci Rev 114(1–4):113–231. https://doi.org/10.1007/s11214-004-1408-2

Walker JD, Chocron S, Waite JH, et al (2015) The vaporization threshold: hypervelocity impacts of ice grains into a titanium Cassini spacecraft instrument chamber. Proc Eng 103:628–635

Weidenschilling SJ, Chapman CR, Davis DR, et al (1984) Ring particles: collisional interactions and physical nature. In: Greenberg R, Brahic A (eds) Planetary rings. University of Arizona Press, Tucson, pp 367–415. https://doi.org/10.1017/9781316286791.010

Weiss JW, Porco CC, Tiscareno MS (2009) Ring edge waves and the masses of nearby satellites. Astron J 138(1):272–286. https://doi.org/10.1088/0004-6256/138/1/272

West R, Lavvas P, Anderson C, et al (2014) Titan's haze. In: Titan, p 285. https://doi.org/10.1017/CBO9780511667398.011

Yoder CF, Synnott SP, Salo H (1989) Orbits and masses of Saturn's co-orbiting satellites, Janus and Epimetheus. Astron J 98:1875. https://doi.org/10.1086/115265

Authors and Affiliations

Mauro Ciarniello[1] · Gianrico Filacchione[1] · Philip D. Nicholson[2] · Matthew M. Hedman[3] · Sebastien Charnoz[4] · Jeffrey N. Cuzzi[5] · Maryame El Moutamid[6] · Amanda R. Hendrix[7] · Nicolas Rambaux[8] · Kelly E. Miller[9] · Olivier Mousis[10,11] · Kevin Baillié[8] · Paul R. Estrada[5] · J. Hunter Waite[9]

✉ M. Ciarniello
mauro.ciarniello@inaf.it

G. Filacchione
gianrico.filacchione@inaf.it

P.D. Nicholson
pdn2@cornell.edu

M.M. Hedman
mhedman@uidaho.edu

S. Charnoz
charnoz@ipgp.fr

J.N. Cuzzi
jeffrey.cuzzi@nasa.gov

M. El Moutamid
maryame.elmoutamid@swri.org

A.R. Hendrix
ahendrix@psi.edu

N. Rambaux
nicolas.rambaux@obspm.fr

K.E. Miller
kelly.miller@swri.org

O. Mousis
olivier.mousis@lam.fr

K. Baillié
kevin.baillie@obspm.fr

P.R. Estrada
paul.r.estrada@nasa.gov

J.H. Waite
hunterwaite@gmail.com

[1] Istituto di Astrofisica e Planetologia Spaziali, Istituto Nazionale di Astrofisica, via del Fosso del Cavaliere, 100, Rome, 00133, Italy

[2] Astronomy Department, Cornell University, Ithaca, 14853, NY, USA

[3] Department of Physics and Astronomy, University of Idaho, Moscow, 83844, ID, USA

[4] Université Paris Cité, Institut de Physique du Globe de Paris, CNRS, Paris F-75005, France

[5] Space Sciences Division, NASA Ames Research Center, MS 245-3, Moffett Field, 94035, CA, USA

[6] Southwest Research Institute, Boulder, CO, USA

[7] Planetary Science Institute, Tucson, AZ, USA

[8] IMCCE, Observatoire de Paris, PSL University, CNRS, Sorbonne Université, Paris, 77 Avenue Denfert-Rochereau, Paris, 75014, France

[9] Southwest Research Institute, San Antonio, TX, USA

[10] Aix-Marseille Université, CNRS, CNES, Institut Origines, LAM, Marseille, France

[11] Institut universitaire de France (IUF), Paris, France

Space Science Reviews (2025) 221:66
https://doi.org/10.1007/s11214-025-01189-z

The Age and Origin of Saturn's Rings

Aurélien Crida[1] · Paul R. Estrada[2] · Philip D. Nicholson[3] · Carl D. Murray[4]

Received: 26 September 2024 / Accepted: 23 June 2025 / Published online: 7 July 2025

Abstract
The origin of Saturn's rings is a long standing mystery in planetary science, for which the age of this system is a critical constraint. After having clarified what the age of the rings may mean, this article reviews several aspects of this question. We discuss the exposure age, which was recently found to be on the order a few 100 Myrs based on the rings' composition and their bombardment rate by micrometeoroids. The derivation of this age is explained, and the possible caveats are discussed, including the idea of a putative phenomenon of cleaning of the rings. We address the age of structures such as the Cassini division, plateaux, ramps, and how they are constrained by interactions with satellites and the effects of bombardment. We then address the dynamical evolution of the rings, due to viscosity, micrometeoroid bombardment, and satellite torques. Initially massive rings are found to viscously spread and lose mass quickly and then converge to the present mass in billions of years. However, the dynamical effects of micrometeoroid bombardment can take the reins over viscosity and may wipe the rings out efficiently. Naturally, consideration of ring origin also motivates us to review the ring formation models existing in the literature, and discuss their implications on the composition and age of the rings, in particular in the frame of large tidal dissipation inside Saturn. Finally, putting together and synthesizing these results, we conclude that a fully consistent picture for the origin and age of Saturn's rings has yet to be established and more work remains to be done to resolve this important question.

Keywords Ring origin · Ring dynamics · Ring age

✉ C.D. Murray
C.D.Murray@qmul.ac.uk

A. Crida
crida@oca.eu

P.R. Estrada
paul.r.estrada@nasa.gov

P.D. Nicholson
nicholso@astro.cornell.edu

[1] Université Côte d'Azur, Observatoire de la Côte d'Azur, CNRS, Laboratoire Lagrange (UMR7293), Boulevard de l'Observatoire, Nice, 06304, France

[2] Space Sciences Div., NASA Ames Research Center, MS 245-3, Moffett Field, 94035, CA, USA

[3] Department of Astronomy, Cornell University, Ithaca, 14853, NY, USA

[4] School of Physical and Chemical Sciences, Queen Mary University of London, London, E1 4NS, UK

'...when we contemplate the Rings from a purely scientific point of view, they become the most remarkable bodies in the heavens... We must either explain its motion on the principles of mechanics, or admit that, in the realms of Saturn, there can be motion regulated by laws which we are unable to explain.'
– James Clerk Maxwell, 1859

1 Introduction and Definitions

Since their discovery in the year 1610 when Galileo first began pointing his telescope to the sky, Saturn's rings have been a source of intrigue and wonder to scientist and layman alike. Galileo himself, due to a lack of magnification of his instrument, mistook the rings to be adjoined moons, "handles" or "lobes" attached to Saturn. It was only later in 1655 that Christiaan Huygens first described them as a disk encircling, but nowhere touching, their parent planet. Jean-Dominique Cassini then found in 1675 the division named after him. Already in 1755, Immanuel Kant described how the inner rings should rotate faster around Saturn than the outer part, and how such a shear should lead to the spreading of the rings, due to the interactions among the particles (what we now call "viscosity"). He assumed that the main rings were divided in individual ringlets to avoid this global evolution. Pierre-Simon de Laplace demonstrated theoretically in 1787 that the rings can not be a solid planar object but must be divided in infinitesimal concentric rings, each orbiting Saturn at its own Keplerian velocity, which was later confirmed by James Keeler in 1895 with spectroscopic observations. James Clerk Maxwell showed in 1859 that uniform solid ringlets are not stable either, so that the rings should be rather 'composed of an indefinite number of unconnected particles, revolving round the planet with different velocities according to their respective distances' (as Kant postulated) and confirmed that the rings dynamically spread inwards and outwards (though he does not use the word 'viscosity'), all without ever having properly observed them. Setting the tone for future times, in the final conclusions of his "On the Stability of the Motion of Saturn's Rings", Maxwell (1859) submits that further investigations should be conducted to determine whether the rings are a permanent fixture, or a transitionary element of our Solar system subject to 'terrestrial corruption and generation, and the old order giving place to new before our own eyes.'

Our interest in the rings since then has only grown, as has our ability to investigate their nature, through ground-based observations, but especially through in-situ robotic missions that have revealed the rings in all their glory. Though all of the giant planets of our Solar System have rings, Saturn's rings are unique in that they are rather massive and composed of $\gtrsim 95\%$ water ice by mass (Doyle et al. 1989; Grossman 1990; Cuzzi and Estrada 1998; Zhang et al. 2017a,b). This is in stark contrast to the other ring systems which are mostly very low mass and composed of ringlets or diaphanous rings that are spectrally dark, where water ice is absent or not the dominant component (Nicholson et al. 1984; Porco et al. 1987, 1995; De Pater et al. 2018). An immediate query arises as to why this is the case. Did those other worlds at some point in the past have massive rings, or have they always been as they are now? If it is the former, then why are Saturn's rings still massive? Are rings that are so massive and so icy offering clues to their evolutionary history? This continues to be a subject of vigorous debate.

The origin and age of the rings is the main question we address in this article: are Saturn's rings ancient, forming in the same epoch as Saturn itself, or could they be much younger, forming recently in the Solar System history? The Cassini spacecraft in its more than a dozen year tour of the Saturn system provided several key observations of ring properties that can help to address this very question. Also, the discovery of large tidal dissipation inside Saturn

(Lainey et al. 2012, 2017, 2020) sheds new light on the dynamics of the whole system, and constrains any global scenario that links the origin of the rings and moons. Our goal here is to *attempt* to construct a potentially consistent picture for whether the rings are young or old. The answer(s) will have bearing on their likely origin.

The debate on ring age centers around understanding the various timescales at play, all of which are important, but may be wildly different from each other. To clarify the debate, it is here useful to first define properly the various types of age for the rings, and what they imply:

Exposure Age For rings, the exposure age refers to the time the rings' "surface" has been bombarded by the interplanetary flux of micrometeoroids. In general though, because exposure to any outside perturbation can range from short periods to the age of a surface, the exposure age is not necessarily the same as the physical age of a body. For example, it is understood that an asteroid's surface is subject to space weathering, which tends to redden a newly exposed surface. From the spectral slope of a feature, one can then estimate how long it has been uncovered; terrains of different exposure ages may be found on the same body. Moreover, active processes of erosion and resurfacing can erase any trace of a surface as old as the body itself, so that in such extreme cases the exposure age does not constrain at all the age of the body (e.g. the Earth).

Structure Age The structure age refers to the age of individual features. Saturn's rings are rich with a variety of structures including the sharp, resonantly-controlled outer edges of the A and B rings, the C ring plateaux and several narrow gaps and ringlets in the Cassini division, and the prominent Cassini division itself. One can also mention undulations in the middle C ring, and the inner edges of the A and B rings and the linear ramp features that connect them to the Cassini division and C ring, respectively. Non-ring examples are chains of mountains on the Earth, or channels carved by ancient rivers on Mars. These generally can only provide a lower limit on the age of a system, but not an upper limit since such structures could have arisen recently (or at any time) in the history of the system (e.g., mountains are created and eroded throughout the geological history of our planet). Thus, structural age is also generally not the same as the physical age.

Dynamical Age The dynamical age of the rings refers to the time they have evolved through physical processes that mostly consist of orbital energy and angular momentum exchange among the ring particles. This leads to modifications of the surface density profile and the total mass of the rings as material falls into Saturn or beyond the Roche radius.[1] Modeling these processes, one could relate the present mass and profile to the past history and, ideally, constrain the age and the initial mass of the rings by integrating their evolution back in time. As we will see below, some degeneracy prevents us from reconstructing the rings' initial state, but dynamics still constrains the evolution of the rings and allows us to estimate something more akin to their true, physical age based on their present mass.

Erosion Age Beside these timescales, there is another key observation which may well influence our determination of the rings' age. Cassini provided several observations during

[1] First examined by Roche (1847), it is a compositionally dependent radial location away from the planet where the distortions from the tidal force are too large for a liquid body to find an equilibrium figure. In the context of rings, planetary tides prevent the coalescence of material into larger bodies or moons (see also Weidenschilling et al. 1984). For solid water ice, the Roche limit lies near the outer edge of the A ring.

its Grand Finale orbits that indicate that the rings are losing mass to the planet (Mitchell et al. 2018; Hsu et al. 2018; Waite et al. 2018; Perry et al. 2018). If the measured rates are characteristic over the rings' lifetime, then it suggests the remarkable outcome that the rings will mostly disappear in a few hundreds of Myr (see also O'Donoghue et al. 2019). These observations introduce the idea that rings may be ephemeral, and that the other giant planets may have already lost theirs.

If the timescale of a major change (e.g. disparition or birth) is of the order of a hundred million years, seeing the rings now in their present state is very fortunate since this time span is only 2% of the age of the Solar System. Note that, using the analogy of Earth's mountains, the presence of a young massive ring currently would not preclude the possibility that Saturn previously had other ring incarnations, or that it had an even more massive ring system early in Solar System history. Indeed, the latter provides an attractive scenario for the formation of the mid-sized, icy moons (Charnoz et al. 2011; Crida and Charnoz 2012; Salmon and Canup 2017, see also Blanc et al. 2025, this collection), not just for Saturn but for Uranus and Neptune as well (Crida and Charnoz 2012, 2014). Together, these smaller satellites and rings represent an intimate, dynamically very rich system deep in the potential well of their parent planets, and as such are more susceptible to stochastic processes from outside (e.g., collisional disruption due to heliocentric interlopers Goldreich et al. 1989; Zahnle et al. 2003; Charnoz et al. 2009b) or from within the planetary system (e.g., dynamical instabilities Wisdom et al. 2022; Teodoro et al. 2023) over their history.

For the remainder of this article, we will address in more detail the question of ring age and present arguments for and against the proposition that the currently-observed rings are young. What we are looking for here is the moment where a huge mass of solids (most probably icy) was deposited inside Saturn's Roche radius, before evolving into the main ring system we observe today. This time would define the rings' *absolute age*. In Sect. 2 we estimate the exposure age for a static ring due to micrometeoroid bombardment, discuss the assumptions that go into this estimate, but also discuss the possibility that bombardment is cleaning the rings rather than darkening them. In Sect. 3 we discuss some small-scale structural features in the rings and the timescales associated with their formation. In Sect. 4 we discuss the long term dynamical evolution of the rings under various effects and compare the implied ages with those associated with exposure and structure. In Sect. 5 we offer a summary of the various origin scenarios for the rings in the literature, and which of these are more likely to apply to the current rings. Finally in Sect. 6 we present a summary and discussion, trying to provide a global picture of the question of Saturn's rings origin and age.

2 Exposure Age

The surface area of Saturn's A and B rings is 2.8×10^{10} km^2 (4.1×10^{10} adding the C-ring and the Cassini Division), which for comparison is $5.8 - 8.3 \times 10^4$ times the surface area of Mimas. They are less massive than Mimas, though. The mass measured by Cassini is $1.54 \pm 0.49 \times 10^{19}$ kg or $0.41 \pm 0.13\ M_{\mathrm{Mimas}}$ (Iess et al. 2019), which is consistent with earlier estimates based on wavelet analysis of several density waves in the massive B ring (Hedman and Nicholson 2016). Owing to their huge surface-area-to-mass ratio (average surface density $\sim 400\ \mathrm{kg\,m^{-2}}$), the rings are highly susceptible to micrometeoroid bombardment (hereafter MB) of exogenic origin (that is: coming from outside the Saturnian system).

Impacts on the rings have long been thought to lead to two effects. The first is that the direct deposition of non-icy material leads to darkening, or "pollution" of the mainly icy rings over time (Doyle et al. 1989; Cuzzi and Estrada 1998; Estrada et al. 2015) The second is that both MB and the ballistic transport (hereafter, BT) of their impact ejecta can lead to structural (and compositional) evolution of the rings through mass and angular momentum transfer (Ip 1984; Cuzzi and Durisen 1990; Durisen et al. 1992; Estrada et al. 2018; Durisen and Estrada 2023; Estrada and Durisen 2023). We will address the effects of the latter in Sects. 3 and 4. In this section, we consider the former, as this is what is used to estimate a "darkening derived" exposure time.

A simple, commonly used formula for the impact mass flux on the rings at a distance r from Saturn is given by:

$$\dot{\sigma}_{\mathrm{im}} \simeq 2F_{\infty}\mathcal{A}(\tau)F_{\mathrm{g}}\left(\frac{r}{r_{\mathrm{B}}}\right)^{-0.8} \tag{1}$$

which is an approximate form fit from the numerical calculations of Cuzzi and Durisen (1990). A new discussion on $\dot{\sigma}_{\mathrm{im}}$ will be provided by Ricerchi & Crida (in prep). Here, F_{∞} is the one-sided, interplanetary flux entering the Saturn system, and F_{g} is the gravitational focusing factor normalized at $r_{\mathrm{B}} = 1.8\ \mathrm{R_S}$ (where $R_{\mathrm{S}} = 60{,}268$ km is Saturn's equatorial radius). The impact flux also depends on the probability of absorption of the exogenic material $\mathcal{A}$ which is a function of the local optical depth τ (Cuzzi and Durisen 1990). In the B ring, at r_{B}, $\tau \gg 1$, and $\mathcal{A} \approx 1$.

The exposure age t_{exp} for a *static* ring can be estimated under the assumption that it began as pure ice and has been steadily polluted to the current state over time. Locally, $t_{\mathrm{exp}} = \Sigma_{\mathrm{pol}}/\eta\dot{\sigma}_{\mathrm{im}}$, where Σ_{pol} is the surface mass density of pollutant (that is non-icy material, supposedly exogenic), and $\eta \leqslant 1$ allows for the possibility that a fraction of the impactor's material does not retain its polluting (darkening) properties or is not fully deposited in the ring particle and may be lost (see Sect. 2.2). Using Eq. (1) with $\mathcal{A} = 1$, the exposure age of the B ring can then be expressed as:

$$t_{\mathrm{exp}} \simeq \frac{\rho_{\mathrm{pol}}}{\rho}\upsilon\frac{\Sigma_{\mathrm{B}}}{2\eta F_{\mathrm{g}}F_{\infty}}, \tag{2}$$

where we have converted from mass to volume fraction $\upsilon \simeq (\rho/\rho_{\mathrm{pol}})(\Sigma_{\mathrm{pol}}/\Sigma_{\mathrm{B}})$, $\rho = \rho_{\mathrm{pol}}\upsilon + \rho_{\mathrm{ice}}(1-\upsilon) \approx \rho_{\mathrm{ice}}$ with $\rho_{\mathrm{ice}} = 900\ \mathrm{kg\,m^{-3}}$, and $\rho_{\mathrm{pol}} = 2800\ \mathrm{kg\,m^{-3}}$ is the pollutant material density which corresponds to a mixture of 30% carbonaceous and 70% silicate (Wooden et al. 2017; Kempf et al. 2023). For the B ring, $\upsilon \sim 0.001 - 0.005$ (e.g., Zhang et al. 2017a, who used RADAR radiometry data and hence scanned the interior of the ring particles and not only the surface). Equation (2) shows that t_{exp} depends on many parameters, which previously were only estimates. The ring mass and the micrometeoroid flux in particular, along with υ, remained uncertain post Voyager but have since been constrained by Cassini observations. We discuss the flux measurement in the next section, and the uncertainty introduced by η in Sect. 2.2 before giving an estimate of the exposure time in Sect. 2.3. Mass outflow from the rings (not taken into account in the above equations) is discussed in Sect. 2.4.

2.1 Micrometeoroid Flux F_{∞} and Gravitational Focusing F_g

The rate $\dot{\sigma}_{\mathrm{im}}$ at which extrinsic micrometeoroids bombard the rings is a key input into the calculation of the rings' exposure age. Several estimates for the flux F_{∞} entering Saturn's

Hill sphere (referred to as "at infinity") had previously been made (e.g., Morfill et al. 1983; Grün et al. 1985; Cuzzi and Durisen 1990). A careful review of these works was done by Cuzzi and Estrada (1998) who re-analyzed Pioneer and Ulysses data to provide a refined estimate and concluded that the flux value of 4.5×10^{-16} kg m^{-2} s^{-1} by Cuzzi and Durisen (1990) was good to within a factor of two or three. The source population of the micrometeoroids was assumed to be Oort cloud projectiles which have very high eccentricities (≥ 0.99) and random inclinations between 0° and 180° and are therefore characterized by very high injection speeds v_∞ into the Saturn system. Such micrometeoroids are only modestly gravitationally focused by the planet, yielding an enhancement factor $F_g \approx 3$ (Cuzzi and Estrada 1998). However, a true estimate of the flux and its dynamical population would have to wait for Cassini.

During its tour of over a dozen years, Cassini's onboard Cosmic Dust Analyzer (CDA) experiment registered millions of impacts with its charge-sensitive entrance grid system (Auer et al. 2002; Srama et al. 2004). Of these, only 163 particles (outside the dust-rich E ring region) were identified to be candidates for an exogenic origin (Kempf et al. 2023). These particles ranged in size from a micron (the detection threshold of the CDA entrance grids) to up to a cutoff of $\sim 60\ \mu$m (which was shown to be statistically significant). Trajectories were derived for all particles that placed them into two categories: *certainly exogenic* in which all possible solutions were unbound to Saturn, and *probably exogenic* in which there was at least one unbound solution. These latter particles are considered likely exogenic due to their high eccentricities and very low injection speeds in the kronian system because small errors in their trajectories can nudge them from bound to unbound. Kempf et al. (2023) discussed the possibility that these particles could also be a result of collisional grinding of Saturn's irregular satellites, which might produce a comparable flux (Bottke et al. 2010), but argued against it (Tamayo et al. 2011). In any case, owing to their very high eccentricities, they would still impact the rings and contribute to $\dot{\sigma}_{\rm im}$.

Kempf et al. (2023) conducted a rigorous analysis of these detections and determined $F_\infty = 6.9 \times 10^{-17}$ kg m^{-2} s^{-1} taking into account only the certainly exogenic particles, and $F_\infty = 2.7 \times 10^{-16}$ kg m^{-2} s^{-1} when all detections are included. The latter value for F_∞ is less than a factor of two lower than the previously accepted number. However, the dynamical properties of these particles were found not to match that of the Oort cloud. The heliocentric orbital elements and injection speeds into Saturn's Hill sphere were derived for all exogenic solutions and were found to be consistent with grains with low to moderate heliocentric inclinations generated by mostly Edgeworth Kuiper Belt (or Jupiter Family Comet) sources (consistent with modeling results of Poppe and Horányi 2012; Poppe 2016; Poppe et al. 2019). A relatively small modal injection speed was found of $v_\infty = 4.3$ km s^{-1} which implies a gravitational enhancement factor of $F_g \approx 30$ at $r_{\rm B}$ (Morfill et al. 1983; Cuzzi and Estrada 1998; Kempf et al. 2023), an order of magnitude larger than for the previously assumed cometary population. One should however note that v_∞ has a large dispersion, so that a unique value for F_g is probably a very crude approximation. Besides, $v_\infty = 4.3$ km s^{-1} does not correspond exactly to $F_g = 30$ so that this value should be taken with care (see Ricerchi & Crida, in prep).

The total micrometeoroid impact rate onto the rings is $\dot{M}_{\rm im} = \int_{R_{\rm min}}^{R_{\rm max}} 2\pi r\, \dot{\sigma}_{\rm im}(r)\, dr$. Using Eq. (1) and the values above to estimate $\dot{\sigma}_{\rm im}(r)$, integrating over the optically thick A and B rings where we assume $\mathcal{A} = 1$, one finds a total rate of incoming material onto the rings $\dot{M}_{\rm im} = 115 - 450$ kg s^{-1}, where the lower bound does not include the *probably exogenic* grains.

2.2 Pollution Efficiency η

A first factor reducing the pollution is of course the ice content of the impactors. Previous models of ring evolution assumed Oort cloud projectiles that were a mixture of carbonaceous material and water ice. On the contrary, particles from the Edgeworth Kuiper Belt are assumed to be composed of a mixture of silicates and carbonaceous materials by the time they arrive at Saturn by photosputtering and sublimation of volatiles (Moro-Martín and Malhotra 2002; Grigorieva et al. 2007), so that there is a priori no need to reduce η for impactor composition reasons.

The η parameter then represents the efficiency with which the extrinsic bombarding material retains its absorptive properties upon impact, either as solid inclusions or potentially as recondensed vapor (e.g., silicates), while a fraction $1-\eta$ of the impacting mass is assumed to be vaporized and recondensed as a non absorbing material (e.g., macromolecular carbon recondensing into CO or CO_2), or lost. The value of η plays a key role in determining the rate at which the rings are polluted, and thus their exposure age.

Several previous papers modeling ring pollution by direct deposition have discussed this parameter. Doyle et al. (1989) assumed a value of $\eta \sim 0.1$ based on the persistence of a certain number of carbon-carbon bonds in particles impacting at speeds of 70 km s^{-1} observed by the Vega spacecraft (Kissel and Krueger 1987). Cuzzi and Estrada (1998) found that the *shape* of the radial profile of ring color is directly tied to the local abundance of pollutant, and was thus related to the value of η. They found that the best fitting values ranged from $\eta \sim 0.05-0.3$. Using lower values of η did not allow for simultaneously explaining the relatively sharp darkening contrast between the low-τ C ring and high-τ inner B ring regions, and the much more gradual color transition that arises from the advection of icier (and spectrally more red) material from the B ring to the C ring due to BT. When allowed to act long enough, BT homogenizes composition regionally, and also globally (independent of η; Cuzzi and Estrada 1998), but it can also eventually erase the contrast in darkening between the two ring regions, because the more polluted C ring becomes the dominant source of darkening of inner B ring material (Estrada et al. 2015, 2018).

Saturn's rings' particles are quite porous ($\gtrsim 70\%$ Zhang et al. 2017a,b, see also French et al. 2007; Morishima et al. 2016), and likely regolith-covered particles to a considerable depth (Salo et al. 2004; Elliott and Esposito 2011; Mondino-Llermanos and Salo 2023). Avdellidou et al. (2017) have studied the fate of mm-sized silicate impactors onto porous water ice targets with porosities $\lesssim 40\%$ using laboratory experiments. Velocities up to a few km s^{-1} were reached, corresponding to a projectile energy density $Q_{\mathrm{p}}=\frac{1}{2}v^2$ up to 5×10^6 J kg^{-1}. They found that for their highest impact speed, the largest fragment was a hundred to a thousand times less massive than the impactor. Impact speeds on the rings are more than ten times faster, and lead to energy densities as high as $Q_{\mathrm{p}} \sim 10^9$ J kg^{-1}, which is 2 to 3 orders of magnitude more than in these experiments. Such energies likely lead to melting and vaporization or even ionization of the whole micrometeorite, possibly suggesting $\eta \ll 1$. Since this regime is not accessible to lab experiments, numerical simulations are necessary.

Hyodo et al. (2025) have modeled impacts of micrometeoroids on ice blocks at 30 km s^{-1} using SPH simulations and they find that the whole impactor is vaporized, together with part of the target. In their model, more than 97% of the vapor then escapes from the ring system, leading to η values as low as $1-3\%$. This number is actually an upper bound, assuming the remaining material keeps its absorptive properties after being vaporized and recondensed, which is not certain. Such a low value for η would suggest then that the rings are resistant to pollution. It would change our view on pollution, but also be a problem to explain the B-C ring transition and homogenization as explained above.

However, the applicability of this result is not straightforward for three reasons. First, the model of Hyodo et al. assumes that the impactor is smaller than any target surface element, and ignores porosity and material strength. Under these assumptions, collisional behavior is size invariant (Melosh 1989). That is, the impact of a micrometeoroid onto a ring particle is no different than a large body onto a planetary surface. These conditions are unlike the high porosity, strength and granularity of a ring particle regolith of significant depth that is composed of grains similar in size to the impactor. Second, these authors only model the highest speed (leading side) impacts, when in fact impact velocities on a ring particle can range from $\sim 5-50$ km s^{-1} (Cuzzi and Durisen 1990). The lower end velocities correspond to a significant fraction of trailing side impacts of which will not lead to vaporization. Third, depending on the location of the target ring particle in the ring and of the location of the impact on the surface of this particle, the vapor cloud may expand towards other ring particles. Some fraction of the escaping material may then condense and be captured on the surface of the neighboring ring particles, increasing the effective η coefficient.

2.3 Estimate of $t_{\rm exp}$

We now can put these pieces together to provide an estimate for the exposure age of the rings from Equation (2) using Cassini observations of the flux, the bulk fraction of non-icy material in the rings, and the ring mass. Iess et al. (2019) report on the total mass of the rings, but also the mass contained in the individual ring segments. For the B ring, this is 0.23 Mimas masses. We determine the average surface mass density $\Sigma_{\rm B} \sim 520$ kg m^{-2} by spreading this mass over the radial extent of the B ring from 92,000 km to 117,500 km. We also take a median pollution value of $\upsilon = 0.003$ as determined by (Zhang et al. 2017a). The exposure age for a constant flux then is

$$t_{\rm exp} \approx 100\,{\rm Myr} \left(\frac{\Sigma_{\rm B}}{520\,{\rm kg\,m^{-2}}}\right)\left(\frac{2.7\times10^{-16}\,{\rm kg\,m^{-2}\,s^{-1}}}{F_\infty}\right)\times \left(\frac{\upsilon}{0.003}\right)\left(\frac{2800\,{\rm kg\,m^{-3}}}{\rho_{\rm pol}}\right)\left(\frac{30}{F_{\rm g}}\right)\left(\frac{0.1}{\eta}\right). \tag{3}$$

If one did not include the *probably exogenic* detected grains (Sect. 2.1), then this timescale would increase to ~ 400 Myr (Kempf et al. 2023). Durisen and Estrada (2023) have done a similar calculation for the C ring where $\upsilon \sim 0.02$ is a typical volume fraction of non-icy pollutant, and mean surface mass densities are much lower, getting a similar value for $t_{\rm exp}$. Note that using $\eta = 0.01$, as suggested by Hyodo et al. (2025) under some conditions, multiplies all these ages by a factor 10, reaching 4 Gyrs.

The actual age of the rings could be smaller than the exposure age if they don't start from a pure water ice composition.[2] Other processes can make the rings older than the exposure age above. Because the rings are losing mass, they were more massive even a few 100 Myr ago which modestly increases these exposure ages (Crida et al. 2019; Estrada and Durisen 2023). Other factors that would increase these ages would be if impactors had significant porosity, or if the micrometeoroid flux has been intermittent over the lifetime of the rings and is exceptionally high today (Crida et al. 2019). However, significant porosity would likely only add another factor of two, while observed structures attributed to MB and BT

[2]For instance, Cuzzi and Estrada (1998) had to assume the presence of a small fraction of intrinsic absorber, like Titan tholen to account for the rings' spectral redness. See Tables 3 and 4 in Miller et al. (2024), this collection.

(discussed in Sect. 3) require a persistent flux to sculpt and maintain. On the other hand, a higher flux in the past, or a larger η would shorten these exposure times.

2.4 Ring Mass Loss and the Possibility of Ring Cleaning

Section 7 of Miller et al. (2024), this collection, presents results on the inflow of small grains and molecules from the rings into Saturn's upper atmosphere. We stress that this mass loss process is different from the viscous spreading of the rings (described in Sect. 4.1), where the ring particles slowly change orbit while staying chemically unaltered. As has been considered for some time (see Sect. 2.2), this inflow of material to the planet is probably due to the meteoritic bombardment itself, with the lost material being a mixture of impactor debris and the high-speed component of the impact ejecta yield (Hsu et al. 2018). [3] We refer the readers to Miller et al. (2024) for more details on the analysis of this inflow (from Saturn's point of view) / outflow (from the rings point of view), and we summarize below the results and discuss their implications for the exposure age of the rings.

The inflow onto Saturn can be split into an equatorial component (close to the ring plane) and a mid-latitude component – also often called "ring rain". The latter source of material emanates from all across the rings while the equatorial inflow, which has been mainly characterized by INMS (Ion and Neutral Mass Spectrometer), originates from the D ring and innermost C ring (Waite et al. 2018). All the analyses of the INMS data are consistent with a Mass Deposition Rate (MDR) of the order of $10^4\ \mathrm{kg\,s^{-1}}$ which is a factor $20 - 100$ times larger than the best estimates of the MB rate $\dot{M}_{\mathrm{im}}$, but is consistent with mass inflow rates that in principle can be produced through ejecta.[4] The composition of the equatorial inflow is very much water depleted compared to the rings, with only $\sim 20\%$ water by mass. Moses et al. (2023) find that the composition of Saturn's upper atmosphere is not consistent with a constant flow of non-icy material of this magnitude. Hence, these authors suggest either that this large flow is a transient and recent phenomenon, possibly linked with recent changes in the D ring, or that a large fraction of the material is inflowing as unablated nanograins $\lesssim 50$ nm in size.

The mid-latitude flow from the rings onto Saturn, on the other hand, has no apparent reason to be transient if MB is a continuous process. But like the equatorial flow, it is relatively water-poor (compared to the rings), with 8 to 30 weight percent of silicate grains reported by Hsu et al. (2018), for an MDR (water+silicates) of $\sim 100 - 370\ \mathrm{kg\,s^{-1}}$, comparable to $\dot{M}_{\mathrm{im}}$ (Sect. 2.1). Interestingly, there is a strong indication of a water inflow as well, since H_3^+ is observed in the atmosphere of Saturn at these latitudes, requiring a MDR of the order of $\sim 430 - 2800\ \mathrm{kg\,s^{-1}}$ (O'Donoghue et al. 2019). This suggests that water may detach from the grains on their way to Saturn (by photosputtering for instance), or that water nanograins arrive smaller than silicate grains and fall below the detection limit of Cassini's CDA. In any case, it shows that after ejection from the rings, water and silicates follow different routes. Hence, the efficiency of their elimination can be different. However, the fluxes of silicates (measured from nano-grains with CDA) and water (inferred from the H_3^+ abundance in the atmosphere of Saturn) are more or less consistent with the global composition of the C ring (6% non-icy by mass) when one adopts the CDA upper bound of 30% and reconstitutes the

[3] Note that the majority of the ejecta though is reabsorbed by the rings and can cause mass inflow through the ring layer from the B to C ring and possibly into the D ring due to BT (Durisen and Estrada 2023, this phenomenon is discussed more in Sect. 4.2).

[4] However, how the material ultimately is transported from ring to planet is not completely understood (Durisen and Estrada 2023).

MDR for the missing water. A similar calculation also yields a composition consistent with the A and B rings ($\sim 1\%$ non-icy by mass) when adopting the lower bound of 8%.[5] This apparent agreement may very well be coincidence, as this ratio could vary considerably within the uncertainties.

The uncertainties in these observations open the possibility that some process may preferentially be removing the dirt from the rings and depositing it in Saturn's atmosphere – a "ring-cleaning" phenomenon, as suggested by Crida et al. (2019). Given that the total flux of non-icy material as measured by CDA (over all latitudes) is comparable to the inflow from MB (if one adopts CDA's upper bound of 30%) such a ring-cleaning mechanism could have a significant influence on the ring composition.[6] One can speculate that the impacts due to MB lifting ring material off of ring particles in the form of nanograins or gas – some of which makes it to the planet, and some of which may escape the system altogether – are somehow preferentially ejecting non-icy material or returning the water to the rings, perhaps as larger nanograins or as recondensing vapor. As we saw above, the infall rate of non-icy nano-grains onto the planet inferred by CDA could be up to $0.3 \times 370 = 111\ \mathrm{kg\,s^{-1}}$, not terribly dissimilar to that being brought in by infalling micrometeoroids ($115 - 450\ \mathrm{kg\,s^{-1}}$), and perhaps this is not an accident. If we entertain for the moment that the non-icy mass leaving the ring is equivalent to that impacting the ring (i.e. that $\eta \equiv 0$) then the exposure age is meaningless. But if in addition the ejecta from the ring particles are depleted in water compared to the rings themselves, the rings are being actively cleaned and could have been more polluted in the past (Crida and Bramardi 2021).

An other way of cleaning the rings has been recently proposed by Esposito et al. (2024). Space weathering by UV radiation from the Sun can alter the organic compounds inside an icy ring particle and locally degrade these complex molecules into CO_2 and other oxygenated products, leaving only a refractory organic residue. Over time, without even physically removing it from the rings (like in the mass loss processes discussed above) the pollutants are chemically altered and might lose their darkening properties. The authors show that this process can be very efficient at Saturn. In the end, a balance between the pollution rate and the destruction rate can be reached. Since pollutants are continually lost by this process, the calculated exposure age is only a lower limit.

A cleaning mechanism faces several challenges. First, the physics of the impacts, ejection, and evacuation of nano-grains and molecules remains to be understood. Second, MB darkens low optical regions like the C ring and Cassini division much faster than the more optically thick A and B rings, a state in which the rings are observed. If a cleaning mechanism is also linked to MB, it should also be more effective in the low optical depth regions, which would mean that they should be cleaner than the A and B rings, and not darker. Regardless, the significant inflow of non-icy ring particles into Saturn shows that we can not exclude the possibility of ring cleaning. If it is actually occurring, then it means we are missing something important, and that our understanding of how the rings evolve compositionally is called into question. More detailed discussion advocating for and against ring cleaning can be found in recent works (Crida et al. 2019; Crida and Bramardi 2021; Durisen and Estrada 2023; Estrada and Durisen 2023).

[5] Applying the upper bound to the A and B rings would imply a much larger flux of "fast" ejecta grains far in excess of that estimated by Hsu et al. (2018), or would require a much higher MB rate which is not observed.

[6] Even though the INMS equatorial flux is much higher than the CDA flux (which is ejected from across the rings), it originates from ring regions so close to the planet that it can hardly be used as an indicator of global ring cleaning.

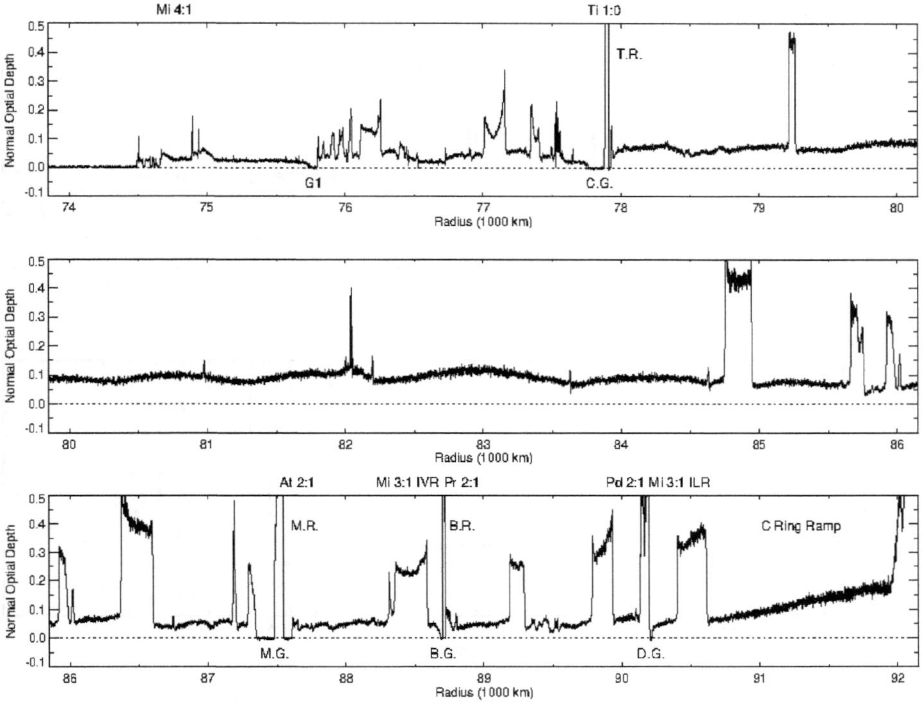

Fig. 1 Normal optical depth profile of the C ring, derived from a Cassini radio occultation experiment at X-band (wavelength = 3.5 cm) on Rev 7. (The original values have been divided by 2 to allow for their exclusion of the forward-scattered signal (French and Nicholson 2000).) Labels below the profile identify the Colombo (C.G.), Maxwell (M.G.), Bond (B.G.) and Dawes gaps (D.G.), while labels above the profile identify the associated Titan (T.R.), Maxwell (M.R.) and Bond ringlets (B.R.). Gap G1 at 75,770 km was identified in Table 13.21 of Colwell et al. (2009). Labels along the upper axis indicate the locations of the important satellite resonances in this region due to Mimas (Mi), Titan (Ti), Atlas (At), and Prometheus (Pr). The inner third of the C ring is dominated by relatively narrow, sharp-edged features with higher than average optical depth known as plateaux, and also contains several narrow gaps, of which the widest is the Colombo Gap. The middle third of the C ring is dominated by gentle undulations in optical depth with wavelengths of $\sim$ 1000 km, while in the outer third we find more plateaux as well as three named gaps (Maxwell, Bond and Dawes). The outermost 1300 km represents a transition region to the more opaque B ring, referred to as the "C ring ramp". Adapted from Fig. 1 in Nicholson et al. (2014b)

3 Structure Age

3.1 Ring-Satellite Interactions

Saturn's rings contain many small-scale features, such as narrow ringlets and sharp-edged gaps, which might, in principle, be used to estimate various 'structure ages', as defined above. These include the outer edges of the A and B rings, which coincide with the strong Janus 7:6 and Mimas 2:1 mean motion resonances, respectively; the Encke and Keeler gaps in the outer A ring, with their embedded satellites, Pan and Daphnis; the Colombo and Maxwell gaps in the C ring, plus a whole series of narrow gaps in the Cassini Division, many of which contain narrow, eccentric ringlets (Colwell et al. 2009). Many of these features are to be found in the lower-optical-depth regions of the rings, especially the innermost or C ring and in the Cassini Division that separates the A and B rings, as illustrated in Figs. 1 and

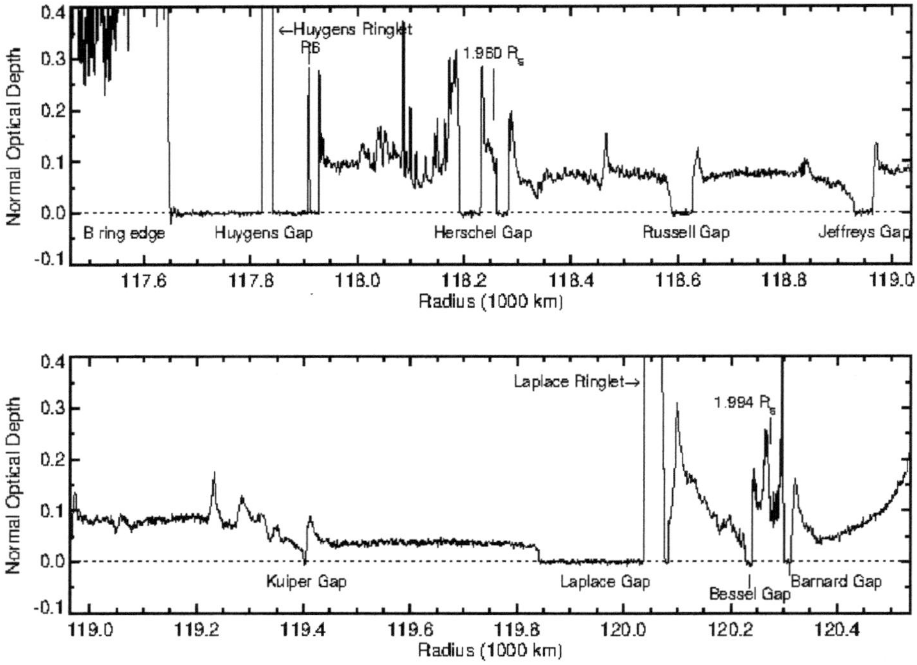

Fig. 2 Normal optical depth profile of the inner part of the Cassini Division, derived from a Cassini radio occultation experiment at X-band (wavelength = 3.5 cm) on Rev 7. (The original values have been divided by 2 to allow for their exclusion of the forward-scattered signal (French and Nicholson 2000).) Labels below the profile identify the eight named gaps in this region, while labels above the profile identify the associated Huygens and Laplace ringlets. The feature at 1.960 R_S is now known as the Herschel ringlet, while that at 1.994 R_S is unnamed. Besides the Mimas 2:1 resonance at the outer edge of the B ring, at 117,570 km, there are no strong satellite resonances in this region, unlike the C ring. This region is dominated by a regular series of relatively narrow, sharp-edged narrow gaps, of which the Huygens and Laplace gaps are the widest. The outer edges of the gaps are, in general, circular while the inner gap edges and those of most of the narrow ringlets are measurably eccentric. As in the C ring, the outermost part of the Cassini Division (not shown here) represents a transition region to the more opaque A ring, referred to as the "Cassini Division ramp". Adapted from Fig. 1 in French et al. (2016)

2. For detailed descriptions of many of these features, the interested reader is referred to the review by Nicholson et al. (2018).

The outer edges of the A and B rings are clearly controlled by the Janus 7:6 and Mimas 2:1 resonances (Porco et al. 1984; Nicholson et al. 2014a; French et al. 2023), while the Encke and Keeler gaps are maintained against viscous closing by the shepherding torques exerted by their embedded satellites (Weiss et al. 2009). But as these particular features are thought to be maintained in a dynamical equilibrium by external satellite resonances or a local shepherding mechanism, they are generally unsuitable for structure age estimates.

How "old" the present configuration of the B ring's outer edge may be depends on how old the Cassini Division itself is. Recent work suggests that it could have formed as recently as ~ 4 Myr ago due to the rings' response to a possible period of inward migration of Mimas, and that it may eventually disappear in the next ~ 40 Myr as Mimas continues to migrate outwards and the A ring to spread inwards (Baillié et al. 2019; Noyelles et al. 2019). Similarly, the present configuration of the A ring's outer edge probably is not long-lived since Janus migrates outwards and the ring itself spreads. However, the details of this evolution are probably unknowable, at least without a much better understanding of resonant confinement and of the accretional histories of the small satellites (Charnoz et al. 2010).

Most of the remaining small-scale structure in the rings is of unknown origin, although it seems likely that many of the gaps that are lacking satellites may be maintained by an as-yet-undiscovered shepherding mechanism involving eccentric ringlets. If this is true, then these gaps are also probably unsuited to estimates of structural age.

One category of small-scale features in the rings that might lend themselves to estimates of structural ages are the dozen or so plateaux in the inner and outer C ring, illustrated in Fig. 1. These mysterious structures are distinguished by widths of 50–150 km, optical depths of ~ 0.4 in a part of the rings where typical values are $\lesssim 0.1$ and relatively sharp edges on one or both sides (Colwell et al. 2009). On the assumption that the plateaux are not somehow dynamically-controlled — by some sort of viscous instability, for example (Schmidt et al. 2009) — but have spread from an initial localized peak in mass density, we can use their widths W to estimate their spreading time due to viscous interactions:

$$t_{\text{spread}} \approx \frac{W^2}{\nu_{\text{C}}} = 5 \times 10^5\ \text{yr} \left(\frac{W}{100\ \text{km}}\right)^2 \left(\frac{6\ \text{cm}^2\,\text{s}^{-1}}{\nu_{\text{C}}}\right), \tag{4}$$

where we have adopted a value for the kinematic viscosity in the C ring ν_{C} based on an assumed ring thickness of ~ 10 m. This timescale would only be the age of the plateaux, perhaps indicating the time at which their mass was introduced at this radius, and not the age of the C ring as a whole. It has also been suggested that the combination of BT and a ballistic transport instability (BTI; Durisen 1995; Latter et al. 2012, 2014a,b) may have played a role in the formation and maintenance of the enigmatic C ring plateaux (Estrada and Durisen 2011; Latter et al. 2012; Estrada et al. 2015, 2018).

A similar estimate may be made for the age of the bimodal structure in the region of the B ring known as B2, between radii of 99,000 km and 106,000 km, where the normal optical depth alternates between $\tau \simeq 1.0$ and $\tau \gtrsim 5$ at intervals of 100 km or so. Assuming that $\nu_{\text{B}} \simeq 30\ \text{cm}^2\,\text{s}^{-1}$ in the B ring (see Sect. 4.3), we obtain a similar estimate of $t_{\text{spread}} \approx 1 \times 10^5$ yr. But the abrupt nature of the observed transitions in optical depth is not very consistent with the idea of viscous spreading so that this time may be irrelevant.

Finally, it has been suggested by Spitale and Porco (2011) that the very narrow Huygens ringlet in the inner Cassini Division (see Fig. 2) may be one of the youngest features in the rings, and perhaps the result of a disrupted moonlet ~ 10 km in size. The ringlet displays a dynamical behavior reminiscent of the B ring edge with a forced $m = 2$ mode at its inner edge, but is also shaped by a strong $m = 1$ (eccentric) mode (French et al. 2016). Since such an unconstrained ringlet would be expected to spread rapidly for any reasonable choice of viscosity, and since there appear to be no shepherd moons to contain it, its average radial width of 18 km suggests an age of perhaps as little as $\sim 10^4$ yr (Spitale and Porco 2011; Spitale and Hahn 2016). [7] However, there are several other such narrow, eccentric ringlets in the C ring, including the Titan, Maxwell and Bond ringlets (see Fig. 1), all of which have quite sharp edges that appear to be inconsistent with unconstrained viscous diffusion but instead suggest some sort of dynamical control (Borderies et al. 1983; Tremaine 2003; Salo and Schmidt 2010).

[7] We note that the same argument applied to the 2 km-wide R6 (or 'Strange') ringlet that also inhabits the Huygens Gap would yield an age of order 100 yr.

3.2 Micrometeoroid Bombardment and Ballistic Transport

In Sect. 2 we pointed out that micrometeoroid bombardment (MB) on the rings leads to two effects, the first of which – enrichment in non-icy pollutant through direct deposition – was discussed therein. The second is that MB and the ballistic transport of the impact ejecta (BT) can lead to the transfer of mass and angular momentum across the rings. This effect is thought to be responsible in great part for the formation, evolution or maintenance of several structures within the rings, as well as the radial compositional evolution of the rings as material is transported between different ring regions. Detailed reviews of the mechanism can be found in Estrada et al. (2018, see also Durisen et al. 1989; Cuzzi and Estrada 1998; Charnoz et al. 2009a; Estrada et al. 2015; Durisen and Estrada 2023).

The primary outcome of a hypervelocity cratering impact on a ring particle is the ejection of a high yield Y (the ratio of ejected mass to the impactor mass) of smaller grains with secondary velocities that range from $\sim 1-100\ \mathrm{m\,s^{-1}}$ (Shu et al. 1985; Cuzzi and Durisen 1990; Tiscareno et al. 2013). Because of aberrations due to the motion of particles about Saturn, and Saturn's motion about the Sun, incoming micrometeoroids predominantly impact on the leading face of ring particles. This means that ejecta have a prograde sense in that they are launched at slightly higher speeds than the local orbital velocity. Ejecta that escape their radius of origin then arc on ballistic elliptical trajectories above the ring plane layer which carry them to ring radii a few km to thousands of km outwards where, depending on the local optical depth some or all may be absorbed by other ring particles. Prograde ejecta carry away more specific angular momentum than needed for a circular orbit at their launch point, but arrive at their new radial location with less specific angular momentum than needed there for a circular orbit. This leads to a radial inward drift at both locations. The divergence of the resultant radial fluxes of material alters the local ring surface densities.

The characteristic timescale associated with BT is called the "gross erosion time" (Durisen 1984)

$$t_G = \frac{\Sigma}{\dot{\sigma}_{\mathrm{ej}}} \approx \frac{\Sigma}{\dot{\sigma}_{\mathrm{im}} Y}, \tag{5}$$

where $\dot{\sigma}_{\mathrm{ej}}$ is the mass flux of ejecta material at a given radius.[8] Since it corresponds to displacement of mostly ring material, t_G is not a measure of exposure age, but of structural age. For our B ring example, $t_G \sim 10^4 - 10^5$ yrs for $Y = 10^4 - 10^5$, but would be much shorter for the less dense C ring.

The left panel of Fig. 3 shows many of the structures that can be seen in the rings, such as the C ring (and Cassini Division) plateaux, undulatory structure in the C ring, and the relatively sharp inner edges of the A and B rings. For the latter, full BT simulations (see Fig. 3, right panel) show that a more-or-less steady impact flux over hundreds to thousands of gross erosion times contributes to shaping the ring edge to its current sharpness, and to form the roughly linear ramp that connects it with the C ring (Estrada et al. 2015, 2018). It is also telling to see a similar ramp structure — with the same radial extent and slope — connecting the inner A ring with the Cassini Division.

It should be noted that the edge itself cannot be created by BT. It had to be somewhat steep to begin with, perhaps as an initial, relatively massive annulus of material whose edge is maintained by the viscosity having a strong dependence on Σ (see Sect. 4; Salmon et al.

[8] Here the approximation is used because $\dot{\sigma}_{\mathrm{ej}}$ depends for any given radius on integration over the full ejecta yield function derived by Cuzzi and Durisen (1990, see, for instance Eq. 21 of Cuzzi and Estrada 1998).

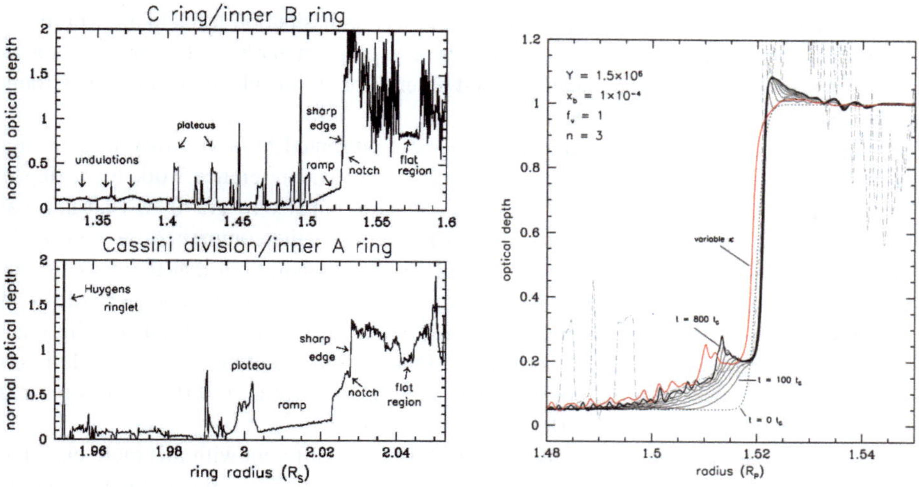

Fig. 3 *Left*: Normal optical depth profiles focusing on the C ring and inner B ring (top), and Cassini Division and inner A ring (bottom). Various structures are labeled. In particular, there is a strong similarity between the inner B ring edge and that of the A ring which has been attributed to ballistic transport (BT). Data from the UVIS α Arae occultation on Rev 32 (Colwell et al. 2009). *Right*: Constant opacity ($\Sigma \propto \tau$) BT simulation of the structural evolution of the inner B ring edge and the development of the C ring ramp up to 800 t_G. The red curve is for a variable opacity model and the dash-dotted curve is the observed optical depth profile. From Estrada et al. (2015, 2018)

2010; Estrada and Durisen 2023). Under these circumstances, BT is particular effective at further sharpening and sculpting the edge (as shown in the right panel of Fig. 3). The structural times necessary to sharpen the edges and form the ramp feature that connects the edge with the C ring are $\sim 10^2 - 10^3$ t_G, or $10^6 - 10^8$ yr, which are similar to our overall exposure time estimate (see Eq. (3)), but these two times are not necessarily the same.

It has been shown that the inner B ring edge can be kept relatively sharp even if the edge migrates inwards as a unit (Latter et al. 2012; Estrada et al. 2018). As demonstrated by Estrada and Durisen (2023), this suggests that it could have started further away from Saturn and has evolved to its current radius of $\sim$ 92,000 km since then. Moreover, Estrada and Durisen also show that the C ring could be formed from material spreading inwards from the inner edge of the annulus to the planet in $\mathcal{O}\left(10^7\right)$ years, setting a lower limit on the dynamical age of the rings due to BT.

4 Dynamical Age

4.1 Viscous Spreading

Like any astrophysical disk in Keplerian rotation, the rings of Saturn spread. Indeed, the angular velocity decreases with the orbital radius r as $\Omega \propto r^{-3/2}$. Hence, any interaction (gravity, collisions, friction...) between two particles orbiting at different radii tends to accelerate the slower, outer one, increasing its angular momentum, while the faster, inner one is slowed down and loses angular momentum. But the specific orbital angular momentum $j = r^2\Omega \propto r^{1/2}$ increases with radius. Consequently, the outer particle moves to a larger orbit while the inner one moves to a closer orbit. At a global scale, the rings spread while orbital angular momentum flows outwards. Meanwhile, dissipation leads to an overall loss of orbital energy so that the net flow of mass is on average inwards, and eventually most of the mass should end up in the planet. A good description of this phenomenon can be found in Lynden-Bell and Pringle (1974).

The exchanges of angular momentum can be represented by a viscosity in a sheared fluid. The key is then to model the viscosity. The standard reference work for a ring of solid particles which interact not only through collisions but also by gravity is Daisaka et al. (2001), but the interested reader may also wish to consult the comprehensive reviews by Schmidt et al. (2009) and Salo et al. (2018). These works point out that the viscosity has three components: translational (or "local", akin to molecular viscosity in a gas), collisional (or "non-local", where momentum is transferred across ring particles in a dense ring), and gravitational. Salmon et al. (2010) performed 1D numerical simulations of the evolution of the radial surface density profile of the rings, using Daisaka et al. (2001)'s viscosity in an analysis similar to that of Lynden-Bell and Pringle (1974).

They find that the behaviour of the rings is drastically different with this more elaborate model of the viscosity, compared to a constant viscosity one. This is due to the fact that the kinematic viscosity depends on the rings' mass density itself. In particular, the gravitational component of the viscosity —which becomes important when the critical dimensionless parameter $Q_{\mathrm{T}} = \kappa c_r / 3.36 G \Sigma$ (Toomre 1964) falls below about 2 and self-gravity wakes form in the rings — depends effectively on the surface mass density of the rings squared. The $Q_T \sim 2$ threshold corresponds roughly to the surface density of the A and B rings (see Salmon et al. 2010, their Fig. 2). As the rings spread radially due to viscosity, they lose mass at their edges: into Saturn when they reach the planet as they drift inwards, and to the formation of moonlets when they cross the Roche radius drifting outwards. The viscous timescale being inversely proportional to the surface mass density of the rings squared for $Q_T \lesssim 2$, the mass evolution can be analytically estimated as (Crida and Charnoz 2014):[9]

$$D(t) \approx \frac{D_0}{\sqrt{1 + 240 D_0^2 (t/T_{\mathrm{R}})}} \underset{t\to\infty}{\sim} \frac{8.40 \cdot 10^{-8}}{\sqrt{t\,[\mathrm{Gyr}]}} \tag{6}$$

where D is the mass of the rings normalized by the planet mass, $D_0 = D(t = 0)$, and T_{R} is the orbital period at the outer edge of the rings (14 h 51.2 min). Remarkably, on long timescales, when rings evolve under viscosity alone the asymptotic ring mass does not depend on the initial mass at all but only on their age, as first noticed in numerical simulations by Salmon et al. (2010). The present mass of the rings, $1.54 \pm 0.49 \cdot 10^{19}$ kg or $D = 2.71 \pm 0.86 \cdot 10^{-8}$ (Iess et al. 2019) is in excellent agreement with the one found by Salmon et al. (2010) and Crida et al. (2019) after 3 to 5 Gyrs.

This agreement can be seen in Fig. 4, which shows simulations similar to those performed by Salmon et al. (2010) for several initial disk masses over 5 Gyr. Saturn's rings' current mass (black line) and error bars (shaded region bracketed by black dotted lines) are indicated (from Crida et al. 2019). For the higher mass disks in particular, evolution is most rapid in the first few 100 Myr, while most of the remaining evolution is the slow decrease given by the asymptotic form of Eq. (6). Interestingly, as pointed out by Crida and Charnoz (2012,

[9] Crida and Charnoz (2014) calculations lead to a coefficient 60 where we have used 240 here. This is because they took the Roche radius for the width of the rings in order to calculate the viscous time, while it should be only half of this, because Saturn extends up to about half the Roche radius. A factor two in the typical length scale yields a factor 4 in the viscous time, hence 240 instead of 60. This actually gives a result closer to the numerical simulations.

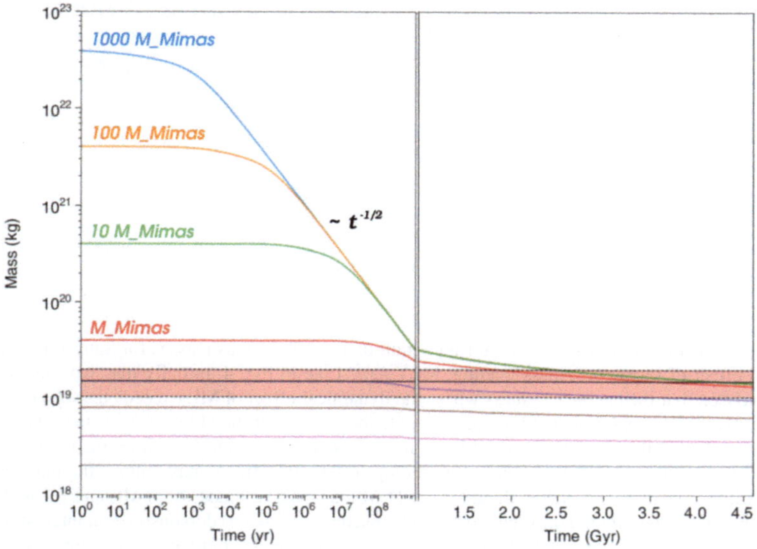

Fig. 4 Viscous evolution models of initial ring annuli for several different masses, up to 1000 Mimas masses (blue curve). As demonstrated by Salmon et al. (2010, see their Fig. 10), initially massive rings for Saturn evolve viscously to an asymptotic mass over the age of the Solar System which is in very good agreement with their current mass of 0.41 ± 0.13 Mimas masses (Iess et al. 2019), given by the black solid line; the shaded region indicates the range of uncertainty in the mass measurement. In the left panel (first Gyr of evolution; log-scale in time), it can be seen that viscous evolution is very rapid early on, so that rings spend the bulk of their lifetime (right panel, 1-4.5 Gyr, linear timescale) at relatively low mass. Rings that start out with masses less than the current mass see little evolution because viscosity is already very weak. Adapted from Crida et al. (2019)

2014), the same scenario apparently does not apply at Uranus and Neptune. The mass of the Uranian rings is likely closer to $\sim 2 \cdot 10^{16}$ kg (Goldreich and Porco 1987; French et al. 1991; Mosqueira and Estrada 2002; Chiang and Culter 2003), making $D \approx 2 \cdot 10^{-10}$, a hundred times below the expected asymptotic value. This suggests either that Uranus never had a massive ring, or that other processes may have taken place that led to such a low mass. In this case, one may wonder what these processes are and why they did not apparently apply to Saturn. Such a possible process is described below.

4.2 Dynamical Effect of Micrometeoroid Bombardment

We have already discussed that MB has the effect of polluting the rings over time with non-icy materials (Sect. 2), and when combined with BT it can lead to the formation and maintenance of structure in the rings (Sect. 3). Most previous papers have mainly been concerned with how ring structure can be formed near ring edges, or due to instabilities (e.g., Durisen et al. 1992; Durisen 1995; Latter et al. 2012; Estrada et al. 2015). However, the observed mass loss rates from Cassini (Hsu et al. 2018; Waite et al. 2018) inspired focus on a more fundamental property of MB and BT, that both mechanisms can drive significant mass inflow and thus, like viscosity, ring dynamical evolution.

Durisen and Estrada (2023) have quantified the magnitude of the mass inflow potentially driven by direct deposition of impactors (referred to as mass loading, ML) and BT using a simplified analysis of the ballistic transport equations similar to that used by Durisen (1995, see also Durisen et al. 1992; Estrada et al. 2018). As explained in Sect. 3.2, the bombardment

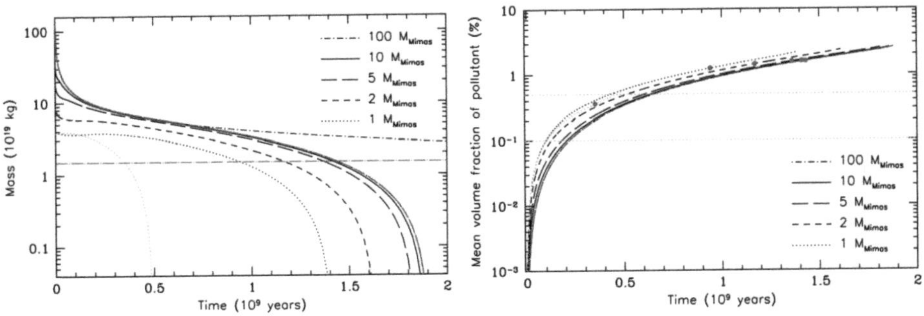

Fig. 5 Viscous evolution models of initial ring annuli of 1 − 100 Mimas masses for Saturn that includes the dynamical effect of impactors (mass loading, ML). *Left:* Like in Fig. 4, initially massive rings evolve very quickly early on, but then achieve a mass where the dynamical effect of ML exceeds that of viscosity which causes the mass to decrease much more quickly (note the top, dot-dashed blue curve is the 100 Mimas mass model that neglects ML from Fig. 4). For ML alone, this occurs at around four times their current mass (red dot-dashed line). Larger initial disk masses lie on top of the 100 Mimas mass curve (the magenta curve is for 1000 Mimas masses). The orange dotted curve (1 Mimas mass) includes both ML and BT with $Y = 10^4$. Smaller initial mass annuli (and thus the most slowly viscously evolving) demonstrate an interesting behavior in gaining mass briefly before reaching a steady mass loss. All models briefly gain mass until the annuli outer edges spread to the Roche limit and begin to lose mass beyond it. However, because the 1 Mimas mass model is already in the ML dominated regime, viscous evolution beyond the Roche limit ceases around ~ 130 Myr, and the rings gain mass again while the inner edge is evolving toward the planet. A steady inwards mass loss rate is then reached around $\gtrsim 230$ Myr. *Right:* The corresponding mean volume fraction of non-icy pollutant for the same models. The green dotted lines bracket the observed non-icy fractions in the A and B rings (Zhang et al. 2017a). The red dots correspond to the time the ring models have achieved the observed ring mass. This suggests that for initially lower mass rings, the observations can be satisfied with the exposure and dynamical age being similar if BT is taken into account

and ejecta both lead to a local loss of angular momentum and therefore a global inward drift of the rings (with a small mass transfer outwards for BT).

Applying this analysis to the B ring, we obtain estimates for the equatorial mass inflow rates there due to BT, ML, and viscosity of:

$$\dot{M}_{\rm BT} \approx 5 \times 10^3 \left(\frac{F_\infty}{2.7 \times 10^{-16}\ {\rm kg\,m^{-2}\,s^{-1}}}\right)\left(\frac{F_{\rm g}}{30}\right)\left(\frac{x_{\rm b}}{10^{-4}}\right)\left(\frac{Y}{10^4}\right)\ {\rm kg\,s^{-1}}; \tag{7}$$

$$\dot{M}_{\rm ML} \approx 3.4 \times 10^3 \left(\frac{F_\infty}{2.7 \times 10^{-16}\ {\rm kg\,m^{-2}\,s^{-1}}}\right)\left(\frac{F_{\rm g}}{30}\right)\ {\rm kg\,s^{-1}}; \tag{8}$$

$$\dot{M}_{\rm visc} \approx 50 \left(\frac{\nu_{\rm B}}{100\ {\rm cm^2\,s^{-1}}}\right)\left(\frac{\Sigma_{\rm B}}{520\ {\rm kg\,m^{-2}}}\right)\ {\rm kg\,s^{-1}}. \tag{9}$$

Despite the uncertainties in the modeling of BT (mainly the value of Y and x_b which is a lower bound for the ejection velocity, see caption of Fig. 3), one can see that the micrometeoroid bombardment can exert considerably more influence on the dynamical evolution of the rings compared to other mechanisms, for the present values of the surface density and viscosity.

Estrada and Durisen (2023) have further used the analysis of Durisen and Estrada (2023) to reexamine viscous evolution models of initially massive ring annuli like those done in Salmon et al. (2010), by adding the dynamical effects of ML and BT. The reader is refered to the paper for more details on the implementation and its necessary caveats. In Fig. 5 (left panel) are shown the evolution models for several initial disk masses as was done in

Fig. 4 (left panel), but here ML is included. Taking the 100 Mimas mass case as an example (black dot-dashed curve), it can be seen that just as before the rings rapidly evolve in the first few 100 Myr, and begin to more gradually decrease in mass after they have shed most of their material to the planet, or beyond the Roche limit. Larger initial mass models follow the same trajectory (the magenta curve corresponds to 1000 Mimas masses). However, since the gravitational viscosity grows as Σ^2, and the dynamical efficiency of MB decreases as Σ^{-1} for optically thick rings, a transition between the regime dominated by viscous evolution and by MB occurs at a threshold mass. Above this mass, viscous spreading dominates the ring evolution and below this mass, MB does. This mass turns out to be of the order of four times the present mass of the rings[10] when one considers ML alone (Estrada and Durisen 2023), as can be seen with the curves corresponding to the initially massive rings diverging away from blue dot-dashed curve (which corresponds to viscosity only) at about 6×10^{19} kg. Initial rings $\gtrsim 100$ Mimas masses eventually reach the current ring mass in ~ 1.5 Gyr, while the smallest case achieves the current mass in $\lesssim 1$ Gyr. Although the dynamical age is still on the order of a billion years, it now seems difficult for a massive ring to reach the age of the Solar System. An additional case for 1 Mimas mass (orange dotted curve) is shown in which BT is also included using an ejecta yield of $Y = 10^4$, which implies that BT would decrease the evolution time, and begin to dominate over viscosity at even larger threshold masses.

In the right panel of Fig. 5 are shown the corresponding evolutions of the non-icy pollutant for the same models. The current fractions as observed for the A and B rings are indicated by the dotted green lines (Zhang et al. 2017a), while the red dots indicate the times at which the current ring mass is achieved for the various models. As seen in the section on the exposure age, all models reach the observed fractions in a few 100 Myr, but they don't all reach the observed ring mass at the same time. By definition, the dynamical age and exposure age in these models are the same, but it is suggested from these simulations that the observations *and* the similar ages can be satisfied if the rings began at lower mass *and* BT plays an important dynamical role. The shortest time to achieve the current ring mass in Fig. 5 occurs around ~ 350 Myr for the initial 1 Mimas mass model that includes BT. A slightly higher yield can lower this age further to ~ 100 Myr or even less. Estrada and Durisen estimate from their models that the initial mass of Saturn's rings would be between $\sim 1-3$ Mimas masses if they formed a few 100 Myr ago for ejecta yields of $10^4 \lesssim Y \lesssim 10^5$.

4.3 Satellite Torques and Ring Lifetimes

As noted in Sect. 3.1, the radii of the outer edges of the A and B rings are currently determined by two strong Lindblad resonances with external satellites: the A ring by the 7:6 ILR with the co-orbital satellite Janus and the B ring by the 2:1 ILR with Mimas. The latter was first discussed quantitatively by Goldreich and Tremaine (1978), who suggested that the (negative) torque exerted on the rings by Mimas at this resonance was strong enough to balance the outward flux of angular momentum through the B ring due to viscosity:

$$F_V(r) = 3\pi \nu_B \Sigma_B \Omega(r) r^2, \tag{10}$$

where $\Omega(r)$ is the local Keplerian angular velocity at radius r, and ν_B the viscosity in the B ring. In this view, the B ring is currently prevented from spreading outward, and its edge

[10] One should note, however, that using the flux value for the certainly exogenic particles leads to ~ 4 times less bombardement, reducing the transition mass.

maintained at the resonance by the transfer of the viscous angular momentum flux (henceforth AMF for short) to Mimas via the gravitational resonance torque. Mimas thus effectively acts to 'anchor' the outer edge of the B ring, whose radial drift is limited by the rate that the satellite's own orbit evolves outwards due to the tidal torque from Saturn plus the reaction torque from the rings.[11]

A similar situation occurs at the outer edge of the A ring, where the gravitational torque associated with the Janus 7:6 ILR was initially believed to balance the outward viscous flux of angular momentum through the A ring (Porco et al. 1984) In this case, however, there are several other satellite resonances of comparable strength in the outer part of the A ring due to Mimas and Janus, plus numerous weaker resonances with the F ring shepherd satellites Prometheus and Pandora and even more numerous (but even weaker) resonances with Atlas, Pan and Daphnis. A careful accounting of the torques from all of these resonances by Tajeddine et al. (2017), using a surface mass density profile obtained from the analysis of density waves by Tiscareno and Harris (2018), showed that the *cumulative* satellite torque is indeed capable of balancing the outward viscous AMF, although the Janus 7:6 resonance is probably too weak to do the job alone.

Although in the above picture the outer edges of the A and B rings are currently almost fixed in radius, except for the slow evolution of their confining satellites' orbits, angular momentum is still being steadily removed from each ring via the resonance torque(s) and transferred to the Janus/Epimetheus coorbital pair and Mimas, respectively.

Tajeddine et al. (2017) also estimated the orbital expansion rates of all the inner satellites from the combined effects of ring and tidal torques, using a constant value of $k_2/Q = 1.59 \times 10^{-4}$ for Saturn from Lainey et al. (2017). The resulting time scales to double their distances from the A ring are listed in Table 3 of Ciarniello et al. (2024), this collection. For Mimas $t_{\mathrm{dbl}} \simeq 1.0 \times 10^9$ yr, for Janus $t_{\mathrm{dbl}} \simeq 3.1 \times 10^8$ yr, and for Pandora and Prometheus, which are less massive but much closer to the rings, it is only $10^7 - 10^8$ yr. So we expect that the outer edge of each ring will move slowly outward, following the appropriate satellite resonance, but that once the A ring edge moves significantly beyond the Roche Limit for icy bodies the outermost ring material will begin to accrete into one or more new satellites, as described by Charnoz et al. (2010). A new A ring edge will then form closer to the Roche Limit, presumably anchored by a different satellite resonance. The evolution of the A ring is thus likely to be discontinuous in nature, with some of its mass being lost outward to new satellites during episodes of spreading, separated by episodes of confinement by different resonances with external satellites, or perhaps by the newly formed satellites themselves. The details of this process, however, are probably unknowable.

At the same time, the cumulative negative torques exerted on the A and B rings by the satellite resonances must augment the gradual inward drift of the ring material due to viscous energy dissipation (see Sect. 4.1). We can make a crude estimate of the timescale of this process for the A and B rings based on their current orbital angular momenta, L_{ring} and the cumulative torque T_{sat} from the satellite resonances currently located within each ring:

$$t_{\mathrm{coll}} = \gamma L_{\mathrm{ring}} / T_{\mathrm{sat}}, \tag{11}$$

where γ is a dimensionless number specifying by what factor the mean angular momentum of the ring has been reduced. Roughly, $\gamma \simeq 1 - (r_f/r_0)^{1/2}$, where r_0 and r_f are the present and final mean mass-weighted ring radii.

[11] The evolution of Mimas' orbit is further complicated by its being in a 4:2 resonance with the larger satellite Tethys. See also Blanc et al. (2025), this collection.

Based on the satellite torques calculated by Tajeddine et al. (2017) and simple analytical estimates of L_A and L_B, we find that the collapse time $t_{coll} \simeq 8.3 \times 10^8$ yr for the A ring and $\sim 2.3 \times 10^9$ yr for the B ring, where we assume that the A ring's mass drifts inward to the current location of the B ring ($\gamma \simeq 0.076$) and the B ring's mass drifts inward to the current location of the C ring ($\gamma \simeq 0.113$). Note that the above estimates are largely independent of the assumed surface mass density profile of each ring, inasmuch as both L_{ring} and T_{sat} are proportional to Σ. These collapse times are quite uncertain due to the unknown future masses and orbital configuration of Saturn's inner moons; besides, when the rings recede, they leave the resonance which was confining them, so that the torques diminish. Perhaps all that can safely be said is that they are consistent with, but do not favor, the idea of an ancient ($\gtrsim 1$ Gyr) ring system.

As noted in Sect. 4.1, the combination of resonant torques, viscous energy dissipation and meteoritic mass loading are expected to lead to a gradual inward collapse of most of the ring material, and to its eventual accretion by Saturn. The details of how this mass is lost from the A and B rings and passes through the Cassini Division and C ring are likely to be bound up with the process of ballistic transport which is believed to sculpt these transition regions, as described in Sect. 3.2 above.

A useful by-product of the analysis of satellite resonant torques is an estimate of the average effective viscosities of the A and B rings, independent of theoretical models and of estimates based on the damping of density and bending waves. On the assumption that the viscous AMF at each radius r in the A or B ring is equal to the cumulative gravitational torque of all the resonances in that ring located external to r, Tajeddine et al. (2017) derived radial profiles of ring viscosity for each ring. For the A ring, the viscosity ν was found to decrease slowly from $\sim 45\ \mathrm{cm^2\,s^{-1}}$ in the inner regions to $15 - 30\ \mathrm{cm^2\,s^{-1}}$ in the region exterior to the Encke Gap, before dropping rapidly to less than $11\ \mathrm{cm^2\,s^{-1}}$ at the outer edge. These estimates are considerably lower than those obtained from weak (and therefore more likely to be linear) density waves, which range from $\nu_A \simeq 30$ to $\sim 300\ \mathrm{cm^2\,s^{-1}}$ (Tiscareno and Harris 2018).

For the B ring, on the other hand, most of the angular momentum is removed at the outer edge by the strong Mimas 2:1 ILR, resulting in a fairly constant value of the AMF across this ring. Assuming a nominal constant value of $\Sigma_B = 750\ \mathrm{kg\,m^{-2}}$, based on the limited number of waves identified in this ring by Hedman and Nicholson (2016), Tajeddine et al. (2017) find that $\nu_B \simeq 30\ \mathrm{cm^2\,s^{-1}}$.

5 Formation Models

In this section, we review various formation models which have been proposed in the literature, loosely ordering them from primordial or ancient scenarios to ones which could explain more recently formed (or rejuvenated) rings. We discuss each model's pros and cons, with a special focus on the predicted age of the rings and on their likely composition.

5.1 Rings as a Remnant of Saturn's Circumplanetary Gas Disk

The simple fact that all of the giant planets in our Solar System have rings, while none of the terrestrial planets do, has long suggested that ring formation may be a by-product of giant planet accretion as unaccreted remnants from their circumplanetary disk (Pollack et al. 1973, 1976). However, it is not sure that the ice giants ever were massive enough to open a gap in the protoplanetary disc and to form their own circumplanetary disc. Besides, the recent discovery of rings around minor bodies like Chariklo (Braga-Ribas et al. 2014; Duffard et al. 2014; Bérard et al. 2017), mostly in the form of narrow ringlets similar to the Uranian and Neptunian rings, suggests that such rings might originate from a different process, common to the ice giants and the centaurs. This being said, ring formation from the circum-planetary disc remains a possible scenario for Saturn.

A major problem with this scenario is how this material could have outlived the gas disk. Taking the present size distribution of particles in Saturn's rings, which extends from mm-size to several meters (Zebker et al. 1985; French and Nicholson 2000; Harbison et al. 2013), it is clear that radial drift due to gas drag in the circumplanetary nebula would pose a problem, sweeping away much if not all this material into the planet's atmosphere in very short order. This drag time will depend on the local gas surface density ($\propto \Sigma^{-1}$) and temperature ($\propto T^{-3/2}$), but the drift time of a meter-size particle would be on the order of tens of years for gas surface densities as low as $\sim 10 - 100$ g cm^{-2} (Mosqueira and Estrada 2003). One possible way to avoid this problem is if an inner cavity near the planet were opened by the planet's magnetic field, but this probably does not apply to Saturn (e.g., Sasaki et al. 2010; Shibaike et al. 2019).

A second big problem is the issue of the temperature close to the planet which must be below the condensation temperature for water ice, whereas all formation models assume the temperature to be significantly higher as one nears the planet due to its high post-accretion luminosity. For example, the compositional gradient apparent in the Galilean moons has been argued to be explained by this (e.g., Lunine and Stevenson 1982; Bierson et al. 2023). Jupiter's diffuse, non-icy gossamer rings would be consistent with this picture, at least compositionally. At Saturn, even if the temperature were cold enough to preserve water in its solid form, there is no obvious reason to expect that the rings made this way would not include silicates and iron, so that a 'cleaning mechanism' acting over the age of the Solar System would need to be invoked. Given the above complications, an origin of the present saturnian rings in the primordial protoplanetary disk no longer seems to be plausible.

5.2 Migration and Tidal Stripping of a Differentiated Satellite

Roche (1847) was the first to suggest that the rings of Saturn may be remnants of a disrupted moon, though it was more than a century later that the workings of such a mechanism began to be explored in more detail (e.g., Pollack et al. 1973; Harris 1984). This ring origin scenario requires that a sufficiently large satellite (or several satellites) be brought inside Saturn's Roche limit to be broken up tidally shortly after. See Sect. 5.4 for an alternate disruption scenario, and Ciarniello et al. (2024), this collection, for the definition and an extended discussion of the Roche Limit.

It is commonly assumed that during their formation in the proto-planetary disk around the star, giant planets as massive as Saturn are surrounded by a circum-planetary disk, in which satellites may form. The reader is referred to Blanc et al. (2025), this collection, for more details on this process. Here, all we need to know is that objects as large as Titan or Jupiter's Galilean moons form in such a circum-planetary disk. These bodies are then subject to migration, just like planets in the proto-planetary disk. With a satellite-to-planet mass ratio of the order of 10^{-4} — equivalent to less than a gas giant relative to the Sun — they would typically be subject to type I migration. The migration is generally inwards, at a speed proportional to the mass of the object (see for instance Baruteau et al. 2014, for a recent review on planetary migration). As a consequence, many such satellites may fall into the planet and are regularly lost, while others are still forming.

Canup (2010) has shown that if a differentiated satellite crosses the Roche limit, its destruction by the tidal forces proceeds by onion-peeling, where first the outer, less dense layers are stripped away by the tides. Losing parts of its light mantle, the satellite becomes denser on average, so that it can resist the tides (the Roche radius decreases with increasing density of the satellite). As it keeps migrating inwards, it loses a new layer of the mantle, and so on. During its migration towards the central planet, the satellite leaves behind debris from its mantle. If the satellite has a similar composition as Ganymede or Titan and is fully differentiated, this mantle is made of almost pure water ice. Canup (2010) suggests that Saturn was large enough at this time ($\sim 90{,}000$ km in radius) for the Roche radius corresponding to silicates to lie within the planet. In this way, the core of the migrating satellite would never be destroyed before being engulfed. Only a disk of water ice from the mantle is left behind.

The mantle debris are pushed outside of the Roche radius by the torques from the infalling satellite, and eventually form a pure ice satellite. This object then migrates inwards either in planetary migration with the gas or, if the subnebula gas has dissipated, tidal interaction with Saturn as it would likely be inside the synchronous orbit at that time[12] (Fortney et al. 2007; Canup 2010), and is then totally destroyed as it enters the Roche limit for ice. This debris then naturally evolves via collisions into a flat ring (perhaps massive enough to resist gas drag if gas is still present).

The rings must be the left-overs from the mantle of the last satellite to have fallen into Saturn, so that they are not destroyed by an other inwards migrating satellite. Hence, this scenario applies towards the end of the life of the circumplanetary disk. The dissipation of the gas soon after this event also helps in the survival of the icy chunks. Canup (2010) also verified that the conditions were cool enough at the end of the circumplanetary disk era for ice to survive. This may not have been the case around Jupiter, which could explain why Jupiter probably never had a massive ring system.

In this model, differentiation of the progenitor allows the ice to be separated from the silicate core, while migration is the solution both to the problem of the entry of the progenitor within the Roche radius and to the ultimate removal of the silicate core (Crida and Charnoz 2010). Because differentiation is needed, the progenitor satellite must be large (which is also necessary for it to migrate fast enough). The resulting rings therefore must be much more massive than the present rings. Canup suggests $\sim 10^{22}$ kg or ~ 250 Mimas masses, which is more than the mass of Rhea. Such a large initial ring mass implies that the rings have lost most of their material as they spread radially (see Sect. 4.1 above), leading to the formation of most of the regular satellites interior to Titan (Canup 2010; Charnoz et al. 2011).

Regarding the timing, rings formed via this satellite migration scenario are necessarily primordial, as this whole process happened while Saturn was still surrounded by a gaseous circumplanetary disk. After dissipation of this disk, one could still have inwards migration due to Saturn's tides of a satellite lying just outside the Roche radius but inside the synchronous orbit in the scenario of Canup (2010, see above). Charnoz et al. (2009a) calculate that it takes less than 1 Gyr for a $\sim 10^{20}$ kg moon ($\sim 3\times$ the mass of Mimas) to migrate inwards from the synchronous orbit to the Roche radius and be destroyed, with a constant tidal quality factor for Saturn $Q \sim 10^5$.

The tidal stripping of a differentiated satellite can explain the initial pure water ice composition of the rings, and implies that they are ancient and were once very massive, having evolved to their present mass over billions of years.

---

[12] A reasonable assumption is that Saturn would have obtained all of its spin angular momentum after gas disk dissipation. However, its radius would be extended at the time (Fortney et al. 2007), so that its spin rate would be slower, thus placing synchronous orbit further out than it is today (Canup 2010).

5.3 Capture and Tidal Disruption of a Passing Centaur

Somewhat in response to the post Voyager paradigm that Saturn's rings may be quite young, Dones (1991) suggested that Saturn's rings could be the product of the tidal disruption of a Saturn crossing comet or Centaur. The idea is similar to a disintegrative capture model proposed for the formation of the Moon (Öpik 1972; Wood and Mitler 1974), in which a body on a parabolic orbit with respect to the planet that passes (well) within its Roche limit experiences a very large gravitational potential difference across its diameter. If the body breaks up into a large number of fragments at closest approach, about half (those on the leading face of the comet) will have velocities less than the local planetary escape speed and will become weakly bound to Saturn with periapses very close to the planet, and apoapses of typically several hundred R_S.

Dones (1991, see also Dones et al. 2007) modeled this process assuming a Centaur-like population with Chiron as an archetype of the typical Saturn crosser with a mean radius of 90 km, mean densities[13] of 1000 $kg\,m^{-3}$ and a distribution of velocities entering Saturn's Hill sphere of $0 < v_\infty < 6\ km\,s^{-1}$. Chiron ($\sim 0.1$ Mimas masses) was the only Saturn crosser known at the time, although about a dozen more have been discovered since,[14] most of which are similar in size or smaller than Chiron (though the largest Centaur known, Chariklo, is roughly 150 km in radius). However, Dones further assumed that the size distribution extended up to 1000 km in size, with a cumulative size distribution described by a power law with an exponent between 2 and 3. His simulations showed that the mean number of close encounters of such objects (*i.e.*, passing within the Roche limit) over the age of the Solar System is $10 - 100$, but that most events leave no debris either because they are moving too fast or do not pass close enough to Saturn.

Dones found that up to $\sim 0.1 - 1$ Mimas masses were captured over 4.5 Gyr using the assumed size distribution, but with effectively all of the mass being captured in single, rare events with the largest bodies that had the closest periapses and lowest injection velocities v_∞. In essence, he found that over the age of the Solar System there would be one potential ring mass capture event with a probability of $\sim 40 - 80\%$. Assuming that such large bodies still exist presently, if such an event were to happen in the last 100 Myr its probability, all other things being equal, would be $\sim 1 - 2\%$. Such events may actually be much rarer since the size-distribution of larger Centaurs appears to be steeper (> 3) than what Dones (1991) assumed (Charnoz et al. 2009a).

Some question as to what happens to the material once it is captured remains, as the detailed process has yet to be modeled (Dones et al. 2007, see also discussion about the fate of debris in Sect. 5.6). Dones (1991) argued that the most likely outcome would be that collisions between fragments and conservation of angular momentum would lead to the formation of an equatorial ring in $\sim 10^5$ yrs. Much of the mass, however, would have a significant probability of impacting Saturn's mid-sized moons (or even Titan), and could help to explain the presence of planetocentric craters on these bodies (e.g., Ferguson et al. 2022). Note that if the incoming Centaur is not differentiated, the composition of the rings would be at most half icy. Similarly as in Canup (2010)'s model described above in Sect. 5.2, differentiation is needed to separate the ice from the rocks and metals in the progenitor. Then, one needs the closest approach to be inside the Roche radius for ice but outside the Roche radius for the core, so that the latter would continue on its hyperbolic or parabolic trajectory and leave the Saturnian system, leaving behind only debris of the icy mantle.

[13] It should be noted that the densities of these objects can vary greatly (Johnston 2018).

[14] See the JPL small bodies database at https://ssd.jpl.nasa.gov/tools/.

Hyodo et al. (2017, see also Charnoz et al. 2009b) modeled this process assuming a population of large differentiated objects (Rhea-like or greater masses of $10^{21}-10^{23}$ kg) scattered from the primordial Kuiper Belt during the LHB using a combination of Smoothed Particle Hydrodynamics (SPH) to model the tidal stripping, and N-body simulations to follow the dynamical evolution of the captured material. They found that $\sim 0.1-10\%$ of the material is captured into planetocentric, highly eccentric orbits and subsequently tidally disrupted into smaller bodies by the planet over time. Eventually due to orbital precession, they expect that encounter velocities will be enhanced such that the bodies would be ground down to cm-to-m sizes. The accompanying dissipation of random velocities between colliding particles would eventually lead to an equatorial and circular ring within Saturn's Roche limit. The composition of the rings may be mostly icy, but some silicates would be expected.

Overall, such a scenario appears most likely to be an option for ancient rings, perhaps delivering enough mass to account for them and the mid-sized satellite system as well. Given that no Centaurs are currently known to be larger than Chariklo (the number of suitable bodies is expected to drop off steeply after the LHB; Zahnle et al. 2003), it is very unlikely that there were any ring mass forming events in the current epoch. Thus the tidal disruption mechanism does not appear to be viable for recent ring formation. In fact, Dones did include a simulation with a maximum size of 90 km, finding that only ~ 0.004 Mimas masses, or $\sim 1.5\times10^{17}$ kg are captured over that time. Coincidentally, this is similar to the estimated $\sim 10^{17}$ kg of *non-icy* material from a captured Centaur thought to be embedded in the middle C ring and accounting for the observed thermal brightness temperature and anomalous opacity there (Zhang et al. 2017b). So capture events of smaller bodies are still likely to be occurring, particularly in the D ring (Hedman et al. 2015; Hedman and Showalter 2016). Naturally, capture of such material would be more easily facilitated if rings are already present.

5.4 Collisional Disruption of a Mimas-Size Moon Due to a Heliocentric Impactor

This ring origin scenario shares some common features with the one discussed in Sect. 5.2 in that it requires that a mid-sized moon(s) (e.g. $\gtrsim$ Mimas) migrates inwards and is left stranded inside or near the Roche limit after the circumplanetary gas disk dissipates. Such a moon can continue to migrate inwards (or outwards) due to planetary tides, but will generally be too small to be broken up by planetary tidal forces unless it migrates very close to the planet (Davidsson 1999). Thus in this scenario, ring formation relies on the collisional disruption of a moon by a sufficiently large Sun-orbiting body.

A heliocentric interloper on the order of $10-20$ km in radius has enough energy to disrupt a Mimas-sized moon (Charnoz et al. 2009b). This would produce debris which, within the Roche radius, could not reaccrete and rather become the rings we see today. Compositionally, this scenario faces the challenge of explaining nearly pure ice rings unless the progenitor itself was predominantly made up of water-ice which, although possible (e.g., Tethys), seems unlikely. It seems more probable that the progenitor moon would be a differentiated body, meaning the rings would be a mixture of ice and significant rock. Charnoz et al. (2009b) suggest that an intact core might be removed through interactions with icy mantle material which was preferentially ejected in the collision, but these authors admit that no obvious way to remove the silicates exists for this scenario.

As for the timing of such an event, in principle the impact could happen at any time in the history of the Solar System. However, two aspects favor an early event (hence old rings). First, for the debris to make up the rings, the progenitor satellite should be just outside of the Roche radius. But we know that it should migrate outwards (or inwards if inside the synchronous orbit as may have been the case up to 1 Gyr, Fortney et al. 2007) quickly due to tidal interactions with Saturn, which makes its presence here unlikely after a few hundred million years. Second, the frequency of available impactors only decreases with time. Charnoz et al. (2009b) suggested that the time of the Late Heavy Bombardment, some ~ 600 Myr after Saturn's formation, is favorable in the frame of the Nice model (Gomes et al. 2005; Tsiganis et al. 2005). This model assumes a global instability in the orbital architecture of the giant planets, destabilizing a massive progenitor Kuiper Belt, leading to a peak of bombardment in the whole Solar System. Charnoz et al. (2009b) find that a Mimas-sized moon would have had a 90% chance of destruction during this epoch. It should be noted that the timing of this instability is now questioned, and that the LHB may not be late anymore (e.g. Morbidelli et al. 2018; Nesvorný et al. 2023). The Nice model global instability (which is not questioned and supported by many other lines of evidence) would then happen shortly after the dissipation of the proto-solar nebula (Griveaud et al. 2024). In this case, rings formed during this phase would be almost as old as Saturn.

More recently, in an effort to account for recent ring formation Dubinski (2019) suggested a modified version of the collisionally disrupted moon scenario in which an Oort comet or Centaur of sufficient size (~ 20 km in radius) impacts a mid-sized moon in orbit essentially straddling the Roche limit of Saturn. In this scenario, a hypothetical, differentiated proto-Mimas-like moon is maintained close to Saturn's Roche zone through an MMR with Enceladus and Dione that may have existed in the past and that would have significantly slowed the moon's outward tidal migration. Note that this model assumes rings are not present before the disruptive collision, so that ring torques (which are likely to dominate over tidal torques at this distance Tajeddine et al. 2017) would not push the moon away. The disrupted material resulting from a sufficiently energetic collision with the heliocentric interloper then spreads along the moon's orbit to form the ring system.

The models conducted by Dubinski place a 2 Mimas mass satellite with a density of $1.03\ \mathrm{g\,cm^{-3}}$ (this is not considerably different than Tethys', which is $0.984\ \mathrm{g\,cm^{-3}}$) at a distance of 140,000 km from Saturn. Such a low density is chosen to match the composition of the rings, while also allowing for the reformation of Mimas from a rocky core remnant and orbiting debris. In this scenario, because the mass of the ring and newly accreted Mimas are comparable, outward migration of Mimas leads to a back reaction from the ring causing material to spread inwards faster than by viscosity alone. Thus, the proposed modified catastrophic collision scenario in this work suggests that the rings and Mimas formed together in the past few hundred million years. Evidence from measurements of Mimas' libration (Tajeddine et al. 2014) and models of orbital and internal evolution of Mimas (Neveu and Rhoden 2017) may be consistent with a recent reaccretion of Mimas from a remnant rocky core such as that proposed in this model. Despite the apparent plausibility of this scenario, it still suffers from the difficulty that such catastrophic events have a very low probability given the current flux of appropriately sized disruption-capable objects with a chance of $\sim 2\%$ in the last 0.5 Gyr (Lissauer et al. 1988; Dones et al. 2009; Charnoz et al. 2009a). Furthermore, maintaining a hypothetical Mimas-sized moon so close to the Roche zone of Saturn in the proposed MMR for most of the age of the solar system seems problematic, especially in light of Saturn's dissipative nature and what would probably be an even larger torque from the nearby rings (if a pre-existing ring were present).

In the end, this model addresses convincingly the composition and mass of the rings, but still strongly favors old rings (in particular if Saturn's tides are efficient at pushing moons away), without excluding that they might be young.

5.5 Recent Destabilization of a Distant Moon

The model presented in this section is the only one which relies on a highly dissipative Saturn. The recently observed fast migration of Titan (Lainey et al. 2020) has prompted the possibility that Saturn's obliquity of $\varepsilon = 27°$ (angle between the spin axis and the normal to the orbital plane around the Sun) could be explained by a substantial migration of this satellite over the age of the Solar System (Saillenfest et al. 2021a).

Obliquity Increase by a Spin-Orbit Resonance A satellite orbiting in the equatorial plane of a planet increases the planet's effective J_2. When the satellite migrates outwards, the latter grows. But the precession frequency of the spin axis of Saturn is $\dot{\Psi} = \alpha(1-e^2)^{3/2}\cos\varepsilon$, where $\alpha \propto J_2$ is the precession constant, e is the eccentricity of Saturn and ε the obliquity. As J_2 and α increase, Saillenfest et al. (2021a) suggest that the precession frequency $\dot{\Psi}$ encountered the Solar System's secular eigen-frequency s_8, corresponding to the precession of the node of Neptune's orbit around the Sun. The planet was then trapped in this secular spin-orbit resonance. As α keeps increasing while the orbits of Neptune and Saturn remain stable, $\cos\varepsilon$ must decrease to keep $\dot{\Psi} = \mid s_8 \mid$, and ε rises.

Saillenfest et al. (2021a,b) show that starting from an obliquity of a few (typically ~ 5) degrees, and assuming the migration rate for Titan given by Lainey et al. (2020), the spin-orbit resonance would have been entered ~ 1 Gyr ago, and the present obliquity can be explained.

Exit of the Spin-Orbit Resonance by the Loss of a Satellite Wisdom et al. (2022), however, contend that Saturn is not currently in the spin-orbit resonance with Neptune's nodal precession, but very close to it. Therefore, they suggest that the above scenario may well have happened, but that the resonance was broken very recently.

A way to break this spin-orbit resonance is to change Saturn's precession constant abruptly, and the authors suggest that this could be done by removing a previous Iapetus-mass satellite, which they call Chrysalis. They show that such a satellite, located initially between Titan and Iapetus, would be caught in the 3:1 MMR with Titan as Titan migrates outwards. This resonance would lead to the destabilisation of the orbit of Chrysalis, whose eccentricity approaches 1 in $\sim 10^5$ yr. The loss of Chrysalis extracts Saturn from the spin-orbit resonance with Neptune. The system then converges to the present situation, provided that this event occurred roughly 100 to 200 Myr ago. As this time is consistent with the exposure age of the rings and the young rings hypothesis, the authors suggest that Chrysalis may have penetrated Saturn's Roche radius, and been disrupted, with a small fraction ($\sim 1\%$) of its debris giving birth to the rings.

Discussion This model is appealing for its consistency, as it explains at the same time Saturn's obliquity, proximity to the spin-orbit resonance with Neptune, and apparently young rings. But one should keep in mind that the first two points do not necessary imply the third one: in Wisdom et al. (2022)'s simulations, there are as many hyperbolic orbits (escape) as Saturn grazing orbits, so that Chrysalis could be lost without ever having the chance to produce a ring system. This may actually be better for the satellite system, since having an Iapetus-mass ($\approx 2 \times 10^{21}$ kg) object repeatedly crossing the orbits of all the mid-sized moons of Saturn may be disastrous.

Moreover, for rings consistent with their current composition to be created by this mechanism, the debris from Chrysalis must be made almost entirely of pure water ice. But for stripped material from Chrysalis to be captured and form a ring, it must have a specific orbital angular momentum low enough that it corresponds to a circular orbit of radius $a_{eq} = a(1 - e^2)$ inside the Roche radius for ice. Since Chrysalis' semi major axis is very large (initially $a \sim 41$ R$_S$), $e \sim 1$ so that $a_{eq} \sim 2a(1 - e) = 2q$ where q is the pericentre distance. Hence $a_{eq} < r_{Roche}$ is equivalent to $q < \frac{1}{2} r_{Roche} \approx 70{,}000$ km. Not only does this leave little space above the surface of Saturn itself (R$_S \approx 60\,000$ km), but this is also inside the Roche limit for rock. So, the debris from Chrysalis would contain silicates as well, if Chrysalis itself is not pure water ice. In a sense, this model may likely only transfer the question of the amazing composition of the rings (which is the main argument for young rings) into that of Chrysalis. No object this size is known to be nearly pure water ice (although Tethys, about 1/3 the mass of Iapetus is quite icy); Iapetus itself is likely about 20% silicates.

Besides, since the mass of Iapetus is more than a hundred times the present mass of the rings, if the rings are made of Chrysalis debris, they must represent only 1% of the mass of this object. Indeed, in the $\lesssim 150$ Myr time-span suggested by Wisdom et al. (2022), the mass of the rings can not decrease significantly as shown in Fig. 5. Even including the effects of BT and extending the time to 400 Myr (which is not compatible with Saturn's current obliquity), one would still need to capture only a few percents of Chrysalis mass. This looks like fine tuning, and questions what happened to the rest of Chrysalis.

5.6 Destabilization of a Pre-Existing Mid-Size Moon System Due to the Evection Resonance

A recent origin scenario for the rings is based on the work of Ćuk et al. (2016) who point out that if the observed mid-sized satellite system of Saturn were ancient in origin, then the outermost moons Tethys, Dione and Rhea should have encountered several key MMRs predicted to be crossed in standard tidal evolution models (Murray and Dermott 1999; Meyer and Wisdom 2007). These would have excited their orbits beyond what we see today. Ćuk et al. argued that because this excitation is not observed, either tidal evolution in the Saturn system is very slow — which would be inconsistent with Lainey et al. (2012, 2017, 2020), the observations that motivated the edition of this collection (see Lainey et al. 2025) — or at least some of the moons are quite young, suggesting that some sort of catastrophic instability occurred within the system in the recent past. They identified the crossing by Rhea of the solar evection resonance[15] as the most likely candidate for the instability. A key argument is that if Rhea had survived crossing the evection resonance, its inclination would be over an order of magnitude larger than today's observed value of 0.35°, and such inclinations are difficult to damp (Murray and Dermott 1999; Chen et al. 2014). Rhea is currently just outside this resonance (located at 8.3 R$_S$) implying that it cannot have formed long ago and tidally migrated outwards to its current location. This appears to be the case even if the currently fast outward migration of Rhea (Lainey et al. 2020) is driven by resonance locking (Fuller et al. 2016) and not by equilibrium tides.

In this scenario, an outwardly-migrating proto-Rhea encounters the evection resonance, and is excited into crossing orbits with a moon immediately interior to it (proto-Dione), leading to a collision between them. This may have initiated a collisional cascade with any remaining moons orbiting progressively closer to the planet. Ćuk et al. hypothesized that subsequent destabilization, destruction and reaccretion of (at least some) of these mid-sized moons could have also led to the formation of rings at the same time, within the last few

[15] The evection resonance is where a moon's apsidal precession period about Saturn is equal to Saturn's orbital period around the Sun.

100 Myr. They further suggested that Titan may have acquired its non-zero eccentricity during this period of instability, and that the present moons interior to Titan would have crater populations characterized by planetocentric rather than heliocentric impactors.

Production of Icy Debris and Its Long-Term Fate Initial examination of the Ćuk et al. (2016) scenario including the long-term evolution of this debris disk was conducted by Hyodo and Charnoz (2017). These authors performed a series of SPH simulations, mirroring the initial conditions[16] of Ćuk et al. (with 2×10^5 particles each, and varying impact geometry). They then dynamically evolved the system by extracting 20K SPH particles for incorporation into an N-body code featuring dissipative hard-sphere collisions (no further fragmentation is considered) and tidally-modified accretion conditions (following the criterion of Canup and Esposito 1995) for the specific case of an impact angle and speed of 45° and $3\ \mathrm{km\,s^{-1}}$. They found that under these assumptions both collisional damping and gravitational reaccumulation were effective. Since all debris orbits intersect close to the original impact point of the two satellites, an intense collision activity takes place there such that even particles put on very eccentric orbits have a high probability to re-collide with the satellite remnants and other local debris in the original region of impact. As a result, mainly re-accumulation into one or two bodies close to the original locations of the disrupted bodies is observed, without the implantation of low-eccentricity particles close to the Roche limit (see Figs. 3 and 4 of Hyodo and Charnoz 2017). Admittedly, the impact simulation did not produce any particle with an orbital angular momentum corresponding to a circular orbit inside the Roche radius (in contrast to Teodoro et al. 2023, see below). Thus, there was little to no possibility of ring formation in the subsequent N-body simulation. One could imagine that the debris would spread inwards, but Hyodo and Charnoz estimated analytically the viscous spreading time and the gravitational reaccumulation time and found that the latter is five order of magnitude shorter than the former.

Recent work by Teodoro et al. (2023) explored the catastrophic collision with much higher-resolution Smoothed Particle Hydrodynamics (SPH) simulations of impacts between two differentiated moons. Figure 6 shows the result of one such calculation for an impact angle of 35° at $3\ \mathrm{km\,s^{-1}}$ and $10^{7.5}$ SPH particles. The collision occurs at around $a_{\mathrm{com}} = 7.3$ $\mathrm{R_S}$ at proto-Dione's apoapsis and descending node, and proto-Rhea's periapsis and ascending node. The collision leads to a significant amount of debris and massive fragments being sent throughout the system, distributed in two branches as seen in the figure (see also Fig. 2 of Hyodo and Charnoz 2017). The material on the inner branch loses energy post-collision and finds itself just after the collision at their apocenter. Material on the outer branch, on the other hand, has gained orbital energy and leaves the collisional debris at the pericenter of their new orbit. On the inner branch, large amounts of debris and fragments (not distinguished here) are placed onto orbits that intersect with those of moons orbiting further inward, as indicated by the dotted lines, and some material even penetrates within Saturn's Roche limit.[17] Because proto-Dione and proto-Rhea are both assumed to have differentiated,

[16]These authors set the radial location of the impact at the evection resonance, but this is Proto-Rhea's semi-major axis at the time. Its $e \sim 0.12$ places the crossing orbit location and subsequent collision with proto-Dione further inward (see Teodoro et al. 2023).

[17]We note that while Teodoro et al. (2023) find a total amount of material crossing the Roche limit of nearly as much as the present ring mass in some cases, Hyodo and Charnoz (2017) did not find any material on their inner branch that came close to crossing the Roche limit (see their Fig. 2), although they had a lower resolution. Teodoro et al. (2023) tried to reproduce Hyodo and Charnoz (2017)'s results and still found material crossing the Roche limit in these cases. There may be several reasons for this discrepancy between the models which the former authors discuss in their paper.

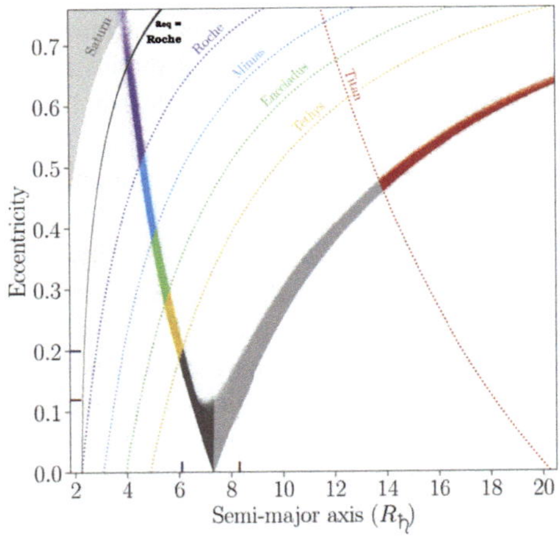

Fig. 6 Eccentricities and semi-major axes of orbiting fragments and debris from a collision between proto-Dione and proto-Rhea at an impact angle of $\beta = 35°$, and an impact velocity of $3\ \mathrm{km\,s^{-1}}$, 9 hours after impact. Each point represents one of the $10^{7.5}$ SPH particles of ($\sim 10^{14}$ kg) each. Large gravitationally bound fragments are not distinguished in this plot. The dotted color curves represent where a fragment's periapse crosses the Roche limit or the orbits of interior moons, or its apoapse (for the outer branch) crosses Titan. The semi-major axes of proto-Dione and proto-Rhea are indicated by the dark red and dark blue tics, respectively. The black solid line on the left indicates where the orbital angular momentum corresponds to a circular orbit at the Roche limit. A small fraction of the ejecta lies between this line and Saturn crossing trajectories (gray shaded region at the top left corner). Adapted from Teodoro et al. (2023)

their largest surviving fragments tend to be core material, while the icy mantle material dominates the component crossing the orbits of the inner moons, the Roche limit and the orbit of Titan. Teodoro et al. (2023) find that for the example shown here, the material delivered interior to the Roche limit is 100% water ice, although if moons progressively further inward are also disrupted, they expect some of their core material could also be delivered (see Fig. 7 of Teodoro et al. 2023, and discussion therein).

However, as discussed above, crossing the Roche limit is not sufficient to lead to ring formation; the material must be placed on circular orbits there. As seen in Sect. 5.5, if angular momentum is conserved, the equivalent circular radius is $a_{\mathrm{eq}} = a(1-e^2)$. To guide the eye, the line $a_{\mathrm{eq}} = r_{\mathrm{Roche}}$ has been added onto the figure from the original paper and is shown as a black solid curve. In the different impact scenarios tested, Teodoro et al. (2023) find between 10^{18} and 10^{19} kg of debris above this line (corresponding to $e \sim 0.7$ in the inner branch). In this case, spreading is almost not necessary. But the evolution and fate of the debris, and where would they eventually be circularized after collisions, possible reaccummulation and encounters is not addressed in this paper.

Summary This new scenario for a recent formation or rejuvenation of the rings clearly requires more detailed modeling. High-resolution simulations of the debris disk's collisional evolution, including gravitational scattering by the inner moons, and impacts of massive fragments with moons orbiting further inward that could potentially lead to a collisional cascade (Ćuk et al. 2016), are necessary in order to constrain the implications for the rings and moons.

Preliminary results show that this model has strong implications for the Saturnian system, associated with short timescales. In particular, the significant amount of debris that crosses the orbit of, and can potentially impact Titan may have implications for the evolution of

Titan's atmosphere. Also, it offers a source of planetocentric impactors which could affect the apparent ages of the surfaces of the satellites – so far estimated to several hundred million years if not billions of years, through arguments which are not only based on cratering rates (e.g., Dalle Ore et al. 2015; Kirchoff et al. 2018; López-Oquendo et al. 2019; Rivera-Valentín et al. 2020), making the recent formation of these objects unlikely. More constraints from the geology would help to determine the maximum amount of resurfacing that they recently suffered. In addition, the interactions of the debris with other moons (e.g., scattering), or with itself (collisional evolution) are likely to be complicated, emphasizing the need for more detailed simulations to determine the ultimate fate of the collisional debris, and whether or not some of that material will form rings.

6 Discussion

In this article we have addressed fundamental questions about the age, origin and future of Saturn's rings. Though all the giant planets have ring systems, Saturn's are by far more massive than the others, and for this alone they demand special scrutiny; but the fact that they are predominantly water ice unlike their other outer solar system counterparts further adds to their apparent uniqueness. Does the relatively massive nature and icy composition of Saturn's rings tell us something about their formation and evolution? These two properties of the rings form the main basis for a vigorous debate about how and when they came to be. We began by establishing that the "when" can be different, just like any planetary surface, depending on whether you consider the age of structures, exposure to erosive mechanisms or the purely dynamical or formation age. For the rings, this refers to the development of structures such as the Cassini division or C ring ramp; the time that meteoroid bombardment has impacted and darkened the rings; or the time over which the rings have evolved due to angular momentum exchange. We concluded that individual structures in the rings are the least likely gauge of the rings' absolute age.

Saturn's rings were initially considered by default to be an eternal ornament of the planet. In the post-Voyager era, the idea emerged that the rings are young, mostly composed of water ice, are on the order of one Mimas mass and experience an (estimated) micrometeoroid flux that suggested a darkening-derived exposure age that was much less than the age of the Solar System. But large uncertainties remained in many key ring parameters which would have to wait until the arrival of Cassini to be clarified. In particular, Cassini measured three key quantities which provide strong constraints on formation models: the composition of the rings (in other words the fraction of non-icy material they contain; Zhang et al. 2017a,b); the bombardment rate of the rings by interplanetary micrometeoroids (Kempf et al. 2023); and the total mass of the rings (Iess et al. 2019). These three quantities allow us to estimate the exposure age, while the mass also constrains the rings' viscous evolution time, or dynamical age. Below we summarize the principal results and our discussion of the various age estimates, before trying to fit things into a consistent global picture.

Exposure Age The time it takes for the rings to reach their present composition, assuming that they start as pure water ice and suffer a constant interplanetary bombardment flux, can be obtained from the three Cassini observations mentioned above. To derive a darkening-derived exposure age one must include the gravitational focusing which greatly increases the flux of micrometeoroids near Saturn, as well as the pollution efficiency η, the fraction of the impacting mass which remains spectrally detectable in the rings as non-icy material.

Adopting what seem to be reasonable values for these parameters, and assuming an unvarying ring mass and impact flux, one obtains an exposure age on the order of $100 - 400$ Myrs for the rings (Kempf et al. 2023). The most uncertain parameter in this age estimate is probably the value of η, as discussed in Sect. 2.2. If η is ten times smaller than generally assumed (as suggested by Hyodo et al. 2025), then these ages should be multiplied by ten, reaching $1 - 4$ Gyrs (however, see discussion at the end of Sect. 2.2). Other factors that can either shorten or lengthen this age estimate are reviewed in Sect. 2.3.

In Sect. 2.4 we point out that the rings are not only being steadily 'polluted' by interplanetary material. They are also losing mass into Saturn in the form of nanograins and individual molecules, at a rate that far exceeds the observed micrometeoroid impact rate of interplanetary material into the entire ring system. This phenomenon potentially affects the ring composition as much as the bombardment, and opens the possibility for ring cleaning. The inflow takes place in the equatorial region of Saturn (INMS and MIMI measurements) – presumably via the D ring – and at the mid-latitudes connected via magnetic field lines to the main rings from charged material being ejected from the ring particles (CDA). The observed equatorial mass influx from INMS into Saturn appears to have a much lower water content than do the main rings themselves (see the extended discussion in Miller et al. 2024, this collection), suggesting a preferential elimination of non-icy material (i.e. cleaning), but it is probably a local (restricted to the D-ring) and possibly transient phenomenon (since it should pollute Saturn's upper atmosphere more than what is observed).

In contrast, the CDA inflow originates from across the rings. It corresponds to an amount of non-icy material that is similar to that entering the rings due to micrometeoroid bombardment (MB). Although the "ring rain" phenomenon may imply a water inflow rate (not observed by CDA) such that in total the composition of the material lost by the rings is similar to that of the C-ring (which is still much richer in non-icy material than the A and B rings), the mere fact that the rate of incoming non-icy material is similar to what is falling into the planet raises the possibility, within the uncertainties, that the influx of interplanetary material into the main rings could be largely balanced by the outflow of nanograins along magnetic field lines into Saturn's atmosphere. This would effectively mean that $\eta \approx 0$ (Sect. 2.2) and that the rings could not be darkened at all. In addition, it has recently been suggested that the organics in the ring particles can be destroyed by UV radiation, and eliminated with time. Thus, even though the color contrast between the optically thin and thick regions of the rings is most easily explained by pollution and darkening due to MB, Cassini's in situ observations prompt us to question the meaning and relevance of the exposure age.

Dynamical Age It is well known that the rings spread due to the various interactions between ring particles, which result in an outward transfer of orbital angular momentum. This can be modeled as a viscosity, a component of which is proportional to the surface mass density of the rings squared (Daisaka et al. 2001). As a result, denser rings spread faster and, as they lose mass into Saturn or beyond the Roche radius, their mass decreases quickly and their evolution catches up with that of less dense rings (see Fig. 4, right panel). The convergence of the mass evolution curves to an asymptotic value close to the present mass over the age of the solar system, irrespective of their initial mass as long as it is appreciably greater than their current mass, appears to be a compelling result, suggesting that the present ring mass is not the product of chance but of a well-understood dynamical evolution (Salmon et al. 2010; Crida and Charnoz 2014). Under such circumstances, one can fairly say that the dynamical age of the rings is a few billion years and that they could be primordial; on the other hand, we can not say with any confidence what their initial mass might have been.

In fact, if the rings were today significantly more massive, they would have to be young. Massive rings can not stay massive for a long time. The dependencies on the ring mass of the dynamical and exposure ages are therefore opposite: the exposure age is directly proportional to the mass of the rings, so that more massive rings are likely to be older from this point of view.

In Sect. 4.2, however, we have shown that the micrometeoroid bombardment can also drive ring dynamical evolution once viscosity weakens sufficiently, or the surface density drops below a threshold value that depends both on the impact rate and on the effectiveness of ballistic transport (BT). With the current interplanetary dust flux, these effects make it impossible for rings to survive for the age of the Solar System. The present ring mass is reached in ~ 1.5 Gyr when one considers only mass loading, whatever the initial mass. Note that this is neither consistent with original rings formed together with Saturn nor with young rings as old as their exposure age. When one also adds the effects of BT, the current ring mass is reached on timescales of a few hundred Myr for impact ejecta yields of $10^4 \lesssim Y \lesssim 10^5$, for an initial ring mass of $\sim 1-3$ Mimas masses. It is therefore possible for the exposure and dynamical ages to be similar within this range of parameters, although the dynamical effects of BT are independent of whether the rings are being polluted or not.

Structures Although it is clear that individual structures can only provide a lower bound to the age of the rings since they could have formed in pre-existing rings (e.g. the Cassini division), the global density profile of the rings provides a strong constraint. In the frame of young rings born recently with the present mass, viscosity has little time to play a significant role so that the initial mass distribution should be consistent with a short time evolution to the present. Salmon et al. (2010) find a qualitative agreement between the global shape of the A and B rings and their models after ~ 150 Myrs of viscous evolution (their Fig. 17) starting with an initial 1 Mimas mass annulus; however, for longer simulations the inner B ring moves inwards past its current location (see their Fig. 18, and also Fig. 13 of Estrada and Durisen 2023). We have seen in Sect. 3.2 that the effects of MB and BT on a pre-existing B ring inner edge can further sharpen it against viscous spreading and produce a linear ramp similar to the B-C transition and an extended C ring analog in about 10 million years. The B ring inner edge will still migrate inward under the dynamical effects of MB, but at a different rate than by viscous spreading. In other words, the inner edge of the B-ring seems to be a young structure.

Global Picture and Conclusion Clearly, the formation age of the rings can not be measured directly. Some aspects of the above discussion suggest old (perhaps primordial) rings — chiefly the purely-viscous evolution models and the greater formation possibilities afforded by conditions in the early solar system — while others argue in favor of young rings, dating back to the time when dinosaurs roamed the Earth ~ 200 Myr ago. The latter include the exposure age of the A and B rings, the structural age of the C ring (if it was formed by BT) as well as the location of the B ring's inner edge (with all the caution that the exposure age and the age of structures may differ from the formation age), and dynamical models that take into account the slowing down of ring particles by micrometeoroids bombardment and a model of the ballistic transport of the ejecta. Depending on which of these arguments they consider more compelling, the reader may prefer one or the other hypothesis.

In either case, to make the story complete, a plausible formation model is needed. In Sect. 5, we have reviewed the various formation models proposed in the literature, with particular attention to whether they favor ancient or young rings.

If the rings are old, several formation models can produce icy rings, possibly much more massive than the present system. In this case, the present rings would be a product of the viscous evolution of the primordial rings, along with the accretion near the Roche limit of most or all of the small and mid-sized moons (Charnoz et al. 2011). In this case, while the ring mass converges to the present mass in a few billion years (but may drop shortly after this under the dynamical effect of the bombardment), the structures like the inner edge of the B ring and the Cassini division must have been carved recently, and the purity of the rings must rely on some cleaning mechanism.

If, on the other hand, the rings are young and their formation age corresponds roughly to their exposure age of a few hundred Myr, then they must still be created from nearly pure water ice. How does one deposit this mass of water ice within the Roche limit of Saturn only a few hundred million years ago? As the high tidal dissipation inside the planet makes it almost impossible to keep a satellite close to the Roche radius for billions of years, one presumably must rely on a catastrophic event happening in the outer saturnian system. The loss of a satellite (e.g., Chrysalis), or a catastrophic collision between the progenitors of Rhea and Dione have been proposed, because the present orbital parameters of the remaining satellites and the obliquity of Saturn suggest that these events may have happened fairly recently. However, the link between these events and the production of icy rings of the right mass remains unclear and needs to be modeled in detail. In particular, the fate of material which by simple energy and angular momentum arguments may have settled into equilibrium circular orbits inside Saturn's Roche limit requires further investigation in order to determine if the rings actually can be formed (or rejuvenated) in this way.

For the present, we must conclude that there is no fully worked-out picture that describes the formation and subsequent evolution of Saturn's rings to their observed state, though there are several promising possibilities. In particular, we still do not know if the rings represent the outcome of a singular event, unlikely to be repeated elsewhere, or of the natural evolution of a giant planet and its retinue of satellites. In this context, we would like to emphasize that, even if the rings we see now are relatively young, this does not exclude the possibility that a more massive ring system existed in the distant past but disappeared a few billion years ago under the dynamical effects of viscous evolution and meteoritic bombardment. We should also consider the possibility that the same processes acting at Uranus and/or Neptune may have led to the formation of primordial massive rings, only the ghostly vestiges of which remain today (Estrada and Durisen 2023). In each case, the regular satellite system not only of Saturn but also of Uranus and/or Neptune may have been formed by the spreading of such a massive, primordial ring, as is suggested today by their similar mass-distance distributions (Crida and Charnoz 2012).

Acknowledgements This paper resulted from the Workshop "New Vision of the Saturnian System in the Context of a Highly Dissipative Saturn" held at the International Space Science Institute (ISSI) in Bern (Switzerland). The authors thank M. Ćuk, B. Noyelles and C. Charnoz for valuable comments. We also thank the two referees for their remarks and suggestions which improved the manuscript.

Declarations

Competing Interests The authors have no competing interests to declare that are relevant to the content of this article.

References

Auer S, Grün E, Srama R, et al (2002) The charge and velocity detector of the cosmic dust analyzer on Cassini. Planet Space Sci 50(7–8):773–779. https://doi.org/10.1016/S0032-0633(02)00019-3

Avdellidou C, Price MC, Delbo M, et al (2017) Survival of the impactor during hypervelocity collisions - II. An analogue for high-porosity targets. Mon Not R Astron Soc 464(1):734–738. https://doi.org/10.1093/mnras/stw2381. arXiv:1612.05060 [astro-ph.EP]

Baillié K, Noyelles B, Lainey V, et al (2019) Formation of the Cassini division - I. Shaping the rings by Mimas inward migration. Mon Not R Astron Soc 486(2):2933–2946. https://doi.org/10.1093/mnras/stz548

Baruteau C, Crida A, Paardekooper SJ, et al (2014) Planet-disk interactions and early evolution of planetary systems. In: Beuther H, Klessen RS, Dullemond CP, et al (eds) Protostars and planets VI, pp 667–689. https://doi.org/10.2458/azu_uapress_9780816531240-ch029. arXiv:1312.4293

Bérard D, Sicardy B, Camargo JIB, et al (2017) The structure of Chariklo's rings from stellar occultations. Astron J 154(4):144. https://doi.org/10.3847/1538-3881/aa830d. arXiv:1706.00207 [astro-ph.EP]

Bierson CJ, Fortney JJ, Trinh KT, et al (2023) The effect of Jupiter's luminosity on the Galilean satellites. In: 54th Lunar and Planetary Science Conference. LPI contributions, p 2263

Blanc M, Crida A, Shibaike Y, et al (2025) Understanding the formation of Saturn's regular moons in the context of giant planet moons formation scenarios. Space Sci Rev 221:35. https://doi.org/10.1007/s11214-025-01156-8

Borderies N, Goldreich P, Tremaine S (1983) Perturbed particle disks. Icarus 55(1):124–132. https://doi.org/10.1016/0019-1035(83)90055-6

Bottke WF, Nesvorný D, Vokrouhlický D, et al (2010) The irregular satellites: the most collisionally evolved populations in the Solar System. Astron J 139(3):994–1014. https://doi.org/10.1088/0004-6256/139/3/994

Braga-Ribas F, Sicardy B, Ortiz JL, et al (2014) A ring system detected around the Centaur (10199) Chariklo. Nature 508(7494):72–75. https://doi.org/10.1038/nature13155. arXiv:1409.7259 [astro-ph.EP]

Canup RM (2010) Origin of Saturn's rings and inner moons by mass removal from a lost Titan-sized satellite. Nature 468:943–946. https://doi.org/10.1038/nature09661

Canup RM, Esposito LW (1995) Accretion in the Roche zone: coexistence of rings and ring moons. Icarus 113(2):331–352. https://doi.org/10.1006/icar.1995.1026

Charnoz S, Dones L, Esposito LW, et al (2009a) Origin and evolution of Saturn's ring system. In: Dougherty MK, Esposito LW, Krimigis SM (eds) Saturn from Cassini-Huygens, p 537. https://doi.org/10.1007/978-1-4020-9217-6_17

Charnoz S, Morbidelli A, Dones L, et al (2009b) Did Saturn's rings form during the Late Heavy Bombardment? Icarus 199(2):413–428. https://doi.org/10.1016/j.icarus.2008.10.019. arXiv:0809.5073 [astro-ph]

Charnoz S, Salmon J, Crida A (2010) The recent formation of Saturn's moonlets from viscous spreading of the main rings. Nature 465(7299):752–754. https://doi.org/10.1038/nature09096

Charnoz S, Crida A, Castillo-Rogez JC, et al (2011) Accretion of Saturn's mid-sized moons during the viscous spreading of young massive rings: solving the paradox of silicate-poor rings versus silicate-rich moons. Icarus 216(2):535–550. https://doi.org/10.1016/j.icarus.2011.09.017. arXiv:1109.3360 [astro-ph.EP]

Chen EMA, Nimmo F, Glatzmaier GA (2014) Tidal heating in icy satellite oceans. Icarus 229:11–30. https://doi.org/10.1016/j.icarus.2013.10.024

Chiang EI, Culter CJ (2003) Three-dimensional dynamics of narrow planetary rings. Astrophys J 599(1):675–685. https://doi.org/10.1086/379151. arXiv:astro-ph/0309248 [astro-ph]

Ciarniello M, Filacchione G, Nicholson PD, et al (2024) The origin and composition of Saturn's ring moons. Space Sci Rev 220:72. https://doi.org/10.1007/s11214-024-01103-z

Colwell JE, Nicholson PD, Tiscareno MS, et al (2009) The structure of Saturn's rings. In: Dougherty MK, Esposito LW, Krimigis SM (eds) Saturn from Cassini-Huygens, p 375. https://doi.org/10.1007/978-1-4020-9217-6_13

Crida A, Bramardi G (2021) Effects of cleaning in Saturn's rings. In: European Planetary Science Congress, EPSC2021-393. https://doi.org/10.5194/epsc2021-393

Crida A, Charnoz S (2010) Solar System: recipe for making Saturn's rings. Nature 468(7326):903–905. https://doi.org/10.1038/nature09738

Crida A, Charnoz S (2012) Formation of regular satellites from ancient massive rings in the Solar System. Science 338(6111):1196. https://doi.org/10.1126/science.1226477. arXiv:1301.3808 [astro-ph.EP]

Crida A, Charnoz S (2014) Complex satellite systems: a general model of formation from rings. In: Complex planetary systems. IAU Symposium, vol S310. Cambridge University Press, pp 182–189. https://doi.org/10.1017/S1743921314008229

Crida A, Charnoz S, Hsu HW, et al (2019) Are Saturn's rings actually young? Nat Astron 3:967–970. https://doi.org/10.1038/s41550-019-0876-y

Ćuk M, Dones L, Nesvorný D (2016) Dynamical evidence for a late formation of Saturn's moons. Astrophys J 820(2):97. https://doi.org/10.3847/0004-637X/820/2/97. arXiv:1603.07071 [astro-ph.EP]
Cuzzi JN, Durisen RH (1990) Bombardment of planetary rings by meteoroids: general formulation and effects of Oort cloud projectiles. Icarus 84(2):467–501. https://doi.org/10.1016/0019-1035(90)90049-F
Cuzzi JN, Estrada PR (1998) Compositional evolution of Saturn's rings due to meteoroid bombardment. Icarus 132(1):1–35. https://doi.org/10.1006/icar.1997.5863
Daisaka H, Tanaka H, Ida S (2001) Viscosity in a dense planetary ring with self-gravitating particles. Icarus 154(2):296–312. https://doi.org/10.1006/icar.2001.6716
Dalle Ore CM, Cruikshank DP, Mastrapa RME, et al (2015) Impact craters: an ice study on Rhea. Icarus 261:80–90. https://doi.org/10.1016/j.icarus.2015.08.008
Davidsson BJR (1999) Tidal splitting and rotational breakup of solid spheres. Icarus 142(2):525–535. https://doi.org/10.1006/icar.1999.6214
De Pater I, Hamilton DP, Showalter MR, et al (2018) The rings of Jupiter. In: Tiscareno MS, Murray CD (eds) Planetary ring systems. Properties, structure, and evolution, pp 125–134. https://doi.org/10.1017/9781316286791.006
Dones L (1991) A recent cometary origin for Saturn's rings? Icarus 92:194–203. https://doi.org/10.1016/0019-1035(91)90045-U
Dones HC, Agnor CB, Asphaug E (2007) Formation of Saturn's rings by tidal disruption of a Centaur. In: AAS/Division for Planetary Sciences Meeting abstracts, vol #39, p 420
Dones L, Chapman CR, McKinnon WB, et al (2009) Icy satellites of Saturn: impact cratering and age determination. In: Dougherty MK, Esposito LW, Krimigis SM (eds) Saturn from Cassini-Huygens. Springer, Dordrecht, pp 613–635. https://doi.org/10.1007/978-1-4020-9217-6_19
Doyle LR, Dones L, Cuzzi JN (1989) Radiative transfer modeling of Saturn's outer B ring. Icarus 80:104–135. https://doi.org/10.1016/0019-1035(89)90163-2
Dubinski J (2019) A recent origin for Saturn's rings from the collisional disruption of an icy moon. Icarus 321:291–306. https://doi.org/10.1016/j.icarus.2018.11.034
Duffard R, Pinilla-Alonso N, Ortiz JL, et al (2014) Photometric and spectroscopic evidence for a dense ring system around Centaur Chariklo. Astron Astrophys 568:A79. https://doi.org/10.1051/0004-6361/201424208. arXiv:1407.4599 [astro-ph.EP]
Durisen RH (1984) Transport effects due to particle erosion mechanisms. In: Greenberg R, Brahic A (eds) Planetary rings. University of Arizona Press, Tucson, pp 416–446
Durisen RH (1995) An instability in planetary rings due to ballistic transport. Icarus 115(1):66–85. https://doi.org/10.1006/icar.1995.1079
Durisen RH, Estrada PR (2023) Large mass inflow rates in Saturn's rings due to ballistic transport and mass loading. Icarus 400:115221. https://doi.org/10.1016/j.icarus.2022.115221. arXiv:2305.13578 [astro-ph.EP]
Durisen RH, Cramer NL, Murphy BW, et al (1989) Ballistic transport in planetary ring systems due to particle erosion mechanisms I. Theory, numerical methods, and illustrative examples. Icarus 80(1):136–166. https://doi.org/10.1016/0019-1035(89)90164-4
Durisen RH, Bode PW, Cuzzi JN, et al (1992) Ballistic transport in planetary ring systems due to particle erosion mechanisms II. Theoretical models for Saturn's A- and B-ring inner edges. Icarus 100(2):364–393. https://doi.org/10.1016/0019-1035(92)90106-H
Elliott JP, Esposito LW (2011) Regolith depth growth on an icy body orbiting Saturn and evolution of bidirectional reflectance due to surface composition changes. Icarus 212(1):268–274. https://doi.org/10.1016/j.icarus.2010.10.031
Esposito LW, Elliott JP, Bradley ET (2024) Space weathering provides a lower limit on the age of Saturn's rings. In: European Planetary Science Congress, EPSC2024-562. https://doi.org/10.5194/epsc2024-562
Estrada PR, Durisen RH (2011) Modeling the long term evolution of the C ring due to meteoroid bombardment. In: Proceedings of the rings 2011 worskhop, Ithaca NY, p 2263
Estrada PR, Durisen RH (2023) Constraints on the initial mass, age and lifetime of Saturn's rings from viscous evolutions that include pollution and transport due to micrometeoroid bombardment. Icarus 400:115296. https://doi.org/10.1016/j.icarus.2022.115296. arXiv:2305.13609 [astro-ph.EP]
Estrada PR, Durisen RH, Cuzzi JN, et al (2015) Combined structural and compositional evolution of planetary rings due to micrometeoroid impacts and ballistic transport. Icarus 252:415–439. https://doi.org/10.1016/j.icarus.2015.02.005
Estrada PR, Durisen RH, Latter HN (2018) Meteoroid bombardment and ballistic transport in planetary rings. In: Tiscareno MS, Murray CD (eds) Planetary ring systems. Properties, structure, and evolution. Cambridge University Press, pp 198–224. https://doi.org/10.1017/9781316286791.009
Ferguson SN, Rhoden AR, Kirchoff MR, et al (2022) A unique Saturnian impactor population from elliptical craters. Earth Planet Sci Lett 593:117652. https://doi.org/10.1016/j.epsl.2022.117652

Fortney JJ, Marley MS, Barnes JW (2007) Planetary radii across five orders of magnitude in mass and stellar insolation: application to transits. Astrophys J 659(2):1661–1672. https://doi.org/10.1086/512120. arXiv:astro-ph/0612671 [astro-ph]

French RG, Nicholson PD (2000) Saturn's rings II. Particle sizes inferred from stellar occultation data. Icarus 145(2):502–523. https://doi.org/10.1006/icar.2000.6357

French RG, Nicholson PD, Porco CC, et al (1991) Dynamics and structure of the Uranian rings. In: Bergstralh JT, Miner ED, Matthews MS (eds) Uranus. University of Arizona Press, Tucson, pp 327–409

French RG, Salo H, McGhee CA, et al (2007) HST observations of azimuthal asymmetry in Saturn's rings. Icarus 189(2):493–522. https://doi.org/10.1016/j.icarus.2007.02.019

French RG, Nicholson PD, McGhee-French CA, et al (2016) Noncircular features in Saturn's rings III: the Cassini division. Icarus 274:131–162. https://doi.org/10.1016/j.icarus.2016.03.017

French RG, McGhee-French CA, Gordon M, et al (2023) Uranus ring occultation observations: 1977-2006. Icarus 395:115474. https://doi.org/10.1016/j.icarus.2023.115474

Fuller J, Luan J, Quataert E (2016) Resonance locking as the source of rapid tidal migration in the Jupiter and Saturn moon systems. Mon Not R Astron Soc 458(4):3867–3879. https://doi.org/10.1093/mnras/stw609. arXiv:1601.05804 [astro-ph.EP]

Goldreich P, Porco CC (1987) Shepherding of the Uranian rings. II. Dynamics. Astron J 93:730. https://doi.org/10.1086/114355

Goldreich P, Tremaine SD (1978) The velocity dispersion in Saturn's rings. Icarus 34(2):227–239. https://doi.org/10.1016/0019-1035(78)90164-1

Goldreich P, Murray N, Longaretti PY, et al (1989) Neptune's story. Science 245(4917):500–504. https://doi.org/10.1126/science.245.4917.500

Gomes R, Levison HF, Tsiganis K, et al (2005) Origin of the cataclysmic Late Heavy Bombardment period of the terrestrial planets. Nature 435(7041):466–469. https://doi.org/10.1038/nature03676

Grigorieva A, Thébault P, Artymowicz P, et al (2007) Survival of icy grains in debris discs. The role of photosputtering. Astron Astrophys 475(2):755–764. https://doi.org/10.1051/0004-6361:20077686. arXiv:0709.0811 [astro-ph]

Griveaud P, Crida A, Petit AC, et al (2024) The Solar System could have formed in a low-viscosity disc: a dynamical study from giant planet migration to the Nice model. Astron Astrophys 688:A202. https://doi.org/10.1051/0004-6361/202450340. arXiv:2406.20075 [astro-ph.EP]

Grossman AW (1990) Microwave imaging of Saturn's deep atmosphere and rings. PhD thesis, California Institute of Technology

Grün E, Zook HA, Fechtig H, et al (1985) Collisional balance of the meteoritic complex. Icarus 62(2):244–272. https://doi.org/10.1016/0019-1035(85)90121-6

Harbison RA, Nicholson PD, Hedman MM (2013) The smallest particles in Saturn's A and C rings. Icarus 226(2):1225–1240. https://doi.org/10.1016/j.icarus.2013.08.015. arXiv:1312.2927 [astro-ph.EP]

Harris AW (1984) The origin and evolution of planetary rings. In: Greenberg R, Brahic A (eds) Planetary rings. University of Arizona Press, Tucson, pp 641–659

Hedman MM, Nicholson PD (2016) The B-ring's surface mass density from hidden density waves: less than meets the eye? Icarus 279:109–124. https://doi.org/10.1016/j.icarus.2016.01.007. arXiv:1601.07955 [astro-ph.EP]

Hedman MM, Showalter MR (2016) A new pattern in Saturn's D ring created in late 2011. Icarus 279:155–165. https://doi.org/10.1016/j.icarus.2015.09.017. arXiv:1509.04755 [astro-ph.EP]

Hedman MM, Burns JA, Showalter MR (2015) Corrugations and eccentric spirals in Saturn's D ring: new insights into what happened at Saturn in 1983. Icarus 248:137–161. https://doi.org/10.1016/j.icarus.2014.10.021. arXiv:1410.3761 [astro-ph.EP]

Hsu HW, Schmidt J, Kempf S, et al (2018) In situ collection of dust grains falling from Saturn's rings into its atmosphere. Science 362(6410):aat3185. https://doi.org/10.1126/science.aat3185

Hyodo R, Charnoz S (2017) Dynamical evolution of the debris disk after a satellite catastrophic disruption around Saturn. Astron J 154(1):34. https://doi.org/10.3847/1538-3881/aa74c9. arXiv:1705.07554 [astro-ph.EP]

Hyodo R, Charnoz S, Ohtsuki K, et al (2017) Ring formation around giant planets by tidal disruption of a single passing large Kuiper Belt object. Icarus 282:195–213. https://doi.org/10.1016/j.icarus.2016.09.012. arXiv:1609.02396 [astro-ph.EP]

Hyodo R, Genda H, Madeira G (2025) Pollution resistance of Saturn's ring particles during micrometeoroid impact. Nat Geosci 18(1):44–49. https://doi.org/10.1038/s41561-024-01598-9

Iess L, Militzer B, Kaspi Y, et al (2019) Measurement and implications of Saturn's gravity field and ring mass. Science 364(6445):aat2965. https://doi.org/10.1126/science.aat2965

Ip WH (1984) Ring torque of Saturn from interplanetary meteoroid impact. Icarus 60(3):547–552. https://doi.org/10.1016/0019-1035(84)90163-5

Johnston WR (2018) TNO and Centaur Diameters, Albedos, and Densities V1.0 This data set is a compilation of published diameters, albedos, and densities for Transneptunian Objects (TNOs) and Centaurs. A total of 194 objects are listed, many with more than one entry. This version covers published values through 31 March 2018. NASA Planetary Data System, p 6. https://doi.org/10.26033/y5sn-4t02

Kempf S, Altobelli N, Schmidt J, et al (2023) Micrometeoroid infall onto Saturn's rings constrains their age to no more than a few hundred million years. Sci Adv 9(19):eadf8537. https://doi.org/10.1126/sciadv.adf8537

Kirchoff MR, Bierhaus EB, Dones L, et al (2018) Cratering histories in the saturnian system. In: Schenk PM, Clark RN, Howett CJA, et al (eds) Enceladus and the icy moons of Saturn, p 267. https://doi.org/10.2458/azu_uapress_9780816537075-ch013

Kissel J, Krueger FR (1987) The organic component in dust from comet Halley as measured by the PUMA mass spectrometer on board Vega 1. Nature 326(6115):755–760. https://doi.org/10.1038/326755a0

Lainey V, Karatekin Ö, Desmars J, et al (2012) Strong tidal dissipation in Saturn and constraints on Enceladus' thermal state from astrometry. Astrophys J 752(1):14. https://doi.org/10.1088/0004-637X/752/1/14. arXiv:1204.0895 [astro-ph.EP]

Lainey V, Jacobson RA, Tajeddine R, et al (2017) New constraints on Saturn's interior from Cassini astrometric data. Icarus 281:286–296. https://doi.org/10.1016/j.icarus.2016.07.014. arXiv:1510.05870 [astro-ph.EP]

Lainey V, Casajus LG, Fuller J, et al (2020) Resonance locking in giant planets indicated by the rapid orbital expansion of Titan. Nat Astron 4:1053–1058. https://doi.org/10.1038/s41550-020-1120-5. arXiv:2006.06854 [astro-ph.EP]

Lainey V, Zannoni M, Robert V, Guillot T (2025) Quantification of tides in giant planets from observations. Space Sci Rev 221

Latter HN, Ogilvie GI, Chupeau M (2012) The ballistic transport instability in Saturn's rings - I. Formalism and linear theory. Mon Not R Astron Soc 427(3):2336–2348. https://doi.org/10.1111/j.1365-2966.2012.22122.x. arXiv:1209.2943 [astro-ph.EP]

Latter HN, Ogilvie GI, Chupeau M (2014a) The ballistic transport instability in Saturn's rings - II. Non-linear wave dynamics. Mon Not R Astron Soc 441(3):2760–2772. https://doi.org/10.1093/mnras/stu737. arXiv:1404.4245 [astro-ph.EP]

Latter HN, Ogilvie GI, Chupeau M (2014b) The ballistic transport instability in Saturn's rings - III. Numerical simulations. Mon Not R Astron Soc 441(3):2773–2781. https://doi.org/10.1093/mnras/stu657. arXiv:1404.4247 [astro-ph.EP]

Lissauer JJ, Squyres SW, Hartmann WK (1988) Bombardment history of the Saturn system. J Geophys Res 93:13776–13804. https://doi.org/10.1029/JB093iB11p13776

López-Oquendo AJ, Rivera-Valentín EG, Dalle-Ore CM, et al (2019) Constraints on crater formation ages on dione from Cassini VIMS and ISS. In: 50th Annual Lunar and Planetary Science Conference, Lunar and Planetary Science Conference, p 2435

Lunine JI, Stevenson DJ (1982) Formation of the Galilean satellites in a gaseous nebula. Icarus 52(1):14–39. https://doi.org/10.1016/0019-1035(82)90166-X

Lynden-Bell D, Pringle JE (1974) The evolution of viscous discs and the origin of the nebular variables. Mon Not R Astron Soc 168:603–637. https://doi.org/10.1093/mnras/168.3.603

Maxwell JC (1859) On the stability of the motion of Saturn's rings. https://doi.org/10.3931/e-rara-244

Melosh HJ (1989). Impact cratering: a geologic process

Meyer J, Wisdom J (2007) Tidal heating in Enceladus. Icarus 188(2):535–539. https://doi.org/10.1016/j.icarus.2007.03.001

Miller KE, Filacchione G, Cuzzi JN, et al (2024) The composition of Saturn's rings. Space Sci Rev 220:70. https://doi.org/10.1007/s11214-024-01104-y

Mitchell DG, Perry ME, Hamilton DC, et al (2018) Dust grains fall from Saturn's D-ring into its equatorial upper atmosphere. Science 362(6410):aat2236. https://doi.org/10.1126/science.aat2236

Mondino-Llermanos AE, Salo H (2023) Axisymmetric viscous overstability in fully self-gravitating systems. Conditions for the onset of overstable oscillations. Mon Not R Astron Soc 521(1):638–661. https://doi.org/10.1093/mnras/stad500

Morbidelli A, Nesvorny D, Laurenz V, et al (2018) The timeline of the lunar bombardment: revisited. Icarus 305:262–276. https://doi.org/10.1016/j.icarus.2017.12.046. arXiv:1801.03756 [astro-ph.EP]

Morfill GE, Fechtig H, Gruen E, et al (1983) Some consequences of meteoroid impacts on Saturn's rings. Icarus 55(3):439–447. https://doi.org/10.1016/0019-1035(83)90114-8

Morishima R, Spilker L, Brooks S, et al (2016) Incomplete cooling down of Saturn's A ring at solar equinox: implication for seasonal thermal inertia and internal structure of ring particles. Icarus 279:2–19. https://doi.org/10.1016/j.icarus.2015.06.025. arXiv:1508.07375 [astro-ph.EP]

Moro-Martín A, Malhotra R (2002) A study of the dynamics of dust from the Kuiper Belt: spatial distribution and spectral energy distribution. Astron J 124(4):2305–2321. https://doi.org/10.1086/342849. arXiv:astro-ph/0207350 [astro-ph]

Moses JI, Brown ZL, Koskinen TT, et al (2023) Saturn's atmospheric response to the large influx of ring material inferred from Cassini INMS measurements. Icarus 391:115328. https://doi.org/10.1016/j.icarus.2022.115328. arXiv:2211.05253 [astro-ph.EP]

Mosqueira I, Estrada PR (2002) Apse alignment of the Uranian rings. Icarus 158(2):545–556. https://doi.org/10.1006/icar.2002.6878

Mosqueira I, Estrada PR (2003) Formation of the regular satellites of giant planets in an extended gaseous nebula I: subnebula model and accretion of satellites. Icarus 163(1):198–231. https://doi.org/10.1016/S0019-1035(03)00076-9

Murray CD, Dermott SF (1999). Solar System dynamics

Nesvorný D, Roig FV, Vokrouhlický D, et al (2023) Early bombardment of the moon: connecting the lunar crater record to the terrestrial planet formation. Icarus 399:115545. https://doi.org/10.1016/j.icarus.2023.115545. arXiv:2303.17736 [astro-ph.EP]

Neveu M, Rhoden AR (2017) The origin and evolution of a differentiated Mimas. Icarus 296:183–196. https://doi.org/10.1016/j.icarus.2017.06.011

Nicholson PD, Jones TJ, Matthews K (1984) Infrared observations of the Uranian rings. In: Brahic A (ed) Planetary rings, p 169

Nicholson PD, French RG, Hedman MM, et al (2014a) Noncircular features in Saturn's rings I: the edge of the B ring. Icarus 227:152–175. https://doi.org/10.1016/j.icarus.2013.09.002

Nicholson PD, French RG, McGhee-French CA, et al (2014b) Noncircular features in Saturn's rings II: the C ring. Icarus 241:373–396. https://doi.org/10.1016/j.icarus.2014.06.024

Nicholson PD, French RG, Spitale JN (2018) Narrow rings, gaps, and sharp edges. In: Tiscareno MS, Murray CD (eds) Planetary ring systems. Properties, structure, and evolution, pp 276–307. https://doi.org/10.1017/9781316286791.011

Noyelles B, Baillié K, Charnoz S, et al (2019) Formation of the Cassini division - II. Possible histories of Mimas and Enceladus. Mon Not R Astron Soc 486(2):2947–2963. https://doi.org/10.1093/mnras/stz445

O'Donoghue J, Moore L, Connerney J, et al (2019) Observations of the chemical and thermal response of 'ring rain' on Saturn's ionosphere. Icarus 322:251–260. https://doi.org/10.1016/j.icarus.2018.10.027

Öpik EJ (1972) Comments on lunar origin. Ir Astron J 10:190

Perry ME, Waite JH, Mitchell DG, et al (2018) Material flux from the rings of Saturn into its atmosphere. Geophys Res Lett 45(19):10093–10100. https://doi.org/10.1029/2018GL078575

Pollack JB, Summers A, Baldwin B (1973) Estimates of the sizes of the particles in the rings of Saturn and their cosmogonic implications. Icarus 20(3):263–278. https://doi.org/10.1016/0019-1035(73)90003-1

Pollack JB, Grossman AS, Moore R, et al (1976) The formation of Saturn's satellites and rings, as influenced by Saturn's contraction history. Icarus 29:35–48. https://doi.org/10.1016/0019-1035(76)90100-7

Poppe AR (2016) An improved model for interplanetary dust fluxes in the outer Solar System. Icarus 264:369–386. https://doi.org/10.1016/j.icarus.2015.10.001

Poppe AR, Horányi M (2012) On the Edgeworth-Kuiper Belt dust flux to Saturn. Geophys Res Lett 39(15):L15104. https://doi.org/10.1029/2012GL052530

Poppe AR, Lisse CM, Piquette M, et al (2019) Constraining the Solar System's debris disk with in situ new horizons measurements from the Edgeworth-Kuiper Belt. Astrophys J Lett 881(1):L12. https://doi.org/10.3847/2041-8213/ab322a

Porco C, Danielson GE, Goldreich P, et al (1984) Saturn's nonaxisymmetric ring edges at 1.95 R_S and 2.27 R_S. Icarus 60(1):17–28. https://doi.org/10.1016/0019-1035(84)90135-0

Porco CC, Cuzzi JN, Ockert ME, et al (1987) The color of the Uranian rings. Icarus 72(1):69–78. https://doi.org/10.1016/0019-1035(87)90120-5

Porco CC, Nicholson PD, Cuzzi JN, et al (1995) Neptune's ring system. In: Neptune and Triton, pp 703–804

Rivera-Valentín EG, López Oquendo AJ, Kirchoff MR, et al (2020) Constraints on the recent cratering rate in the Saturn system from water ice crystallinity derived crater formation ages on dione. In: 51st Annual Lunar and Planetary Science Conference, Lunar and Planetary Science Conference, p 2839

Roche ME (1847) Memoire sur la figure d'une masse fluide, soumise à l'attraction d'un point éloigné. Acad Sci Montp 1:232–262

Saillenfest M, Lari G, Boué G (2021a) The large obliquity of Saturn explained by the fast migration of Titan. Nat Astron 5:345–349. https://doi.org/10.1038/s41550-020-01284-x. arXiv:2110.04104 [astro-ph.EP]

Saillenfest M, Lari G, Boué G, et al (2021b) The past and future obliquity of Saturn as Titan migrates. Astron Astrophys 647:A92. https://doi.org/10.1051/0004-6361/202039891. arXiv:2101.10366 [astro-ph.EP]

Salmon J, Canup RM (2017) Accretion of Saturn's inner mid-sized moons from a massive primordial ice ring. Astrophys J 836(1):109. https://doi.org/10.3847/1538-4357/836/1/109. arXiv:1702.04385 [astro-ph.EP]

Salmon J, Charnoz S, Crida A, et al (2010) Long-term and large-scale viscous evolution of dense planetary rings. Icarus 209(2):771–785. https://doi.org/10.1016/j.icarus.2010.05.030. arXiv:1006.0633 [astro-ph.EP]

Salo H, Schmidt J (2010) N-body simulations of viscous instability of planetary rings. Icarus 206(2):390–409. https://doi.org/10.1016/j.icarus.2009.07.038
Salo H, Karjalainen R, French RG (2004) Photometric modeling of Saturn's rings. II. Azimuthal asymmetry in reflected and transmitted light. Icarus 170(1):70–90. https://doi.org/10.1016/j.icarus.2004.03.012
Salo H, Ohtsuki K, Lewis MC (2018) Computer simulations of planetary rings. In: Tiscareno MS, Murray CD (eds) Planetary ring systems. Properties, structure, and evolution, pp 434–493. https://doi.org/10.1017/9781316286791.016
Sasaki T, Stewart GR, Ida S (2010) Origin of the different architectures of the Jovian and saturnian satellite systems. Astrophys J 714(2):1052–1064. https://doi.org/10.1088/0004-637X/714/2/1052. arXiv:1003.5737 [astro-ph.EP]
Schmidt J, Ohtsuki K, Rappaport N, et al (2009) Dynamics of Saturn's dense rings. In: Dougherty MK, Esposito LW, Krimigis SM (eds) Saturn from Cassini-Huygens, p 413. https://doi.org/10.1007/978-1-4020-9217-6_14
Shibaike Y, Ormel CW, Ida S, et al (2019) The Galilean satellites formed slowly from pebbles. Astrophys J 885(1):79. https://doi.org/10.3847/1538-4357/ab46a7. arXiv:1909.00285 [astro-ph.EP]
Shu FH, Dones L, Lissauer JJ, et al (1985) Nonlinear spiral density waves - viscous damping. Astrophys J 299:542–573. https://doi.org/10.1086/163722
Spitale JN, Hahn JM (2016) The shape of Saturn's Huygens ringlet viewed by Cassini ISS. Icarus 279:141–154. https://doi.org/10.1016/j.icarus.2016.04.035. arXiv:1509.07228 [astro-ph.EP]
Spitale JN, Porco CC (2011) Forced and free distortions in Saturn's Huygens ringlet. In: AAS/Division of Dynamical Astronomy Meeting, vol #42, p 8.05
Srama R, Ahrens TJ, Altobelli N, et al (2004) The Cassini cosmic dust analyzer. Space Sci Rev 114(1–4):465–518. https://doi.org/10.1007/s11214-004-1435-z
Tajeddine R, Rambaux N, Lainey V, et al (2014) Constraints on Mimas' interior from Cassini ISS libration measurements. Science 346(6207):322–324. https://doi.org/10.1126/science.1255299
Tajeddine R, Nicholson PD, Longaretti PY, et al (2017) What confines the rings of Saturn? Astrophys J Suppl Ser 232(2):28. https://doi.org/10.3847/1538-4365/aa8c09. arXiv:1710.08462 [astro-ph.EP]
Tamayo D, Burns JA, Hamilton DP, et al (2011) Finding the trigger to Iapetus' odd global albedo pattern: dynamics of dust from Saturn's irregular satellites. Icarus 215(1):260–278. https://doi.org/10.1016/j.icarus.2011.06.027. arXiv:1106.1893 [astro-ph.EP]
Teodoro LFA, Kegerreis JA, Estrada PR, et al (2023) A recent impact origin of Saturn's rings and mid-sized moons. Astrophys J 955(2):137. https://doi.org/10.3847/1538-4357/acf4ed. arXiv:2309.15156 [astro-ph.EP]
Tiscareno MS, Harris BE (2018) Mapping spiral waves and other radial features in Saturn's rings. Icarus 312:157–171. https://doi.org/10.1016/j.icarus.2018.04.023. arXiv:1708.03702 [astro-ph.EP]
Tiscareno MS, Mitchell CJ, Murray CD, et al (2013) Observations of ejecta clouds produced by impacts onto Saturn's rings. Science 340(6131):460–464. https://doi.org/10.1126/science.1233524
Toomre A (1964) On the gravitational stability of a disk of stars. Astrophys J 139:1217–1238. https://doi.org/10.1086/147861
Tremaine S (2003) On the origin of irregular structure in Saturn's rings. Astron J 125(2):894–901. https://doi.org/10.1086/345963. arXiv:astro-ph/0211149 [astro-ph]
Tsiganis K, Gomes R, Morbidelli A, et al (2005) Origin of the orbital architecture of the giant planets of the Solar System. Nature 435(7041):459–461. https://doi.org/10.1038/nature03539
Waite JH, Perryman RS, Perry ME, et al (2018) Chemical interactions between Saturn's atmosphere and its rings. Science 362(6410):aat2382. https://doi.org/10.1126/science.aat2382
Weidenschilling SJ, Chapman CR, Davis DR, et al (1984) Ring particles: collisional interactions and physical nature. In: Greenberg R, Brahic A (eds) Planetary rings. University of Arizona Press, Tucson, pp 367–415. https://doi.org/10.1017/9781316286791.010
Weiss JW, Porco CC, Tiscareno MS (2009) Ring edge waves and the masses of nearby satellites. Astron J 138(1):272–286. https://doi.org/10.1088/0004-6256/138/1/272
Wisdom J, Dbouk R, Militzer B, et al (2022) Loss of a satellite could explain Saturn's obliquity and young rings. Science 377(6612):1285–1289. https://doi.org/10.1126/science.abn1234
Wood JA, Mitler HE (1974) Origin of the moon by a modified capture mechanism, or half a loaf is better than a whole one. In: Lunar and Planetary Science Conference, p 851
Wooden DH, Ishii HA, Zolensky ME (2017) Cometary dust: the diversity of primitive refractory grains. Philos Trans R Soc Lond Ser A 375(2097):20160260. https://doi.org/10.1098/rsta.2016.0260
Zahnle K, Schenk P, Levison H, et al (2003) Cratering rates in the outer Solar System. Icarus 163(2):263–289. https://doi.org/10.1016/S0019-1035(03)00048-4
Zebker HA, Marouf EA, Tyler GL (1985) Saturn's rings: particle size distributions for thin layer models. Icarus 64(3):531–548. https://doi.org/10.1016/0019-1035(85)90074-0

Zhang Z, Hayes AG, Janssen MA, et al (2017a) Exposure age of Saturn's A and B rings, and the Cassini division as suggested by their non-icy material content. Icarus 294:14–42. https://doi.org/10.1016/j.icarus.2017.04.008

Zhang Z, Hayes AG, Janssen MA, et al (2017b) Cassini microwave observations provide clues to the origin of Saturn's C ring. Icarus 281:297–321. https://doi.org/10.1016/j.icarus.2016.07.020

Publisher's Note Springer Nature remains neutral with regard to jurisdictional claims in published maps and institutional affiliations.

Space Science Reviews (2025) 221:35
https://doi.org/10.1007/s11214-025-01156-8

Understanding the Formation of Saturn's Regular Moons in the Context of Giant Planet Moons Formation Scenarios

Michel Blanc[1,2] · Aurélien Crida[3] · Yuhito Shibaike[4,5] · Sebastien Charnoz[6] · Maryame El Moutamid[7] · Paul Estrada[8] · Olivier Mousis[2] · Julien Salmon[9] · Antoine Schneeberger[2] · Pierre Vernazza[2]

Received: 20 March 2024 / Accepted: 26 February 2025 / Published online: 20 March 2025

Abstract
This article explores the different formation scenarios of the Kronian moons system in the context of a highly dissipative Saturn, with the objective of identifying the most likely of these scenarios. First, we review the diversity of objects – moons and rings – orbiting solar system giant planets, and the diversity of their architectures, which formation scenarios must reproduce. We then identify in this broader context the specific features of the Saturn system, such as the particularly large spectrum of its moon masses, the uniqueness of Titan and the presence of both dense and tenuous rings, before discussing the applicability of the different giant planet moon formation scenarios to the Saturn case. We discuss each of the most relevant scenarios and their respective merits. Finally, we tentatively propose a "favorite" scenario and we identify the key observations to be made by future space missions and/or Earth-based telescopic observations to validate this scenario or possibly alternative ones.

Keywords Moons of Saturn · Rings of Saturn · Formation of moons · Migration of moons

1 Introduction and Objective

It is generally accepted that giant planet moons are a by-product of the formation of their parent planets, similar in a way to our understanding of planets as a by-product of the formation of their parent star. Beyond this general statement, the different theories that explain the formation of giant planet moons differ by the way they describe the relationships and relative timing between the formation of the planet in the Solar Nebula (SN), the formation of its circumplanetary disk (CPD), the capture or accretion of satellitesimals in the CPD, the generation of rings, and the capture of planetesimals from the SN or, after its dissipation, from solid body reservoirs of the early Solar System.

This diversity of formation processes is not really surprising, knowing the large differences existing in the physical, chemical and dynamical characteristics of moons, as well as in their arrangement in space – the moon system architectures – among the only four giant planet systems of the Solar System. Among them, the Saturn system offers the broadest diversity of objects: in increasing order or radial distance, dense rings and tenuous rings, small moons inside or near the rings that are usually called "ring-moons", small and medium size

Extended author information available on the last page of the article

regular moons outside the rings region, one large moon (Titan), and finally a large number of irregular moons. Thanks to the Cassini-Huygens mission, this system is also the one that has been best documented by close-in spacecraft observations that covered all families of objects, from small inner moons embedded in the rings to large irregular ones.

Attempting to trace back from these detailed observations of the contemporary Saturnian system the mechanisms and series of events that led to its assembly is therefore particularly relevant, because of the significant number of observational constraints that can be compared to model predictions. The main purpose of this article is to address this challenge by confronting predictions of these different formation models to observations. Since the Saturnian system is only one of the four giant planet systems of the Solar System, while the diverse formation mechanisms proposed may apply to different extents to all four of them, we place our study of the formation scenario of Saturn's moons in the broader context of the diversity of formation scenarios proposed for the four giant planet systems.

This article is organized to follow this logic.

Section 2 is devoted to providing the broader context of the diversity of moons and moon system architectures.

We first review the variety of observational characteristics of individual moons and of moon systems architectures in the light of past and current Earth-based telescopic and space mission observations. On this basis, we identify the specific features of the different categories of Saturnian moons that have to be explained by candidate formation scenarios.

In Sect. 3, we focus on the formation of the Saturnian moons system.

As an introduction, we review the different types of moon formation scenarios that have been proposed for the four giant planet systems: result of a giant impact; formation by accretion in a circum-planetary disk (CPD); formation by viscous expansion from a massive disk; capture... For each category of formation scenario, we discuss its relevance to the formation of the different giant planet systems and draw some conclusions on the key characteristics that each type of formation scenario tends to reproduce best.

Then we analyze the relevance of these different formation scenarios to the different categories of Saturnian moons and to the moon system as a whole. We discuss the merits of the different formation scenarios on the basis of available observations and in the context of the fast migration of moons which is the main subject of this Topical Collection.

In Sect. 4, we draw some tentative conclusions on the relative success of the different models in reproducing the different categories of moons and tentatively propose a "favorite" scenario. We identify additional observations, laboratory data and simulation studies that are needed, first to discriminate among the different scenarios, and then to elaborate a more comprehensive scenario that will make it possible to assemble all the pieces of the moons formation puzzle.

2 Characteristics of Giant Planet Moons and Architectures of Their Systems

2.1 Introduction

The moon systems of the four giants can be divided into three families which generally occupy three domains of radial distance. From inside out:

- A first family of small, prograde inner moons, generally in close dynamical coupling to their planet's rings and occupying near-circular orbits in the equatorial plane, are generally referred to as "ring-moons". They mainly reside within the Roche limit for ice, or extend only slightly beyond. Their sizes vary from a km to a hundred km at most;

- A second family of small-to-large-size regular prograde moons resides on near-circular orbits in the equatorial plane at distances of a few to a few tens of planetary radii. They are mostly decoupled from the rings, though a few of these objects display weak resonances with them. Their sizes range from a few 100 to a few 1000 km;
- Finally, a third family of small irregular moons is found at all four giant planet systems on mid-to-high-inclination, eccentric orbits extending to radial distances encompassing a significant fraction of the Hill radius of their parent planet. A majority of the smaller size component of this population occupy retrograde orbits, while a smaller number of larger size objects, above about 50 km diameter, dominate on prograde orbits.

We successively describe the observational characterizing of these three families, ending up with the regular moons which are the main focus of this article.

2.2 Small Inner Moon Systems Coupled to the Rings

Figure 1 illustrates the intricate relative geometry of the radial distributions of small inner moons and rings in the four giant planet systems, relative to their planet's radius. Moons are indicated as black dots while the different rings are represented as gray areas, with increasing intensities of gray representing increasing densities of ring components.

Direct inspection of this figure reveals the extreme diversity of these small moons and of their relation to rings at the four planets.

Jupiter hosts only four small regular moons, i.e. in order of increasing planetocentric distance Metis and Adrastea (on very close orbits), then Amalthea and Thebe. The orbits of these moons coincide with sharp boundaries in Jupiter's tenuous dust rings: the main ring for Metis and Adrastea, and the Gossamer rings for Amalthea and Thebe. This has been interpreted as a result of the generation of an extended dust ring along these moons orbits by meteorite impact on their surfaces, which subsequently drifts inward under the combined effects of dust charging and motion in the Jovian magnetic and electric fields. A comprehensive description of the Jovian rings-small moons system and ring-moon interactions within has been given by Burns et al. (2004).

The ring-small moons systems of ice giants have recently been reviewed (Showalter 2020). Uranus is host to a total of 13 small inner moons, whose radial distances range from the edges of the main ring to nearly the orbit of Miranda, the innermost "classical" moon. From inside out, one finds first Cordelia and Ophelia, separated by the inner edge of the main rings. Then a group of 9 moons, as close to one another as a 3 percent fraction of their semi-major axis, forms the most densely packed set of moons in the Solar System. Finally, Puck and Mab orbit between this group and Miranda, surrounding a tenuous ring called the μ ring.

Taken together, the total mass of these small moons would form a dense ring, if fragmented. Their packed configuration is the seat of many resonances and instabilities, to the point that its lifetime is likely not larger than one million years. This made Showalter (Showalter 2020) conjecture about the relationship between the lifetime and the age of this system, in an apparent paradox: "The inner satellite system of Uranus presents us with a challenging puzzle. On the one hand, we have numerous indications that the system is unstable, with catastrophic collisions expected over time scales of approximately a million years or less. Cometary impacts are also capable of breaking apart small moons. On the other hand, we would expect the system to have formed, along with Uranus itself, over 4 billion years ago. How can we reconcile these two ideas? Perhaps the material that makes up these moons and rings is primordial, but its current configuration into this specific set of moons and rings is not." While the main point of this analysis still holds, it has been moderated

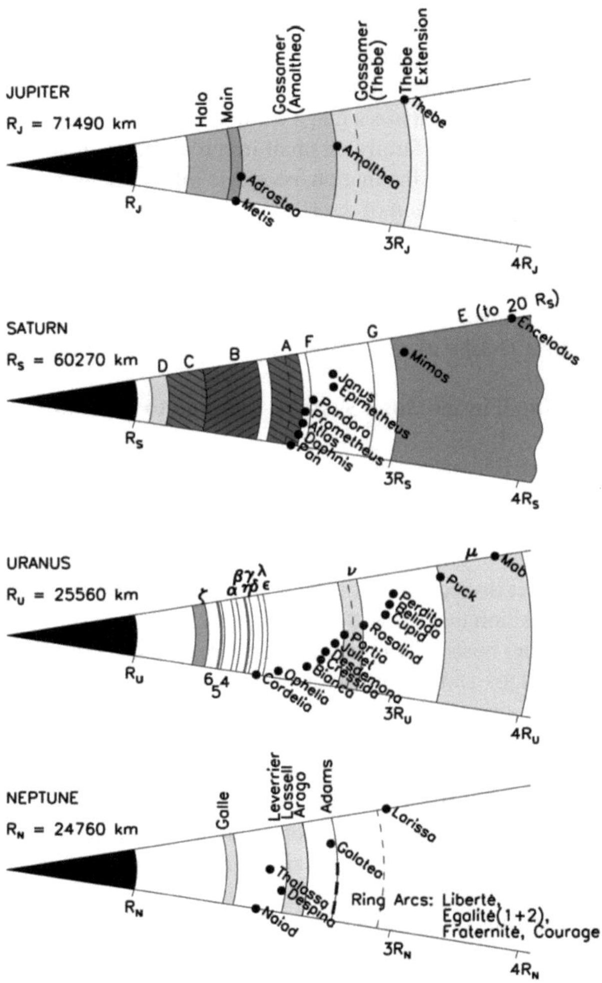

Fig. 1 Schematic comparison of the ring-moon systems of giant planets. The Roche radius for ice is shown as a dashed line. From Blanc et al. (2021)

by Ćuk et al. (2022) who revisited the stability of the inner moons of Uranus using updated orbital elements and considering tidal dissipation. They found that the Belinda group can be stable on 10 8 yr timescales due to an orbital resonance between Belinda and Perdita and that, while tidal evolution cannot form the Belinda–Perdita resonance, convergent migration could contribute to the long-term instability of the Portia group. Additionally, recent modeling of the bombardment history of the giant planets (Bottke et al. 2024) suggests that these inner moons might not have experienced impacts that would result in catastrophic disruption, at least for the largest of them. Finally, in the event a moon would experience significant disruption, the debris would likely re-accumulate rapidly into a new moon (or possibly several smaller moons), and/or be absorbed by nearby other moons, depending on the degree of radial dispersion of the debris. These analyses leave us with the idea that, while the material forming the small moons may be dating back to the giant collision that tilted Uranus' spin axis, their current configuration may be permanently evolving, redistributing material between objects and in radial distance over a broad range of timescales.

Neptune's six small moons offer a slightly different picture. While the four innermost ones orbit close to several of Neptune's narrow rings, the two outermost ones orbit outside of them. Being less packed than the small Uranian moons, these moons do not offer the same puzzle as Uranus, it has been suggested that they may be the result of another catastrophic event in the early solar system: according to the dominant theory, any primordial Neptune moon system may have been disrupted by the capture of Triton on a retrograde, highly eccentric orbit. Triton's orbit may have been progressively circularized afterwards, partly via tidal interactions with the planet, and partly by collisions with and accretion of the preexisting moons system Triton's capture contributed to disrupt, as suggested by Rufu and Canup (2017). In this scenario, the tiny moon system we see today may have formed by re-accretion of the material of the putative primordial moons after completion of this circularization (Agnor and Hamilton 2006). It should be noted, however, that there is no definite proof of this scenario. From the fact that Neptune's small moons' sizes are comparable to those of Uranus, their formation via a scenario similar to the case of the small moons of Uranus and Saturn should not be totally discarded.

Saturn's stands out in this family portrait by the diversity of its moons: excluding the thousands of very small moons embedded into the main rings and the 7 major regular moons, one can count 17 small regular moons. A large fraction of them occupy special orbit configurations; four on a Trojan orbit with larger moons, three of them are in co-rotation eccentricity resonance with Mimas (Hedman et al. 2010), two moons shepherding Saturn's F-ring, two mutually co-orbital moons, and two more orbiting within gaps in Saturn's rings. Hyperion is locked in an orbital resonance with Titan. All other small regular moons orbit near the outer edge of the dense A Ring (as best seen in Fig. 1). The diversity of these configurations points to a diversity of formation scenarios, which will be extensively discussed in Sect. 3.

2.3 Irregular Moon Systems

While most regular satellites occupy near-circular orbits deep inside the potential wells of their host planet, a population of small irregular satellites can be found around all four giants at larger distances, up to one half or two thirds of the Hill radius (see Nicholson et al. (2008) and Jewitt and Sheppard (2005) for comprehensive reviews). Figure 2, adapted from Sheppard et al. (2023), shows the main characteristics of the distribution of orbital parameters of these populations, using a different color coding for each planet: preferred inclinations mainly between 30° and 60° (prograde) and between 130° and 170° (retrograde); rather large eccentricities spanning from 0.1 to 0.5, and strong tendency for a clustering in compact dynamical groups, such as the Ananke group for Jupiter or the Norse group for Saturn. The distribution of these irregular moons between prograde and retrograde orbits is also worth noticing. Figure 2 shows that a majority of the population is retrograde, but Bottke et al. (2010), studying this distribution as a function of moon size, showed that retrograde moons are statistically smaller, while prograde orbits dominate among the largest of these objects, at and above 50 km diameter. They used a dynamical model of moon populations to show that this intriguing size distribution may result from the collisional re-shaping of an original population of objects similar to the Trojans captured during the chaotic episode of giant planets migration described by the Nice model.

Saturn again occupies a special place in the inventory. First, the Cassini exploration of the Saturn system allowed an inventory and characterization of a fraction of the irregular moons family: using the time available between observations of the main mission targets, a systematic photometric study of irregular moons provided light curves over a broad range of phase angles for 25 of these previously poorly known objects (Denk and Mottola 2019).

Among them, Phoebe, the first irregular moon of Saturn, discovered as early as in 1899, was the object of a targeted flyby by the spacecraft that provided a 90 percent imaging coverage of its surface, revealing a dark (albedo 6 per cent), heavily cratered and near-spherical

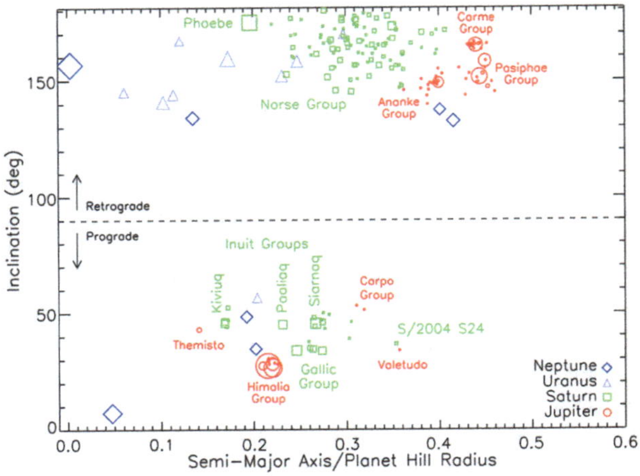

Fig. 2 Distribution of irregular satellites at Jupiter, Saturn, Uranus, and Neptune in a plane in which the horizontal axis is the semi major axis expressed in units of the planet's Hill radius, and the vertical axis is orbital inclination. Retrograde satellites appear in the upper part at inclinations above 90 degrees. In the top left corner one can see Phoebe, the only object that has been resolved and studied in detail by Cassini, and Triton, the only large moon of Neptune. The figure clearly shows the grouping of a large fraction of these objects in compact dynamical families (adapted from Sheppard et al. 2023)

body; spectroscopic IR studies of its composition revealed the presence of water ice, carbon dioxide, phyllo-silicates, organics and possibly iron-bearing minerals at its surface (Clark et al. 2005). Analysis of its density, indirectly related to its porosity, and of its implications on the history of its accretion, led to question the origin of this moon as a captured KBO (Castillo-Rogez et al. 2019) and to suggest that it could rather have formed from the same reservoir that gave birth to C-type asteroids.

Furthermore, thanks to spectacular progress made in recent years in telescope surveys of its Hill sphere in search for new moons, particularly at the CFH telescope on Mauna Kea (Denk and Mottola 2019), Saturn is today the planet with the largest number of known moons, with a total of 146 including 122 irregular moons discovered as of 2023 (see Sheppard et al. 2023 and the "green" squares of Fig. 2). These irregular moons generally have featureless visible and near infrared spectra dominated by water absorption bands. They are neutral or moderately red in color, pointing to some similarity with C-type, P-type or D-type asteroids, as well as with Jupiter Trojans. This similarity arguably points to a commonality in the capture mechanism for all of those populations, such as proposed by the giant planet instability scenario of the early Solar System.

2.4 Regular Moon Systems

A notable characteristics of giant planets of our Solar System is the presence of systems of regular moons orbiting each of them. To analyze them first individually, Fig. 3 shows the main giant planet moons placed in the broader context of the moons of all solar system planets, with Earth to scale. The figure reveals that each system displays deeply different characteristics. Jupiter stands out with its four large Galilean moons with comparable sizes not very different from the Earth's moon size. In contrast, Saturn has only one large moon, Titan, which gathers just by itself 96% of the total cumulative mass of its 146+ moons. Then Uranus has no large moon but instead 5 medium-size moons with diameters similar

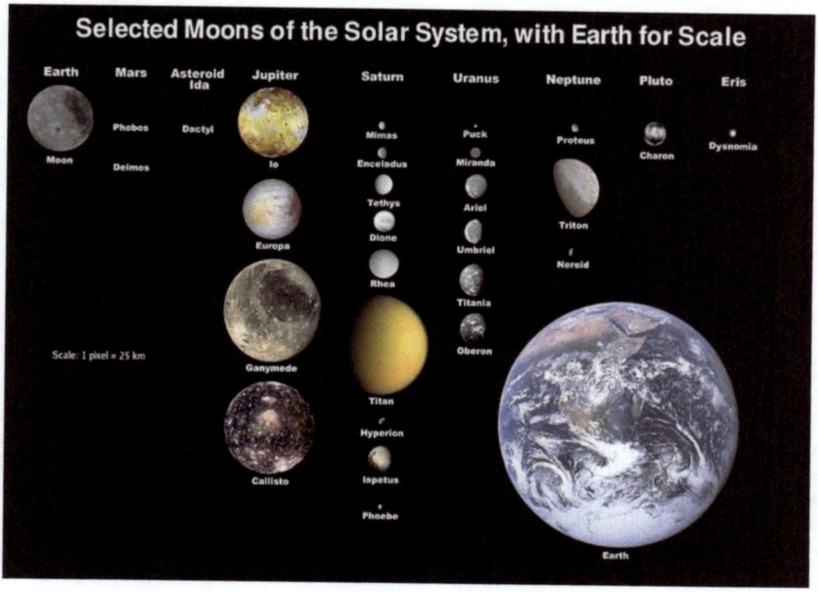

Fig. 3 Main moons in the solar system, with their sizes to scale

to those of Saturn. Finally, while all other major moons occupy low-eccentricity prograde orbits close to the equatorial plane, Neptune's Triton, which represents just by itself 99.7% of the mass of the Neptunian moon system ($2.1 \cdot 10^{-4}$ Neptune mass), resides on a nearly circular but retrograde orbit with 156° inclination best explained by a capture origin, hence its classification in the family of irregular moons (see previous paragraph).

As long as these moons are thought to have primarily formed in situ, their ubiquity has often been considered as indicative of their status of by-products of the formation of their parent planet (e.g., Canup and Ward 2006). However, we will see in this article that in situ formation scenarios themselves cover different paths, and that capture is possible for some of them. Some additional observational characteristics of moons and their systems are particularly important to help distinguish between the different scenarios.

First, there are important characteristics to note in the overall architecture and physical parameters of each system of regular moons. As emphasized by Canup and Ward (2006), a common characteristic among Jupiter, Saturn and Uranus is the range of the total mass of regular moon systems, always between 1 and 3 times 10^{-4} the mass of their parent planet. In contrast, the individual masses of moons span a broader range, from tiny objects to 10^{-4} times their planet mass, the record being for Titan with $2.4 \cdot 10^{-4}$ times the mass of Saturn.

Another characteristic of the architecture of these regular moon systems is the variation of their mass with their distance to the planet, presented in Fig. 4 in units of planetary radius and mass. This diagram reveals a systematic increase of mass with radial distances at lower masses, which works particularly well for Saturn, starting from 0 at $\sim 2-3\ R_p$. This tendency reaches a plateau at large masses, in the mass range where both Titan and the four Galilean moons reside. The common trend observed in particular for the three outermost planets suggests a common process at play.

Finally, the density-radius relationship of the group of large moons (Fig. 5) provides strong constraints on their bulk density and on their composition. For sizes below about 100 km radius, they are less dense than water ice, suggesting a high degree of porosity as

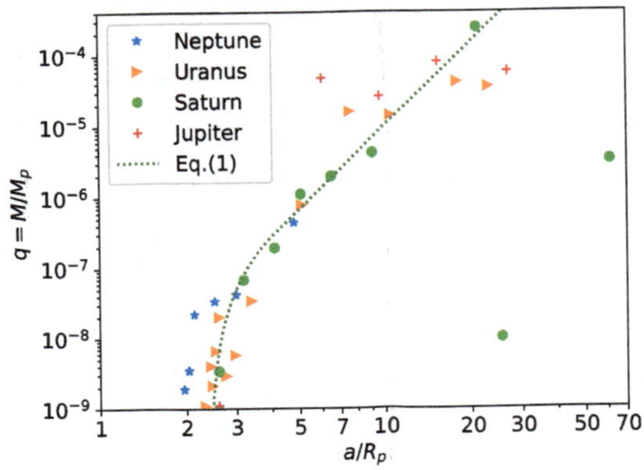

Fig. 4 Architecture of the regular moon systems of the four giant planets (Triton is not considered regular and thus not shown). Shown are the masses of the moons (normalized to that of their parent planet) against their semi-major axis (normalized to the radius of their parent planet). The dotted line corresponds to Eq. (1) (Crida and Charnoz 2012, see Sect. 3.2.3), with a coefficient of proportionality equal to $2.3 \cdot 10^{-7}$ and a Roche radius at 2.4 R_p (140,000 km for Saturn)

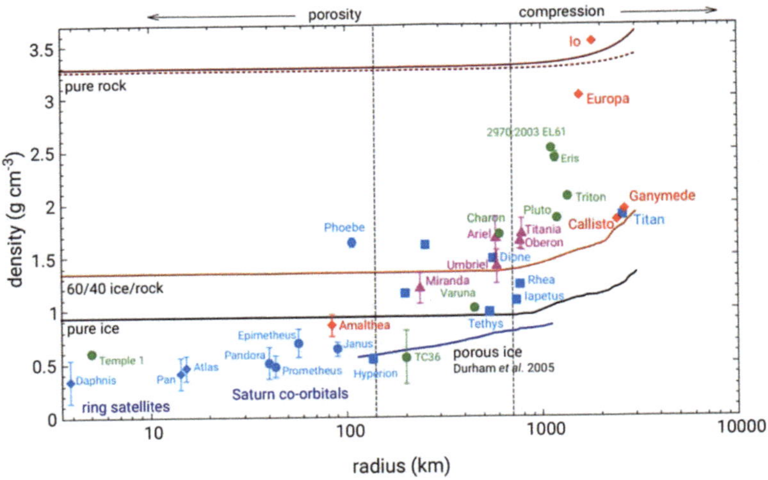

Fig. 5 Variation with radius of the density of outer solar system regular moons. Most of the density estimates used here are from spacecraft observations of radius and mass derived from gravity science analyses of flybys. Data for Ganymede and Callisto (orange diamonds), Uranus satellites (purple triangles), Triton (green circle) and Titan (blue square) are from Yoder (1995). Amalthea data are from Anderson et al. (2005). Data for mid-sized and small Saturn moons (radii $\sim 10-1000$ km, blue squares) are from Jacobson (2004), Jacobson et al. (2006), Porco et al. (2005b,a, 2007), Thomas et al. (2007). Pluto, Charon and KBO (green circles) data are, for Pluto/Charon from Buie et al. (2006), Eris (Brown and Schaller 2007), Varuna (Jewitt and Sheppard 2002), 2003EL61 (Rabinowitz et al. 2006), 1999TC36 (Stansberry et al. 2006). The curves labeled "60/40 ice/rock" (by mass) and "Pure Ice" are theoretical calculations of densities of a sphere with a rock-ice ('rock' density $= 3662\ \mathrm{kg\,m^{-3}}$) mixture and pure water ice composition, taking into account the presence of high density phases of water ice in the interior of a larger body (Lupo and Lewis 1979). The curve labeled "Porous Ice" represents the density of an ice sphere with a porosity throughout equal to that allowed at the central pressure of the object, based on laboratory compression experiments with cold ice compaction (Durham et al. 2005). Variation of rock density with moon size for the "pure rock" case is expected to be very small; it was approximated using two EOS, one from Mueller and McKinnon (1988), and one for Poirier (2000), for a rock composition corresponding to a 3.2 to 3.3 g/cc density. Adapted from Johnson and Estrada (2009)

well as a large water content. For increasing sizes, their representative densities reach the density for water, then of rock-water mixes, suggesting not only a composition change but also a decrease of porosity with increasing mass and gravity. The large moons Io and Europa, among those with radii larger than 1000 km, stand out with densities near pure rock.

The concurrent variations of the density and of the mass of regular moons with distance to the planet provide important constraints on their formation (in situ or by capture) and on their evolutionary path, including the evolution of their volatile-to-heavy elements ratio from the time of their differentiation to contemporary times, which will be discussed in detail in Sect. 3 for the case of Saturn.

2.5 Specific Features of Saturn's Moons in the Context of Giant Planets Systems

General reviews of the moons of the outer Solar System have been presented among others by Johnson (2005) at the eve of the Cassini mission, and just after by Dougherty and Spilker (2018), who reviewed the tremendous breakthrough in our knowledge of Saturn's moon system accomplished during the 13 years in orbit of this historic mission between 2004 and 2017. Saturn's rings and moons system clearly stands out by the large number of its objects and their broad diversity in mass, size, radial distance and composition. This system extends in radial distance from regions where moonlets accretion and disruption is still at play, inside the Roche radius for ice in the main rings, through intermediate distances dominated by gravitational dynamics, orbital evolution and resonances, to distant regions, a fraction of the Hill radius, where one finds the largest population of known irregular moons in the Solar System. Let us focus on regular moons, which offer the greatest challenge to our understanding of their formation.

First, inside Saturn's dense rings, one finds a large and diverse population of embedded moons, revealed by Cassini, such as the two small moons Pan and Daphnis which orbit inside two gaps of the main rings, respectively the Encke and Keeler gaps. Then, in-between rings and irregular satellites, 20 regular moons strike by their diversity in sizes, radial distances and dynamics. In order of increasing radial distances, one first finds five small moons: Atlas, Prometheus, Pandora, Janus and Epimetheus; the two last ones, located on very close orbits, are dynamically coupled to the point that they exchange their positions every fourth year. Then five mid-sized moons (Mimas, Enceladus, Tethys, Dione and Rhea) orbit inside of Titan, while two others (Hyperion and Iapetus) orbit outside Titan. Hyperion is locked in a 4:3 mean motion resonance with Titan. All these moons nicely align on the mass-distance average trend revealed by Fig. 4, except for the two outer ones Hyperion and Iapetus which can be seen as the two green dots well below the main trend line in that figure. Finally, to be comprehensive, one finds an additional family or small moons orbiting in the region of mid-sized inner moons. The triad of Alkyonide moons (Methone, Anthe and Pallene) orbit between Mimas and Enceladus, and four additional small moons are Trojan companions of Tethys (Telesto and Calypso) and of Dione (Helen and Polydeuces). Finally, in-between Rhea and Hyperion stands Titan, Saturn's unique large moon and also the only moon in the Solar System with a dense atmosphere.

Despite this intricate superposition of small and mid-sized regular moons, the same general trend already noticed in Fig. 4 emerges from their mass vs. radial distance relationship, as illustrated in Fig. 6: a systematic increase of moons mass with planetocentric distance, with two different characteristic slopes for small inner moons (blue trend line) and for mid-sized ones (red trend line).

Taken together, there are three main features that moon formation models have to reproduce at Saturn:

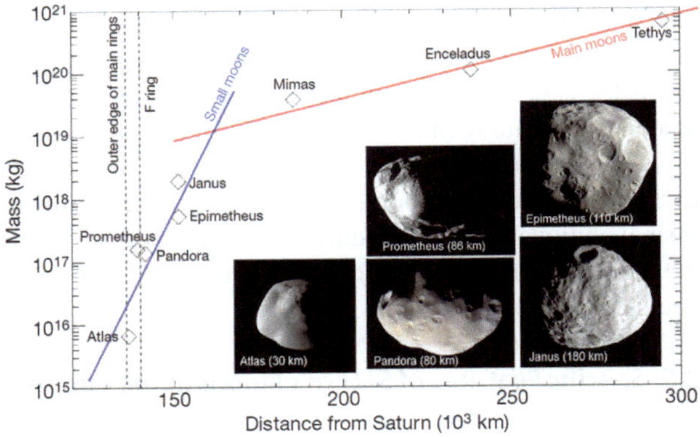

Fig. 6 Mass vs distance relationship for Saturn inner moons. One can note the marked change of slope between small moons (from Atlas to Janus, blue regression line) and mid-sized ones (from Mimas to Tethys and beyond, red regression line). From Charnoz et al. (2010). Courtesy NASA-JPL-SSI

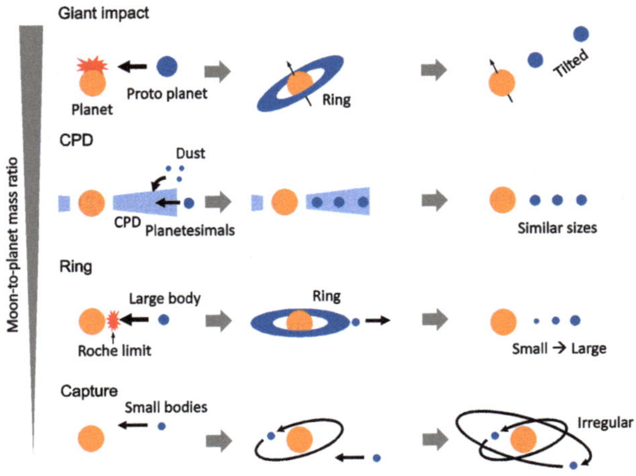

Fig. 7 An illustration of the four main mechanisms of moon formation showing their typical characteristics. The moons in the solar system were likely formed by one or combinations of these four elementary mechanisms, which are shown from top to bottom in the order of the typical moon-to-mass ratios of the systems that they can form

- the systematic trend found on the mass-distance relationship of small inner moons and mid-sized moons;
- the presence of one single large Moon, Titan;
- and finally the presence of two mid-sized moons outside Titan's orbit.

In the next section, we are going to explore the broad diversity of proposed formation mechanisms to try and assemble the puzzle of the Saturn moons system.

3 How Did Saturn's Moons Form?

3.1 Introduction

In order to best understand the specific case of Saturn and identify the most credible scenario(s) for the formation of its moons, we start by reviewing the diversity of moon formation scenarios currently considered for Earth, Mars and the four giant planet systems. The main four families of scenarios are illustrated in Fig. 7, from top to bottom in the order of the typical moon-to-mass ratios of the systems that they can form.

As one will see, the age of moons, when it is known, can be a strong discriminant between the different scenarios, since it points directly to the period of their formation. Information about moons can be partly derived from their cratering record, with due consideration of the evolution of their surface after formation. In this respect the article by Bottke et al. (2024) which describes a comprehensive scenario for the bombardment history of the giant planet satellites, and references therein, provide important constraints for the ages of their different moon systems which are very complementary to the dynamical considerations we primarily develop in this section. In particular, the formation scenarios of mid-sized moons must take into account that their surface ages are on the order of 4 Billion Years old according to the cratering records, provided most craters are made by heliocentric projectiles.

- formation by accretion in a debris disk formed by a giant impact or another type of catastrophic event: this is currently the favored scenario for the formation of Earth's Moon, recently reviewed by Canup et al. (2023). It was also proposed for the two moons of Mars, Phobos and Deimos (Rosenblatt and Charnoz 2012), and in a different way for Uranus by Morbidelli et al. (2012) to explain the formation of its system of mid-sized, prograde regular moons. These authors conjectured that the current Uranian moon system could result from the re-accretion of debris of a primordial moon system (similar to the Jovian one) disrupted by the giant impact(s) that likely tilted the spin axis of the planet. This scenario was later analysed in detail by Salmon and Canup (2022), who found that a set of specific conditions are needed, as well as some ad-hoc assumptions, in order for this scenario to work at Uranus. We will see in Sect. 3.2.1 that this mechanism, i.e. re-accretion of a "new generation" of moons from a debris disk, has also been considered for Saturn's mid-sized moons.
- Formation from a "primordial" circumplanetary disk (CPD) is the favorite candidate for the Jovian system, dominated by its four large Galilean moons of similar sizes (Mosqueira and Estrada 2003a; Canup and Ward 2006). This generic CPD-related model may be applied to the primordial formation of prograde, regular moons occupying orbits near the equatorial plane of their host planet around a planet. It may be that it applies only to Jupiter in the Solar System, as the discussion of its applicability to Saturn will suggest in Sect. 3.2.2.
- Formation/migration from a massive ring: this scenario, initially proposed for Saturn owing to its large dissipation (Charnoz et al. 2011), before being generalised to the ice giants and the Earth (Crida and Charnoz 2012), will be discussed in detail in Sect. 3.2.3.
- Capture of small bodies from Sun-centered orbits has been proposed by Pollack et al. (1979) and Colombo and Franklin (1971) to explain the population of irregular moons of all four giant planets. For a capture to occur, a collision or a close encounter with a third body must be involved, such as the close encounters between giant planets proposed in the Nice model (Nesvorný et al. 2007a). Another possibility is to assume tidal disruption of a binary TNO by the capturing giant planet (Agnor and Hamilton 2006). For the Neptune system, current models propose that Triton's capture from the population of Trans-Neptunian Objects (TNO) and its subsequent circularization may have destroyed any primordial moon system. Thus the capture hypothesis for Triton concurrently explains the absence of other large or mid-sized moons around Neptune. Note however that capture of the small irregular moons of Jupiter, Saturn and Uranus does not seem to have perturbed their regular moons systems.

In the following sections, we are going to successively discuss how these different formation scenarios have been applied to Saturn's regular and irregular moons. These scenarios describe the resulting moon systems, but also introduce constraints on the locations where these moons have been assembled or delivered. In the context of a "highly dissipative Saturn" (Lainey et al. 2020), these locations differ significantly from their current locations

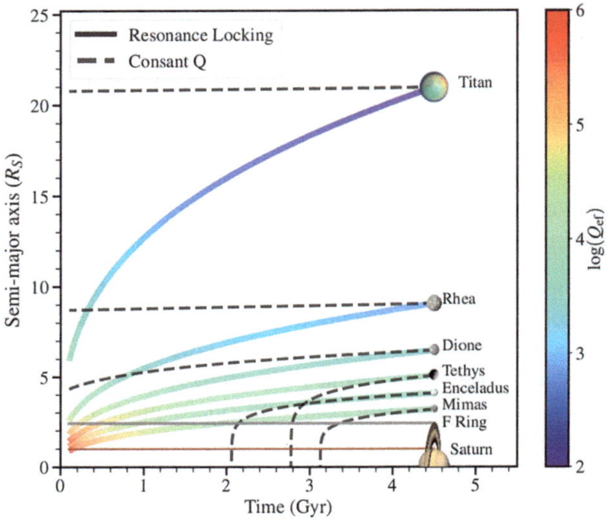

Fig. 8 This figure, originally from Fuller et al. (2016), displays the expected radial migration trajectories over the age of the Solar System for two different types of tidal interactions driving migration. In the "classical" tidal interaction theory, the Q factor is basically constant and moon migration slows down quickly with increasing planetocentric distance (dashed curves); in contrast, tidal migration driven by propagting tidal waves and resonant locking (thick continuous curves) leads to similar migration time scales, on the order of the age of the Solar System, for all moons. See Lainey et al. (2020) for a more detailed analysis

as a result of fast orbital migration. Determining the migration rate of each moon thus becomes a key element to retrieve the time and place of its formation. This migration rate theoretically depends on the Q factor associated to the tidal interaction mode responsible for orbital migration. In the "classical" tidal interaction theory, each moon interacts with the stationary tidal bulge it generates in the atmosphere and interior of its host planet. For a basically constant Q, the migration rate decreases with increasing radial distance, implying that distant moons like Titan experienced very limited migration, while moons orbiting closer to the planet experienced large ones. This model has recently been challenged by Fuller et al. (2016) who noticed that, if one considers all possible oscillations of the planet excited by tidal interaction with a moon, certain propagating tidal modes may produce a "resonance locking" between moons and planet: with this "resonant locking" mechanism, concurrent temporal evolution of the radial migration of the moon and of the characteristic resonance modes of the planet while it cools down may produce a migration regime which is independent of radial distance and equally well applies to distant moons like Titan and to inner moons. The corresponding migration timescales are on the order of the age of the solar system for all moons, as shown in Fig. 8: starting from the current location of moons and moving back in time, the "classical" tidal theory (dashed curves) predicts that Titan must have formed close to its current location whereas inner moons may have formed significantly later than the Solar System. In contrast, assuming resonant locking in tidal migration (thick continuous curves), all moons migrate at comparable rates over the age of the solar system. Recent measurements of Titan's orbit expansion rate by two independent methods (Lainey et al. 2020) tend to validate the predictions of the resonant locking model. Even so, further measurements are needed to ultimately discriminate between the different types of tidal interaction and orbital migration illustrated in Fig. 8 and to draw consequences on the validity of the different formation scenarios.

The reader should note that, while Fig. 8 captures the main point of resonance locking, (1) it simplifies it somewhat, as for instance the Mimas-Tethys and Enceladus-Dione pairs must evolve convergently to be in their present resonances, and (2) the actual migration rates traced by the solid curves of 8 are functions of the model of Saturn interior used, and may be significantly different if one chooses another interior model. A more detailed description of

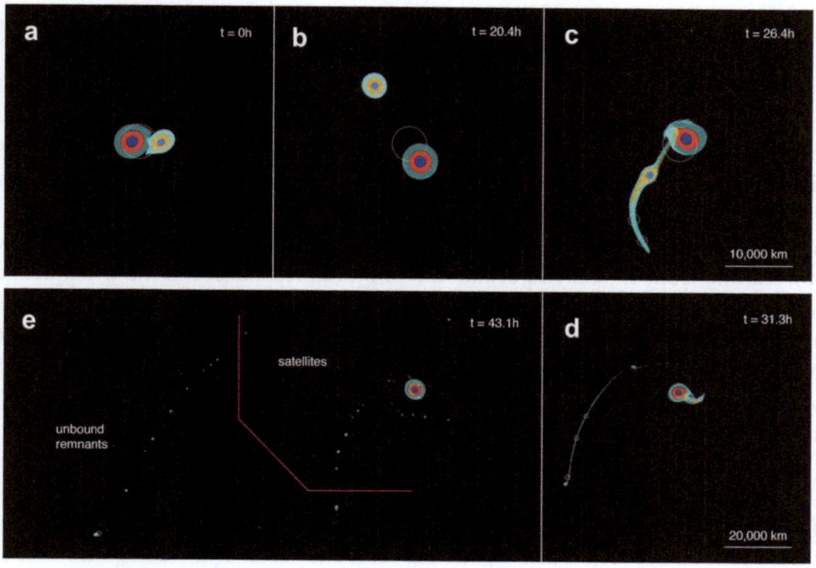

Fig. 9 Example of one of the giant collisions simulated by Asphaug and Reufer (2013) to simultaneously reproduce Titan (from merging of several smaller-sized bodies in successive collisions) and the mid-sized moons (from clumps of one or several of these collisions). Here five consecutive snapshots of a 1:3 mass ratio collision between a satellite of a mass comparable to Titan and a smaller impactor are shown clockwise from a to e. While a series of clumps escape the system, a few of them (right of the red broken line in panel e) remain in Saturn-centered orbit and form MSM's. In this scenario, Titan's formation bears some similarity with the oligarchic growth of terrestrial planets in the inner Solar System. See Asphaug and Reufer (2013) for more details

the latest orbital migration theories can be found in Chap. 4 of this Topical Collection (Ćuk et al. 2024).

3.2 Formation of the Regular Moons of Saturn

In this section we successively review the different scenarios that may apply to the formation of Saturn's regular moons and discuss their respective merits.

3.2.1 Late Formation of Mid-Sized Moons After Disruption and Re-Accretion of a Previous Moon System

This first family of scenarios proposes that Saturn's moons formed from the collisional disruption and re-accretion of a previous moon system. A first study along this line (Asphaug and Reufer 2013) started from the initial assumption that Saturn's CPD, just like Jupiter's, gave birth to a moon system similar to the Galilean moons that was later disrupted during an episode of global instability. Following a detailed critical analysis of the physics of collisions between moons and the clumps of ice-rich material produced in these collisions, this study performed a full simulation of the moon formation sequence using a 3-D Smooth Particle Hydrodynamics (SPH) code, together with an equation of state describing the colliding differentiated bodies. In this simulation, Titan is formed at the same time as the unique large-size end-product of a series of mergers, in a way similar to the oligarchic growth of planetesimals that likely led to the formation of inner planets. Figure 9 shows a time series of the events occurring in one of these collisions. A merit of this scenario is that it simultaneously reproduces Titan and the mid-sized moons as the products of the same series of collision-disruption-merger events. However, the assumptions made on the properties of collisions and on the rheologic behavior of colliding moons and fragments are not easy to validate by current observations, and the model does not explain the mass-distance relationship of small moons and the circularization of the orbits of new-born moons after their re-accretion.

Another scenario of moon re-accretion from an earlier system was proposed by Ćuk et al. (2016), starting from an analysis of the role of orbital resonances in the dynamical evolution of mid-sized moons. Using two different numerical integrators, the authors studied two resonance crossings that should have happened in the past based on the current orbits: the 3:2 MMR of Tethys and Dione, and the 5:3 MMR of Dione and Rhea which should have happened later. Their simulations suggested that the Tethys-Dione 3:2 resonance crossing always produces excessive final inclinations for these moons which are incompatible with their present values. On the other hand, simulations of the Dione-Rhea 5:3 resonance reasonably reproduce current eccentricities.

If only the second resonance crossing happened, but not the first one, Dione and Rhea must have formed between the two potential resonance crossings. Taking into account the fast tidal migration observed by Lainey et al. (2012), this implies that formation of this moons was recent, maybe as young as 100 Myr. Building on this "young moons" hypothesis, Ćuk et al. (2016) proposed that Saturn's mid-sized moons formed from a disk of solids that briefly existed late in the history of the Saturn system from the catastrophic disruption of a previous moon system. In this scenario, this disk extending from inside to outside of the Roche limit could have been the progenitor of the current rings. Like the previous one, this model does not explain the circularization of small moons orbits and the mass-distance relation shown in Fig. 4.

As noted by the authors, further validation of this scenario and of other similar "young moons" scenarios rests on three main types of evidence: 1) a better estimate of the tidal heating rate of Enceladus; 2) a possible post-Cassini reconsideration/validation of the fast tidal migration rates of Saturn's moons observed by Lainey et al. (2012), in particular because in a typical resonance locking mechanism the period ratios do not evolve, preventing resonance crossing (see Fig. 8), and 3) a careful analysis of the cratering records of icy moons, a critical element to date the age of their surfaces that requires taking into account the distribution between impactors in planetary and in sun-centered orbits.

3.2.2 Formation from a CPD

Regular moons of giant planets, particularly those of considerable size, are thought to originate within circumplanetary disks (CPDs), akin to how planets form within protoplanetary disks (PPDs). In our solar system, this phenomenon is evident with the Galilean moons of Jupiter, and possibly with Titan and Iapetus around Saturn. CPDs emerge as byproducts of the gas accretion process of planets, as the accreting gas carries angular momentum, allowing planets more massive than Saturn to carve gaps within the PPD (e.g., Crida et al. 2006). It is improbable that Uranus and Neptune harbored their own CPDs, as they likely remained engulfed within the PPD (Crida et al. 2006). In this section, we present an overview of proposed scenarios regarding the formation of large moons within CPDs. The dimensions of CPDs typically span approximately 0.3 Hill radius of their host planets (approximately 0.1 AU for both Jupiter and Saturn) (e.g., Lubow et al. 1999). Moons are theorized to originate within the high-density regions situated close to the centers of these disks, typically at radial distances smaller than 0.03 to 0.1 Hill radius, as described in the majority of formation models.

Two classical types of formation scenarios have been delineated in the literature: the minimum mass model and the gas-starved disk model. In the first model, CPDs are segregated from their parent PPDs. The solid mass utilized for moon formation within these disks approximates their current total mass in the Jupiter, Saturn, and Uranus systems, denoted as $M_s \sim 10^{-4}\ M_p$. Within the Saturnian system, the mass contrast between Titan and Iapetus finds explanation in the distribution of solid masses; the majority of mass is concentrated within the centrifugal radius situated around Titan's current orbit, with only a marginal quantity of solids near Iapetus's current orbital position contributing to its formation (Mosqueira and Estrada 2003a,b).

The other classical scenario is the gas-starved disk model (Canup and Ward 2002). In this framework, gas accretion onto the CPD persists during moon formation, albeit at a rate modest enough to maintain the disk temperature below the sublimation point of ice across most regions of the CPDs, thereby facilitating the formation of icy moons. Solid material is consistently delivered to CPDs from their parent PPDs, with varying accretion rates assumed in hydrodynamic (HD) and magneto-hydrodynamic (MHD) simulations (e.g., Gressel et al. 2013; Kanagawa et al. 2015; Müller et al. 2018).

Recent numerical simulations and observations of planets embedded in gaps in PPDs have revealed that modest gas accretion persists even after embedded planets have carved gas gaps along their orbits (Haffert et al. 2019; Szulágyi et al. 2022). This observation lends stronger support to the gas-starved disk model over the minimum mass model. Consequently, in the following we delve into recent models grounded in the gas-starved disk concept. A range of concepts rooted in this framework have been investigated to replicate the typical (total) satellite-planet mass ratio, $M_s/M_p \sim 10^{-4}$, along with other characteristics of large moons.

In the original concept of the gas-starved disk model, multiple generations of large moons form successively within the CPDs. N-body simulations suggest that the satellite-planet mass ratio, $M_s/M_p \sim 10^{-4}$, can be broadly replicated through a balance between their loss via Type I migration toward the planets and their formation through the collisional growth of satellitesimals (Canup and Ward 2006), which are kilometer-sized building blocks of moons (Shibaike et al. 2017). Some subsequent studies proposed that the orbital positions of Io, Europa, and Ganymede, which are locked in the Laplace resonance, could be reproduced if the strong magnetic field of young Jupiter generated an inner cavity within the CPD, thus halting moon migration (refer to the upper panels of Fig. 10) (Sasaki et al. 2010; Ogihara and Ida 2012). Conversely, young Saturn, being smaller than young Jupiter, likely lacked the capacity to generate a sufficiently robust magnetic field to carve out a wide inner cavity at the CPD's center. This scenario would have led to the survival of only one large moon and perhaps the existence of massive ice rings (see Sect. 3.4).

However, a critical challenge with this scenario lies in the formation of satellitesimals. As dust particles supplied to the CPDs increase in size through mutual collisions and adhesion, they tend to drift towards the planets as centimeter to meter-sized pebbles before reaching the size of satellitesimals, akin to the movement of dust particles in PPDs towards stars before they can evolve into planetesimals (Shibaike et al. 2017; Zhu et al. 2018). Three primary solutions have been proposed to address this issue: 1) outward gas flows interacting with dust and pebbles in specific regions facilitate the formation of satellitesimals via the streaming instability, 2) planetesimals formed within PPDs are captured by CPDs and assume the role of satellitesimals, and 3) dust particles efficiently grow within laminar magnetic wind-driven CPDs.

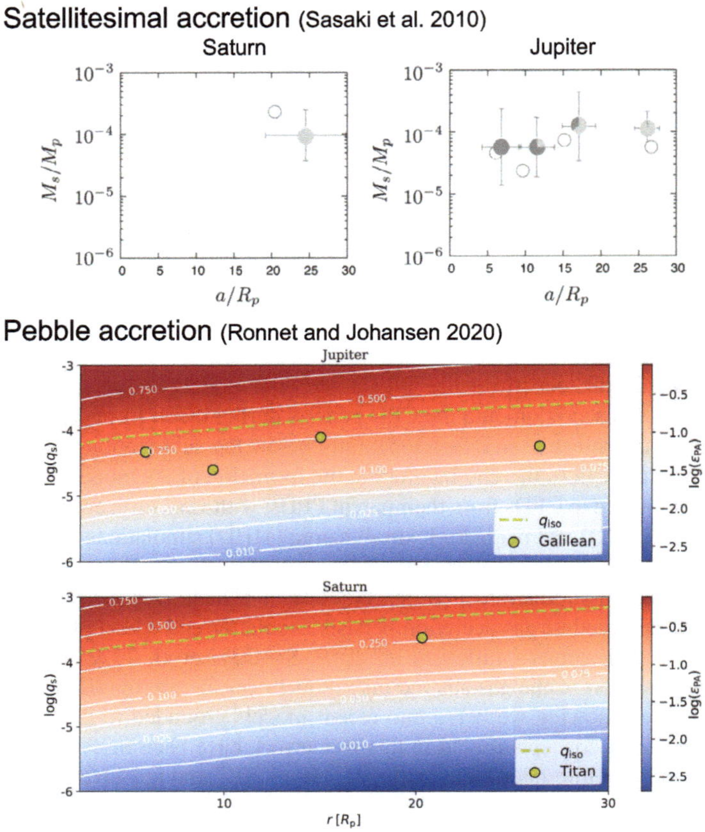

Fig. 10 Formation of the moons around Saturn and Jupiter in the classical satellitesimal accretion and recent pebble accretion scenarios. The upper panels show that the mass and orbits of both Saturnian and Jovian moons are reproduced by population synthesis simulations in Sasaki et al. (2010). The thickness of gray represents the rock-to-ice mass ratios, which are also reproduced well. However, the plenty of satellitesimals assumed to exist in the CPDs in the simulations could be problematic, as we explain in the text. The lower panels show that the current mass of moons (yellow circles) is lower than the pebble isolation mass (i.e. the satellite mass where pebble accretion stops by curving gaps around the satellites; yellow dashed curves), meaning that the moons can be formed by pebble accretion if their orbits are fixed. The color maps show the pebble accretion efficiency (the ratio of the mass of accreting pebbles to the mass of the satellites growing out of the drifting pebbles). See Sasaki et al. (2010), Ronnet and Johansen (2020) for more details. Reproduced with permission ©AAS and ©ESO

The operational mechanisms and characteristics of outward gas flows within CPDs remain a subject of contention. According to HD and non-ideal MHD simulations, such flows may manifest in the midplane of CPDs, particularly in the outer regions of the satellite formation zone, contingent upon specific viscosity and opacity conditions (Tanigawa et al. 2012; Gressel et al. 2013; Szulágyi et al. 2014; Szulágyi 2017; Schulik et al. 2020). Outflows could aggregate particles at the orbital location of approximately 0.1 Hill radius or beyond, fostering a consistent generation of satellitesimals via the streaming and/or gravitational instability, potentially resulting in successive generations of moons (Drążkowska and Szulágyi 2018; Cilibrasi et al. 2018; Batygin and Morbidelli 2020).

The capture of planetesimals has also been explored in various studies (e.g., Ruskol 2006; Suetsugu and Ohtsuki 2017). Gas pressure bumps emerging at the outer perimeters of gaps carved by young Jupiter/Saturn within PPDs could serve as efficient sites for planetesimal formation (e.g., Lyra et al. 2009; Shibaike and Alibert 2020). Conversely, since planetesimals are also pushed away from the vicinity of planets due to gravitational interactions as gap structures form within the gas, capturing planetesimals by CPDs necessitates high eccentricity for them to approach planets closely enough (Fujita et al. 2013). Furthermore, the migration of an outer planet can perturb planetesimals around the inner planet, potentially explaining why Jupiter, affected by Saturn's migration, boasts four large satellites while Saturn has only one (Ronnet et al. 2018; Eriksson et al. 2022).

Recent non-ideal MHD simulations have also demonstrated that CPDs can function as laminar magnetic wind-driven disks. Within such laminar disks, dust particles can settle onto the midplane more efficiently compared to the turbulent CPDs assumed in prior studies, allowing for more efficient growth through mutual collisions. Consequently, satellitesimals can form before dust descends towards the central planet (Shibaike and Mori 2023).

Pebble accretion presents an alternative mechanism for moon formation. Initially proposed in the context of planet formation within PPDs, pebble accretion utilizes drifting dust directly as the building material for planets (Ormel and Klahr 2010; Lambrechts and Johansen 2012). Here, *pebbles* are conceptualized as objects (dust aggregates) large enough to decouple from the gas (unlike dust grains) but small enough to be significantly affected by gas friction (unlike planetesimals).[1] Pebbles typically range from centimeters to meters in size, depending on the density of the surrounding gas. Planetary embryos efficiently accrete pebbles drifting towards the central stars due to gravitational focusing effects and aerodynamic drag, which influences pebbles more strongly than planetesimals. Consequently, the feeding zone for pebbles is considerably larger than that for planetesimals. This mechanism operates analogously for the formation of large moons if their seeds are supplied to CPDs and their migration is halted by the inner cavity (refer to the lower panels of Fig. 10) (Shibaike et al. 2019; Ronnet and Johansen 2020). Captured planetesimals ablated by CPD gas can form pebbles through recondensation, and large planetesimals can also serve as seeds for moon formation (Ronnet and Johansen 2020).

Two potential challenges arise when applying pebble accretion to the case of Saturn. Firstly, the ablated material from which Saturnian moons accreted might lack ice, which contradicts their observed low densities. Nonetheless, if this material was abundant in hydrated minerals, the dehydration of particles and outward diffusion of released water vapor could have facilitated the recondensation of substantial amounts of ice within the formation region of the moons (Ronnet and Johansen 2020).

The other issue pertains to the necessity of inner cavities to impede rapid inward migration (Ronnet and Johansen 2020). Given that this condition likely wasn't met at Saturn, the formation of Titan through pebble accretion becomes challenging, or at best, it necessitates that this moon formed in the very inner region of the CPD before migrating outward. This scenario aligns with the recent observation of the current rapid outward orbital evolution of Titan: Lainey et al. (2020) demonstrated that Saturn's moons migrate outward due to Saturn's tides at such a swift rate that it would take less than the age of the Solar System to move from the Roche radius to their current positions. This observation opens the possibility for a more intricate and dynamic history than previous formation scenarios, as outlined in Sect. 3.4.

[1] Given that this definition corresponds to a Stokes number S_t (friction stopping time multiplied by orbital frequency) on the order of *one* regardless of size, s_t *one* may be a more suitable term than *pebble* (Crida 2023).

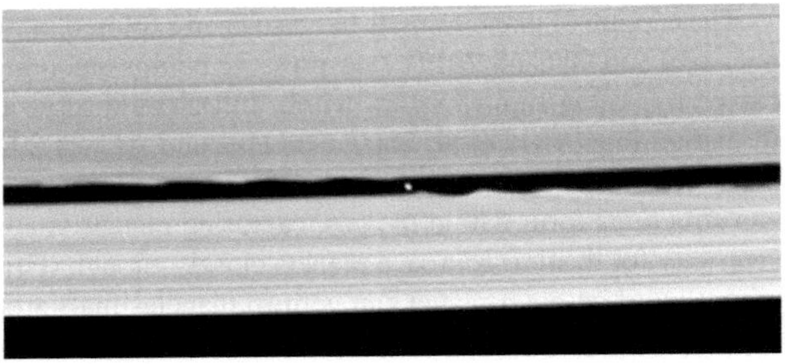

Fig. 11 Famous Cassini picture of Daphnis in the Keeler gap at the outer edge of the A-ring. The gravitational effect of the moonlet on the rings leads to the waves which can be seen trailing behind Daphnis outside its orbit, and leading ahead of Daphnis inside its orbit

3.2.3 Formation/Migration from a Massive Disk

We call *tidal disk* a disk of condensate debris which is dynamically cold, but prevented from coalescence into self-gravitating objects by the tidal forces from the central planet. In other words, a tidal disk is a debris disk inside the Roche radius. Saturn's main rings are the best example of such a tidal disk, but the moon-forming disk of molten lava which probably surrounded the Earth is an other case.

Interactions among the disk material yield a radial spreading of the tidal disk. When a dynamically cold disk of solids (like Saturn's rings) spreads, self-gravity dominates over tides beyond the Roche limit of the planet, and solids aggregate into elongated objects. As they orbit around the planet, these objects can keep accreting material from the rings, since the Roche radius depends only on the density of the considered material, not on the size of the moonlet.

However, moonlets also migrate outwards. Indeed, their orbital velocity is smaller than that of ring particles, so that the gravitational perturbation caused by a moonlet on the orbits of ring particles decreases the angular momentum of the latter, and increases that of the former. Ring particles lose angular momentum by gaining eccentricity, but are quickly circularised by collisions with other ring particles. Such a phenomenon can be observed in the vicinity of the ring moons Pan and Daphnis, and explains how these two satellites maintain their gaps open (see Fig. 11). Accumulating interactions with many ring particles, the moonlet gains angular momentum, and therefore is transferred onto a circular orbit of larger radius. The migration rate is proportional to the masses of the moonlet and of the rings, and to the distance to the edge of the rings to the power -3 (see for instance Lin and Papaloizou 1986).

Depending on the outcome of the competition between the disk spreading rate and the moon migration rate, one could have a single moon absorbing all the mass from the tidal disk as it migrates away (which would be the Earth-Moon case) or a series of moons regularly popped out of the rings and migrating away at constant mass; in this case, since their migration rate decreases with distance to the rings, the separation between moons decreases and they suffer collisions and merge, leading to a mass increase with orbital distance.

This process has been modeled numerically by Charnoz et al. (2010) around Saturn's present rings and nicely reproduces the handful of small moons Janus, Epimetheus, Prometheus, Pandora and Atlas, which share the same spectrum as the rings and a low bulk density, strongly suggesting that they are made of the same material. Besides, the migration rate due to their interactions with the rings is such that it is impossible for them to have been on their orbit for billions of years. A numerical application shows that Janus should have been inside the Roche radius just a few hundred million years ago. There is now a consensus that these small moons are born recently, from the spreading of the present rings beyond the Roche radius.

This model has then been generalized to all the mid-sized moons of Saturn, using two key recent results. First, ring formation and evolution models (Canup 2010; Salmon et al. 2010) suggest that Saturn's rings could be primordial but have been much more massive in the past. In this case, they could have lost a mass equivalent to most of the mid-sized moons over the age of the Solar System by spreading beyond the Roche radius. Second, the strong dissipation measured inside Saturn (Lainey et al. 2020) allows for a migration of the mid-sized moons to their present orbit from the Roche radius over the age of the Solar System (see Appendix of Charnoz et al. 2011). This is a key element in the discussion since migration due to the interaction with the rings vanishes past the location of the 2:1 mean motion resonance (MMR) with the Roche radius, that is 222,000 km from Saturn, between the orbits of Mimas and Enceladus. Hence, with the previously estimated quality factor for tides inside Saturn which makes migration over the age of the solar system negligible, it would be impossible to form the mid-sized moons by this process. Conversely, with the strong dissipation measured about 15 years ago, Saturn's tide can take over the interaction with the rings and keep pushing the satellites outwards. Starting from massive rings, numerical simulations reproduce quite nicely the distribution of all mid-sized moons up to Rhea (Charnoz et al. 2011). Besides, this model offers an explanation for the presence and variety of cores among these 5 moons, whose silicate to ice ratio varies from 7% to 70%. In this model, silicates originally present in the rings would have aggregated into a few large objects within the rings (because the Roche radius for silicates is only $\sim 90{,}000$ km). This handful of proto-cores of random masses would then have been covered with ice during the process of ring spreading, providing the seeds for Mimas, Enceladus, Tethys, Dione, Rhea. In this model, the 5 aforementioned moons were born in sequence, the further and more massive the earlier (thus the older the object is now), but they grew through multiple collisions after having left the rings.

Using an analytical description of this model, Crida and Charnoz (2012) demonstrated that the spreading of a tidal disk beyond the Roche radius, in the case when it does not yield one single large moon, leads to the formation of a series of moons with a precise mass - distance relation (where the mass q is expressed in units of the central planet mass, and Δ is the distance to the Roche radius, normalised by the latter):

$$q \propto \begin{cases} (\Delta/\Delta_{2:1})^{9/5} & \text{for } \Delta < \Delta_{2:1} \\ \left(\frac{\Delta+1}{\Delta_{2:1}+1}\right)^{3.9} & \text{for } \Delta > \Delta_{2:1} \end{cases} \tag{1}$$

where $\Delta_{2:1} \approx 0.58$ is the location of the 2:1 MMR with the Roche radius. This analytical relation nicely fits the one observed in the Saturn system, from Pandora to Rhea and even Titan; the observed masses are within a factor 60% of the analytic relation across 4 orders of magnitude in distance (see Fig. 4).

It is interesting to note that each of these three results from around 2010 – (i) initial massive rings which spread and lose mass over the age of the solar system, (ii) formation of satellites at the Roche radius followed by migration to their present position and (iii) strong tidal dissipation inside Saturn – are, alone, inconsistent with the present system. However, taken together, they draw a new consistent picture for the origin of the Saturnian ring and satellite system. But each relies on the other two. For instance, the strong tidal dissipation is not possible if the satellites have been here since Saturn formed. But reciprocally, the satellites can not have been formed in sequence from the spreading of the rings if they don't migrate fast. And it is absurd to consider initially massive rings if their mass can not have been evacuated, or to imagine that the moons come from the rings if the latter can not efficiently spread and send mass beyond the Roche radius.

In this frame, the newest results on Saturn's tides, suggesting that instead of the constant Q paradigm, the resonant lock mechanism may prevail (Lainey et al. 2020), can modify the picture. Indeed, Eq. (1) relies on the constant Q for $\Delta > \Delta_{2:1}$. But with the resonant lock, satellites migrate faster as they go further and do not merge anymore. The mass–distance relation would then be the consequence of a relation between the mass at $\Delta_{2:1}$ and the age, which remains to be established. On the other hand, the observed fast migration of Titan offers the possibility to include Titan in the process (Lainey et al. 2020), while it was impossible to bring it to its position in 5 Gyrs with the constant Q paradigm. So some work remains to be done to update this satellite formation model.

Even more interestingly, the relation (1) matches the mass–distance distribution of the regular moons of Uranus and Neptune, suggesting that ice giants also had massive rings in the past (Crida and Charnoz 2012), which should then have been later completely eroded by meteoritic bombardment. In that sense, the Saturnian system would be an illustration of a more general process: see Crida et al. (2025, this collection), for a more detailed description of this mechanism. As one can see from Fig. 4, this same relation does not work for Jupiter and the Galilean moons. But the case of Mars has also been shown to be consistent with this model, although indirectly: in one of their formation scenarios, Phobos and Deimos would be formed from debris accumulated in the MMRs with the most massive satellite coming out of the rings. All the satellites directly formed from the rings would then have already fallen back onto Mars due to tides (Rosenblatt et al. 2016). However alternative scenarios have been proposed, such as for instance the direct formation of Deimos from a disk of giant impact debris that extended beyond the synchronous orbit (e.g., Canup and Salmon 2018).

Finally, let us note that the model described here does not address the origin of the tidal disk. In the case of the Earth, Mars, and maybe of Uranus, a giant impact is the favored hypothesis. In the case of Saturn, the ring formation model which best fits the scenario described here is based on the tidal stripping of the icy mantle of a Titan-mass satellite which migrated inside the Roche radius inside Saturn's CPD (Canup 2010). Hence, this satellite formation model actually relies on a first generation of satellites formed in the CPD as described in the previous section. We will propose a possible formation scenario of the mid-sized moons and Titan based on this idea in Sect. 3.4. The reader is referred to the review of the origin of the rings presented in Crida et al. (2025, this collection) for more discussion on this topic.

3.3 Origin of Irregular Moons

Most of the irregular moons orbits have large inclinations to their planet's equator and a majority of them are retrograde (see Fig. 2), suggesting that they originate from Sun-orbiting small bodies that have been captured by Saturn during their crossing of its Hill sphere (see also Fig. 7). Indeed, irregular moons may have been captured from previously heliocentric orbits by gas drag (Pollack et al. 1979), collisions during the Solar Nebula period (Colombo and Franklin 1971), or three-body interactions during the final stages of the large-scale dynamical instability described in the Nice model (Nesvorny et al. 2007b). These different scenarios involve a diversity of capture mechanisms.

(1) "pull-down" capture during the short phase of rapid growth of a body like Jupiter due to the run-away accretion of its gas envelope (Heppenheimer and Porco 1977): the resulting fast expansion of its Hill sphere may result in the entry and subsequent capture of small bodies into this sphere; this mechanism might also work for Saturn.

(2) Gas-drag capture during the Solar Nebula period: energy dissipation of planetesimals due to friction with gas from the SN or CPD may transfer an object initially on a hyperbolic

orbit onto a closed one (Pollack et al. 1979). Ćuk and Burns (2004) performed a dynamical simulation to show that this mechanism could have worked for the Himalia group of Jupiter and conjectured that a similar mechanism could explain the capture of Phoebe.

(3) Collisions and three-body interactions are also a way of exchanging energy between bodies in a hyperbolic orbit to transfer one of them onto a closed orbit (Colombo and Franklin 1971). Nesvorny et al. (2007b) proposed that irregular moons of giant planets were captured from the planetesimal disk during the encounters between these planets described in the Nice model. They showed via numerical simulations that nearby planetesimals can be deflected into planet-bound orbits during planetary encounters with an efficiency large enough to produce the observed populations of irregular satellites at Saturn, Uranus and Neptune. Jupiter, however, which does not experience encounters with other planets in the Nice model, would have acquired its irregular satellites by a different mechanism. The collision and three-body interaction mechanisms have been explored specifically for Saturn by Turrini et al. (2009). They solved numerically the inverse capture problem for several of the irregular moons Albiorix and Siarnaq for the prograde group and Mundilfari and Phoebe for the retrograde one, and found that the current population of these moons is compatible with a comet or centaur origin. Based on the analysis of Cassini images of crater impacts at the surface of Phoebe, they also concluded that single-collision capture was possible but of low probability for this moon, thus favoring multiple collision scenarios.

It remains possible that the most massive irregular moons ($\geq 10^{-7} M_J$) formed within CPDs (Cilibrasi et al. 2021). This scenario conjectures that they might have been excited by interaction with the other forming satellitesimals prior to settling in the residence domain of irregular moons. However the probability of such an event was found to be low, typically $\sim 1\%$ in the case of the Jupiter system (Cilibrasi et al. 2021).

Close-in exploration of irregular moons systems, still poorly known, by space probes and astrometric studies, revealing their mass, shape, density and surface state, appears to be the only way to discriminate between their different origins and formation scenarios, at Saturn just as at the other giant planet systems.

3.4 A Tentative Scenario

Our review of the different families of formation scenarios that can be applied to Saturn's moons, illustrated in Fig. 7, first confirms that capture is the scenario that applies to irregular moons at Saturn, just as it does at other giant planets. Discussion of the merits and difficulties of the three models proposed for the formation of mid-sized and large moons is more complex. To summarize:

- formation by solid re-accretion after a giant impact or the chaotic disruption of a primordial moon system has been proposed in at least two studies. Both studies have difficulties in explaining (1) their current water-rich composition, and (2) their mass-distance relationship. While additional measurements, EOS determinations and dynamical simulations have been identified that could possibly validate a scenario of this family, this first scenario does not appear as the most promising avenue to resolve the puzzle of the formation of the regular moons system;
- formation from a CPD, the best candidate for Galilean moons, has a major problem when applied to Saturn. As shown in Fig. 8, tracing back in time the current location of mid-size moons to their potential formation site in a primordial CPD takes them inside the Roche limit. The implication is that, if formed in the CPD, they would have been lost into Saturn soon after their accretion. Only Titan could possibly escape this primordial loss, if driven outward fast enough by a tidal resonance locking soon after its accretion; this

negative inference however needs to be taken with due caution, since (1) one can imagine that the Q factor characterizing the efficiency of tidal coupling of moons to Saturn may have evolved in the course of Saturn's life, and (2) that Q also depends on the internal structure, which is still poorly known.

- finally, formation of small and mid-sized moons by viscous expansion of a massive disk appears as a consistent scenario for the following two reasons: (1) the outward edge of Saturn's primordial magnetic cavity (if any) was likely inside the Roche radius; (2) tracing back in time the radial position of these moons (Fig. 8) strongly suggests they have been inside or at the edge of the Roche limit at the time of their formation. Of course one should keep in mind that this reconstruction of the moons' past orbital distances that go back to the Roche limit ignores the satellites' mutual interactions, including orbital resonances, for the sake of simplification: it is therefore for the time being a promising theoretical concept, still to be evolved into a detailed model. This scenario is also the one that provides a natural explanation for the mass-distance relationship illustrated in Figs. 4 and 6. Note that whether the current rings are the remnants of this massive disk (Salmon et al. 2010; Crida et al. 2019) or a recent addition to the Saturn system (Iess et al. 2019) does not matter here: the massive disk could have disappeared completely before new young rings form (Estrada and Durisen 2023).

Building on these elements, we propose a tentative scenario that can be applied in the same terms to the formation of the Saturnian and Jovian systems, and validate this scenario by a quantitative assessment of the relative magnitudes of the different "critical radii" of the problem: planet radius, Roche limit and size of the primordial magnetospheric cavity.

To set this scenario, let us start from a quantitative evaluation of the truncation radii associated with the magnetospheric cavities of young Saturn and Jupiter. The radius of the inner cavity formed by the pole-aligned dipole magnetic field of the planet is.

$$R_{\mathrm{cav}} = \left(\frac{\mathcal{M}^4}{4GM_\mathrm{p}\dot{M}_\mathrm{g}^2} \right)^{1/7} \tag{2}$$

where $\mathcal{M} = B_\mathrm{s} R_\mathrm{p}^3$ (Takasao et al. 2022). At the time when gas accretion was ongoing and moons formed in their CPD, planet radii, R_p, were approximately 1.5 times larger than current ones (Fortney et al. 2011). The strength of magnetic fields on the surfaces of planets, B_s, were also about 10 times larger than the current ones (Christensen et al. 2009; Reiners and Christensen 2010). Figure 12 shows that the predicted truncation radius by the magnetospheric cavity of young Saturn is smaller than the planetary radius when $B_\mathrm{s} = 2.0$ G, 10 times larger than the current value (Connerney 1993), and $\dot{M}_\mathrm{g} = 0.006 - 0.06\ M_\mathrm{J}\ \mathrm{Myr}^{-1}$, consistent with the gas-starved disk model (Canup and Ward 2002). This means that there was no inner cavity during moon formation and the CPD had a direct connection with the surface of Saturn. In contrast, the predicted truncation radius of young Jupiter is larger than the planetary radius when $B_\mathrm{s} = 42$ G (10 times larger than the current value (Connerney 1993)) and $\dot{M}_\mathrm{g} = 0.02 - 0.2\ M_\mathrm{J}\ \mathrm{Myr}^{-1}$ (consistent with the gas-starved disk model and the ice mass fractions of the Galilean moons (Canup and Ward 2002; Shibaike et al. 2017, 2019)).

The Roche limits of young Saturn and Jupiter are

$$R_{\mathrm{Roche}} = 2.456 \left(\frac{\rho_\mathrm{p}}{\rho_\mathrm{m}} \right)^{1/3} R_\mathrm{p}, \tag{3}$$

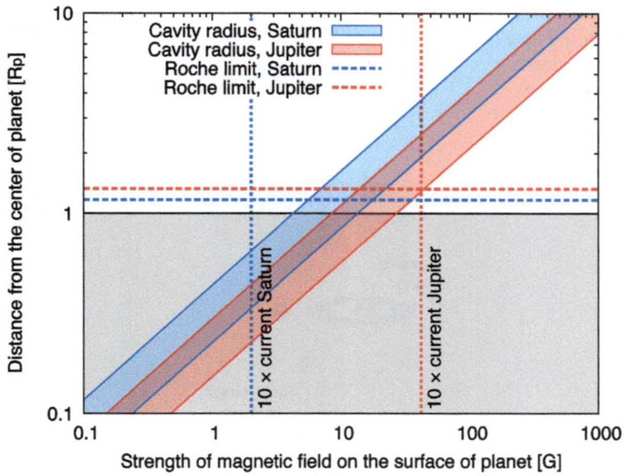

Fig. 12 Normalized radii of the disk inner cavities truncated by the magnetic fields of young Saturn and Jupiter predicted by Eq. (2) (shaded blue and red areas, respectively). We assume gas accretion rates of young Saturn and Jupiter as $\dot{M}_g = 0.006 - 0.06$ and $0.02 - 0.2\ M_J\ \mathrm{Myr}^{-1}$, respectively, consistent with the typical assumptions based on the gas-starved disk model (Canup and Ward 2002; Shibaike et al. 2017, 2019). The blue and red dashed lines are the Roche limits of young Saturn and Jupiter, respectively, where the planet density is $\rho_p = 0.20$ and $0.39\ \mathrm{g\,cm^{-3}}$ (assuming 1.5 times larger radii than the current (Fortney et al. 2011)), and the moon density is $\rho_m = 1.9$ and $2.5\ \mathrm{g\,cm^{-3}}$ (Titan and Io; see Fig. 5). The blue and red vertical dotted lines represent the strengths of the magnetic fields of young Saturn and Jupiter at their surfaces, assumed to be 10 times larger than their current value, $B_s = 2.0$ and 42 G, respectively (Connerney 1993; Reiners and Christensen 2010). The black horizontal line represents the surfaces of the planets (i.e. the planet radii)

where ρ_p and ρ_m are the densities of planets and moons, respectively. Figure 12 shows that the Roche limits of both Jupiter and Saturn were larger than the surfaces of the planets, but only the cavity radius of Jupiter was larger than the Roche limit. Note however that the evolution of planetary atmospheres is not well understood, so that the Roche limits could have been smaller than the "swollen" planet radii (Fortney et al. 2011).

Based on these estimates, we propose a dual formation scenario for the Saturnian and Jovian systems. The main elements of this scenario are schematically presented in Fig. 13. Let us consider Saturn first, and a situation in which two large moons first form in its CPD. These large moons should have similar masses, according to both the pebble and satellitesimal accretion scenarios, because their mass at formation is determined by the Type I migration timescale or by the pebble isolation mass (Canup and Ward 2006; Ronnet and Johansen 2020). Let us assume that both moons have Titan's current mass. Similarly to the scenario proposed by Canup (2010), the inner moon (Moon 1) crosses the Roche limit during its Type I inward migration and its ice-rich mantle (if it is already differentiated) is stripped off, delivering ice fragments to a forming disk, while its rocky core resists the tidal force and falls into Saturn. Then, following dissipation of the gas disk, a massive icy ring forms out of the fragments of Moon 1, as predicted by the SPH simulation in Canup (2010). Mid-sized moons could then form by viscous expansion of this massive ring, as described in Sect. 3.2.3 above. A constraint placed on this scenario by the current silicate content of Saturn's mid-size moons (i.e., Fig. 5) is that the primordial rings born out of the stripping off of Moon 1 should contain enough silicates to provide them to these moons – or that the current rings, with their high ice content, were formed later after erosion of a primordial ring. This connects us to the issue of the age of the rings, addressed by Crida et al. (2025, this collection).

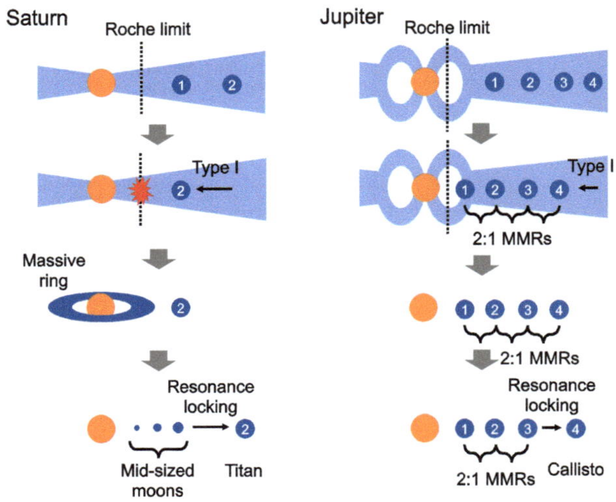

Fig. 13 Schematic picture of the tentative scenario proposed here for the formation of mid-sized and large moons of Saturn and Jupiter

At the same time, the outer moon (Moon 2) could be at the origin of Titan, but for this to be possible it must have halted its inward migration during the gas disk phase before falling inside the Roche limit. This requires that Moon 2 formed late, just before the CPD dissipated, so that its migration was slower and short. Alternatively, powerful resonance locking tides from Saturn may have captured Moon 2 and pushed it outwards as it was migrating in due to the CPD. Keeping Moon 2 is a bit tricky, but in any case the fast migration of Titan observed by Lainey et al. (2020) allows for more inwards migration of this Moon 2 than in previous models.

While Iapetus is not directly addressed in this scenario, it leaves room for Iapetus to have been able to form at a larger distance in the same CPD that gave birth to moons 1 and 2, as a third object outside the orbit of Moon 2. Its small mass would have avoided its significant Type I migration from the time of its origin.

A strong point of this scenario is that it also matches well the formation of Jovian moons: first, the four large Galilean moons form in the CPD by pebble or satellitesimal accretion. Then they move inward by Type I migration, and the innermost one (Moon 1) stops at the inner cavity, which resides outside the Roche limit because of the stronger magnetic field of Jupiter (Shibaike et al. 2019). Then the other three moons continue to drift inward until they are captured into the 2:1 MMR one by one, forming the Laplace resonance, just like giant planets around stars in low viscosity disks (Griveaud et al. 2023)). After dissipation of the CPD gas, the outermost moon (Moon 4, i.e., Callisto) may have left the resonance and moved outward under the effect of resonance locking, as also suggested by orbital dynamical simulations (Lari et al. 2023).

4 Conclusions and Directions for Future Work

In this article, we explored the different formation scenarios of the Kronian moons system in the context of a highly dissipative Saturn, with the objective of identifying at least one tentative scenario consistent with both observations and current theories and models. In order to do so, we first reviewed the diversity of moons and moon system architectures existing among the four giant planet systems of our Solar System, in the light of past and current Earth-based telescopic and space mission observations. This allowed us to identify the specific features of the different categories of Saturnian moons (small, mid-sized and large) that have to be explained by candidate formation scenarios.

We then reviewed the different types of moon formation scenarios that have been proposed for the four giant planet systems: result of a giant impact; formation by accretion in a circumplanetary disk (CPD); formation by viscous expansion from a massive disk; capture of small bodies from Sun-centered orbits. For each category of formation scenario, we discussed its relevance to the formation of Saturnian moons and of the Saturn system as a whole. We discussed the merits and difficulties of the different formation scenarios on the basis of available observations and in the context of the observed fast migration of moons. Finally, we proposed one tentative formation scenario that can be applied in the same terms to the cases of Saturn and Jupiter, starting from the assumptions that moons formed inside CPDs and migrated inwards and that the strength of the magnetic fields, and thus the radius of the inner magnetic cavity, was different between the two planets.

This scenario, like alternative ones, will need further observations, theories and simulations to be fully validated: all CPD-based scenarios strongly depend on knowledge of the structure of the CPD, which is unfortunately still controversial. There will be two ways to investigate CPDs in more detail in the near future: 1) numerical simulations with higher resolution and 2) observations of CPDs around young exoplanets.

Current numerical simulations still do not have enough spatial resolution to reveal the detailed gas structures and flows from inside the satellite formation region (inner than 0.01 Hill radius) to the entire CPDs. High resolution MHD simulations including the whole process of gas accretion on planets are also necessary.

Observations of CPDs around forming exoplanets will be another strong contribution to revealing the conditions prevailing in CPDs. There has already been one detection of a CPD around a gas-accreting exoplanet: dust continuum at (sub)millimeter wavelength with ALMA from PDS 70 c (Isella et al. 2019; Benisty et al. 2021). A few candidate CPDs around planets embedded in gas gaps have also been identified by their signature in the dust continuum and in the gas velocity distortion (Andrews et al. 2021; Bae et al. 2022). Such observations can constrain the properties of CPDs as well as those of their host planets (Shibaike and Mordasini 2024). As the number of observations will progressively increase, statistical information will also progressively become available.

Progress in our knowledge of the interiors of giant planets, as recently accomplished for Jupiter with the Radio Science experiment on board NASA's Juno, is also greatly needed to better quantify the efficiency of tidal coupling between moons and Jupiter (the Q factor) and the actual rate of tidal migration of moons. In the framework of this Topical Collection dedicated to "a highly dissipative Saturn", we did not explore all the possible complexities related to the Q factor. No doubt, this will be an important avenue for future research, heavily dependent on improved knowledge of Saturn's and more generally giant planets interiors.

Finally, new and additional information on the icy material content of giant planets regular moons will also provide still missing ground truth for formation scenarios. Detailed analysis of the cratering records and related ages of moons surfaces, recently reviewed in detail by Bottke et al. (2024), provides important constraints on the different formation scenarios, in particular by favoring formation of mid-size moons very early in the History of the Solar System, provided that most of the colliders were in heliocentric orbit and given we know the primordial population of impactors. In addition, determining the chemical and isotopic compositions of the surfaces and plumes of icy moons will provide critical information on the thermal history of the reservoirs of icy materials which have existed in protoplanetary and circumplanetary disks (Horner et al. 2008; Bierson and Nimmo 2020). There is no doubt that ESA's JUICE mission, as well as NASA's Europa clipper mission, which are both heading to the icy moons of Jupiter, will return to Earth critical new observations of the plumes and atmospheres of the Jovian icy moons in the next decade, likely to be followed by missions to Titan and Enceladus in the Saturn system, as described in the Horizon 2061 Planetary Exploration foresight exercise (Blanc et al. 2023). This new wave of missions to the gas giants of our solar system is our best chance to far better constrain the formation scenarios of their moons in the coming decades.

Acknowledgements We thank William F. Bottke and the other anonymous reviewer for the very valuable comments. This paper resulted from the Workshop "New Vision of the Saturnian System in the Context of a Highly Dissipative Saturn" held at the International Space Science Institute (ISSI) in Bern (Switzerland).

Author Contribution All authors have contributed to this review article.

Funding Information Y.S. contributed to this work within the framework of the NCCR PlanetS supported by the Swiss National Science Foundation under grants 51NF40_182901 and 51NF40_205606, and acknowledges support from the Swiss National Science Foundation under grant 200021_204847 'Planets In Time'. This work was supported by JSPS KAKENHI Grant Numbers JP22H01274, JP23K22545, and JP24K22907.

Data Availability Not applicable.

Code Availability Not applicable.

Declarations

Competing Interests The authors have no relevant financial or non-financial interests to disclose.

References

Agnor CB, Hamilton DP (2006) Neptune's capture of its moon Triton in a binary-planet gravitational encounter. Nature 441(7090):192–194. https://doi.org/10.1038/nature04792

Anderson JD, Johnson TV, Schubert G, et al (2005) Amalthea's density is less than that of water. Science 308(5726):1291–1293. https://doi.org/10.1126/science.1110422

Andrews SM, Elder W, Zhang S, et al (2021) Limits on millimeter continuum emission from circumplanetary material in the DSHARP disks. Astrophys J 916(1):51. https://doi.org/10.3847/1538-4357/ac00b9. arXiv:2105.08821 [astro-ph.EP]

Asphaug E, Reufer A (2013) Late origin of the Saturn system. Icarus 223(1):544–565. https://doi.org/10.1016/j.icarus.2012.12.009

Bae J, Teague R, Andrews SM, et al (2022) Molecules with ALMA at Planet-forming Scales (MAPS): a circumplanetary disk candidate in molecular-line emission in the AS 209 disk. Astrophys J Lett 934(2):L20. https://doi.org/10.3847/2041-8213/ac7fa3. arXiv:2207.05923 [astro-ph.EP]

Batygin K, Morbidelli A (2020) Formation of giant planet satellites. Astrophys J 894(2):143. https://doi.org/10.3847/1538-4357/ab8937. arXiv:2005.08330 [astro-ph.EP]

Benisty M, Bae J, Facchini S, et al (2021) A circumplanetary disk around PDS70c. Astrophys J Lett 916(1):L2. https://doi.org/10.3847/2041-8213/ac0f83. arXiv:2108.07123 [astro-ph.EP]

Bierson CJ, Nimmo F (2020) Explaining the Galilean satellites' density gradient by hydrodynamic escape. Astrophys J Lett 897(2):L43. https://doi.org/10.3847/2041-8213/aba11a

Blanc M, Mandt K, Mousis O, et al (2021) Science goals and mission objectives for the future exploration of ice giants systems: a Horizon 2061 perspective. Space Sci Rev 217(1):3. https://doi.org/10.1007/s11214-020-00769-5

Blanc M, Ammannito E, Bousquet P, et al (2023) Planetary Exploration Horizon 2061: a long-term perspective for planetary exploration. Elsevier, Amsterdam. https://doi.org/10.1016/C2020-0-01626-9

Bottke WF, Nesvorný D, Vokrouhlický D, et al (2010) The irregular satellites: the most collisionally evolved populations in the Solar System. Astron J 139(3):994–1014. https://doi.org/10.1088/0004-6256/139/3/994

Bottke WF, Vokrouhlický D, Nesvorný D, et al (2024) The bombardment history of the giant planet satellites. Planet Sci J 5(4):88. https://doi.org/10.3847/psj/ad29f4

Brown ME, Schaller EL (2007) The mass of dwarf planet Eris. Science 316(5831):1585. https://doi.org/10.1126/science.1139415

Buie MW, Grundy WM, Young EF, et al (2006) Orbits and photometry of Pluto's satellites: Charon, S/2005 P1, and S/2005 P2. Astron J 132(1):290–298. https://doi.org/10.1086/504422. arXiv:astro-ph/0512491 [astro-ph]

Burns JA, Simonelli DP, Showalter MR, et al (2004) Jupiter's ring-moon system. In: Bagenal F, Dowling TE, McKinnon WB (eds) Jupiter: the planet, satellites and magnetosphere. Cambridge University Press, pp 241–262

Canup RM (2010) Origin of Saturn's rings and inner moons by mass removal from a lost Titan-sized satellite. Nature 468(7326):943–946. https://doi.org/10.1038/nature09661

Canup R, Salmon J (2018) Origin of Phobos and Deimos by the impact of a Vesta-to-Ceres sized body with Mars. Sci Adv 4(4):eaar6887. https://doi.org/10.1126/sciadv.aar6887

Canup RM, Ward WR (2002) Formation of the Galilean satellites: conditions of accretion. Astron J 124(6):3404–3423. https://doi.org/10.1086/344684

Canup RM, Ward WR (2006) A common mass scaling for satellite systems of gaseous planets. Nature 441(7095):834–839. https://doi.org/10.1038/nature04860

Canup RM, Righter K, Dauphas N, et al (2023) Origin of the moon. Rev Mineral Geochem 89(1):53–102. https://doi.org/10.2138/rmg.2023.89.02

Castillo-Rogez J, Vernazza P, Walsh K (2019) Geophysical evidence that Saturn's moon Phoebe originated from a C-type asteroid reservoir. Mon Not R Astron Soc Lett 486(1):538–543. https://doi.org/10.1093/mnras/stz786

Charnoz S, Salmon J, Crida A (2010) The recent formation of Saturn's moonlets from viscous spreading of the main rings. Nature 465(7299):752–754. https://doi.org/10.1038/nature09096

Charnoz S, Crida A, Castillo-Rogez JC, et al (2011) Accretion of Saturn's mid-sized moons during the viscous spreading of young massive rings: solving the paradox of silicate-poor rings versus silicate-rich moons. Icarus 216(2):535–550. https://doi.org/10.1016/j.icarus.2011.09.017. arXiv:1109.3360 [astro-ph.EP]

Christensen UR, Holzwarth V, Reiners A (2009) Energy flux determines magnetic field strength of planets and stars. Nature 457(7226):167–169. https://doi.org/10.1038/nature07626

Cilibrasi M, Szulágyi J, Mayer L, et al (2018) Satellites form fast & late: a population synthesis for the Galilean moons. Mon Not R Astron Soc Lett 480(4):4355–4368. https://doi.org/10.1093/mnras/sty2163. arXiv:1801.06094 [astro-ph.EP]

Cilibrasi M, Szulágyi J, Grimm SL, et al (2021) An N-body population synthesis framework for the formation of moons around Jupiter-like planets. Mon Not R Astron Soc Lett 504(4):5455–5474. https://doi.org/10.1093/mnras/stab1179. arXiv:2011.11513 [astro-ph.EP]

Clark RN, Brown RH, Jaumann R, et al (2005) Compositional maps of Saturn's moon Phoebe from imaging spectroscopy. Nature 435(7038):66–69

Colombo G, Franklin F (1971) On the formation of the outer satellite groups of Jupiter. Icarus 15:186–189. https://doi.org/10.1016/0019-1035(71)90073-x

Connerney JEP (1993) Magnetic fields of the outer planets. J Geophys Res 98(E10):18659–18680. https://doi.org/10.1029/93JE00980

Crida A (2023) Planetary formation and early phases. C R Phys 24(S2):1–16. https://doi.org/10.5802/crphys.161

Crida A, Charnoz S (2012) Formation of regular satellites from ancient massive rings in the Solar System. Science 338(6111):1196. https://doi.org/10.1126/science.1226477. arXiv:1301.3808 [astro-ph.EP]

Crida A, Morbidelli A, Masset F (2006) On the width and shape of gaps in protoplanetary disks. Icarus 181(2):587–604. https://doi.org/10.1016/j.icarus.2005.10.007. arXiv:astro-ph/0511082 [astro-ph]

Crida A, Charnoz S, Hsu HW, et al (2019) Are Saturn's rings actually young? Nat Astron 3:967–970. https://doi.org/10.1038/s41550-019-0876-y

Crida A, Estrada PR, Nicholson PD, Murray CD (2025) The age and origin of Saturn's rings Space Sci Rev 221

Ćuk M, Burns JA (2004) Gas-drag-assisted capture of himalia's family. Icarus 167(2):369–381. https://doi.org/10.1016/j.icarus.2003.09.026
Ćuk M, Dones L, Nesvorný D (2016) Dynamical evidence for a late formation of Saturn's moons. Astrophys J 820(2):97. https://doi.org/10.3847/0004-637x/820/2/97
Ćuk M, French RS, Showalter MR, et al (2022) Cupid is not doomed yet: on the stability of the inner moons of Uranus. Astron J 164(2):38. https://doi.org/10.3847/1538-3881/ac745d
Ćuk M, El Moutamid M, Lari G, et al (2024) Long-term evolution of the Saturnian system. Space Sci Rev 220:20. https://doi.org/10.1007/s11214-024-01049-2
Denk T, Mottola S (2019) Studies of irregular satellites: I. Lightcurves and rotation periods of 25 Saturnian moons from Cassini observations. Icarus 322:80–102. https://doi.org/10.1016/j.icarus.2018.12.040
Dougherty MK, Spilker LJ (2018) Review of Saturn's icy moons following the Cassini mission. Rep Prog Phys 81(6):065901. https://doi.org/10.1088/1361-6633/aabdfb
Drażkowska J, Szulágyi J (2018) Dust evolution and satellitesimal formation in circumplanetary disks. Astrophys J 866(2):142. https://doi.org/10.3847/1538-4357/aae0fd. arXiv:1807.02638 [astro-ph.EP]
Durham WB, McKinnon WB, Stern LA (2005) Cold compaction of water ice. Geophys Res Lett 32(18):L18202. https://doi.org/10.1029/2005GL023484
Eriksson LEJ, Ronnet T, Johansen A, et al (2022) A low accretion efficiency of planetesimals formed at planetary gap edges. Astron Astrophys 661:A73. https://doi.org/10.1051/0004-6361/202142391. arXiv:2202.12723 [astro-ph.EP]
Estrada PR, Durisen RH (2023) Constraints on the initial mass, age and lifetime of Saturn's rings from viscous evolutions that include pollution and transport due to micrometeoroid bombardment. Icarus 400:115296. https://doi.org/10.1016/j.icarus.2022.115296. arXiv:2305.13609 [astro-ph.EP]
Fortney JJ, Ikoma M, Nettelmann N, et al (2011) Self-consistent model atmospheres and the cooling of the Solar System's giant planets. Astrophys J 729(1):32. https://doi.org/10.1088/0004-637X/729/1/32. arXiv:1101.0606 [astro-ph.EP]
Fujita T, Ohtsuki K, Tanigawa T, et al (2013) Capture of planetesimals by gas drag from circumplanetary disks. Astron J 146(6):140. https://doi.org/10.1088/0004-6256/146/6/140
Fuller J, Luan J, Quataert E (2016) Resonance locking as the source of rapid tidal migration in the Jupiter and Saturn moon systems. Mon Not R Astron Soc Lett 458(4):3867–3879. https://doi.org/10.1093/mnras/stw609
Gressel O, Nelson RP, Turner NJ, et al (2013) Global hydromagnetic simulations of a planet embedded in a dead zone: gap opening, gas accretion, and formation of a protoplanetary jet. Astrophys J 779(1):59. https://doi.org/10.1088/0004-637X/779/1/59. arXiv:1309.2871 [astro-ph.EP]
Griveaud P, Crida A, Lega E (2023) Migration of pairs of giant planets in low-viscosity discs. Astron Astrophys 672:A190. https://doi.org/10.1051/0004-6361/202245208. arXiv:2303.04652 [astro-ph.EP]
Haffert SY, Bohn AJ, de Boer J, et al (2019) Two accreting protoplanets around the young star PDS 70. Nat Astron 3:749–754. https://doi.org/10.1038/s41550-019-0780-5. arXiv:1906.01486 [astro-ph.EP]
Hedman MM, Cooper NJ, Murray CD, et al (2010) Aegaeon (Saturn LIII), a G-ring object. Icarus 207(1):433–447. https://doi.org/10.1016/j.icarus.2009.10.024. arXiv:0911.0171 [astro-ph.EP]
Heppenheimer TA, Porco C (1977) New contributions to the problem of capture. Icarus 30:385–401. https://doi.org/10.1016/0019-1035(77)90173-7
Horner J, Mousis O, Alibert Y, et al (2008) Constraints from deuterium on the formation of icy bodies in the Jovian system and beyond. Planet Space Sci 56(12):1585–1595. https://doi.org/10.1016/j.pss.2008.04.010. arXiv:0808.3476 [astro-ph]
Iess L, Militzer B, Kaspi Y, et al (2019) Measurement and implications of Saturn's gravity field and ring mass. Science 364(6445):aat2965. https://doi.org/10.1126/science.aat2965
Isella A, Benisty M, Teague R, et al (2019) Detection of continuum submillimeter emission associated with candidate protoplanets. Astrophys J Lett 879(2):L25. https://doi.org/10.3847/2041-8213/ab2a12. arXiv:1906.06308 [astro-ph.EP]
Jacobson RA (2004) The orbits of the major Saturnian satellites and the gravity field of Saturn from spacecraft and Earth-based observations. Astron J 128(1):492–501. https://doi.org/10.1086/421738
Jacobson RA, Spitale J, Porco CC, et al (2006) The GM values of Mimas and Tethys and the libration of Methone. Astron J 132(2):711–713. https://doi.org/10.1086/505209
Jewitt DC, Sheppard SS (2002) Physical properties of trans-Neptunian object (20000) Varuna. Astron J 123(4):2110–2120. https://doi.org/10.1086/339557. arXiv:astro-ph/0201082 [astro-ph]
Jewitt D, Sheppard S (2005) Irregular satellites in the context of planet formation. Space Sci Rev 116(1–2):441–455. https://doi.org/10.1007/s11214-005-1965-z
Johnson TV (2005) Geology of the icy satellites. Space Sci Rev 116(1–2):401–420. https://doi.org/10.1007/s11214-005-1963-1
Johnson TV, Estrada PR (2009) Origin of the Saturn system. In: Dougherty MK, Esposito LW, Krimigis SM (eds) Saturn from Cassini-Huygens, pp 55. https://doi.org/10.1007/978-1-4020-9217-6_3

Kanagawa KD, Muto T, Tanaka H, et al (2015) Mass estimates of a giant planet in a protoplanetary disk from the gap structures. Astrophys J Lett 806(1):L15. https://doi.org/10.1088/2041-8205/806/1/L15. arXiv:1505.04482 [astro-ph.EP]
Lainey V, Karatekin O, Desmars J, et al (2012) Strong tidal dissipation in Saturn and constraints on Enceladus' thermal state from astrometry. Astrophys J 752(1):14. https://doi.org/10.1088/0004-637x/752/1/14
Lainey V, Casajus LG, Fuller J, et al (2020) Resonance locking in giant planets indicated by the rapid orbital expansion of Titan. Nat Astron 4:1053–1058. https://doi.org/10.1038/s41550-020-1120-5. arXiv:2006.06854 [astro-ph.EP]
Lambrechts M, Johansen A (2012) Rapid growth of gas-giant cores by pebble accretion. Astron Astrophys 544:A32. https://doi.org/10.1051/0004-6361/201219127. arXiv:1205.3030 [astro-ph.EP]
Lari G, Saillenfest M, Grassi C (2023) Dynamical history of the Galilean satellites for a fast migration of Callisto. Mon Not R Astron Soc Lett 518(2):3023–3035. https://doi.org/10.1093/mnras/stac3299. arXiv:2210.00424 [astro-ph.EP]
Lin DNC, Papaloizou J (1986) On the tidal interaction between protoplanets and the protoplanetary disk. III. Orbital migration of protoplanets. Astrophys J 309:846. https://doi.org/10.1086/164653
Lubow SH, Seibert M, Artymowicz P (1999) Disk accretion onto high-mass planets. Astrophys J 526(2):1001–1012. https://doi.org/10.1086/308045. arXiv:astro-ph/9910404 [astro-ph]
Lupo MJ, Lewis JS (1979) Mass-radius relationships in icy satellites. Icarus 40(2):157–170. https://doi.org/10.1016/0019-1035(79)90061-7
Lyra W, Johansen A, Klahr H, et al (2009) Standing on the shoulders of giants. Trojan Earths and vortex trapping in low mass self-gravitating protoplanetary disks of gas and solids. Astron Astrophys 493(3):1125–1139. https://doi.org/10.1051/0004-6361:200810797. arXiv:0810.3192 [astro-ph]
Morbidelli A, Tsiganis K, Batygin K, et al (2012) Explaining why the Uranian satellites have equatorial prograde orbits despite the large planetary obliquity. Icarus 219(2):737–740. https://doi.org/10.1016/j.icarus.2012.03.025
Mosqueira I, Estrada PR (2003a) Formation of the regular satellites of giant planets in an extended gaseous nebula I: subnebula model and accretion of satellites. Icarus 163(1):198–231. https://doi.org/10.1016/S0019-1035(03)00076-9
Mosqueira I, Estrada PR (2003b) Formation of the regular satellites of giant planets in an extended gaseous nebula II: satellite migration and survival. Icarus 163(1):232–255. https://doi.org/10.1016/S0019-1035(03)00077-0
Mueller S, McKinnon WB (1988) Three-layered models of Ganymede and Callisto: compositions, structures, and aspects of evolution. Icarus 76(3):437–464. https://doi.org/10.1016/0019-1035(88)90014-0
Müller A, Keppler M, Henning T, et al (2018) Orbital and atmospheric characterization of the planet within the gap of the PDS 70 transition disk. Astron Astrophys 617:L2. https://doi.org/10.1051/0004-6361/201833584. arXiv:1806.11567 [astro-ph.EP]
Nesvorný D, Vokrouhlický D, Morbidelli A (2007a) Capture of irregular satellites during planetary encounters. Astron J 133(5):1962–1976. https://doi.org/10.1086/512850
Nesvorny D, Vokrouhlicky D, Morbidelli A (2007b) Capture of Irregular Satellites During Planetary Encounters. AAS/Division for Planetary Sciences Meeting Abstracts #39 p 32.11
Nicholson PD, Cuk M, Sheppard SS, et al (2008) Irregular satellites of the giant planets. In: Barucci MA, Boehnhardt H, Cruikshank DP, et al (eds) The Solar System beyond Neptune, pp 411–424
Ogihara M, Ida S (2012) N-body simulations of satellite formation around giant planets: origin of orbital configuration of the Galilean moons. Astrophys J 753(1):60. https://doi.org/10.1088/0004-637X/753/1/60. arXiv:1205.0301 [astro-ph.EP]
Ormel CW, Klahr HH (2010) The effect of gas drag on the growth of protoplanets. Analytical expressions for the accretion of small bodies in laminar disks. Astron Astrophys 520:A43. https://doi.org/10.1051/0004-6361/201014903. arXiv:1007.0916 [astro-ph.EP]
Poirier JP (2000) Introduction to the physics of the Earth's interior. Cambridge University Press, Cambridge
Pollack JB, Burns JA, Tauber ME (1979) Gas drag in primordial circumplanetary envelopes: a mechanism for satellite capture. Icarus 37:587–611. https://doi.org/10.1016/0019-1035(79)90016-2
Porco CC, Baker E, Barbara J, et al (2005a) Cassini imaging science: initial results on Saturn's rings and small satellites. Science 307(5713):1226–1236. https://doi.org/10.1126/science.1108056
Porco CC, Baker E, Barbara J, et al (2005b) Cassini imaging science: initial results on Phoebe and Iapetus. Science 307(5713):1237–1242. https://doi.org/10.1126/science.1107981
Porco CC, Thomas PC, Weiss JW, et al (2007) Saturn's small inner satellites: clues to their origins. Science 318(5856):1602. https://doi.org/10.1126/science.1143977
Rabinowitz DL, Barkume K, Brown ME, et al (2006) Photometric observations constraining the size, shape, and albedo of 2003 EL61, a rapidly rotating, Pluto-sized object in the Kuiper Belt. Astrophys J 639(2):1238–1251. https://doi.org/10.1086/499575. arXiv:astro-ph/0509401 [astro-ph]

Reiners A, Christensen UR (2010) A magnetic field evolution scenario for Brown dwarfs and giant planets. Astron Astrophys 522:A13. https://doi.org/10.1051/0004-6361/201014251. arXiv:1007.1514 [astro-ph.EP]

Ronnet T, Johansen A (2020) Formation of moon systems around giant planets. Capture and ablation of planetesimals as foundation for a pebble accretion scenario. Astron Astrophys 633:A93. https://doi.org/10.1051/0004-6361/201936804. arXiv:1912.04765 [astro-ph.EP]

Ronnet T, Mousis O, Vernazza P, et al (2018) Saturn's formation and early evolution at the origin of Jupiter's massive moons. Astron J 155(5):224. https://doi.org/10.3847/1538-3881/aabcc7. arXiv:1804.02892 [astro-ph.EP]

Rosenblatt P, Charnoz S (2012) On the formation of the Martian moons from a circum-Martian accretion disk. Icarus 221(2):806–815. https://doi.org/10.1016/j.icarus.2012.09.009

Rosenblatt P, Charnoz S, Dunseath KM, et al (2016) Accretion of Phobos and Deimos in an extended debris disc stirred by transient moons. Nat Geosci 9(8):581–583. https://doi.org/10.1038/ngeo2742

Rufu R, Canup RM (2017) Triton's evolution with a primordial Neptunian satellite system. Astron J 154(5):208

Ruskol EL (2006) The origin of Jovian and Saturnian satellites in accretion disks. Sol Syst Res 40(6):456–461. https://doi.org/10.1134/S0038094606060025

Salmon J, Canup RM (2022) Co-accretion + giant impact origin of the Uranus system: post-impact evolution. Astrophys J 924(1):6. https://doi.org/10.3847/1538-4357/ac300e

Salmon J, Charnoz S, Crida A, et al (2010) Long-term and large-scale viscous evolution of dense planetary rings. Icarus 209(2):771–785. https://doi.org/10.1016/j.icarus.2010.05.030. arXiv:1006.0633 [astro-ph.EP]

Sasaki T, Stewart GR, Ida S (2010) Origin of the different architectures of the Jovian and Saturnian satellite systems. Astrophys J 714(2):1052–1064. https://doi.org/10.1088/0004-637X/714/2/1052. arXiv:1003.5737 [astro-ph.EP]

Schulik M, Johansen A, Bitsch B, et al (2020) On the structure and mass delivery towards circumplanetary discs. Astron Astrophys 642:A187. https://doi.org/10.1051/0004-6361/202037556. arXiv:2003.13398 [astro-ph.EP]

Sheppard SS, Tholen DJ, Alexandersen M, et al (2023) New Jupiter and Saturn satellites reveal new moon dynamical families. Res Notes Amer Astron Soc 7(5):100. https://doi.org/10.3847/2515-5172/acd766

Shibaike Y, Alibert Y (2020) Planetesimal formation at the gas pressure bump following a migrating planet. I. Basic characteristics of the new formation model. Astron Astrophys 644:A81. https://doi.org/10.1051/0004-6361/202039086. arXiv:2010.10594 [astro-ph.EP]

Shibaike Y, Mordasini C (2024) Constraints on PDS 70 b and c from the dust continuum emission of the circumplanetary discs considering in situ dust evolution. Astron Astrophys 687:A166. https://doi.org/10.1051/0004-6361/202449522

Shibaike Y, Mori S (2023) Effective dust growth in laminar circumplanetary discs with magnetic wind-driven accretion. Mon Not R Astron Soc Lett 518(4):5444–5456. https://doi.org/10.1093/mnras/stac3428. arXiv:2211.08947 [astro-ph.EP]

Shibaike Y, Okuzumi S, Sasaki T, et al (2017) Satellitesimal formation via collisional dust growth in steady circumplanetary disks. Astrophys J 846(1):81. https://doi.org/10.3847/1538-4357/aa8454. arXiv:1708.01080 [astro-ph.EP]

Shibaike Y, Ormel CW, Ida S, et al (2019) The Galilean satellites formed slowly from pebbles. Astrophys J 885(1):79. https://doi.org/10.3847/1538-4357/ab46a7. arXiv:1909.00285 [astro-ph.EP]

Showalter MR (2020) The rings and small moons of Uranus and Neptune. Philos Trans R Soc Lond Ser A 378(2187):20190482. https://doi.org/10.1098/rsta.2019.0482

Stansberry JA, Grundy WM, Margot JL, et al (2006) The albedo, size, and density of binary Kuiper Belt object (47171) 1999 TC_{36}. Astrophys J 643(1):556–566. https://doi.org/10.1086/502674. arXiv:astro-ph/0602316 [astro-ph]

Suetsugu R, Ohtsuki K (2017) Distribution of captured planetesimals in circumplanetary gas disks and implications for accretion of regular satellites. Astrophys J 839(1):66. https://doi.org/10.3847/1538-4357/aa692e. arXiv:1703.07917 [astro-ph.EP]

Szulágyi J (2017) Effects of the planetary temperature on the circumplanetary disk and on the gap. Astrophys J 842(2):103. https://doi.org/10.3847/1538-4357/aa7515. arXiv:1705.08444 [astro-ph.EP]

Szulágyi J, Morbidelli A, Crida A, et al (2014) Accretion of Jupiter-mass planets in the limit of vanishing viscosity. Astrophys J 782(2):65. https://doi.org/10.1088/0004-637X/782/2/65. arXiv:1312.6302 [astro-ph.EP]

Szulágyi J, Binkert F, Surville C (2022) Meridional circulation of dust and gas in the circumstellar disk: delivery of solids onto the circumplanetary region. Astrophys J 924(1):1. https://doi.org/10.3847/1538-4357/ac32d1. arXiv:2103.12128 [astro-ph.EP]

Takasao S, Tomida K, Iwasaki K, et al (2022) Three-dimensional simulations of magnetospheric accretion in a T Tauri star: accretion and wind structures just around the star. Astrophys J 941(1):73. https://doi.org/10.3847/1538-4357/ac9eb1. arXiv:2211.01072 [astro-ph.SR]
Tanigawa T, Ohtsuki K, Machida MN (2012) Distribution of accreting gas and angular momentum onto circumplanetary disks. Astrophys J 747(1):47. https://doi.org/10.1088/0004-637X/747/1/47. arXiv:1112.3706 [astro-ph.EP]
Thomas PC, Burns JA, Helfenstein P, et al (2007) Shapes of the Saturnian icy satellites and their significance. Icarus 190(2):573–584. https://doi.org/10.1016/j.icarus.2007.03.012
Turrini D, Marzari F, Tosi F (2009) A new perspective on the irregular satellites of Saturn - ii. Dynamical and physical origin. Mon Not R Astron Soc Lett 392(1):455–474. https://doi.org/10.1111/j.1365-2966.2008.14100.x
Yoder CF (1995) Astrometric and geodetic properties of Earth and the Solar System. In: Ahrens TJ (ed) Global Earth physics: a handbook of physical constants. AGU, Washington DC, p 1–31.
Zhu Z, Andrews SM, Isella A (2018) On the radio detectability of circumplanetary discs. Mon Not R Astron Soc Lett 479(2):1850–1865. https://doi.org/10.1093/mnras/sty1503. arXiv:1708.07287 [astro-ph.EP]

Authors and Affiliations

Michel Blanc[1,2] · Aurélien Crida[3] · Yuhito Shibaike[4,5] · Sebastien Charnoz[6] · Maryame El Moutamid[7] · Paul Estrada[8] · Olivier Mousis[2] · Julien Salmon[9] · Antoine Schneeberger[2] · Pierre Vernazza[2]

✉ Y. Shibaike
yuhito.shibaike@nao.ac.jp

M. Blanc
michel.blanc@irap.omp.eu

A. Crida
aurelien.crida@oca.eu

S. Charnoz
charnoz@ipgp.fr

M. El Moutamid
maryame@astro.cornell.edu

P. Estrada
paul.r.estrada@nasa.gov

O. Mousis
olivier.mousis@lam.fr

J. Salmon
julien@boulder.swri.edu

A. Schneeberger
antoine.schneeberger@lam.fr

P. Vernazza
pierre.vernazza@lam.fr

1 Research Institute in Astrophysics and Planetology, Toulouse, France

2 Laboratoire d'Astrophysique de Marseille, Marseille, France

3 Université Côte d'Azur, Observatoire de la Côte d'Azur, Nice, France

4 National Astronomical Observatory of Japan, Tokyo, Japan

[5] University of Bern, Bern, Switzerland

[6] Institut de Physique du Globe de Paris, Paris, France

[7] Department of Astronomy, Cornell University, New York, USA

[8] NASA Ames Research Center, California, USA

[9] IMCCE, Observatoire de Paris, Paris, France

Space Science Reviews (2026) 222:20
https://doi.org/10.1007/s11214-026-01275-w

Paving the Way for Future Space Missions in the Context of High Tidal Dissipation in the Saturnian System

Valéry Lainey[1] · Aurélien Crida[2] · Matija Cuk[3] · Jeffrey N. Cuzzi[4] · Dominic Dirkx[5] · Gianrico Filacchione[6] · James Fuller[7] · Carly J.A. Howett[8] · Kelly Miller[9] · Francis Nimmo[10] · Nicolas Rambaux[1] · Marco Zannoni[11]

Received: 4 September 2025 / Accepted: 2 February 2026 / Published online: 13 February 2026

Abstract
The recent discovery of strong tidal dissipation in Saturn's interior has radically changed our view of the Saturnian system. While some questions are naturally answered by the new paradigm, others are emerging and require further measurement. This article presents the next key questions to be addressed by future space missions and analysis. Suggestions for space measurements to discriminate between different scenarios concerning the formation, evolution and internal state of the Saturnian system are given.

Keywords Saturnian system · Future space missions · Tides · Dynamics · Cassini mission

1 What Physical Mechanisms Drive the High Dissipation Observed in Saturn?

One of the great surprises of the Cassini mission was certainly the discovery of the intense tidal energy dissipation in Saturn (Lainey et al. 2012, 2017). Successive measurements revealed that the frequency dependence of the amount of energy dissipated was very different from the usual models that were based on a constant quality factor Q or delay time Δt. As described in Fuller et al. (2024) and summarized below, while there are several mechanisms that could lead to significant tidal loss of mechanical energy (Ogilvie and Lin 2004; Remus et al. 2012; Terquem 2021; Barker and Astoul 2021; Dewberry 2024), the mechanism proposed by Fuller et al. (2016) is currently the only one to explain the observed frequency variation of Q (Lainey et al. 2025). This mechanism is based on the excitation of the planet's internal eigenmodes by the tidal forces associated with the moons. When a resonance is encountered, a large amount of mechanical energy is dissipated as heat via turbulence in the fluid part of the planet. As the planet cools and its structure and rotation rate evolve, the frequencies of the eigenmodes vary with time. As a result, the moons all have a high probability of encountering resonances with the planet's modes. Hence, because of the intense internal friction generated, the moons are then pushed outwards from the planet via an exchange of angular momentum between Saturn's rotation and the moons' orbits. The result is a dynamic equilibrium in which each moon migrates outwards as its resonant eigenmode evolves. Thus, according to the mechanism developed by Fuller et al. (2016), the commonly

Extended author information available on the last page of the article

used physical parameter Q must be replaced by analysis of the parameter t_{tide} equal to the ratio $\frac{a}{\dot{a}}$. According to the resonance locking model, each moon raises a tidal wave whose frequency is resonant with a normal mode, causing the moon to migrate outwards at a rate dictated by t_{tide}.

However, several questions remain to be addressed. Firstly, Jacobson (2022) obtained an orbital expansion for Titan that disagrees with that published by Lainey et al. (2020). Specifically, when combining the radiometric and astrometric data, Jacobson (2022) obtain a tidal Q of Saturn at Titan's forcing frequency that is about an order of magnitude larger than both results (one from astrometric data, one from radiometric data) of Lainey et al. (2020), while they obtain a statistically consistent result when omitting the Voyager and Cassini tracking data from their analysis. A fuller discussion of this is described in Fuller et al. (2024). As Titan's orbital expansion is, along with that of Rhea, the most important piece of information for asserting the existence of a resonance locking mechanism between the satellites' orbital evolution and Saturn's evolving internal oscillation modes (Fuller et al. 2016), this conflict between different results needs to be resolved. Since the solution for the orbital evolution depends on a combination of astrometric and radiometric data, a global re-analysis is necessary, with a particular focus on how uncertainties in the data and models propagate to uncertainties in Saturn's dissipation. Ideally, an independent team should reanalyze the radio-science data from Cassini's flybys of Titan, and help to confirm or deny the presence of such a rapid orbital expansion (11 ± 2 cm/year) of Titan's orbit (Lainey et al. 2020). Such an analysis is now underway (Hener et al. 2025)

Another open question is that of dissipation within Saturn's core. Work focusing on this question has suggested that a significant fraction of the system's mechanical energy may be dissipated in Saturn's core (Remus et al. 2012). In theory, accurate measurements of the inner moons' migration rates may allow us to separate the fraction of energy dissipated in the core by viscoelastic dissipation from that dissipated by resonant modes in the fluid envelope. In practice, however, orbital expansion measurements are still subject to large error bars. Moreover, the exchange of angular momentum between Mimas and the Saturnian rings may be significant due to the 2:1 resonance (Lissauer and Cuzzi 1982), inducing an orbital expansion of Mimas, similarly to tides. Last, Mimas' orbital expansion is very weakly constrained (Lainey et al. 2020) with a da/dt of 25.3 ± 20.9 cm/year (at 3-sigma) leaving only five truly constraining measurements, associated with the moons Enceladus, Tethys, Dione, Rhea and Titan. Thus, it is desirable that future space missions add further constraints on the orbital expansions of these moons via astrometry and radio-science during future flybys, orbit phases or landers. Measurements of the orbital expansion of Mimas will especially be important.

Another important point is the amplitude of tidal effects in the moons themselves. As the dissipated energy is taken from the orbital energy, the tides in the moons decrease their eccentricities and semi-major axes. The orbital expansion measurements obtained by Lainey et al. (2020) assumed that tidal dissipation in Enceladus was in the range 3-55 GW. This is consistent with observationally constrained upper limit of 54 GW (Miller et al. 2024). Because of the degeneracy between tidal migration driven by Saturn and that driven by dissipation within Enceladus, error bars on dissipation in Saturn at the frequency of Enceladus and Dione are very large.

Furthermore, no significant dissipation in the other moons was modeled, on the assumption that this would be negligible within the error bars of the measurements. Nevertheless, Lainey et al. (2024) recently showed that Mimas could be the center of intense tidal effects due to the inferred presence of a young ocean beneath its ice pack. This suggests that the interpretation of orbital expansion measurements could be revised, at least for Mimas, and potentially for other moons (see Sect. 5).

Finally, testing the resonance locking mechanism within a system other than Saturn is more than desirable. For example, Fuller et al. (2016) predicted that if this mechanism were to take place within Jupiter, Callisto should exhibit a significant orbital expansion. Such a measurement will be conceivable with the arrival in the Jovian system of the next generation of space missions, such as JUICE and Europa Clipper (Magnanini et al. 2024; Fayolle et al. 2024).

2 Determining the Formation Process and the Age of the Main Moons

With the observed fast outwards migration of Saturn's regular satellites, the question of their origin must be addressed from a new perspective. In particular, it is not clear that the currently observed moons formed in a circum-planetary disk, close to where they currently orbit. The highly variable composition of the mid-sized moons, from 7 to 70% silicates with no apparent radial gradient, also suggests some stochasticity in their formation process. Meanwhile the low density and pure water ice composition of the small moons suggests a connection to the rings. Therefore, the idea has emerged that the regular satellites might come from the spreading of the rings beyond the Roche radius, where the material naturally agglomerates under the effects of self-gravity (Charnoz et al. 2010), and then migrated outwards by interaction with the rings and with Saturn itself. The fast outwards migration, with a constant Q model, allows to form by this mechanism all the mid-sized moons inside the orbit of Titan within the age of the Solar System (Charnoz et al. 2011). Moving to the resonance-locking mechanism makes it possible to add Titan in the list of the satellites formed by the spreading of a massive primordial ring beyond the Roche radius (Crida 2020). However, Iapetus does not fit in this frame, and most likely was formed in the circum-planetary disk. So was the progenitor of the massive rings in the model by Canup (2010), leading to a hybrid satellite formation scenario around Saturn: a primordial satellite is the progenitor of the rings, who are the progenitors of most satellites observed today. The theories on the formation of Saturn's satellites are reviewed in Blanc et al. (2025).

Details in this global picture are still to be understood however. It is expected that merger events occur in the so-called pyramidal regime (Crida and Charnoz 2012), so that the age of an object may not be so clearly defined. The age of the surfaces imposes a lower limit of the time of the last major collision these moons suffered. Impacts with debris from a catastrophic collision in the system may perturb the age estimated from the cratering rate (Asphaug and Reufer 2013). Other constraints can potentially be derived by the surface ice amorphous/crystalline ratio (Dalle Ore et al. 2015).

The dynamical interactions between the moons and their exact orbital parameters place constraints on their relative migration history. Whether the moons migrate outwards following a classical constant Q model or following a resonance locking model completely changes their relative migration velocity, and hence their resonant interactions. The presence of current mean-motion resonances between three pairs of moons (Mimas-Tethys, Enceladus-Dione and Titan-Hyperion) requires current (or recent) convergent migration in each of these pairs. While convergent migration is possible in resonance locking, especially when only the inner moons are caught in resonance locks, divergent migration is also possible and the dynamics can be quite complex. More detailed planetary and tidal evolution models are needed to predict how often convergent vs. divergent migration occurs. Additionally, classical tides are a stronger function of distance than the resonance-lock tides, so, depending on the satellite masses and the Q factor at relevant frequencies, classical tides may even dominate migration of sizable moons close to the planet.

While the ring-spreading satellite-formation model that implicitly assumes the fragmentation of a big moon opens the way for a highly variable density of the produced moons due to the stochastic nature of the formation of silicate chunks in the rings and of the accretion history (Charnoz et al. 2011, Sect. 4), it does not address the short tidal evolution timescales obtained if Q is assumed to be constant. However, in the resonance-locking model, Q is not constant and ancient formation of the midsize moons is fully compatible with their present-day orbital expansion (Fuller et al. 2016). As a consequence, models for the early formation of the midsize moons that can date back to a primordial disk are not ruled out. Still, it should be kept in mind that there are several other comparably-short-timescale indications associated even with Titan itself (such as its eccentricity damping timescale and lifetime of atmospheric CH_4) which need more study (Nixon et al. 2018). One possibility is that the system was dynamically excited recently, for instance by a mid-sized satellite becoming destabilized and ultimately colliding with Saturn (Wisdom et al. 2022).

The age of the surfaces of the midsize moons probably contains the best clues to the time and mode of their formation (or re-formation). Unfortunately, obtaining absolute ages for satellite surfaces depends on what model of impactor flux is used (Zahnle et al. 2003), so they cannot be used as a test of such models. Traditional assumptions of heliocentric impactors lead to ancient ages but may not be fully consistent with the observations in several ways (Rhoden et al. 2024). Indeed, some of the observations are better fit by a Saturn-unique planetocentric population operating over a shorter period of time. Certainly, more effort along the lines of crater analyses is extremely important (Bottke et al. 2024). Other constraints may be derived from the surface ice amorphous/crystalline ratio; here, more laboratory or experimental work may be needed to characterize the conversion rates. It was also recently proposed that hypervelocity dust impacts may efficiently clean icy surfaces (Hyodo et al. 2025), potentially explaining young-looking terrains without requiring recent formation.

3 Constraining the Evolution and Age of the Ring System

The origin and age of Saturn's ring system is the prominent topic of this special issue. Crida et al. (2025) discuss the hypotheses behind the "primordial" and the "recent" ring formation scenarios that are, so far, the two main families of formation models proposed to explain a range of observations, including several critical new ones by Cassini: the rings' bulk composition, the interplanetary micrometeoroid bombardment rate, and the rings' mass.

In the "primordial" age scenarios the rings are billions of years old and are either the in-situ remnants of the planetary subnebula from which the entire planetary system formed, or the product of the destruction of a large close-in differentiated moon - either by tides or by late-heavy-bombardment interlopers. Since then, the dissipative dynamical processes occurring within the disk generated the regular icy satellites, which were sequentially expelled from the disk itself, causing its progressive loss of mass and explaining the regular moons' mass-distance distribution (Charnoz et al. 2011; Crida and Charnoz 2012). The last satellites to be formed were the small ring satellites that share a composition similar to the main ring particles (Charnoz et al. 2010; Ciarniello et al. 2024), and the rings naturally reach the present mass in $\sim$4 Gyrs of evolution (Salmon et al. 2010).

Conversely, the "recent" age scenarios are motivated by a combination of compositional and structural arguments suggesting an age of a few hundred million years. The rings could have formed recently in two different dynamical scenarios. First, the eccentricity of an outer moon may increase following its capture in resonance with the rapidly outward migrating

Titan. By passing through pericentre within the planet's Roche limit, this moon may have undergone a rapid disintegration caused by tidal forces before escaping, and its remnants are assumed to have formed the ring disk (Wisdom et al. 2022). Second, inner moons and the rings could have originated or been reconfigured in a collision between moons interior to Titan that was induced by the evection resonance between a moon and the Sun (Cuk et al. 2016; Teodoro et al. 2023).

Crida et al. (2025) review the pros and cons of these two types of scenarios. The temporal evolution of the Saturn system is discussed in order to determine the ring age with respect to the *exposure* to the micrometeoroids bombardment flux, to the stability of the ring dynamical *structures* (e.g. orbital resonances) and to the *dynamical* evolution of the disk through orbital energy and angular momentum conservation laws. Reflecting the open debate in the community, the authors do not agree on some of the implications, so the interested reader is encouraged to read the chapter carefully and make up their own mind based on the balance of the evidence.

In the future, further efforts are necessary to better describe the current rings environment within the two proposed scenarios. More studies of crater and impact statistics across the Saturn system (compared to other outer planet systems) will be useful. The comparison with the other Outer Planets ring systems is one necessary advance because some common mechanisms may drive the origin and successive evolution of planetary rings. Why are Saturn's rings so massive and extended with respect to the ones encircling Jupiter, Uranus and Neptune? What can we learn from recently discovered rings around the Centaur (10199) Chariklo (Braga-Ribas et al. 2014), and trans-Neptunian objects (136108) Haumea (Ortiz et al. 2017), and (50000) Quaoar (Morgado et al. 2023; Pereira et al. 2023)? Additional efforts must also resolve the compositional differences within Saturn's system of rings and satellites. Why are the rings and inner satellites almost depleted in organic matter while the satellites orbiting beyond Titan organic-rich (Filacchione et al. 2010)? Is this difference inherited from a radial compositional gradient within the circumplanetary disk from which they formed? Does it reflect a successive evolution? Or is it a strong indication that the inner satellites formed from the spreading of the rings, while the ones beyond Titan were born in the circum-planetary disk? Early intense bombardment and satellite disruption could have mixed or buried early-delivered organics, complicating interpretations of present-day surface compositions. Resolving the entire ring formation and evolution scenario will remain a challenging task for future planetary scientists.

4 When Did the Ocean of Enceladus Form?

This is a hard question to answer. Currently, there are really only two constraints: one is from orbital dynamics, which might bound the age of Enceladus or the 2:1 Enceladus:Dione resonance; the other is from a combination of crater counts and heat flux estimates. Neither gives an answer in which one can be very confident.

The current orbital migration rate of Enceladus is known (e.g. Lainey et al. 2020). In principle, one could track Enceladus's position back in time based on the current rate, and derive an age. If one assumes that Saturn's dissipative properties (as parameterized by the dissipation factor Q) are constant, one gets an Enceladus age of order 1 Gyr. However, there is no reason to think that Q is constant. On the contrary, the resonance locking mechanism (Sect. 1; Fuller et al. 2016, 2024) has observational support, and implies that Q is time-variable and is potentially consistent with a several Gyr age for Enceladus. The issue of resonance-locking is discussed in some detail in Nimmo et al. (2023) and Ćuk et al. (2024).

An important consequence of the resonance locking mechanism is that it implies roughly constant tidal heating inside Enceladus for as long as the Enceladus-Dione resonance has existed (Nimmo et al. 2023). Furthermore, although the magnitude of this heating is uncertain, the best guess is close to the estimated global heat loss rate derived from shell thickness measurements (Hemingway and Mittal 2019). This means that Enceladus has likely been in steady-state for as long as the resonance has lasted, which also means that the present-day ocean is a long-lived feature (Nimmo et al. 2023).

The orbital evolution of the Saturnian satellites prior to the present day is poorly understood (Ćuk et al. 2024); there are no orbital constraints on the age of the Enceladus-Dione resonance. However, there are hints that it is old based on heat flux inferences and crater counts, even though recent analyses make this far from certain (Rhoden et al. 2024).

Some geological features on Enceladus, such as flexural uplift at rifts and viscously-relaxed craters, allow the heat flux at the time of deformation to be inferred (Nimmo et al. 2023). These heat fluxes are typically high (of order 100 mW/m^2), comparable to the present-day heat fluxes at the South Polar Terrain. However, the geological features are located in other, older regions of Enceladus, suggesting that high heat fluxes have been a prominent feature of Enceladus's history. Since tidal heating is the only plausible energy source, these older features suggest that either the Enceladus-Dione resonance or some prior resonance was operating in Enceladus's past. However, estimating the actual age of these features is not easy (Rhoden et al. 2024).

Enceladus possesses regions with large variations in crater densities, some of which (e.g. high northern latitudes) are heavily cratered and appear "old" (Patterson et al. 2018). Converting crater densities to actual ages requires a model of the impact flux, which has already been partly constrained by New Horizons observations (Doner et al. 2024) and will be further refined by JUICE and Europa Clipper. Assuming heliocentric impactors, both Zahnle et al. (2003) and Bottke et al. (2024) find some Enceladus regions are older than 4 Gyr, and Giese et al. (2008) obtained an age of 3.5 Gyr for a flexurally-uplifted rift. At face value, these derived ages support the argument that Enceladus itself is old and that tidal heating is also ancient. However, one complication noted by Bottke et al. (2024) (see also Rhoden et al. 2024) is that the impact flux in the Saturnian system may include planetocentric impactors (Ferguson et al. 2022). If this is the case, then the current age estimates are likely too old.

Of course, tidal heating might be episodic (e.g. driven by different resonances). The freezing timescale for an ocean that extends to depths of $\sim$80 km is about 200 Myr, so one could in principle imagine a situation in which multiple oceans are created, each one during a different pulse of tidal heating. Given the fragmentary record of tidal heating on Enceladus, episodic heating is possible but not required. However, once a subsurface ocean fully freezes, restarting it through tidal heating alone is difficult, somewhat favoring scenarios with long-lived or persistent oceans.

At present, the answer that requires fewest deviations from the inferred histories of other satellite systems appears to be that Enceladus is old ($\sim$ 4 Gyr) and its ocean equally old. Tidal heating has probably been operating at least episodically since early in Enceladus's history, and the power supplied by tidal heating via resonance locking is sufficient to maintain an ocean over 4 Gyr. However, these conclusions are not secure. In particular, episodic heating intervals with periods of a mostly- or totally-frozen Enceladus in between are conceivable, as is a relatively recent resetting event; and the impact fluxes on which absolute age determinations are based could be incorrect.

Despite the successes of astrometry, dynamical models or observations are unlikely to solve this particular problem. The real key is to develop better impact flux models. A good start would be estimates of the present-day outer solar system impact flux, which will probably be provided by JUICE and Europa Clipper. However, such constraints may only be of limited use if the Saturnian system experienced impacts from a population of planetocentric bodies that no longer exists.

5 Probing the Presence of Other Possible "Ocean World" Moons

To date, three of Saturn's moons are thought to host a global ocean beneath their surface: Titan, Enceladus and perhaps Mimas (Iess et al. 2012; Thomas et al. 2016; Lainey et al. 2024).

In the case of Titan, the original arguments for an ocean on the basis of the tidal response (Iess et al. 2012) and obliquity (Baland et al. 2011) have recently been re-examined and the data re-interpreted as indicating an *absence* of a subsurface ocean (Petricca et al. 2025). This is a surprising result and will certainly lead to further analyses of the available observations.

In the case of the latter two moons, their small size seemed to be a considerable challenge to the existence of an ocean present beneath the entire surface. Indeed, the surface-to-volume ratio of these small moons implies that internal heat is released very rapidly, with the expected corollary that any ocean must freeze quickly if not maintained by active heating. From this point of view, the presence of a global ocean on Enceladus was already unexpected; it has been subsequently explained by the discovery of intense tides in Saturn that maintain Enceladus' eccentricity over a long period of time (e.g. Fuller et al. 2016).

The apparent discovery of an ocean beneath the entire surface of Mimas is even more astonishing, given the absence of resurfacing. However, it should be noted that evidence for this claim is not definitive. In particular, the requirement that the ocean form within the last few tens of Myr requires some kind of recent heating episode of unknown origin. Second, formation of such an ocean should result in large global compressive stresses, of which there is no evidence (McKinnon and Schenk 2025).

If it is actually the case that Mimas possesses a subsurface ocean, this raises the question of whether other moons might conceal a global ocean beneath their surface. Radio-science measurements associated with flybys of Dione by the Cassini probe raise the possibility that this moon too could hide a global water ocean (Zannoni et al. 2020). Indeed, the combined analysis of Dione's gravitational field and its topography reveals a significant deviation from hydrostatic equilibrium. One possible interpretation is the presence of a thin global ocean beneath a thick layer of ice. Could the other medium-sized moons be hiding global oceans? For the time being, radio-science data have not yielded gravitational fields precise enough to tell us more. Interestingly, isostatic modeling minimizing crustal stresses supports the presence of an ocean for Dione as well (Beuthe et al. 2016).

As a result, future space missions will have to consider close flybys of these moons, with one of their objectives being to test the possibility of the presence of a global ocean beneath their surface. Experience has shown that the absence of resurfacing does not always indicate the absence of liquid water beneath the surface, the type example being Callisto, although the presence of an ocean there is not definitively established either (e.g. Cochrane et al. 2025).

In practice, the presence of an ocean can be determined by geophysical measurements such as topography, gravity and magnetic fields, including tidal Love numbers, or derived from rotational dynamics, including obliquity and physical librations. Future measurements at target satellites should include precise radio science tracking of the spacecraft, magnetometer, altimeter and a camera to characterize satellite rotation (or a radar for Titan). The global inversion of all available data would then allow us to precisely characterize the interior of the moon, including depth and density of a liquid ocean layer.

6 Conclusion

Far from existing exclusively in the Galilean system, ocean moons are just as present in the Saturnian system. While this conclusion would not have been possible without the data from the Cassini mission (including the discovery of high tidal dissipation inside Saturn), many questions remain as to their number, formation and evolution. Similarly, the age of Saturn's moons and rings remains uncertain in spite of multiple indications of short timescales for the rings and Titan, at least. Finally, the physical mechanism responsible for the intense tides in Saturn has yet to be confirmed. For this reason, further work, including measurements from a future space mission, are required. In the present article, we have attempted to put together a set of thoughts that could feed into future space projects, hoping this will inspire future space mission proposals.

Acknowledgements This paper resulted from the Workshop "New Vision of the Saturnian System in the Context of a Highly Dissipative Saturn" held at the International Space Science Institute (ISSI) in Bern (Switzerland).

Declarations

Competing Interests The authors declare no competing interests.

References

Asphaug E, Reufer A (2013) Late origin of the Saturn system. Icarus 223(1):544–565

Baland RM, Van Hoolst T, Yseboodt M, Karatekin Ö (2011) Titan's obliquity as evidence of a subsurface ocean? Astron Astrophys 530:A141

Barker AJ, Astoul AAV (2021) On the interaction between fast tides and convection. Mon Not R Astron Soc 506:L69–L73. https://doi.org/10.1093/mnrasl/slab077

Beuthe M, Rivoldini A, Trinh A (2016) Enceladus's and Dione's floating ice shells supported by minimum stress isostasy. Geophys Res Lett 43:10–088

Blanc M, Crida A, Shibaike Y, et al (2025) Understanding the formation of Saturn's regular moons in the context of giant planet moons formation scenarios. Space Sci Rev 221(3):35. https://doi.org/10.1007/s11214-025-01156-8

Bottke WF Vokrouhlický D Nesvorný D, et al. (2024) The Bombardment History of the Giant Planet Satellites. Planet Sci J 5. https://doi.org/10.3847/PSJ/ad29f4

Braga-Ribas F, Sicardy B Ortiz J, et al (2014) A ring system detected around the Centaur (10199) Chariklo. Nature 508:72–75. https://doi.org/10.1038/nature13155

Canup RM (2010) Origin of Saturn's rings and inner moons by mass removal from a lost Titan-sized satellite. Nature 468:943–946

Charnoz S, Salmon J, Crida A (2010) The recent formation of Saturn's moonlets from viscous spreading of the main rings. Nature 465(7299):752–754. https://doi.org/10.1038/nature09096

Charnoz S, Crida A, Castillo-Rogez JC, et al (2011) Accretion of Saturn's mid-sized moons during the viscous spreading of young massive rings: solving the paradox of silicate-poor rings versus silicate-rich moons. Icarus 216:535–550

Ciarniello M, Filacchione G Nicholson PD, et al (2024) The origin and composition of Saturn's ring moons. Space Sci Rev 220:72. https://doi.org/10.1007/s11214-024-01103-z

Cochrane CJ, Vance SD, Castillo-Rogez JC, Styczinski MJ, Liuzzo L (2025) Stronger evidence of a subsurface ocean within Callisto from a multifrequency investigation of its induced magnetic field. AGU Adv 6(1):e2024AV001237

Crida A (2020) Large tides alert on Saturn. Nat Astron 4:1024–1025. https://doi.org/10.1038/s41550-020-1109-0

Crida A, Charnoz S (2012) Formation of regular satellites from ancient massive rings in the Solar System. Science 338(6111):1196. https://doi.org/10.1126/science.1226477

Crida A, Estrada PR, Nicholson P, Murray C (2025) The age and origin of Saturn's rings. Space Sci Rev 221(5):66. https://doi.org/10.1007/s11214-025-01189-z

Cuk M, Dones L, Nesvorny D (2016) Dynamical evidence for a late formation of Saturn's moons. Astrophys J 820:97

Ćuk M, El Moutamid M Lari G, et al (2024) Long-term evolution of the Saturnian system. Space Sci Rev 220:20. https://doi.org/10.1007/s11214-024-01049-2

Dalle Ore CM, Cruikshank DP, Mastrapa RME, Lewis E, White OL (2015) Impact craters: an ice study on Rhea. Icarus 261:80–90. https://doi.org/10.1016/j.icarus.2015.08.008

Dewberry JW (2024) Tidal Synchronization Trapping in Stars and Planets with Convective Envelopes. Astrophys J 966. https://doi.org/10.3847/1538-4357/ad344d

Doner A, Horányi M Bagenal F, et al (2024) New Horizons Venetia Burney Student Dust Counter observes higher than expected fluxes approaching 60 au. Astrophys J Lett 961(2):L38

Fayolle MS, Dirkx D, Cimo G, Gurvits LI, Lainey V, Visser PN (2024) Contribution of PRIDE VLBI products to the joint JUICE-Europa Clipper moons' ephemerides solution. Icarus 416:116101

Ferguson SN, Rhoden AR, Kirchoff MR, Salmon JJ (2022) A unique Saturnian impactor population from elliptical craters. Earth Planet Sci Lett 593. https://doi.org/10.1016/j.epsl.2022.117652

Filacchione G, Capaccioni F, Clark RN, Cuzzi JN, Cruikshank DP, Coradini A, Stephan K (2010) Saturn's icy satellites investigated by Cassini–VIMS: II. Results at the end of nominal mission. Icarus 206(2):507–523

Fuller J, Luan J, Quataert E (2016) Resonance locking as the source of rapid tidal migration in the Jupiter and Saturn moon systems. Mon Not R Astron Soc 458:3867–3879

Fuller J, Guillot T, Mathis S, Murray C (2024) Tidal dissipation in giant planets. Space Sci Rev 220:22. https://doi.org/10.1007/s11214-024-01055-4

Giese B, 6 colleagues (2008) Enceladus: an estimate of heat flux and lithospheric thickness from flexurally supported topography. Geophys Res Lett 35. https://doi.org/10.1029/2008GL036149

Hemingway DJ, Mittal T (2019) Enceladus's ice shell structure as a window on internal heat production. Icarus 332:111–131

Hener J, Dirkx D, Fayolle S, et al. (2025) Re-visiting determination of Saturn-Titan tidal interaction parameters using the open-source Tudat software. EPSC-DPS Joint Meeting 2025, Helsinki, Finland, 7–12 Sep 2025, EPSC-DPS2025-1567. https://doi.org/10.5194/epsc-dps2025-1567

Hyodo R, Genda H, Madeira G (2025) Pollution resistance of Saturn's ring particles during micrometeoroid impact. Nat Geosci 18(1):44–49

Iess L, Jacobson RA, Ducci M, et al (2012) The tides of Titan. Science 337:457. https://doi.org/10.1126/science.1219631

Jacobson RA (2022) The orbits of the main Saturnian satellites, the Saturnian system gravity field, and the orientation of Saturn's pole. Astron J 164. https://doi.org/10.3847/1538-3881/ac90c9

Lainey V, 10 colleagues (2012) Strong tidal dissipation in Saturn and constraints on Enceladus' thermal state from astrometry. Astrophys J 752:14

Lainey V, 10 colleagues (2017) New constraints on Saturn's interior from Cassini astrometric data. Icarus 281:286–296

Lainey V, Casajus LG Fuller J, et al (2020) Resonance locking in giant planets indicated by the rapid orbital expansion of Titan. Nat Astron 4:1053–1058. https://doi.org/10.1038/s41550-020-1120-5

Lainey V, Rambaux N Tobie G, et al (2024) A recently formed ocean inside Saturn's moon Mimas. Nature 626:280–282. https://doi.org/10.1038/s41586-023-06975-9

Lainey V, Zannoni M, Robert V, et al (2025) Quantification of tides in giant planets from observations. Space Sci Rev 221(7):93. https://doi.org/10.1007/s11214-025-01209-y

Lissauer JJ, Cuzzi JN (1982) Resonances in Saturn's rings. Astron J 87:1051–1058. https://doi.org/10.1086/113189

Magnanini A, Zannoni M, Casajus LG, Tortora P, Lainey V, Mazarico E, Iess L (2024) Joint analysis of JUICE and Europa Clipper tracking data to study the Jovian system ephemerides and dissipative parameters. Astron Astrophys 687:A132

McKinnon WB, Schenk PM (2025) Is Mimas a Dyson satellite? The fate of small melting moons. Lunar Planet Sci Conf 56:2897

Miller KE, Filacchione G Cuzzi JN, et al (2024) The composition of Saturn's rings. Space Sci Rev 220:70. https://doi.org/10.1007/s11214-024-01104-y

Morgado BE, 58 colleagues (2023) A dense ring of the trans-Neptunian object Quaoar outside its Roche limit. Nature 614:239–243. https://doi.org/10.1038/s41586-022-05629-6

Nimmo F, Neveu M, Howett C (2023) Origin and evolution of Enceladus's tidal dissipation. Space Sci Rev 219:57. https://doi.org/10.1007/s11214-023-01007-4

Nixon CA, Lorenz RD Achterberg RK, et al (2018) Titan's cold case files - outstanding questions after Cassini-Huygens. Planet Space Sci 155:50–72. https://doi.org/10.1016/j.pss.2018.02.009

Ogilvie GI, Lin DNC (2004) Tidal dissipation in rotating giant planets. Astrophys J 610:477–509. https://doi.org/10.1086/421454

Ortiz JL, Santos-Sanz P Sicardy B, et al (2017) The size, shape, density and ring of the dwarf planet Haumea from a stellar occultation. Nature 550:219–223. https://doi.org/10.1038/nature24051

Patterson GW, 4 colleagues (2018) The geology of Enceladus. Enceladus Icy Moons Saturn 95

Pereira CL, 68 colleagues (2023) The two rings of (50000) Quaoar. Astron Astrophys 673. https://doi.org/10.1051/0004-6361/202346365

Petricca F, Vance SD, Parisi M, Buccino D, Cascioli G, Castillo-Rogez J, Downey BG, Nimmo F, Tobie G, Journaux B, Magnanini A (2025) Titan's strong tidal dissipation precludes a subsurface ocean. Nature 648(8094):556–561

Remus F, Mathis S, Zahn J-P, Lainey V (2012) Anelastic tidal dissipation in multi-layer planets. Astron Astrophys 541:A165. https://doi.org/10.1051/0004-6361/201118595

Rhoden AR, Ferguson SN Bottke W, et al (2024) Geologic constraints on the formation and evolution of Saturn's mid-sized moons. Space Sci Rev 220:55. https://doi.org/10.1007/s11214-024-01084-z

Salmon J, Charnoz S, Crida A, Brahic A (2010) Long-term and large-scale viscous evolution of dense planetary rings. Icarus 209:771–785

Teodoro LFA, Kegerreis JA, Estrada PR, Cuk M, Eke VR, Cuzzi JN, Massey RJ, Sandnes TD (2023) A recent impact origin of Saturn's rings and mid-sized moons. Astrophys J 955:137

Terquem C (2021) On a new formulation for energy transfer between convection and fast tides with application to giant planets and solar type stars. Mon Not R Astron Soc 503:5789–5806. https://doi.org/10.1093/mnras/stab224

Thomas PC, Tajeddine R, Tiscareno MS, et al (2016) Enceladus's measured physical libration requires a global subsurface ocean. Icarus 264:37–47. https://doi.org/10.1016/j.icarus.2015.08.037

Wisdom J, Dbouk R, Militzer B, Hubbard WB, Nimmo F, Downey BG, French RG (2022) Loss of a satellite could explain Saturn's obliquity and young rings. Science 377:1285–1289

Zahnle K, Schenk P, Levison H, Dones L (2003) Cratering rates in the outer Solar System. Icarus 163:263–289. https://doi.org/10.1016/S0019-1035(03)00048-4

Zannoni M, Hemingway D, Gomez Casajus L, Tortora P (2020) The gravity field and interior structure of Dione. Icarus 345:113713. https://doi.org/10.1016/j.icarus.2020.113713

Publisher's Note Springer Nature remains neutral with regard to jurisdictional claims in published maps and institutional affiliations.

Authors and Affiliations

Valéry Lainey[1] · Aurélien Crida[2] · Matija Cuk[3] · Jeffrey N. Cuzzi[4] · Dominic Dirkx[5] · Gianrico Filacchione[6] · James Fuller[7] · Carly J.A. Howett[8] · Kelly Miller[9] · Francis Nimmo[10] · Nicolas Rambaux[1] · Marco Zannoni[11]

✉ D. Dirkx
d.dirkx@tudelft.nl

1 LTE, Observatoire de Paris, PSL Research University, Sorbonne Université, CNRS, Univ. Lille, 77 avenue Denfert Rochereau, 75014 Paris, France

[2] Laboratoire Lagrange (UMR7293), Université Côte d'Azur / Observatoire de la Côte d'Azur, Boulevard de l'Observatoire, CS 34229, 06300 Nice, France

[3] SETI Institute, Mountain View, CA, USA

[4] Space Sciences Division, NASA Ames Research Center, MS 245-3, Moffett Field, 94035, CA, USA

[5] Delft University of Technology, Kluyverweg 1, 2629 HS Delft, The Netherlands

[6] IAPS-INAF Institute for Space Astrophysics and Planetology via del Fosso del Cavaliere, 100, 00133, Rome, Italy

[7] TAPIR, Mailcode 350-17, California Institute of Technology, Pasadena, CA, 91125, USA

[8] Clarendon Labs, Atmospheric, Oceanic and Planetary Physics Department, University of Oxford, Parks Road, Oxford, Oxfordshire, OX1 3PU, UK

[9] Southwest Research Institute, San Antonio, TX, USA

[10] Department of Earth and Planetary Sciences, University of California Santa Cruz, Santa Cruz CA, 95064, USA

[11] Dipartimento di Ingegneria Industriale, Università di Bologna, Forlì, Italy

MIX
Papier aus verantwortungsvollen Quellen
Paper from responsible sources
FSC® C105338

If you have any concerns about our products,
you can contact us on
ProductSafety@springernature.com

In case Publisher is established outside the EU,
the EU authorized representative is:
Springer Nature Customer Service Center GmbH
Europaplatz 3, 69115 Heidelberg, Germany

Printed by Libri Plureos GmbH
in Hamburg, Germany